Television and Video Engineering
Second Edition

Television and Video Engineering

Second Edition

Late Arvind M Dhake

Formerly, Professor of Electrical Communication
College of Engineering
Pune-5

McGraw Hill Education (India) Private Limited
NEW DELHI

McGraw Hill Education Offices
New Delhi New York St Louis San Francisco Auckland Bogotá Caracas
Kuala Lumpur Lisbon London Madrid Mexico City Milan Montreal
San Juan Santiago Singapore Sydney Tokyo Toronto

McGraw Hill Education (India) Private Limited

Copyright © 1995, 1979, by McGraw Hill Education (India) Private Limited.

37th reprint 2014
RYCCRRYURBADX

ISBN (13 digit): 978-0-07-460105-1
ISBN (10 digit): 0-07-460105-9

Published by McGraw Hill Education (India) Private Limited,
P-24, Green Park Extension, New Delhi 110 016 typeset in Times Roman at
Replika Press Pvt. Ltd., and printed at Avon Printers, Main Loni Road,
Jawahar Nagar Industrial Area, Shahdara, Delhi 110094

Dedicated to

My Students and TV Enthusiasts

Preface to the Second Edition

The last decade has seen a proliferation of television and video systems in India, and perhaps the world over. Television in all its colour and glamour has gained an edge over the print media or radio, because of the instant visual impact it can provide through satellites spreading images. Television technology has now become a vital tool to the information revolution that is sweeping across the countries of the world. The low cost video systems, VCR-cameras or camcorders have brought about a video revolution in the field of home entertainment, education, training, advertising and electronic news gathering. Dramatic developments in flat panel displays, reduction in the cost of image scanning systems, LCD displays and integrated subsystems is going to affect our communication capabilities and life-style in a big way. This revision had to take into account all these wider implications.

The revised second edition deals with all the new developments in television engineering. It maintains a wide angle perspective to all the issues involved in television and video engineering, zooming on to specific topics to explain the details that are of common interest. With the expanding horizons and improvements in technology, it is indeed a difficult task to strike a balance, but I hope the readers would appreciate the approach adopted. By introducing colour television fundamentals in early chapters, the emphasis has been shifted to colour television systems and associated integrated circuits. The upcoming digital television technology, video information services, advanced television technology evolving through high definition TV, MAC signals used in direct broadcasting satellites, are all dealt with in fair detail. These topics highlight the intense development activities going on to sharpen the TV images and bring up the quality of video program production and distribution. With the microprocessors, frame grabbers, digital signal processing chips and holography in the fray, films and video enter into the domains of advanced TV research dealing in digital video graphics with spatial imaging, 3-dimensional pictures and animation.

The book has been updated to include the latest developments in television and video hardware and renamed *Television and Video Engineering* as such, to signify the wider coverage of the video.

Chapter 1 reviews the historical development and the present scope of television systems in a perspective. The chapter may be understood better after the reader becomes familiar with the operating principles of television discussed in the early chapters.

Chapter 2 and 3 deal with the basic principles of television and the standards used for transmission and reception of television signals for those not initiated into television fundamentals.

Chapter 4 describes the principles of television cameras, lens systems, and details of image pickup tubes and devices including the CCD imagers. Camera signal processing and camera chain is explained briefly.

Chapter 5 explains the fundamentals of colour, colour television cameras, basic principles of compatible colour television transmission and the standard colour encoding systems.

Chapter 6 discusses the requirements of television studio, TV studio equipment and explains in depth principles of video recording. Television transmission requirements and design principles of television transmitters, antenna and relay systems including the satellite relays are dealt with in chapter 7.

Chapter 8 and 9 discuss the propagation phenomena and antenna systems with special reference to television reception.

Chapter 10 to 19 discuss the TV receiver circuits stagewise in detail, to bring out the requirements and design considerations of each important block in the television receiver. Chapter 16 describes the large screen projection techniques and flat panel technologies including the LCD and the plasma displays that have acquired importance in small portable TV sets and laptop computers on the one hand and large screen HDTV on the other. Chapter 19 discusses miscellaneous topics like SMPS, remote control and highlights the impact of the ICs on TV receiver circuits. Single chip integrated circuits for economy and high performance sets are presented.

Chapter 20 discusses detailed features of the common ICs used in a typical integrated colour TV receiver, to explain the function of complete IC colour TV receiver.

Chapter 21 discusses, TV test instruments and their use in TV receiver testing and alignment, and draws out a systematic procedure for servicing a TV receiver, ending up with a fault localization table that should help quick trouble shooting.

Chapter 22 deals with system design features of closed circuit television and the modern colour video systems in detail. Single tube colour camera and signal processing features are explained. This is followed by features of the commonly used VCR systems and camcorders including the U-matic, VHS, Betacam and their improved derivatives. Video disc recording including the much publicized Laservision is also discussed. With the compact CCD imagers and the ever expanding capacity of IC memories, their use in video imaging and storage has exciting possibilities as evinced by the IC card camera discussed at the end.

Chapter 23 discusses the TV program distribution via the media of cable TV networks and satellites, both of which have recently gained a lot of ground because of the thirst of discerning video files for choice programs and global circuits.

Chapter 24 describes the information age applications of TV and video systems, visual information services like teletext pioneered by the BBC and now broadened by the computer based viewdata services at home. This leads us to the video display units. CRT terminals and graphic interfaces with personal computers, which are explained in detail to elucidate the graphic adapter cards like CGA, EGA and VGA.

Chapter 25 deals with digital television in depth. Principles and standards for digitization are discussed, followed by attempts to improve quality of picture by use of digital signal processing techniques and microcomputer control in the TV receiver.

Finally, chapter 26 exposes the frontier technologies in television, the Multiplex Analog Components packets signals to improve the picture quality in satellite systems and high definition television—a challenge of the video technology to the photographic films and print media. The impact of the lead shown by the Japanese HDTV technology which has sent tremors in the US and Europe is highlighted. The methodologies being attempted by the European consortium countries to catch up through digital compatible route and MAC are discussed.

The new edition should meet the requirements of a full fledged course in Television Engineering or Video Systems at the undergraduate and advanced diploma level. The book should be useful reference to practicing engineers and technicians for widening knowledge of the subject. I am sure that it will provide a lot of interesting information on television to satisfy any curious reader, whether a student, a professional or a TV enthusiast.

Arvind M Dhake

Preface to the First Edition

Although a wide range of books is available on television systems, the with need for a modern book based on the continental circuits and technology CCIR system-B standards commonly used in many countries, including India, has been generally felt. This book is intended to meet the requirements of a modern textbook on television engineering suitable for engineering students at the degree, diploma and the technician's level. It has grown out of my teaching courses in the subject to the students of electronics and telecommunication over a number of years. The book should also prove highly informative to practising engineers and service technicians in the field.

The book aims at a comprehensive coverage of television systems including principles of television studio and broadcasting equipment, antennas and propagation, television receiver circuit techniques and CCTV systems. Emphasis has naturally been laid on television receiver systems and circuits that are of wider interest. The approach assumes only some preliminary knowledge of radio and electronics and progressively elucidates the operating principles of the television systems to bring out the design requirements of the various circuits, particularly in TV receivers. Currently popular circuits are discussed to include both tube and transistor circuits. There has been a clear shift now towards the use of solid state devices and integrated circuits in the new designs which have been discussed in detail. A great deal of application information is given wherever possible.

The systems and circuits include in the book are primarily black and white, considering the current practice in the country. Principles of colour television have been explained in the last chapter to provide a basic understanding of the colour TV systems in use.

Chapter 1 reviews the historical development and the present scope of television systems including modern trends in the field. This chapter may be understood better after the reader becomes familiar with the operating principles of television discussed in later chapters.

Chapters 2, 3 and 4 deal with the basic principles of television and the standards used for transmission and reception of television signals.

Chapters 5, 6 and 7 describe the television cameras, television studio equipment and transmission principles, including satellite broadcasting. These chapters may be skipped in initial reading by those not very familiar with communication circuits, and have primary interest in TV receivers.

Chapters 8, 9 and 10 discuss the propagation phenomena and antenna systems with special reference to TV reception.

Chapters 11 to 22 explain the TV receiver circuits stagewise in detail to bring out the requirements and design considerations of each important block in the receiver including practical circuits in use. Chapter 22 highlights the impact of ICs in TV receiver circuits and also discusses associated topics like ultrasonic remote control and TV games.

Chapter 23 discusses the TV test instruments and their use in TV receiver testing and alignment; and draws out a systematic procedure for servicing a TV receiver, ending up with a fault localisation table that should help quick fault finding.

Chapter 24 describes the system design features of closed circuit television, block schematics of the camera control circuits and video monitors, and a portable video tape recorder commonly used with CCTV systems.

Chapter 25 explains the fundamentals of colour, colour television camera and the basic principles of transmission and reception of colour TV in the NTSC, SECAM and PAL systems and the colour picture tubes.

The book should meet the requirements of a full fledged course in television engineering at the undergraduate and diploma level. The book should be useful to practising engineers and technicians in widening their knowledge of the subject and modern solid state circuits. The somewhat involved analytical sections of the chapters on deflection coils and video circuits and amplifiers may be omitted when using the book at the diploma or technician level.

ARVIND M DHAKE

Acknowledgements

A book of such wide coverage cannot be prepared without help from numerous sources and fellow professionals.

I am indeed grateful to the numerous sources, the various companies and publishers who have permitted use of their literature in preparing the book. I specially thank:

Philips Components, Peico Electronics & Electrical Ltd. Bombay.
Philips Components, Philips International B.V. Endhoven.
Centrex Publishing Co. Eindhoven.
INTERMETALL, Halbleiterwerk der Deutsche ITT Industries GmbH.
Bharat Electronics Limited, Bangalore.

Institution of Electrical and Electronics Engineers Inc. New York.
Film and Television Institute of India, Pune.
Director General, Doordarshan, New Delhi.
Director General, Indian Standards Institution, New Delhi.

Franzis Verlag, Munchen.
McGraw-Hill Publishing Company, New York.

For details of the Indian standards on Television indicated in summary extract form, reference should be made to these listed in the references and available from Indian Standards Institution, New Delhi and their branch offices.

I am indebted to the authors and publishers of the various references listed at the end of the book.

It a great relief and pleasure for me to release this elaborately revised edition of the popular text book, and dedicate it to my students and the numerous TV fans who were the moving spirit behind this effort, pressing for up-to-date information on the subject.

I have to place on record my sincere thanks to Prof. M. Murugesan of IIT Bombay; Mr. M.Y. Thote, formerly Chief Engineer Doordarshan, Mr. S.P. Bhatikar, Chief Engineer Doordarshan and staff of Doordarshan and the Film Institute of India, Pune; Mr V.J. Mulay, for their interest and useful suggestions in preparing the book. I am grateful to the Principal College of Engineering Pune, and the Director of Technical Education for permission to publish this book.

Last but not the least, I record, my appreciation of the help and encouragement from my wife Vimal, children Preeta and Ajay during preparation of the manuscript, and forbearing the loss of togetherness of many evenings and weekends.

Author

Professor Arvind M Dhake passed away soon after completing the final draft manuscript of this major substantially revised second edition.

In order to complete the editorial process, we sought the help of Professor R.B. Joshi, Department of Electrical Engineering, IIT Bombay who resolved many of the technical and copyediting queries related to this project. We are grateful to him for his time and expertise.

The Publishers

Contents

1

Introduction

1.1 THE HISTORICAL DEVELOPMENT OF TELEVISION

For ages man has dreamt of transmitting sight and sound from one place to another. The isolation of selenium by the Swedish scientist, Berzelius, in 1817 and the discovery of the light-sensitive properties of selenium by May, in 1873, heralded the possibility of converting light from pictures into electrical signals. It was only after 1892, when Elster and Geitel devised a photoelectric cell based on this property, that its application was made. Inventors could see the selenium cell as a counterpart of the carbon microphone which altered its electrical resistance according to the sound pressures impinging on it. Like sound converted into electrical signals by the microphone, the selenium cell could possibly send picture signals over wires by transforming them into electrical counterparts. The transmission of a picture required the simultaneous sensing of the light intensity at the various elements in the picture by thousands of selenium cells at the sending end, and transmitting the signals over an equal number of wires to an equal number of reproducing devices. This simultaneous method, although tried by Renaux and Fournier in 1906, was obviously uneconomical and is so even today. Modern fibre optics has some potential of doing this optically by guiding light from the picture elements along a bunch of glass fibres over limited lengths.

Another breakthrough in television was made by Paul Nipkow, a German experimenter, when he invented the scanning disc in 1884. It was Nipkow's idea to change the picture into electrical bits at the sending end by means of spirally arranged holes in a rotating disc in front of the picture. The sequential transmission of these bits over a single wire, and reconstruction by a similar scanning technique at the receiving end, serves as the basic principle of present-day television.

The actual television was first demonstrated in 1925–27, by J.L. Baird in London and by C.F. Jenkins in Washington, both working independently of each other, by using the technique of mechanical scanning, employing the Nipkow rotating disc. The disc had small apertures spirally located around the periphery and these apertures carried lenses associated with a photocell. Light reflected from the scene was made to pass through the apertures of the disc and resultant light was collected by a photocell which produced a proportional current. The monitor display employed a similar Nipkow disc with apertures, rotating at the same speed as the transmitting disc and neon gas discharge lamp as the light source which produced light proportional to the current fed into it. The scanning thus took place along arcs of circles traced by aperture holes in the discs. The mechanical scanner operated on instantaneous or non-storage principles allowing only limited light from each scanned spot to pass through the aperture holes for a short scanning time. The mechanical system of scanning had limitations of scanning speed and picture size also.

As electronic vacuum tubes were developed, use of the cathode ray tube for electronic scanning or writing purposes was thought of by B.L. Rosing, who demonstrated the capability of the cathode ray tube at the receiving end in a television system in 1907. In 1908, A.A. Campbell Swinton proposed an all-electronic television system based on a camera tube containing a multiplicity of photoelectric cells in parallel so that each cell could store a charge until scanned by an electron beam. Based on this principle, various designs were proposed between 1920 and 1930. In 1923, V.K. Zworykin filed a U.S. patent for a camera tube known as the 'Iconoscope'. The all-electronic television system employing an iconoscope and an electrostatic cathode ray tube was demonstrated by Zworykin at the Westinghouse Research Laboratories in 1924–29, and a second demonstration employing an image dissector for the pick-up and a magnetically deflected and focused cathode ray tube for display was made by P. Farnsworth in 1929. By 1930, most of the work on scanning had changed to electronic systems that had a far greater promise. By 1935, necessary tubes, devices and ancillary circuits, such as for deflection and video amplification, had been developed and the need for setting up scanning and modulating standards became obvious.

Regular television broadcasts were introduced in the late thirties. In 1936, Britain became the first country to provide regular TV programmes, followed by the USA in 1939. But it was only after the end of World War II that television broadcasting grew into popular, commercial entertainment and news media, and receivers with important circuit features like flyback voltage supply for the picture tube, horizontal AFC to improve the horizontal synchronisation, etc., were developed. The intercarrier sound system, linking sound and picture carriers together, was introduced to enable easy tuning, particularly on UHF channels which were allocated for TV broadcasting in 1952.

Colour Television

A colour sequential system developed by the Columbia Broadcasting Service (CBS) was adopted in the United States of America in 1950, for colour TV transmission. This system employed a rapidly rotating disc containing filters of the primary colours, viz. red, green and blue in sequence, interposed between the lens and the camera tube at the transmitter and also a similar disc between the receiving picture tube and the viewer, both discs rotating at the same synchronized speed. This system, although a breakthrough in colour television, suffered from many drawbacks, such as loss of luminosity due to filters, large size of the discs compared to actual display, besides the mechanical problems of the discs. It was given up soon in favour of an all-electronic system developed by the Radio Corporation of America (RCA).

The RCA system compatible with the monochrome system was experimentally vindicated in early 1949, but was accepted by the Federal Communication Committee (FCC), USA, for colour transmissions in 1953, after being standardized as the National Television Subcommittee (NTSC) system, in the USA. Modified systems, viz. Sequential Colour and Memory (SECAM) in France and Phase Alternation Line (PAL) in West Germany were introduced later to overcome problems of phase error in transmission paths, in the NTSC system. After long controversies over the relative merits of these systems and their derivatives, all the three colour TV systems have survived to be accepted in different countries. The choice has been often affected by the monochrome system standard prevalent in the particular country.

1.2 TELEVISION BROADCASTING

Television with its real time image transmission has an instant impact on the viewer, intensely engaging the mind. It is almost an extension of the human eye in modern life, in countries with extensive TV networks using Electronic News Gathering (ENG) techniques with an on-the-spot TV camera linked to the studio with microwaves. The impact of television is widespread. It caters to millions of viewers when used for public broadcasting, and has opened new avenues in the field of public entertainment, techniques of news dissemination and social education. Television viewing influences people in their social behaviour, political views, buying habits and so on. It is widely acknowledged as a powerful mass communication medium even though some of its undesirable effects, like addiction to viewing the 'idiot box' hours on end, are also manifest. One may evident compare its influence with that of literature, stage or the movies. The great civilizing effect of all literature is that it takes people's vision beyond the immediate. Literature enables them to make the philosophical leap that Jean-Paul Sartre described as 'seeing the other side of self'. Television simply does this more effectively, than any other kind of artform that went before. Unlike the stage and the movies, the episode TV serial does not end in catharsis. Week after week the characters come back and evolve at the slow pace of day-to-day life, exposing themselves to the viewers more intimately than most relatives and friends. Television serials become an ongoing part of life, and for many susceptible people, can be barely distinguishable from real life.

Today, it is hard to imagine a world without television, harder still to imagine what the world of the last half a century would have been, without the flickering images from BBC, NBC in the '40s, and all that followed. We might have had a less violent society, because the typical child would not have been exposed to thousands of actual and simulated violent crimes on the news and entertainment. We might have been a healthier society, in which the children played outdoors instead of watching the TV hour after hour. We might have been a society in which people still felt respect for the establishment— institutions and their leaders instead of the TV-bred skepticism of these. There would have been perhaps less consumerism in soaps, shampoos and the like, and the meals cooked at home would not be outdistanced by snacks and fast foods, so enticingly advertised on TV. But we might also have been a less alert world, one in which citizens were not so widely informed about economy, socio-political matters, foreign military adventures that run the risk of war, one in which starvation in Ethiopia could never inspire Live Aid. We would have certainly missed the live broadcasts of the breathtaking space adventures of man, Neil Armstrong's small step on the moon as 'a giant leap for mankind', the world of sports events and the Olympic Games we almost take for granted. The world we live in today would have been very different without television.

Television has spread its influence among peoples of all countries, much faster than the earlier landmarks in the history of science. Television transports us around the world far faster than any vehicle or jet for that matter, and links together humankind the world over, through a flotilla of satellites. It has almost replaced print as the primary means of communication. The capital costs and the operational expenses in the production and broadcasting of TV programs are quite high. Yet, its importance as a mass communication medium for political image building and propagating social objectives like family planning, health care, social education and improved agricultural techniques has been recognized in all developing countries also. Television is now commonly used as a valuable teaching aid, and video instructional programs are being produced in all subjects at all levels of education.

1.3 EXTENDED COVERAGE OF TELEVISION

Owing to the use of VHF-UHF frequencies for television broadcasting, direct reception of TV signals is limited to line-of-sight distances—usually ranging from 75 to 150 km. These are dependent on the heights of the transmitter and receiver antennas and the terrain in between. In fringe areas the TV programs can be extended to specific subscribers by a cable TV system relying on good high gain aerials mounted at favourable heights. For extending the broadcast service, relay stations that receive the signal via microwave links or coaxial cable and rebroadcast it to the extended regions are used. A chain of such terrestrial relay stations can cover wide areas.

Thanks to the rapid strides made in satellite launching and space communication technology, TV has become a wide ranging and powerful medium of mass communication in this ever-shrinking world. It has now become possible to have international programs with global coverage by linking national TV systems through satellites. Communication Satellites in geostationary orbits have opened new vistas for global television broadcasting. In 1962, world-wide television was made possible by the use of the TELSTAR satellite. Now there are a number of INTELSAT Geostationary satellites, making international, global or space television relay possible. Direct domestic broadcasting satellites for television and telecommunications have been launched by several countries. These satellites emit higher power radiations—enough to be received directly through low-cost, medium- to small-size dish antennas on conventional television receivers augmented by a front-end converter.

The Satellite Instructional Television Experiment (SITE) was the first attempt in this direction, using the NASA ATS-6 satellite made available to India in 1975–76. Direct reception through low-cost medium-size, 3-metre dish antennas was possible on conventional TV receivers augmented by a front-end converter. With higher powers from the satellites launched subsequently by various countries, direct satellite TV reception has become easier with 1 to 1.6-metre dish antennas and reduced cost of the front-end microwave converters. These satellites include the Eutelsat in Europe, TV-sat D3 of FRG, UNISAT of Britain, AUSSAT of Australia, BS-2 of Japan, PALAPA in Indonesia, the INSAT-1 series by India, and the communication satellites like INTELSAT-V, which have made international TV directly accessible from space, provided suitable standard receivers or converters are available.

Television broadcasting equipment is usually quite sophisticated and expensive. This is so because the technical capabilities of the system have to be considered to ensure aesthetically pleasing pictures. Television studios employ extensive lighting control facilities and equipment with special effect capabilities that lend gloss and glamour to the televised image. TV cameras used in broadcast studios are high performance types with special facilities and circuitry not required in less professional closed circuit television applications. Mobile trolley mounting, panning and tilting facilities along with built-in video monitor and camera lens controls, make it more bulky.

A wide variety of equipment is used to support television cameras in a broadcast studio. Video tape recorders, telecine film cameras for display of movie films and slides, slow motion disc recorder facilities, video switching consoles, pulse distributors, audio mixer systems, etc. provide special effects and professional production techniques. With the introduction of digital techniques and high definition TV in the studio equipment the quality of production has remarkably improved; and the elaborate editing facilities with special video effects provide unprecedented flexibility for program producers.

1.4 TECHNOLOGY TRENDS

In the sixties, the use of transistors and integrated circuits, along with the availability of small size TV camera tubes like vidicon and its derivatives like plumbicon and saticon having improved characteristics and ease of operation, had made the video camera and video recorders compact and portable, besides increasing their reliability and reducing the cost. Single-chip ICs integrating all small signal functions in a TV receiver have already been announced, minimizing the discrete components and service adjustments. With solid state circuitry employing integrated circuits, and incorporating more complex circuit functions, the trend is to make the systems of the modular or single board mounting style. The solid state Charge Coupled Device (CCD) sensors and the Liquid Crystal Displays brought in the solid state technology in the video camera and the flat panel LCD TV display. LCD pocket TV in monochrome and colour too, are already available. Instead of phosphors, the LCD set has molecular gates that open to expose different colours. These are improved every year and refined to challenge the supremacy of the electron beam in image sensing and reproduction. LCD colour displays of up to 5 in. are available at reasonable prices, while active matrix colour displays of up to 14 in. have been demonstrated recently. The technology to build large screen is within the reach but the cost is still prohibitive. The last bastion of electron tubes in consumer electronics is showing signs of crumbling.

In developing as well as advanced industrialized countries, colour television is now commonplace. With colour television reaching its peak, new products such as camcorders, optical video disc recorders, large flat screen displays, high definition television are under development. Techniques for bandwidth reduction, to allow the high resolution channels in the available channels are under study.

Recent technological efforts towards this end have shown encouraging results. The bandwidth required for television with these techniques can be drastically reduced while the picture quality can be good. A recent technique for this is the pseudo-random scan. If the picture elements are illuminated with a well organized randomness, the slow scan produces good results even with moving objects. Such bandwidth reduction techniques have been employed in high definition systems under development.

1.5 CLOSED CIRCUIT TELEVISION SYSTEMS (CCTVs)

In closed circuit television systems, camera signals are generally supplied through coaxial cables or low power wireless links to TV monitors or receivers located at not-too-distant locations. Since the signal is not radiated into space, these systems may employ arbitrary and more elaborate high resolution standards or simpler sync standards. Large RF powers are not required, and hence the system costs are lower. Compact, high quality cameras using small camera tubes or CCD image sensors and VCRs of reliable solid state circuitry have been developed to suit various applications. These are now available at much lower costs in recent years, and hence CCTV and video systems have found wide-ranging applications in education, industry, medicine, aerospace, surveillance, traffic control, security monitors, data transmission, etc. In general, CCTV systems are a valuable aid in remote monitoring, or surveillance in places of hazardous surroundings where human monitoring may be risky or impossible. CCTV systems are also particularly useful in educational demonstrations of scientific experiments, surgical operations, and in the viewing of microscope plates by many students or classes at a time. Compact video recording facilities are also now available for recording the signal and replay it as many times as required.

CATV Cable Television (CATV)

Cable TV was started to provide strong TV signals to communities in shadow zones where broadcast signals were weak or in city areas where TV signals were prone to ghost interference from reflected signals from high rise buildings and structures. Supplying stronger signals for better reception and a wide choice of the number of available channels, cablevision TV systems have become very popular in developed countries. Cable TV programmes on standard channels are made available to subscribers via coaxial cables much the same way as a telephone service. TV broadcasting employs RF carrier frequencies in the VHF and UHF ranges which have the limitation of line-of-sight propagation.

In remote suburban or rural areas far from a transmitter, the signal is weak, suffering from noise and channel interference. Even in big cities, in areas near the transmitter, strong signals produce severe ghost images on the screen due to multipath reception of TV signals by direct path, and reflections from tall buildings. In both situations, the solution of feeding the signal to individual subscribers over cables has been found convenient. The signals on standard channels are distributed via trunk amplifiers which boost the signals to a high level. The boosted signals are then distributed through cables and distribution amplifiers to subscribers paying monthly charges. The signal can originate from a CATV studio or can be derived from a broadcast program received by a good receiving antenna system. The latter type or system is on a smaller scale, in apartment buildings which have a common master antenna on top and an amplifier feeding a number of TV receivers in the building. The arrangement is known as the Master Antenna Television (MATV) system.

Although broadcast television is considered a high quality professional system, its ability to produce pictures of high resolution have been limited by the restricted bandwidth allowed for broadcast television signals on most systems standardized by International Radio Consultative Committee (CCIR) and other bodies. The number of channels available for broadcasting are limited. Cable Television (CATV) systems have no such restriction and may have very high resolution capabilities because of the use of a wider bandwidth.

1.6 BROADCAST INFORMATION SERVICES

Advances in microelectronics have accelerated the developments and introduction of new communication methods for visual information. Existing television channels are now being used in many countries to provide visual text or graphic information of public interest, directly in unused channels or during the vertical blanking interval without interfering with the normal TV program. These *view data* or *videotext* information systems have different versions in different countries—Teletext, Ceefax, Oracle, Videotext, Prestel, Antiope and so on, providing one-way or interactive information exchange.

With the widespread use of computers for information processing, the TV receiver or monitor screen has been most suitable for temporary display of the information for viewing and editing, before taking a hard copy on paper in a printer. In low cost home computers, the video information is modulated on a standard channel to suit the antenna input of a TV receiver. A TV monitor with enhanced video bandwidth is used in professional computers and PCs to improve resolution to serve as the CRT terminal or video display unit (VDU).

Teletext service is *a one-way* system based on broadcast or cable TV system, while *videotext* or *viewdata* as often called, is based on a *two-way* switched telecommunication network. In teletext, screen-fulls of video data are transmitted as coded digital data to suitably equipped television receivers. A large number of such frames are continuously transmitted in a cyclic sequence from central information storage facility, during the vertical blanking interval of the television channel, or in an entire unused television channel. The frames desired to be viewed can be selected by the viewer by entering frame identification numbers on a key pad accessory.

Videotext or viewdata is a *two-way* information service using colour TV receiver as video display unit. The information is transmitted bidirectionally as modulated data via a telephone line system, from the information source, which is a *computer data base*. The information can be displayed on modified television receivers or CRT terminals/VDUs. A subscriber to the database can select the desired information via the keypad accessory for the adapted TV set, or keyboard of the video monitor. The system is interactive, and has the possibility of user-to-user communication. There is instant communication with the data base but it has limited simultaneous use capability. *Prestel* was the first videotex service launched in UK, and is emulated by some countries while others are developing their own, with their proprietary names— Teletel in France, Teledon in Canada, Viewtron in USA, Captain in Japan and so on.

1.7 HIGH DEFINITION TELEVISION (HDTV)

Since the late seventies, considerable research interest has been created in developing a television system that could compete with the high definition and the quality of cinematographic images produced by the 35 mm colour motion picture film, by the pioneering efforts of Japan's Broadcasting company, NHK. HDTV uses nearly twice as many scan lines as standard TV and about 5 times the picture details in a wide-screen format. The advantages of video equipment over photographic cine-film equipment in producing and editing programs have given a big boost to HDTV systems for studio use. In the USA and Canada, the road to broadcasting HDTV is seen as an evolutionary process, in gradual steps of improved definition television (IDTV) and extended definition television (EDTV).

In 1988, the Seoul Olympic games were broadcast throughout Japan in HDTV using INTELSAT-V and the direct broadcasting satellite BS-2b. Regular direct broadcast satellite programmes in HDTV are slated for 1990, while attempts to set international HDTV standards have not been successful, because of the 30 v. 25 frame rate difference in the US-Japanese and European standards and the desire to provide compatibility with existing systems. A variety of such systems proposed are under study. The Japanese will send one signal for the old sets and another HDTV signal for the new. HDTV requires a broadcast space five times the ordinary TV channels, making it uneconomic. Some proposals are based on the concept of using an additional channel for HDTV signal details. Compressed bandwidth HDTV systems are under study to make the bandwidth requirements less stringent. HDTV promises the biggest changes in the media in the coming decade. In this decade wide-band fibre optics technology could be used to transmit an unlimited number of channels on laser-produced light via optical fibres and finely threaded glass cabling, offering an unlimited array of programmes to the viewer. According to Steven Benton of MIT, stereoscopic 3-D TV, using holographic techniques, could also be a reality in the coming decade.

Large Screen TV

Large screen television systems have been developed, based mostly on the projection technique, that serve to expand images to life-size or greater. Present systems offer a resolution of 650 TV lines or 1000 lines on RGB for computer applications on screen sizes over 3 m. Flat screen high resolution CRTs have been specially designed to large sizes up to 50×50 cm. Other flat panel displays using other technologies like electroluminescence, gas discharge plasma, light emitting diodes and liquid crystal displays based on pixel excitation, are also being developed. The costs of all these are yet too high for home use. The development of a high resolution large screen display of about 1 m at an affordable price is essential for the development of HDTV broadcasting.

1.8 DIGITAL TELEVISION

As VLSI circuits incorporating more and more complex signal processing functions are being made available, digital technology is now being increasingly applied to video systems, offering several technical and economic advantages. In digital TV, the analog picture signal is converted into digital form by sampling it at a high rate of 3 or 4 times the colour subcarrier frequency and converting it into digital form by an analog-to-digital converter. Typically, this is done by quantizing the signal in 256 levels, each sampled signal being represented by an equivalent 8-bit binary number. A chain of such 1/0 bit streams represents the video signal in digital form. At the receiver the analog form is recovered from the digital by a suitable digital-to-analog converter.

As the digital signal is in a 1 or 0 bit form, there is no degradation of the video signal due to noise or distortion while processing it in the studio chain, VCRs or in the receiver. Many repeaters can be introduced in the path or video tapes can be replicated in succession without visible deterioration of the picture. The digital bit stream can be stored as long as necessary to introduce signal corrections for stable synchronisation or for TV standards-conversion of the signal.

The bandwidth requirement of the digitized video signal is much larger, as the bit rate is 8 times the sampling rate used, which is 3 or 4 times the colour subcarrier frequency. This comes to about 50 MHz, which is too large for a ground based broadcast system. The sampling rate is very high requiring fast analog-to-digital converters. Satellite based systems or optical fibre based systems could possibly afford the large bandwidth requirement of digital video or audio transmission as they do not have the bandwidth limitation. Digital techniques are however used in various forms in the video equipments in the TV studios and also in TV receivers for improved signal processing. These provide easier time base correction, subtitling, special effects, noise reduction and give a better picture quality.

MAC Encoding

In the new broadcast media, such as cable TV or satellite TV, the restriction on the bandwidth is not as serious as in terrestrial VHF/UHF broadcasts. New encoding systems have been developed to provide higher quality of picture and get over some of the problems of spectrum sharing between the luminance and the colour components creating problems like cross colour or cross luminance. These include extended PAL, Multiplexed Analog Components (C-MAC, D-MAC, D2-MAC) systems developed in Europe for satellite broadcasting. MAC technique time compresses luminance, chrominance and sound signals, and then buffers and transmits them consecutively within the same line width, eliminating the cross signal interference.

Television is thus poised for many revolutionary technological changes, and as the most powerful medium of mass communication, would be the mainstay of the several innovations in consumer electronics.

REVIEW QUESTIONS

1.1 Draw the schematic of an elementary television system based on mechanical scanning. Draw the scanning pattern that ensues.
1.2 What is the principle of the colour sequential system? Why was it not accepted for the colour television system? Does it have any advantages?
1.3 Enumerate the various applications of CCTV systems, bringing out special advantages in their use.
1.4 Why is cable TV increasingly popular in advanced countries?
1.5 What is the impact of satellite communication on television broadcasting?
1.6 Explain the techniques of large screen television display systems, bringing out their limitations.
1.7 Explain the use of TV for display of public utility information through broadcasts or computer database.
1.8 What is the basic principle of digital television? Indicate the merits of digital TV?

2

The Basic Television System and Scanning Principles

The basic technique of converting an optical image into electrical signals, sending them over wires or through space and receiving them to reproduce the image on a screen is discussed in this chapter. The conversion technique is illustrated by the construction of the iconoscope, a camera pick-up tube which was commonly used in TV studies for over 20 years in the early period of the development of television. A modern solid state image sensor is also described. This has an altogether different scanning technique. marking elimination of 'tubes' from the TV camera.

2.1 SOUND AND PICTURE TRANSMISSION

In television, the basic problem is to devise a system by which one can transmit both sound and pictures and receive them over long distances. The technique for transmission of sound is rather simple. Sound waves are mechanical waves produced in the air by vibrating elements like reeds in the voice box of the speaker, or vibrating diaphragms, strings, etc., and are sensed by ears as audio signals varying with respect to time in the frequency range of 20 to 20,000 Hz. For transmission of sound over long distances, the sound waves are converted with the aid of a microphone into electrical signals so that they can be sent over wires as in a telephone. If these audio signals are translated into radio frequencies by modulation, they can be radiated into space by feeding the RF energy to suitable aerials. At the receiver, the RF signal is amplified and the low frequency audio signals are separated by demodulation. These are further amplified to drive a sound reproducing device like a loudspeaker, which converts them back into sound.

Transmitting a picture by converting it into electrical signals is not so simple because the picture has both space and time variations of brightness information contained in it. It has a different optical brightness at each different point in it, and this goes on varying with respect to time, in live scenes or in movies. The human eye can sense all the elements of the picture projected on its retina with the aid of the many cone and rod sensors and the optical nerves communicating with the brain. Simultaneous transmission of the brightness information existing all over the points in the picture and receiving it on an equal number of point light sources is very difficult, practically impossible. Such a parallel television system would require as many sensors to sense brightness levels at the various points and seed the information in parallel to the reproducing elements, via as many electrical wires or channels. In order to simplify the process, the sequential 'scanning' technique is used to convert the picture information into a single valued function of time. This makes it possible to transmit the picture information over a single pair of wires serving as a channel.

2.2 THE SCANNING PROCESS

The scanning technique is a process similar to the reading of written information on a page, starting at the top left, and progressing line by line downwards to the end at the bottom right. The picture is not instantaneously read off; it takes a finite though small scanning time. The brightness information at the picture elements in the picture is converted into electrical signals at a photosensitive surface exposed to the optical image of the picture on one side and scanned by an electron beam on the other side.

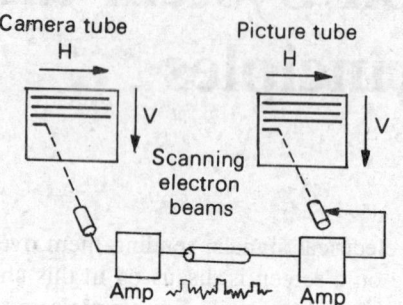

Fig. 2.1 Scanning process

The scanning is done line by line, horizontally from left to right at a fast rate, and vertically from top to bottom at a slower rate as illustrated in Fig. 2.1. The retrace of the beam is very fast compared to the forward scan, and scanning during retrace or flyback is blanked by cutting off the beam during the horizontal and vertical flyback intervals. The picture is thus scanned for its brightness information over a finite number of lines in it. The larger the number of lines, the greater are the details of the picture scanned.

During reproduction, an exactly similar scanning process is used synchronous with that at the sending end. The scanning electron beam in the picture tube 'paints' the television picture on the viewing screen by producing proportional glow at the corresponding points. The scanning rate is fast enough to create an illusion of continuity because of the persistence of vision of the human eye.

2.3 CAMERA PICK-UP DEVICES

The scene or the picture to be televised is focused with the help of a lens system onto a photosensitive target near the faceplate of a camera pick-up tube to form an accurate, well-defined image on it. The surface of the target plate may be regarded as a mosaic formed by a large number of elementary areas. As the surface is photosensitive, the electrical state of each minute area varies in accordance with the intensity of light falling on it at a particular instant. The minute photoelectrical elements on the surface of the target plate produce an electrical pattern corresponding to the picture illumination.

The electrical response of each minute element is read off with the help of an electron beam circuit to produce electrical impulses. The elements are scanned in a predetermined sequence and speed so that an electrical waveform with respect to time is generated to represent the brightness information at various points in the picture. The target plate is held at a positive potential with respect to the cathode of the pick-up tube and the actual beam current varies in accordance with the electrical state of the picture element (pixel) being scanned at a given instant, thus producing a varying voltage across a resistor through which the beam current is made to pass.

The beam scans the image horizontally by means of a magnetic field set up by the horizontal deflection coils which are supplied with the required sawtooth currents for linear deflection. Simultaneously the beam proceeds vertically downwards slowly by means of the magnetic field of the vertical deflection coils which are also supplied suitable sawtooth currents. The scanning must be done at a fast speed over and over again so that the changing or moving pictures can be considered as steady for the scanning period.

Photoelectric Conversion

The photosensitive target plate converts the variations in the intensity of light into variations in electrical signals utilizing: (i) a photoemissive surface, or (ii) a photoconductive coating or layer.

(i) Photoemissive surfaces Light is in the form of bundles or packets of electromagnetic energy called photons that have energy inversely proportional to the wavelength. The photons of light have energy adequate to dislodge electrons from surfaces made of certain metals like caesium silver, bismuth, lithium, etc. belonging to the alkali group. Photoemissive surfaces of this type are used in some camera tubes, viz. the iconoscope and the image orthicon. The photoelectric efficiency of such surfaces expressed as the ratio of the number of photoelectrons emitted to the number of incident photons is referred to as 'quantum efficiency', a parameter useful for comparing of photosensitivities of different surfaces.

(ii) Photoconductive coatings or layers The resistance of semiconducting materials like selenium, antimony trisulphide and lead oxides is affected by the light incident on it. The photons of light are absorbed by the semiconducting materials to generate electron-hole pairs as charge carriers. This reduces the resistance or increases the conductivity of the material, proportional to the incident light. Photoconductive coatings of these types are used in targets of camera tubes like vidicon and plumbicon.

To illustrate the basic construction of a camera pick-up tube, the iconoscope pick-up tube is discussed here, although the tube is now only of historical significance. This is followed by a description of the solid state image sensor that represents a breakthrough in solid state pick-up devices.

Iconoscope

This was one of the first electronic pick-up tubes developed by Zworykin in the late 20's for use in the all-electronic television system. It made use of electron beam scanning of a photoemissive mosaic surface which had charge storage properties. The schematic construction of the tube is shown in Fig. 2.2.

The mosaic surface consists of a thin sheet of mica, one side of which is made photosensitive by being embedded with minute separate globules of caesium-silver compound. The other side is coated by a conducting metal film, so that each tiny globule forms a small capacitor with the metal film, the mica sheet forming the dielectric in between. The image is focused upon the mosaic surface by means of the optical lens. Light falling in the mosaic globules causes them to emit electrons,

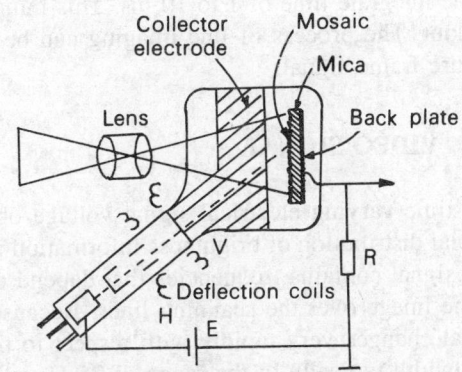

Fig. 2.2 Iconoscope pick-up tube

most of which are drawn away by the collector electrodes. This leaves a positive charge pattern on the mosaic proportional to the light intensity distribution. Since the globules are discrete (separated from each other), the charge distribution stores itself across the small globule capacitors. A high velocity electron beam from the electron gun scans the mosaic, impinging on each globule to restore the charge lost by photoelectric emission. In this process, is sends a current pulse in the beam circuit as each globule is capacitively coupled to the back electrode that gives out a signal voltage across resistance R.

In practice, the uneven secondary emission due to the high velocity scanning beam, and the charge leakage between adjacent globules tend to make the iconoscope response nonlinear and limit its sensitivity. This led to the development of low velocity scanning tubes like the orthicon and the image orthicon with considerable improvement in sensitivity and characteristics. However, in the early years of television, until the early 40's, the iconoscope was the most widely used pick-up tube for TV cameras.

Solid State Image Sensor

The most recent that is poised to get rid of the TV camera 'tube' and the electron beam scanning is the solid state image sensor. The device uses a charge coupling principle for transfer of charge from one picture element to the next with the help of a series of electrodes, and is hence called a charge coupled device (CCD) image sensor. This consists of a Metal Oxide Semiconductor (MOS) transistor family type silicon chip with an array of metal electrodes in groups of three electrodes that are provided with three-phase clock pulses for shifting the charge packets in the silicon substrate, proportional to the illumination at the surface.

As the optical image is focused on the front face of the CCD silicon chip, a corresponding charge pattern is formed in the substrate due to electrons produced by the light collecting beneath the central electrodes that are the most positive of the three sets of electrodes. The spot under each triplet set of electrodes forms a picture element. Scanning is done by shifting the charge packets by phasing the clock pulses to the three electrodes to be in succession. The charge pattern is thus clocked out as a series of pulses of varying amplitude according to the original optical image.

The process of scanning takes place in two steps, an integrate period and a readout period. During the integrate period, one set of electrodes is maintained at a constant voltage creating a depletion region under each electrode. Electron-hole pairs are generated within the depletion region under each electrode. Electron-hole pairs generated within the depletion region are collected under each electrode in the integrate period and are read out by the phased clock pulses. The readout time is around 100 μs, short compared to the integrate time of 1 to 10 ms. This minimizes smear effects due to changes in illumination during readout. The process of line imaging can be extended to the frame imaging system for forming a TV picture frame signal.

2.4 VIDEO SIGNAL

The time varying electrical signal voltage obtained from the TV camera tube circuit representing the special distribution of brightness information in the image is called the picture signal or the video signal. The signal contains frequencies that depend on the scanning speed and the variations in the brightness of the image over the scanning lines. Because of the fast scanning rate used in television systems, the signal changes very rapidly with respect to time and hence contains a wide range of frequencies with bandwidth typically in the range of 25 Hz to 5 MHz. The weak signal obtained from the camera pick-up tube is amplified in video amplifiers which are RC-coupled amplifiers with extended frequency response.

The video signal must, in addition, contain synchronising pulses as signals that identify the retrace or flyback of the scanning beam and can be used to control the scanning rate at the receiver so that the scanning at the receiver and the transmitter takes place in synchronism. The retrace should no be visible on the receiver screen and is hence blanked by pulses that cut off the electron beam in the picture tube as well as in the camera pick-up tube. The blanking and sync pulses are thus inserted during the retrace period of the scanning to form a composite video signal.

2.5 TRANSMISSION AND RECEPTION OF VIDEO SIGNALS

The camera video signal thus amplified and mixed with sync and blanking pulses can be transmitted to the receiving end over a cable in closed circuit television or is used to amplitude modulate a radio frequency (RF) carrier which can be radiated into space with the help of an antenna for transmission over long distances. The large bandwidth of about 5 MHz of the video signal necessitates the use of very high frequencies in the very high frequency (VHF) and the ultra high frequency (UHF) ranges since the carrier frequencies must be greater than the modulating 5 MHz signals. The bandwidth occupied by each TV channel is also consequently large, greater than 5 MHz. Propagation at the VHF and UHF range frequencies is normally restricted to line-of-sight distance by the space waves and hence the range of TV reception is also limited.

In television transmission, along with the picture, sound also is to be transmitted. Another RF carrier is used for transmitting the sound. Frequency modulation (FM) is more commonly employed for sound carriers as it offers noise-free reception with much less power than in amplitude modulation (AM), and can provide high fidelity up to 15 kHz if adequate bandwidth is used. Frequency modulation is not suitable for video signals as the ghost interference due to multipath reception of FM TV signals is more disturbing and annoying than in AM, besides having a larger bandwidth requirement.

In FM, the beat between two frequencies of the same signal arriving *via* two paths with different propagation delays produce very objectionable interference patterns in the form of multiple ghosts of Moire patterns. Spacing of the Moire or the frequency of repetition of ghosts depends upon the contrast ratio between adjacent areas, thus varying with the movements or changes in the scene.

FM has, however, the important advantage of noise interference reduction, and is hence employed in microwave relay or satellite links for TV broadcasts where multiple path reception is avoided by highly directional antennas for transmission and reception.

The block diagram of a basic television system is given in Fig. 2.3.

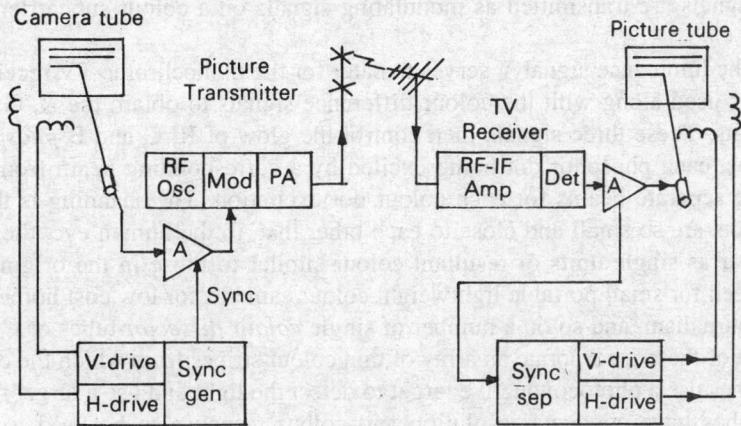

Fig. 2.3 Basic television system

The video signal from the camera pick-up tube, amplified and mixed with the blanking and sync signals from the sync pulse generator, is supplied to the AM visual or picture transmitter. The transmitter does not transmit both the sidebands as usual. In order to keep down the bandwidth of the channel, it transmits one sideband fully, while the other sideband is transmitted partially as a 'vestigial sideband'

The sound section of the TV system is separate. The audio signals from the microphone are amplified suitably and are used to frequency modulate the carrier of the aural or sound transmitter. The power outputs of the picture and sound transmitters are combined in a diplexer and fed to a common transmitting antenna system to be radiated together.

Reception

At the receiving end, the picture is reproduced on a cathode ray picture tube screen with the help of the focused electron beam of varying intensity falling on the screen to produce a proportional glow at the various positions on the phosphor coating of the screen.

The beam intensity is controlled by the picture video signal brought over cables in the CCTV or obtained by the demodulation of the RF waves received from the antenna and amplified by RF/IF amplifiers. The beam also scans the picture raster in exact synchronism with the beam at the camera tube so that the intensity of the glow produced at the various positions tallies with that of the corresponding picture element on the picture scanned at the camera. The scanning by the beam spot is so fast that the human eye is unable to follow the spot movement and because of persistence of vision, views the picture as a continuous one.

2.6 PRINCIPLE AND WORKING OF COLOUR TELEVISION

If colour reproduction of a picture is desired, the colour information at each picture element (in the form of the relative content of the component primary colours red, green and blue) must be found and proportional electrical signals sent as chrominance or colour information. In modern colour television systems, the same scene is supplied to three camera tubes, each made responsive to one of the three primary colours, by passing the optical rays through corresponding filters before they strike the target plates. The electrical signals proportional to the red, green and blue colours have characteristic amplitudes and phases, and are combined to produce: (i) a *luminance* signal Y, proportional to the monochrome brightness signal, and (ii) two colour-difference signals (R–Y) and (B–Y), called the *chrominance* signals. The chrominance signals are transmitted as modulating signals on a colour subcarrier within the video band itself.

At the receiver, the luminance signal Y serves to cater for the monochrome TV receivers, while in the colour receivers it is used along with the colour difference signals to obtain the R, G and B signals by suitable adder circuits. These three signals then control the glow of R, G and B *phosphor dot-triads* in a colour picture tube, each phosphor dot being excited by a corresponding beam from a triple electron beam gun providing separate beams for each colour dot excitation. The scanning is thus a triple beam scanning, and the dots are so small and close to each other that, to the human eye, the tricolour units of phosphor dots appear as single units of resultant colour similar to those in the original scene.

Because of the need for small portable lightweight colour cameras for low cost home use, professional use for electronic journalism, and so on a number of single *colour dissector* tubes or CCD cameras have been designed. Most of these incorporate an array of thin colour stripes from which the colour information is derived. These tubes use a photoconductive target to detect the light and use a target readout technique. Modern cameras tubes have adequate resolution and colour response to be used to image the scene through proper *colour stripe filter array*, which rearranges this image onto the faceplate target of the tube. Typically this array consists of two sets of different colour stripes inclined to each other so that two different colour carrier frequencies are developed as the beam scans across the periodic array structure, and three colour information can be derived.

2.7 CHARACTERISTICS OF THE HUMAN EYE

In formulating the requirements of camera tube scanning and reproducing systems, the characteristics of the human eye must be taken into consideration. These characteristics are: (i) visual acuity—the ability to resolve finer details in a picture; (ii) persistence of vision; and (iii) brightness and colour sensation. Visual acuity or resolution determines the number of scanning lines, persistence of vision determines the rate of scanning, and the brightness and colour sensation characteristics affect the camera tube and the picture tube light transfer and spectral response requirements. Colour television systems in particular have been formulated taking into account the colour sensation characteristics and limitations of the human eye in colour perception.

Resolution

As shown in Fig. 2.4, the lens system of the eye focuses the optical image of the object being observed on the retina which contains light sensitive cellular structures of two kinds, the rods and the cones. It is believed that the rods sense primarily the brightness levels, including very faint impressions, while the cones are mainly responsible for colour perception. There are some 6,500,000 cones and about 100,000,000 rods connected to the brain through some 800,000 optic nerve fibres. Sharpest vision is obtained in the small central region in the retina called *fovea*, just opposite the lens. Within the fovea and the area around it, called *macula lutea*, the predominant light sensitive cells are the cones while the rest of the peripheral area contains mostly rods. Within the fovea, each individual cone has independent transmission paths to the optic nerves and the brain. Away from the fovea, however, each optic nerve fibre is connected to two or three rods or cones. At low brightness levels, the stimulus given to the single rod or cone is not sufficient to produce a measurable sensation, in which case the sensation from a group of cells is combined to produce effective sensation but, naturally, with loss of detail.

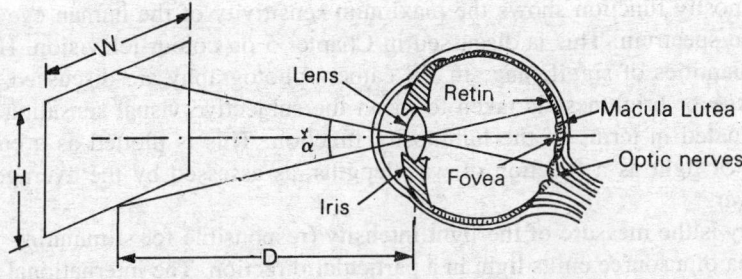

Fig. 2.4 Human eyes optics

The resolving power of the human eye depends upon where the image is formed on the retina, being highest at the fovea where the cones have individual fibre paths. The ultimate limit of resolution of details in the image, called acuity of vision, corresponds, therefore, to the separation between the cones and rod elements. This distance of separation subtends an angle of about 1 minute at the centre of the lens which then is the ultimate limit of resolution of the human eye. This means a resolution of 1 mm at a distance of about 3 metres.

There is also a limit to the resolving power of the eye due to the diffraction effects caused by the finite sized aperture (iris opening D) and the wavelength of light λ. The minimum viewing angle of resolution is given by 1.22 λ/D radians. This limit, however, is lower than that due to the rod-cone structure of the eye.

The minimum resolvable angle of the eye depends upon the scene brightness and the contrast ratio in the scene, with the limiting resolving angle of about 1 minute. Contrast ratio is the ratio of the differential brightness of a small gray object to its surrounding background brightness level. The contrast ratio C, the scene brightness B, and the minimum resolvable angle α_0 are found to be related by the equation:

$$B = \text{Constant}/C^2\alpha_0^2 \tag{2.1}$$

Field of Vision

As the density of nerve fibres falls off, as one moves away from the fovea, the resolving power in the peripheral region decreases proportionally. The decrease is quite rapid, falling off to about 20% at an angle of $\pm 10°$ with the main axis. Objects outside this appear blurred. The main field of vision of the eye is defined by the vertical and horizontal viewing angles of 30 to 40°, beyond which the visual acuity falls off to more than 15% of the maximum at the fovea. The extended field of vision beyond these angles has very poor resolution and serves primarily as a view finder. It is also useful in dark adapted vision, and in augmenting colour impressions of a scene.

2.8 BRIGHTNESS PERCEPTION AND PHOTOMETRIC QUANTITIES

Visible light is the range of the electromagnetic spectrum of radiant energy that can be perceived by the human eye. This has a range of wavelengths from about 0.4 to 0.7 μm. The effect of visible (luminous) energy on the eye depends upon the wavelength and is expressed as the relative luminosity of monochrome radiation. This spectral response curve varies from observer to observer, but for dealing with photometric quantities, the response has been standardized for an 'average' observer in the form of the *luminosity function* based on results of tests by the Commission Internationale de 1' Eclairage (CIE).

The relative luminosity function shows the maximum sensitivity of the human eye in the green and yellow regions of the spectrum. This is discussed in Chapter 5 on colour television. Here, some of the basic photometric quantities of significance in TV camera photography are discussed.

The term *luminosity* or brightness is taken to mean the subjective visual sensation of brightness of light or colour, evaluated in terms of the luminosity function. This is plotted as a curve representing relative impressions of light as a function of wavelengths, as assessed by the average observer, for a constant radiant power.

Luminous intensity is the measure of the light intensity (responsible for stimulating visual sensation) at which the total area of a source emits light in a particular direction. The international unit of luminous intensity is the *candela* (cd) which is equivalent to 1/60th of the luminous intensity radiated by one centimetre of a black body (full radiator) at the temperature of solidification of platinum (1774°C), at right angles to its surface. This quantity gives no information about the total amount of light flux emitted by a source in all directions. Light intensity of a source may be different in different directions depending upon the area of the source, reflector employed, etc.

Luminous flux is the radiated luminous power (power of visible light) expressed in terms of its effect on the average or normal human eye. The unit of luminous flux is *lumen* (lm) and is the luminous flux emitted per unit solid angle (steradian) by a uniform point source of one candela.

Luminance is the quantity of light intensity emitted per square centimetre of an illuminated area. Luminance is thus an attribute of a surface emitting or reflecting it (in contrast to the term illuminance which expresses light received by the surface). The international unit of luminance is the stilb (sb) which is the luminance of a surface having luminous intensity of one candela per square centimetre. A smaller unit is the *nit* which is the luminance of a surface emitting one candela of luminous intensity per square metre. One stilb is thus equal to 10,000 nits. Some countries use the unit of foot-lambert which is equivalent to 3.43 nits. A foot-lambert is the luminance of a surface emitting or reflecting light at the rate of one lumen per square foot.

Illumination or *illuminance* is the average luminous flux incident onto a surface. Its unit is *lux* (lx), defined as the illuminance of a surface area of 1 square metre, over which a luminous flux of 1 lumen is uniformly distributed. It is also expressed in foot-candles as the illuminance of a surface area of one square foot, over which a luminous flux of one lumen is uniformly distributed. One foot-candle is equivalent to 10 lx.

Illuminance is not directly related to visual sensation due to the brightness or luminance of the illuminated surface, which depends upon the reflectivity of the surface. Illumination is the only quantity in light engineering that can be measured directly by means of photoelectric meters. A summary of the commonly used photometric equivalents is given in Table 2.1.

Table 2.1 Photometric Equivalents

Quantity	*Photometric unit*	*Equivalent unit based on lumen as unit of flux*
Luminous intensity	1 candela (cd)	1 lm/ster
Surface luminance	1 stilb = 1 cd/cm^2	1 $lm/ster/cm^2$
	= 10^4 cd/m^2	
	1 nit = 1 cd/m^2	10^{-4} $lm/ster/cm^2$
	1 ft-lambert	$1/\pi$ $lm/ster/ft^2$
	= 3.43 nits	
Illuminance of a surface	1 lux	1 lm/m^2
	1 ft-candle = 10 lx	1 lm/ft^2
Luminous energy	1 talbot	1 lm sec

Typical values of some photometric quantities are given in Table 2.2, to give a visual idea of the units employed.

Table 2.2 Typical Values of Photometric Units

Example	*Illuminance (lx)*	*Luminance (nits)*
Motion picture screen	40 to 200	10 to 30
Picture tube screen	—	40 to 80
TV studio	750 to 2000	—
Reading a page	20	—
Lamp filament	—	10^7

Picture Contrast and Tonal Gradations

In nature the range of brightness levels is very large, almost infinite from pitch darkness of a closed cave to the brilliance of sunlight. By adoption of the iris, the human eye can adapt to a very wide range of luminance levels, the upper limit being up to about 50,000 nits. The lowest discernible luminance is about 0.03 nits ordinarily, although in dark-adapted vision the eye can discern levels below 10^{-5} nits.

The visual sensation of brightness to the human eye is a logarithmic function of the picture luminance over the restricted range adaptable to the iris adjustment, as expressed by the *Weber-Fechner law*:

$$\Delta S = K \log B_1/B_2 \tag{2.2}$$

where B_1 and B_2 are two luminance levels and ΔS is the incremental visual sensation due to them. If a value of ΔB is established for which the brightnesses of two adjacent areas are just noticeably different, the ratio $\Delta B/B$ is known as *Weber's fraction*. Actually the fraction is dependent on the brightness level and on the viewer's state of adaptation to it. It has a nearly constant value of about 0.02 over a brightness range of about 1 to 300 nits, while at very low levels of brightness, the Weber's fraction is relatively larger.

The contrast in a picture or scene is expressed as a ratio of the maximum to the minimum luminance B_{max}/B_{min}. The *contrast ratio* discernible to the eye depends upon the level of ambient light. In bright daylight illumination, the eye can accommodate a contrast range of up to 1000 : 1, while at lower levels of illumination it can make out a range of about 10 : 1. The darkest part in a picture cannot be darker than the ambient luminance from the unenergized screen, due to ambient light externally falling on it and the internal scattered light. The reflectance of the TV screen may over 25%. Such ambient light is always present and is in fact recommended to reduce eye strain. In a TV picture the contrast ratios are less, in the range of 10 to 40, while in a well designed dark theatre a contrast range of over 100 can be obtained. For the normal illuminance of the TV screen, of 1 to 40 nits, the incremental contrast sensitivity of the eye is around 0.03. With contrast ratios of 10 to 40, the eye can notice a maximum number of tonal gradations of 80 to 130, 100 on the average.

2.9 ASPECT RATIO AND RECTANGULAR SCANNING

Subjective tests have indicated that best viewing comfort, *panoramic effect* and artistic appreciation are obtained when the picture raster has a rectangular format with an aspect ratio, i.e. width-to-height ratio, of 4 : 3. This is because of the binocular vision due to the pair of eyes in the horizontal plane as compared to that in the vertical plane. Also the fovea, the region of maximum resolution at the centre of the retina, has greater area along the width than along the height. A larger width of the raster ensures a more efficient use of the area of the fovea. In motion picture industry, the aspect ratio of 4 : 3 had been accepted, until the advent of the cinerama and stereophony technique. In television systems also, the picture format with an aspect ratio of 4 : 3 is commonly employed as it is most pleasing aesthetically and less fatiguing to the eye.

For *High Definition Televisions* intended to give better quality wide screen pictures a larger aspect ratio has been favoured. The Japanese Broadcasting Corporation NHK who pioneered work in HDTV, conducted psychological experiments on the size of the screen, the aspect ratio, the angle of vision, for giving the sense of impact and reality. A wider field of vision was required to evoke the visual psychological effects of high level viewing, especially during large screen projection. They found a larger aspect ratio desirable and selected 5 : 3.

An aspect ratio of 16 : 9 is favoured to give greater flexibility in shooting and releasing TV programs. By using a *shoot and protect* scheme with an aspect ratio of 16 : 9, releases could be made in any aspect ratio between 4 : 3 and 2.35 : 1, as shown in Fig. 2.5. If the master shooting has a 16 : 9 aspect ratio, a 4 : 3 aspect ratio release uses the full height of the master and appropriate width. A release with 2.31 : 1 aspect ratio can use full width of the master and appropriate height. Other releases can use either full height or full width. Critical portions are contained in the inner rectangular while the extended rectangle forms the 16 : 9 aspect ratio master, as seen in Fig. 2.5.

Rectangular Scanning

For picture raster of rectangular shape, rectilinear scanning is most convenient. There are two scanning procedures taking place simultaneously, one moving the beam horizontally from left to right at a fast rate

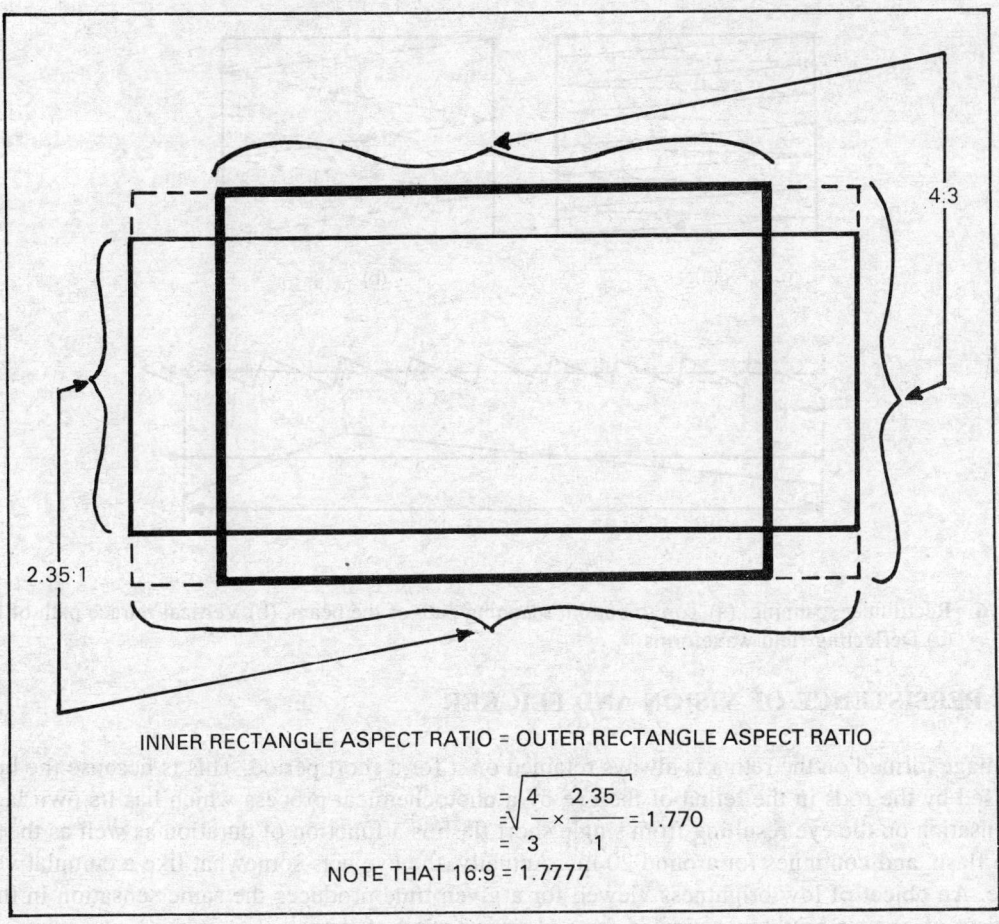

INNER RECTANGLE ASPECT RATIO = OUTER RECTANGLE ASPECT RATIO

$$= \sqrt{\frac{4}{3} \times \frac{2.35}{1}} = 1.770$$

NOTE THAT 16:9 = 1.7777

Fig. 2.5   Shoot and protect scheme with 16 : 9 aspect ratio

and the other moving the beam vertically downwards at a slower rate. Movements of the beam are at constant speeds during the forward and the downward scans and the scanning is thus linear in both directions.

Figure 2.6 illustrates the beam paths and deflection fields for reproducing the picture on the picture tube screen. The beam travels in both horizontal and vertical directions as shown by the continuous lines. The broken lines indicate the retrace which occurs rapidly with respect to the forward scan and is normally blanked by cutting off the beam in the picture tube as well as the camera tube, both during the horizontal and vertical retrace.

For each vertical scan, the picture is assumed stationary, and is scanned at a fast rate along a number of horizontally slanting lines. For good resolution, the number of lines scanned per picture must be large enough. The scanning lines must not be disturbingly visible to an average viewer at a distance of about five to six times the picture height. A relatively low vertical scanning rate is desirable because it allows the beam to scan a large number of horizontal lines without reaching an inordinately high horizontal scanning frequency. The lower limit to the vertical scanning rate is set by the limit of persistence of vision of the human eye.

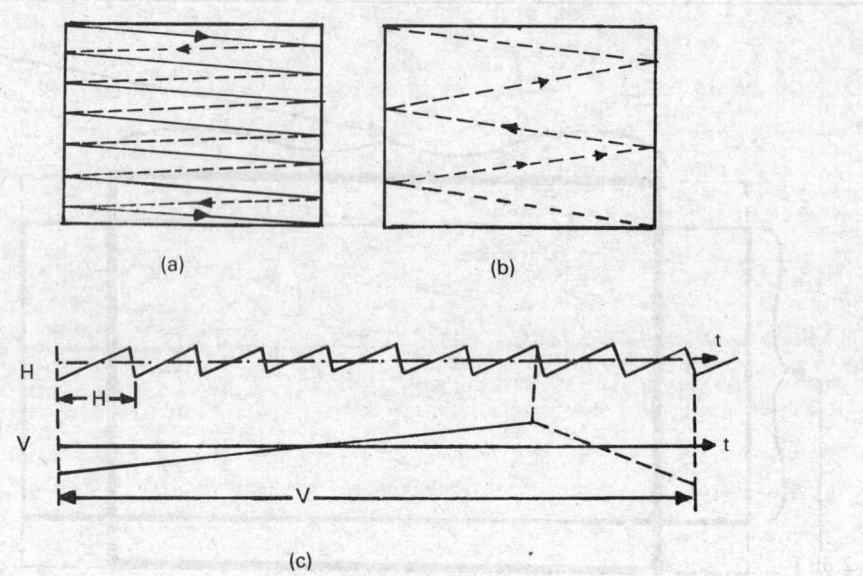

Fig. 2.6   Rectilinear scanning: (a) Top to bottom scanning path of the beam, (b) Vertical retrace path of the beam, (c) Deflecting field waveforms

2.10   PERSISTENCE OF VISION AND FLICKER

The image formed on the retina is always retained on it for a short period. This is because the brightness is sensed by the rods in the retina of the eye by a photochemical process which has its own lag. Hence the sensation on the eye resulting from single short flash is a function of duration as well as the intensity of the flash, and continues for around 20 ms. Actually, the eye acts somewhat like a cumulative storage device. An object of low brightness viewed for a given time produces the same sensation in the eye as an object of greater brightness viewed for a shorter period of time.

For intermittent flashes of light incident on the retina, a vigorous photochemical process continues for a fraction of a second (around 20 ms) even after the stimulus has disappeared. The continuation of the photochemical process means continuation of brightness impression in the visual centre of the brain and is called 'persistence of vision'. This is used in cinema and television in obtaining the illusion of continuity by means of rapidly flashing picture frames. If the flashing is fast enough, the flicker is not observed and the flashes appear continuous. The repetition rate of the flashes at and above which the flicker effect disappears is called the 'critical flicker frequency' (CFF). This is dependent on the brightness level and the colour spectrum of the light source. Because the human eye has the greatest sensitivity to yellow-green light, the flicker effect is maximum in that region.

In *cinema*, the classical example of this effect, a film speed of 16 frames per second was used in earlier films to obtain the illusion of movement. Lack of smooth movement was noticeable in these films. The present-day standard for movie film speed is 24 frames per second and at this speed, these effects are very much reduced. The flicker problem is further reduced in modern projectors by causing each frame to be illuminated twice during the interval it is shown, by means of fan blades. The resulting flicker rate is quite acceptable for cine-screen projection, because it is viewed in subdued light and a wide angle display area.

In *television*, the field rate is concerned with (i) large area flicker, (ii) smoothness of motion, and (iii)

motion blur in the reproduced picture. As the field race is increased, these parameters show improvements but tend to saturate beyond 60 Hz. Further increase does not pay off, and increases the bandwidth. Hence the picture field scanning is generally done at the same rate as the mains power supply frequency, which conveniently happens to be 50 or 60 Hz for the same reasons of reducing illumination flicker from electric lamps. At 60 Hz, the flicker is practically absent, while at 50 Hz, a certain amount of borderline flicker may be noticed at high brightness levels used to overcome surrounding ambient light conditions. Use of the frequency of the power mains for vertical scanning reduces possible effects like supply ripple and 50 Hz magnetic fields, in the reproduced picture.

2.11 VERTICAL RESOLUTION

The ability of a scanning system to resolve vertical details in a scene depends upon the number of scanning lines used per frame. This is generally referred to as the scan ratio of a television system. The realistic limit to the number of lines is set by the resolving capability of the human eye, viz. about one minute of visual angle. For comfortable viewing, an angle of about 10 to 15° can be taken as the optimum visual angle. Hence the best viewing distance for watching television is about 4 to 8 times the height of the picture, i.e. a visual angle of about 10° as shown in Fig. 2.7.

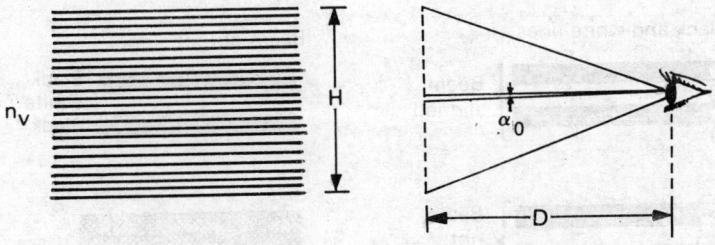

Fig. 2.7 Minimum number of scanning lines

The maximum number of dark and white elements which can be resolved by the human eye in the vertical direction in a screen of height H is given by n_v, the number of lines of vertical resolution, according to the relation

$$\frac{H}{D} = n_v \times \alpha_0$$

where n_v is the number of black and white lines of resolution, α_0 the minimum angle of resolution in radians, and D the distance of the viewer.

Thus, for $D/H = 6$, i.e. visual angle about 10°

$$n_v = \frac{H}{D\alpha_0}$$

$$= \frac{1}{6} \times \frac{60}{1} \times \frac{180}{\pi}$$

$$\approx 600 \text{ lines}$$

and for $D/H = 4$. i.e. visual angle about 15°

$$n_v = \frac{1}{4} \times \frac{60}{1} \times \frac{180}{\pi}$$

$$\approx 900 \text{ lines}$$

The maximum number of alternate dark and white elementary horizontal lines which can be resolved by the eye are thus 600 for a 10° visual angle, or 900 for a 15° visual angle. There has been a large difference of opinion about what constitutes a sufficient number of lines because of the subjective assessment involved. This has been made further difficult to assess because of the effects of the finite size of the scanning beam spot.

2.12   THE KELL FACTOR

In a practical scanning systems, the maximum vertical resolution obtainable is less than the active number of lines available for scanning. This is because of the finite beam size and its alignment not coinciding with the elementary resolution lines. Consider a finite size of a beam spot scanning a series of closely spaced horizontal black and white lines of minimum resolvable thickness, when the beam spot also is nearly as much in size as the thickness of the line, as shown in Fig. 2.8.

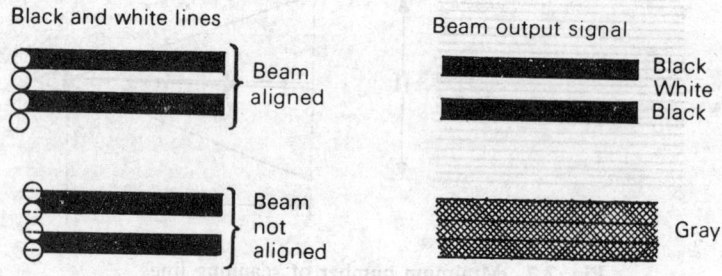

Fig. 2.8   Effect of beam spot size on vertical resolution

If the beam is in perfect alignment, the output will be exactly following the lines as black or white levels, as shown in the figure. If, however, the beam spot is shifted slightly and happens to align on the junction of the black and white lines, it actually senses both black and white areas simultaneously. Hence, it integrates the effects of both areas to give a resultant gray level output, in between the black and white levels. This happens for all scanning positions and the output as well as the reproduced picture will be a continuous gray without any vertical resolution at all. In positions of intermediate alignment of the beam, it will be more on one line than on the adjacent one and the output lines will be reproduced with diminished contrast.

This indicates that there is a degradation in vertical resolution due to finite beam size. Statistical analysis as well as subjective tests based on a bar pattern consisting of tapered wedges of almost horizontal, converging, alternate black and white bars, have indicated that the average number of effective lines is of the order of 0.7 times the total active scan lines present. This factor indicating the reduction in effective number of lines is called the 'Kell factor'. It is obviously not a precisely determined quantity, and values from 0.64 to 0.85 are ascribed to it.

What the Kell factor indicates is that it is unrealistic to state that the vertical resolution is equal to the number of active lines. In a picture, not all lines or parts of lines are fully effective at all times. The number of active lines multiplied by the Kell factor leads to a smaller figure for a more realistic assessment of available vertical resolution.

In the 625-line system, the number of active lines left after deducting lines lost in vertical blanking (as explained in Sec. 2.13) are (625 – 40 =) 585 lines. With a Kell factor of 0.7, the vertical resolution is (0.7 × 585 =) 409.5 lines. The horizontal resolution may not exceed this value, multiplied by the aspect ratio.

2.13 HORIZONTAL RESOLUTION AND VIDEO BANDWIDTH

The horizontal resolution of a television system is the ability of the scanning system to resolve the horizontal details, i.e. changes in brightness levels of elements along a horizontal scanning line. Since such changes represent vertical edges of picture detail, it follows that the horizontal resolution can be expressed as a measure of the ability to reproduce vertical information or lines of resolution as shown in Fig. 2.9.

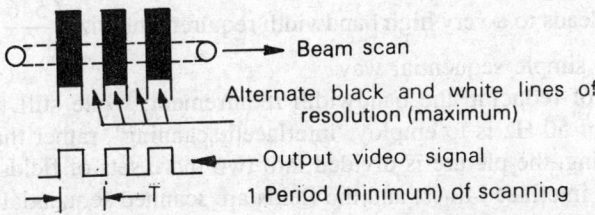

Fig. 2.9 Horizontal resolution and video bandwidth requirement

The horizontal resolution in a scanning system depends upon the rate at which the scanning spot is able to change brightness level as it passes through a horizontal line across the vertical lines of resolution shown in the figure.

In a 625-line system, there are effectively about 410 lines of vertical resolution. The horizontal resolution should be of the same order. Because of the aspect ratio of 4 : 3, the number of vertical lines for equivalent horizontal resolution will be $\left(410 \times \dfrac{4}{3} \approx\right)$ 546 black and white alternate lines, which means (546 × 1/2 =) 273 cycles of black and white alternations of elementary areas. For the 625-line system, the horizontal scan or line frequency f_H is given by:

$$f_H = \text{number of lines per picture} \times \text{picture scan rate}$$
$$= 625 \times 25 = 15625 \text{ Hz}$$

as each picture line is scanned 25 times in one second. The total line period is thus

$$T_H = \frac{1}{f_H} = \frac{1}{15625} \text{ sec} = 64 \ \mu\text{s}$$

Of this period, 12 μs are used for the blanking of the flyback retrace. Thus the 546 black and white alternations, i.e. 273 cycles of complete square waves, are scanned along a horizontal raster line during the forward scan time of (64 – 12 =) 52 μs. The period of this square wave is 52/273 $\approx$ 0.2 μs, giving the highest fundamental frequency of 5 MHz, which is adequate as the highest video frequency in the signal.

The highest fundamental video frequency in a scanning system is thus given by

$$f_{max} = \frac{\text{active lines} \times \text{Kell factor} \times \text{aspect ratio}}{2 \times \text{line forward scan period}}$$

$$= \frac{N_a \times K \times a}{2 \times t_{fH}} \tag{2.4}$$

where t_{fH} is the horizontal-line forward scan period.

2.14 INTERLACED SCANNING

From considerations of flicker, it has been found that 50 picture frames per second is the minimum requirement in television scanning. For a 625-line system, this means that the horizontal line scanning frequency should be 31,250 Hz, with the line period of 32 μs. For a desired resolution of 546/2 alternations in the horizontal line, this leads to a very high bandwidth requirement, viz. $\left(\dfrac{546}{2} \times \dfrac{1}{(32-6)} \approx\right) 10$ MHz, if the line scanning is the simple sequential way.

An ingeneous method of reducing the bandwidth requirement, while still maintaining an effective vertical picture scan rate of 50 Hz is to employ 'interlaced scanning', rather than the simple sequential raster. In interlaced scanning, the picture is divided into two more sets of fields each containing half or other fractional number of interlaced lines, and the fields are scanned sequentially. In 2 : 1 interlace, the 625 lines are divided into two sets of $312\frac{1}{2}$ lines each.

The first set of $312\frac{1}{2}$ odd number lines in the 625 lines, called the first field or the *odd field*, are first scanned sequentially. Halfway through the 313th line, the spot is returned to the top of the screen and the remaining $312\frac{1}{2}$ even number lines, called the second field or the *even field* are then traced interleaved between the lines of the first set as shown in Fig. 2.10.

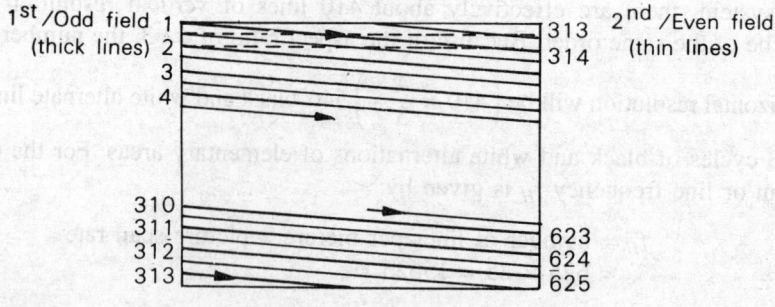

Fig. 2.10   Interlaced raster

This is done by operating the vertical field scan at 50 Hz so that the two successive interlaced scans, each at a 25 Hz rate, make up the complete picture frame. This keeps the line scanning speed down, as only $312\frac{1}{2}$ lines are scanned in 1/50 second. The 625 lines of the full picture are scanned in 1/25 second, thus keeping down the bandwidth requirement.

Here, though the picture is scanned 25 times per second, the area of the screen is covered in an interlaced fashion at twice the rate, viz. 50 times per second. A close examination may reveal the small area 'interline flicker', as actually each individual line repeats only 25 times per second. But this is tolerable and the overall effect is closer to that of a 50 Hz scanning. The flicker becomes noticeable at high brightness levels only.

In practice, the flyback from the bottom to the top is not instantaneous and takes a finite time equal to several line periods. Up to 20 lines are allowed for vertical flyback after each of the two fields that make a complete picture. This means that out of 625 lines, only (625 – 40 =) 585 lines actually bear picture information. These are called the active lines.

For interlaced scanning, the number of lines for the picture must be odd. It is then necessary to supply two regularly spaced synchronizing pulses to the field time base for every picture frame. One of the pulses must be sent at the middle of the last line in the odd field, while the other must be sent at the end of the last full line of the even field. The odd-line field starts at the top left and ends at the bottom centre, while the even-line field begins at the top centre and ends at the bottom right. This interleaves the two fields exactly, giving a complete scanning. By doubling the vertical deflection speed, the apparent picture frequency is doubled. The lines now slant twice as much but this is not noticeable. Vertical flyback occupies, in practice, a period equivalent to the time required to trace 5 to 10 lines. The path of the return line becomes a multiple zigzag under the influence of horizontal as well as vertical deflection.

The flyback times must be exactly the same, following both odd and even fields, and the size and position of both should repeat exactly. A difference of as little as 0.1 μs in the flyback time following the fields, causes the two to be dislocated by about 10% of the line length, enough to cause noticeable 'pairing of lines' reducing vertical resolution significantly.

REVIEW QUESTIONS

2.1 Explain the terms photoemission and photoconductivity. State which materials display these properties, and which camera tubes are based on these.

2.2 Explain the basic requirements of a pick-up tube for producing television signals. Indicate how these are met with in an iconoscope camera tube.

2.3 Compare the photoelectric conversion and scanning process in an iconoscope and in a CCD solid state image sensor.

2.4 Give reasons for the following:
   (a) The range of direct TV reception is limited to line-of-sight.
   (b) It is necessary to use VHF or higher frequencies for transmitting TV signals over radio waves.
   (c) A TV signal is more complicated than a radio signal.
   (d) In colour television, colour difference signals are transmitted rather than the three primary colour signals.
   (e) Frequency modulation is not used for video signaling television broadcasting.
   (f) Frequency modulation is generally used for sound transmission in television broadcasting.

2.5 Discuss the visual characteristics of a human eye. Explain how these affect the television system standards for scanning.

2.6 Write notes on: (a) Kell factor, (b) persistence of vision, and (c) field of vision.

2.7 Why is an odd number of lines used for scanning? How does interlaced scanning help reduce the bandwidth of the video signal?

2.8 If an n-line television system sends f pictures per second, the aspect ratio is a line blanking period is a fraction l of each line period, field blanking is a fraction m of each field period and K the Kell factor, show that the bandwidth required for video signal is given by $n^2(1 - m)afK/2(1 - l)$.

2.9 A television standard has 819 scan lines and picture scan rate of 50 Hz with 2 : 1 interlace. Assuming 15% as blanking time, find the video bandwidth requirement of the system.

2.10 A television receiver in the 625/50 system shows a resolution of 300 horizontal lines and 400 vertical lines. What does this mean in terms of receiver performance?

2.11 A facsimile-picture transmission system scans a picture wrapped around a drum revolving at 50 rpm, by means of a light spot and photocell moving axially to scan the picture fully in 5 min. The picture of 4 : 3 aspect ratio is wrapped with its height along the drum axis. Estimate the minimum bandwidth requirement of the amplifier to amplify the video signal from the photocell.

2.12 Explain the difference between the terms (a) luminosity and luminous intensity, and (b) luminance and illumination.

MULTIPLE-CHOICE QUESTIONS

2.1 The density of light emitted from a TV screen is called: (a) illuminance, (b) luminance, (c) luminosity, (d) lumens.

2.2 Lux is a unit of (a) luminous flux, (b) illuminance, (c) luminous intensity, (d) a and b.

2.3 For a screen at a distance of thrice its height, the physiological limit to observable resolution is: (a) 600 lines, (b) 800 lines, (c) 1000 lines, (d) 1200 lines.

2.4 If the number of lines in the picture is doubled, the video bandwidth needs to be: (a) doubled, (b) quadrupled (c) unaffected, (d) halved.

2.5 Compared to progressive scanning, the interlacing technique reduces the bandwidth because: (a) the picture scanning rate is increased, (b) the picture scanning rate is reduced, (c) the effective picture scanning rate is kept same while the pixel scanning rate is halved, (d) b and c.

2.6 Kell factor indicates: (a) reduction in contrast ratio in a picture due to beam spot size, (b) reduction in vertical resolution due to beam spot size, (c) reduction in horizontal resolution in picture due to bandwidth.

2.7 The effective line resolution observed due to Kell factor is reduced by a factor of approximately: (a) 0.5, (b) 0.6, (c) 0.7, (d) 0.9.

2.8 The aspect ratio used for picture in high definition television is: (a) 6 : 5, (b) 4 : 3, (c) 16 : 9, (d) 5 : 3.

2.9 Weber's fraction is: (a) the ratio of B_{min}/B_{max} in a picture, (b) $\Delta B/B$ for adjacent areas, (c) $\Delta B/B$ for incremental eye sensitivity of brightness for eye, (d) $(B_{max} - B_{min})/B_{max}$ ratio.

3

Composite Video Signal and Television Standards

The composite video signal is formed by the electrical signal corresponding to the picture information in the lines scanned in the TV camera pick-up tube and the synchronizing signals introduced in it. It is important to preserve its waveform as any distortion of the video signal will affect the reproduced picture, while a distortion in the sync pulses will affect synchronization resulting in an unstable picture. The signal is, therefore, monitored with the help of an oscilloscope, at various stages in the transmission path to conform with the standards. In receivers, observation of the video signal waveform can provide valuable clues to circuit faults and malfunctions.

Television signal standards specify complete technical details of the video and audio signal waveforms and the modulation characteristics in the television broadcast system. These include the number of lines, scanning frequencies, interlace, the amplitude and time dimensions of the composite video waveform, characteristics of vision and sound modulation channel bandwidth, etc. In this chapter, the CCIR system-B standards for scanning, composite video waveform and modulation are first discussed, followed by a review of the various international standards approved by CCIR.

3.1 LINES AND SCANNING

The CCIR system-B employs 625 lines with 2 : 1 interlace, i.e. it uses two fields sequentially scanned but interleaved to form a complete picture. The field frequency is 50 Hz so that the picture frequency is 25 Hz. The horizontal line frequency is thus $625 \times 25 = 15,625$ Hz, maximum permissible deviation being 0.1%.

The aspect ratio, which is the ratio of the width to the height of the TV picture, is 4 : 3. In practice, the display area of the picture tube has an aspect ratio nearer to 5 : 4. This limitation due to the problems of manufacturing rectangular screen tubes is, by the way, useful for overscan that eliminates vertical stripes due to ringing oscillations in the line time base at the end of flyback.

3.2 VIDEO SIGNAL COMPONENTS

The video carrying picture and sync information signal consists of three parts: (i) the video signal corresponding to the picture information in the lines scanned at the camera, (ii) the synchronizing pulses to synchronize the horizontal and vertical scanning at the transmitter and the receiver, and (iii) the blanking pulses to make the retrace invisible. The picture information and the sync pulses are not required simultaneously. The sync pulses are needed at the end of the horizontal and vertical scans, when flyback retrace is desired. During the forward scan, the video signal level changes between black and white levels occupying intermediate gray levels also, depending upon the picture brightness in the scanned areas. The synchronizing pulses occupy a different amplitude level corresponding to the blacker than black level in the video signal, so that it can be separated at the receiver easily, using simple amplitude separator circuits. The ratio of the picture detail to the sync pulse amplitude is of the order of 7 : 3, as shown in Fig. 3.1.

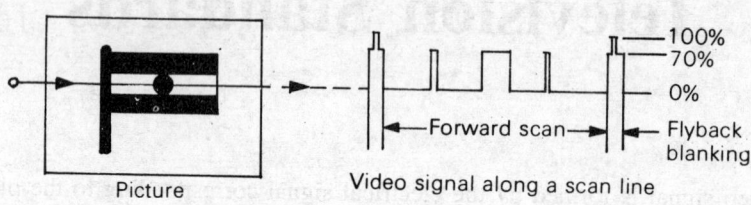

Fig. 3.1 Picture-sync ratio: (a) picture, (b) video signal

If the picture content is increased at the expense of sync pulse amplitude, then the sync pulse amplitude is insufficient to keep the picture locked in when the signal to noise (S/N) ratio in the received signal falls. On the other hand, if the sync pulse amplitude is increased at the expense of the picture signal, then the picture may lock better under low signal to noise (S/N) ratio conditions, but the picture signal content is of too low an amplitude to set up a picture with good contrast. A *picture-to-sync ratio* of 7 : 3 is about the optimum, so that when the S/N ratio reaches a certain low level, the sync fails at the same time as the picture ceases to be satisfactory.

Blanking Signal

After each line is scanned by the forward motion of the beam from left to right, return movement to the left must be as quick as possible, and must be prevented from reproducing unwanted luminous traces on the screen. Though actual return takes a small fraction of this line, the damped oscillatory currents in the deflection circuits show up as light and dark stripes at the left of the screen. Both these effects are suppressed by inserting the horizontal blanking (HB) at the black level extended to about 19% of the line period, as shown in Fig. 3.2.

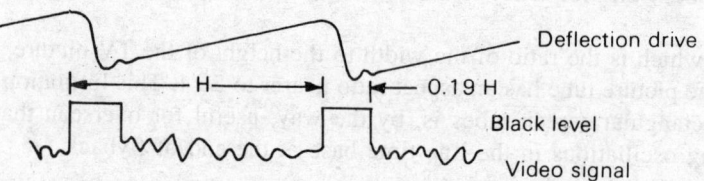

Fig. 3.2 Blanking signal

Similarly, after the vertical scan from the top to the bottom is completed, the retrace from the bottom to the top, and oscillatory conditions in the frame and the line time bases are also blanked by a vertical blanking (VB) pulse covering 20 line periods after each field, i.e. 6.5% of the field period.

3.3 HORIZONTAL SYNC AND BLANKING STANDARDS

The amplitude level and time period distributions in the composite video and RF signal during the horizontal period H (= 64 μs), as per CCIR system-B standards are shown in Fig. 3.3.

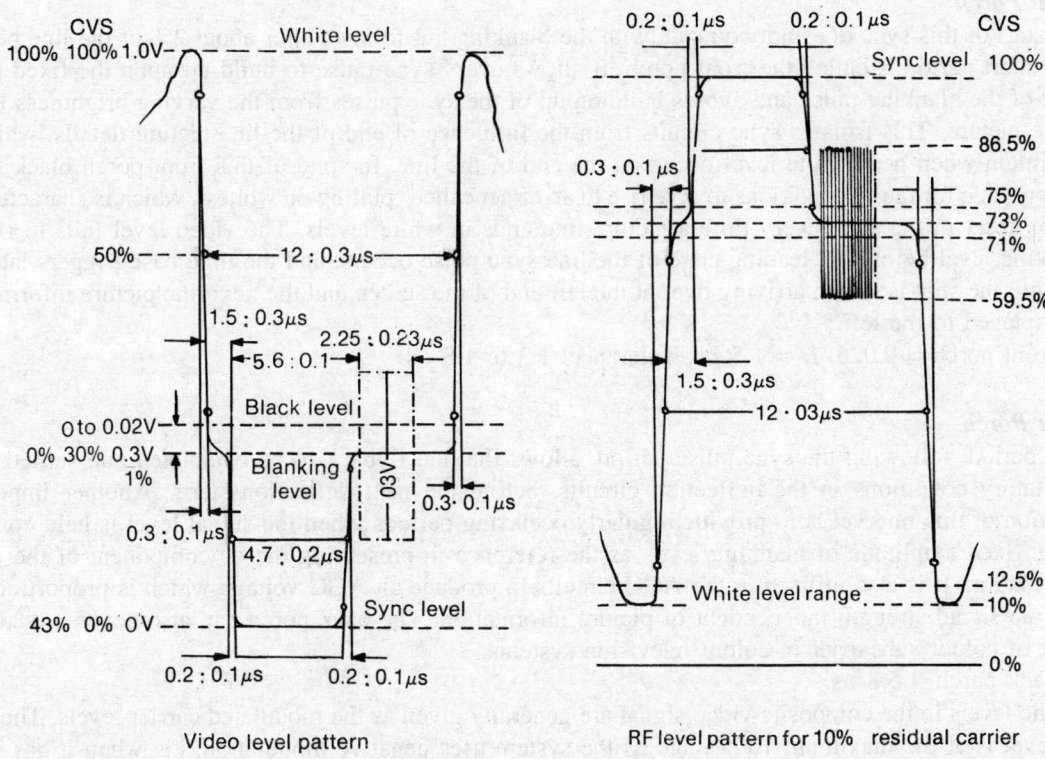

Video level pattern

RF level pattern for 10% residual carrier

Fig. 3.3 Horizontal sync and blanking signal standards

Line Period H
This is the total duration of one complete line. The line frequency f_H is $625 \times 25 = 15{,}625$ lines per second, so that the period

$$H = \frac{1}{f_H} = \frac{1}{15625} = 64 \ \mu s$$

Line Blanking (LB) Period
This is the part of each line during which the line sync pulse is inserted. During this period, the beam flyback is initiated and completed and the beam cut off by the black level amplitude of the video signal. HB or $LB = 0.19H = 12$ μs. Tolerance: 11.8 to 12.3 μs.

The line blanking period is divided into three sections because of the sync pulses mounted on it midway.

Line Sync Pulse

This is the important short pulse sent from the transmitter to correct the horizontal scanning rate at the receiver if it deviates from the desired rate. Its width $HS = 0.075H = 4.7$ μs. Tolerance: 4.5 to 4.9 μs. Its rise time at the transmitter is better than 0.25 μs.

Front Porch

The start of this sync does not coincide with the blanking but follows after about 2% of the line period. This short period is called the front porch. It allows every sync pulse to build up upon the fixed black level of the blanking pulse and avoids building up of the sync pulses from the varying brightness levels in the picture. This isolates sync circuits from the influence of end of the line picture details, which is maximum when peak white level occurs at the end of the line. In spite of this front porch black level, it is possible for faulty conditions to give rise to an effect called 'pulling on whites', which is characterized by an upset in the sync every time a picture line ends in white levels. The video level fails to rise to blanking level before the leading edge of the line sync pulse occurs, and the time base triggers late. As a result, the spot is late in arriving over at the left end of the screen and the next line picture information is displaced to the left.

Front porch = 0.025, $H \approx 1.5$ μs, Tolerance: 1.3 to 1.8 μs.

Back Porch

This period, following the sync pulse period, allows the line flyback to be completed and settled from oscillatory conditions in the deflection circuits, before the next deflection starts. Another important function of this interval is to provide regularly occurring periods when the signal level is held constant at the fixed amplitude of blanking level, as the reference in preserving the dc component of the video information. It is also utilized in the AGC circuits to produce an AGC voltage which is proportional to the true signal strength independent of picture information. The back porch can also accommodate the burst of colour subcarrier in colour television systems.

Back porch ≈ 5.8 μs.

The levels in the composite video signal are generally given as the modulated carrier levels. The sync tip level is at the maximum 100%, and as the system uses negative modulation, i.e. when it has white levels going negative, the blanking level is at 75% and the peak white level limited to 10%. The minimum level of 10% is necessary in view of the intercarrier system which requires some minimum picture carrier to beat with the sound carrier to produce the intercarrier.

Pedestal

In some systems, the blanking level is blacker than black. The difference between the two, if any, is called the 'pedestal' or 'set-up'. It is generally kept at a minimum, as close to zero as possible, unless otherwise specified.

3.4 VERTICAL SYNC AND BLANKING STANDARDS

At the end of each field scanning $312\frac{1}{2}$ lines, a vertical field synchronising waveform is inserted in the composite video signal. The waveform is shown in Fig. 3.4, for the odd and even fields.

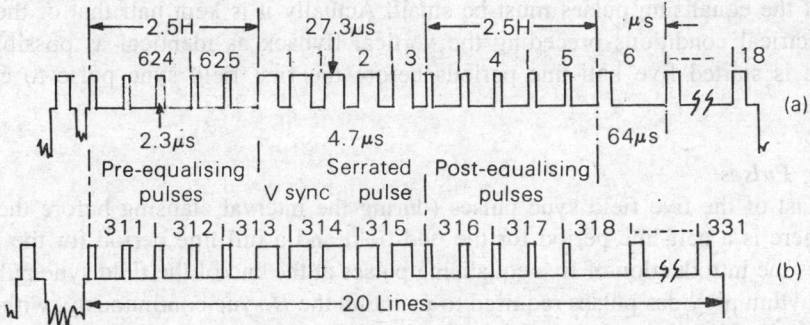

Fig. 3.4 Field sync and blanking signal standards: (a) for odd field, and (b) for even field

Field Blanking Period

This is the period during which the picture information is entirely suppressed, the flyblack retrace of the field timebase initiated and completed, while the beam is cut off by the black level. The duration of the field by vertical blanking period is equal to 20 line periods.

$$VB = 20H = 1280 \ \mu s$$

Field Synchronizing Pulse

This pulse is of much longer duration in order to make it distinguishable from the H or line sync pulses. It is a little less than $(5/2)$ H line periods. In order to avoid the interruption of line synchronization during the time the field sync pulse is operating, it is split by serrations, each 4.7 μs wide, into five half-line rhythm narrower pulses, each $(32 - 4.7 =)$ 27.3 μs wide. The front edges of these correspond exactly with the line sync pulses they replace. The extra half-line sync pulse edges that occur between each pair of pulse edges actually required for line synchronization, are unnecessary. However, they do not affect the line time base as it is insensitive to pulses in these intermediate times. The half-line rhythm is necessary because, with interlaced scanning, the field sync pulses must occur at the middle of the lowest 313th line of the first or odd-field, and at the end of the lowest 625th line of the second or even-field.

Equalising Pulses

The receiver deflection circuits have to separate the vertical field sync pulses from the line sync pulses and process-integrate the serrated vertical sync pulse to produce a composite single field pulse which is used to trigger and synchronize the vertical oscillator. For achieving a perfect interlace, the pulse produced must be in exactly similar conditions on both odd and even fields. Because of the unequal line periods preceding the field sync pulses for odd and even fields, viz. half line period before the odd field pulse and full line period before the even field pulse, the electrical conditions in the processing circuits are different. This effect of uneven line period can be reduced simply by increasing the interval between the preceding line pulse and the field sync pulse. This interval has been fixed at five half-line periods. During this period, the horizontal time base must not, however, be deprived of the horizontal sync pulses. For this purpose, five narrow pulses, 2.3 μs each, also occurring at half-line rhythm, are inserted in this five-half-line period before the field sync pulses. These are called the pre-equalising pulses. The half-line rhythm of equalising pulses ensures line synchronization during both the fields, the first, third and fifth equalising pulses synchronizing the line oscillator during the odd field ending, while the second and fourth pulses do so during the even field ending. The synchronizing rising edges are drawn as thick edges of the pulses as shown in Fig. 3.4.

The width of the equalising pulses must be small. Actually it is kept half that of the *H* sync pulses to keep the electrical conditions preceding the vertical flyback as identical as possible. The vertical blanking period is started five half-line periods before the first field sync pulse to cover these five equalising pulses.

Post Equalising Pulses

Following the last of the five field sync pulses (during the interval elapsing before the next line sync pulse occurs) there is a half-line period for the odd field and a full line period for the next even field. This necessitates the introduction of five equalising pulses at the end of the field sync pulses also, so that their half-line rhythm provides pulses required to maintain the *H* sync continuously, without any missing gap. For accurate interlaced spacing of lines, the vertical flyback must not only start regularly but its duration also should be regular, that is, the instant at which its downward motion is started should repeat regularly. Though this instant is decided by the receiver time base circuitry, the circuit must be kept free from any interference. The sudden change-over from the normal full-line rhythm to the half-line rhythm occurring irregularly between the odd and even fields, may sometimes cause the horizontal time base to fall out of step.

The extra equalising pulses have an effect of moving the half-line and full line discrepancies away from the field pulses, and are useful in producing identical electrical conditions in the receiver field pulse processing circuitry, on odd and even fields, both before and after the arrival of the five field sync pulses. They allow high quality, well-interlaced pictures to be reproduced even from receivers with simpler deflecting circuits.

3.5   VIDEO MODULATION AND VESTIGIAL SIDEBAND SIGNAL

Amplitude modulation is used for the video signal. The composite video signal waveform has distinct polarity for black and white levels in the picture. Hence two types of modulations are possible.

(i) *Positive modulation*   In this, the increase of brightness towards white causes increases of the carrier amplitude. The peak white has 100% modulation while the black and sync levels have lower and minimum modulation respectively.

(ii) *Negative modulation*   In this, the sync tip levels are at the 100% modulation level. Blanking level corresponding to black is at 75%, and increase in brightness towards white causes a decrease in carrier amplitude down to the minimum of 10% at peak white.

The CCIR system-B uses negative modulation. In such modulation, the impulsive interference can be more troublesome because it causes a sudden random increase in the signal level which is similar to the existing sync pulses, and it is more difficult to eliminate the effects of the interference pulses on the sync system. However, the visible interference on the picture tube screen produces black spots which are less annoying than the white blobs in positive modulation systems.

Video Bandwidth

The composite video signal is allowed a 5 MHz bandwidth with the highest video frequency of 5 MHz. The low frequency content is right down to dc or zero frequency if the steady background brightness level, which may remain constant or change from scene to scene, is considered. The lowest frequency must be the vertical picture scan rate 25 Hz.

This large video bandwidth necessitates the use of VHF-UHF frequencies as RF carrier frequencies. Amplitude modulation of the carrier with the video signal produces two sidebands, each 5 MHz wide.

This double sideband vision signal will occupy a 10 MHz frequency-range spectrum with the carrier at the centre. The actual band space allocated to the TV channel will have to be greaer still to allow for an attenuation slope of 0.5 MHz on either side as the bandwidth cannot be abruptly restricted to 5 MHz.

Vestigial Sideband Transmission Signal

Because of the extensive bandwidth requirements of the video signal, it is desirable to make use of a bandwidth saving technique. The signal information is fully contained in each of the two sidebands of the modulated carrier; and provided the carrier is present, one sideband may be suppressed altogether. The signal sideband transmission technique can reduce the bandwidth requirement to half, viz. 5 MHz. However, it is not possible to do this in the case of a television signal because a television signal also contains very low frequencies including even the dc information. It is impossible to design a filter which will cut out the unwanted band while passing the carrier frequencies and low frequency components of the other sideband, without objectionable phase distortion.

As a compromise, a part of only one sideband is suppressed. The transmitted signal consists of one complete sideband together with the carrier, and a vestige of the partially suppressed sideband as shown in Fig. 3.5(a). In the vestigial sidebnd transmission system, there is a saving in the bandwidth required, and the filtering required is not so difficult to achieve.

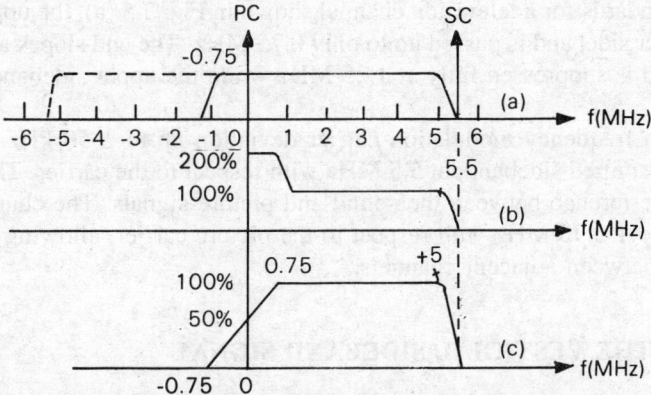

Fig. 3.5 Vestigial sideband channel characteristics: (a) VSB channel, (b) Video power content, (c) Receiver response.

3.6 SOUND MODULATION AND THE INTERCARRIER SYSTEM

In addition to the vision band, each television channel has its own associated sound signal, the carrier and the sidebands of which are positioned just outside the upper or lower limit of the vision signal. Frequency modulation is used for sound transmission as it offers noise-free reception and higher fidelity. In the intercarrier sound technique, the sound intermediate frequency (IF) and the sound demodulator at the receiver are tuned to the beat frequency produced by the picture and sound carriers at the video detector. The beat frequency is called the intercarrier sound IF. The system has the advantage that the sound IF is always correct, free from the local oscillator drifts in the receiver. Also, because the picture is an amplitude modulated signal and the sound is frequency modulated, cross-modulation between them is readily avoided.

Pre-emphasis

At the transmitter, the higher frequencies are given a boosted characteristic in order to enhance the signal-to-noise ratio at the higher frequencies at which the audio power is much less, making them more susceptible to noise interference. This boost given to high frequencies prior to modulation is called the pre-emphasis. It is given to the signal by a highpass RC filter of a time-constant of 50 μs.

At the receiver, the demodulated audio signal is provided *de-emphasis* by a low-pass filter of the same time-constant of 50 μs in order to bring down the boost at the high frequencies.

Carrier Deviation

The maximum FM deviation of the sound carrier by the peak level audio signal is restricted to ± 50 kHz.

In the intercarrier, the difference between the picture and the sound carriers is 5.5 MHz which serves the sound IF in the receiver sound section. This intercarrier with sound modulation is just outside the passband of the video amplifier and, therefore, has to be rejected by a suitable trap circuit in the video path to the picture tube.

3.7   STANDARD CHANNEL CHARACTERISTICS

In the CCIR system-B standards for a television channel shown in Fig. 3.5 (a), the upper sideband is fully transmitted while the lower sideband is passed up to only 0.75 MHz. The end slopes are allowed 0.5 MHz so that the lower sideband is suppressed fully at 1.25 MHz while the upper sideband is fully attenuated at 5.5 MHz.

The sound carrier with frequency modulation carrier deviation up to ± 50 kHz is positioned at the extremity of the fully transmitted sideband, at 5.5 MHz with respect to the carrier. This is a logical place for it to minimize the interference between the sound and picture signals. The channel width is thus 7 MHz, from – 1.25 MHz to + 5.75 MHz with respect to the picture carrier, allowing for a guard band of 0.25 MHz as separation between adjacent channels.

3.8   RECEPTION OF THE VESTIGIAL SIDEBAND SIGNAL

In the vestigial sideband signal, the lower modulation frequencies from 0 to 0.75 MHz are present in both sidebands while the rest of the frequencies from 0.75 to 5 MHz are present only in one sideband. If all the channel frequenies are equally amplified in the RF and IF amplifiers of the TV receiver and then led to the video detector, the double sideband frequencies will give rise to a total of twice the voltage output relative to the higher frequencies from 0.75 to 5 MHz that have only a single sideband component, as shown in Fig. 3.5(b).

In order to make for this eventuality, it is necessary to shape the picture IF response curve of the receiver so that the frequencies in the double sideband range are provided less amplification than the rest. This is arranged by placing the picture carrier halfway down the slope of a response curve that slopes linearly over the double sideband frequency range from – 0.75 to + 0.75 MHz relative to picture carrier as shown in Fig. 3.5(c). The width of the sloping edge is twice that of the vestigial sideband. It will be observed that the sum of output contributions from the two sidebands together make up the same output at all the frequencies.

3.9 TELEVISION BROADCAST CHANNELS

For television broadcasting, channels have been assigned in the VHF and UHF ranges. The allocated frequencies are:

Lower VHF range	Band I	41–68 MHz
Upper VHF range	Band III	174–230 MHz
UHF range	Band IV	470–582 MHz
UHF range	Band V	606–790 MHz

(Band II 88–108 MHz is allotted for FM broadcasting.)

The channel allocations in band I and band III are given in Table 3.1. There are four channels in band I, of which channel 1 (6 MHz) is no longer used for TV broadcasting, being assigned to other services.

Table 3.1 Television Channel Allocations CCIR-B

Band	Channel	Frequency range, MHz	Picture carrier, MHz	Sound carrier, MHz
I	1	41– 47	Not used for TV	
(41–68 MHz)	2	47– 54	48.25	53.75
	3	54– 61	55.25	60.75
	4	61– 68	62.25	67.75
III	5	174–181	175.25	180.75
(174–223 MHz)	6	181–188	182.25	187.75
	7	188–195	189.25	194.75
	8	195–202	196.25	201.75
	9	202–209	203.25	208.75
	10	209–216	210.25	215.75
	11	216–223	217.25	222.75
Addl. chl.	12	223–230	224.25	229.75

Channels 2 to 8 are illustrated in Fig. 3.6 in respect of their frequency characteristics.

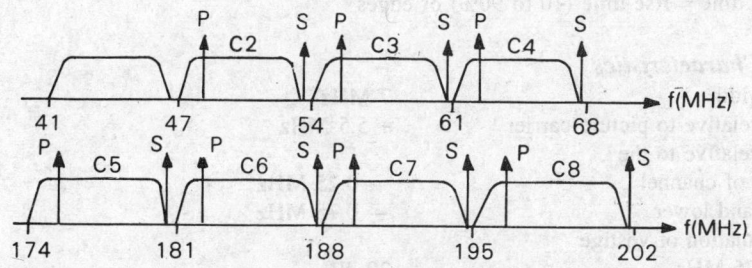

Fig. 3.6 CCIR channels 2 to 8

Carrier stability has to be within ± 1 kHz. When transmitters operating on the same channel operating at different place causes co-channel interference due to freak distant reception, the transmitters may have 'off-set' operation by operating their picture carriers shifted + 10.5 kHz and – 10.5 kHz with respect to the assigned frequencies, in order to reduce the interference effects.

3.10   CONSOLIDATED CCIR SYSTEM-B STANDARDS*

The main characteristics of the CCIR system-B for monochrome television, adopted in India are given below.

Video Characteristics

Number of lines per picture	625
Interlace ratio	2 : 1
Field frequency	50 fields/sec
Picture frequency	25 pictures/sec
Line frequency tolerance	15625 lines/sec
	± 0.1%
Aspect ratio	4 : 3
Scanning sequence for lines	Left to right
Scanning sequence for fields	Top to bottom
Video bandwidth	5 MHz
Approximate gamma of picture signal	0.5

Composite Video Signal Characteristics

Line period H	64 μs
Line blanking period HB	18.5 to 19.2% of H, 11.8 to 12.3 μs
Front porch FP	2 to 2.8% of H, 1.3 to 1.8 μs
Line sync pulse HS	7 to 7.7% of H, 4.5 to 4.9 μs
Build-up time of line sync pulse	0.31 to 0.62% of H, 0.2 to 0.4 μs
Field period V	20 ms
Field blanking period FB	[(18 to 22) H + 12] μs, 1164 to 1420 μs
Duration of pre-equalising pulse sequence	2.5 H, 160 μs
Duration of vertical pulse sequence	2.5 H, 160 μs
Duration of post equalising pulse sequence	2.5 H, 160 μs
Interval between field sync pulses (serrations)	7 to 7.7% of H, 4.5 to 4.9 μs
Duration of equalising pulse	3.4 to 3.75% of H, 2.2 to 2.4 μs
Build-up time field sync pulse	0.31 to 0.62% of H, 0.2 to 0.4 μs
Build-up time of field blanking pulse	6 μs

Note: Build-up time = rise-time (10 to 90%) of edges

Radio Frequency Characteristics

Channel bandwidth	7 MHz
Sound carrier relative to picture carrier	+ 5.5 MHz
Sound carrier relative to the nearest edge of channel	— 0.25 MHz
Vestigial sideband lower	— 0.75 MHz
Minimum attenuation of vestige beyond – 1.25 MHz	20 dB
Type and polarity of video modulation	A5, AM negative
Sync level as percentage of peak carrier	100%
Blanking level as percentage of peak carrier	72.5 to 77.5%
Difference between black level and blanking level as percentage of peak carrier	0 to 2%

---

*Based on Appendix A of IS:4545—1968.

Peak white level as percentage of peak carrier	10 to 12.5%
Sound modulation	
Type	F3 FM
Carrier deviation	± 50 kHz at 100% modulation
Pre-emphasis	50 μs
Ratio of radiated powers of vision and sound	5 : 1

3.11 VARIOUS TELEVISION BROADCAST SYSTEMS

As television systems developed rather independently in various countries there have been a number of television standards in use all over the world. However, of late there have been concerted efforts to reduce these because of the need for exchange of TV programs.

All television broadcast systems use a 4 : 3 aspect ratio, 2 : 1 interlace and a scanning sequence of left to right and top to bottom. The field scanning frequency is usually the same as the power supply frequency, except in some countries like Canada, Japan, and some South and Central American states that have a 50 Hz supply frequency and yet use a scan rate of 60 fields per second. Modern low ripple receiver designs do not involve any great problem of hum-bar or rolling ripples in the picture. The 60 Hz scanning systems provide a greater freedom from flicker at high brightness levels than the 50 Hz systems. High resolution systems, such as the French 819-line system, could not be fully exploited economically in the allocated UHF and VHF bands, because of large bandwidth requirements. There is a revived interest in high resolution high fidelity television service in HDTV, for cine film grade picture, if not better. All systems, except the old British 405–line system, employ horizontal polarization because of reduction of interference from man-made sources, such as cars, in the horizontal polarization mode.

With growing international communication and exchange of programs, the various television systems have been standardized as CCIR systems, designated as systems A, B, ..., N; system B is the popularly known CCIR system. The salient characteristics of the various systems are given in Fig. 3.7 and Table 3.2.

Systems A, B, C, D, E and F have been in use in Europe for band I and band III transmission. Of these, systems A, C, E and F have not been recommended for further services. Systems G, H, I, K and L are employed for transmission in the UHF bands IV and band V only. Systems M and N are used for all VHF and UHF bands.

Although the FCC approved NTSC standards for monochrome compatible *colour TV standards* were introduced in the United States in 1956 and subsequently in a number of other countries, the European countries delayed the adoption in favour of modified systems—SECAM developed in France, and PAL developed in Germany. These systems aimed at overcoming the implementation problems of NTSC which was prone to colour distortions due to phase errors in the transmission path. The SECAM was adopted in France, USSR and the East European countries, while the PAL system has been adopted by numerous countries in Europe as well as in the United Kingdom. Two basic standards have been adopted for the international exchange of TV programs:

	FCC standard	CCIR standard
Lines/picture	525	625
Fields/s	60	50
Colour system	NTSC	PAL/SECAM
Video bandwidth	4.2 MHz	5/5.5/6 MHz
Colour subcarrier	3.58 MHz	4.43 MHz

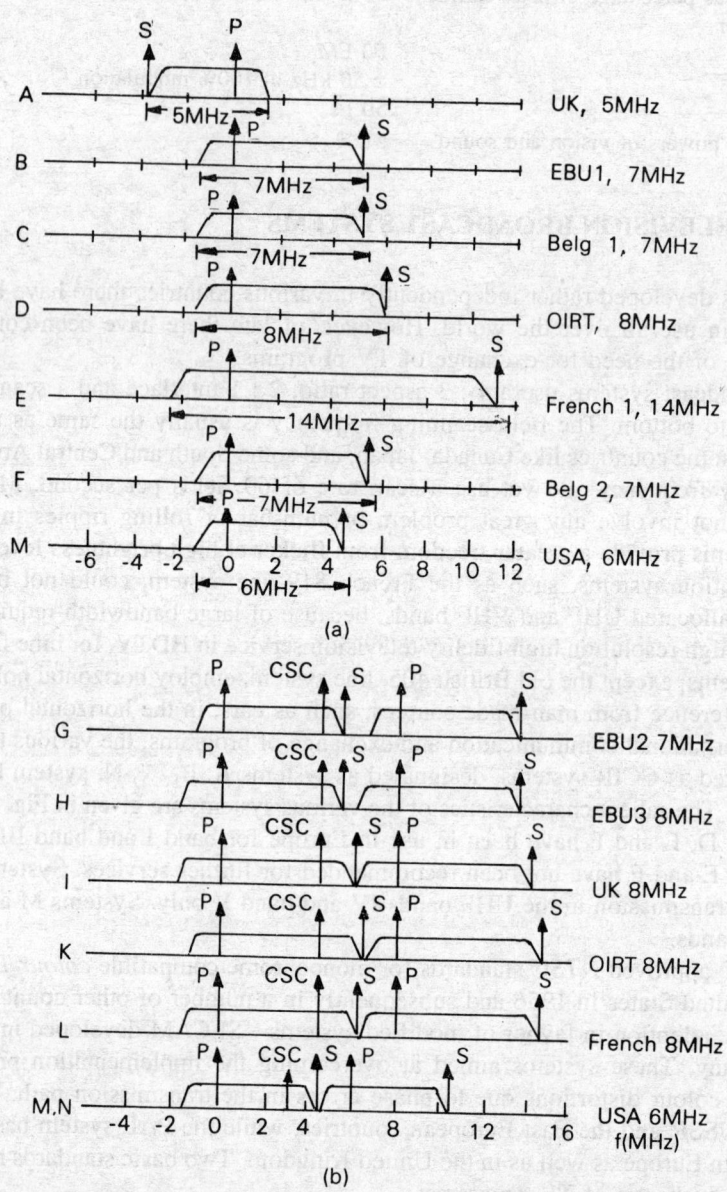

Fig. 3.7 Channel characteristics of various television systems: (a) For VHF channels, (b) For UHF channels

The main problem of standards conversion is the conversion of field frequency from 50 Hz to 60 Hz and vice versa. For this purpose, the picture information must be stored and then scanned at the new frequency. The *electro-optical analog standards converter* uses the screen of a high-resolution display tube of suitable persistence. The display is picked up like an "open scene" in the output standard by means of a camera tube. A *digital standards converter* converts the picture signal information from analog into digital form, reads it into a digital memory, reads it out a new scanning rate and reconverts it into analog form.

Table 3.2 International Television Standards—CCIR Systems

Standards	A (Engl. 1)	B (EBU 1) CCIR	C (Belg. 1) Luscom	D (OIRT)	E (French 1)	F (Belg. 2)	M (USA) FCC
Lines	405	625	625	625	819	819	525
Fields/sec	50	50	50	50	50	50	60
Pictures/sec	25	25	25	25	25	25	30
Lines/sec	10125	15625	15625	15625	20475	20475	15750
Line period	98.7 μs	64 μs	64 μs	64 μs	48.84 μs	48.84 μs	63.5 μs
Video BW (MHz)	3	5	5	6	10	5	4.2
Gamma	0.4 to 0.5	0.5	0.5	0.5	0.6	0.5	0.45
Vestigial side band (MHz)	+ 0.75	− 0.75	− 0.75	− 0.75	− 2	− 0.75	− 0.75
Sound carrier relative to picture (MHz)	− 3.5	+ 5.5	+ 5.5	+ 6.5	+ 11.15	+ 5.5	+ 4.5
Video modulation AM	+	−	+	−	+	+	−
Sound modulation	AM	FM	AM	FM	AM	AM	FM
Sound carrier deviation		± 50 kHz			± 50 kHz		± 25 kHz
Pre-emphasis		50 μs			50 μs		75 μs
Channel bandwidth (MHz)	5	7	7	8	14	7	6
Picture to sound power	4 : 1	5 : 1	4 : 1	5 : 1	4 : 1	4 : 1	10 : 1
Carrier levels −% of peaks							
Black	35%	75%	25%	75%	25%	25%	75%
Blanking	30%	75%	25%	75%	25%	25%	75%
White peaks	10%	100%	100%	10%	100%	100%	12.5%

Systems for UHF

Standard lines 625	G	H (Belgium)	I (Engl. 2)	K (OIRT)	L (French 2)	N (South America)
Fields 50/s						
Video BW (MHz)	5	5	5.5	6	6	4.2
Sound rel. to picture (MHz)	5.5	5.5	6	6.5	6.5	4.5
Picture modulation AM	—	—	—	—	+	—
Sound modulation	FM	FM	FM	FM	AM	FM
Picture to sound power	10 : 1	10 : 1	10 : 1	10 : 1	10 : 1	10 : 1
Vestigial SB (MHz)	0.75	1.25	1.25	0.75	1.25	0.75
Colour system	PAL/ SECAM	PAL/ SECAM	PAL	SECAM	SECAM	SECAM

In the standards converter for *colour television*, the incoming signal must be divided into its luminance and chrominance components, decoded and remodulated onto the other colour carrier. If only the colour system is to be converted, e.g. PAL into SECAM, the nuber of lines and the fields frequency being equal, no picture memory is required. It then suffices to separate and transcode the chrominance signal and to modulate the new carrier as required (transcoder principle).

Broadcasting of TV Programs

The public television service is operated by broadcasting picture and sound from picture transmitters and associated sound transmitters in three main frequency ranges in the VHF and UHF bands. By international ruling of the ITU, these ranges are exclusively allocated to television broadcasting. Subdivision into operating channels and their assignment by location are also ruled by international regional agreement. The continental standards are valid as per the CCIR 1961 Stockholm plan. The details of the various system parameters are as follows.

Band	Frequency	Channel	Bandwidth
I	(41) 47 to 68 MHz	2 to 4	7 MHz
II	87.5 (88) to 108 MHz	VHF FM sound	
III	174 to 223 (230) MHz	5 to 11(12)	7 MHz
IV	470 to 582 MHz	21 to 27	8 MHz
V	582 to 790 (860) MHz	28 to 60 (69)	8 MHz
VI	11.7 to 12.5 GHz	superseded by satellite	
Special	68 to 82 (89) MHz	2 (3) S	7 MHz
channels	104 to 174 and	channels	
Cable TV	230 to 300 MHz	S1 to S20	7 MHz

Types of Modulation
Vision: C3F (vestigial sideband AM)

Vestigial sideband ratios:
0.75 MHz/4.2 MHz = 1 : 5.6   for system M 525/60, 6 MHz
0.75 MHz/5.0 MHz = 1 : 6.7   for system B 625/50, 7 MHz
1.25 MHz/5.5 MHz = 1 : 4.4   for system I 625/50,   8 MHz

The saving of frequency band is about 40%; the polarity is negative because of the susceptibility to interference of the synchronizing circuits of early TV receivers (exception: positive modulation); residual carrier with negative modulation 10% (exception 20%).

Sound: F3E; FM for better separation from vision signal in the receiver (exception: AM).

Sound carrier above vision carrier within RF channel, inversion at IF; (exception: standards A, E and, in part, L).

Dual-sound-carrier Systems
Details are discussed in section 15.6.

System parameters (for standards B/G)

	Channel-1	Channel-2
RF sound carriers		
If Frequencies	fvision + 5.5 MHz (+/–500 Hz), eqvt. to 352fh	fvision + 5.7421875 MHz (+/–500 Hz),
Vision/sound power ratio	13 dB	20 dB
Modulation	FM	FM
Frequency deviation Max.	< +/–50 kHz	< +/–50 kHz
nominal value	+/–30 kHz	+/–30 kHz
Preemphasis	50 μs	50 μs
AF bandwith	40 to 15000 Hz	40 to 15000 Hz
Sound Modulation		
Mono	mono	mono
Stereo	(L + R)/2 = M	R
Dual sound mono	channel A	channel B
Identification		
Pilot carrier frequency	—	54.6875 kHz (+/–5Hz) (eqvt to 3.5fh)
Modulation	—	AM (with identification frequency)
Modulation depth	—	50%
Identification signal		
mono		none
stereo		(fh/133 =) 117.5 Hz
dual sound		(fh/57 =) 274.1 Hz
Frequency deviation of transmitter carrier (due to pilot tone)		+/– (2.5 kHz +/–0.5 kHz)
Synchronization		pilot carrier and identification frequencies phase-locked with fh

The two *sound channels* arrive from the studio via radio link with 15 KHz bandwidth at the TV transmitter, where matrixing is performed for compatibility; (L + R)/2 for channel-1, R for channel-2. An additional sound modulator is used to modulate the second sound carrier with sound channel–2 and with the AM modulated pilot carrier.

The mode *identification* is transmitted in (data) line 16 (329) of a normal TV picture from the studio to the dual-sound coder of the TV transmitter via the conventional TV lines (i.e. not the sound lines) From the 13 usable words of this data line the first two bits of word 5 are provided for mode identification in bi-phase code as follows:

Identification	Bit 1	Bit 2
Stereo	1	0
Mono	0	1
Dual sound	1	1
Fault	0	0

Vision/sound power ratio is 3 : 1 / 4 : 1 / 10 : 1 / 20 : 1, depending on the standard. Ratios of 5 : 1 and 10 : 1 are conventionally used; 20 : 1 is used in Germany, its advantage being energy saving and low

intermodulation distortion in TV transposers and TV transmitters with common vision-sound amplification and in Cable Television; 20 : 1 : 0.2 for dual-sound broadcasts in the B/G standard.
Channel bandwidths are 5/6/7/8/14 MHz, depending on standard; conventional values are 6/7/8 MHz and 14 MHz are still valid for a certain transition period.

TV Broadcasting Standards

Four broadcasting standards are in most common use at present (see also the standards table for monochrome television):

Standard code	FCC M	CCIR B/G	British I	OIRT D/K
Number of lines	525	625	625	625
Field frequency	60 Hz	50 Hz	50 Hz	50 Hz
Channel width	6 MHz	7/8 MHz	8 MHz	8 MHz
Vision/sound carrier spacing	4.5 MHz	5.5 MHz, 5.74 MHz	6 MHz	6.5 MHz
Vestigial sideband	0.75 MHz	0.75 MHz	1.25 MHz	0.75 MHz (1.25 MHz)
Vision IF	45.75 MHz	38.9 MHz	39.5 MHz	38.9 MHz (38.0 MHz)
Vision/sound ratio	5 : 1	10 : 1 20 : 1 20 : 1 : 0.2 [dual sound]	5 : 1	10 : 1

The *British modification* I of the 625 line CCIR standard, being one of the last systems adopted, represents today the best compromise providing:

    1. optimum utilization of 8 MHz channel width,

    2. increase of vestigial sideband from 0.75 MHz to 1.25 MHz, consequently a broadening the Nyquist slope from 1.5 MHz to 2.5 MHz and reducing group-delay error near the carrier to about 60 ns (half-correction); precorrection in the TV transmitter is thus not necessary in this part of the video band,

    3. bandwidth of upper sideband of chrominance signal increased from 0.57 MHz to 1.07 MHz, thus no "second" vestigial-sideband system is necessary for colour transmission,

    4. increase of residual carrier from 10% to 20% so that highly saturated colours (yellow) with modulation of the TV transmitter to ultra-white are better reproduced, involving, however, a loss in useful-signal level of 1.5 dB, which can only be compensated for by a 40% increase of transmitter power (e.g. from 10 kW to 14 kW).

Digital Coding of Colour TV Video Signals and Sound Signals

Detailed aspects of digital television and other advanced aspects are discussed in the later chapters. National and international organizations are attempting at present to establish a uniform digital coding standard or at least an optimal compromise for the TV studio and transmission on the basis of CCIR 601 recommendation for digital interfaces. Some aspects of these standards and other broadcast services are given here together for completing the information on standards, though these will be better appreciated after going through Chapters 24, 25 and 26.

TV Studio

The (West European) EBU has prepared the following *Digital coding standard* for video signals:

- Component coding (Y signal and two colour-difference signals);
- Sampling frequencies f sample in the ratio 4 : 2 : 2 with 13.5 MHz ($3 \times f$ chrominance) for the luminance component and 6.75 MHz ($2 \times f$ chrominance) for each chrominance component;
- Quantization q is 8 bits/amplitude value;
- Data flow/channel
  13.5×10^6 values/s $\times$ 8 bits/amplitude value
  $= 2\ (6.75 \times 10^6$ values/s $\times$ 8 bits/amplitude value)
  $= 108$ Mbits/s.
  108 Mbits/s + 2×54 Mbis/s = 216 Mbits/s.
  i.e. the required bandwidth is approximately 100 MHz.

Transmission

This high channel capacity can only be achieved with internal studio links via coaxial cables or fibre optics. In public communications networks of present-day technology, the limits per channel lie at the hierarchical step of 34 Mbits/s for microwave links, later 140 Mbits/s. Therefore great attempts are being made at reducing the bit rate with the aim of achieving satisfactory picture quality with 34 Mbits/s per channel.

Terrestrial TV transmitters and coaxial copper cable networks are unsuitable for digital transmissions. Satellites with carrier frequencies of about 20 GHz and above may be used.

The digital coding of sound signals for *satellite sound broadcasting* and for the *digital sound studio* is more elaborate with respect to quantizing than for video signals.

A quantization q of 16 Bits/amplitude value is required to obtain a quantizing signal-to-noise ratio S/Nq of 98 dB,

$$[S/Nq = 96 + 2)\ dB].$$

The sampling frequency must follow the sampling theorem f sample $=/> 2 \times f_{max}$, where f_{max} is the maximum frequency of the baseband.

Planned Sound Coding Standard

	f-sample	Quantization	q Data flow/chl
Direct satellite sound broadcasting with 16 stereo channels	32 kHz	16 bits	512 kbits/s
Digital sound studio	up to 48 kHz	16 bits	768 kbits/s

Broadcast of Special Services

Using television screen display, text communication systems known as *teletext/videotext* services are now on trial worldwide under various names. These are discussed in chapter 26.

Teletext

Preliminary standards typical for 625-line system B/G (FRG) and I (UK) are summarized below:

Clock frequency	6.9375 MHz, eqvt. to 444 fh
Half-amplitude duration	144.14 ns per bit

Data signal	H: 0.462 Vpp = 66% picture
	L: 0 V (blanking level)
Coding	8 bits/word incl. 1 parity bit
Code	NRZ (nonreturn to zero)
Words	per line 45; incl. 2 for clock run-in,
	1 for framing code, 2 for address code
Transmission time per line	45 words $\times$ 8 bits $\times$ 0.144 us = 51.89 us
	(TV line without blanking interval = 52 us)
Transmission time per page	$\dfrac{24 \text{ text lines}}{4 \text{ TV lines/picture}} = 6 \times \dfrac{1}{25} \text{ s} = 0.24 \text{ s}$
Wait time, max.	75 text pages 0.24 s = 18 s
average	9 s approx.
Lines occupied	1st field: 20/21, and field: 333/334

Viewdata/Bildschirmtext

Preliminary standard with page dialled from subscriber's telephone set and displayed on domestic TV receiver:

Text format	1 page of 24 lines of 40 characters
Coding	10 bits/character (word), incl. 1 each parity start and stop bit
Data load	24 lines $\times$ 40 characters $\times$ 10 bits
	= 9600 bits/page
Data flow	1200 bits/s
Transmission time	$\dfrac{9600 \text{ bits/page}}{1200 \text{ bits/s}} = 8 \text{ s/page}$
Modulation	F1B [FSK]
in data channel	H: 1300 Hz, L: 2100 Hz,
in return channel	H: 390 Hz L: 450 Hz

Channels and IF carriers The VHF/UHF channel frequency definitions for the various systems differ depending on the channel bandwidth and other factors. The IF carriers used in the receivers also vary from country to country as seen in the Table 3.3.

System M Features

FCC allocations for TV channels in the system M commonly used in the USA and other countries, are somewhat different from those in system B, besides the smaller channel width (6 MHz), and other differences listed in Table 3.4. The channel frequencies as per FCC standards are given in Table 3.4.

TV receivers for this system commonly employ vision IF of 45.75 MHz and sound IF of 41.25 MHz, with an intercarrier sound IF of 4.5 MHz. The field scanning rate is 60 Hz against 50 Hz used in system B, and the line scanning rate is 15,750 Hz, fairly close to the 15,625 Hz in system B. The system uses six pre-equalising and six post-equalising pulses with six serrations in the vertical sync pulse of $3H$ duration. Some channels in the two systems overlap. For example, channels 3 and 7 in this system overlap to some extent with channels 4 and 5, respectively, in system B. Hence, the pictures on the respective channels can be received and locked with some manipulation of the field and line oscillators. The sound, however, cannot be received because of the difference in the intercarriers, unless the intercarrier IF circuits and the video IF amplifier response is modified properly.

Table 3.3 If carriers used in the various systems

Standard	Channel width (MHz)	Vision IF (MHz)	Sound IF (MHz)	Intercarrier (MHz)
B, C/G, H	7/8	38.9	33.4	5.5
D/K OIRT	8	38.9	32.4	6.5
D/K China	8	37.0	30.5	6.5
E/L France	13.5/8	28.05	39.2	11.15
I	8	38.9	32.9	6.0
M, N	6	45.75	41.25	4.5

Table 3.4 The FCC Channels Frequencies in System M (VHF bands)

Channel number	Frequency range (MHz)	Picture carrier (MHz)	Sound carrier (MHz)
2	54–60	55.25	59.75
3	60–66	61.25	65.75
4	66–72	67.25	71.75
5	76–82	77.25	81.75
6	82–88	83.25	87.75
7	174–180	175.25	179.75
8	180–186	181.25	185.75
9	186–192	187.25	191.75
10	192–198	193.25	197.55
11	198–204	199.25	203.75
12	204–210	205.25	209.75
13	210–216	211.25	215.75

REVIEW QUESTIONS

3.1 Explain the basis for keeping the picture-sync ratio at about 7 : 3.

3.2 Sketch the CCIR B standard composite video waveform: (i) locked to H rate, (ii) locked to V rate on an oscilloscope, corresponding to the following patterns: horizontal bars, vertical bars, chessboard, horizontal gradation and vertical gradation bars.

3.3 What will be the nature of the pattern on the TV screen if the video signal contains (i) a 50 Hz sinewave pick-up sigal, (ii) a 100 Hz pick-up signal, (iii) a 15,625 Hz pulsed signal, (iv) a 31,250 Hz pulse signal, (v) a 15,625 Hz sawtooth signal, and (vi) a 5.5 MHz sinewave signal?

3.4 Define and explain the following terms: pedestal, gamma, pre-emphasis and de-emphasis, and intercarrier sound signal.

3.5 State the advantages and disadvantages of VSB modulation. Explain how the VSB reception of TV signal is compensated for its frequency response.

MULTIPLE-CHOICE QUESTIONS

3.1 The horizontal resolution of a picture depends upon
   (a) the number of horizontal lines
   (b) the horizontal scanning rate

(c) the vertical scanning rate

(d) (a) and (b)

3.2 VSB modulation is preferred for TV transmission because

(a) it reduces the bandwidth required to half

(b) it avoids the phase distortion problems at low frequencies

(c) it is less critical than SSB modulation

(d) b and c

3.3 The vertical resolution in a TV picture depends upon

(a) the vertical scanning rate

(b) the number of horizontal lines

(c) the focusing of the picture tube beam

(d) b and c

3.4 In the standard TV signal a 1.5 MHz video signal will be present at

(a) only the upper sideband frequency

(b) only the lower band frequency

(c) both sideband frequencies

(d) attenuated lower sideband frequency.

3.5 In CCIR B 525/50 standards the equalising pulses

(a) occur at double the line rate and have same width as the sync pulses

(b) occur at the line rate and have width equal to half the sync pulses

(c) occur at double the line rate before the field sync pulse and have a width equal to half the sync pulses

(d) occur before and after the field sync pulse and have a width equal to half the line sync pulse

3.6 The front porch is kept ahead of the line sync pulse

(a) to black out the line before the sync initiates the flyback

(b) to allow the video signal to settle to a constant amplitude before the line sync is initiated

(c) to blank out the transient oscillations in the horizontal deflection

(d) to prevent effect of video amplitude on the synchronization

(e) b and d.

3.7 The higher power in the lower sidebands of the VSB TV signal is compensated in the TV receiver by proper frequency response in

(a) the RF tuner

(b) the video IF amplifier

(c) the video amplifier

(d) a and b.

4

Television Pick-up Devices and Cameras

A television camera chain comprises an optical lens system, a camera pick-up tube and video processing circuits. The construction and characteristics of camera lenses and camera tubes are described in detail. This is followed by a discussion on video processing of the pick-up tube signal in the studio camera chain.

4.1 CAMERA LENSES

To televise a picture or a scene, it is first necessary to focus the light emitted or reflected from the scene accurately on the photosensitive surface of a camera tube by means of an optical lens system. Intelligent selection of lenses can considerably improve the quality of the image produced. It is, therefore, very useful to know the characteristics and properties of various lenses and the practical manner in which they are used in conjunction with the TV camera tube controls.

Focal Length
The focal length of a lens is the distance between the centre of the lens and the image formed by it of an object at infinity. Lenses of longer focal length are physically of greater dimension, and offer an effective magnification, compared to lenses of shorter focal length. This is illustrated in Fig. 4.1, where the image is focused on the face plate of a camera tube, with the help of a longer focal length in Fig. 4.1(a), and a smaller focal length in Fig. 4.1(b).

Magnification obtained with the longer focal length lens is obvious, as the image occupies a larger area of the face plate, and may fall outside the limits of the face plate area with a still longer focal length

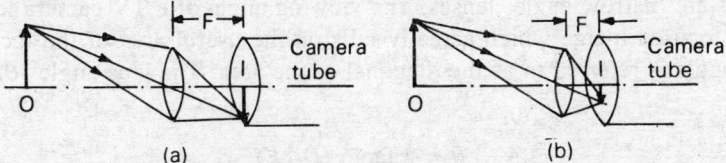

Fig. 4.1 Effect of focal length on viewing angle: (a) Longer focal length gives a larger image and a narrow viewing angle, (b) Shorter focal length gives a smaller image and a wide viewing angle

lens. It is, therefore, useful to employ a lens of longer focal length when observing distant objects in detail. Almost all TV cameras employ lenses that have focal lengths slightly adjustable by movement of the front element of the lens located on the barrel of the lens. Telephoto lenses of large focal lengths provide close-ups of distant objects by the magnification effect of the large focal lengths.

Focal Power

The focal power (ϕ) of a lens, expressed in diopters, is the reciprocal of the focal length f in metres. The shorter the focal length, the greater is the focal power. Convex lenses are assumed to have a positive focal power, while concave lenses have a negative focal power. The optical power of a system made up of two thin lenses, each of focal power ϕ_1 and ϕ_2, and separated by a small distance d, is defined as:

$$\phi = \phi_1 + \phi_2 - d\phi_1\phi_2 \tag{4.1}$$

Aberrations Due to Lenses

Failure of light rays emerging from the same point but taking different paths through the lens to converge to a single point, after passing through a lens, is called *spherical aberration*. As a result, the image shows a low contrast, highlights getting surrounded by a halo. The diameter of the halo of a point image (d_h) is:

$$d_h \propto h^3/f^2 \tag{4.2}$$

where h is the effective radius of the lens, and f the focal length.

Spherical aberration is obviously reduced by narrowing down the effective radius of the lens by placing an aperture stop or iris in the beam path. This, however, reduces the amount of light entering the lens and, hence, the light efficiency of the lens. Another method, which does not reduce the light aperture, employs a combination of several lens elements with two or more convex or concave lens elements. The focal power of such a lens system is diminished but the aberrations are considerably reduced.

Compound Lenses

Lenses used in photographic cameras generally consist of a combination of simple lens elements matched for radii of curvature, thickness, refractive indices and separation between them to minimise spherical aberrations and other optical distortions like coma, astigmatism, chromatic aberrations, etc. Light efficiency of the lens in producing a sufficiently bright and large image depends upon the speed of the lens, the reciprocal of the f-stop number defined in the following paragraphs.

Viewing Angle of Lens

It is obvious from Fig. 4.1 that a lens of shorter focal length views a wider area than one of a longer focal length. This implies that lenses of shorter focal lengths are 'wide angle' lenses while those with longer focal lengths are 'narrow angle' lenses. The viewing angle of a TV camera lens usually refers to that portion of the focused image which actually falls on the useful area of the face plate of the pick-up tube. The lens angle is referred to as the diagonal of the area. The lens angle (θ) is obviously given by:

$$\theta = 2 \tan^{-1} (D/2F) \tag{4.3}$$

where D is the diagonal and the F focal length. Horizontal (θ_H) and vertical (θ_V) viewing angles can also be expressed in terms of width W and height H, respectively, as given by the relations:

$$\theta_H = 2 \tan^{-1} (W/2F); \quad \theta_V = 2 \tan^{-1} (H/2F)$$

Field of View

This describes the width and height of the scene viewed and is determined by the focal length, lens-to-object distance and the width and height of the scanned area. Obviously, by geometry,

$$\text{Field of view (width or height)} = \text{scanned width (or height)} \times \frac{\text{lens-to-object distance}}{\text{focal length}}$$

Lens Speed

The speed of a lens is determined by the amount of light it allows to pass through it. Lenses of greater diameters are thus faster. Most camera lenses have a variable opening called *iris* located in the lens assembly that can alter the lens speed. It can be used to compensate for varying conditions of scene illumination, camera sensitivity, etc.

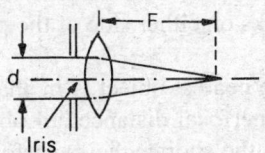

Fig. 4.2   Lens speed = $f/(F/d)$

Lens speed is defined as the ratio between the aperture, i.e. the lens opening d and the focal length F, viz. $d/F = 1/(F/d)$ as indicated in Fig. 4.2. The lens speed is expressed as (f-stop number) = $f/(F/d)$, where F/d is referred to as a 'stop number' marked on the iris or aperture ring of the lens barrel as $f/1$, 1.4, 2, 2.8, 4, 5.6, 8, 11, 16, 22, etc. When the aperture is adjusted to a higher stop, the amount of light is typically reduced by the square of the ratio of the stop numbers.

Depth of Field

When a lens is focused on an object at a certain distance, objects nearer and also those farther away may not appear in focus. 'Depth of field' is the distance between the nearest object in focus and the farthest object in focus, when the lens is nominally focused on a given point. This is the distance on the object side of the lens. 'Depth of focus' is the distance between the nearest in-focus image behind the lens and the farthest in-focus image behind the lens when the lens is nominally focused on a given point. This distance is on the image side of the lens.

The depth of field depends upon a number of factors like the focal length, aperture and the lens-to-object distance. Variation in these factors causes defocusing to some extent. This may not be apparent to the human eye up to a certain limit decided by the visual acuity or resolving power of the human eye.

A point object observed at a distance may be defocused without noticeable distortion or blurring. This limiting area over which the image of the object may be blurred or diffused without being noticed is called the *circle of confusion* of the human eye.

TV camera tubes also have a circle-of-confusion characteristic due to the resolution limit set by the mosaic structure of photosensitive elements and the finite scanning beam spot size.

A smaller aperture, i.e. larger stop number, gives a greater depth of field because the light rays from the outer edge of the lens are cut off, narrowing the maximum convergence angle at the focal point as shown in Fig. 4.3.

This reduces the defocused image area to a smaller circle. The circle of confusion must be, in practice, kept smaller than a scanning line width, unnoticeable by a television viewer.

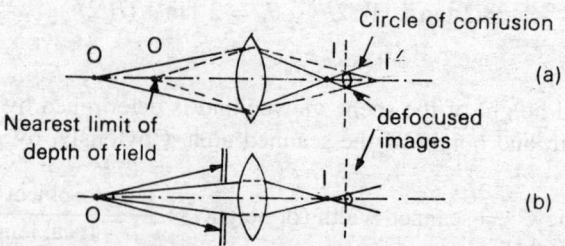

Fig. 4.3   Circle of confusion: (a) Larger aperture gives a larger circle of defocusing, (b) Smaller aperture gives a smaller circle of defocusing.

The wide angle, i.e. large focal length, lens exhibits a larger depth of field for a particular distance than a narrow angle lens because of the smaller shift in the focal point in a wide angle lens with respect to change in distance of the object. Often, lenses are calibrated by manufacturers along the focal ring to show the approximate depth of field with respect to the lens speed in the stop number setting. With the focal pointer at a particular distance in focus, the stop number marks on either side of the pointer indicate the depth of field limits on the focal distance ring.

The 'hyperfocal distance' of a lens is the distance at which the nearest object is in sharp focus when the lens is focused on to infinity. If the lens is focused on the hyperfocal distance, all objects from one half the focussed distance to infinity will be in sharp focus. Thus the shorter the hyperfocal distance of a lens, the greater is the depth of field.

Hyperfocal distance H is given by:

$$H = \frac{1}{d_c} \cdot \frac{F^2}{f} \tag{4.4}$$

where d_c is the diameter of circle of confusion, F the focal length, and f the stop number.

When the object-to-lens distance is D and the hyperfocal distance is H, the limits of the depth of field are given by:

$$\text{Nearest limit} = \frac{H \times D}{H + D}$$

and

$$\text{Farthest limit} = \frac{H \times D}{H - D}$$

Example: Consider a TV camera using a 50 mm lens at $f/8$ stop, height of scan area 24 mm and 625-line scanning.

$$\text{'Circle of confusion' diameter } (d_c) = \frac{\text{height of scan}}{\text{active lines}}$$

$$= \frac{24}{600} = \frac{1}{25} \text{ mm}$$

Hence, the hyperfocal distance $H = \dfrac{25 \times 50^2}{8}$ mm

$$= 7.5 \text{ m}$$

If the lens is focused at 7.5 m, all objects from half this distance, viz. 3.75 m to infinity, will appear in focus.

If the lens is focused at 1 m,

$$\text{Nearest limit in focus} = \frac{7.5 \times 1}{7.5 + 1} = 0.89 \text{ m}$$

$$\text{Farthest limit in focus} = \frac{7.5 \times 1}{7.5 - 1} = 1.15 \text{ m}$$

It will be observed that the depth of field increases if a lens is focused farther away, while it is reduced considerably when the lens with the same stop number is focused for a close-up.

Lens Turret

Television cameras can produce images to different scales according to the viewing angle or the focal length of the lens employed. Narrow angle (less than 20°) lenses are suitable for telephoto or close-ups. Medium angle (20° to 60°) lenses are suitable for commonly televised scenes and are called universal lenses. Angles over 60° are used in wide angle lenses required for wide angle location shots.

The lens complement of a television camera is mounted on a turret. Use of turret-mounted lenses with different viewing angles enables a quick change of lenses in operation. The less turret is screwed on to the front of the camera, and rotation of the turret brings in the desired lens in front of the camera tube. An image orthicon turret typically holds four lenses of different focal lengths, 35 mm, 50 mm, 150 mm and a zoom of 40 to 400 mm with facility of adaptor up to 1200 mm.

Zoom Lenses

In modern cameras, particularly colour cameras, the turret lens system has been completely eliminated by the use of a zoom lens with a variable focal length of range 10 : 1 or ever greater. The zoom lens offers wide lens capabilities. The variation of viewing angle and field of view without loss of focus provides a dramatic close-up control as indicated in Fig. 4.4(a).

This appears to the viewer as if he is approaching or receding from the scene. The variable focal length of a zoom lens is obtained by moving individual lens elements of a compound lens assembly. If spacing d between two lens elements of focal powers ϕ_1 and ϕ_2 (reciprocals of focal lengths) is changed, the focal power of the combination given by $\phi_{tot} = \phi_1 + \phi_2 - \phi_1 \phi_2 d$ will also change, enabling the focal length or viewing angle to be controlled. This affects the distance of the image also, requiring movement of one more lens element.

An important feature of a variable focal length zoom system is to keep the focused image plane fixed relative to the stationary element of the system while the focal length is varied throughout the range provided. This feature can be provided by two types of arrangements as shown in Fig. 4.4(b) and (c). In one type of zoom lens shown in Fig. 4.4(b), there are two movable lens elements. Movement of the concave element B produces a change in the focal length, while a smaller movement of element A gives mechanical image shift compensation to hold the image plane stationary, relative to the rear fixed element C. In this type of arrangement, at least one of the lens movements requires use of some kind to non-linear linkage or cam control.

Another type of arrangement for variable focal length consists of a sequence of alternate stationary and movable elements as shown in Fig. 4.4(c). The relative focal powers and positions of all the elements are chosen to permit the two movements to be identical so that they can be linked together mechanically as a single linear movement. This arrangement can provide exact compensation for the image plane shifting at a number of points in the range of the focal length, leaving small finite errors of focusing in the intermediate ranges between the corrected points. This is shown by the dotted line of image shift in

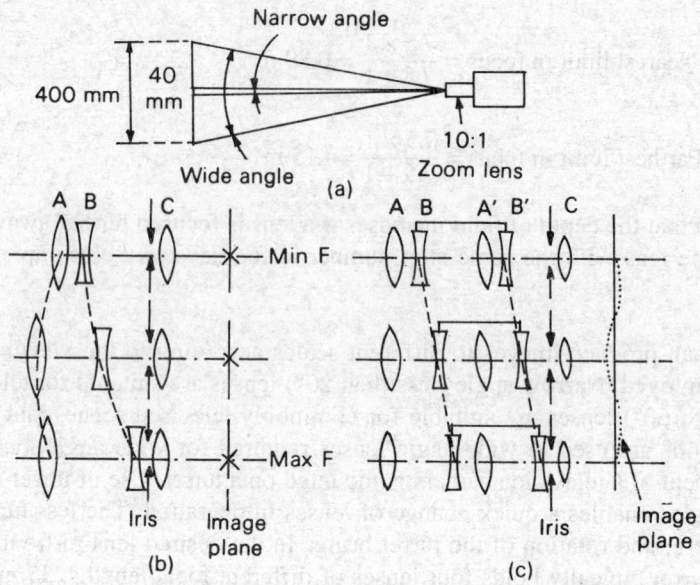

Fig. 4.4(a) Field of view of a 10 : 1 zoom lens, (b) Arrangement using mechanical compensation of image shift, and (c) Arrangement for optical compensation of image shift

Fig. 4.4(c). The method relies broadly on the optical compensation for the image shift. The iris or aperture of the system is generally situated near the rear stationary element C, and is also controlled automatically to maintain a constant *f*-stop number of the lens and, hence, the illumination on the face plate of the pick-up camera tube.

The zoom lens can in principle simulate any fixed lens that has a focal length within its zoom range. It should be remembered, however, that the zoom lens is not a fast lens. Under poor lighting conditions, faster fixed focus lenses mounted on a turret are preferred. Zoom lenses are subject to greater aberrations because of the difficulty in compensating them with so many movable lenses.

Lens Mounts

A wide range of special camera lenses are available to match the various applications for which a camera can be used. Hence the lens mounting system has been standardized to enable maximum possible applications by interchanging the lens system. Surveillance cameras require wide angle lenses to permit large areas in their field of view. Telephoto lenses permit a close observation of distant objects. Microscopic adapters and fibre optic systems are useful in observations of close-ups. The so-called *C-mount*, derived from the 16 mm film cameras, which uses a threaded cylinder about 25 mm in diameter, has been standardized for CCTV cameras also. In professional cameras the *bayonet mount* is more common, in which the lens uses index tabs that match the recesses in the cameras for mounting and holding the lens. A locking ring secures the lens in place by a partial turn.

For establishing the correct focus, the practical reference is the flange at the back of the lens which comes in contact with the front of the lens mount. When the camera is first set up, the pick-up tube is racked back and forth until the correct optical focus is obtained. This adjustment is often called the *flange-back or back-focus* adjustment.

4.2 AUTO-FOCUS SYSTEMS

Focusing has been made automatic in modern cameras, relieving the chore of adjusting optical focus every time the camera is moved to point at different objects. By auto-focus systems, operator adjustment errors are also eliminated. Automatic focus employs a servo system to control the drive motor for the mechanical movement of the lens focus ring, in accordance with the actual measure of distance between the lens flange and the object. The distance is measured by a sonar type ultrasonic system, or by an infrared system using the triangulation technique.

Ultrasonic Sonar Technique

In this method, an ultrasonic pulse of energy of short duration (typically 70 kHz, 380 μs), is sent towards the object by means of a piezoelectric ceramic transducer, and the time taken for receiving the reflected pulse is measured to represent the distance. The lens focus ring is adjusted by its motor drive through a microprocessor controlled amplifier such that the distance reading on the lens ring tallies with the distance measured in terms of the clock count between the transmitted and the received pulse. The clock reading is latched and compared with a digital readout of a digital coding ring for generating the correction signal.

The transmit pulse is short, timed during the vertical blanking interval to avoid interference in picture. The system has to be receptive for a large variation in the strength of the received echo and should be insensitive to stray sonics. This is achieved by creating a time-varying reference voltage profile for expected echo-strength from human objects, and allowing only those echos that exceed this to be fed through.

Infrared Triangulation Technique

In this method, a beam of infrared light from an infrared LED is directed along the optical axis of the camera lens. The IR light reflected from the object intercepting the beam is received and focused by a receiving lens into a detector block, moving across through a servo motor and mechanically linked with the camera lens focus ring, as shown in Fig. 4.5.

Light reflected from a distant object falls to the left, while light reflected from a near object falls to the right of the detector block. The detector block moves across the optic axis and rotates the focus ring along with it, through a servo motor. The servo motor is driven by a servo amplifier fed by an error signal obtained as differential input from a pair of infrared sensitive photodiodes mounted adjacently on the detector block. These photodiodes can provide a differential error signal, because of the different angles at which the reflected echo is received from objects at different distances. The error signal is used to set the correct focus by rotating the motor in the proper direction to stabilize when the diode outputs are equal.

4.3 TELEVISION CAMERA PICK-UPS

A television camera pick-up tube/device serves as an eye of the television system, consisting of a light sensitive target and scanning or sensing system to generate electrical signals. It must have characteristics similar to the human eye, viz. sensitivity to the visible spectrum, a wide dynamic range with respect to light intensity and good resolution capability. Several types of camera pick-up devices have been developed employing different target materials to meet the needs of operating characteristics, the physical size, the cost factors, operating conditions, etc. The targets of these devices basically utilize either (i) the photoemission

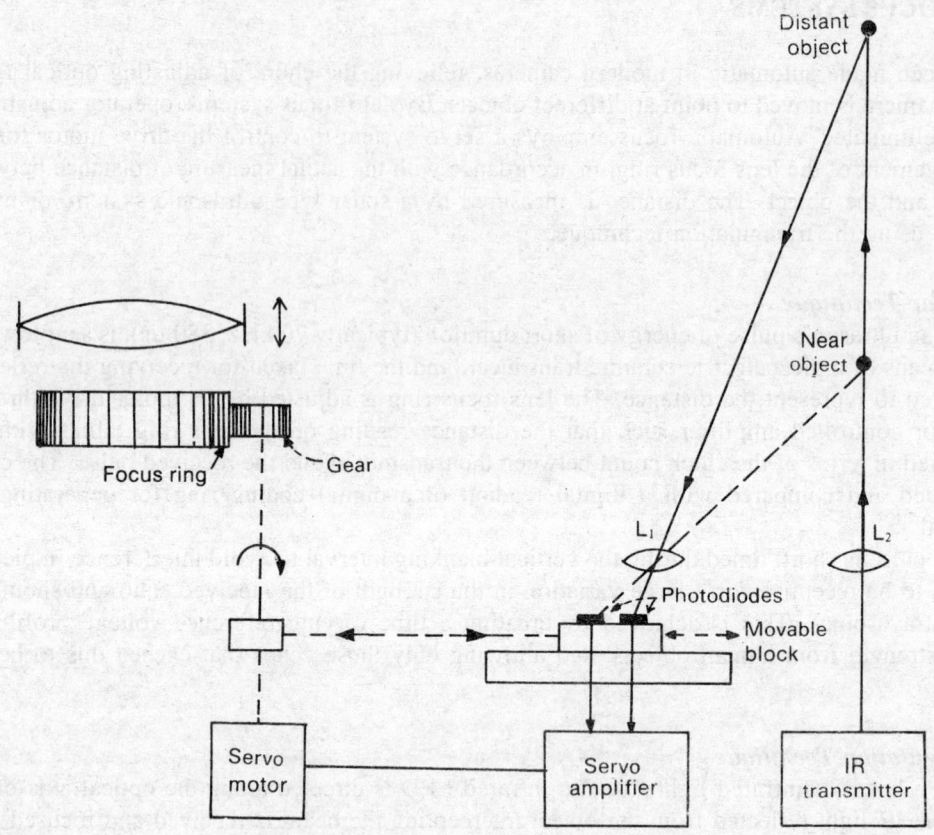

Fig. 4.5 Infrared autofocus triangulation

principle as in image orthicon; or (ii) the photoconduction process in bulk semiconductors as in vidicon, or junction semiconductors as in plumbicon, newvicon tubes, or completely solid state CCD sensors.

Camera Pick-up Characteristic

The performance of a camera pick-up is given by the following main characteristics and properties:

(i) *Light transfer characteristics* This is the plot of output current against the face plate illumination, on a log-log scale. The slope of this plot is defined as the 'gamma' of the pick-up.

(ii) *Spectral response* For black and white television, the pick-up should have the same spectral response as that of the eye so that colours are rendered in their proper gray tones. Pick-ups for colour TV have a greater response to the respective primary colour each pick-up handles. Special tubes with spectral response in infrared, ultraviolet or even x-rays are also available that surpass the vision capability of the human eye.

(iii) *Sensitivity* This is the output photo-signal current per lumen, measured with a standard light source—a tungsten filament lamp operating at a colour temperature of 2870 °K.

(iv) *Dark current* Even if there is no illumination on the face plate of a camera pick-up tube, there is a small amount of signal current flowing in the output circuit. This current is called the dark current.

It is generated by electron hole charge carriers in the photosensitive surface due to thermal and other energies, and may limit the low illumination sensitivity of the pick-up tube.

(v) *Lag characteristics* This refers to the inability of the photosensitive layer in the pick-up to follow faster changes in illumination. After removal of high light illumination, the photoelectrons and the holes takes a finite time to disappear, leaving a fading memory of the previous scene and cause comet tails or smears in the moving objects.

(vi) *Resolving power* It is found that as the number of black and white lines of resolution in the picture is increased, the signal current produced is not able to attain full changes in the black and white levels. Consider a pattern consisting of different widths of black and white strips projected on the face plate of the camera pick-up, as shown in Fig. 4.6(a).

Some parts of the pattern contain wide black and white stripes, such that say 40 of them will fill the photosensitive screen on the face plate of the camera tube. At other places, the stripes are narrow, such that say 400 of them will fill the width of the screen.

When an electron beam scans the picture, the corresponding charge image on the photosensitive screen produces signal currents with a fundamental frequency having a period equal to the time required for scanning of the wide stripes. Current changes during the scanning of the wide stripes will yield a peak-to-peak value of b, while during the scanning of narrow stripes, current changes for the same levels of dark and white are much less, corresponding to a as shown in Fig. 4.6(a), the alternations being much faster.

The ratio a/b expressed as the percentage modulation depth is commonly adopted as a measure of the resolving power of a camera tube. The percentage modulation depth a/b is reduced as the number of lines of resolution is increased due to cross modulation and lags between adjacent picture elements (pixels) and a certain amount of scattering of light. Besides this, there will be lag processes in the scanning circuit to affect it. A typical plot of percentage modulation with the number of lines of resolution for a plumbicon camera tube is shown in Fig. 4.6(b).

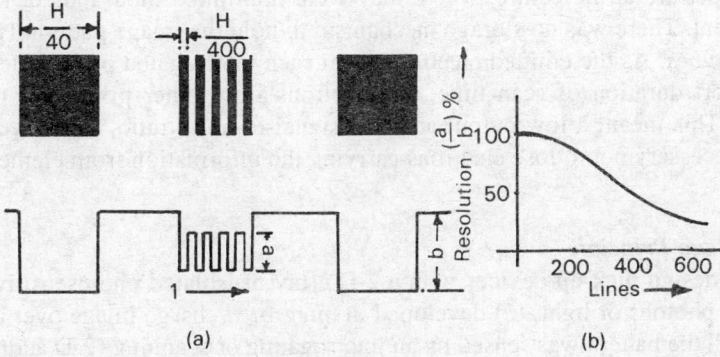

(a)   (b)

Fig. 4.6 Resolving power of a camera pick-up: (a) Single current modulation, and (b) Percentage modulation

Types of Camera Pick-up Devices

TV camera pick-up devices use a photosensitive surface for conversion of optical image into electrical signal. Depending on the type of material used for this surface the tubes can be classified as photoemissive o photoconductive camera tubes. Of the *photoemissive camera tubes*, image orthicon has been the most common, primarily used for broadcast studio applications because of its high sensitivity, stability and high quality picture. The vidicon because of its small size, simplicity and low cost has been popularly used for industrial, educational and aerospace applications besides its special use in broadcasting of stills and telecine projections.

Because of its large size and complex construction and set-up procedure, the image orthicon tube was not quite suitable for colour cameras and hence was gradually replaced by the *photoconductive camera tubes* like Vidicon, Plumbicon, Saticon, Newvicon and other derivatives with different types of photoconductive target layers. In such tubes, a photoconductive layer on the faceplate converts the image focused on it into a charge distribution which is then scanned by an electron beam and transformed into an electrical signal. These tubes are compact and simpler in construction suitable for multitube colour TV cameras. Variations of vidicon with modified targets like 'silicon multidiode vidicon', are used in special industrial applications. For compact portable colour TV cameras, *single tube colour dissectors* have been developed to separate the primary colours at the target itself through a set of integrated colour stripe filters, also referred to as *Filter Integrated Colour (FIC) vidicons*. These are discussed in Chapter 24, along with portable CCTV camera systems.

CCD image sensors are based on solid state integrated circuit technology. They trap the photon generated electron holes at each photosite pixel and scan these by suitable charge transfer techniques. Such sensors are the latest addition to camera pick-up devices, and in the package of a standard IC, are most suitable for compact, lightweight minicameras for consumer as well as military and space surveillance applications.

4.4 EARLY CAMERA TUBES AND THE STORAGE PRINCIPLE

All the early camera pick-up devices employed the photoemitter targets. The first one was the *image dissector* in which the electrons emitted from the target plate on one face of the tube were accelerated and brought to focus on a plane on the opposite side of the tube. The focused streams of photoemitted electrons were directly used to obtain the picture signal. They were deflected by horizontal and vertical magnetic fields, to enter a narrow aperture at the centre, where they were multiplied thousands of times to provide the scanned signal current. There was no storage mechanism to hold the image pattern. Hence such tubes are called *non-storage tubes*. As the emitted electrons from each interrogated picture element, were directly used for a very short duration of scan time, the electrons from other pixels not passing through the aperture were lost. This meant a low output and low signal-to-noise ratio, unless very high light levels were used. It was necessary not to lose electrons carrying the information from elements not interrogated or scanned.

Charge Image Storage Principle
It was necessary to design pick-up devices with a 2-D array of isolated photosensitive picture elements, which (1) *absorbed* photons of light, (2) developed in *integrated* charge image over a image scan period and (3) *stored it* until the pattern was sensed by an interrogating or scanning (2-D addressing) mechanism. In the camera pick-up tubes, a scanning electron beam was used to sense the amount of charge stored at each pixel to develop a fully time domain signal whose waveform corresponded to the brightness at each point in succession as the scanning proceeded. In storage type pick-ups, the device integrates and stores the photon generated charge on each pixel obviating the loss of potential information during the interval (about 20 ms) between successive scans at that pixel. The scanning in CCD solid state sensors is by two dimensional pixel addressing by suitable pulses applied to the integrated cells.

As photoemitters are by themselves good conductors of electricity, continuous films of photoemitters could not be directly used as storage planes for the charge image. Early camera tubes used photoemissive material in discrete grannular arrays on insulating substrates. This was done in the *iconoscope,* which was the first storage type tube developed (see Sec. 2.3 for construction). A mosaic of isolated photoemissive globules of caesium-silver compound was constructed on an insulated mica substrate. Light incident on the mosaic was absorbed to emit electrons that were attracted by the surrounding collector, creating a positive *charge pattern on the mosaic* that was scanned by electron-beam scanning the surface from the same side. Exposing the mosaic plate to light and scanning it by electrons from the same side presented problems in construction and scanning. The *high velocity scanning* employed created secondary electrons which along with the photoelectrons raining back to the positive charge image gave an unstable output.

Low Velocity Scanning

The scanning beam is used to interrogate the pixels and sense the amount of charge generated by the incident light. The beam deposits electrons on the positively charged pixels to develop corresponding signal. As the electrons in the beam approach the target, they are decelerated to land at the target pixels at a very low (near zero) velocity, averting secondary emission. With low velocity scanning it was found convenient to utilize the return beam method of signal generation. Electrons that are not used to recharge the target pixels are slowed to stop in front of it and returned towards the gun, carrying equal but opposite polarity of information. The return beam can be directed to an electron multiplier at the cathode to improve signal-to-noise ratio before amplification.

In order to simplify scanning by electron beam from the mosaic side, the back plate of the insulated mosaic was made a thin transparent signal electrode, and a thin 2-D array of photoemissive material was placed on transparent mica or glass insulator as in a family of *low velocity* electron beam tubes called *orthicons* and CPS (cathode potential stabilized) *emitrons*. The magnetic focusing aided perpendicular *orthogonal* landing on the target. Though these tubes gave good resolution with precise grey scales, they showed instability on highlights, and picked-up stray signals from ac fluorescent lighting. The magnetic focusing of the beam in multiple nodes helped orthogonal landing on the target at low velocity.

Return Beam Signal

When a near-zero velocity beam is used for scanning the target, it can be conveniently used as a return beam to carry the pixel information in inverted form. Electrons not used to neutralize the positive charge on the pixels are slowed down in front of the pixels to stop are returned towards the gun following a path alongside the forward beam. The return beam carrying information of opposite polarity is guided into an electron multiplier to improve the signal level and sensitivity in a *multiplier orthicon*.

Image Orthicon

In an image orthicon the two-sided target structure was further developed to enable use of a more efficient continuous film of semitransparent photocathode. By formation of the charge image on a target plate separate and away from the photocathode, the problems of signal instability on highlights and stray signals due to fluorescent lighting were also eliminated. First produced in 1945, image orthicon became popular in broadcast studios replacing the earlier tubes like iconoscope, orthicon, and image dissector. It could operate over a wide range of illumination levels, from bright sunlight to dark shadows and produce high quality pictures in various lighting conditions including outdoor scenes of high contrast, to be accepted as 'workhorse' of the television studio until its place was taken over in the late 70's by improved versions of the plumbicon, which had the compactness required in colour cameras. The image orthicon is now of only historical significance. The schematic diagram of the image orthicone is shown in Fig. 4.7. The tube consisted of three sections:

(i) *Image section*: This is where the optical image on the photocathode is converted into a charge image and is then transferred on to the target plate, as shown in Fig. 4.8.

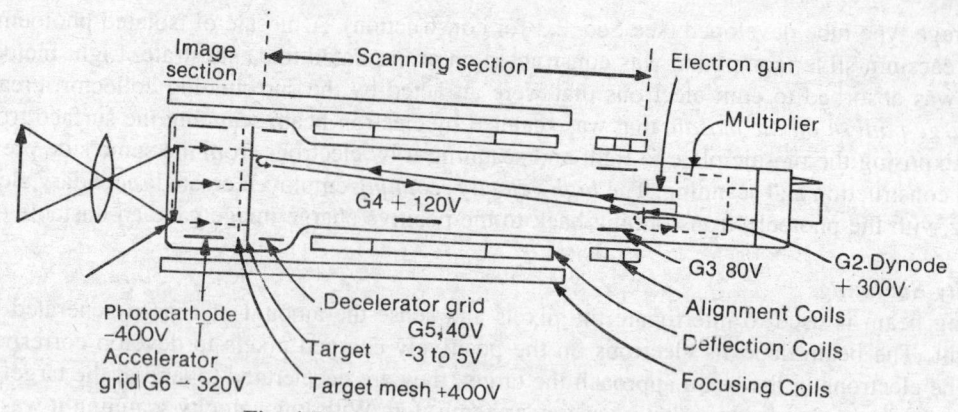

Fig. 4.7 Schematic diagram of the image orthicon

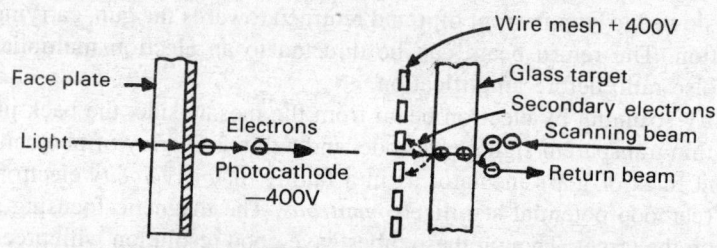

Fig. 4.8 Formation of charge pattern on the target

Proportional to the light incident on the transparent photocathode, photoelectrons are emitted from each pixel on the cathode held at about – 400 V relative to target plate. These are focused on to the target at corresponding points, accelerated to the thin glass target plate to produce secondary emission from it. The secondary electrons are collected by a fine mesh very close to it, leaving electron deficiencies or a positive-charge image on the target. Electrons at the back of the thin semiconductor glass target leak through the thin section to transfer the charge image to the back, so that it can be scanned by the electron beam in the scanning section.

(ii) *Scanning section* Here, the high resolution beam from the electron gun scans with the help of deflection coils, the charge image on the target at a low velocity, as the beam is decelerated by grid G3. The scanning beam is maintained in focus by the focusing coil and electric field of G4. The return beam, deficient of electrons proportional to the positive charges on the back of the target, is amplitude modulated proportional to light variations in the scene. It is returned towards the cathode alongside the forward beam, but is guided into the five-stage electron multiplier.

(iii) *Multiplier section* The dynodes of the multiplier are coated with low work function material capable of high secondary emission. The returning electrons strike the first dynode surface of grid G2 that surrounds the cathode. Return beam electrons are multiplied as they strike the dynodes and further multiply as they accelerate by progressively higher potential of the multiplier dynodes and finally to the anode. The overall magnification is of about 1000, giving a signal of 500 mV across 20 k.

Performance of IO
With a variety of bi-alkali and multi-alkali materials used in the photocathodes, the image orthicon could provide a variety of spectral responses, from the one close to human eye response to extremely high response in the infrared region. Sensitivity obtainable was high, equivalent to 5000 ASA or more.

The light transfer characteristic is linear in the dark region with a gamma of unity, bending into a knee on highlights and stabilizing at higher lighting levels. The knee is reached when the illumination causes the target to be fully charged with respect to the mesh which attracts no more secondaries. With extreme highlight sensitivity and *self-adjusting knee* in exposure transfer characteristics, the image orthicon could produce high quality pictures over a wide lighting range including outdoor scenes of high contrast. Operated on higher illuminations, it had average gamma is 0.5 to 0.6, complementary to picture tube performance.

4.5 VIDICON

The vidicon was the first tube based on the photoconductive principle, developed by RCA in the early 1950's. Because of low cost, small size (1-inch diameter), simplicity and ease of operation, it became at once popular in CCTV applications. With advances in vidicon technology, its characteristics were improved greatly to make it attractive in some TV applications. Use of different types of photoconductive materials as target layers led to the development of other compact camera tubes with highly improved characteristics suitable for colour TV broadcast applications. These derivatives of vidicon include the *Plumbicon, Saticon, Newvicon and Chalnicon*, and their basic arrangement of the electron gun, electrodes and coils is similar to that of the vidicon, with some improvements in the gun design.

Construction

Figure 4.9 illustrates the general construction of a photoconductive tube like vidicon or plumbicon. An electron gun produces the scanning electron beam, which is directed by the *focusing and deflection coils* to land upon and scan the target containing a photoconductive layer. The electron gun comprises an indirectly heated cathode (0 V), control grid G1 (– 100 V) and the accelerating grid G3 (275 to 300 V). The voltage on G1 controls the beam current, and Grid 2 (first anode) accelerates the electrons, which subsequently pass through the cylindrical electrode G3 and a fine mesh G4. The electrostatic field of the grid and the magnetic field of the external focusing coil are both used to focus the electron beam on the target. The fine wire mesh usually connected to G3 serves to provide a uniform decelerating field between the fine wire mesh and the target so that the electron beam lands on the target perpendicularly, at a low velocity without producing secondary emission. Two sets of *alignment coils* produce an adjustable transverse magnetic field, ensuring that the beam is aligned parallel to the tube axis, so that it lands perpendicularly on the target.

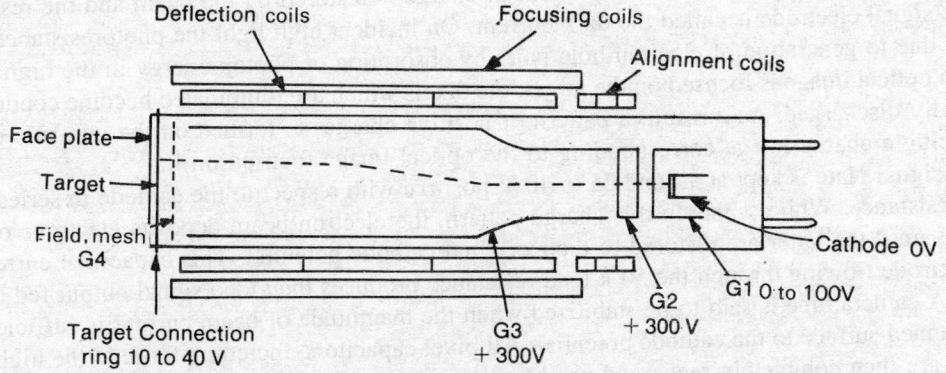

Fig. 4.9 Construction of a photoconductive camera tube like vidicon

The *target* section consists of an optically flat faceplate, on the inside of which a transparent conducting film of tin oxide, also known as *Nesa glass*, is deposited. The conducting tin oxide layer forms the electrical contact with the target connecting to the metal target electrode through a ring of indium. A thin (2- to 3-μm) layer of photoconductive material such as selenium or antimony trisulphide (Sb2 S3), is vapour-deposited on the conducting film usually in two to four sublayers. Its properties are dependent upon the antimony-sulphur ratios and the porosities of the sublayers. Two types of layers are in use: type A for standard industrial and educational applications and type B for x-ray image intensifier applications in medicine.

Photoconductive principle   The basic principle of operation of a vidicon is illustrated in Fig. 4.10.

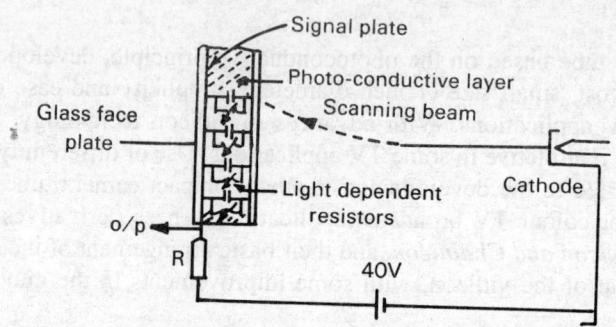

Fig. 4.10   Circuit for signal production in vidicon

The photoconductive target material in vidicon is an intrinsic semiconductor which has a very high resistivity in darkness, capable of decreasing with increasing illumination. In a small area on the surface, it can store a charge without appreciable lateral spread. The layer can therefore be considered as a mosaic structure of isolated picture elements (pixels) each equivalent to a light dependent resistor (LDR) in parallel with a pixel capacitor. When the target is scanned, each pixel capacitance of it gets charged by the beam electrons approaching the target at a low velocity; and continue to land until the scanned surface is stabilized at the cathode potential approximately. Thus a potential difference is established across the layer, with each pixel capacitor charged to nearly the same potential as that applied to the signal electrode.

Charge Image

In the dark, the photoconductive layer has a high resistance of about 20 M, so that very little charge leaks away between successive scans. This small amount of charge is restored by the beam and the resulting current to the signal electrode is called the dark current. On incident high light the photoresistance falls to about 2 M due to generation of electron-hole pairs by absorption of photon energy in the high light. Thus when an optical image is focused on the target, the pixels which are illuminated become conductive and are partially discharged. As a result, a pattern of positive charges is formed on the gun side of the target, producing a charge image corresponding to the optical image focused on it.

The target signal plate is kept at around 25 to 40 V positive with respect to the cathode in series with a 50 k load resistance. While scanning this charge pattern, the electron beam deposits electrons on the positive pixels until the latter are restored to their original cathode potential. This capacitive current to the signal electrode flowing through the 50 k load resistance produces the video signal output fed to the preamplifier. A camera tube is said to be stabilized when the magnitude of beam current is sufficient to restore the scanned surface to the cathode potential. All pixel capacitors, including those at the highlight of the image, are then completely recharged by the scanning beam.

Typical illumination can produce signal peak currents of 300 to 400 nA, which across 50 k load resistance produce 15 to 20 mV signal voltage, with a dark current of about 20 nA.

Storage Property

Each of the pixels in the target layer is sampled repetitively during the scanning at the frame rate. Hence the change in the photoresistance due to current carriers released at a given location due to the incident light integrated over a full period of one frame, sets the charge pattern in readiness for the next scan. The light information at each pixel is thus integrated over the time elapsed between the samples, and stored in the form of a light-dependent positive charge image pattern on the cathode side of the pixels. This storage property of the capacitive pixels of the photosensitive target accounts for the high sensitivity of vidicon-like tubes.

Dark Current

Even when there is no illumination on the face plate, there is a slight discharge of the pixel capacitors owing to the finite value of the resistance of the photoconductive layer. This produces some recharging signal current in dark conditions also, known as the dark current. Vidicon dark current is due to electron-hole pairs generated by thermal energy and increases with temperature and applied target voltage. As the target is very thin (2–4 μm), the electric field gradient across the photoconductive layer is very high. Increase in the target voltage increases the field gradient and the probability of recombination of the light generated electron hole carriers reduces. This is reflected in increase of the dark current and the sensitivity.

Image Lag

The time delay in establishing a new signal current in the camera to follow the rapid changes in the target illumination is called 'image lag' or simply 'lag'. In the photoconductive camera tubes this occurs in two forms: (i) *photoconductive lag* determined by the properties of the target materials, and (ii) *capacitive lag* or *beam lag* attributed to the storage effect of the pixel capacitance and the beam resistance. The image lag causes *smear or comet tails* following fast-moving objects in the scene, and the prolonged exposure of a bright stationary object results in a slow decaying after image of x-ray type appearance. This long-term but faint after-image is called *burn-in* or picture sticking.

Photoconductive Lag

This is due to the sluggishness of the photoconductive target elements in following the brightness changes. In vidicon, the Sb2S3 used as the semiconductor target material is an *n*-type semiconductor shows considerable lag and burn-in. In the *n*-type semiconductor, the minority carriers—holes—continue to arrive at the scanned surface for a considerable time even after high light has been removed. These carriers reach the negative side directly, or by a process of being captured and released by the doped *n*-type (sulphur) impurities. There is constant interchange between the *holes trapped in the impurities* and the holes actually flowing through the crystal. Materials like antimony trisulphide or selenium have many traps accentuating lag effects at high target voltages, where the vidicon has higher sensitivity required at low light levels. It is better to work with high light levels with low contrasts, as is possible in telecine and studio lighting conditions.

Capacitive Lag

The time for recharging of the target elements by the electron beam during each scan is determined by the pixel capacitance and the beam resistance time constant $C_t R_b$, where the beam resistance R_b is given by the scanning face potential divided by landed beam current. For typical values of face potential of about 0.2 to 4 V with respect to cathode, and a beam current of 0.3 µA, the beam resistance is in the range of 1 to 10 MΩ.

If the time constant $R_b C_t$ is large, the recharging is not complete in one scan and for moving objects, incomplete erasure will produce a smearing tail behind the moving object. High illumination of bright areas tend to stick. The beam lag is kept low by reducing the beam resistance by providing sufficient excess beam current that will charge the discharged high light pixel capacitors fast in one scan only. However, too much beam intensity (adjustable by the grid cathode bias) increases spot size and the resolution will be reduced.

For measurement purposes, this lag can be defined in two forms, *decay or discharge lag* occurring at the transitions from light to dark and *build up lag* occurring at the transitions from dark to light. The decay lag is expressed as a percentage ratio of the residual signal current measured 60 ms and 200 ms (at 50 Hz) after removal of illuminance to the initial highlight signal current stabilized over 5 ms exposure. The build up lag is measured after 10 ms of darkness expressed as the percentage ratio of the intermediate signal current, measured 60 ms and 20 ms (at 50 Hz) after restoring the high light, to the final current.

The vidicon lag characteristics of vidicon and plumbicon are shown in Fig. 4.11. The lag is dependent on the signal current, dark current and temperature. At low signal currents, discharge lag predominates whereas at high signal currents photoconductive lag is preponderant. A typical lag 200 ms after removal of an illumination giving a signal of 200 nA, for the 1-in vidicon Philips type 1240 with standard layer type A, at a dark current of 20 nA is 8% (1.6 nA).

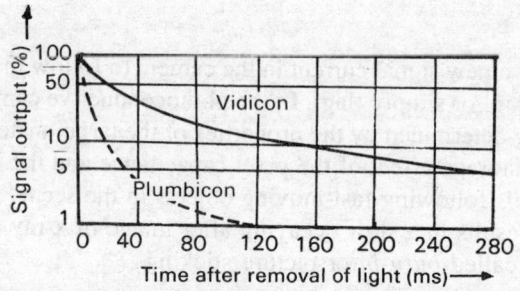

Fig. 4.11 Lag characteristics of vidicon and plumbicon

Light Transfer Characteristics

This is plotted for a vidicon as signal output current vs. illumination on the tube face with dark current, i.e. target voltage as the parameter as shown in Fig. 4.12.

The gamma or the slope of the characteristic plotted on a log-log scale, varies from 0.4 to 0.6 for low target voltages and high illumination, and from 0.8 to 0.9 for high target voltages and high illumination. Average gamma is approximately 0.65. Compared to linear gamma, such characteristics are non-critical of input lighting, i.e. say, a change of input illumination by a factor of two, will raise the signal by only 50%. Such characteristics also improve the visibility of low light details.

It is noteworthy that the gamma of a vidicon is approximately the complement of that of the receiving picture tube, and the two combined produce light tone rendition without any gamma correcting circuitry. The nonlinearity of a vidicon is opposite to that of the picture tube, and the two tend to cancel out to make an overall linear system.

Target Voltage Setting

An important feature of a vidicon is its ability to operate at different levels of sensitivity, by simply

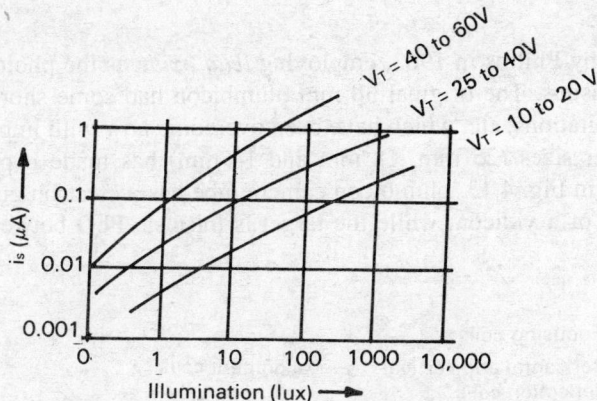

Fig. 4.12 Light transfer characteristics of vidicon

varying the target voltage (i.e. also the dark current) with respect to the high light illumination on the face plate. The sensitivity varies approximately as the 2.5th power of the change of the target potential. Its operating condition, i.e. the target voltage, can be varied to suit the requirement of various fields in which the tube is employed, as follows:

(i) Low illumination: High target voltage
$$VT = 40–60 \text{ V, for industrial use}$$
$$\text{Gamma} = 0.8–0.9$$

(ii) Medium illumination: Medium target voltage
$$VT = 25–40 \text{ V, for studio broadcast}$$
$$\text{Gamma} = 0.6–0.8$$

(iii) High illumination: Low target voltage
$$VT = 10–20 \text{ V, for telecine}$$
$$\text{Gamma} = 0.4–0.6$$

Automatic Gain or Sensitivity Control

This can be easily provided in vidicons by feedback circuits which increase the target voltage as the incident light decreases. A very simple way to do this is to include a high, AC bypassed resistance in series with the target voltage source.

Spectral Response

The spectral response of the vidicon is generally quite close to that of the human eye, with a peak in the yellow green region. This can be modified by proper choice of target material (using the so-called Type B layer) to shift towards the ultraviolet region, for applications in x-ray image intensifiers equipped with P11 or P20 output phosphors.

Resolution

The resolution capability of a vidicon varies over a wide range. As resolution elements are made smaller the signal output from the tube decreases, since the beam spot size is then larger and recharges a number of elements instead of one, so that there is an averaging of output. Typically it has 400 lines of resolution for 55% modulation.

4.6 PLUMBICON

Plumbicon was introduced by Philips in 1963, employing *lead oxide* as the photoconductive layer, with several improved characteristics. The original 30 mm plumbicon had some shortcomings like poor red response, contrast range limitations, etc. which have been overcome now with improved layer techniques. Also, availability in smaller sizes (25 mm, 18 mm and 14 mm) has made it popular in smaller high quality cameras. As shown in Fig. 4.13, plumbicon camera tube has a electron gun and scanning system essentially similar to those of a vidicon, while the target is intrinsic PbO between *p*-doped PbO and *n* layers.

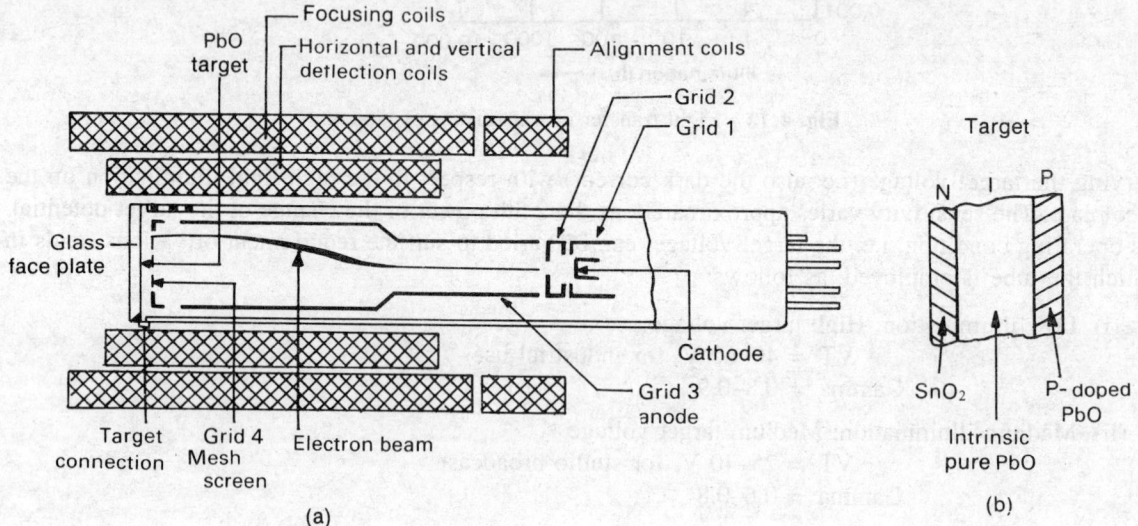

Fig. 4.13 Plumbicon target and camera tube construction (*Courtesy*: ELCOMA PHILIPS)

Target

Plumbicon uses a lead oxide (PbO) target which has a lower dark current, improved sensitivity and lower photoconductive lag as seen in Fig. 4.14. To form the target, pure lead monoxide (PbO) is vapour deposited on a thin conducting film of tin oxide (Nesa glass) formed on the inside of the face plate. This film serves as the signal electrode and forms *n*-type semiconductor in the PbO layer. The pure *intrinsic* PbO layer is doped suitably to form a *p*-type semi-conductor layer on the beam side of the PbO surface. Thus an intrinsic layer of PbO is sandwiched between thin *n*-type and *p*-type layers. These three layers (10 to 18 μm thick) form a continuous array of pin diodes that are reverse biased by the target voltage set at 40 to 45 V. Because of relatively large band gap of the thick intrinsic layer of PbO, few electron hole pairs are generated at normal operating temperature. This gives extremely low dark current (1 to 3 nA), and linear transfer characteristics. With light, the diode current increases due to the photon generated electron-hole pairs. Signal currents around 300 nA are obtained for high lights.

The layer is porous, built up of crystallites of very small dimensions compared with the scan line spacing, giving excellent resolution and low burn-in, and make plumbicon eminently suitable for colour TV cameras.

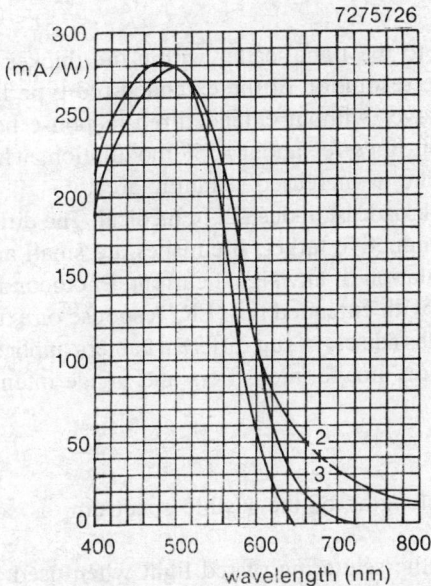

7275726

Fig. 4.14 Spectral response of 30 mm Plumbicon: with (1) High resolution layer (2) Extended red layer, (3) IR filter and antihalation disc (*Courtesy*: Elcoma Philips)

Lag

In plumbicon, the predominant PbO layer of intrinsic layer has few traps in it and the photoconductive lag is negligible. Lag is primarily due to the low p-n junction capacitance, as the photoconductive layer in the tubes is relatively thick (10 to 18 microns). Plumbicon shows very little discharge lag at normal signal currents. Discharge lag becomes apparent under low key lighting conditions, when the signal currents are small. The lag depends on the junction layer capacitance and the beam resistance.

The discharge lag is reduced by lowering the target capacitance and the effective beam resistance at low currents, by applying light bias and using a diode gun. In low capacitance plumbicon tubes, a target pin or tab, rather than a target ring, is used to connect with the SnO_2 conducting layer to reduce the surface area of target connection and hence overall capacitance. Low output capacitance of the tube increases the S/N ratio. To reduce lag at low lights, provision is made in some plumbicon tubes for applying light bias in the pin base. Light from a small lamp falls on the pumping stem of the tube and is conveyed by a forked glass light pipe into the collector space. It then falls directly or via reflection against the collector wall on the target. The lamp fits into the metal sleeve of the pumping stem.

Diode Gun and Dynamic Beam Control

In the conventional triode gun, grid 1 and the anode potentials converge the electrons emitted by the cathode to produce crossover in the electron beam. Electron interaction in the beam, in the region of crossover, increases the differential beam resistance increasing the lag. In the diode gun design, *grid 1 is made positive relative to the cathode*. This reduces beam convergence and so eliminates the crossover. The result is reduced differential beam resistance and a larger beam reserve. The consequent reduction in lag permits use of thinner photoconductive layers to improve resolution particularly in smaller tubes. With larger beam reserve of the diode gun, excessive highlights can be handled using dynamic beam control. When the beam encounters a highlight, the sharp rise in signal current is detected by a feedback network which then increases the control grid voltage, thus raising the beam current to read out the highlight.

Sensitivity

The sensitivity of the plumbicon is better than that of the vidicon, because of all the liberated carriers that contribute to the photocurrent, typically 400 μA/lumen as against 100 to 150 μA/lumen for vidicon at the colour temperature of 2856 K. The plumbicon and similar tubes forming a reverse biased p-n junction *do not permit automatic sensitivity control* (ASC) by regulating the target potential. Such tubes cannot be retrofitted in cameras that employ variable target voltage for ASC purposes. Adequate control is therefore achieved by iris control and neutral density filters.

Resolution

The resolution of the plumbicon is somewhat poorer due to the light scatter within the thicker target. Thinning of the target layer worsens the red response. The resolution of the extended-red type layer is higher than that of the standard layer. A high-resolution layer without extended-red response has also been developed. Typical resolution for 25 mm tubes is 400 lines for about 45% modulation, while for 30 mm tubes it is 400 lines for around 65% modulation due to increased scanning area.

Plumbicon tubes are available in different quality grades—broadcast, industrial or medical. The difference is in the degree of freedom from blemishes on the photoconductive target. Blemishes are small areas in the form of sharp spots or smudges producing uneven modulation of any signal current. Photoconductive layers of these have standard (SR) or high resolution (HR), with extended red (ER) response or extended red response and IR reflecting filter (ER-F) on anti-halation glass discs, to suit different camera applications, viz. B/W, luminance, red, green, blue channels, medical x-ray image intensifiers and image intensifiers used for scientific purposes or surveillance.

Spectral Response

This shows a peak in the response near the shorter wavelengths of the visible spectrum, as seen in Fig. 4.14.

Earlier types had poor sensitivity and resolution due to the scattering of red light when used in red channel of colour cameras. The deficiency is due to the wide band gap of 2 eV of PbO, corresponding to a cut-off of 620 nm. PbO does not absorb different kinds of light to the same degree. Shorter wavelengths are absorbed at low penetrations; blue light is almost completely absorbed after penetrating 5 μm of the PbO layer, but quite a high portion of red light passes through the target. Spectral sensitivity of the target can be varied between wide limits, by the thickness of the i layer of PbO. A higher red sensitivity can be obtained by making the i layer thicker, while a higher blue sensitivity can be achieved by making the n-type layer as thin as possible, and ensuring an adequate electrostatic field adjoining the n-contact. Improved red response is obtained by using the PbO, PbS mixture, by doping the lead monoxide layer with sulphur, which have bandgaps less than 2 eV.

Anti-Comet-Tail (ACT) Gun

The transfer characteristics of a plumbicon tube are linear up to a point determined by the available beam current. In broadcast television, the beam current is usually limited to a level which is sufficient to stabilize to twice the normal peak white signal output current to be expected from the scene. This limits the dynamic range, and local highlights in excess of the expected may cause blooming due to beam pulling and loss of detail. In extreme cases, the target is unstabilized locally and the pixels lose more charge between successive scans than the beam can replenish. 'Comet tails' following moving highlights are observed.

These effects can be reduced by limiting the maximum potential excursions on the target by lower target potential of about 20 V in place of 45 V. However the lower field strength across the target causes a reduction in sensitivity and slight tendency to burn-in. The anti-comet-tail (ACT) gun was designed to reduce these effects. In plumbicon with such a gun the beam current is strongly increased during line flyback, and most of the recharging of the target element capacitors in areas of extreme highlight occurs in the flyback period.

The first anode is split into two parts: the anode a_1 and the limiter a_2, electrically interconnected together. An additional electrode g_3 is placed between them. The limiter cathode and the limiting aperture are both increased. During the normal read-out scan the potential of g_3 is maintained near a_1a_2, bringing the scanning beam into *focus at the target*, as shown in Fig. 4.15(a).

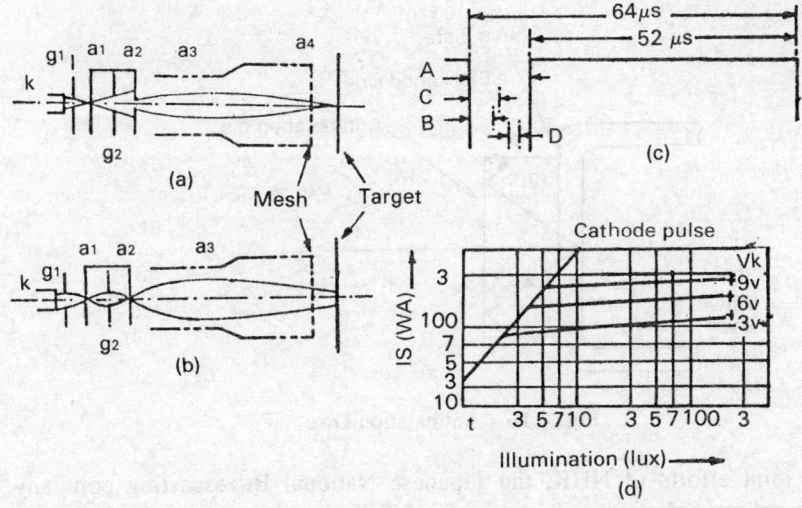

Fig. 4.15   ACT gun operation in plumbicon: (a) Normal readout, (b) High intensity defocussed beam during flyback, (c) Pulse timing of ACT gun, and (d) Transfer characteristics. (*Courtesy*: Mullard)

During line flyback, a negative going pulse (210–250, 5 μs) is applied to g_2 to focus the scanning beam on to the aperture in the limiter g_4, as shown in Fig. 4.15(b). Simultaneously the beam current is strongly increased by a positive going pulse (20–30 V, 6 μs) on g_1. Thirdly, a positive going pulse (0–15 V) is applied to the cathode during flyback. In this way a defocused beam carrying large current as great as 100 μA scans the surface of the photoconductive layer during line flyback. This beam contains sufficient current to replenish even the extreme highlight areas, bringing the surface here to the cathode potential during flyback. Potential levels below this contain the picture information and are not affected. Consequently, during the normal read-out, the scanning beam does not encounter target potentials higher than the cathode potential. Stabilization is therefore possible everywhere, and blooming and comet tails are significantly reduced.

The effective light transfer characteristics, when ACT is operating, are shown in Fig. 4.15(d). Up to the knee, the linear unity gamma characteristic is retained, while after the knee has been reached, the characteristic still exhibits a grey scale although it is greatly compressed. The knee level can be adjusted by varying the amplitude of the cathode pulse. Highlights up to and over five lens stops in excess of normal picture are handled without blooming, comet-tail and other distortions appearing in the picture.

Antihalation Disc

The plumbicon target is a light tan in colour as it reflects the red part of the spectral range of light. Diffuse reflected light from the target can be caught in the face plate of the tube by total internal reflection as shown in Fig. 4.16, and cause stray light 'halation'. A thick antihalation glass disc cemented on the face plate causes the internally reflected rays to strike the side wall, which is made optically black to absorb the stray light. Further reduction of stray light halation can be achieved by fitting a mask on the antihalation disc, with an aperture slightly larger than the useful scanning area of the target.

4.7   SATICON

Saticon was originally developed by Hitachi in 1973, as a low cost pick-up tube with broadcast standard

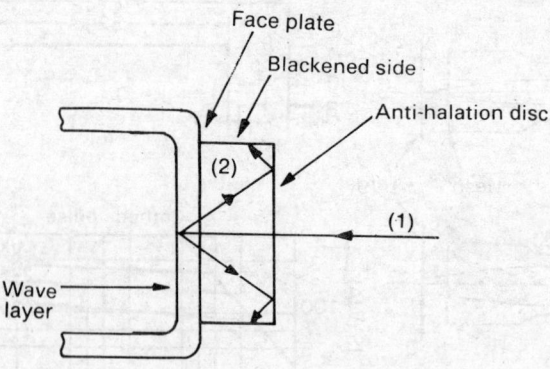

Fig. 4.16 Antihalation Disc

characteristics, with joint efforts of NHK, the Japanese National Broadcasting company. The target material used was amorphous Selenium with Arsenic and Tellurium incorporated in trace amounts, which gave the tube the name S-A-T-icon. Selenium has bandgap of 2 eV, limiting sensitivity beyond 620 nm, which is improved by addition of tellurium. Addition of arsenic in trace amounts prevents the tendency of amorphous selenium to crystallize, maintaining it glassy, while retaining the high resistivity of the photoconductor. Because of high resistivity, and high field of 125 kV/cm through the layer, there is little lateral diffusion of charge carriers. The addition of tellurium in the front portion of the layer near the target corrects the relative poor sensitivity towards the red end of the visible spectrum. The Saticon target construction is shown in Fig. 4.17.

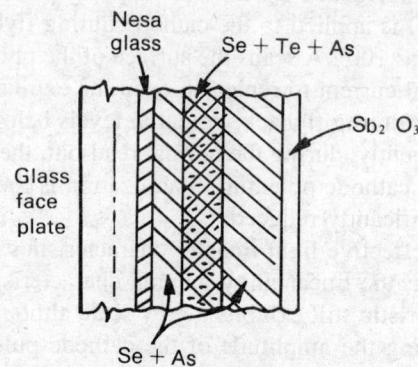

Fig. 4.17 Saticon target

The target end of the saticon is sealed with an optical glass faceplate, on the inside surface of which is a thin layer of tin oxide as the conducting material. This layer, also known as *Nesa* glass, is electrically conducting besides being transparent. A ring of indium serves to join the faceplate to the cylindrical envelope, also forming the electrical contact to the Nesa layer. A surrounding metallic ring provides the target contact generally available through a deflection yoke assembly.

Tellurium is incorporated into the target in a narrow band somewhat set back from the Nesa glass-selenium-arsenic junction. A pn-junction is formed at the boundary of the p-type selenium-arsenic and n-type tin oxide Nesa layer. With the positive target voltage applied to the target, the Saticon behaves like a *reverse biased* junction. The dark current is quite low, of the order of 0.25 nA per sq cm. Another layer of p-type antimony trisulphide is deposited on the gun side of the target to suppress secondary emission of electrons.

The addition of *tellurium layer* to the target gives a *nearly black* appearance. Hence there is little scattering of light within the target. This helps preserve high picture contrast and tonal quality with very high resolution. With hardly any reflections from the black target layer back towards the lens, there is no need for an antihalation disc for flare correction. The sensitivity of saticon is somewhat lower than that of plumbicon, while its spectral response is similar to it, lower than the extended red plumbicon.

The lag due to the junction capacitance and the high beam resistance is reduced by what is called *bias lighting*. The faceplate of the saticon is kept uniformly illuminated by a low level LED light source assembly built ino the camera optical system. The bias lighting establishes enough static signal current to reduce the beam resistance and permit rapid charging and discharging of the target capacitance. The static bias value is subtracted from the video signal to restore correct black level in the signal and obtain the full swing between black and peak white.

The saticon has very uniform characteristics, yielding a gamma of almost unity, and provides excellent resolution, high sensitivity, good spectral response and low lag when used with bias lighting. It is widely used in studio broadcast colour cameras, telecine and also in single tube colourplexer cameras using stripe filters.

4.8 NEWVICON

Newvicon, featuring high resolution and very high sensitivity in the entire visible spectrum extending into the near infrared region, was developed by Matsushita Electric Corp. in 1974. It employs a *heterojunction photoconductive layer* consisting of a sublayer of zinc selenide (ZnSe) and a sublayer formed by a mixture of zinc telluride (ZnTe) and cadmium telluride (CdTe). The multi-layer target structure is as shown in Fig. 4.18a. On the inside surface of the Nesa glass, a layer of n-type zinc selenide is deposited. A further layer of p-type cadmium telluride, a compound of zinc, cadmium and tellurium in amorphous form, is deposited on this. The ZnSe layer acts as a light transmission barrier to prevent crystallization of the amorphous CdTe layer. This is followed by a final layer of antimony trisulphide, to suppress the secondary electron emission on the gun-side surface. A pn-junction is formed at the boundary of zinc selenide and amorphous Zn-Cd-Te layer.

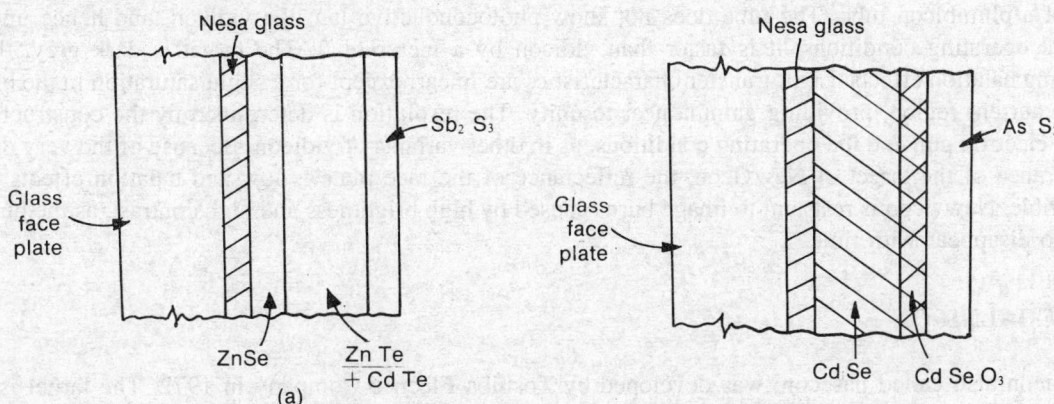

Fig. 4.18 (a) Newvicon target, (b) Chalnicon target

The target voltage setting is somewhat critical. Too high a setting results in appearance of spots and higher dark current, which is 5 to 10 nA, at 30° C. Too low a voltage setting results in poor lag characteristic and high light burns. Newvicon tube does not permit automatic sensitivity control (ASC) by means of adjustments in target voltage. Adequate control has to be achieved by other means like iris control and neutral density filters. Newvicon has the highest sensitivity of all the camera tubes, three to four times as high as that of plumbicon, with a very wide spectral response extending well into the infrared region, as seen in Fig. 4.19.

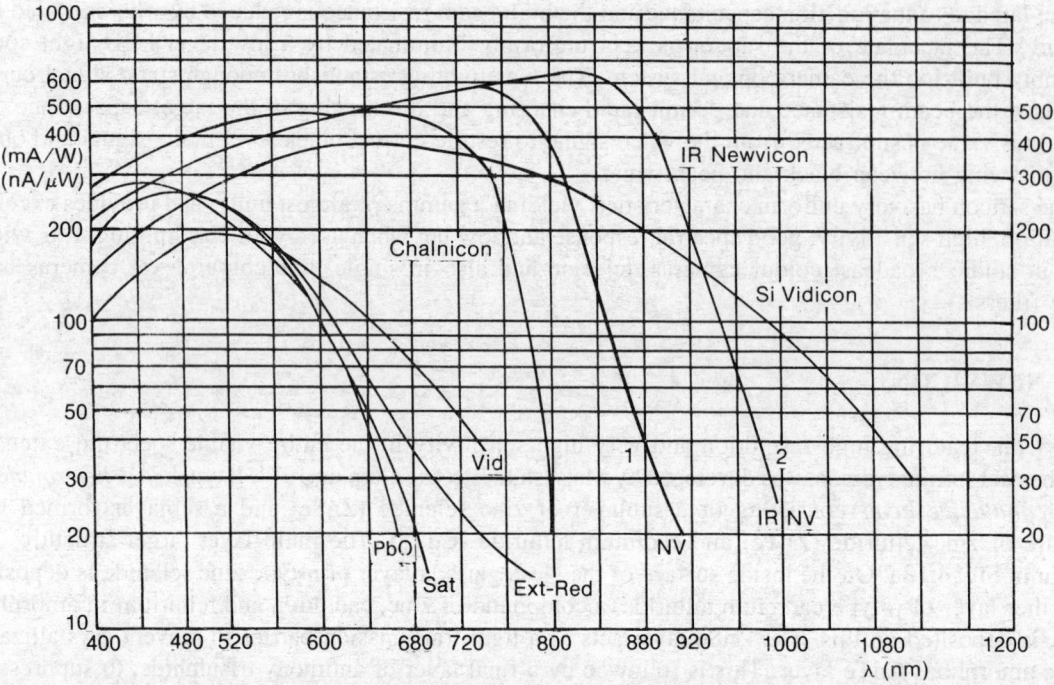

Fig. 4.19 Typical spectral response of Newvicon and other photoconductive tubes (ELCOMA PHILIPS)

Because of its much larger target capacitance, the lag of a Newvicon tube is significantly higher than that of a plumbicon tube. The tube does not show photoconductive lag like vidicon, and hence under normal operating conditions, it is faster than vidicon by a factor of 2. The target is dark grey, thus reducing halation effects. Light transfer characteristics are linear, except for a slight saturation in the high signal current region, providing gamma near to unity. The resolution is determined by the construction of the electron gun and the operating conditions, as in other variants of vidicon. Because of the very dark appearance of the target of Newvicon, the reflectance of the face plate is low and halation effects are negligible. Newvicon is resistant to image burns caused by high brightness and high contrast, as the burns tend to disappear with time.

4.9 CHALNICON

Chalnicon also called pasecon, was developed by Toshiba Electric Company in 1972. The target is of multilayer construction with cadmium selenide as the photoconducting layer, as shown in Fig 4.18(b). The cadmium selenide layer is deposited by vapour deposition, after which a backing layer of cadmium selenium trioxide ($CdSeO_3$) is formed at a higher temperature. This is followed by arsenic trisulphide (As_2S_3) deposited on the scanned surface, on the gun side, to eliminate secondary electron emission. Cadmium selenide gives an efficiency of almost 100%, an electron hole pair produced by every incident photon absorbed. Cadmium selenide has a bandgap of 1.7 eV, which limits the response beyond 720 nm. Doping with cadmium telluride improves the red and infrared sensitivity in special versions of chalnicon. The addition of a porous layer of arsenic selenide (As_2Se_3) porous layer on the scanned side reduces the lag by reducing the capacitance.

The target voltages are somewhat critical, 20 to 50 V, depending on temperature; low voltages cause image lag, while excessive voltage produces white spots. The lag is due to junction capacitance and beam resistance, and is more objectionable at low lights where bias lighting becomes necessary. Dark current is low, 1 to 2 nA. Spectral response is wide, extending well into the red. Hence the achronym pasecon, from *panchromatic selenium vidicon*. It is primarily used in cameras meant for surveillance.

4.10 SILICON DIODE ARRAY VIDICON/SILICON VIDICON

Multidiode silicon vidicon is a variation of the vidicon in which a new type of target comprising a silicon diode array is used. This type of target is advantageous in environments of high illumination that bring in burn and lag problems with ordinary vidicons.

The target plate in the silicon vidicon consists of a very thin, (10–12 μm thick) wafer of n-type silicon over which an array of silicon photodiodes is formed. To form this, a thin SiO_2 insulating layer is first formed to passivate the surface. By photomasking and etching processes, an array of fine opening is made in the oxide layer to expose the n-type silicon islands, and a very thin layer of p-type silicon is formed. A fine layer of gold is deposited on each p-type opening thus, as shown in Fig. 4.20.

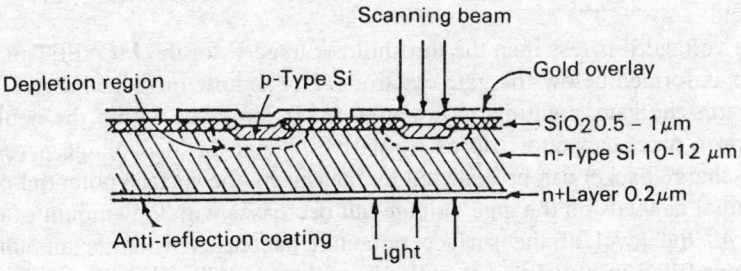

Fig. 4.20   Silicon diode array plate

An array of photodiodes, each about 8 μm in diameter, is thus formed, typically containing 540 × 540 diodes. The diodes are reverse biased by applying +10 V to the n-type overlay side. The substrate is illuminated by light from the image focussed onto it. The incident photons of light generate electron-hole pairs which diffuse to the p-type regions by the depletion field of the diodes. The n-type of substrate material has poor mobility for p-type carrier holes. The hole mobility and lifetime are sufficiently large for most of them to survive long enough to diffuse through the substrate and enter the depletion region. The holes are here swept over to the p-type region by the depletion field, establishing a positive charge on the gold overlay contact pad of the diodes. This reduces the reverse bias potential of the diodes. When the electron beam scans the surface subsequently, it deposits electrons and returns the diode side of the target to the cathode potential. The increase in the target current thus caused by the incident electrons at the respective diodes represents the video signal. For high lights, currents of 0.7–1.3 μA per ft-candle are obtained. Signal output is linear with gamma of 1.0.

Silicon vidicon is less susceptible to damage or burns due to excessive high lights. It has a lower lag and high sensitivity which can be extended to the infrared regions. It is used for industrial applications requiring different operating conditions.

4.11   CCD IMAGE SENSORS

As mentioned earlier, the new generation of image pick-ups based on charge transfer devices (CTDs) are increasingly used as solid state image sensors. The signal is represented by isolated packets of charge stored on elementary capacitors formed in adjacent arrays and can be transferred between the capacitors in a controlled manner. Devices using discrete MOS transistors for charge transfer between capacitors are known as bucket brigade devices (BBD). Devices in which the charge transfer is arranged by fabricating the elementary capacitors close together on a chip are known as charge coupled devices (CCD). The stored charge on elementary MOS capacitors is proportional to the incident light, each charge packet representing a sample of the pixel information to be processed.

The image sensors consist of monolithic large scale integrated circuits employing a large number arrays of charge coupled devices (CCD) or charge injection devices (CID), serving as photo sites or pixel capacitors. In these devices, light energy incident on each photo site is absorbed to liberate a proportional number of electrons and holes. In the p-type semiconductor, the holes are repelled out of the depletion region to the p-type substrate, and because of the reverse bias applied to the metal oxide interface on the substrate, the electron minority carriers are attracted towards the gate electrode. These are trapped near the metal oxide layer, forming an inversion layer of charge packet proportional to the incident light. These minority charge carriers trapped in the so-called potential wells, are tapped and scanned once in each frame or field.

When the gate voltage V is less than the threshold voltage V_T of the MOSFET pixel, a depletion layer devoid of carriers is formed below the gate electrode. If V is now increased above V_T, minority carriers are attracted towards the gate, creating an potential inversion layer within the depletion region. As the charge injected on a pixel capacitor causes a relative change in the voltage across it, changes in the magnitude of the charge packet can be detected by measuring the surface potential of the semiconductor. The surface potential depends on the gate voltage but decreases with the amount of charge trapped in the inversion layer. As the level of the surface potential decreases with the amount of charge trapped (analogous to water filled in a well), it is called a potential well.

Advanced MOS technology is employed to fabricate the closely spaced multiple MOS capacitors as imaging pixels. Arranged in linear or area configuration with appropriate on-chip scanning circuit and low-noise preamplifier, these constitute the focal plane image sensor in a compact video camera. Resolution is determined by the number of pixels on the target area available on the silicon substrate. This area as available now is similar to that of the common 18 mm pick-up tube, providing typically some 196,096 CCD pixels, in a 512×383 format. Special higher pixel resolution CCD cameras for HDTV use are also available.

CCD Imaging

The photogenerated charge packet in each pixel is transferred to the adjacent pixel by means of sequentially clocked voltage pulses applied to adjacent pixels creating potential wells. The moving charge packets are transferred to an output diode that has a capacitance of a fraction of 1 pF. This gives a high speed charge transfer with practically no lag and high *S/N* ratio for the video signal. There are two methods of the charge transport, one using a surface channel conduction device (SCCD), and the other using buried or bull channel conduction device (BCCD). In the SCCD, the conduction channel is at the silicon oxide interface owing to a positive bias applied to the electrodes fabricated over the photosites. This allows the charge packets to interact with the interface states, lowering the efficiency due to recombinations. In the BCCD, the conduction channel is ion-implanted inside the bulk semiconductor, some 0.5 micron from the silicon oxide interface, a layer of n-type silicon separating the interface from the p-type substrate. The charge packets interact with an insignificant number of bulk states, and therefore noise contribution is not significant. The charge transfer efficiency is also better, because of larger fringing fields between the pixels in the channel.

Charge Transfer Device Architecture

Each photosite accumulates charges proportional to the light incident on it, integrated over the 1/50th second period between samples of scanning. The charge packets are transferred in time sequence using digital techniques. In the CID area imager, the individual photosites are addressed by x-y co-ordinates as an XY matrix of MOS capacitor pairs, each constituting a pixel. The photogenerated charge packets are not transferred to common output—charge transfer occurs only between the capacitor pairs within a pixel. The charge packets are stored by biasing at least one of the capacitors. When the capacitor pair is pulsed to zero, the potential well collapses and the stored minority carrier charge is injected into the substrate where most of the carriers recombine, and the injected charge provides the video signal.

In the CID imagers, a column at a time is allowed to float and the change in the potential after a row transfer under the column provides the pixel read-out signal. Area imaging is obtained by suitable scanning of the horizontal and vertical shift registers. The effective output capacitance of a CID is much higher than that of the CCD, a disadvantage in low light imaging. The XY addressed random access feature, and freedom from optical smearing and blooming effects, are however useful in some applications.

CCD imagers, used in TV cameras, are fabricated in two basic configurations, one using the frame storage mode and the other an interline transfer mode. RCA and some European manufacturers use the frame transfer arrays while most Japanese CCD cameras now use a third technique, as a hybrid of the two, viz. the frame interline transfer.

In the frame storage method, a charge is integrated in the imaging section and the packets of each of the photosites are rapidly transferred to the digital frame storage section during the vertical blanking interval for subsequent read-out through a serial shift register, as shown in Fig. 4.21(a), to generate continuous video signal.

The frame storage configuration has practically all the sensitive imaging area with transparent electrodes available, and both front as well as back illumination are possible. Each cell has a combined dual function of optical sensing and shift register. During the transfer process each cell has charge passed through it from the cell above. If the cells are exposed to light during this process, additional charges will be generated, especially on bright scenes. To avoid the resulting blurring effect, mechanical shutter is required to cut off light during the shifting process. An interlaced read-out is obtained by imaging with alternate electrode sets, giving read-outs of field 1 and field 2 alternately.

In the interline transfer method, the charge packet for every alternate line of the image in the first field is transferred to the vertical columns, as shown in Fig. 4.21(b). The selection of the line to be transferred to the vertical opaque column for subsequent transfer to the serial shift register is made by the vertical shift registers shown to the right side, which turns on each horizontal row of switches sequentially during the blanking period of each line. The interlace lines are switched in sequence for the next field. As the horizontal row of switches is turned on, the charge collected by the vertical columns is stored temporarily in the horizontal shift register and buffer shown at the bottom. Thus during each blanking, a line pattern of charges is read into the horizontal register in parallel. During the active line scan time a clock signal transfers this pattern out of the H shift register in serial form, repeating the transfer process line by line. The discrete stepped waveform is smoothed by a low pass filter to get the analog video waveform. The vertical and horizontal registers are masked by opaque layers of aluminium, leaving only the photosites exposed to light. There is a tendency to streak vertically from bright objects due to leakage from adjacent vertical shift registers. Yet no shutter mechanism is required for interruption of light during shift interval.

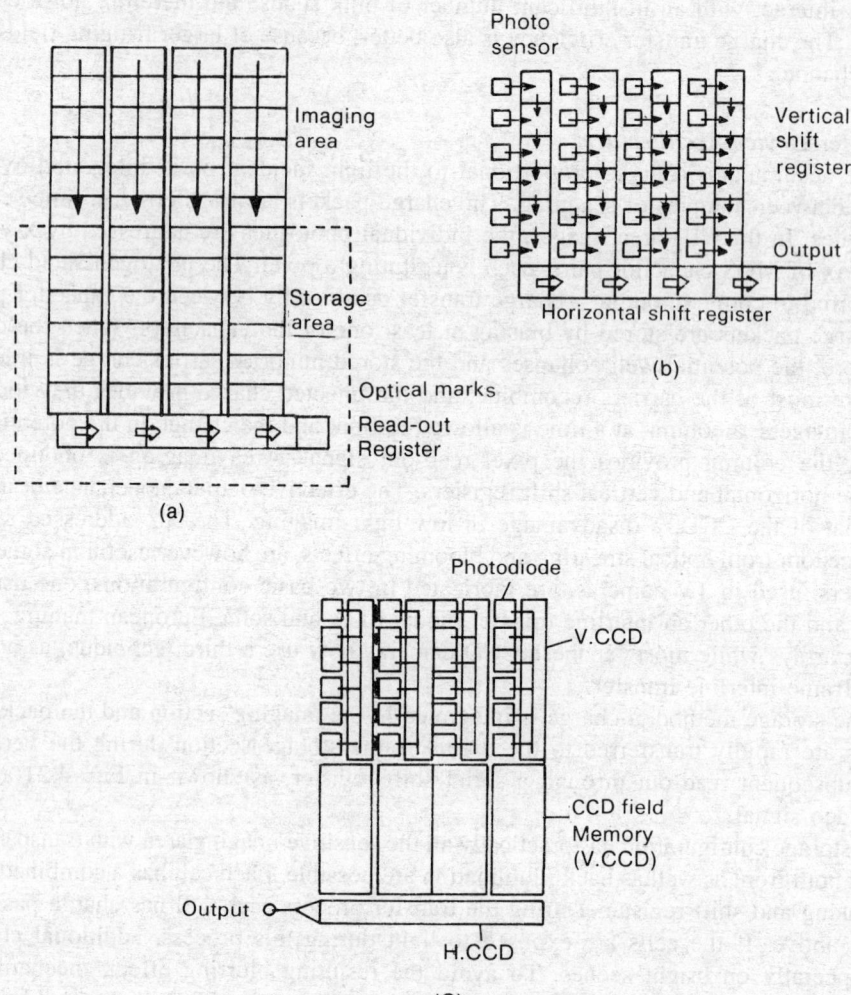

Fig. 4.21 (a) Frame storage transfer (Sony), (b) Interline charge transfer (Sony), (c) Frame interline charge transfer (NEC)

Frame Interline Transfer

This is a hybrid technique used by most Japanese manufacturers in premium cameras. As shown in Fig. 4.21(c), it is an interline array with an attached 1-field (312.5 lines) memory, a concept borrowed from the frame transfer architecture. This enables a high speed shift of the cell charges (over a 100 times that of interline transfer) into the protected memory area reducing streaking from sources of bright light, like the sun or headlights.

Details of the charge coupling technique in the CCDs is illustrated in Figs 4.22 (a–b), where depletion bias on a MOS junction due to voltage V, is shown. The surface electrodes of polysilicon are translucent so that light can enter from the top side. Charge proportional to the incident light accumulates in the depletion well, during the time between two scans, and is moved to the neighbouring site by creating a deeper well on that site while forming a higher well on the opposite side. This method can be repeated

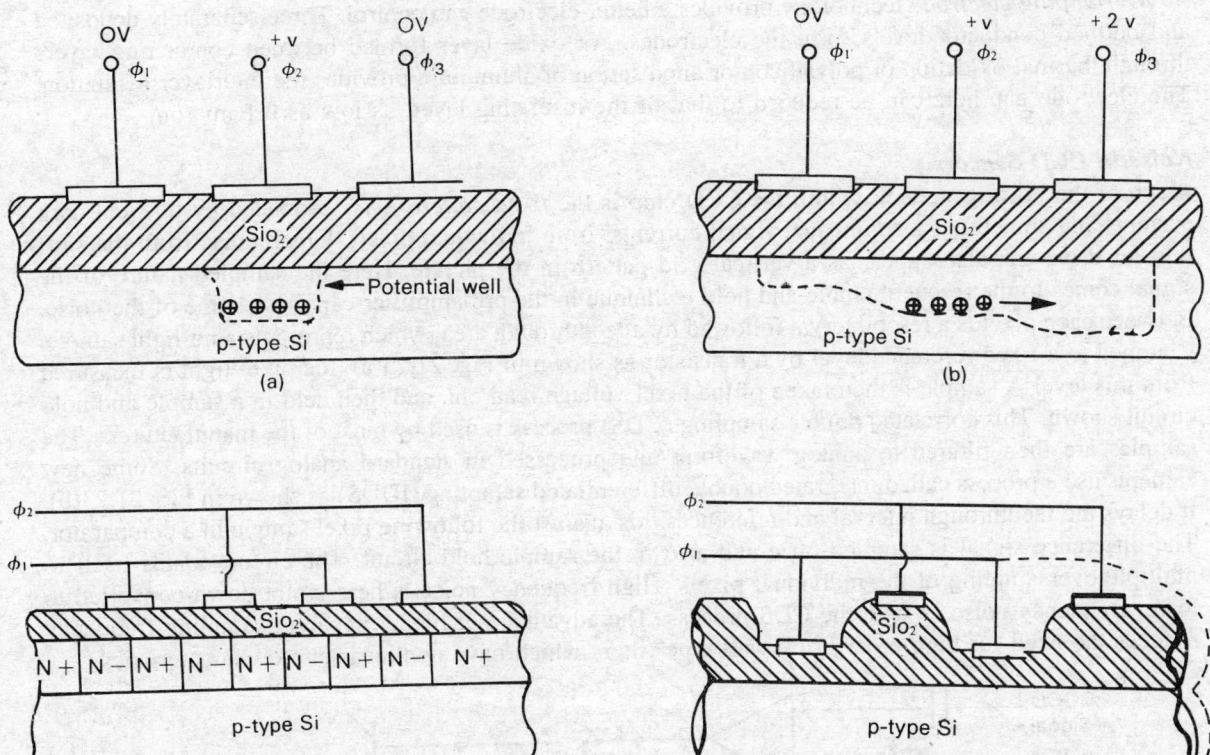

Fig. 4.22 Charge coupling techniques in CCDs: (a–b) Depletion well and 3-phase clocking, (c–d) 2-phase clocking for charge movement
(a) Depletion bias and potential well; (b) Charge movement by 3-phase clocking; (c) Unequal doping; (d) Unequal thickness

to move the charge by one more site ahead, until the end of the resister is reached. Here, an integral floating gate MOSFET reads out the line to the external circuits.

If the fabrication of photosites is identical and symmetrical a three phase clocking system is required to ensure the movement of the charge in the accumulated in the photosites in the desired direction. If an unsymmetry in adjacent sites is built-in, a two phase clocking is also able to achieve the desired direction of charge movement. This is possible (i) by unequal doping of alternate sites to obtain a built-in field gradient in the direction as shown in Fig. 4.22(c), or (ii) by unequal thickness of the silicon dioxide layer between the polysilicon electrodes and the substrate, as in Fig. 4.22(d).

Fabrication Technology
Manufacturing processes for CCD imagers need processes capable of creating the electrode gaps. Aluminium gate technology was the first technology used to fabricate CCD devices. In the shadow mask deposition process fine electrode gaps are produced by a process involving overetching of a thick aluminium layer coated on silicon and depositing a thin layer in between the gaps. Overetching allows fine gaps between the old and new deposits because of the *shadow mask deposition* through the overhanging photoresist on the old deposits. The photoresist and aluminium layer on it can be removed by float-off technique.

Overlapping electrode technology provides a better electrode gap control. Three separately deposited and defined conductor levels form the electrodes. An oxide layer formed between conducting layers through thermal oxidation of polysilicon or anodisation of aluminium provides the interlayer insulation. The electrode gap here can be reduced to that of the insulating layer, as low as 0.1 micron.

Noise in CCD Sensors

Much of the noise generated within the CCD chip is the *fixed pattern noise* produced by unequal dark currents and sometimes also unequal signal currents, from individual pixels. Clock noise from the shift register drive can also appear as a vertical grid pattern in the picture. Here the sampled nature of the signal comes to the rescue. Sample and hold technique in the pre-amplifier can mask some of the noise. Between each pixel is a reset interval followed by a feedthrough area, which represents a no-light sample. Clamped to a fixed reference level by a transistor as shown in Fig. 21.23(a), the pixel light is measured from this level A sample is then taken of the pixel voltage read out, and then held in a sample and hold circuit shown. This correlated double sampling (CDS) process is used by most of the manufacturers. The samples are then filtered to analog waveform and processed in standard analog circuits. Some new cameras use a process called integrated double differentiated sampling (IDDS) as shown in Fig. 21.23(b). It delays the feedthrough interval and references this against the following pixel sample in a comparator. The difference signal is sampled once and also in the sample hold circuit. The circuit adapts itself to multiple oversampling of the individual pixels. High frequency noise is here is not down-converted to lower frequency noise as with the CDS process. The advantage of higher apparent sampling frequency reduces the need for high order brick wall type filters which have nonlinear phase characteristics.

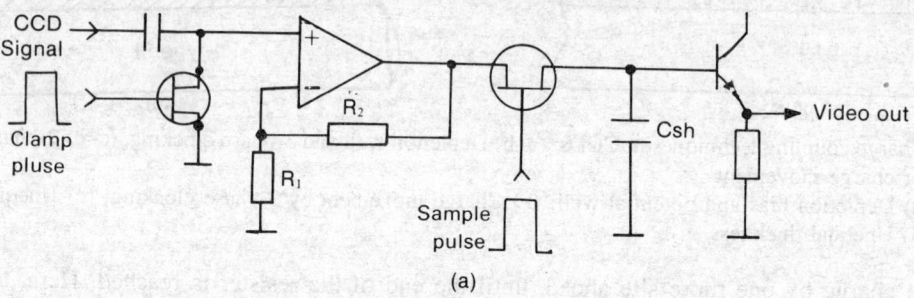

(a)

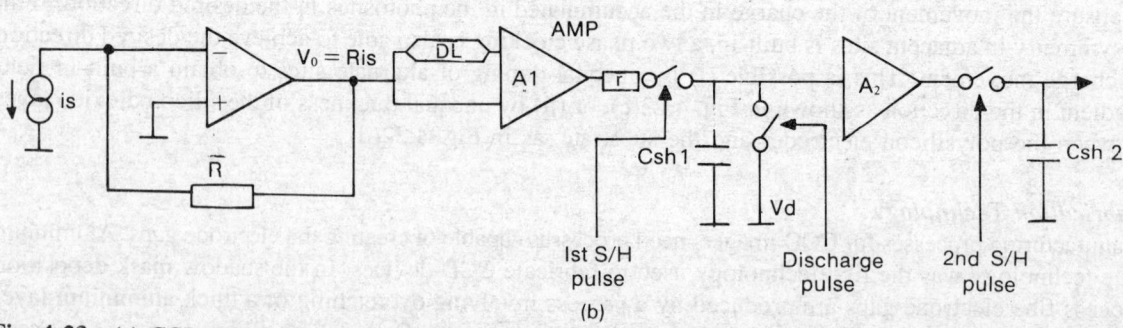

(b)

Fig. 4.23   (a) CCD preamplifier with correlated double sampling (CDS), (b) CCD preamplifier with integrated double differentiated sampling (IDDS)

Resolution CCD sensors developed for broadcast cameras have 750×493 pixel resolution, allowing more than 550 TV lines. Better horizontal resolution is possible at the expense of sensitivity because of the smaller size of the pixel. For better resolution in colour, red and blue CCDs are mounted horizontally offset from the green by half the pixel distance. Fine details in the form of high frequency information from red and green channel CCDs are combined in the horizontal detail ciruit of the camera. High resolution on white edges is obtained by adding this information back to the R, G and B camera channels. Information from the blue channel CCD is omitted, because of low sensitivity of CCD to blue light and the attendent noise problems.

Many broadcast cameras employ the vertical detail circuit (also called contouring circuit), that uses two H delay lines circuit to process the green channel CCD signal. The green channel represents information closest to the B/W luminance information. Vertical details are extracted by adding half of the non-delayed video to half of the 2H delayed video, and subtracting the resultant value from the full amount of the H delayed video. The currently employed glass delay lines are mechanically sensitive to shock and vibrations. The delays could be built into the CCD structure by triple recursive horizontal shift registers. This would enables separate delays in the R, G and B channels and the vertical detail signal can be derived from the true B/W video signal.

Advantages of CCD Sensors

The sensitivity of the CCD sensors is high, comparable to that of common pick-up tubes. An illumination of 10 lux can give an acceptable picture. Spectral response is also good, extending into the infrared region. Image lag is practically absent because the charge transfer mechanism is instantaneous. The sensor can withstand excessive illumination without permanent damage of burns. The CCD elements are not scanned by electron beam. Hence there are no geometric distortions, nonlinearities, and so on, in the raster but for the distortion due to lens optics. A CCD camera has better vertical detail capability than a tube camera. Due to excessive aliasing this appears as an excessively sharp edged picture rather than the more natural video from tube cameras. The higher resolution is most useful for computer vision and pattern recognition purposes, where the aliasing effects can be ignored by computer. In fact digital image processing can enable high quality origination of video by modern CCD cameras. Solid state cameras are lightweight and operate at low power usually requiring less than 100 mW at low voltage. Colour cameras using single colour CCD pick-up devices have been developed using a colour filter mosaic to provide colour dissection also.

4.12 COMPARISON OF TV CAMERA PICK-UP DEVICES

A broad comparison of the various pick-up devices used in video cameras is given in Table 4.1. Actual performance of individual tube varies considerably with different makes, broadcast or industrial quality type and operating conditions.

Vidicon has been common in compact low cost compact CCTV cameras and telecine applications where large contrast ratios have to be handled. Its variable sensitivity feature offers advantages. Plumbicon has improved performance because of low lag and dark current as needed in professional cameras. The spectral response limitations and highlight problems have been improved with spectral targets and construction. Of the new vidicon derived tubes, Newvicon hs the highest sensitivity, good spectral response extendable to the infrared range, very high resolution and shows negligible burn-in. The dark current is temperature sensitive, doubling every 7–8 °C.

The demand for smaller portable colour video cameras for on site ENG has led to development of

Table 4.1 Comparison of camera pick-up devices

Device	Vidicon	Plumbicon	Saticon	Newvicon	Chalnicon	CCD imager
Devel. by	RCA	Philips	NHK-Hitachi	Matsushita	Toshiba	RCA
Year	1951	1963	1973	1974	1972	?
Sizes	←————————— 14,18,25 mm diam —————————————→					IC chip
		30/45 mm				
Scan-area	←————————— 9.6 × 12.8 for 25 mm size ——————————→					8.8×6.6 mm
Target	←————————————— photoconductive ——————————————→					
	Sb_2S_3	PbO pin diodes–Sb_2S_3	SeAsTe ZnCdTe –Sb_2S_3	ZnSe- CdSeO$_3$ –As_2S_3	CdSe- Si-photo- diode CCDs	polysilicon,
Sensi- tivity	300 nA 140?	400 μA/lm 280?	300 nA 60? 280?	240 nA/lx 500?	200 ? 450?	130 nA/lx 300–400
Illumi- nation	10– 100 lx	10 lx	0.1–20	0.5–1 lx	?	10 lx
Dark current	20 nA	3 nA	0.25 nA	10 nA	1 nA	Small, reduced by cooling
Av. gamma	0.74	0.95	1	1	1	1
Lag after 60 ms	21%	5%	3% with bias light	17%	10%	Lag-free over- load blooming
Lim. Reso- lution	500– 800	40–50% at 400		650		550–700
Spectral resp. max. cutoff	550 nm	500 nm 650–850	550 nm 750 nm	750 nm 900 nm	700 nm 800 nm	850 nm > 1000 nm
Special features	Variable sensitivity, lag large	spl. target for red response	Temp. sensitive	No image burn, highest sensitivity	Sensitive	Rugged Sensi- tive Low power Integrated lin. scan Busy pict.

single-tube colour dissectors which are discussed in Chapter 24. Compact lightweight CCD imagers have robustness, low power requirements and ability to have independent control over the integration time of the image, which can be as low as 1 ms. The readout time can also be controlled to produce a high resolution freeze frame of fast moving objects. As the performance of CCD imager improves they are bound to replace the present tubes in ENG telecine and studio cameras as well.

4.13 PICK-UP TUBE DEFLECTION UNIT

The camera pick-up tube mounts itself inside a deflection coil unit which contains the focusing coil, the horizontal and vertical deflection coils, and the alignment coils or magnets. The focusing coil surrounds the entire tube extending from the electron gun to the face plate of the tube. It produces an axial magnetic field because of a dc current passing through it, as shown in Fig. 4.24(a).

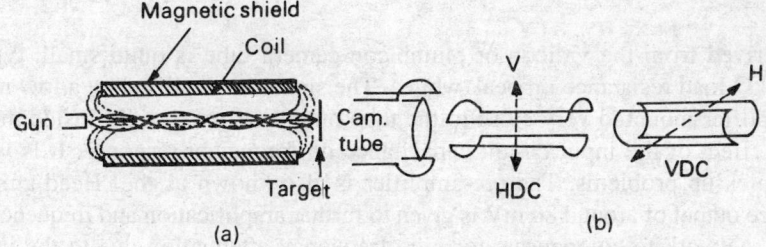

Fig. 4.24   Camera tube deflection unit: (a) Focusing coil, (b) Deflection coils

If the motion of the electrons in the beam is along the axial magnetic field, they proceed straight along the axis without any interaction with the focusing field. If the divergent electrons enter the field at a small angle with it, they experience a force at right angles to the field and their motion, so that they move along the focusing field in a spiral-cycloidal motion to converge at a certain distance depending upon the intensity of the magnetic field. The focusing coil current adjustment can thus serve as the beam focus adjustment. The beam may have a number of focus nodes or lobes (4 in an I.O.) between the cathode and the target. The diameter of successive nodes tends to increase due to mutual repulsion of electrons.

The horizontal and vertical deflection coils are a pair of coils each in the shape of yokes to mount on the cylindrical portion of the pick-up tube, as shown in Fig. 4.24(b). The horizontal deflection coils produce a vertical field and the vertical deflection coils produce a horizontal field. The field strength of the deflecting magnetic field is about 1/10th of the focusing coil. The electron beam is deflected horizontally and vertically by passing sawtooth currents in the coils at the rate required for standard rectilinear scanning. The required currents have to be supplied by the deflection drive circuits of the camera chain.

The alignment coils are a pair of coils positioned just outside the limiting aperture that produce a magnetic field at right angles to the tube axis to enable alignment of the beam scanning on the target on the face plate of the tube. The coil assembly as a whole is enclosed in a mumetal shield to prevent effect of any external field on the deflection scanning.

Typical deflection coil unit data and drive requirement for a 1 in. vidicon type AT 1102/01 are as follows.

Line deflection coil $L = 0.75$ mH, $R = 2.5$ Ω
$\qquad\qquad\qquad I_{pp} = 170$ mA
Field deflection coil $L = 23$ mH, $R = 80$ Ω
$\qquad\qquad\qquad I_{pp} = 24$ mA
Focusing coil resistance $= 4220$ Ω
Focusing coil current $= 17$ mA

4.14   VIDEO PROCESSING OF CAMERA PICK-UP SIGNAL

The video signal obtained from the target plate of a camera pick-up tube is quite small and requires amplification and some signal processing before it can be fed to video monitors or to the transmitter chain. Processing of the video signal includes amplification of the signal at a wide bandwidth, frequency response compensation, camera cable equalisation, aperture correction, gamma correction, dc clamping of the sync and blanking waveforms, white clipping, shading correction, and so on. For this processing, the signal is passed through the following stages.

Pre-amplifier

The video signal derived from the vidicon or plumbicon camera tube is quite small, typically 300 nA flowing through 50 kΩ load resistance on peak whites. The signal is amplified by a low noise high input impedance FET amplifier mounted very close to the tube output terminal, right inside the camera head. This minimizes the effect of the input circuit capacitance on frequency response. It is well shielded to reduce RF or hum pick-up problems. The pre-amplifier is also known as the 'Head amplifier'.

The low impedance output of around 80 mV is given to further amplification and frequency compensation by high peaking *RC* network to compensate for high frequency attenuation due to the input distributed capacitance and source capacitance of the target plate. The output from the head amplifier is sent to the video processing amplifier through the camera cable.

Cable Equaliser   High frequencies suffer attenuation due to the capacitance of the coaxial cable carrying the video signal from the camera unit to the processing circuitry in a central location. The loss of high frequencies in the camera cable is compensated for by a passive equalizer network. This can be adjusted to suit the length of the cable is use.

Aperture Correction and Image Enhancement

The camera lens optics, the beam splitter for RGB colours in colour TV cameras, and the finite size of the scanning beam, all contribute to loss of resolution and high frequency content in the signal output. Only an ideal point size beam will give sharp transitions between black and white shades to produce a well contrasted square wave signal. As illustrated in Fig. 4.25(a) a finite diameter beam spot tends to blur or round off the output signal produced at the black and white crossings. The finite beam spot provides an integrated output of the black and white areas it crosses during the transitions. This results in a somewhat sin *x/x* (phaseless) type of loss in the output signal.

The quality of picture is improved in broadcast calibre cameras by aperture correction and image enhancement techniques. Boosting the high frequency content in the signal provides electronic correction that has an effect similar to reduction of beam size, as if the beam has been passed through a small aperture within the camera. One method is to employ the second derivative compensating signal. The process involves double differentiation of the signal and subtracting it from the original signal as shown in Fig. 4.25(b)..

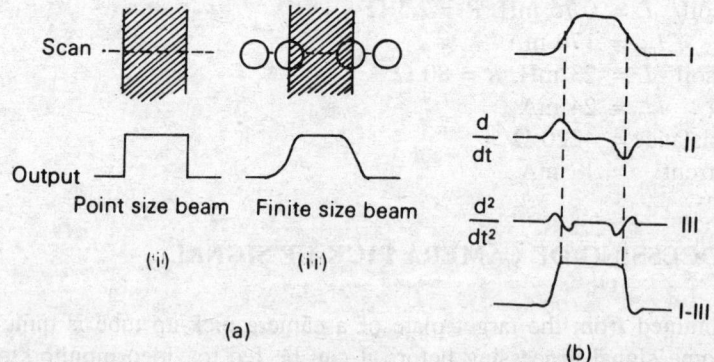

Fig. 4.25   (a) Effect of finite beam size, (b) Aperture correction

The aperture correction is applied to restore the depth of modulation at the high end 400 lines·to that at low end 40 line resolution. Another commonly used correction for image enhancement is to employ a delay line transversal filter which also provides a high frequency boost without phase shift, and compensates the sin x/x phaseless (without ringing) loss. Such a corrector produces a correction signal with both the anticipatory and a following overshoot around the transions in the picture, and combines them in op-amps and summing amplifier. The corrections should not be overdone, lest the image should be unnaturally sharp-edged and busy. Because the human eye is most sensitive to details in the mid-grey-scale, it is beneficial to modulate the aperture and enhancement detail correction signals as a function of brightness.

Gamma Correction It has been observed in the discussion on camera tubes that the electrical output of the camera pick-up tube is not exactly a linear function of the light input. Again, the light output of the picture tube is also not a linear function of the electrical input. The exponent of the transfer function is called the *gamma* of the system. The output v. input transfer characteristics may be plotted on a linear or a logarithmic scale, as illustrated in Fig. 4.26.

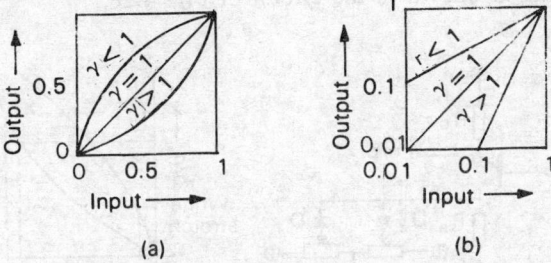

(a)   (b)

Fig. 4.26 Gamma correction: (a) Linear scale plot, (b) log-log scale plot of the transfer characteristics

If the system is linear, the gamma of the system is unity. The transfer characteristic is a straight line on the linear plot, and has unity slope on the plot on a log-log scale. If the gamma of the system is less than 1, the characteristics from a bow upwards, while if the gamma is greater than 1, they form a low downwards. On the log-log plot, the slope of the linear characteristics obtained equals the gamma of the system. The gammas of the various equipments in the television camera chain are indicated in Fig. 4.27.

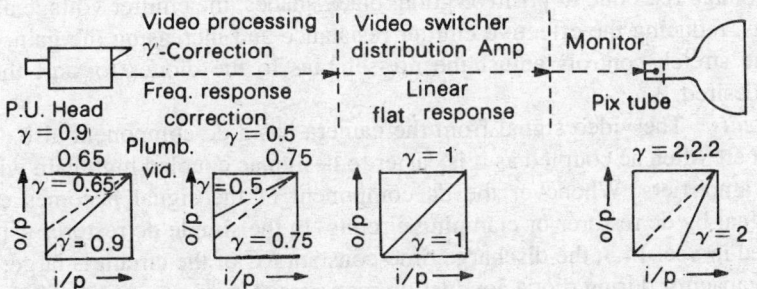

Fig. 4.27 Gamma of television system equipment

The overall transfer characteristics from the pick-up tube to the receiver picture tube should be unity or 1.2. The latter value gives a slightly more pleasing picture under normal lighting conditions by accentuating the tonal gradations in the white region. A monochrome picture tube has a gamma of about 2, stretching the whites and compressing the black. Colour tubes have a gamma about 2.2. To allow for this, the gamma of the transmission chain should be 1/2 to 1/2.2. Most television systems employ a gamma of 0.5; some a gamma of 0.45.

The camera pick-up tubes have different values of gamma for I.O., vidicon and plumbicon. The I.O.

has somewhat nonlinear characteristics; the gamma is dependent on the lighting conditions (high key or low key), 0.5 at peak whites in high key lighting and unity at dark grays in low key lighting. This stretches the blacks and compresses the whites.

The gamma of plumbicon is somewhat less than 1, typically 0.9. The gamma correction network introduces a correction of 0.5/0.9 = 0.55 ≈ 0.5. In special cases like low key photography, the gamma correction is reduced to 0.3, which stretches the black and brings out the black-dark grey details.

If the vidicon camera is used, its average gamma is 0.65, so that the uncorrected system gamma will be 0.65 × 2 = 1.3. A gamma correction of 1/1.3 ≈ 0.75 is, therefore, necessary.

The gamma correction is made in the processing amplifier after the video signal is raised to an adequately high level. The gamma correction network works basically on diode conduction threshold adjustments by initial biasing as shown in the circuit of Fig. 4.28.

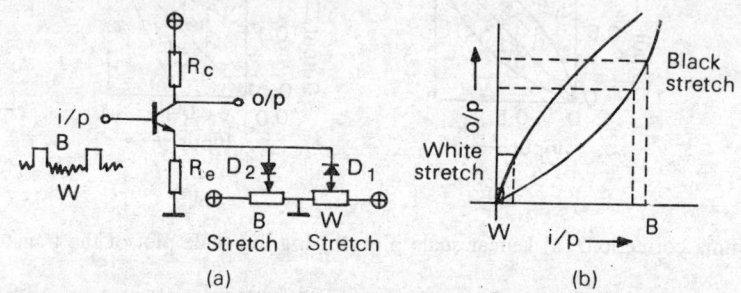

Fig. 4.28 Gamma correction circuit and its transfer characteristics

The signal is direct-coupled to the amplifier so that the emitter voltage changes with the input levels corresponding to the black and white. When the emitter voltage falls due to negative-going white levels, diode D_1 conducts and shunts the emitter resistance R_e, increasing the gain and stretching the whites. When the input voltage rises due to positive-going black shades, the emitter voltage also rises so that the diode D_2 conducts, reducing the effective emitter resistance and increasing the gain, thus stretching the black shades. The stretch controls adjust the pre-set bias to the diodes to start the stretch action at particular levels desired.

Clamping Circuits   The video signal from the camera has a dc component also. The various stages in the video chain are often ac coupled as it is easier to design ac coupled high gain wide band amplifiers than dc coupled amplifiers. Whenever the dc component of the signal becomes essential, the dc is restored to the signal by dc restorer or clamping circuits. In the simple dc restorer using a diode and *RC* circuit as described in Sec. 14.3, the discharge time-constant *RC* of the circuit is large, and the dc voltage generated on the capacitor during diode conduction may persist over several lines before acquiring lower dc levels that may follow.

Gated or keyed clamping has the advantage of faster response to both increasing and decreasing dc levels due to mean brightness variations in the picture; and is, therefore, most commonly used in the camera and studio equipment. In this process, the sync tips (or rather the back porches of the blanking pulses) are clamped to a fixed voltage or grounded by solid state transistor switches by keying or gating pulses. The keyed clamping is done line by line by means of clamping pulses derived from the sync generator.

The dc restoration must be effected by the clamping circuits before the signal is given to gamma correction, blanking insertion or white clip and black clip circuits. The dc restoration or keyed clamping of the black level reduces hum and low frequency distortion that may be incidentally introduced in the video signal. The sync tips get clamped to a reference level eliminating the superposed hum and low frequency levels.

White Clip

A video signal in the camera chain has a normal amplitude of 1 V peak-to-peak, of which 0.7 V is the picture signal and 0.3 V is the sync signal. The composite video signal waveform is adjusted on the video waveform monitors on an amplitude scale after the IRE* scale of units. The video-picture portion of 0.7 V is set to occupy 100 IRE units while the sync pulse portion of 0.3 V is set to occupy 40 IRE units.

Although the set lighting and the camera controls are adjusted so as to avoid levels more than 100 IEEE units, high light speckles from shiny surfaces of musical instruments, ornaments, etc. cannot be avoided. It is not desirable to reduce the overall level to keep these tiny speckles under the 100% limit, at the camera control. Allowing them to go beyond 100% would overload and saturate the output, which may cause streaking. In order to avoid this, the peak whites are limited or clipped at 104 IRE units, by suitable clipping circuits. In the camera chain, this limiting is done initially at 150% and then at 104% at later stages.

Black Clipper   The black clipper clips the picture signal black level and establishes a reference black level. A black clip always succeeds a blanking adder.

In a *blanking adder*, a large negative-going pulse of the duration of the blanking pulse is added to the video signal and then the combined signal is clipped at the reference black level in the black clipper as shown in Fig. 4.29. This ensures that there is no noise signal present during the blanking period.

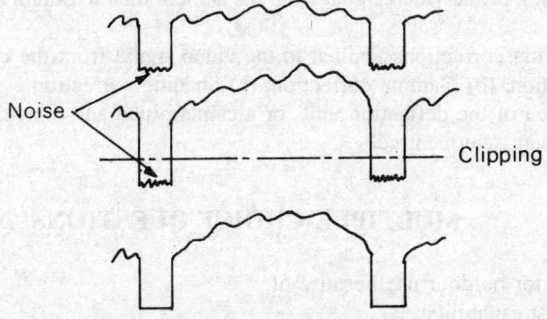

Fig. 4.29   Black clipper

Sync Adder   After the video signal is blanked and clipped to a reference level, the sync pulses are added to the combined signal in a sync mixer or adder circuit.

Shading Correctors   It will be observed with some cameras that there is a gradual and slight fall in the video level from one end to the other either horizontally or vertically even though the scene is uniformly lit and the camera is optimally aligned. This happens especially in I.O. cameras. It may result from variations in the tube sensitivity or from the deflection circuitry or optics. To offset this, horizontal and vertical shading correctors are provided. These correctors add sawtooth signals to the video signals to raise the average level at the end which is darker.

Polarity Reversal   Polarity reversal circuits convert the positive picture into negative by inverting the black and white levels so that the peak white becomes peak black and the peak black becomes peak white. The clamping circuit requires to be suitably modified to clamp the black porch which was originally the reference black, at the peak white level.

---

*The short form IRE is still commonly employed as the original name for IEEE

REVIEW QUESTIONS

4.1 Define the terms: Visual angle of a lens, Depth of field, Circle of confusion, Hyperfocal distance of a lens and Resolving power of a lens.

4.2 What are the types of target plates used in image orthicon, vidicon, plumbicon and multi-diode vidicon? Compare their properties.

4.3 Compare the construction and characteristics of the image orthicon and vidicon camera tubes.

4.4 State the disadvantages of vidicon camera tube explaining how these are caused. Indicate how the limitations are overcome in other variant of vidicon.

4.5 What are the shortcomings in the earlier plumbicons? How are these removed in the modern plumbicon?

4.6 Explain the following phenomena occurring in the camera tubes:
(a) I.O. ghost, (b) Halo effect, (c) Photoconductive lag, (d) Beam lag, (e) Discharge lag in plumbicons.

4.7 Discuss in detail the lag phenomena in photoconductive camera tubes and the ways and means of reducing the lag.

4.8 What is meant by the resolving power of a camera tube? How is it specified?

4.9 What is meant by the gamma of a camera tube? Explain how the gamma of the camera tube, the camera signal chain and the picture tube are matched to give a overall faithful reproduction of the picture.

4.10 What are the special features of the targets and the characteristics of Newvicon, Saticon, and Chalnicon?

4.11 Explain the following: (a) Antihalation disc, (b) Diode gun (c) Light pipe, (d) ACT gun.

4.12 Explain the construction of a CCD imager. What are the methods of charge transfers? Explain one in detail.

4.13 What are the charge coupling techniques used in CCD for generating the video signal?

4.14 How does CID imager differ from CCD? Discuss the advantages of CCD imager in TV cameras compared to other pick-up devices.

4.15 Discuss the processing of the video signal from the camera tube to obtain a composite video signal from the camera tube.

4.16 How are the following corrections applied to the video signal from the camera tube?
(a) Aperture correction, (b) Gamma correction, (c) Shading correction.

4.17 Give the construction of the deflection unit for a camera tube and discuss the action of the focusing coils in it. How is the beam scan centred?

MULTIPLE-CHOICE QUESTIONS

4.1 Vidicon is suitable for outdoor use because of:
(a) its high contrast capability.
(b) its variable gamma characteristics
(c) its self adjustable sensitivity
(d) its good spectral response

4.2 Bias lighting is useful in:
(a) reducing beam lag
(b) improving linearity
(c) reducing halation
(d) (a) and (c)

4.3 Automatic sensitivity control by target voltage control is readily possible in:
(a) Vidicon
(b) Newvicon
(c) Plumbicon
(d) Chalnicon

4.4 Plumbicon has gamma of:
(a) near unity
(b) 0.7

   (c) 0.5
   (d) 0.6 to 0.9
4.5 Major advantage of Plumbicon is:
   (a) better spectral response,
   (b) near unity gamma
   (c) reduced lag
   (d) small dark current
4.6 Operation of CCD image sensors is based on:
   (a) photoconductivity
   (b) photovoltaic principle
   (c) photoemissivity
   (d) none of these
4.7 Silicon vidicon is based on:
   (a) photoconductivity
   (b) photoemissivity
   (c) photovoltaic principle
   (d) none of these
4.8 Silicon vidicon is most useful when:
   (a) damage due to excessive high light is likely
   (b) high sensitivity extended to IR region is required
   (c) very high resolution is required
   (d) (a) and (b)
4.9 The gamma of a TV picture tube is complementary to that of the:
   (a) vidicon
   (b) plumbicon
   (c) image orthicon
   (d) newvicon
4.10 Dark current is proportional to target voltage in:
   (a) plumbicon
   (b) vidicon
   (c) newvicon
   (d) (b) and (c)
4.11 Beam lag in a camera tube is due to:
   (a) pixel capacitance
   (b) beam resistance
   (c) photoconductive effects in pixels
   (d) (a) and (b)
4.12 Photoconductive lag is due to:
   (a) traps in the impurities in the pixels
   (b) slow decay of the photon generated carriers
   (c) pixel capacitance
   (d) (a) and (b)
4.13 Pick the wrong statement:
   (a) Aperture correction is applied to the video signal by boosting the HF response
   (b) Gamma correction is necessary to ensure linearity of the camera studio chain
   (c) the white clip limits the high light speckles from reflections
   (d) the black clipper adds the blanking pulses to the video signal.
4.14 CCD imagers present no lag because:
   (a) there is no charge storage
   (b) the capacitance of pixel capacitors is very low
   (c) the charge transfer is fast
   (d) none of these

5

Colour Television Signals
and Systems

The black and white or monochrome television system reproduces an image that is very different from the original coloured scene, but has for long been accepted to be of sufficient information and entertainment value, as borne out by the success of black and white motion pictures. The appeal of colour television lies in its greater naturalness which is much more informative and entertaining in many situations. For example, the use of colour television for medical education in demonstration of surgical operation or in art appreciation programmes has obvious advantages. In general, colour heightens the contrast and appears to add a third dimension to the picture. Hence colour televisions have rapidly replaced black and white televisions throughout the world, although colour systems are more expensive.

Use of solid state devices and integrated circuits of greater complexities and capabilities unthinkable some years ago, have replaced the earlier electron tube circuits. These solid state devices coupled with SAW filters, availability of low cost delay lines and a new range of more efficient colour picture tubes with simpler deflection assembly have lowered the production as well as maintenance costs, by reducing critical adjustments for the beam convergence and IF alignments. Besides reducing the cost and consumption of colour television systems, this has resulted in greater reliability along with more compact and portable designs. Colour TV cameras, and video cassette recorders along with TV receiver monitor are now available at modest costs as consumer durables. The use of colour TV receiver as a terminal for home computers and for video games has made increased demands on its performance. Fast developments in CCD cameras and colour LCD TV displays indicate further proliferation of colour video systems in the consumer market in future.

5.1 COLOUR FUNDAMENTALS

It is well known that light is a form of electromagnetic energy consisting of a spectrum of frequencies of wavelengths ranging from about 7000 Å for red light to 4000 Å for violet light, visible to human eye. Each wavelength in this range is perceived by the eye as a certain tint or hue. Practically all such hues are present in white sunlight besides energies at other wavelengths beyond this range, viz. infrared on the high side and ultraviolet on the low side as shown in Fig. 5.1.

The infinitely large number of wavelengths in the spectrum produces a gradation of continuous hues from violet to red. Each distinct hue is called a spectral colour that is perceived by the eye as the purest highly 'saturated' colour, though the eye cannot distinguish them as closely. Thus, when a glass prism

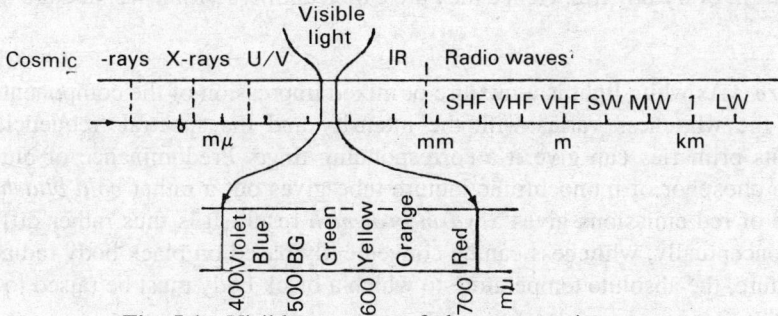

Fig. 5.1 Visible spectrum of electromagnetic waves

separates white sunlight into pure spectral colours, only six fairly distinct colour are perceived by the eye, viz. red, orange, yellow, green, blue and violet. Spectrally pure saturated colours occur in nature rarely, e.g. in light emitted by incandescent vapour or gases and in purer form in laser beams.

Apart from the spectrally pure saturated colours, there are an infinite variety of non-saturated colours or mixed colours which may consist of (i) a fairly *wide continuous group of* wavelengths giving impressions of tempered colours, (ii) a *whole spectrum* of wavelengths with only a group of *wavelengths accentuated,* giving the impression of a soft drab colour, (iii) two or more *independent contiguous groups* of wavelengths, giving the impression of single tempered colour in the spectrum or a new colour not present in the spectrum at all. The Third variety implying mixing or colours, forms the basis of painting, colour printing and also colour television.

The colour of an object is a function of the unabsorbed wavelengths of light reflected from it. That is why an object may appear to possess a different colour under artificial light from an electric bulb, from when viewed in bright sunlight. An object illuminated by white light containing all wavelengths absorbs some wavelengths while reflecting or transmitting others. The colour of the object is decided by the reflected colours for an opaque object, while for a transparent object the wavelengths transmitted through it determine the colour of the object, which acts as an optical bandpass filter. White objects reflect all colour wavelengths and transparent objects like glass transmit all colour wavelengths through themselves. A white screen is, therefore, an essential requirement for colour slide film projection.

5.2 MIXING OF COLOURS AND COLOUR PERCEPTION

Mixing of colours can take place in two ways: (i) by subtractive mixing, and (ii) by additive mixing. In *subtractive mixing*, reflecting properties of pigments are used. These pigments absorb all wavelengths but for their characteristic colour wavelength. When pigments of the same or more colours are mixed they reflect wavelengths that are common to both. Since pigments are not quite saturated they reflect a fairly wide band of wavelengths. This type of mixing takes place in painting or colour printing. For this purpose, appropriate mixtures of primary colours, red, yellow and blue (more precisely, magenta, yellow and cyan) can produce any desired colour sensation. This is shown in Fig. 5.2.

In additive mixing, light from two or more colours from independent sources or obtained through different filters can create a combined sensation of a different colour. This can happen when tiny light emitting dots are very close to each other, giving a common impression or when filtered light of different colours falls on the same place on a white screen which can reflect all of them together. The additive mixture of the three primary colours—red, green and blue in adjustable intensities can create most of the colours encountered in every day life. Hence they are called *additive primaries,* and are used in television as basic colours.

Colour temperature   As white light is a mixture or mixed impression of the component primaries in the visible spectrum, the whiteness varies with the intensity and the spectral frequencies in each. The predominance of its primaries can give it a corresponding tinge. Predominance of blue emissions, for example, from the phosphor of monochrome picture tube gives out a rather *cold bluish* rendition while the preponderance of red emissions gives a *warm brownish* result. It is thus rather difficult to define a standard white. Conceptually, whiteness can be conveniently based on black body radiation, referred by its colour temperature, the absolute temperature to which a black body must be raised to render a desired standard white light.

Reddish light from a candle measures to a colour temperature of 1000 K, and a photo flash gives a whiter light at a temperature of 3000 K. An incandescent tungsten lamp produces light at around 3200 K while quartz halogen lamps gives whiter light above 5000 K. Sunshine is pure white at colour temperature of 6000 K, rising to higher values of up to 10,000 K, if skylight blue is to be effected. In television systems, this has been fixed for standard daylight at 6500 K, corresponding to what is known as illuminent D or D 6500. There have been other standards like the standard white A (incandescent tungsten lamp), white B (sunlight at noon), standard white C (somewhat bluer normal light).

The addition or mixture of the three colours—red, green and blue—in adjustable intensities can create most of the colours encountered in everyday life. Hence red, green and blue are called *additive* or *primary* colours, and are used in television as basic colours. The different colours produced by additive mixing of these three colours is shown in Fig. 5.2(a). Addition of two of the primary colours produces the *secondary* colours: cyan, yellow and magenta. Mixing appropriate proportions of the three primaries, or a secondary with the opposite primary, produces white light. The effect caused by the mixing of subtractive or secondary colours is also shown in Fig. 5.2(b). Mixing proper proportions of the three primary-additive colours produces white, whereas mixing of the three secondaries-subtractive pigments or a secondary with its opposite primary produces black.

It is thus possible to simulate white colour also be pairs of two different colours in certain proportions. Such pairs of *opposite colours* (e.g. Cyan + Red, Magenta + Green, and Yellow + Blue) which, when additively mixed, appear white to the eye, are known as *complementary* colours. *Cyan, magenta* and *yellow* are the complementary colours to the additive colour primaries red, green and blue, and are also called minus red, minus green and minus blue, respectively. They are called *secondary* or *subtractive* colours. Cyan is a greenish blue colour and magenta is purplish red. These can be readily observed in colour photographic negatives. Hence in a subtractive colour system, blue, red and yellow are generally mistaken as the subtractive colours in painting or printing technology, where the mixture of pigments applied on a white paper absorbs (subtracts) colours in the incident light and reflects the resultant colours. In processing colour photographic film or paper, emulsions layers sensitive to red, green and blue produce dye images in complementary colours: cyan, magenta and yellow. A correct mix of cyan, magenta yellow can be used to match any colour in nature. White is the absence of any of the three, while black is a mixture of these in saturated colours.

Colour Perception

The physiological basis for the mixing process is not very clear. One theory formulated by Helmholtz is that the cones of the retina which are responsible for colour perception, are of three kinds responsive to the primary colours red, green and blue, their sensitivity being spread over a broad rather than a

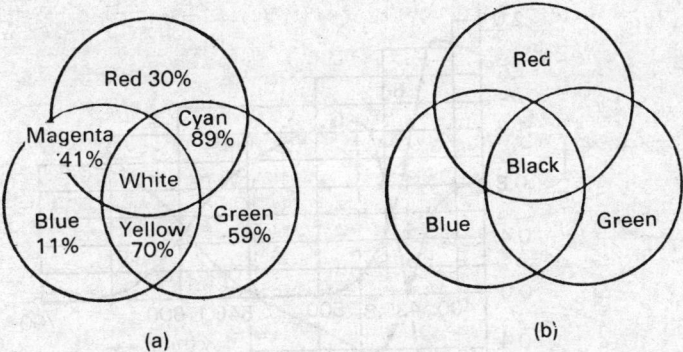

(a) (b)

Fig. 5.2 Mixture of additive and subtractive colours

(a) Additive or primary colours: *R, G* and *B*

$R + G$ = Cyan, $G + B$ = Yellow, $B + R$ = Magenta

$R + C = W$, $G + M = W$, $B + Y = W$, and $R + G + B = W$

(b) Subtractive or secondary colours: *C, M,* and *Y*

$W - R$ = Cyan, $W - G$ = Magenta, $W - B$ = Yellow

$C + M$ = Blue, $M + Y$ = Red, $Y + C$ = Green

$C + M + Y$ = black

$W - R - G$ = Blue, $W - G - B$ = Red, $W - R - B$ = Green

narrow, band of frequencies. As a result of extensive tests with hundreds of observers, the primary spectral colours and their intensities required to produce the greatest number of spectral colours have been standardized by the Commission International d'Eclairage-Illumination (CIE) based on the principles of colour matching in colorimetry. The CIE primaries are Red (7000 nm), Green (546.1 nm) and Blue (435.8 nm). The three colour component values called the tristimulus values of the various spectral colours have been standardized and tabulated. These are represented graphically in the form shown in Fig. 5.3.

White light is produced when the intensities of the primaries are in the proportion—red—30%, green—59% and blue—11%. Colour vision depends greatly on the illumination. In a brightly illuminated scene, the colours are better resolved. Some colours are more easily seen than others. This is indicated in the relative luminosity curve showing the relative sensation of brightness produced by individual spectral colours radiated at a constant energy level, as indicated in Fig. 5.4.

The human eye is most sensitive to green at a wavelength of about 555 nm. Its ability to distinguish between shades of similar colour is limited. Because of this limitation of the eye, satisfactory colour reproduction is possible even though some of the true spectral shades are omitted. Resolution of colour is difficult in objects of small size which appear as objects of black and white shades only. Hence it is logical to transmit high resolution details as luminous information only. The eye tends to integrate or add on the effect of small units of colour very close to one another. In order to define coloured light, the following terms are used:

   (i) *Brightness* or *Luminance*: This is the amount of light intensity or energy received by the eye regardless of colour.

   (ii) *Hue*: This is the predominant spectral purity of the colour light.

   (iii) *Saturation*: This indicates the spectral purity of colour light. Since single frequency colours rarely occur alone, this indicates the amounts of other colours present, i.e. the amount of pastel shading.

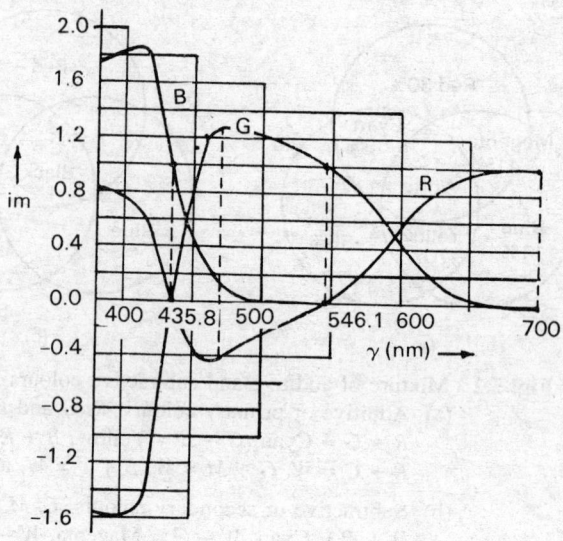

Fig. 5.3 Tristimulus values of spectral colours (reproduced from *colour TV explained* by Holm; *Courtesy:* Centrex Publications)

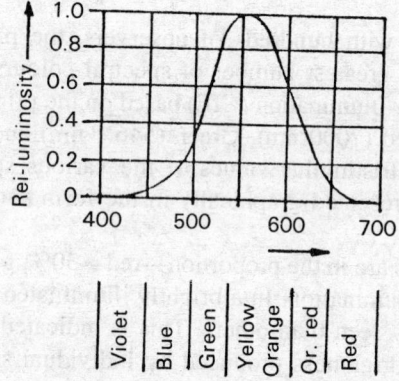

Fig. 5.4 Relative luminosity response of human eye (reproduced from Colour TV explained by Holm; *Courtesy:* Centrex Publications)

5.3 CHROMATICITY DIAGRAM

A chromaticity diagram is a convenient space coordinate representation of all the spectral colours and their mixtures. based on the results of colorimetry referred to above. It has been noted that all the spectral colours can be made up from the three components red, green and blue with tristimulus values of each. Any other mixed colour can also be composed from appropriate modified components of R, G and B taking into account the relative luminosity curve of the eye for the spectral colours. These can be transformed as coordinates in space to represent all the various colours, with their hue and saturation in the $x–y$ plane and brightness along the z-axis.

Without going into the mathematical transformation, we note that such a diagram is formed by arranging the colours of the rainbow along a horseshoe-shaped triangular curve as shown in Fig. 5.5.

The various saturated pure spectral colours are represented along the perimeter of the curve, the corners representing the three primary colours red, green and blue. Such deep saturated colours occur rarely in everyday life. As the central area of the triangle is approached, the colours become pale and less saturated, representing mixed colours. The white lies at the central point W with coordinates $x = 0.31$ and $y = 0.32$. Actually there is no specific white light; sunlight, skylight and daylight are all forms of white light, with component colours of each differing considerably. The point W in effect is a small area, inside which the standard white B (sunlight at noon), standard white C (somewhat bluer normal daylight), the standard illuminent D (daylight with north skylight) used for television purposes and the standard white A (for incandescent tungsten lamp), lie. Along the base line of the triangle lie purple colours that

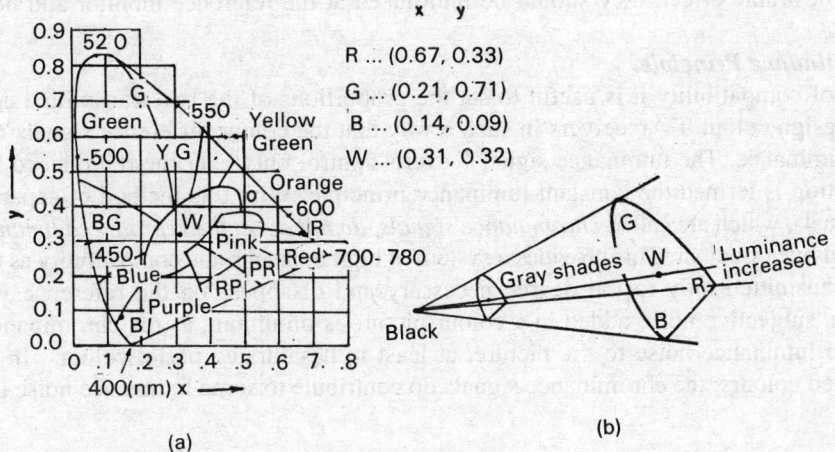

Fig. 5.5   (a) Chromaticity diagram (b) Colour pyramid

are mixtures of red and violet. These are not present in the natural rainbow spectrum and occur only as' visual sensations.

A practical advantage of the chromaticity diagram is that, in a particular arrangement of x and y coordinates, it is possible to determine the *result of additive mixing* of any two or more colours by joining the relevant coordinates by a straight line. The point dividing the line in the ratio of the intensities of the two colours represents the resultant colour of intensity equal to the sum of the two, and the dominant wavelength pointed by the line joining the point with the point W.

The *hue* determined by the spectral colour at the radius from the white point W to the outer periphery, can be specified by the angular measure of the radius with a reference line say, red (700 nm). The *saturation* can be specified by stating the distance from the white.

The colour diagram contains all colours of equal brightness. If the brightness of colour is represented by the z-axis at right angles to the x–y plane, all colours of all brightness must appear as a solid pyramid as shown in the Fig. 5.5(b). As brightness increases, the colour diagram becomes larger and more colour details are visible. At low levels of brightness, the colours become less distinct and the diagram shrinks to a point zenith, where the brightness is zero and no colour is seen. This point represents black.

The colour pyramid represents colours and their brightnesses as the eye sees them. The effect of brightness variation on the white centre of the colour diagram is a narrow white cone going from black at its zenith through all shades of gray to white of the brightest colour plane. The *cone* axis is called the Y or brightness axis, since it represents along it the brightness levels.

In monochrome television, only the y-axis variations are transmitted. In colour television, additional signals specifying the hue and saturation must be transmitted. This coordinate system was used in formulation of the colour television standards as the *Y-coordinate is completely independent of colour* and can be used by the monochrome or black and white television receiver to make the systems compatible with each other, without need for conversion. Conversely the monochrome Y-signal can be received by the colour television receiver, producing a black and white picture in the absence of the colour signals, providing reverse compatibility.

Colour Representation

Although it is not known how much accuracy of colorimetric reproduction is desirable, it is believed that *colorimetric distortions* are not desirable in the transmission process. If at all distortions are employed to produce a desirable effect, they should be intorduced at the reference monitor and not subsequently.

Constant-luminance Principle

For the sake of compatibility it is useful to set the proportions of the two transmitted colour difference signals and design colour TV receivers in such a way that the colour difference signals do not affect the reproduced luminance. The luminance signal Y exerts control only over the reproduced luminance. This basic assumption is termed the constant luminance principle. With this method of separation the colour difference signals, which are called *chrominance signals, do not affect the reproduced luminance* but control the reproduced hue or colour. This provides a system of high monochrome compatibility as the chrominance signals are transmitted only to the degree necessary and disappear on the reference white. With this proportioning, subjective noise added in a colour picture is minimum, as the chrominance channel does not contribute luminance noise to the picture, at least in desaturated pastel colours. In the presence of highly saturated colours, the chrominance signals do contribute to some luminance noise in the luminance channel.

Colour Circle

In colour television, the triangular chromaticity diagram is converted into a colour circle or colour wheel. The R, G and B primaries are the radial vectors, phased about 120 deg. apart. The degree of saturation of each colour is measured along the vector from the centre to the circumference, and the hue can be specified by the phase angle of the colour vector relative to a reference axis V corresponding to the $(R-Y)$ vector. The colour circle diagram is shown in Fig. 5.6, and colour plate I.

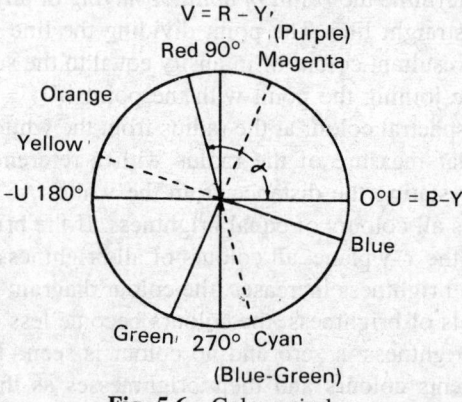

Fig. 5.6   Colour circle

5.4   COLOUR TELEVISION CAMERA

A simple colour television camera block schematic is shown in Fig. 5.7. It is essentially composed of three monochrome camera tube systems, in which each camera tube receives selectively filtered primary colours to produce an electrical signal proportional to the respective colour. Light from the scene processed by the objective lens is directed at an arrangement or mirrors that *splits the light into the three basic colours red,* green and blue. The mirrors are coated with special dichroic material so that they reflect a specific colour and allow other spectral frequencies to pass through.

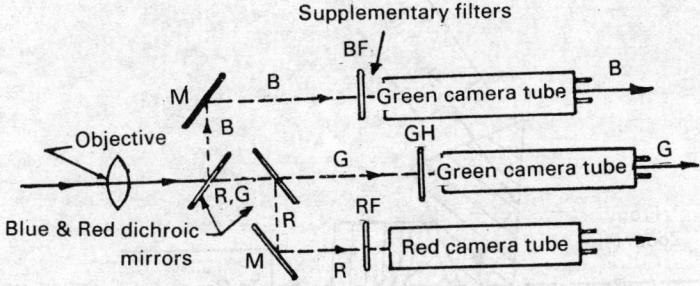

Fig. 5.7   Colour splitter—dichroic mirror system

The action of *dichroic mirrors* is based on the interference produced by light waves in a very thin optical medium as for example caused and observed in the form of the variety of colours in soap bubbles or in thin oil films on water. In dichroic mirrors, a highly polished glass plate is coated with a *thin translucent dichroic film* of materials like zinc sulphide and cryolite. As a ray of light falls on the mirror, it is partly reflected at the glass-film boundary and also from film-air boundary. The *reflected rays interfere* and may cancel each other or add together depending upon the wavelength, and the reflected ray will be thus of a selected colour. The coatings can be substantially loss-free and using several coatings of materials with high and low refractive indices, almost any desired colour characteristics can be obtained.

In Fig. 5.7, mirror B reflects blue light and allows the green and red constituents to pass through. The blue constituent is reflected to the front surface mirror which directs it to the blue camera tube through a supplementary filter BF. The red and green lights passing through B fall on another dichroic mirror R which *reflects the red* light and passes green component through to the green camera tube. The reflected red light is guided by a reflective mirror to the red camera tube through the additional red filter RF. Thus the colour splitter dichroic mirrors split the original multicoloured image into its red, green and blue constituent shades to focus them on to the face plates of three plumbicon or other derivatives of vidicon camera tubes, made responsive to the respective colours by means of supplementary filters.

In recent systems, use has been made of glass prisms cemented together in place of the individual mirrors. Such an arrangement is shown in Fig. 5.8. The *dichroic layers are applied to the mating surfaces* which not only seal the sensitive interface layers from the atmosphere and dust, but also reduce the lens-to-tube distance. A plumbicon camera with prismatic colours splitting system is relatively easy to operate, small in weight and size, besides recording superior colour pictures.

The *scanning* of the camera tubes is accomplished simultaneously by a master deflection oscillator that drives all the three tubes. The three video signals represent the red, green and blue primaries of the colour diagram and by selective use of these, almost any colour in the visible spectrum can be reproduced. This can be done in a tricolour picture tube that has three electron guns controlling the luminescence of closely spaced triangular groups of *three colour phosphor dots,* on the screen of the tube. The dots are so tiny and so closely spaced that the eye cannot see them in individual colour but perceives the integrated or additive colour.

5.5   COLOUR TV SIGNALS

The three primary colour signals produced by the three camera tubes represent the proportions of red, green and blue light components of the pixel being scanned. These signal voltages *R, G* and *B* representing the respective colours are corrected for the differences in the nonlinearity of response of the camera tubes

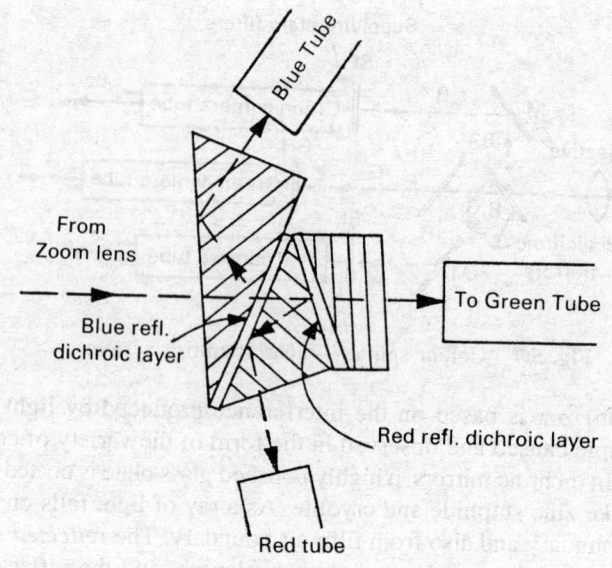

Fig. 5.8 Prismatic colour splitting

and the picture tube to provide the *gamma corrected R', G' and B'* signals. For simplicity of notation we shall hereafter refer to these gamma corrected signals by symbols *R*, *G* and *B* only. These gamma corrected RGB signals are adjusted to an arbitrary value of 1 V each, for the standard white of maximum, i.e. 100% intensity.

These signals could be directly used for transmission and reproduction of picture on a three beam colour picture tube with very fine RGB phosphor dots or stripes deposited sequentially along the screen. However for making the colour video signal compatible for reception on a monochrome TV system, the *R*, *G* and *B* signals are combined to form the luminance signal *Y* which serves the requirements of monochrome TV signal. The colour components are represented by the colour difference signals *R-Y* and *B-Y*, forming the chrominance signal which can be decoded in a colour TV receiver to obtain the *R*, *G* and *B* signals required to drive a colour picture tube for colour picture reproduction.

Luminance Signal

The luminance signal is obtained by adding the primary *R*, *G* and *B* colour signals in the proportion, red 30%, green 59% and blue 11%, taking into account the difference in sensitivity of the human eye to these colours, according to the relative luminosity response curve of Fig. 5.4. The outputs adjusted to a uniform maximum level of 1 V for the standard white, are attenuated proportionally and added in a resistive matrix network, to obtain:

$$Y = 0.30\ R + 0.59\ G + 0.11\ B$$

where *Y* represents the luminance signal and *R*, *G* and *B* represent the red, green and blue signal voltages. This is the signal containing the gray scale information of the picture. *Y* has the maximum value of 1 V for peak white, as it includes all the *R*, *G* and *B* signals. For other colours *Y* is the sum of the luminosity contribution of the primary components, as follows:

Colour	White	Yellow	Cyan	Green	Magenta	Red	Blue	Black
	$R+G+B$	$R+G$	$G+B$	G	$R+B$	R	B	—
Y	1.0	0.89	0.7	0.59	0.41	0.3	0.11	0

Chrominance Signals

Instead of sending three separate RGB signals besides the luminanace signal Y, only two colour signals in the form of colour difference signals can be transmitted as the four are related by a simple mathematical equation. In practice, for compatibility we have to transmit Y for catering to the monochrome receivers. Hence the other chrominance information is transmitted as $R - Y$ and $B - Y$ colour difference signals obtained by subtracting Y from R and B signals. There is no need to send $G - Y$, as this can be derived from Y, $R - Y$ and $B - Y$ signals. We have

$$0.30\,R + 0.59\,G + 0.11\,B = Y, \text{ which can be split as } (0.30 + 0.59 + 0.11)Y$$

This gives

$$G - Y = \frac{-0.30(R - Y) - 0.11(B - Y)}{0.59}$$

$$= -0.51(R - Y) - 0.186(B - Y)$$

It would have been possible to transmit any two of the three colour difference signals along with the luminance signal and derive the third. However $G - Y$ is not chosen for transmission because G is on the average the largest amplitude colour signal so that $G - Y$ is also on the average the smallest. As such, it would be more vulnerable to noise interference in the transmission path. Derivation of the other colour difference signals from $G - Y$ also requires a knowledge of the gain multiplier instead of the attenuation factor required in the above relation for $G - Y$, as seen from the following relations.

$$R - Y = \frac{0.59(G - Y) - 0.11(B - Y)}{0.3} = 1.97(G - Y) - 0.37(B - Y)$$

$$B - Y = \frac{0.59(G - Y) - 0.30(R - Y)}{0.11} = 5.40(G - Y) - 2.70(B - Y)$$

It is interesting to note here that when the televised scene is devoid of colour, i.e. when only luminance grey shades are transmitted, colour difference signals fall to zero amplitude.

For peak white, when $R = 1$, $G = 1$ and $B = 1$

$$Y = 0.30(1) + 0.59(1) + 0.11\,(1) = 1$$
$$R - Y = 1 - 1 = 0$$
and also $\quad B - Y = 1 - 1 = 0$

On grey shades the *RGB* signals from the camera tubes are less than 1 V, but still equal, and hence the colour difference signals are zero. On colour scenes, the *RGB* signals are not equal and hence colour difference signals are produced. Colours below full saturation mean that white has been added to the predominant hue, in terms of the RGB primaries in proportion required for luminance signal. Thus when the camera is in the process of scanning a coloured scene, the *R*, *G* and *B* components become unequal; but the *Y* signal still retains 30% of red, 59% of green and 11% of red. The *luminance* of a given picture area is to be essentially *unaffected by signals carrying colour information* for that area (constant luminance principle).

For the standard vertical *colour bar pattern* consisting of a white bar followed by 100% saturated bars of yellow, cyan, green, magenta, red and blue from left to right, ending with a black bar of zero luminance and zero colour, is widely used in colour testing. The relative values of luminance Y, the colour difference signals and U and V, the *weighted,* that is, reduced, magnitudes of colour difference signals (as used in PAL system), for these colour bars, are given in Table 5.1. The magnitude C and the phase angle 0 of these colours are also given in the table. The vectorial representation of these colours on the colour circle is shown in Fig. 5.6.

Table 5.1   Relative values of luminance and chrominance signals for 100% saturated colours

Colour:	White	Yellow	Cyan	Green	Magenta	Red	Blue	Black
Y	1.0	0.89	0.7	0.59	0.41	0.3	0.11	0
$B - Y$	0	− 0.89	+ 0.3	− 0.59	+ 0.59	− 0.3	+ 0.89	0
$R - Y$	0	+ 0.11	− 0.7	− 0.59	+ 0.59	+ 0.7	− 0.11	0
$G - Y$	0	+ 0.11	+ 0.3	+ 0.41	− 0.41	− 0.3	− 0.11	0
C_{uw}	0	0.9	0.76	0.83	0.76	0.76	0.9	0
$Y + C_{uw}$	1	1.79	1.46	1.42	1.17	1.06	1.01	0
U	0	− 0.439	+ 0.148	− 0.291	− 0.291	− 0.148	+ 0.439	0
V	0	+ 0.097	− 0.614	− 0.517	+ 0.517	− 0.614	− 0.097	0
C_0	0	0.44	0.63	0.59	0.59	0.63	0.44	0
ϕ	—	167	283	241	61	103	347	—

Chrominance Signal Bandwidth

The human eye cannot resolve fine details in colour. In colour television transmission, it is not necessary to transmit any colour information that lies above 1.3 or at the most 1.5 MHz, because the eye cannot recognize colours of objects below a certain finite size. These pixel areas are interpreted as luminance information with only grey shades.

For complex colours formed from *combinations of the three primaries,* perception by the human eye is limited to relatively larger area pixels which produce video frequencies *below* 0.5 MHz. For less complex colours formed out of *two primaries* the perception extends to finer pixels generating video spectrum between 0.5 and 1.3 MHz. Hence there is little point in transmitting high resolution chroma signals, which would be difficult to accommodate in the available channel because of cross talk interference with the luminance signal and would be wasted anyway, on reception. The luminance signal Y carries all the fine details in the picture, and the chroma signal is overlaid on the display in a much coarser form.

A video bandwidth of 4 MHz represents a horizontal resolution of about 400 lines. A chroma bandwidth of 0.5 MHz would provide 1/8th of this resolution, viz. 50 lines or strips. On a picture tube of 50 cm width this would mean loss of horizontal colour resolution details below 1 cm. The higher resolution is carried by luminance signal only as gray shades. The vertical colour resolution is no problem since vertical scanning produces lower video frequencies that are readily accommodated in the chroma signal bandwidth. The subjective viewing of the combination of high definition luminance and rather woolly chrominance obtained with a chroma signal bandwidth of about 1.3 MHz, is found to be in practice quite satisfactory.

5.6   COLOUR TV SIGNAL TRANSMISSION

Compatibility

The present day television systems are fundamentally based on the work of the National Television Systems Committee (NTSC) in the USA in the years 1951–53. The committee formed by some of the leading research laboratories had the aim of evolving a colour television system that would be compatible with the existing black and white television system. It was important that the existing black and white monochrome TV receivers were able to receive the colour transmissions directly and that a colour TV receiver could produce a black and white picture on its colour picture tube, for reverse compatibility. This required that the additional colour information sent in the colour TV system be accommodated in the same channel bandwidth carrying the black and white information, maintaining the same carrier and scanning frequency standards. The black and white picture should not be affected by the colour contents and vice versa, avoiding cross luminance and cross colour effects.

In the NTSC system these requirements were ingeniously fulfilled, albeit in a complicated way. That is why some parts of the system appear rather circumstantial. The ideas to make the system compatible are indeed ingenious and fascinating even today, although with modern high performance systems the *limitations of compatibility* are seen in the form of cross luminance and cross colour effects due to sharing of the bandwidth. It may be noted that if television had started with colour right from the beginning, the system would have been much simpler. For nomochrome compatibility in the NTSC system, the three primary colour signals R, G and B from the colour camera are combined to obtain the luminance signal and chrominance signal I and Q derived from colour difference signals $R-Y$ and $B-Y$. The two are sent on a single colour subcarrier by *quadrature amplitude modulation*. The magnitude of the resultant subcarrier vector C represents the saturation of the colour while its phase represents the hue.

Chrominance Modulation

The colour subcarrier has to convey chrominance information by two colour difference signals $R-Y$ and $B-Y$ (their weighted values V and U) in the PAL and SECAM system, or their derivatives I and Q signals in the NTSC system. The two signals are sent simultaneously on one subcarrier by *quadrature amplitude modulation in the NTSC and PAL* systems, while the $R-Y$ and $B-Y$ signals are sent on alternate lines in sequence, on a *frequency modulated subcarrier in the SECAM* system. For Quadrature Amplitude Modulation in NTSC or PAL, the two subcarriers are derived from the same source but phased 90 deg apart to carry independent information without affecting each other, as shown in Fig. 5.9.

Combined Video Signal and Weighting Factors

The resultant of the two quadrature phase subcarriers, the subcarrier vector C, vectorially represents the colour information, saturation by its amplitude, given by the square root of the sum of their squares, and the phase given by the $\tan^{-1}$ of the ratio of the carriers. When the peak-to-peak amplitude of chrominance vector C for saturated colours is added to the corresponding luminance signal Y to form the combined video signal, the peak-to-peak amplitude of this colourplexed video signal $(Y + C)$ *can exceed the permissible range* of the video signal. This is calculated for the colour bar signals in Table 5.1.

For example, for 100% saturated yellow C, the total amplitude $Y + C$ varies between $(0.89 + /- 0.9)$, i.e. 1.79 and -0.01. The increase in the total *peak-to-peak amplitude* of the combined signal would overload the video circuits and also cause *over-modulation* at the transmitter, unless the whole of the signal is reduced to prevent this. But this would reduce the available signal at the receiver reducing the service area of the transmitter. A better solution which provides a compromise between the luminance and chrominance signal deterioration is to *reduce the $R-Y$ signal by a factor of 0.877 and $B-Y$ signal by a factor of 0.493* separately, in order to exploit the full permissible range of the combined waveform. This may still result in up to 33% *overloading* beyond black and white levels for 100% saturated yellow and cyan of full amplitude, as shown in Fig. 5.10. But this occasional overloading can be permitted without incurring serious distortion.

The colour vector is thus given by

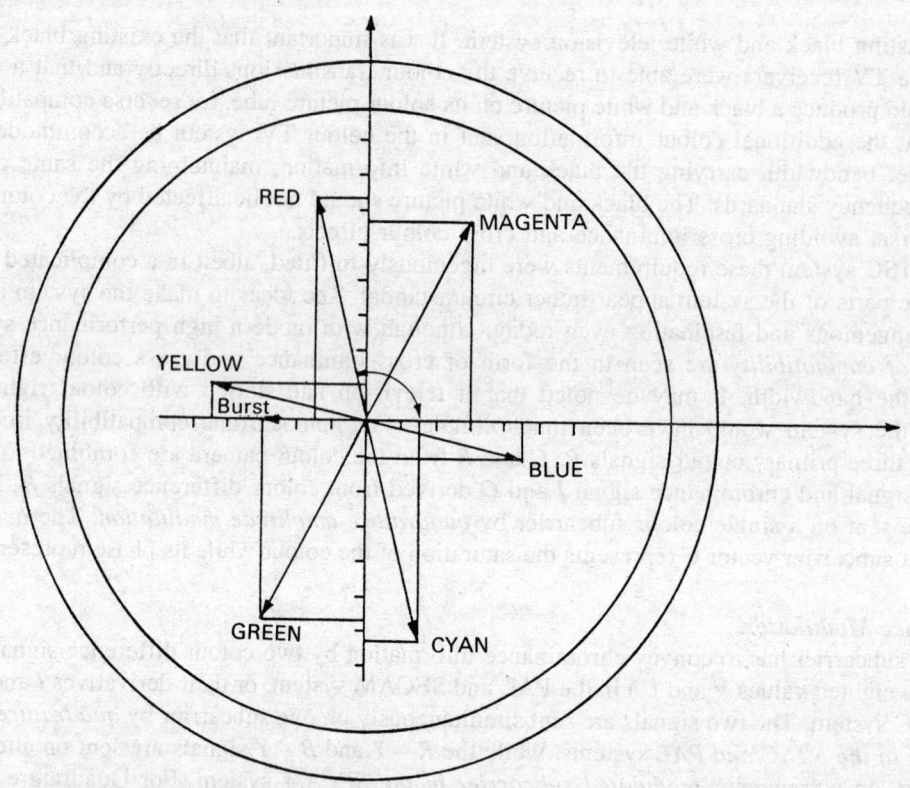

Fig. 5.9 Vector representation of chroma signals
Red $= -0.15\ U + 0.63\ V$, Green $= -0.29\ U - 0,\ 52\ V$, Blue $= 0.43\ U - 0.1\ V$

$$C_{uw} = \sqrt{(B-Y)^2 + (R-Y)^2} \qquad \text{for unweighted signals}$$

or

$$C = \sqrt{U^2 + V^2} \qquad \text{for weighted signals}$$

and the phase angle given by:

$$Q_{uw} = \tan^{-1}\ (R-Y)/(B-Y) \qquad \text{for unweighted signals}$$

or

$$\phi = \tan^{-1}\ (V/U) \qquad \text{for weighted components}$$

The resultant colour vector is located in the four quadrants depending upon the hue represented by the phase angle. For the standard colour bar signal the amplitude C and the phase angle ϕ with the weighted quadrature components U and V used in PAL system, are given in Table 5.1. The phasers are vectorially located in the various quadrants on the vector diagram as shown in Fig. 5.11.

Chrominance Bandwidth

The chrominance signals with bandwidth *limited to* 1.3 MHz modulate an auxiliary carrier near the upper end of the video band. This subcarrier is so chosen that the spectral energy of the sidebands of the carrier can be interleaved in the gaps existing in the black and white video spectrum. The combined colourplexed video signal is then transmitted on standard TV channels. At the receiver, the chrominance components are separated and decoded by suitable demodulators to get the RGB signals that control the beams of a tricolour picture tube.

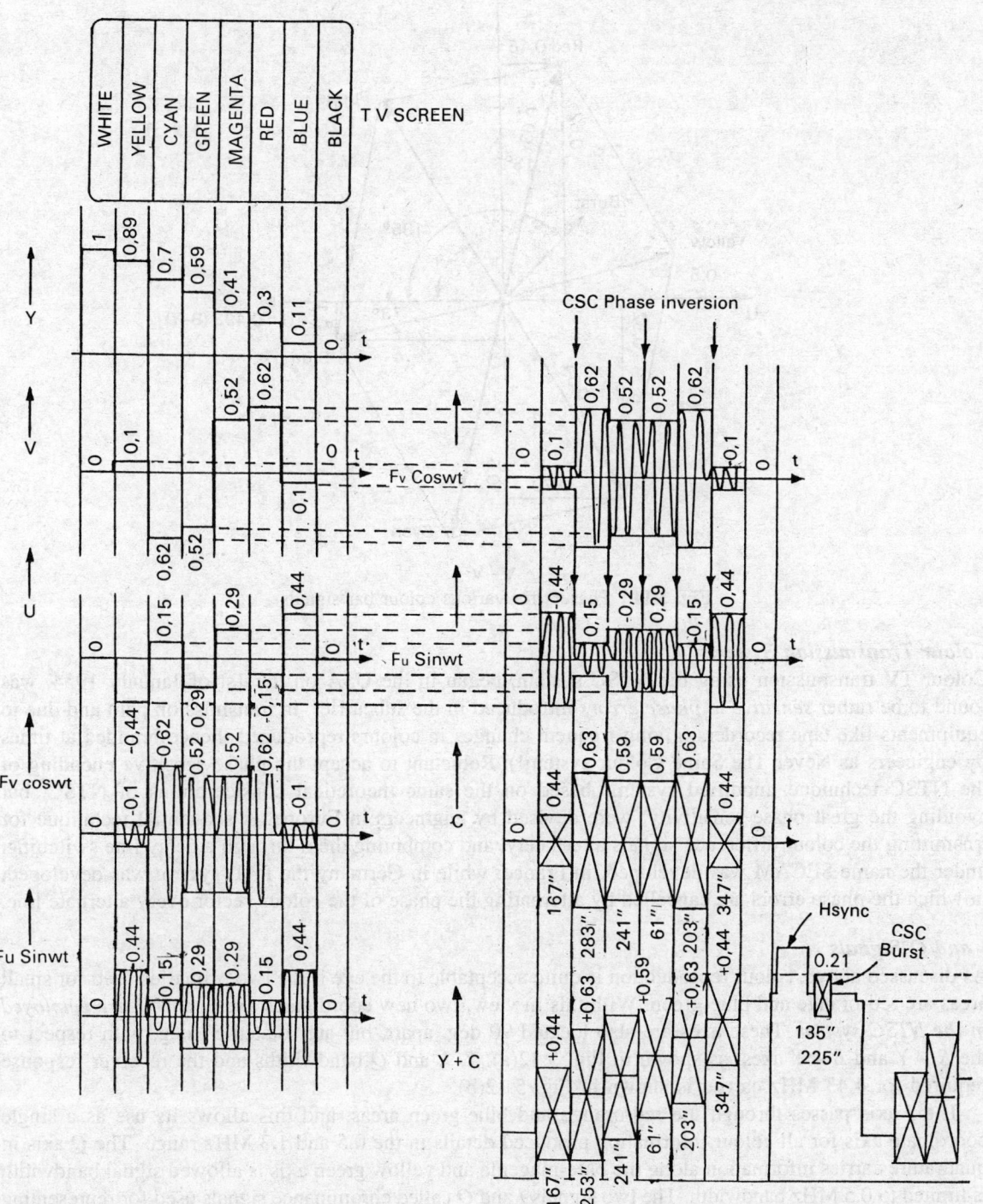

Fig. 5.10 Weighted chrominance components of the colour bar colourplexed video signal [2.8 PAL-sims/grob 8.17]

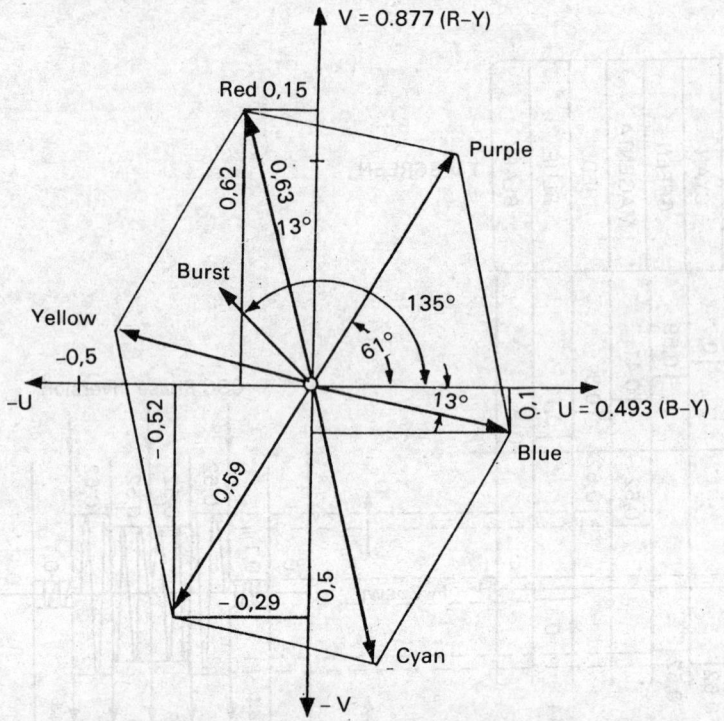

Fig. 5.11 Phasers for various colour bar signals

Colour Transmission Systems

Colour TV transmission using the NTSC system, began in the USA on the 1st of January 1954, was found to be rather *sensitive to phase errors* introduced in the subcarrier, in transmission path and due to equipments like tape recorders, giving frequent changes in colours reproduced (hence branded at times by engineers as Never The Same Colour system!). Reluctant to accept the phase sensitive encoding of the NTSC technique, modified systems based on the same theoretical considerations of NTSC, but avoiding the great phase sensitivity, were devised by engineers in Europe. A sequential technique for trnsmitting the colour difference signals alternately, and combining them through a delay line switching, under the name SECAM was developed, in France: while in Germany the PAL system was developed, in which the phase errors are cancelled by alternating the phase of the colour vector every alternate line.

I and Q Signals

As discussed above, colour reproduction is quite acceptable to the eye if the two primaries used for small areas are red-orange and blue-green. With this in view, two new coordinates axes *I* and *Q are employed in the NTSC system.* These axes are also located 90 deg. apart, but are rotated 57 deg. with respect to the $R - Y$ and $B - Y$ axes, as shown in Fig. 5.12(a). *Y, I,* and *Q* bandwidths and the receiver response required for 4.43 MHz carrier is shown in Fig. 5.12(b).

The *I* axis passes through the red-orange and blue-green areas, and this allows its use as a single coordinate axis for all colour information produced details in the 0.5 and 1.3 MHz range. The *Q* axis in quadrature carries information along the blue-magenta and yellow green axis is allowed signal bandwidth is limited to 0.5 MHz bandwidth. The two signals *I* and *Q* called chrominance signals used for representing the hue and saturation of the colour, are derived from the colour difference signals $R - Y$ and $B - Y$, as follows:

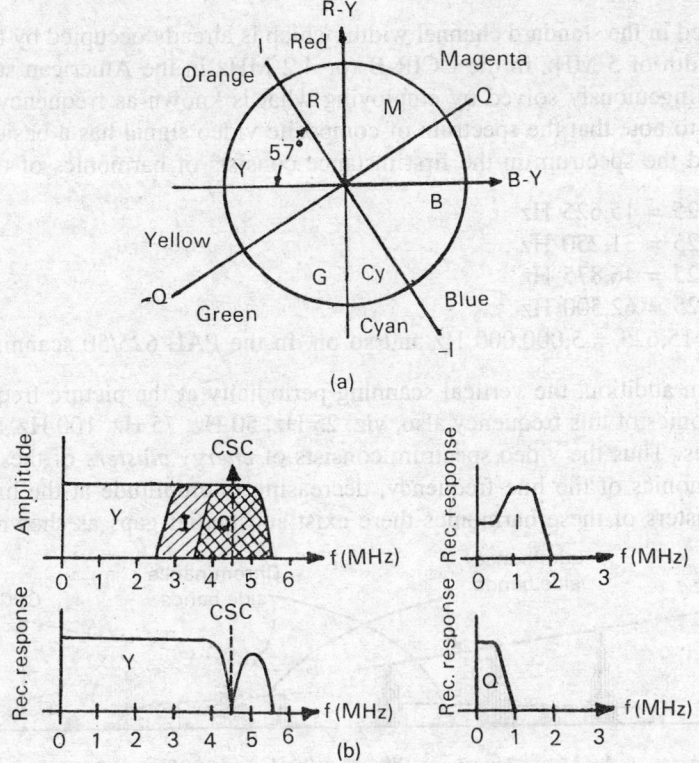

Fig. 5.12 (a) *I* and *Q* axes relative to *R − Y* and *B − Y* axes (b) *Y, I,* and *Q* bandwidth and the receiver response

$$I = 0.74(R - Y) - 0.27 (B - Y)$$
$$= 0.60 R - 0.28 G - 0.32 B$$

$$Q = 0.48(R - Y) + 0.41 (B - Y)$$
$$= 0.21 R - 0.52 G + 0.31 B$$

In the *PAL colour TV system,* which was designed as a variant of the NTSC system, to eliminate its susceptibility to phase errors, the colour difference signals *R − Y* and *B − Y are directly used* to carry the colour information. Before modulation, these signals are *weighted* by factors of 0.877 and 0.493 to produce *V* and *U* chroma signals. Weighting reduces the possibility of overmodulation of the transmitter carrier in the presence of highly saturated colours during very light or very dark parts of the picture. They are both allowed a bandwidth of 1.3 MHz, which gives somewhat better colour reproduction in PAL system.

These chrominance signals with limited bandwidth are used to modulate an auxiliary OSC/104 is so chosen that the spectral energy of its sidebands can be interleaved in the gaps existing in the spectrum of the black and white TV signal. The combined video signal is then transmitted on standard TV channels. At the receiver, a corresponding reverse process takes place to obtain the RGB signals that control the beam currents of a tricolour picture tube.

Frequency Interleaving

The colour subcarrier and the sidebands produced by its modulation with the chrominance signals have to be accommodated in the standard channel width, which is already occupied by the monochrome signal *Y*, with its bandwidth of 5 MHz in the CCIR-B (or 4.2 MHz in the American standard CCIR-M). The problem has been ingeniously solved by employing what is known as frequency interleaving.

It is interesting to note that the spectrum of composite video signal has a basic periodicity of the TV line frequency, and the spectrum in the first instance consists of harmonics of the line frequency, viz.

$1 f_h = 1 \times 15,625 = 15,625$ Hz
$2 f_h = 2 \times 15,625 = 31,250$ Hz
$3 f_h = 3 \times 15,625 = 46,875$ Hz
$4 f_h = 4 \times 15,625 = 62,500$ Hz
$320 f_h = 320 \times 15,625 = 5,000,000$ Hz and so on, in the PAL 625/50 scanning.

Since there is, in addition, the vertical scanning periodicity at the picture frequency of 25 Hz, there is a group of harmonics of this frequency also, viz. 25 Hz, 50 Hz, 75 Hz, 100 Hz, and so on, around each of these frequencies. Thus the video spectrum consists of *energy clusters* of these harmonics occurring nearabout the harmonics of the line frequency, decreasing in amplitude at the higher order harmonics. In between the clusters of these harmonics there exist substantial gaps as shown in Fig. 5.13.

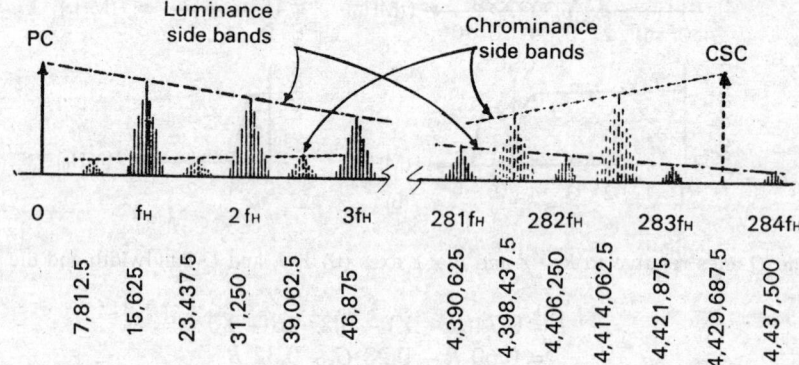

Fig. 5.13 Frequency spectra of video signals and interleaved chrominance frequencies

These gaps occur at *odd multiples of half line frequency*, and can be used to accommodate another series of harmonics of the line frequency. If the colour subcarrier is chosen as some odd multiple of one-half the line frequency, the subcarrier sidebands can lie in the gaps existing between the consecutive harmonics of the line frequency. To avoid interference with the monochrome signal, it is chosen on the higher side of the video band, 567 times the half-line frequency in the PAL system.

$$f_{sc} = 567 \times 15,625/2 = 4,429,687.5 \text{ Hz}$$
$$= 4.43 \text{ MHz approx.}$$

Hence the maximum amplitude of the sidebands of the colour subcarrier occur where the monochrome sidebands are small. The higher order colour sidebands nearer to the monochrome picture carrier are smaller in amplitude and produce minimum interference in that region where the monochrome sidebands are larger in amplitude. This is illustrated in Fig. 5.14.

Colour Subcarrier Frequency
When a monochrome TV receiver receives colour transmission, dot pattern interference can appear along each raster line due to the colour subcarrier frequency lying in the video passband. As the colour

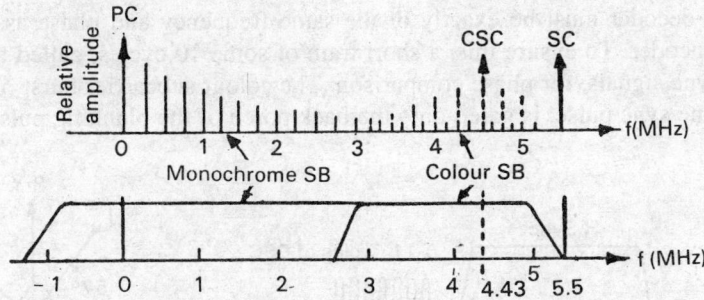

Fig. 5.14 Frequency interleaving and colour sidebands

subcarrier is chosen to be an exact multiple of half line frequency, the interference pattern produced is of opposite polarity for successive odd and even fields. This produces dark black and white dots on the screen alternating on consecutive scans, resulting in overall cancellation effect on the eye of the observer. In the colour TV receiver, this is further reduced by a notch filter tuned to the colour subcarrier, in the Y channel.

Another objectionable interference is likely to be produced in the picture due to the *beat between the sound IF and the colour subcarrier,* e.g. at $4.50 - 3.58 = 0.92$ MHz in the NTSC system. This can be reduced if the subcarrier is an exact multiple of even harmonic of the line frequency. For the sound IF of 4.5 MHz in the NTSC system, the colour subcarrier could be chosen as an even multiple nearest to the prevalent line frequency. The subcarrier can be 4.5 MHz/286 = 15,734.5 Hz, which is quite close to the line frequency of 15,750 Hz used in the monochrome systems and hence within the catching range of the horizontal AFC. For the line frequency of 15,734.5 Hz, the field frequency for interlaced 525/2 lines per field is by division 59.94 Hz, which is also quite close to 60 Hz and can trigger and synchronize the vertical oscillator readily.

In order to *reduce the dot pattern interference,* the subcarrier in the PAL system is made equal to an odd multiple of quarter of the line frequency plus an offset of half the vertical frequency. This provides a phase reversal for the dots on successive fields and minimizes the visibility of the *slow crawling* dot pattern on the screen.

The PAL subcarrier is thus chosen at

$$f_{sc} = \frac{f_h}{4}\,(2 \times 567 + 1) + \frac{f_v}{2} = \frac{f_h}{4}\,(1135) + \frac{f_v}{2}$$

$$\simeq f_h \left(284 - \frac{1}{4}\right) + \frac{f_v}{2}$$

As chrominance modulating information is effectively present in the sidebands of the colour subcarrier, the *subcarrier is suppressed* at the encoder of the transmitter. This reduces the subcarrier interference dot pattern in the received picture, although the subcarrier with correct phase has to be regenerated for synchronous demodulation.

Colour Burst

In order to minimize the subcarrier interference in the reproduced picture, the colour subcarrier is totally suppressed in the chroma balanced modulator which gives out only the chroma sidebands. For demodulation of the chroma signals, it is necessary to regenerate it in the receiver. The subcarrier regenerated at the receiver chroma decoder must be exactly of the same frequency and phase as that of the subcarrier at the transmitter encoder. To ensure this, a short train of some 10 cycles, called the 'colour burst', is sent along with the sync signals, for phase comparison. The colour subcarrier burst of peak to peak amplitude equal to that of the sync pulse, is gated onto the back porch of the blanking pulse, as shown in Fig. 5.15.

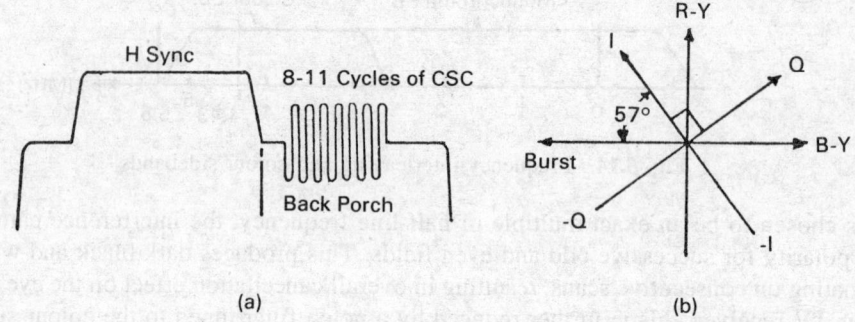

(a)   (b)

Fig. 5.15   Colour subcarrier burst signal and its phase relation with *I* and *Q* axis

The burst signal, in conjunction with a phase comparator AFC circuit is used to lock the local crystal oscillator to the frequency as well the phase of the subcarrier at the transmitter. This provides accurate phase timing reference for the chroma axes. The burst vector is oriented along the $-(B - Y)$ axis and hence the *I* axis lags behind it by 57°, while the $(R - Y)$ axis is a further 90° behind. As the burst signal must maintain a constant phase relationship with the scanning signals to ensure proper frequency interleaving, the horizontal and vertical sync pulses are also derived from the subcarrier through frequency dividers.

5.7   NTSC SYSTEM

NTSC Encoder
In the NTSC system, the *I* signal is oriented along the axis along red-orange and blue-green phasers, while the *Q* signal lies along the blue-magenta yellow-green axis at right angles to it. These signals are obtained in the matrixing circuits from the gamma corrected values *R, G* and *B* from the camera, as shown in Fig. 5.16.

As already seen, the *Y, I* and *Q* signals are related to the colour difference signals or *RGB* signals as follows:

$$Y = 0.30\ R + 0.58\ G + 0.11\ R$$
$$I = 0.74(R - Y) - 0.27(B - Y) = 0.60\ R - 0.28\ G - 0.32\ B$$
$$Q = 0.48(R - Y) + 0.41(B - Y) = 0.21\ R - 0.52\ G + 0.31\ B$$

The luminance signal *Y* with full bandwidth of 4.2 MHz and the chrominance signals *I* and *Q*, with reduced bandwidths of around 1.3 and 0.5 MHz respectively (< 2 dB at 1.3 MHz for *I* and < 2, dB at 0.4 MHz for *Q*), as explained above, are then used to produce colourplexed video signal. The bandwidths of the *Y, I* and *Q* signals are shown in Fig. 5.17, along with the corresponding receiver decoder response required.

The bandwidths are limited in order to reduce the crosstalk interference between the monochrome and colour signals, and at the same time provide adequate colour resolution perceptible to the human eye. The filter stages that limit these bandwidths imply *propagation delays* that vary with the bandwidths, being

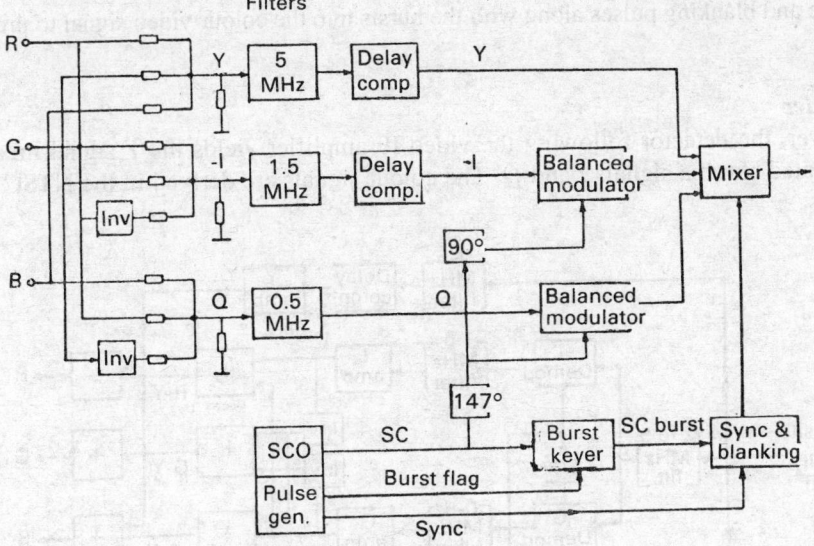

Fig. 5.16 NTSC encoder

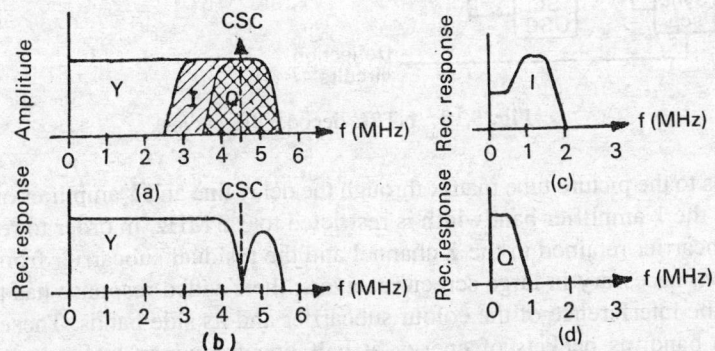

Fig. 5.17 *Y, I* and *Q* bandwidths and the NTSC receiver response

greater for filters with smaller bandwidth, e.g. 1 μs for *Q* and 0.1 μs for *Y* channel. These differences in delays are compensated for by delay lines in the *I* and *Q* channels, in order to avoid mismatch of the vertical contours of the picture.

The colour subcarrier of 3.579545 MHz is Quadrature Amplitude Modulated by the *I* and *Q* chroma signals in the *two balanced modulators*. The I modulator gets a subcarrier lagging 57° behind the colour burst reference signal, sent from the subcarrier oscillator directly to the sync and blanking mixer-adder. The *Q* modulator gets its subcarrier with an additional 90° lag, to provide quadrature phase axis. The modulated *I* and *Q* signals with suppressed subcarrier are added to the sync signals and the *Y* signal to produce the colourplexed video signal, which can be espressed as:

$$M = Y + I \cos (\omega t + 33°) + Q \sin (\omega t + 33°)$$

The subcarrier reference signal from the subcarrier oscillator is gated in by the burst gate flag pulses to feed subcarrier bursts of a minimum of eight cycles to the sync and blanking adder. This mixes the standard sync and blanking pulses along with the bursts into the colour video signal to produce colourplexed video signal.

NTSC Decoder

At the receiver, the detector following the video IF amplifier yields the Y signal mixed with the quadrature modulated chroma signals I and Q. The colour signals are derived in the NTSC decoder as shown in Fig. 5.18.

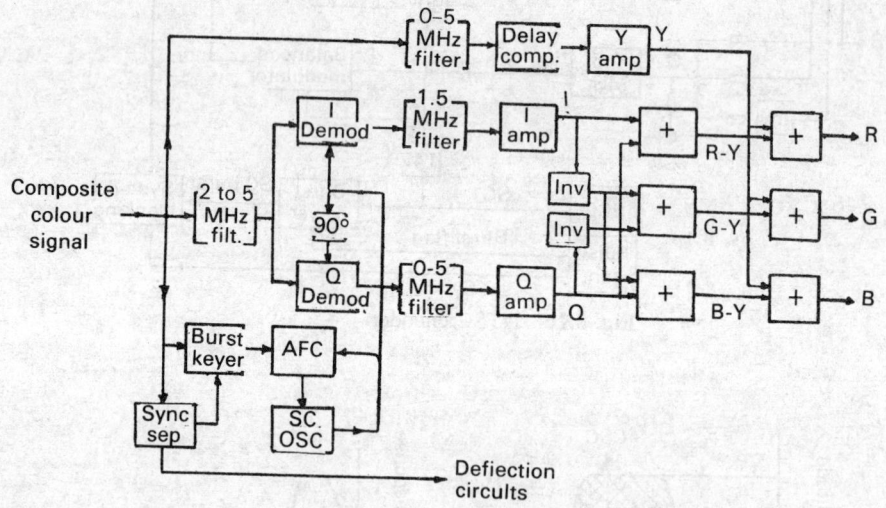

Fig. 5.18 NTSC decoder (Receiver)

The Y signal goes to the picture tube matrix through the delay line and Y amplifier of 4.2 MHz bandwidth. In *simple receivers* the Y amplifier bandwidth is restricted to 3.2 MHz, in order to reduce the interference from the colour subcarrier retained in the Y channel and the residual subcarrier from the chroma section. For better resolution, necessary in large screen receivers, the Y video response has to be extended to full 4.2 MHz, without the interference of the colour subcarrier and its side bands. These chroma components occur in the video band, as packets of energy at half line frequency intervals in energy gaps of the monochrome signal. Hence they can be removed by inserting, in the luminance channel, a *comb filter* which rejects the colour subcarrier and its sidebands, thus reducing the interference in the picture caused due to them.

The detected video signal is passed through a *chroma bandpass* filter amplifier tuned to 3.58 MHz and to allow 2.5 to 4.1 MHz chrominance frequencies to the colour demodulators. The filters serve to reduce the strong lower sidebands of the Y signal, which may otherwise cause severe interference in the colour demodulators. The two chrominance components I and Q are separated by two synchronous demodulators, which are supplied along with the modulated I and Q signals, locally generated subcarrier 90° phase shifted with respect to each other. The subcarrier oscillator is *frequency and phase locked* to the subcarrier burst signal separated from the detected signal by the sync separator and burst keyer, with the help of the AFC circuit.

Synchronous Demodulation

Synchronous detection provides a vector product of the reference subcarrier signal which defines the axis along which the colour signal is detected, and the colour subcarrier. Consider the quadrature modulation of the colour subcarrier of angular frequency ω. The in-phase carrier is given by:

Synchronous Demodulation   Let us consider the quadrature modulation of the colour subcarrier. The in-phase subcarrier is given by

$$v_1 = a_0 \cos \omega_0 t, \text{ which carries the } I \text{ modulating signal}$$
$$v_1 = a_i \cos \omega_i t.$$

In the balanced modulator, the output is given by

$$v_0 = a_0 \cos \omega_0 t \cdot a_i \cos \omega_i t$$
$$= a_0 a_i \cos \omega_0 t \cos \omega_i t = A_1 \cos \omega_0 t \cos \omega_i t \qquad (5.1)$$
$$[= \frac{1}{2} A_1 \{\cos (\omega_0 + \omega_i)t + \cos (\omega_0 - \omega_i)t\}]$$

Similarly for the quadrature subcarrier,

$$v_2 = a_0 \sin \omega_0 t, \text{ carries the } Q \text{ modulating signal}$$
$$v_q = a_q \cos \omega_q t;$$

and the balanced modulator output is given by

$$v_0 = a_0 a_q \sin \omega_0 t \cos \omega_q t$$
$$= A_2 \sin \omega_0 t \cos \omega_q t, \text{ where } A_2 = a_0 a_q \qquad (5.2)$$
$$[= \frac{1}{2} A_2 \{\sin (\omega_0 + \omega_q)t + \sin (\omega_0 - \omega_q)t\}]$$

Thus the output at the video detector consists of the modulated subcarrier signal.

$$v_{sc} = A_1 \cos \omega_0 t \cos \omega_i t + A_2 \sin \omega_0 t \cos \omega_q t \qquad (5.3)$$

In the synchronous demodulator which also consists of a balanced configuration, this is multiplied with the locally generated subcarrier

$$v_{LO} = a_L \cos (\omega_0 t + \phi)$$

The demodulator output is then,

$$v_d = v_{sc} \cdot v_{LO}$$
$$= A_1 a_L \cos \omega_0 t \cos \omega_i t \cos (\omega_0 t + \phi)$$
$$\quad + A_2 a_L \sin \omega_0 t \cos \omega_q t \cos (\omega_0 t + \phi)$$
$$= \frac{1}{2} A_1 a_L \cos \omega_i t \{\cos (2\omega_0 t + \phi) + \cos \phi\}$$
$$\quad + \frac{1}{2} A_2 a_L \cos \omega_q t \{\sin (2\omega_0 t + \phi) - \sin \phi\}$$
$$= \frac{1}{2} A_1 a_L \cos \omega_i t' \cos \phi - \frac{1}{2} A_2 a_L \cos \omega_q t \sin \phi + \text{HF terms.}$$

If the HF terms are filtered out by a low-pass filter,

$$v_d = \frac{1}{2} A_1 a_L \cos \omega_i t, \quad \text{if } \phi = 0 \qquad (5.4)$$

and $\qquad v_d = \frac{1}{2} A_2 a_L \cos \omega_q t, \quad \text{if } \phi = 90°$ $\qquad\qquad\qquad\qquad$ (5.5)

There are two ways of obtaining the vector products required in the preceeding (i) In one method the large signal of the reference phase is added to the subcarrier colour signal. When the reference signal is much greater than the colour signal, the components which are in phase with the reference signal add directly to its amplitude, while the quadrature components add as square root of the sum of the squares, which remains constant as the reference signal predominates. If the combined signal is applied to an envelope detector, the output is proportional to the modulation in the colour subcarrier along the reference axis. A balanced diode synchronous detector of this type is indicated in Fig. 5.19(a).

(ii) In *product demodulation* a current proportional to one of the signals, either the colour subcarrier or the reference signal, is modulated with the second signal by a suitable circuit. Figure 5.19(b) shows two difference amplifiers, where TR1 and TR2 generate constant currents, proportional to the chrominance modulated subcarrier voltages applied at their bases. The currents are switched between the differentially connected transistor pairs TR3–TR4 and TR5–TR6, by means of the reference oscillator voltage across their bases. The switched output across the collectors of the transistors includes the demodulated colour difference signal. The input signal and the reference oscillator may be applied single ended with fixed voltage at the other bases, and output also taken single ended at one of the collectors.

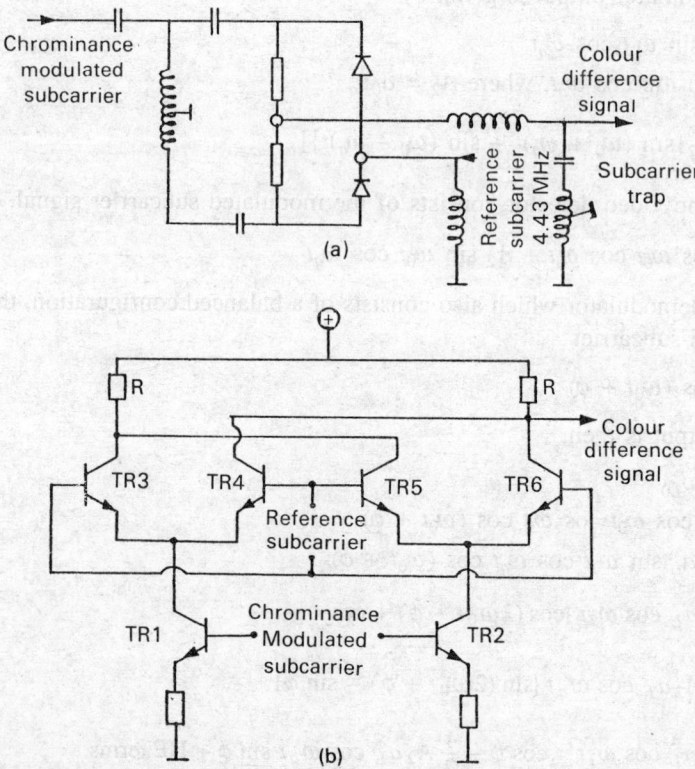

Fig. 5.19 (a) Diode synchronous detector, (b) Transistor synchronous detector

Chroma Signal Processing

The *I* demodulator gets its subcarrier lagging 57° behind the burst phase, and the *Q* demodulator gets it with an additional 90° lag, thus providing the exact phase required for detection of *I* and *Q* signals. These are used in the colour matrix circuit to produce colour difference signals, with deweighting of $R - Y$, by $1/0.877 = 1.144$ and $B - Y$ by $1/0.493 = 2.023$, to restore their original proportion. The $G - Y$ signal is obtaied from these so that the colour difference signals along with the Y signal can be readily matrixed to provide the RGB drives to the colour picture tube.

When receiving monochrome transmission, the chroma bandpass amplifier and colour processing circuitry should be inactive, to prevent coloured interference patterns on the screen. A *colour killer* circuit disables the chroma amplifiers by detecting the absence of colour subcarrier burst, which indicates that the transmission is monochrome. During reception of colour TV transmission, the presence of subcarrier biases the chroma amplifier on so as to process the chroma signal for obtaining the RGB drives.

Differential Phase Errors

As the chroma signal C represents the hue information by its phase relationship with respect to the subcarrier, any deviation in the phase between the subcarrier at the local receiver and the subcarrier at the transmitter encoder as represented by the colour burst, results in incorrect colour hues in the reproduced picture. So long as the phase timing errors introduced in the transmission path on the burst subcarrier and chroma signals are equal, their mutual phase relationship remains constant. If, however, there are nonlinearities in the signal path, these phase relationships are no more constant, and incorrect hues are produced.

As the chroma signal is superposed and rides on the luminance signal, it can lie anywhere in the dark black and peak white levels. The colour subcarrier burst is always at the constant black level. Any nonlinearity in the system introduces level dependent phase timing errors, affecting the colour burst and chroma signals differently. This leads to wrong directions of the chroma vectors and hence errors in reproduction of hues.

In the transmission path such errors can be introduced in a number of ways. These may arise from differential phase distortion in the studio equipments, along the transmission links, due to multipath reception, mistuned or misaligned RF-IF circuits and VTR speed variations. In the NTSC receiver, a Hue or Tint control is essentially provided to correct these phase errors, and needs adjustments occasionally, e.g. when a channel or the program source is changed.

In order to avoid or overcome the problems arising out of these phase errors, engineers in Europe developed modified techniques such as SECAM and PAL, and their variants. This incidentally led to long controversies over their relative merits and demerits affecting the choice of colour TV systems for adoption by different countries. A system which combined all features of all three systems was developed by French and Russian engineers, as SECAM IV and V at NIR, the Russian National Institute of Research.

5.8   SECAM SYSTEM

In order to avoid the problem of phase errors on account of the subcarrier carrying both the chrominance signals simultaneously, H.de France proposed the SECAM (sequential colour a memoire) system in France in 1958. The system rested on the same theoretical principals of the NTSC system, but the two chrominance signals are transmitted on *alternate lines in sequence*. As a result, half of the colour information is lost on each line. This *reduces vertical colour resolution*. It is however not of much consequence as far as the eye is concerned since it cannot distinguish colours in so much detail. In the NTSC system, although the horizontal resolution is reduced by limiting the bandwidth of I and Q signals, the vertical resolution is maintained to the full. SECAM went through progressive developments and versions, of which SECAM III is accepted standard for systems D, K, K_1 (OIRT) and L (France).

In the SECAM system the colour difference signals $R - Y$ and $B - Y$ bandlimited to 1.5 MHz are applied to the FM modulator alternately, via an electronic commutator switching at line frequency rate, as shown in Fig. 5.20.

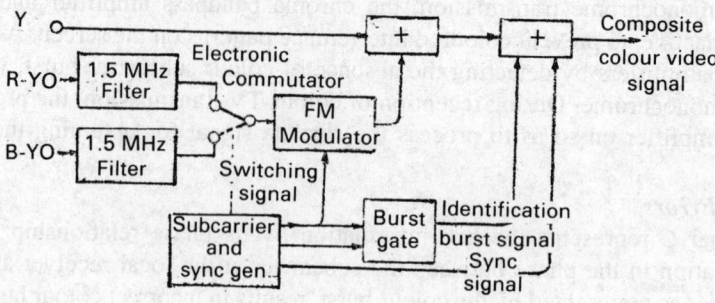

Fig. 5.20 SECAM encoder

The colour subcarrier is thus modulated alternately with the two chrominance signals from successive lines. Use of frequency modulation makes the system less sensitive to pulse interference. Hence phase errors in video recording and transmission are ignored. The channel bandwidth is 8 MHz with 625 line 50 Hz scanning, with the video bandwidth of 6 MHz and sound carrier at 6.5 MHz with amplitude modulation. The chrominance signals DR and DB are weighted colour difference signals obtained from gamma corrected primaries as

$$DR = -1.9 (R - Y) \quad \text{and} \quad DB = -1.5 (B - Y)$$

The nominal subcarrier frequency is 4.4286 MHz. However for suppressing the dot pattern interference due to residual subcarriers, *two separate subcarriers* f_0R and f_0B are used for the DR and DB signals, viz.

$$f_0R = 4.40625 \text{ MHz} \quad \text{and} \quad f_0B = 4.250 \text{ MHz}$$

The negative values of $R - Y$ are required to give positive frequency deviations of the subcarrier to keep them away from the upper end of the video band.

The modulated subcarrier is then combined with the luminance channel signal, the sync and blanking signal and burst or switching pulse signal to form a composite colour signal. The burst signal or switching pulse signal is sent every alternate line and is used as switch identification pulse for phase synchronisation of the electronic commutator at the receiver decoder, to ensure correct switching.

In the SECAM receiver, the decoder transforms these sequential colour difference signals into simultaneous ones. A simplified block diagram of the SECAM decoder is shown in Fig. 5.21. The chrominance amplifier passes through the FM band of about 1 MHz around the subcarrier. The chroma signal is passed along, a path directly and along another path via a 64 μs delay line to the two poles of an electronic commutator switch. The alternating *DR and BR signals* from successive lines are thus *switched along with the one line period delayed version,* to the cross connected output poles of the commutator. This makes the two colour difference signals effectively simultaneous, although they are actually from successive lines.

The two outputs of the electronic commutator lead to the colour difference channels that contain FM demodulator and amplifier stages feeding the $R - Y$ and $B - Y$ signals to the picture tube matrix. The

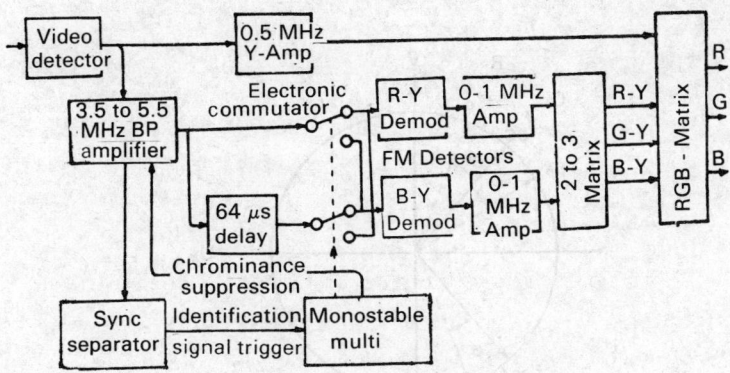

Fig. 5.21 SECAM decoder

identification signal is separated from the video signal and is used to trigger a monostable which controls the electronic commutator to switch on it in a definite sequence.

The SECAM system is free from problems of hue and saturation due to colour difference signals being sent separately on the FM subcarrier. There is no need for the generation of a local reference subcarrier. Hue and saturation controls are not required. Brightness and contrast controls are adequate, the latter serving as saturation control.

5.9 PAL SYSTEM

This system of Phase Alternation by line (PAL), put forward by Prof. Walter Bruch of Telefunken in Germany, is very similar to the NTSC, except that simple colour difference signals $R - Y$ and $B - Y$ called V and U after weighting, are used as chroma signals and the colour subcarrier phase is reversed on every other line. These modifications make the PAL system less sensitive to differential phase errors that can occur at a number of places in the studio chain and the transmission path. Since colours are represented by the phase angles of the colour subcarrier vector, the phase errors or unequal phase rotations of individual frequencies of the colour subcarrier spectrum, relative to the phase of the subcarrier transmitted in the form of the burst, cause colour shifts in the reproduced picture. When the phase error is constant it can be corrected by a phase adjustment, as can be done in the NTSC system by providing the hue or tint control in the decoder at the receiver.

In the PAL system, the U and V signals are transmitted as simultaneous pairs of chrominance components in the form of amplitude modulated sidebands of a pair of suppressed subcarriers in quadrature, as in the NTSC system. The phase of the V signal is, however, reversed on alternating lines. Phase errors that may be present in one line are *counteracted by equal and opposite errors produced in the next line*. These opposite phase errors in the colour subcarrier vectors for every alternate line tend to cancel out the effect of colour shifts on the human eye.

Consider an orange colour being sent as indicated on the colour circle of Fig. 5.22, by the vector f_{sc} sent on say the Nth line. During the process of transmission, a phase error lag by an angle ϕ in the vector f_0 will present a red colour rather than orange. In the PAL system, the V signal of colour vector is phase alternated from line to line through 180° by menas of an electronic switch at the transmitter encoder. This means that the subcarrier transmitted in every alternate $N + 1$ st line is $f_{sc'}$, the mirrored subcarrier which is the complex conjugate of the original f_{sc}, sent on the Nth line. The mirrored subcarrier also undergoes a phase error lag by the same angle ϕ to be shifted to the position $f_{0'}$.

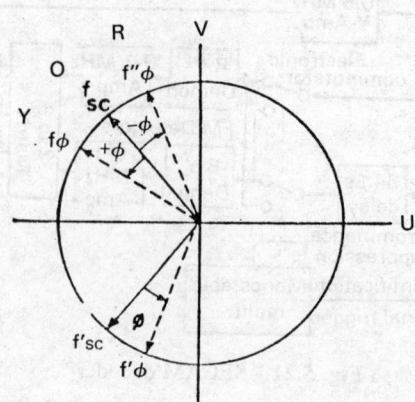

Fig. 5.22 Cancellation of phase errors in PAL system

At the receiver decoder the subcarrier is now again mirrored in the U axis by reversing the V component of the colour vector. With this, the phase angle of the mirrored subcarrier $f_{0''}$ undergoing second mirroring at the decoder of the receiver, gets reversed for the next line, to $+\phi$. The vector $f_{0''}$ with a phase error lead, corresponds to yellow colour as seen on the colour circle. By displaying the two colours red and yellow produced by the negative and positive phase errors in quick succession on every alternate line, the impression on the human eye due to persistence of vision and averaging effect, is that of the intermediate colour, viz. orange in this case. The resultant observed colour will be desaturated. As the phase error increases the saturation is progressively reduced as it corresponds to the vectorial sum of the signals on adjacent line. This technique was referred to as simple PAL.

Hanover Bars

Though the human eye interprets incorrect hues on adjacent lines as a mixture to simulate the correct hues, there are brightness variations associated with the display of wrong hues because of nonlinearities in the system. Especially on saturated colours the eye can distinguish the brightness variations in the wrong hues and horizontal patterning becomes apparent. Due to the *strobing effect of interlace,* two adjacent lines carry a hue error, then the next two carry a complementary hue error, and so on. The coarse hue patterns appear to move and are referred to as 'Hanover bars'. They become increasingly objectionable as the phase errors increase, saturation decreases and vertical colour resolution is reduced.

Delay Line PAL

Instead of relying on the human eye to combine the hues on adjacent lines, misplaced in opposite directions by any differential phase error, the chroma signals may be *combined electronically by means of a delay line* and additional circuits in the receiver, when it is referred to as PAL-D, now considered to be the standard PAL. Hanover bars are not apparent as in simple PAL unless there is maladjustment in the receiver decoder.

Swinging Burst

At the PAL receiver the $R - Y$ output from the V demodulator must reverse its phase. This can be achieved by inverting the phase of the reference subcarrier every alternate line in synchronism with the transmitted signal. An identification signal is needed to synchronise the PAL decoder switch. This synchronising signal is conveyed by the colour bursts contained in the colourplexed signal. The PAL colour bursts are made to swing in phase 45° on either side of $-U$ axis, in synchronism with the phase alternations of the V subcarrier as shown in Fig. 5.23.

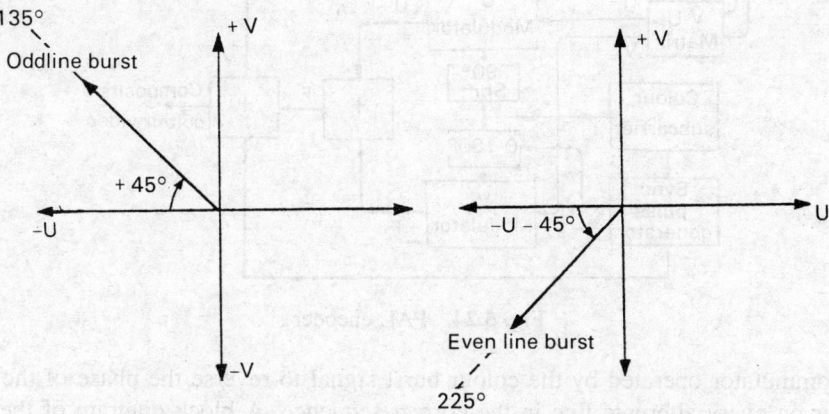

Fig. 5.23 Swinging burst to synchronise phase alternation

For the odd numbered lines with normal V signal, the burst phase lags 45° behind the $-U$ axis, while for the odd lines carrying the inverted V signal the burst phase leads the $-U$ axis by 45°. The average phase of the subcarrier bursts coincides with the $-U$ axis to which the local reference oscillator locks. As the PAL burst switching is done on every alternate line, the switch is operated at the half-line rate of about 7.8 kHz. An indentical frequency is generated in correct phase at the decoder by a bistable flip-flop toggled at the line rate and locked in phase by the 7.8 kHz identical signal separated from the composite video.

PAL Encoder

The block schematic of a PAL encoder is shown in Fig. 5.24. The gamma corrected *RGB* signals are combined in the Y matrix to form the Y signal. The $U - V$ matrix combines the R, B and $- Y$ signals to obtain $R - Y$ and $B - Y$, which are weighted to obtain U and V signals. Weighting by the factor 0.477 for $R - Y$, and 0.895 for $B - Y$ prevents overmodulation on saturated colours. This gives:

$$Y = 0.30\ R + 0.59\ R + 0.11\ B$$
$$U = 0.477\ (R - Y)$$
$$V = 0.895\ (B - Y)$$

The colour subcarrier is modulated in quadrature by the U and V signals, band limited to 1.2 MHz. The subcarrier is fed to the U modulator over a 90° phase shift network while the V modulator gets the subcarrier over a PAL switch to obtain phase inversion i.e. 0–180° deg phase shift on alternate lines. The two modulated chrominance carrier signals F_u and $+/- F_v$ are combined in the adder stage and are further mixed with the sync and blanking signals, as also the swinging burst obtained from the sync and blanking generator. The swinging burst requires a burst gating and ident switch before mixing into the composite signal. The Y signal is provided delay compensation and added together with the chroma signals to form the composite colour video signal.

PAL Decoder

The decoder for the PAL system is similar to the NTSC decoder supplemented by a 64 μs delay line and

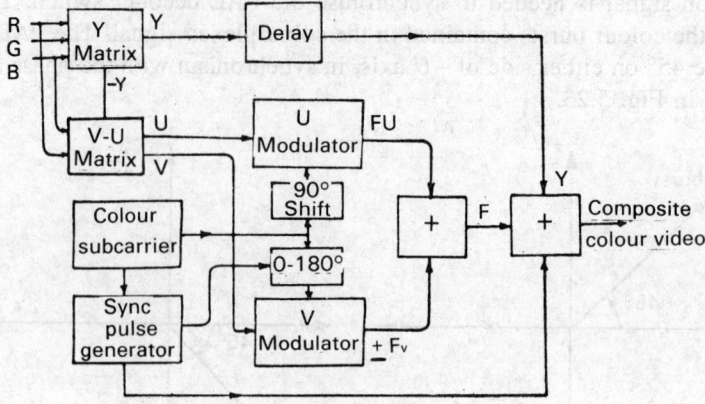

Fig. 5.24 PAL encoder

an electronic commutator operated by the colour burst signal to reverse the phase of the subcarrier for the *V* modulator on every alternate line in the correct sequence. A block diagram of the PAL decoder is shown in Fig. 5.25.

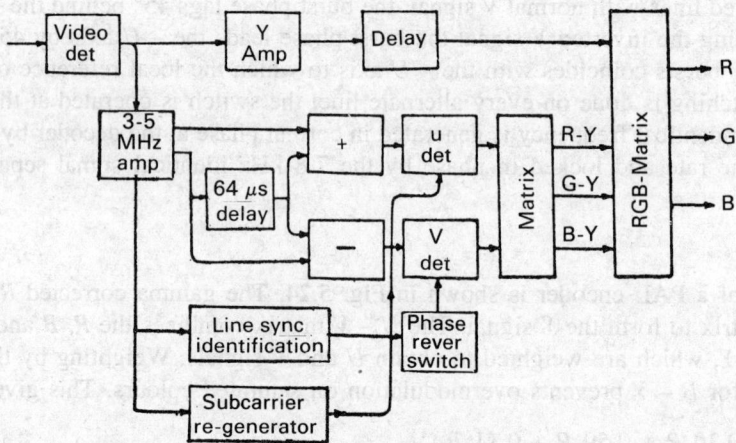

Fig. 5.25 PAL decoder

The delay line is typically an ultrasonic delay line fitted with two piezoelectric ceramic transducers, one acting as the input driven by the chroma signal while the other gives out a chroma signal delayed by one line period. The *U* and *V* signals are separated by combining the direct signal and the delayed output in an add and subtract network, a combination refered to as delay line matrix. Here it must be remembered that the *V* signal is phase reversed every alternate line, so that if line one has chroma signal $U + jV$, the next line has $U - jV$.

When the output of the delay line is added to the direct signal, the output will be $(U + jV) + (U - jV) = 2U$. This enables us to separate the subcarrier component *U*. When the output of the delay line is *subtracted* from the direct signal, the output is $(U + jV) - (U - jV) = 2jV$. Thus we obtain the *V* signal free from the *U* components. The regenerated subcarrier is phase reversed on every alternate line after identification to produce correctly phased *V* signals. These *V*, *U* signals and the *Y* signals are then matrixed together to produce equivalent *RGB* colour primaries that feed to tri-colour picture tube.

REVIEW QUESTIONS

5.1 Discuss the main characteristics of the human eye with regard to perception of colour. What is the difference between additive and subtractive mixing of colours?

5.2 Explain the terms: complementary colours, primary colours, hue, saturation, colour circle, dichroic prisms, and colour temperature.

5.3 How are the colours represented on the chromaticity diagram? What is the significance of the diagram in colour television?

5.4 Describe the construction of a colour television camera and its optical system using dichroic mirrors or prisms.

5.5 Deduce the luminance and chrominance vector amplitudes for the standard colour bar pattern having 100% saturated colours. Explain from the above why it is necessary to weight down the chrominance signals before modulating them on colour subcarrier.

5.6 What are the features of colour television systems that make them compatible with monochrome systems? Under which circumstances can a system be partially compatible or fully compatible with monochrome systems. What are the constraints imposed on performance due to compatibility?

5.7 Show how it is enough to send two colour difference signals only along with the luma signal to obtain the three primary signals.

5.8 How are the chrominance signals transmitted on one carrier in the (a) NTSC, (b) PAL and (C) SECAM systems?

5.9 Explain how synchronous demodulation is used to derive the chroma components from quadrature modulated colour subcarrier.

5.10 Explain the following terms: (a) differential phase errors, (b) Hanover bars, (c) swinging burst , and (d) delay line matrix.

5.11 Explain the basis of the SECAM colour television systems bringing out its features distinct from the NTSC systems.

5.12 Give the block diagrams to explain the working of:
(a) SECAM encoder and decoder, and (b) PAL encoder and decoder.

5.13 Compare the performance and the complexity of the NTSC, PAL and SECAM systems.

5.14 Give reasons for the following:
(a) The primary colours in television are different from the primary colours used in painting.
(b) The bandwidth of the *I* and *Q* signals in the NTSC system is unequal and less than that for the monochrome luminance signal.
(c) A burst of subcarrier is sent along with the sync and blanking pulses.
(d) *G* – *Y* is not chosen as one of the chroma signals for transmission
(e) NTSC decoder is usually provided with tint control.

MULTIPLE-CHOICE QUESTIONS

5.1 The primary pigent colours are:
(a) red, green and blue
(b) red, yellow and blue
(c) magenta, cyan and yellow
(d) none of these

5.2 The signals *V* and *U* are:
(a) same as *I* and *Q*
(b) same as colour difference components $R - Y$ and $B - Y$

(c) modified I and Q

(d) modified colour difference components

5.3 Addition of two complementary equal intensity colours in television would appear to human eye as:

(a) white

(b) black

(c) unsaturated white or black

(d) grey

5.4 Colour burst is used in PAL to:

(a) synchronise subcarrier phase

(b) to identify the phase alternation line

(c) detect presence or absence of colour in the transmission

(d) all of these

5.5 Chroma signals in PAL consist of:

(a) I and Q signals

(b) $R - Y$, and $B - Y$ signals

(c) $R - Y$ and $G - Y$ signals

(d) V and U signals

5.6 The vertical colour resolution in SECAM system is:

(a) one half that of the NTSC system

(b) same as that of PAL

(c) half of that in PAL system

(d) (b) and (c)

5.7 Delay line matrix is used in a PAL receiver:

(a) to derive R, G and B colours

(b) to separate the colour difference components

(c) cancel the phase errors

(d) (b) and (c)

5.8 A colour subcarrier is so chosen that:

(a) it is a multiple of the line frequency

(b) it is an odd multiple of the half line frequency

(c) it gives the least annoying interference dot pattern

(d) (b) and (c)

5.9 Hue complementary to magenta is:

(a) yellow

(b) blue

(c) cyan

(d) green

5.10 In PAL system, the colour subcarrier burst is phased at:

(a) blue

(b) yellow

(c) cyan

(d) none of these

5.11 Pick the false alternative in the NTSC system:

(a) bandwidth for I is 1.3 MHz

(b) bandwidth for Q is 0.5 MHz

(c) Hues for I are red-orange and blue-cyan

(d) hues for Q are blue and yellow

5.12 The average magnitude of the 4.43 MHz modulated colour signal is:

(a) saturation value of the colours

(b) brightness value of the colours

(c) zero for all colours

(d) none of these

5.13 Constant luminance principle requires that:

(a) luminance contribution of the two colour difference signals be equal

(b) luminance contribution of chrominance and luminance signals together be constant

(c) colour difference signals not affect luminance in picture

(d) none of these

5.14 Colour difference signals $R - Y$ and $B - Y$ are reduced in amplitude before chrominance modulation in order to:

(a) prevent overloading of the colour subcarrier

(b) prevent overmodulation of the picture carrier at the transmitter

(c) reduce cross colour interference on the luma signal

(d) (b) and (c)

5.15 The phase of the colour subcarrier burst in PAL is along the axis at:

(a) $+ 180°$

(b) $0 \pm 45°$

(c) $180 \pm 45°$

Studio Equipment, Organization and Control

6.1 TECHNICAL FACILITIES IN TELEVISION STUDIOS

Broadcast television programs are generated in a television studio complex that has a number of studios having sophisticated technical facilities for production and control of programs. These programs high professional standards of picture quality. The studios have, therefore, extensive lighting control and equipment facilities besides the usual acoustic treatment and air conditioning.

Television programs generally originate in a TV studio or from a photographic cine-film projector optically linked to a TV camera. In order to facilitate editing, captioning and rebroadcasting when required, elaborate video recording facilities are essentially provided. Thus the sources of television programs may be: (i) television camera, (ii) video tape recorder, (iii) telecine, or (iv) external signal.

(i) Studio cameras   Some programs are telecast live from the TV cameras in the studios. This includes news reading, panel discussions, etc. and are done in a smaller presentation studio. Programs like dramas, varieties entertainments, etc. are produced in larger studios with at least three cameras for different shots and are generally recorded for convenience of editing, captioning, etc. I.O. cameras have been used for long but are now being gradually replaced by plumbicon cameras.

(ii) Video tape recorders (VTRs)   High quality video tape recording facilities are provided in a separate dust-free VTR room, housing two or more broadcasting standard video tape recorders. Two inch tape quadruplex head recording machines have been standardized for this purpose. One or half inch helical scan tape recorders have been used for outdoor field recording, where less rigorous quality standards can be tolerated; but these are not very stable and require time-base correctors to be put through into the studio telecast chain.

(iii) Telecine   A photographic cine-film projector or a slide projector optically linked with a vidicon forms a telecine program chain. Generally, at least, two telecine chains are provided.

(iv) External signal   Live outside broadcasts from special functions, sports stadia, etc. are telecast by linking the field TV cameras with the studio chain via a microwave relay link.

A wide range of equipment is used to support the cameras in the TV studios which require a centralized control and switching facilities. The video signal from the various sources must have common station sync-genlock pulses, video signal amplification, and processing and distribution equipment. Facilities for mixing and routing of programs from various sources to specific destinations, talk-back facilities between the operators and program producers, special effects, audio mixer consoles and associated equipment, make the system fairly complex. These facilities are distributed in the studios, the production control room (PCR), VTR room, telecine room, and the master control room or the central apparatus room in the studio complex.

6.2   TELEVISION STUDIOS

Acoustic treatment is provided with a perforated hardboard which has internal glass wool backing for reverberation time control. The background for the scene to be televised is provided by a cyke-cyclorama screen of special perforated plastic cloth covering nearly three sides of the studio by its U-shaped mounting fixture. The light blue gray shade provides a background of about third to fourth gray in the black-to-white ten-step gray scale. It prevents the merging of the normal shades with the background.

Lighting

The lighting in a TV studio has to be very elaborate as it can tremendously affect the artform of the reproduced picture. A number of light fittings distributed over the studio are necessary.

Basically the fittings employ incandescent lamps and quartz iodine lamps at appropriate colour temperatures. Quartz iodine lamps are also incandescent lamps with quartz glass envelope and iodine vapour atmosphere inside. These lamps are more stable in operation and colour temperature with respect to aging. The lamp fittings generally comprise spot lights of 0.5 kW and 1 kW and broads of 1 kW, 2 kW and 5 kW. Quartz iodine lamps of 1 kW provide flood lights. A number of these fittings are suspended from the top so that they can be adjusted unseen. The adjustments for raising and lowering can be done by (i) hand operation for smaller suspensions, (ii) winch motor operated controls for greater mechanical loads of batten suspensions carrying a number of light fittings, (iii) unipole suspensions carrying wells of light fittings manipulated from a catwalk of steel structure at the top ceiling where the poles carrying these are clamped.

As many as 100 to 120 fittings are typically employed in bigger studios, 28 battens each carrying four light fittings. The lighting is controlled by varying the effective current flow through the lamps by means of silicon controlled rectifier (SCR) dimmers. These enable the angle of current flow to be continuously varied by suitable gate-triggering signals. The lighting patch panels and SCR dimmer controls for the lights are provided in a separate room. The lighting is energized and controlled by switches and faders on the dimmer console in the PCR, from the technical presentation panel. The lighting has to prevent shadows and produce desired contrast effects. Following are some of the terms used in lighting.

High key lighting is the lighting that gives a picture that has gradations that fall between gray shades and white, confining dark gray and black to few areas as in news reading, panel discussions, etc.

Low key lighting is the lighting that gives picture having gradations falling from gray to black with few areas of light gray or white.

Key light is the principal source of direct illumination often with hinged panels or shutters to control the spread of the light beam.

Fill light is the supplementary soft light to fill details to reduce shadow contrast range.

Back light is the illumination from behind the subject in the plane of camera optical axis, to provide 'accent lighting' to bring out the subject against the background of the scene.

A view of TV studio lighting, camera and microphone boom operator is shown in Fig. 6.1.

Camera Units

In order to command different views of the program scenes, three I.O. or plumbicon camera units are generally used in bigger studios. Two of these are usually mounted on ground-operated dollies, and the third one is mounted on a crane for dramatic shots. Each camera unit is mounted on a friction head with

Fig. 6.1 A view of TV studio lighting, camera and microphone boom operator (*Courtesy*: Film and Television Institute of India)

panning handle that moves the camera left or right for pans, or up and down for tilts. The friction head is fixed on a mobile dolly or pedestal to permit a relatively free movement over the studio floor. It also facilitates tracking for the desired type of camera shots and dramatic effects in conjunction with the camera lens optical adjustments.

The camera unit consists of a pick-up head and a view finder located on top of the pick-up head to see the covered field and help setting of the picture. The pick-up head comprises the camera lens system, the pick-up tube along with associated pre-amplifier and sweep deflection circuits. It also houses intercommunication and signalling facilities with the camera control unit (CCU) in the production control room, to provide signalling and communication between the cameraman and the CCU engineer and the program producer.

Audio Pick-up

For sound arrangement, the microphone placement technique depends upon the type of program. In some cases, e.g. discussions, news and musical programs, the mikes may be visible to the viewers and these can be put on a desk or mounted on floor stands. In other programs, for instance, dramas , the mikes must be out of view. Such programs require hidden microphones or a boom-mounted mike with a boom operator. A unidirectional microphone mounted on the boom arm, high enough to be out of sight, is desirable here. The boom operator must manipulate the boom properly. Lavaliere microphones and hidden microphones are also useful in such programs.

In a television studio, there is considerable ambient noise resulting from off-the-camera activity, hence directional mikes are frequently used. The studio walls and ceilings are treated with sound absorbing material to make them as dead as possible. Artificial reverberation is then required to achieve proper audio quality.

6.3 PRODUCTION CONTROL ROOM (PCR) FACILITIES

A major objective of TV program control facilities is to maintain a smooth continuous flow of program material. The overall control of the program is done in the production control room by the producer with the help of a production assistant, a CCU engineer and an engineer at the vision mixer as seen in Fig. 6.2. They have, in front of them, the switching panel of the vision mixer console and a stack of monitors for the individual cameras, preview monitors for the VTRs or telecine and a transmission monitor for displaying the switched output, with the aid of which the program is edited.

Fig. 6.2   A view of a production control room (*Courtesy*: Film and Television Institute of India)

The producer and the program assistant have in front of them the talk-back control panel for giving instructions to the cameramen, boom operator, audio engineer, floor manager and the floor assistant. The producer can speak to the studio floor and also to the selected VTR or Telecine operator, over the intercommunication system. During rehearsals, the producer can employ a public address system to give instructions to people on the studio floor and other rooms, but during actual programs, the communication is done over the intercom system on headphones.

Camera Control Unit (CCU)

The studio cameras are connected to the video distribution amplifier and mixer via the camera control unit. The CCU contains the controls for the aperture, optical focus, zoom of the lens system and the beam focus and brightness control of the camera tube. The CCU engineer has to be alert to adjust the picture video signal—BAS level to a constant 1 V level, for possible variations in the lighting conditions. For this purpose, precision A-grade picture and video monitors are provided at the CCU, to which the selected signal can be connected. The picture, blanking and sync (BAS) levels are made to conform to the standard specifications.

The lighting is controlled by the lighting director at the technical presentation panel by the side of the CCU, located next to the vision mixer. This enables the producer to select the desired source or sources in order to compose his program. The switching panel of the vision mixer has its switching panel located in the PCR, while the associated amplifier units are located in the master control room (MCR). The interconnections between the units, the sources and the various areas are made by 75 ohm multicore coaxial cables.

Vision Mixer

A vision mixer or video switcher, as it is also called, enables the program producer to select the desired sources or a combination of the sources in order to compose the program. The vision mixer is typically a 10×6 or 20×10 crossbar switcher selecting anyone of the 10 or 20 input sources to 6 or 10 different output lines. The input sources include: Camera-1, Camera-2, Camera-3, Telecine-1, Telecine-2, VTR-1, VTR-2, Test signal, etc. Some of these sources that have their sync coincident with the station sync are called synchronous, while others having their own independent sync are called non-synchronous.

The vision mixer provides for the following operational facilities for editing of TV programs:

(i) Take—selection of any input source, or cut—switching cleanly from one source to another.

(ii) Dissolve—fading in or fading out.

(iii) Lap dissolve—dissolving from one source to another with an overlap or mixing.

(iv) Superposition of two sources—keyed caption when the selected inlay is superposed on the background picture.

(v) Special effects—a choice of a number of wipe patterns for split screen or wipe effects.

The selected output can be monitored in the corresponding preview monitor of the monitor stack. All the picture sources are available on monitors. The preview monitors can be used for previewing the telecine, VTR, test signals and transmission outputs. The camera signals can be previewed on these monitors as well as on the individual picture monitors. This can be done by switching at the crosspoints of the source lines and monitor lines coming at the video switcher. The switcher also provides cue facilities to switch camera transmission lights, etc.

In the process of mixing and superposing, the switcher and the associated amplifier units must satisfy the following technical requirements:

(i) There should be no visible disturbance of sync when switching between synchronous sources.

(ii) Disturbance of sync pulses when cutting to a non-synchronous source should not exceed 1 ms.

(iii) The output BAS signal should be standard 1 V_{pp}, and the peak whites should be properly clipped.

(iv) The fading circuit should not introduce excessive distortion during fading, it should be typically less than 2% at full fade in, 5% at mid position and 15% at other positions.

There are three types of video switchers:

(i) *Mechanical pushbutton switcher*   This has the video signals on the actual switch contacts. The blanks of switches is interlocked to prevent simultaneous punching up. This type of switcher is used primarily for portable field units.

(ii) *Relay switcher*   This employs rack-mounted reed relays for cross point switching. It has magnetically activated reed switch-contacts, in an evacuated glass envelope, that operates within 1 ms. The switching action is overlapped for about 1 ms, preventing loss of sync during switching. This is called D-switching.

(iii) *Vertical interval switcher or electronic crossbar*   This is an all electronic solid state switcher that switches one source to another during the vertical blanking interval following the vertical sync in a matter of a few microseconds. After the cut button is pressed, a memory holds the information until the next field blanking period when the switching takes place.

The simplified schematic diagram of a typical video switcher is shown in Fig. 6.3. The system consists of a bank of cutting switches, an A-B mixing amplifier, special effects equipment, sync adder-stabilizing amplifier and monitors. Each vertical row of switching buttons on a video switching panel corresponds to a single picture source and each horizontal row corresponds to an output bus. The buttons operate the electronic switches at the cross-points of the input and output lines during the vertical blanking interval.

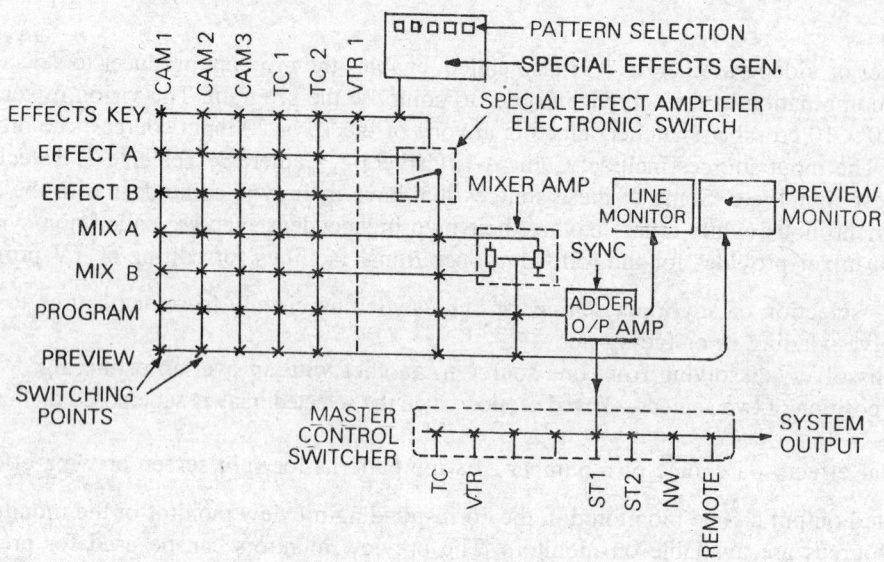

Fig. 6.3   Simplified schematic diagram of a video switcher

One output line or bus feeds to the outgoing transmitter line and is called the 'program bus'. The other buses provide signals for special effects equipment, A-B mixer and preview monitors. The output buses are interlocked so that at a time only one picture source can appear at a time, although a given input signal may be connected simultaneously to several output buses. The preview buses and the associated preview monitors permit the producer to check the incoming program before actually switching to it.

An *A-B mixer* that can superpose two video input signals is formed by combining the outputs of two fading circuits as shown in Fig. 6.4. Each fading circuit employs a difference amplifier with the input to the common emitter and output at the collector of one transistor which is turned on to off by its base bias which fades out the signal. Mixing is done by having a common collector load so that the output is an addition of the two input signals.

The two mix buses A and B in Fig. 6.1 and the variable gain mixing amplifier facilitate dissolve, lap dissolve and superpose operations. For lap dissolve, the two picture signals involved are punched up on the mix buses A and B, and the joystick levers of the two variable gain stages are locked in such a way that as the level of one is increased, that of the other is decreased by a corresponding amount. One may separate the two control levers and reduce the level of one signal to zero before increasing the gain control and hence the level of the next picture. For superposition, one may also separate the control levers to adjust the relative levels of the two picture signals to be combined.

Special Effects

The vision mixer provides a choice of a number of wipe patterns for split screen or wipe effects between sources on two rows on the video switching panel. This is done with the help of a waveform module that generates keying waveforms used to drive an electronic switch that switches the output between the two input signals. The waveforms provide means for blanking out one or more parts of the areas of a signal picture, inserting another signal into these areas, and changing the separation boundary in any desired shape. This technique is also called *inlay* or *external key* which replaces parts of the TV picture by another.

The simple special effect of *horizontal wipe* can be obtained by using only the line-rate keying pulses.

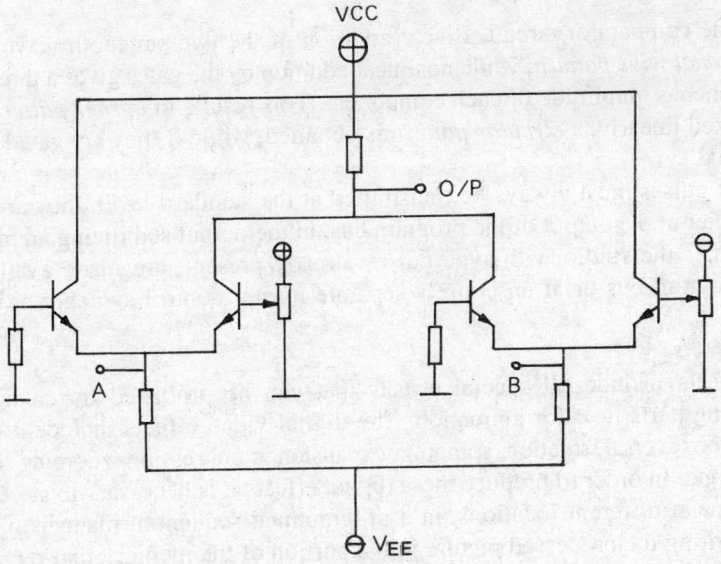

Fig. 6.4   A-B mixer

These key the signals from two camera sources alternately, by means of an electronic switch. This produces a vertical split screen display as shown in Fig. 6.5.

For obtaining the wipe effect, a line frequency sawtooth is fed to a comparator circuit. This compares its amplitude with a variable dc bias from a fader which in turn alters the on-off duty cycle of the output keying waveform given to the switch. This varies the on-off duration of the line interval for the two sources in a complementary way so that at limiting position of the fader, only one picture occupies the screen and the movement of the fader wipes one picture with the other horizontally.

A vertical wipe effect by a horizontal split can be similarly obtained by using field frequency sawtooth for the comparator. The dc bias control now produces a variable on-off cycle at the vertical rate and vertical wipe is produced, as indicated in Fig. 6.5(b).

For obtaining various other wipe patterns, sawtooth waveforms, triangular waveforms at the line and field rate and parabolic waveforms obtained from these by integration are given to the pattern selection

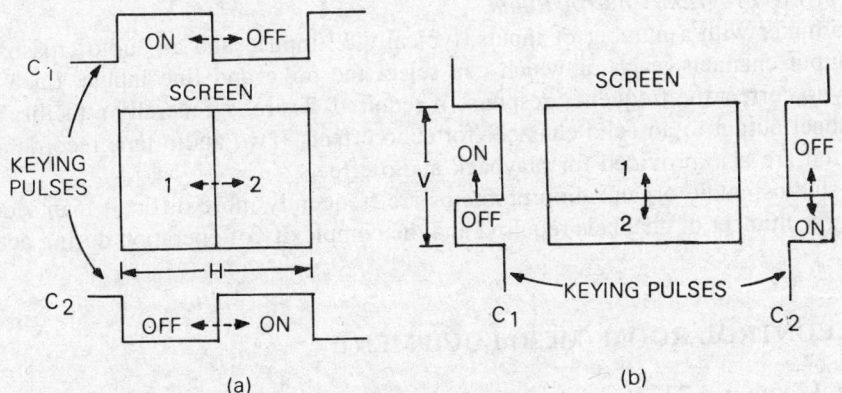

Fig. 6.5   (a) Horizontal wipe effect, (b) Vertical wipe effect

matrix feeding the comparator circuit. Linear addition of the two sawtooth waveforms at line and field rates gives a *diagonal wipe pattern*; while nonlinear addition by the gates gives a threshold action depending upon the instantaneous amplitude of each component. This results in *corner patterns*. If the line and field parabolas are mixed linearly, a *circular pattern* is obtained; while if they are gated together a *rectangular pattern* is obtained.

Since the sync pulses must always be transmitted at the standard level, they are normally added after the mixing amplifier at the output of the program bus, in the output stabilizing amplifier. Remote-external signals, which enter the studio with sync pulses already present, are made available in the secondary portion of the program bus or at an entirely separate master control switcher as shown in Fig. 6.3.

Digital Video Effects

Digital video effects include all special effects that can be produced by changing the picture size, zooming or moving parts in it for animation. The digital video effects include effects known as freeze frame, splitting, line reversal, rotation, spinning, expansion, compression, zooming in and out, perspective, resolution, decay, etc. In order to produce these digital effects it is necessary to store the pixel information and make it appear at different locations, in a programmed sequential manner to achieve motion. This can be done by writing the processed picture into a portion of the memory that represents a normal raster and reading it later, by use of digital technology and microprocessor control. The digital effect unit can be separate, or an integral part of the vision mixer.

The video signal which has a horizontal resolution of over 350 lines, is digitized into 8-bit data, sampled at 2×350 samples per line (see Section 25). This allows 256 levels and resolution for normal broadcast quality picture. With about 585 active lines per picture, this is equivalent to $25 \times 585 \times 700 \cong 10$ M bits of data. The information is stored in a frame memory, from where it can be fast accessed and processed by a microprocessor to produce the required effect. The digital effects controller is a microprocessor based unit that provides required control information for digital video processor. This includes control panel data, cross point select data, key and wipe signals, effects processor data and so on. The digital effects processor uses these signals to digitally process and manipulate the signal for producing the desired effect with the help of subroutine programs. The digital effects controller provides operational control serving as the human interface. Depending on the operational environments and use, the digital effects unit can be a separate stand-alone unit, or can be an integral part of the video switcher to give more creative flexibility in mix effects with more sources.

Audio Console of the Program Control Room

This is an audio mixer with a number of inputs, typically 32 inputs, and an audio crossbar distributor providing 10 output channels, each of which can select the mike and line inputs. Each channel has equaliser facility to correct the frequency response if required. There is generally a facility provided for feeding any channel output to an echo chamber for echo effects. Two audio tape recorders and a turn-table playback unit are also provided for playback audio effects.

In television studios, audio organization problems are frequently more difficult than video problems because of the large number of channels required and the complexity of operation during actual program time.[14]

6.4 MASTER CONTROL ROOM (MCR) EQUIPMENT

The master control room in a TV studio complex houses a master routing switcher and electronic video equipment associated with each studio on separate racks as seen in Fig. 6.6. It also houses the VTR

Fig. 6.6 Master control room equipment (*Courtesy*: Film and Television Institute of India)

switching trays and video jack fields for assigning VTR and telecine to particular studios. Centralized video equipments that include sync pulse generators, pulse distributors, master routing switcher equipments, and test equipments, are also present.

A special video oscilloscope-line monitor is provided for measurement of pulse tips, rise-time and other parameters of the line signal. The video monitors provided are precision grade class-A type, employing 90° deflection picture tubes rather than the common 110° picture tube, for better linearity. Good linearity and contrast are to be ensured before the signal goes to the transmitter.

Sync separator and regenerator is often used for reshaping the sync pulses to eliminate the hum that may have got into the signal in the process of routing. This ensures a clean signal to the transmitter. The picture blanking sync (BAS) levels are also controlled at the master control. These levels must be constant within tolerances allowed. This is facilitated by waveform monitors with suitable tolerance graticules.

Master routing switcher is typically a 10 × 6 cross bar switcher capable of switching or selecting any of the 10 inputs sources, viz. test signals (normally sawtooth—shading signal with 1 kHz tone), black, telecine, VTRs, studios, etc., going to different areas, viz. transmitter, VTR room, preview monitor to MCR, etc.

Video distribution amplifiers are power amplifiers providing five or six outputs at the standard 1-V, 75-ohm impedance for feeding a video signal via coaxial cables to various places in the studios.

Pulse distribution amplifiers provide the station sync at the H and V rate from a crystal controlled source, at a standard amplitude of -4 V, typically. These pulses from the crystal controlled sync pulse generator are also called genlock or sync lock pulses.

Test generators provide various electronic patterns for video testing. These include:

   (i) ramp or sawtooth providing a shading pattern.
  (ii) grill providing electronic cross hatch line pattern for adjusting linearity of monitors and cameras.
 (iii) gray shades providing ten shades of gray for testing gradation capability of the studio chain.
 (iv) $\sin^2$ pulse and bar test (PBT) pattern for video testing of the equipments. $\sin^2$ pulse enables evaluation of the HF response, and the bar enables LF response evaluation as described later in Section 14.8.
  (v) Picture line-up generating equipment (PLUGE) generates waveforms for rapid and accurate adjustment of the operational controls of a picture monitor. This is possible in one form by providing two adjacent

vertical bars with slight difference of + 3% and – 3% with respect to the background at the pedestal level. The monitor brightness control should be so adjusted that the – 3% bar merges into the background leaving the + 3% bar clearly visible. An additional pulse associated with the bars provides for peak white signal.

The equipment in MCR need not be clustered at one place. The pulse generators, pulse amplifiers, distribution amplifiers and test generators are sometimes kept on racks in a separate central apparatus room (CAR), where these can be extended according to requirement. *Earthing* of the video equipments is very important to prevent ground loops. For racks, the power earth and video earth are kept separate in order to prevent ground looping that may introduce power frequency hum into the signals

Pulse Processing

The synchronisation of the deflection and scanning systems in the various cameras and other video equipment is ensured in the studio chain by providing a common 'station sync' from a crystal controlled sync pulse generator (SPG) The camera chain pulse processing has the following functions:

(i) Providing pulses for scanning generators that generate necessary drive waveforms for the horizontal and vertical deflection coils, and

(ii) Introducing the blanking and sync pulses into the camera signal while also blanking the camera, during flyback or retrace.

The deflection of electron beam in the pick-up tube of a camera is required horizontally at the rate of 15625 times/s (the *H* rate or the line rate). The vertical deflection from top to bottom is required at the rate of 50 times/second (the *V* rate or the field rate). The deflections are effected electromagnetically by sending sawtooth currents through the deflection coils that produce required deflecting fields

At a time there may be two or three cameras shooting the same scene at different angles and shots The signals from all these cameras, as also the signals from other video sources that may be occasionally brought in, must have their scanning sequence in phase or time coincident. All the sawtooth drive currents in the deflection coils of the various cameras should start at exactly the same moment. The common 15,625 Hz synchronising pulse signal needed for the horizontal scanning generators is called the *H* drive or the line drive (LD) The common 50 Hz synchronising signal required for the vertical scanning generators is called the *V* drive or field drive (FD).

As the beam scans from left to right, video signal information is picked up; but as the beam quickly flies back, the unwanted video information during flyback needs to be blanked out. This is done by injecting blanking pulses during both the *H* and *V* retrace period, to the pick-up tube to disable it from developing any information. The blanking done at the camera tube is called the camera blanking. In addition to this, system blanking is introduced later in the signal with correct blanking intervals and levels. The system blanking pulses at the *H* and *V* rate are called mixed blanking (MB) or Austastung (A) pulses.

At the TV monitors and receivers, it is necessary to ensure that the picture tube beam deflection takes place in step with the camera tube beam deflection. For this purpose, synchronising signals at both *H* and *V* rate are added to the video signal to control the scanning rate in he picture tube. The combined sync pulses are referred to as mixed syc (MS) or *S* pulses.

The basic pulses required in the camera chain are as follows:

(1) line drive or *H* drive, (2) field drive or *V* drive, (3) system blanking at the *H* and *V* rate, (4) synchronising pulses at the *H* and *V* rate, and (5) camera blanking at the *H* and *V* rate.

The first four of these are developed by the sync pulse generator (SPG). They are designated as LD, FD MB and MS, or as *H, V, A,* and *S* pulses respectively. Camera blanking pulses are generated from the *H* and *V* drive pulses. In some camera chains, the MS pulses are used to develop the line and field drive pulses and these are sent to the camera head.

The basic block schematic diagram of a camera pulse processing chain is shown in Fig. 6.7.

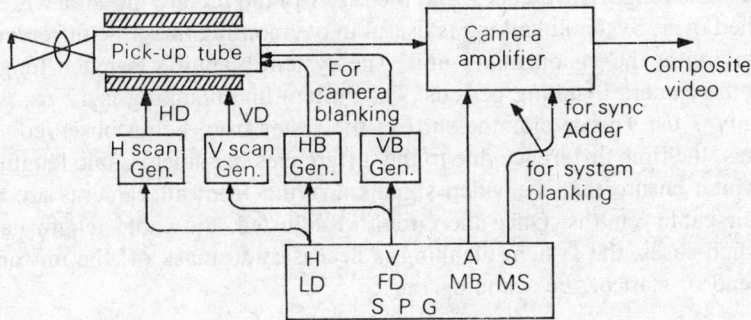

Fig. 6.7 Camera pulse processing

The schematic uses the LD and FD pulses of the SPG for triggering the *H* and *V* scanning generators and the blanking generators. The MB pulses are given to the camera amplifier for system blanking and the MB pulses are given to the sync adder.

Sync Pulse Generator

A sync pulse generator or a sync signal generator as it is often called, comprises: (i) crystal controlled or a mains locked timing system, (ii) pulse shapers that generate required trains for blanking, synchronisation and deflection drives, and (iii) amplifier distributors that supply these pulses to various studio sources in a studio complex.

The timing unit in the sync pulse generator has a master oscillator at a frequency of about 2*H*, that can be synchronised by: (i) a crystal oscillator, at 2*H* (31,250 Hz) exactly, (ii) an external 2*H* frequency source, or (iii) the ac mains frequency with the help of a phase detector and an AFC circuit that compares the 50 Hz vertical frequency rate with the mains frequency. This is shown in Fig. 6.8.

The required pulse timings at *H* and *V* rate are derived from the 2*H* master oscillator through frequency dividers as shown in the figure. The blanking and sync pulses are derived from the 2*H*, *H* and *V* pulses employing suitable pulse shapers and pulse adders or logic gates.

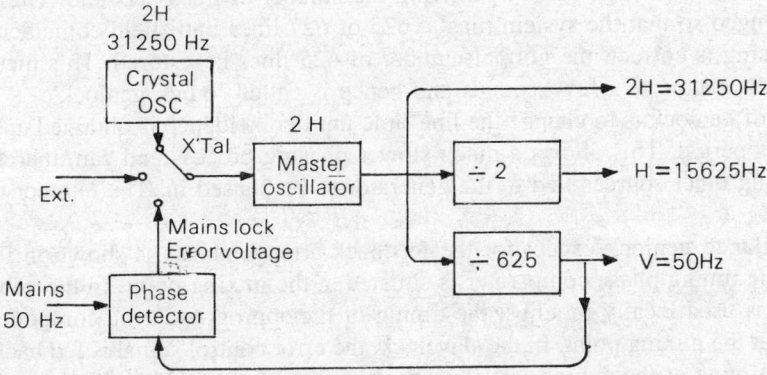

Fig. 6.8 Timing unit of a sync pulse generator

System Blanking

When cameras are placed at different locations, they may require different camera lengths and hence the line drive pulses applied to the cameras may be unequally delayed by the propagation delay of the cable, which is around 0.15 μs/100 ft of the cable. This can cause a time difference between the cameras proportional to the cable length differences, and the raster in the picture monitor will shift slightly as the cameras are switched over. System blanking is useful in overcoming this time difference between the two camera signals arriving at the vision mixer unit. The system blanking is much longer in duration and encompasses both the camera blanking periods. The system line blanking is 12 μs, whereas the camera line blanking is only 7 μs. This avoids the shift in the raster from being observed.

In recent cameras, the time difference due to the differences in camera cable lengths is offset by auto-phasing circuits which ensure that the video signals arriving from all cameras are all time-coincident irrespective of their cable lengths. Once the circuit is adjusted, the cable length has no effect on the timings. Even in such cases, the system blanking is necessary to mask off the unwanted oscillations or distortions at the end or start of the scanning line.

Principle of Genlock and Slavelock Techniques

In a studio complex, the various program sources can be mixed, superposed or inlaid if they are synchronous, with their line and field sync in phase at the mixing point. If the sources are driven from the same SPG, it is relatively easy to make them synchronous. This can be done by applying either the 'genlock' or the 'slavelock' or by building out the necessary delay from the SPG.

Non-synchronous sources driven from separate SPG's cannot be mixed or superposed. They can only be switched by cutting the picture and sync signals together. Switching between two non-synchronous sources leads to a sudden change in the transmission sync pulses causing a temporary loss of sync in the driven monitor time bases. The line time base may recover fairly quickly, but the field time base may take several seconds to recover if the phase change is large resulting in picture roll. This can be avoided if the field sync components of the two sources are brought approximately in phase by suitable phase shifting networks at the moment of switching.

In the genlock process, the line and field components of the local SPG are locked in frequency and phase to the line and field components of a remote incoming video signal without producing any visible disturbance in the monitor. The line and field sync components of the incoming composite video signal are separated and are used to lock the local SPG master oscillator through a timing phase comparator.

In order to bring the local line and field sync components in phase with the remote ones, the local line or field frequency is changed for some time. Field phasing is achieved automatically by deviating the field frequency from its normal value by altering the number of lines per field. The field frequency divider count is changed so that the system runs at 623 or 627 lines until field coincidence is achieved. When the field phasing is correct, the normal number of 625 lines is restored. This method can give a fairly rapid lock, in a matter of a few seconds and hence is called 'quick genlock'.

Another method of genlock is to change the line-time until coincidence is obtained and then resetting it to the usual 64 μs period. This allows a much slower phasing process and can, therefore, be carried out without disturbing other sources tied to the generator to be phased in. The number of lines in each field is constant.

Slavelock is similar to genlock except for the feedback arrangements, as shown in Fig. 6.9.

In both modes, the timing phase comparator is situated at the mixing point. In the genlock mode, the error control signal is used locally to adjust the timing of the appropriate sync component of (the SPG to) the video signal at the mixing point. In the slavelock, the error control signal is fed back in an inverted sense to correct the timing of the appropriate synchronising component of (the SPG of) the contribution from the incoming video signal.

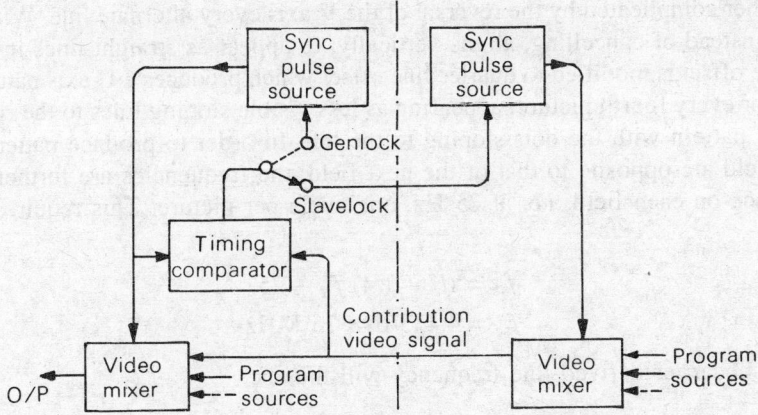

Fig. 6.9   Genlock and Slavelock arrangement

Any system capable of slavelock operation can also be used for genlock operation. But a genlock system cannot operate for slavelock unless the error signal is suitable for other generators and the feedback delay is tolerated by the system for maintaining stability.

Colour Sync Pulse Generators

Older monochrome video source equipment used four-line standard pulses to the equipment, viz. MS, LD, FD and MB pulses. A limited number of colour TV equipment used these four sets of pulses plus the colour subcarrier CSC, the PAL indent flag and the colour burst gate. The next generation colour equipment of solid state design, produced three line distribution, viz. MS, MB and the CSC. Modern equipment employing LSI circuits use self-contained sync generators that require only a single reference pulse for operation. The colour-black signal with the black burst is taken as the de facto standard for single-line distribution. The sync and the subcarrier must be carefully separated from video in order to maintain the exact timing.

PAL Sync Pulse Generator Design

The functions and design of sync pulse generators have changed over the years. They are required as master or slave sync generators, or as source oriented sync generators. In the master sync generator, a master crystal clock feeds the LSI circuit which contains all the counter-dividers and decoders needed to generate all the line and field standard pulses. The clock frequency should be chosen to be a harmonic of the subcarrier, so that the edges of the clock pulses can form the edges of the SPG signals. These edges are therefore critical and should have rise and fall times of 150 ns +/– 50 ns. The PAL subcarrier is required at 4.3361875 MHz within +/ – 1 Hz. In addition, a burst gate signal that allows the subcarrier to ride on the back porch of the HB pulses, and a PAL ident signal to synchronise the V signal switching, are required.

Colour Subcarrier and Horizontal Frequency

In the NTSC system, the subcarrier is chosen to be an odd multiple of half line frequency, to fit in the chroma signals in the gaps of the luma energy spectrum. The phase of the subcarrier thereby reverses every next frame, and the effect of the subcarrier sneaking through Y amplifier to he CRT is more or less cancelled out. For saturated colours, the cancellation is not fully effective because of nonlinearity of CRT transfer characteristics and chroma phase modulation, when the dots tend to move. In PAL system, this is further complicated by the reversal of the V axis every alternate line. With half line offset, the dot structure, instead of cancelling, aligns vertically, to appear as straight lines in exactly the same position. Hence the offset is modified to quarter-line offset which produces a U axis pattern which repeats every eight fields or every fourth picture, appearing as less visible sloping lines to the right. The V signal produces a similar pattern with the dots sloping to the left. In order to produce patterns where the dot positions of one field are opposite to that of the next field, the frequencies are further modified with a half cycle difference on each field, i.e. at 25 Hz, one cycle per picture. This requires the PAL colour subcarrier to be:

$$f_{sc} = (N - 1/4) f_H + 25$$

For $N = 284$,

$$f_{sc} = 4.433\ 618\ 75\ \text{MHz}$$

For the SPG, the subcarrier derived line frequency will thus be:

$$fH = (f_{sc} - 25)/(284 - 1/4)$$

With this 1/4 line offset in the subcarrier frequency, there is no practical frequency which is harmonically related to both sync and the subcarrier. Phase locked loop is utilized in the SPG to obtain these, as shown in Fig. 6.10.

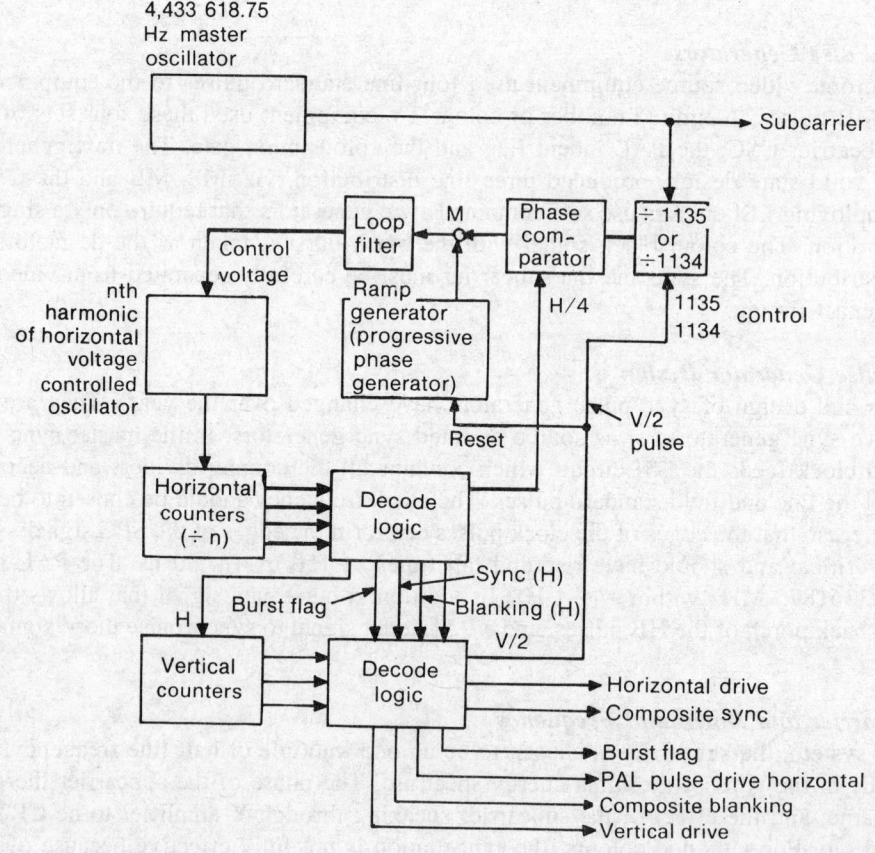

Fig. 6.10   PAL sync pulse generator block diagram. (Reproducd from *Handbook of TV Eng.,* Benson, McGraw-Hill, NY)

The 25 Hz offset between the line sync f_H and subcarrier is obtained by adding the phase advancing term into the phase locked loop and changing the counter by one count (1135/1134) every $V/2$, coincident with the reset of phase advancing ramp. The phase comparison is done at $H/4$ rate. For certain purposes, the SPG design may have the facility to operate in non-offset mode or in the genlock mode indicated earlier. The outputs of the SPG are fed to pulse distribution amplifiers (PDA), usually at $2V$ level. These pulse amplifiers supply to various equipments and areas in the studio complex.

6.5 TELECINE EQUIPMENT

Cine films still form a significant part of broadcast programs and commercials on TV. Films have the advantage of worldwide acceptance on TV as well as cinema circuits, because they can be played with telecine equipment to any TV standard, without the need for complex standards conversion. The films can also be easily duplicated. The quality of cine film images is better than present day TV systems. Until HDTV VTR standard finds a worldwide acceptance and is available at economic prices, the film would continue as an important medium for recording and storing pictures and sound. Hence telecine equipment has kept abreast of modern technological developments for improved performance and facilities.

Many television programs originate from photographic cine films, usually of the standard 35 mm and 16 mm. Slides are often required to be used in television programs. The telecine film camera chain, which couples film projectors or slide projectors to the television camera, therefore, forms an important program source in the TV studios. Telecine is also needed in CCTV systems and post production processes of simple editing to insert programs or add visual effects. For converting the film images into video signals, three types of equipments are used:

  (i) Projector telecine,
 (ii) Flying spot scanner,
(iii) Digital CCD telecine.

Projector telecine and flying spot scanners have been used since long. The recently introduced digital CCD telecine finds increasing use because of the several advantages of its solid state technology.

The sound track for film is generally recorded on the edge, as a combined optical (COMOPT) variable area audio track. Light shining through the clear area of the track on a photocell varies and reproduces proportional sound signal. When additional language sound track is used, separate magnetic (SEPMAG) or separate optical (SEPOPT) audio track is added at the centre or edge of a separate sound film.

(i) Projector Camera
This is the simplest form of telecine equipment. In this the slide projector is coupled to a Vidicon Camera tube to form a telecine film camera chain. For telecine, a source of high intensity of illumination of about 1000 lux is used to illuminate the face plate of a Vidicon Camera tube. With such a source the camera tube can be operated with a low target voltage (about 20 V) without posing a problem of lag or *smear*. Accurate colour registration can be a problem in the three tube camera system, especially for prolonged working. Adjustments for gamma, white and black levels controls are provided to take care of films of variable characteristics.

Multiplexer Systems For high utilization of the film camera chain it is convenient to use a multiplex type of system where one single television camera is used for three or four separate film sources viz. 35 mm film 16 mm film and slide projectors. The selection of film source in multiplex systems is made by use of optical multiplexer. Such multiplexed telecine setups are called film islands. In these setups, the film projectors may have 'reverse run' facility, and for sound reproduction side-optical or side-magnetic break playback facility is available.

The selection of film sources is made in two ways:

(i) by using semi-mirrors or prisms, or by means of rotation or moving mirrors.

A semi-mirror passes part of the incident light and reflects the rest of it, due to the front coated reflecting layer. An arrangement of this type employing semi-mirrors and prisms is shown in Fig. 6.10.

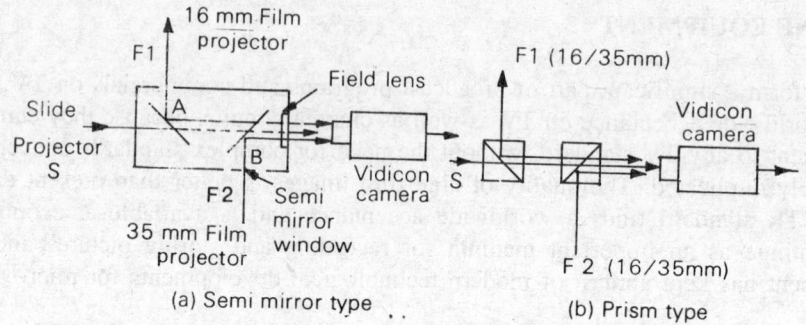

Fig. 6.11 Optical multiplexers: (a) Semi-mirror type, (b) Prism type

Light from the slide projector passing through the mirrors or prisms in straight paths, as also lights from the 16 mm/35 mm motion film projectors reflected by semi-mirrors A and B are focused on the field lens. Same optical path length is maintained for both the projectors with appropriate spacing. The field lens collects light from various film sources in the form of a real image at the centre plane of the lens. The multiplexer may be used with full light in the telecine room if a cover is used for the unit to guard against dust as well as stray light. In order to obtain equal light levels on the field lens from each projector, the window in the cover adjacent to the mirror B for the projector subject to a single reflection, is a semi-mirror. The semi-mirror or prism multiplexer has the advantage of simplicity of adjustment, no problems of moving parts and instantaneous selection of the source.

In a *movable mirror multiplexer*, front coated opaque mirrors are moved back or forward by means of small motors to bring the required one in the optical path depending upon the projector used. The movement of the mirrors may be arranged to be vertical, horizontal or swivel. For each projector, one mirror is used while the other is moved out of the straight optical path. A movable mirror-type multiplex is potentially capable of handling more projectors on a single camera than the fixed-mirror-type. The use of movable mirrors in the multiplexer may eliminate the need for dousers on individual projectors, since mirrors can be moved or turned to douse a particular projector.

Modern motion picture films are taken at the rate of 24 frames/sec. In order to avoid flicker that may be observed at this rate during projection, the illuminating flux of the film is interrupted by a shutter to give a projection rate of 48 frames/sec. In telecine projectors, there are additional problems because of the difference in the frame rate of TV fields and the films. The frame time of film is 1/24 sec, equal to 41.7 ms, the picture being flashed twice during this period. The pull-down time of the film is 25% of the frame time, equal to 10.4 m sec achieved during the shutter blanking. The TV picture frame time is 40 m sec with two interlaced fields of 20 ms each, with retrace time of less than 9% of the field scan period. The difference of one picture frame/sec in the film projection and TV scanning would cause a rolling bar in the picture and a reduced signal output in parts of the image on the camera tube scanned in the dark.

In order to overcome this, the film is pulled down at the rate of 25 frames/sec by a suitable speed correction to the drive mechanism. This increases the pull-down rate of the film by 4% and results in a little faster movements; but the distortion is hardly noticeable, nor can the slight rise in the pitch of the sound become appreciable. The film motion is thus made coincident with the TV frame rate. The dark (memory) scanning of one field still continues; but with the photoconductive lag of the vidicon camera tube commonly used in telecine projectors, the signals from the dark and illuminated fields are nearly of the same order.

In the 60/30 fields/frames/sec TV systems, it is necessary to convert the 24 frames/sec rate into 30 frames/sec rate in order to avoid presentation of incomplete information during some scanning periods. One way to do this is to project one frame three times and the next frame two times, and repeat so alternately by means of an intermittant mechanism of 3 : 2 pull-down cycle for the film. The pull-down is covered by 60 Hz shutter. The first set of alternate 12 frames are thus projected three times and the remaining set of alternate 12 frames are projected two times giving $(12 \times 3 + 12 \times 2 =)$ 60 fields/sec from the 24 frames/sec.

Video Tape Recording (VTR) Although magnetic tape sound recording has long since (1936) been in use, it was not until 1956 when magnetic tape recorders to suit the requirements of video recording were introduced. The requirements are:

(i) a very high writing speed to cover the 5 MHz video bandwidth,
(ii) some form of octave band compression to accommodate the wide 18 octave signal range, in the 6 dB/octave characteristics of a recording system,
(iii) high order of tape speed or time base stability.

The gap in the magnetic head must have a certain minimum physical size to place adequate field strength on the tape. If the tape speed is held constant and the frequency is increased, the recorded wavelength approaches the physical size of the head-gap as shown in Fig. 6.13(a). This results in cancellation of field and zero output at that frequency. The high frequency limit is thus set by the head-gap size and the tape speed. During playback, the strength of the induced signal in the playback head depends upon the rate of change of magnetic field recorded on the magnetic tape. Hence for a constant recorded amplitude, the induced output voltage is proportional to the frequency. The output is doubled if the frequency is increased by a factor of two. The output voltage thus increases at a rate of 6 dB/octave.

At still higher frequencies, the output starts again falling because of losses in tape and cancellation of the energy as the wavelength becomes comparable to the gap-width of the head. The fall in response is quite rapid as shown in Fig. 6.11(b), where the dotted line near the base indicates the noise level. Consequently, the signal-to-noise ratio at high frequencies becomes poor. The magnetic recording response is compensated by the equalisation curve as shown in Fig. 6.11(b). For the smallest practical gap of 0.1 mil or 100 μ in, the required tape velocity will be $v = \lambda \times f$, where λ must be at least twice the gap width to avoid cancellation of signal, and f is the maximum video frequency.

$\therefore$ Tape velocity $(v) \geq 2 \times 100 \ \mu \times 5 \ M \geq 1000$ in/s.

This means a very large footage of tape with uneconomic tape usage and unwieldy tape spools. The control mechanism for tape transport at high speeds of this order is impractical. In order to overcome the tape speed problem, modern video tape recorders use rotating heads moving across a tape running at a practical speed of 15 in/sec. The resultant head-to-tape velocity is quite high.

The 6 dB/octave characteristic of the tape recording is overcome by frequency.

(ii) Flying Spot Scanner
In this system, the cine film passes continuously in front of a special low persistence high brightness CRT

(Cathode Ray Tube) screen. A high intensity flying spot of light produced on the phosphor screen of the CRT scans the film image. The light transmitted through the film and colour splitter arrangement, is collected by corresponding pickup photomultipliers to generate RGB video signals. The signal generated is instantaneous a since camera pickup tubes do not employ charge storage principle as is used in usual camera tubes. In order to compensate for the after glow of the phosphor, high peaking is applied to boost HF response. In recent machines, a laser beam is used for scanning, which is accomplished by rotating prisms to deflect the beam dynamically.

The film movement and spot scanning on the screen must be coordinated to generate properly interlaced fields. In recent flying spot scanners, jump scan, progressive scan and digital store techniques are used, where complicated optical-mechanical transport is simplified by resorting to complex electronics which is now easier to implement. The film is moved continuously at a constant velocity past the raster by a capston servo. In jump scan, the raster is moved faster than the film to produce scan for the first field; the raster then jumps back to produce the second field. This hopping action has to be accurate to ensure perfect interlace. In progressive scan and digital store, all the 625 lines of the image are scanned progressively and the information is stored in a frame memory, from where it is read out from alternate lines to give out an interlaced-scan video signal.

(iii) CCD Film Scanner

For post-production work, flying spot telecine is used. This has the advantages of full resolution zooming, colour grading adjustments, and wide contrast ranges in both negative and positive film stock. In TV broadcast operations, the CCD telecine-film scanner is more economical and also offers certain advantages. Such as scanner is simple in construction and opertion because of continuous vertical optical scanning, rather than jump scan or fast pull down along with the twin optics of the conventional telecine. The CCD sensors have long reliable life, with no burn-in, field lag or afterglow problems. Each frame is scanned only once, line-sequentially and without conventional deflection voltages. As such there is no field flicker and the vertical resolution is absolutely constant. A high quality picture with high resolution, excellent S/N ratio and brilliant colour rendition is possible even in red hues. Digital frame store (DFS) can be provided for stills and slow or fast operations. New machines offer features like scratch concealment and electronic multiplexing.

6.6 CCD DIGITAL TELECINE

Digital telecine has been developed using CCD linear array sensors. The cine film is illuminated by a slit of light from a lamp to get the signals representing a TV line. The slit of light passing through the film is split into RGB primary colours by dichroic prismatic splitters and the separate RGB colours are then focused on three individual arrays of CCD sensors. Neutral density filters are necessary to compensate for the different RGB sensitivities of the three CCD arrays.

Each linear array has typically 1024 CCD elements and the charges representing a particular line on the film are clocked out at 19.6 MHz. A complete frame is scanned by moving the film across the line. The CCD arrays can provide high signal of over 1 V, without the need for video preamplifiers. The signal is filtered to remove the CCD read-out clock frequency components and then processed in usual analog form for gamma correction, black level clamp and white clipping requirements.

In order to produce conventional interlaced fields, the signal is stored in a picture frame store in digitized Y, U, V component form (component coding). The luminance and chrominance components are digitized separately and stored line-by-line in addressable video RAM. Alternate lines from this are read out and stored into field 1 and field 2 memories under control of a microcompuer, so that interlaced signal is obtained by reading out the field memories alternately. The exact addressing of all lines makes possible different readout programs for stills and forward/reverse operation in slow motion, shuttle, several ixed and variable speeds. The field interlaced digital signal readout is converted into analog components and encoded into the standard colour composite video signal.

It is easily possible to freeze-frame on the screen as the complete frame is stored in the video memory. This permits an instant start picture, by storing the required frame and displaying it as a frozen frame, backing up the machine and then arranging transmission only after the restarted picture matches with the frozen frame. A wide range of forward and reverse speeds is also possible to achieve by controlling the film transport speed electronically. Some systems use a fourth CCD sensor in addition to the RGB sensors, to sense infrared signals. This is used with electronic circuitry to detect scratches or dirt and replace the dropout signal by adjacent line information.

Some systems are designed to be completely digital in signal processing, eliminating setup controls required in analog processing and produce stable signals using microprocessor control for automatic programmed adjustments. In order to be able to adjust colour setup on a freeze frame, an additional digital store is placed before the signal processing circuitry. Video signal processing of a completely digital telecine is shown in Fig. 6.12.

While eight-bit coding is normally satisfactory for the TV signal, an 11-bit coding is used to provide adequate coding levels in the low range of the input signal, and enable proper gamma correction in digital form. Since 11-bit high speed flash A-D converters are not available, two eight-bit ADCs are used in combination to produce 11-bit (or more bit) resolution. One ADC provides the eight most significant bits, while the other ADC, fed via an eight-times gain amplifier provides the eight least significant bits. The digital signal is then numerically processed into logarithmic form for gamma correction by means of a microcomputer resident program. Gamma correction and masking colour correction are applied by digital multipliers and adders in the RGB channels. The logarithmic signal is multiplied by the gamma and its antilog is determined by an exponential converter, in order to enable black level adjustments. The signals are brought in the eight-bit code to pass through the remaining linear processors.

The chroma signals R-Y, B-Y and luma signal Y, are obtained by suitably matrixing the RGB signals. The R-Y and B-Y signals are filtered digitally to reduce their bandwidth to half, so that they can be multiplexed to form a chroma data stream. The three digital signals are then individually stored in separate field store memories, and read alternately to give interlaced signal. Aperture correction is provided to the Y signal using digital delay lines. Special amplifier circuits are provided to suppress low level noise in the signal due to film grain, and the signals are finally converted into analog Y and RGB signls for encoding into PAL/NTSC format.

6.7 TELEVISION RECORDING SYSTEMS

Television recording is as old as television itself, though there is a vast difference between the first video recording demonstrated in 1926 by J. Baird on 78 rpm wax phonographs of Edison's design and present-day high-quality video tape and disc recorders. Kinescope recording was used in the early days to make a photographic cinefilm recording of the visual images on a special kinescope—a high-intensity picture tube monitor. It was in 1956 that the first video tape recorder went on the air with its ability to store video programs and reproduce at quality far better than kinescope recording. Since then, the transition from the large bulky equipment to compact portables has been quite remarkable, and the video tape recorders have made a great impact on broadcasting, education and home entertainment.

Other forms like the electronic video recorders (EVR) using photographic film or the disc recorders also employing optical methods, have not proved quite popular. These are well suited for replication type

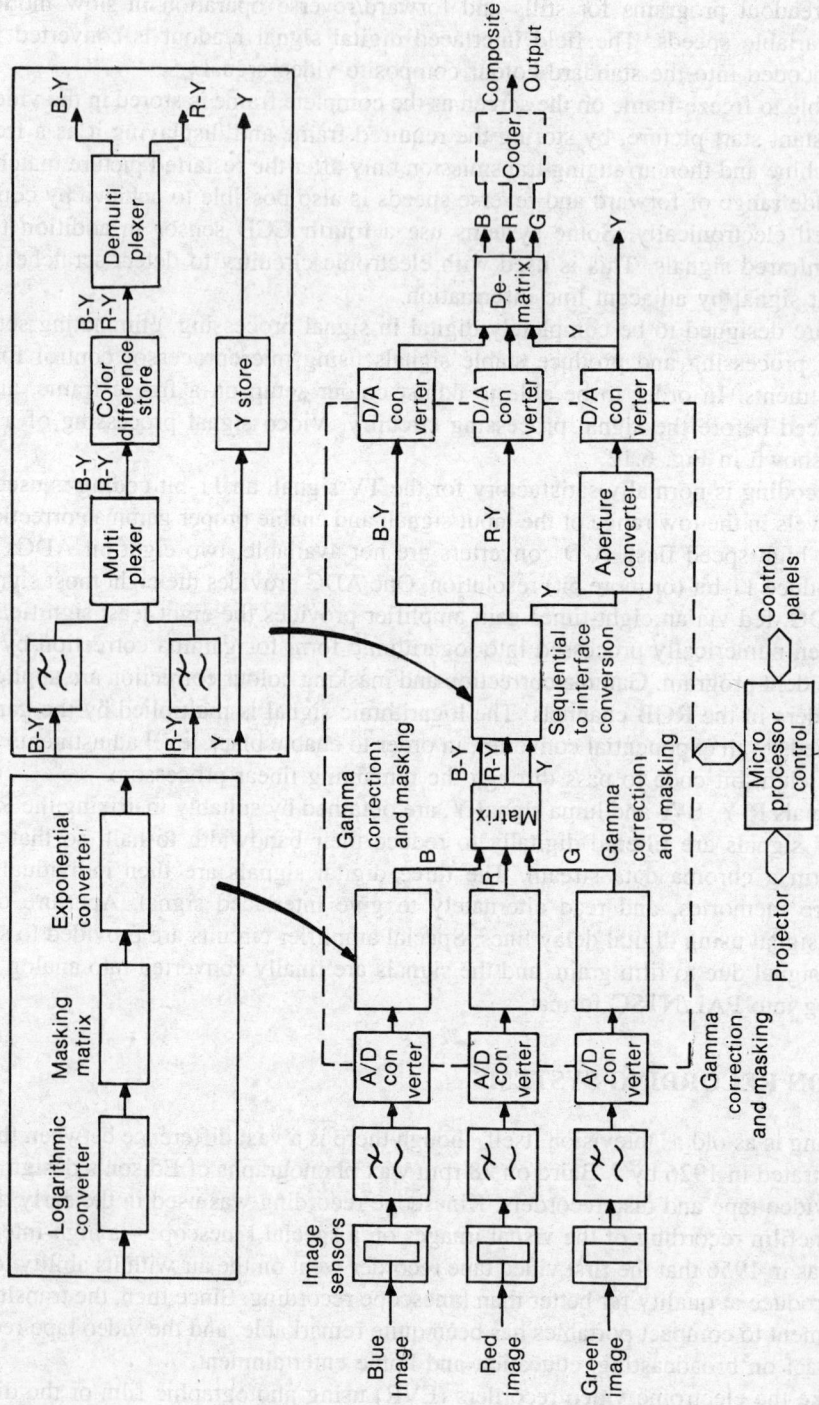

Fig. 6.12 Block diagram of video signal processing for digital telecine. Marconi B3410 (*Courtesy:* Marconi Communication Systems Ltd.)

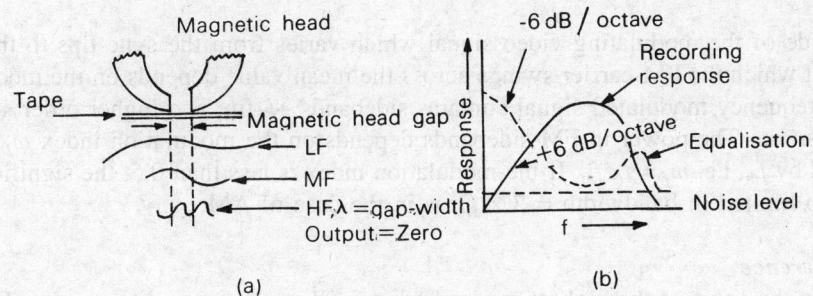

Fig. 6.13 (a) Effect of head gap-width on frequency, (b) Frequency response of magnetic tape recorder and equalisation curve

programs for replay only, but viewing only the same program is not a very attractive proposition for the viewers. The recording process is too elaborate and costly for the limited demand for copies of the video programs. The holographic system developed by the RCA is noteworthy in that the technique may some time reappear for 3-D stereoscopic TV systems!

Holographic Recording System

Hologram is a recorded interference pattern between light from an object and a reference beam, both derived from the same coherent light laser source. If a beam similar to the reference beam falls on the hologram, it is diffracted to form an image beam which appears to diverge from a reconstructed image located in the object position. Depending on the type of recording medium and the relative convergence of the object beam and the reference beam, holograms of different types are obtained. Phase holograms are formed in transparent photoresist or thermoplastics. These only change the phase of the reconstructing light wave by a variation in the refractive index of the hologram medium or by a variation of the hologram thickness. These surface relief holograms find application in video recording as they can be embossed on inexpensive plastic PVC films. If a lens is interposed between the object and the hologram film, and the object is placed in the focal plane of the lens, a Fraunhöfer hologram is obtained. During reconstruction these holograms can be translated in their own plane without displacing or distorting the image.

In holographic recording, a sequence of Fraunhöfer holograms of successive frames is recorded on a photoresist-coated PVC film tape. The exposed hologram tape is developed and the master PVC tape is nickel coated by a metal plating process. The PVC is stripped away to form the metal master. Replication of the metal master is done on inexpensive thin plastic tape by pressing them out together through a pair of rollers. Duplication is thus simple and speedy.

During playback, reconstruction of the encoded image is done with a low power laser beam similar to the reference beam, shot on the hologram recorded on the tape film. The reconstruction of the image is made on a vidicon camera target as in telecine system so that the image is picked up by the vidicon scanning system as a video signal, and is reproduced through RF modulator on a conventional TV receiver.

Holographic recording is thus a method of permanent video recording. It offers significant advantages in economy of cheap recording material, easy and fast duplication and simplicity of playback, although the master recording process is fairly complicated.

Television Recording Using FM Signal Carrier

A video signal occupies almost 18 octaves, if the lowest frequency is taken as 25 Hz and the highest as 5 MHz. This is reduced by frequency modulation of a carrier. The deviation of the frequency depends upon the amplitude of the modulating video signal which varies from the sync tips to the peak white levels. The rate at which the FM carrier swings across the mean value depends on the modulating video frequency. The frequency modulated signal contains sidebands +/–fm, and higher order sidebands +/ – $2f_m$ +/–3 f_m and so on. The power in FM sidebands depends on the modulation index m_f, given by the deviation divided by f_m, i.e. $m_f = f_c/f_m$. If the modulation index is less than 0.5, the significant power of FM signal is also limited in bandwidth = $2 \times f_m$ as in the case of AM.

Sidebands Interference

In video recording the value of the highest carrier frequency that can be used as restricted by the design of the head and the magnetic tape response. Hence the lowest sideband frequency $(f_c - f_m)$, corresponding to the highest video frequencies, can be very near zero frequency, and the sidebands due to harmonics, viz. $(f_c - 2f_m)$, $(f_c - 3f_m)$, etc. are in fact folded back into the first order sideband range of the video signal creating in-band interference as shown in Fig. 6.14(a). Distortion in modulator circuits may produce second harmonic of the carrier frequency $2f_c$, which gets modulated with f_m to produce interference due to $(2f_c - f_m)$ as shown in Fig. 6.14(b).

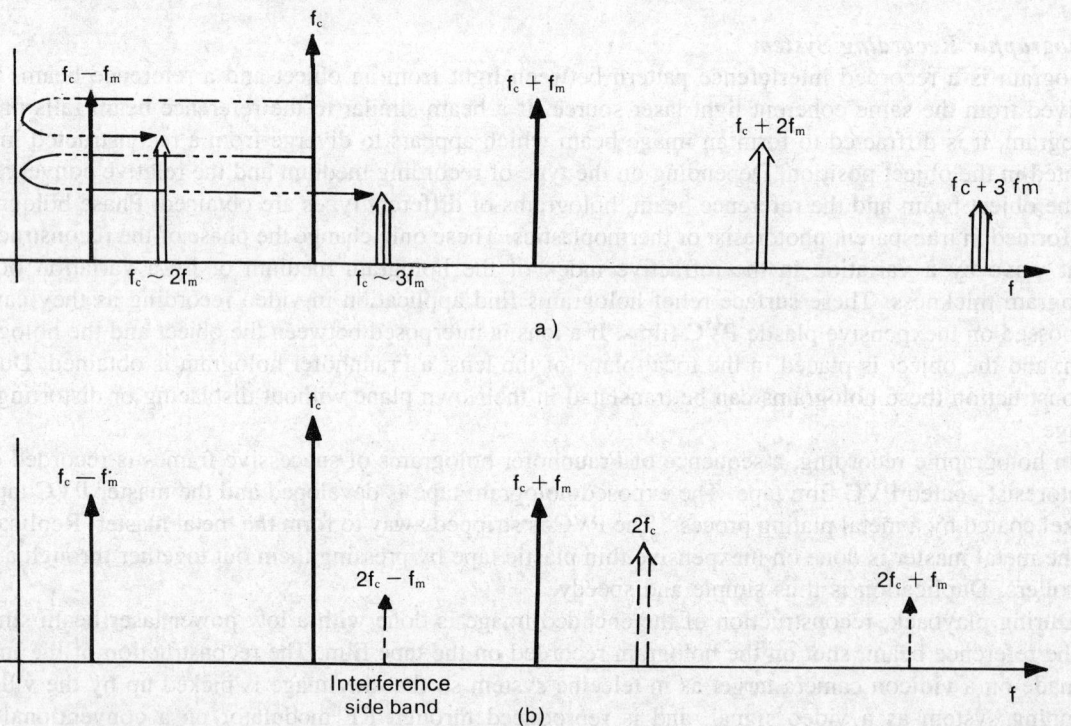

Fig. 6.14 In-band interference due to: (a) folded back sidebands, (b) sidebands created due to harmonics of carrier

In colour TV recording this can be a serious problem as the video signal contains colour signals at the high end of the spectrum. Making the FM carrier higher helps in reducing this interference, because only the higher order sidebands which have lower power, can cause inband interference. The bandwidth of video is also restricted to reduce the interfering sidebands. The FM carrier frequency has to be properly chosen to minimize the interfering signals produced by the higher sideband power contained in the video signal and due to the 2nd and 3rd harmonics of the carrier generated by distortion. Choice of a lower carrier means lower deviation and poor S/N ratio. While a higher frequency carrier allows greater frequency deviation, the signal recovery from tape can be weaker. Moreover, large deviations at low frequencies can also cause problems of differential gain and phase.

In an FM system, high frequencies which normally contain small amounts of the total energy are boosted up to improve the S/N ratio. In the colourplexed video signal, the colour signal occupies the high end of the frequency spectrum and contains a large amount of energy. This limits the amout of pre-emphasis that can be given to the high frequencies.

The highest FM carrier that can be used depends on the recorder design. While earlier low band VTRs used lower frequencies around 5.79/5.54 MHz at the blanking level, later high band models could use 7.9/7.8 MHz. The latest super high-band broadcast recorders can use a still higher carrier of 11.35 MHz.

Two types of VTR systems are in use: (a) transverse scanning, often called Quadruplex head systems, and (b) helical scanning or slant-track recording systems.

Quadruplex-head magnetic tape systems   In these systems, an assembly of four quadruplex heads records on and reproduce from a transverse-laterally scanned tracks on a magnetic tape. Systems of this type have been widely used for professional TV broadcasting and have been standardized for a 2 in tape format shown in Fig. 6.15.

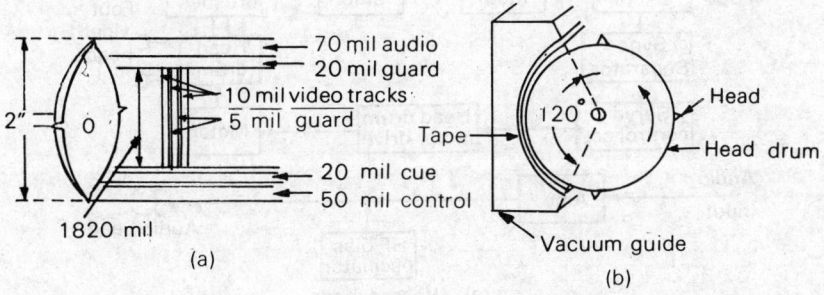

Fig. 6.15   (a) 2 in tape format, (b) Head drum with tape guide

The tape format has a 70 mil sound track at the top, a 50 mil control track at the bottom, a second sound track of 20 mil above the control track and the remaining 1820 mil are filled with 10 mil transverse video tracks, separated by 5 mil guard bands. The primary nominal tape transportation speed is 15.625 in/s and a secondary speed of 7.81 in/s. The four heads are mounted accurately 90° apart on a 2 in drum or headwheel rotating at 15,000 rpm, giving a writing speed of 1500 in/s. Elaborate servo systems have to be employed to control the tape speed, capston motor, the head drum motor and the vacuum tape guide so that a high time base stability of the order of 100 ns, is attained.

The tape is formed into a concave surface by special tape guides as shown in Fig. 6.15(b), which ensure that each head will maintain a proper fixed pressure contact with the oxide coating of the tape through an arc of about 120°. Since the head drum rotates at 250 rps, the contact time is $\frac{1}{3} \times \frac{1}{250} = \frac{1}{750}$ s. In this much time, the tape advances by $15.625 \times \frac{1000}{750} = 20$ mil. The bottom of each track is thus displaced longitudinally by 20 mil. One vertical field is covered by $\frac{1}{1/250} \times \frac{1}{50} = 5$ complete rotations of the head drum; and as the drum has four heads, each picture frame contains 20 track lines or each picture contains 40 tracks occupying about $\frac{1}{25} \times 15 = 0.6$ in length of the tape.

During recording, the composite video signal is supplied to the four rotating heads simultaneously in the form of a frequency-modulated signal. With the carrier frequency of 5 MHz clamped at the composite video blanking level (for low band operation), the peak whites cause an upward frequency deviation of 1.8 MHz, while the sync tips cause a downward deviation of 0.7 MHz. This means a total deviation of 2.5 MHz, from 4.3 MHz at the sync tips to 6.8 MHz at the peak whites. For high band operation, the total deviation is about 3 MHz with the peak white at 10 MHz, sync tip at 7.06 MHz and the blanking level clamped at 7.9 MHz.

During playback, the video head outputs are first amplified and then fed to an electronic switcher that selects the signal from the head that is in contact with the tape. The selection takes place during horizontal retrace blanking so that the switching transients are not visible. The output from the electronic switcher is demodulated to get back the video information from the RF carrier.

For synchronisation, the sync from the incoming video signal is used to obtain a 250 Hz signal which is amplified to sufficient power to drive the head drum. This 250 Hz signal is also recorded on the control track of the tape, for replay synchronisation servo.

The basic arrangements for recording and playback in a quadruplex head tape recorder are shown in Fig. 6.16.

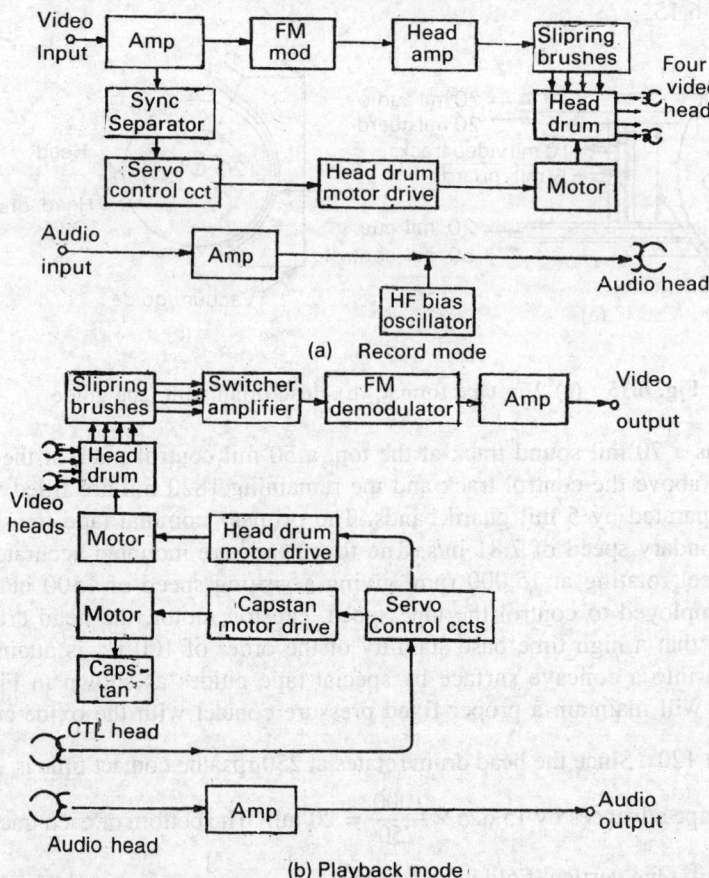

Fig. 6.16 Simplified block diagram of a quadruplex head video tape recording system: (a) Basic arrangement in record mode, (b) Basic arrangement in playback mode

6.8 HELICAL SCAN RECORDING SYSTEMS

The 2 in quadruplex head machines introduced in 1956 dominated the TV broadcast industry for over two decades. Built to tightly controlled professional broadcast standards, they were bulky and costly. The need for less sophisticated, cheaper and compact VTRs had long since been felt. Introduction of helical scan or slant track recording was an important step in this direction. These machines were considered in the mid-50s' and became practical only in early 60's when Sony introduced the first helical scan' recorder, much cheaper than the quadruplex 'Ampex' recorder. In these recorders the tape is wrapped about the scanning rotating head drum assembly in a helix at a small angle with the plane of rotation of the head drum.

As the head drum is big, a few centimetres in diameter, and the recording is along a lengthwise slanting track, the wow and flutter caused by tape speed and tension variations become appreciable and the time base stability suffers considerably. Hence their use in broadcasting was extremely limited until the introduction of digital time base correctors in 1973. These made it possible to incorporate wider windows in the time base correctors than in analog correctors, to control the excessive timing errors.

In the helical scan format, a single recording and playback track could be used to record and replay a whole field, eliminating troublesome banding problems in quadruplex machines. Recording a single field per head had a further advantage in that the tape could be moved at other variable speeds producing slow/fast motion or freeze picture, useful for editing purposes.

The helical machines be classified as full wrap using omega loop with a single head, or half wrap with two heads for scanning, as shown in Fig. 6.17. They can also be classified as field per scan or segmented scan machines. In the latter, some five to six head passes are required for one field scan. A number of manufacturers have developed their own brand of machines with different tape widths (1 in, 1/2 in, 1/4 in, 8 mm), tape speeds (14.29 cm/s, 9.53 cm/s … 1.87 cm/s) and helix angles (3 … 20°).

In order to facilitate interchangeability of programs, common 1-in. tape helical VTR standards appropriate for the broadcast quality and professional needs have been set up by the SMPTE Working Groups. The Ampex format having one field per scan, 1-in tape 360-deg omega wrap, being in production for a number of years became SMPTE type A format. With the improved versions coming in from Sony, broadcast studio quality VTRs were standardized to SMPTE type C format for professional use.

Helical 1-in SMPTE Type B Format

The segmented scan format with 180 deg angle half wrap was introduced from Bosch-Fernseh, originally meant for military applications. It is a cross between the transverse and helical machines, having advantages of mechanical ruggedness and flexibility in writing speeds desirable for PAL recording. Slow motion effects are, however, complex to execute. The FM carrier is 7.40 at blanking deviation ranging from 6.76 MHz (sync tip) to 8.90 MHz (Peak white). Since 1978, the broadcast version of the 1-in segmented field VTRs have been conforming to the type B format, shown in Fig. 6.18.

Studio Quality 1-in SMPTE Type C Format

The 1-in machines having one field per head scan and 360-deg omega-wrap machines have been conformed to this type C format set up by the SMPTE working Group for interchangeability of broadcast programmes. The standard has major advantages in capabilities for slow motion and easy editing, because in any field-per-scan machine the pictures can be reproduced at speeds faster than real time or slower down-to-stop motion as each scan has its picture identity. Broadcast quality can be obtained for pictures from reverse motion to fast forward motion by use of AST technique for the head track alignment. The FM carrier is 7.68 MHz at blanking, deviating from 7.16 MHz (sync tip) to 8.90 (peak white). Important features of the type C format are shown in Fig. 6.19.

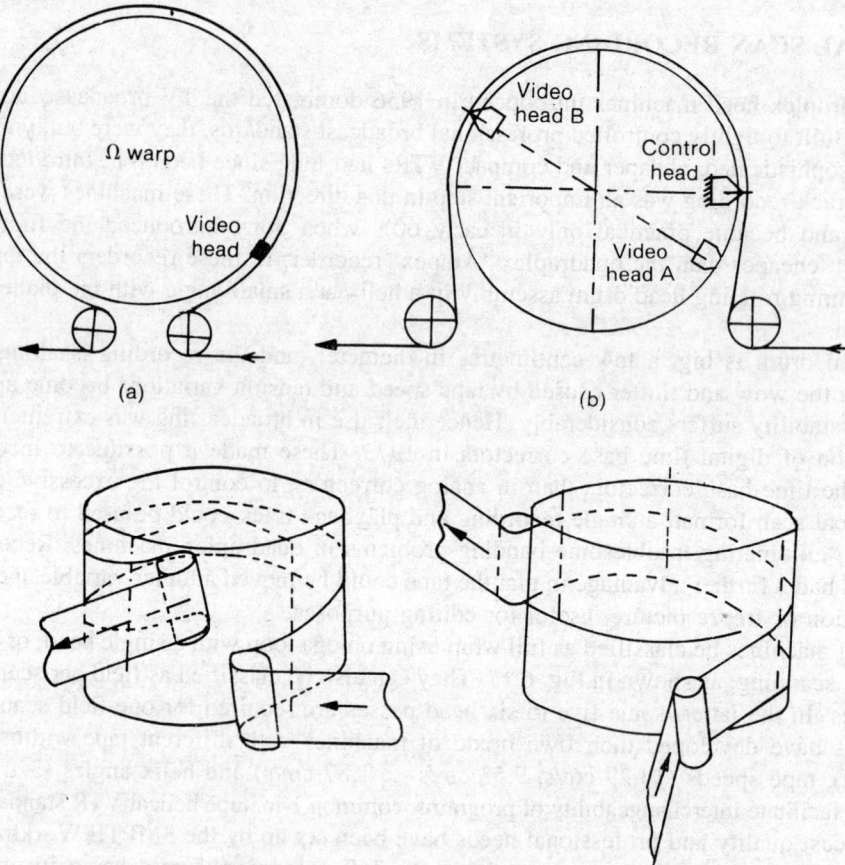

Fig. 6.17 Helical scan systems: (a) Full wrap, and (b) Half wrap

The head configuration in this format is such that the tape can be *monitored while recording* to have a check on the recorded video. Being a single head machine with only one channel for the video, the video electronics is simple without problems of matching of multiple channels as in segmented scan recorders.

6.9 AUTOMATIC SCAN TRACKING (AST)

This is a recently developed technique of automatically aligning the video head tip along the video track path during scanning for replay. It is of great help in improving signals available from helical scan recorders. In this technique, the video tip is mounted at the end of a piezoelectric bimorph strip fixed to the drum at the other end. A bimorph is the bonding together of two planes of piezoelectric ceramics, such that, with application of electric potential, one of these planes expands and the other contracts, giving high deflection sensitivity. By application of the correcting voltage to the bimorph through a servo, the video tip can thus move at right angles to the scanning direction, so as to position it correctly on the track centre. The correcting motion of the piezoelectric bimorph can be fast enough to allow the head jump one track pitch, during the vertical blanking period, eliminating the interruptions in the signals due to skewing across the tracks at any speeds.

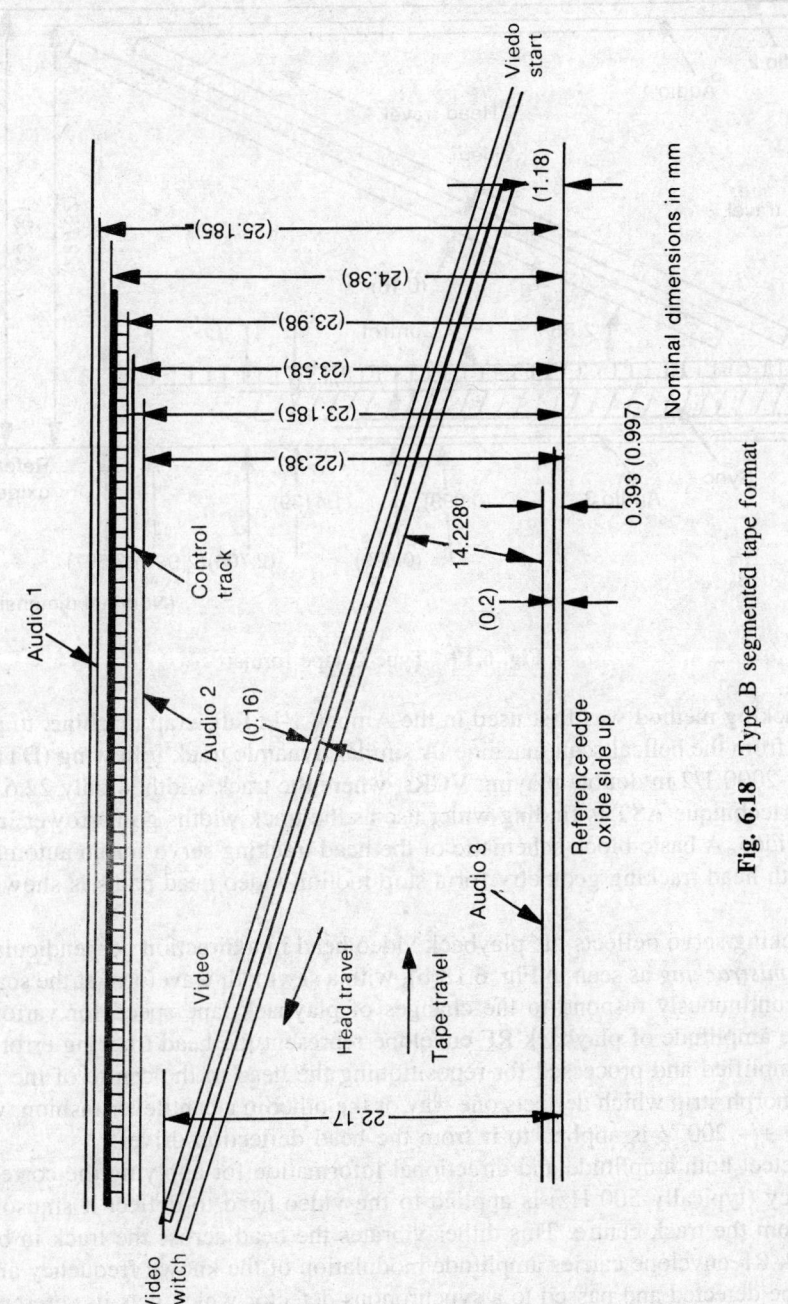

Video start

(1.18)

(25.185)

(24.38)

(23.98)

(23.58)

(23.185)

(22.38)

0.393 (0.997)

14.228^0

(0.2)

22.17

Nominal dimensions in mm

Audio 1

Control track

Audio 2

(0.16)

Video

Head travel

Tape travel

Audio 3

Reference edge oxide side up

Video switch

Fig. 6.18 Type B segmented tape format

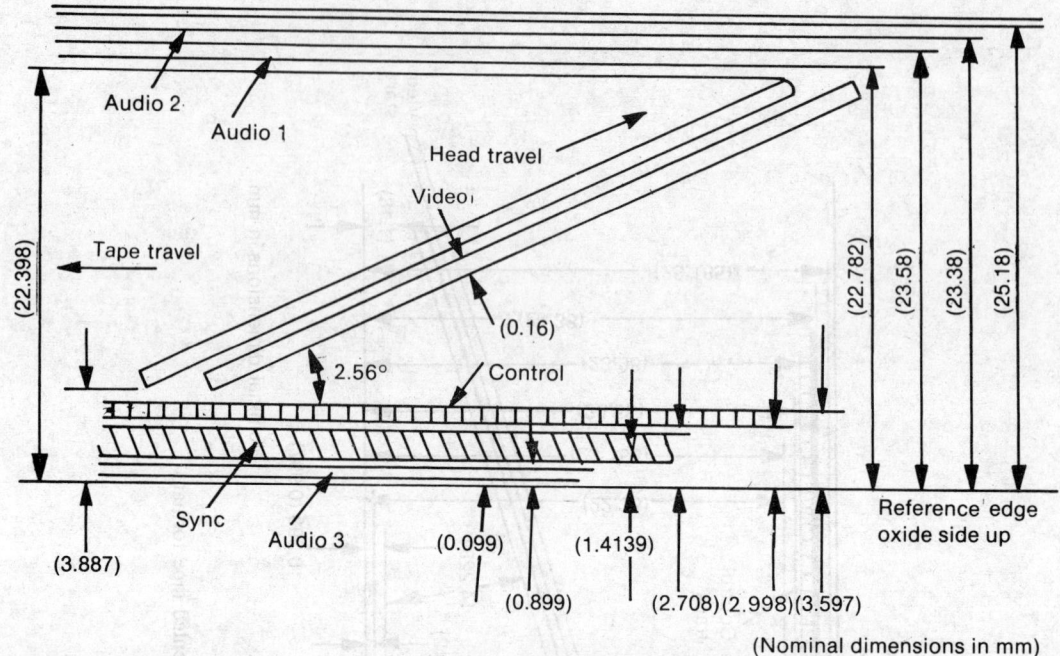

Fig. 6.19 Type C tape format

This head tracking method was first used in the Ampex 1-in full wrap machine, to provide broadcast quality pictures from the helical scan machine. A similar dynamic track following (DTF) method is used in the Philips V-2000 1/2 in double playing VCRs, where the track width is only 22.6 μm (0.9 mil). As a very powerful technique AST is finding wider use as the track widths go narrower in newer machines like the 8 mm video. A basic block schematic of the head tracking servo for an automatic scan tracking (AST) along with head tracking geometry for a stop motion video head paths is shown in Figs. 6.20(a) and (b).

The head tracking servo deflects the playback video head in a direction perpendicular to the recorded track to *prevent mistracking* as seen in Fig. 6.18(b), with a sawtooth waveform at the scanning frequency, and also must continuously respond to the changes of playback tape speed for various slow motions. Variations in the amplitude of playback RF envelope represent the head tracking errors. These tracking variations are amplified and processed for repositioning the head to the centre of the track through the piezoelectric bimorph strip which deflects one way or the other in a cantilever fashion, when a correcting voltage of up to +/– 200 V is applied to it from the head deflection driver.

In order to detect both amplitude and directional information for applying the correction, an error or 'dither' frequency (typically 500 Hz) is applied to the video head to deflect it sinusoidally by a small amount away from the track centre. This dither vibrates the head across the track in both directions so that the playback RF envelope carries amplitude modulation of the known frequency and phase. The RF signal is envelope detected and passed to a synchronous detector which gets its reference from a square wave in phase with the RF dither signal. This synchronously detects any small errors in track curvature or skewing when operating at a different speed, and provides correction signal to the piezoelectric bimorph in the appropriate direction.

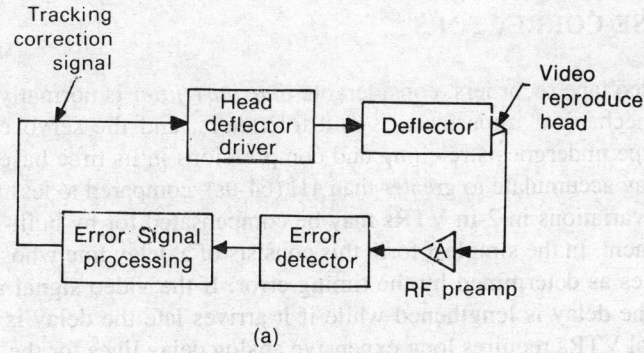

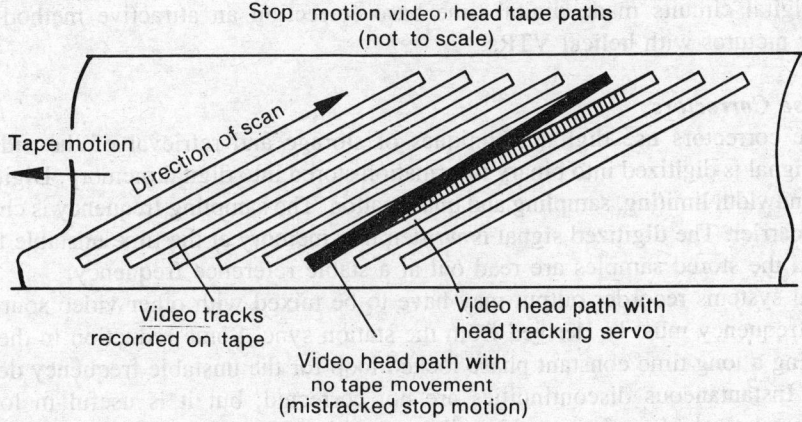

Fig. 6.20 (a) Block schematic of AST servo, and (b) Head tracking (Reproduced from *TV Engg. Handbook*, Benson, *Courtesy:* McGraw-Hill, NY)

6.10 DROP-OUT COMPENSATION

The loss of signal or significant decrease in the output of a recorded signal on replay by more than 12 dB for over 5 μs is considered as a drop-out. Although frequency and severity of dropouts are minimized in developing new tapes and VTRs, in practice some droputs do occur, increasing with use as the tape wears out and gets contaminated with dust and other matter. A speck of dust of diameter 1 μm is enough to cause a significant dropout as the bouncing action removes the contact between the tape and the head.

Drop-out compensation is a method for detecting the portions of signal where dropouts occur and of replacing these by video signals without dropouts. The replaced video has to be a good estimate of the missing video to minimize the visibility of dropouts. The amplitude of the FM video signal delivered by the head is detected by the dropout detector, which generates a pulse if it falls below a preset value. The detection pulse operates an electronic switch, so that the direct video signal containing the drop-out is removed from the output and is substituted by the video content of the previous line.

Direct substitution video from the adjacent line is not suitable for colour TV system because of the colour encoding structure being a different for adjacent lines. The simplest solution to this is in the use of a substitute video from the nearest scan line that has the same encoding structure This becomes more complicated in PAL system where the colour subcarrier of the adjacent line is 90 degrees phase shifted and the V signal is phase reversed too.

6.11 TIME BASE CORRECTORS

In helical scan video tape recorders, considerable *time base jitter* is normally present. Due to tape dimensional changes, mechanical imperfections and tolerances, and the servo circuit limitations, the video output from the tape undergoes stretching and compressions in its time base. For the helical scan VTRs, these variations may accumulate to greater than 1H (64 μs), compared to less than 1 μs for 2-in quadruplex VTRs. The small variations in 2-in VTRs may be compensated for by built-in electrically variable delay lines in the equipment. In the simplest form this consists of a delay line whose effective length is adjusted by switching diodes as determined by the timing error. If the video signal arrives early as compared to the station sync, the delay is lengthened while if it arrives late the delay is shortened. The large elastic variations in helical VTRs requires long expensive analog delay lines for the video signals. An alternative approach is use digital memory for storage and readout with necessary time base correction. Decrease in the cost of digital circuits made digital time base correction an attractive method for producing broadcast-quality pictures with helical VTRs.

Digital Time Base Correctors
Modern-day base correctors use digital techniques of storage and retrieval of the video signal. The incoming video signal is digitized into binary information stored into digital memory. Digitization implies dc-restoration, bandwidth limiting, sampling and quantization. The sampling frequency is chosen a multiple of the colour subcarrier. The digitized signal is written into memory at the time unstable frequency from the playback, and the stored samples are read out at a stable reference frequency.

In professional systems recorder output may have to be mixed with other video sources. Hence the stable reference frequency must be derived from the station sync. Some reduction in the velocity error is possible by using a long time constant phase locked loop for the unstable frequency derived from the playback signal. Instantaneous discontinuities are not corrected; but it is useful in low cost helical recorders that are relatively free from sudden discontinuities in active picture. Professional time base correctors may offer this mode as an option for use in processing the playback of recorders that have no servo system.

A basic block diagram of a digital time base corrector is given in Fig. 6.21. The playback video is band limited through the input low pass filter and is sampled at three or four times f_{sc} and converted into eight-bit digital signal. Three to four samples of this signal are temporarily stored in a high speed memory, to form a 24- or 32 bit word, before they are strobed and multiplexed into the main memory once in each cycle of f_{sc}, in order to extend the read/write time.

The sampling frequency is obtained from an oscillator operating at three or four times the subcarrier f_{sc}, phase locked to the play back video colour sync. The output of he oscillator is adjusted in phase such that after division by three or four the resulting f_{sc} is phase synchronised with the playback burst throughout the line scanning.

The output readout clock is the same multiple of the stable reference subcarrier. Its output is used to recall and demultiplex the words stored in the main memory. The word is then commutated at the sampling rate to retrieve the original stream of digital samples which are converted into analog signal in the DAC. This is followed by the resampler operating at three or four times the f_{sc} to the eliminate transients, and finally the LP filter to remove the out of band harmonics.

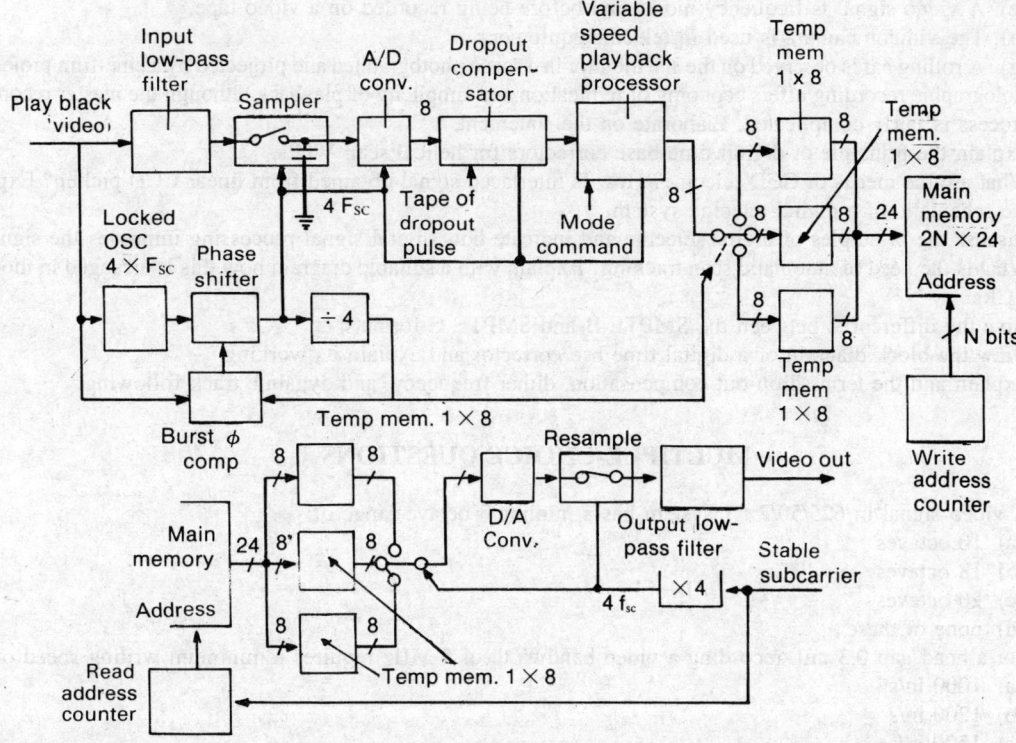

Fig. 6.21 Block diagram of a digital time base corrector (Reproduced from *TV Engineering Handbook*, Benson, *Courtesy*: McGraw-Hill, NY)

REVIEW QUESTIONS

6.1 Describe the facilities for progrm production and editing in a modern TV studio complex.

6.2 Discuss how the video signal from the various program sources is routed to the transmitter.

6.3 What are the methods of lighting adjustments and control of illumination in a TV studio? Explain the terms: high key lighting, low key lighting, key light, back light.

6.4 What are the facilities provided in a vision mixer for basic editing, operations? Indicate how special effects for wipes are produced.

6.5 What is meant by the following: Station sync, system blanking, camera blanking, sync regenerator, genlock, slavelock?

6.6 State the different methods of television picture recording. Explain the principles of any two of them.

6.7 What are the potential advantage and limitations of photographic recording?

6.8 Explain how the harmonics of an FM carrier causes in-band interference in video recording. How is it reduced?

6.9 Sketch the typical tape format of recording in a quadruplex head and a helical scan tape recorder.

6.10 How are slow motion and stop action effects obtained in a TV program with the help of recorders?

6.11 Give reasons for the following:
   (a) System blanking pulses are essential in forming a composite video signal.
   (b) It is necessary to use rotating heads in video recording.
   (c) Servo control, speed control and speed control circuits are essential for tape and head drum drives in a VTR.
   (d) Helical scan tape recorders require time-base correctors before being used as program sources in a studio chain.

    (e) A video signal is frequency modulated before being recorded on a video tape.

    (f) The vidicon camera is used in telecine equipment.

    (g) A rolling bar is observed on the TV monitor in a scene photographed and projected by a cine-film projector.

6.12 Holographic recording offers economy of replication and simplicity of playback although the master recording process is fairly complicated. Elaborate on the statement.

6.13 Explain the principle of digital time-base correctors for helical scan VTRs.

6.14 What are the merits of CCD telecine? How is interlaced signal obtained from linear CCD pickup? Explain the schematic of a digital telecine system.

6.15 Discuss the principles of digital telecine and indicate how digital signal processing improves the signal.

6.16 What is the need of automatic scan tracking? Explain with a suitable diagram how this is arranged in modern VCRs.

6.17 Give the differences between the SMPTE B and SMPTE C formats.

6.18 Draw the block diagram of a digital time bse corrector and explain its working.

6.19 Explain and the terms drop-out compensation, dither frequency, and dynamic track following.

MULTIPLE-CHOICE QUESTIONS

6.1 A video signal in 625/50/2 : 1 system has a minimum octave range of:
    (a) 10 octaves
    (b) 18 octaves
    (c) 20 octaves
    (d) none of these

6.2 For a head gap 0.3 mil, recording a video bandwidth of 5 MHz requires a minimum writing speed of:
    (a) 1000 in/s
    (b) 1200 in/s
    (c) 1500 in/s
    (d) 300 in/s

6.3 Frequency modulation is used for recording video signal because:
    (a) it gives high fidelity
    (b) it improves S/N ratio
    (c) it enables octave band compression of recorded signal
    (d) (b) and (c)

6.4 Dither frequency is applied to the video head to:
    (a) frequency modulate the recorded signal
    (b) amplitude modulate the playback output
    (c) average out the pickup during playback due to mistracking
    (d) (b) and (c)

6.5 High-band video recording:
    (a) improves S/N ratio
    (b) reduces interference from harmonic sidebands
    (c) enables larger frequency deviation
    (d) all of these

6.6 Field per scan helical recorders are:
    (a) more rugged
    (b) easy to edit for slow motion
    (c) reduce tape usage
    (d) (a) and (b)

6.7 Automatic scan tracking is ensured by:
    (a) head drum servo
    (b) head tracking servo
    (c) tape capston servo
    (d) all of these

7

Television Transmission and Relay Systems

7.1 REQUIREMENTS OF TV BROADCAST TRANSMISSION

The requirements of television signal modulation and the various standards for television transmission have been considered in Chapter 4. It may be recalled here that standard channels of 7 MHz bandwidth have been provided in bands I and III in the VHF range and bands IV and V in the UHF range, in the CCIR System-B. At these frequencies, the propagation takes place by space waves limited by the maximum line-of-sight distance between the transmitting and receiving aerials. The signal strength available from a transmitter is proportional to the square root of the transmitter power and varies inversely as the distance from the transmitter. The signal strength at any place in the service area must be large enough to overcome the noise at that place and provide a satisfactory picture. In the VHF bands, the man-made noise generated by vehicle ignition systems, and other sparking is quite a predominant factor in determining minimum field strength requirements.

In television broadcasting, the *primary area* served by a TV transmitter is classified into various grades depending upon the signal field strength contours. *Grade A* service is the area within the signal strength contour which gives a satisfactory noise-free picture on the average receiver. *Grade B* service is the area within the field intensity contour that may result in intermittent noise in the picture on the average receiver. The local or principal community being served should get a signal better than twice that in the grade A service. The required field intensities in the different bands are indicated in Table 7.1. The signal strengths are expressed in dBμ, referred to 1 μV/m, i.e. dB$\mu = 20 \log_{10} \dfrac{E}{1}$, where E is the signal strength in μV/m.

Location of the antenna at a high point of elevation is desirable from the point of view of wider coverage and reduction of shadow effects due to hills and tall buildings which may reduce the signal intensity considerably. The height of the transmitting antenna is very important. Increase in the height is of greater advantage than increase in the transmitter power. For example, doubling the antenna height is nearly equivalent to increase in the transmitted power by a factor of about five. In general, the transmitting antenna should be located at the most central point and height available. When the shape of the desired service area and population distribution make the central location difficult, a directional transmitting antenna system may be used.

Table 7.1 TV Signal Strength Requirements*

Band	Grade A	Grade B	Local community
I	68 dBμ = 2510 μV/m	47 dBμ = 224 μV/m	74 dBμ = 5010 μV/m
III	71 dBμ = 3550 μV/m	56 dBμ = 631 μV/m	77 dBμ = 7100 μV/m
IV & V	74 dBμ = 5010 μV/m	64 dBμ = 1585 μV/m	80 dBμ = 10000 μV/m

*These are based on FCC regulations in USA. The values may differ in other countries depending upon the level of man-made and other noise interference. CCIR has recommended a minimum signal strength of 55 dBμ in band III.

Transmitter Power

The power radiated by a transmitter is usually expressed as the Effective Isotropic Radiated Power (EIRP). The EIRP is a measure of the actual signal power radiated by the transmitter in the direction of the receiving area. The transmitter output power is partly reduced by losses in the transmission line and coupling network, but is provided a boost in the direction of the horizontal radiation because of the directional gain provided by the transmitting antenna. The EIRP is, therefore, defined as the transmitter power multiplied by the loss factor for the transmission line and the coupling network, and the antenna directive gain over an isotropic radiator radiating equally in all directions.

A television transmitter comprises a visual transmitter for the picture and an aural transmitter for the sound. For the visual transmitter, the EIRP is given in terms of the peak carrier corresponding to the sync peaks modulated at 100%. The aural transmitter power is rated in terms of the rms value and is 10 or 20% of the visual power. Because of frequency modulation, the range of the aural transmission, with even 10% power, is satisfactory from the point of view of having noise-free reception over the same range as the visual transmitter using amplitude modulation. Greater powers are not only uneconomical, but also worsen problems of sound carrier beats with the colour subcarrier in colour TV transmissions.

In order to avoid co-channel and adjacent channel interference, the channel assignment has to be done with care so that such stations are as far apart as possible. The maximum powers of TV stations are limited by these considerations, depending upon the transmitting antenna heights and the band. For antenna heights of up to 300 m, the powers are generally limited to 20 dBk (100 kW) in band I, 25 dBk (316 kW) in band III and 37 dBk (5000 kW) in band IV and V.

If the EIRP from a TV station is P watts, free space intensity, E, at a distance d in metres is given by:

$$E = \frac{\sqrt{30\,P}}{d} \text{ V/m} \tag{7.1}$$

In practice, the ground reflections modify this as discussed later in Sec. 8.5. Taking into account the effect of the ground, the net field strength at small angles of phase difference is given by:

$$E = \frac{120\,h_t\,h_r\,\sqrt{P}}{\lambda\,d^2} \text{ V/m} \tag{7.2}$$

where h_t, h_r are antenna heights, d is the distance and λ is the wavelength in metres.

The net field strength is thus inversely proportional to the wavelength and varies inversely as the square of the distance.

7.2 DESIGN PRINCIPLES OF TV TRANSMITTERS

In a TV transmitter, the amplitude modulation of the picture carrier by video signal can be carried out at a high level or a low level. In early transmitter designs, direct modulation was used. The picture carrier was directly modulated by the video signal. This can be done at a high level in the final power amplifier or at a low level RF driving amplifier. In modern transmitters, IF modulation at low level, is commonly used. The IF modulated transmitter has many advantages, including simplicity of design, ease of operation and superior performance.

High level modulation has the following disadvantages:

(1) The video modulator has to supply high video voltages of up to several hundred volts into highly capacitive loads taking amperes of currents at the video frequencies, and, therefore, employs the transmitting type of tubes.

(2) The grid modulated stage presents a varying load to its driver.

(3) The vestigial sideband characteristics are obtained here by a filter that must handle high power at the operating channel. In high level modulation, the linearity and high video driving power present design problems. The VSB filter, to filter out the lower side band partially, must handle the total transmitter output power.

Low Level Modulation

In this method, one or more stages of linear amplifiers follow the modulated stage. These stages must be designed for linear and wide band operation. The VSB filter can be located immediately after the modulated amplifier and hence can be operated at a much lower power. The linear amplifiers that follow the modulated stage have to be tuned with care and skill with the help of a special apparatus. Earlier power tubes used for linear amplification suffered from considerable nonlinearity, low gain and bandwidth. Improvements in vacuum tubes have made available a new family of high power ceramic tetrodes with high gain and bandwidth capability at VHF frequencies with excellent linearity. These improvements in RF power tubes and swept frequency techniques for tuning adjustments have made possible better linear amplifier designs. High power RF transmitters can now be designed with one or more RF linear amplifiers following the low level modulated stage.

7.3 IF MODULATION

This is a low level modulation in which the video signal modulation is carried out at an intermediate frequency, typically the standard 38.9 MHz video IF used in TV receivers. The modulated IF is then converted to the channel frequency by heterodyning it with a suitable oscillator frequency. This method has a number of advantages although the circuitry becomes more complex and the stages that follow must have a linearity of high order.

IF modulation has the following advantages:

(i) Since the modulation takes place at the IF at a low power level, the modulator section and the visual exciter can be built with solid state devices only, giving greater efficiency and reliability.

(ii) It is possible to introduce the VSB filter at a low power level just after the IF modulator and can be designed with lumped components, not only to shape the lower side band but also the upper side band response.

(iii) The subsequent stages following the VSB filter can be tuned permanently for wide band linear operation, simplifying tuning procedure and operation.

(iv) Group delay equalisation can be carried out at the IF under optimum conditions. Delay compensation requires only three sections whereas when done at video frequencies, seven or eight sections are required to compensate the same VSB filter.

(v) Visual excitors for all channels in the VHF or UHF bands are identical in IF modulation systems. This results in considerable economy.

In the IF modulation system, two highly stable oscillators are required, one at the IF carrier and the other at the sum of the radiated visual carrier and the IF. The heterodyne oscillator chain (including the multipliers) and the mixer have to be provided to translate the visual IF modulated signal to the radiated channel carrier frequency. Solid state linear amplifiers can be provided to raise the power from the mixer to a level sufficient to drive the final tube amplifier. At the VHF band frequencies, linear solid state amplifiers delivering up to 60 W of peak power are currently available.[16]

7.4 POWER OUTPUT STAGES

The power output of VHF transmitters generally employ triodes or tetrodes in grounded grid configuration, and quarter-wave mode re-entrant cavities. In the VHF range, the lumped components, because of their small size, cannot handle large RF currents and voltages while the dimensions of the normal cavity chambers are too large for practical designs. Transmission line circuits also present their peculiar problems of coupling, etc. Hence ceramic tetrodes with quarter-wave mode re-entrant cavities are conveniently used to overcome the problems.

Output stages of the UHF band transmitters employ tetrodes or klystrons. Tetrodes provide greater efficiency but their power gains are less and tube failure risk is greater. The life of a klystron is greater than the present day available triodes or tetrodes. Klystron provides a gain almost three to four times that of the tetrodes. This reduces the driving power requirement drastically to enable direct drive from semiconductor stages. Because of the lower efficiency of the klystron amplifier, the cooling system of the klystron transmitter must handle a larger dissipation. The power supply voltages required are also considerably higher.

Modern UHF transmitters use linear output amplifiers employing multi-cavity klystrons or travelling wave tubes that have a high power gain. The requirements of the modulator drive are just a few watts that can be readily obtained from solid state modulators. Although the modulator output power required is low and the RF drive required is also quite small, the incidental phase modulation introduced by solid state modulators may be of sufficient magnitude to require a pre-phase modulation of the RF drive signal to cancel the incidental phase modulation.

7.5 BLOCK DIAGRAMS OF TV TRANSMITTERS

A television transmitter consists of the visual transmitter for the picture and aural transmitter for sound. Outputs from both the transmitters are connected to a common transmission line feeder to the transmitting antenna through a combining unit called the diplexer.

In high level modulation, the vestigial sideband filter is also located at the output so that the filter and the diplexer can be integrated into a single unit called the filterplexer.

The block diagram of a typical 10 kW television transmitter with high level modulation is shown in Fig. 7.1.

The channel picture carrier is obtained from a crystal oscillator through a chain of multipliers and is raised to the required power level by the drivers and final power amplifier. The 10 kW final power

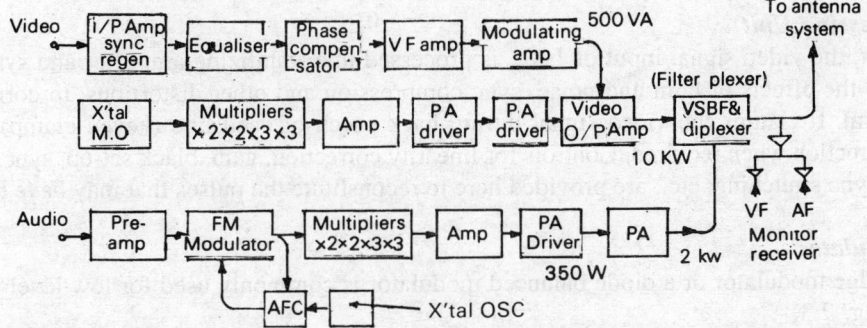

Fig. 7.1 Block diagram of a 10 kW VHF transmitter using high level modulation

amplifier is grid-modulated by the video signal from the modulating video amplifier that supplies the necessary 500 VA of video power. The incoming video signal is processed in the *visual exciter* and amplified to this level. In the *aural exciter*, the incoming audio signal frequency modulates a crystal-controlled oscillator which drives a chain of multipliers to raise the FM carrier frequency to the channel sound carrier required. The aural exciter drives the power amplifiers to produce the required 2 kW power which combines in the diplexer with the visual power after the latter has been passed through the VSB filter.

The block diagram of an IF modulated TV transmitter is shown in Fig. 7.2.

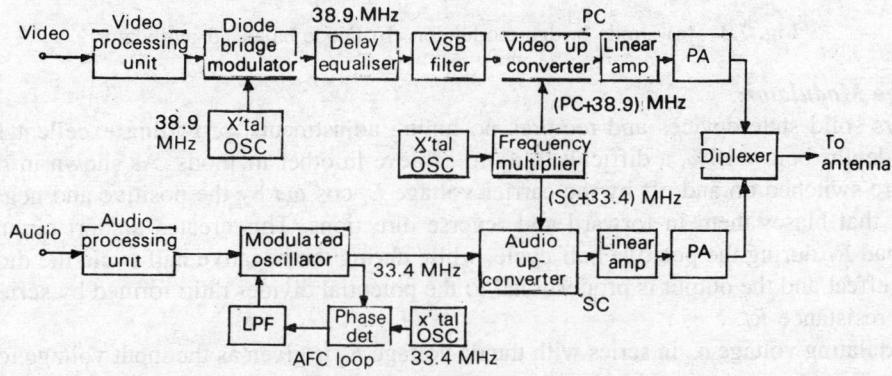

Fig. 7.2 Block diagram of IF modulated TV transmitter

As explained in Sec. 7.3, the modulation in this type of transmitter is carried out at the standard video IF and sound IF which are upconverted to the required picture carrier and sound carrier by heterodyning them with another crystal oscillator frequency. The VSB filter is included at the low IF level. The upconverted carriers are provided power amplification in linear amplifiers and combined in the diplexer unit.

Details of the important blocks of Fig. 7.2 are discussed in the subsequent sections.

7.6 VISUAL EXCITER

This consists of a video processing unit, a visual, modulator, VSB filter and phase compensator or delay equaliser and frequency converter.

Video Processing Unit

In this block, the video signal input of 1 V_{pp} is processed in a stabilizing amplifier and sync regenerator to minimize the effects of hum and noise, sync compression and other distortions, to correct it into the standard form. It clamps the video signal during back porch of each line (keyed clamp) and provides linearity correction when required. Controls for linearity correction, gain, black set-up, sync level, picture-sync ratio, sync stretching, etc., are provided here to reconstitute the pulses that may have been distorted.

Visual Modulator

A diode bridge modulator or a diode balanced modulator is commonly used for low level modulation as shown in Fig. 7.3.

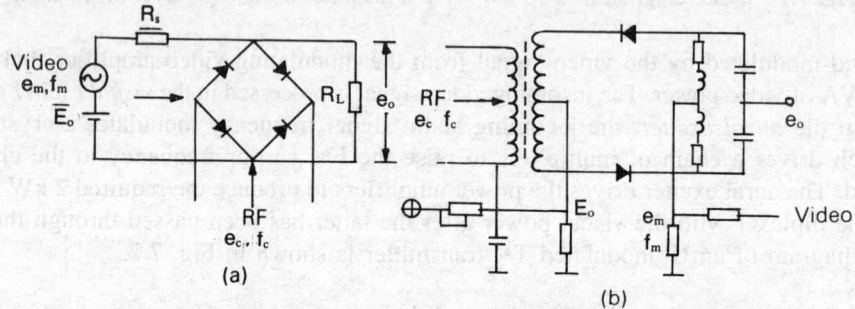

Fig. 7.3   (a) Diode bridge modulator, (b) Diode balanced modulator

Diode Bridge Modulator

This employs solid state devices and requires no tuning adjustments, providing excellent linearity for modulation depths below 10%, a difficult thing to achieve in other methods. As shown in the diagram, the diodes are switched on and off by the carrier voltage $E_c \cos \omega_c t$ by the positive and negative halves of its cycle, that biases them in forward and reverse directions. This creates a short circuit condition across the load R_L during the positive half cycle, while during the negative half cycle the diodes present no shunting effect and the output is proportional to the potential divider ratio formed by series resistance R_s and load resistance R_L.

If the modulating voltage e_m in series with the dc voltage E_0 is given as the input voltage to the circuit, the output is a square wave type with the chopped input signal. This contains the fundamental amplitude modulated carrier f_c and its harmonics. The instantaneous amplitude of the fundamental is given by:

$$e_0 = \frac{R_L}{R_s + R_L} (E_0 + E_m \sin \omega_m t) \cos \omega_c t$$

$$= A_c (1 + m \sin \omega_m t) \cos \omega_c t$$

where $A_c = \dfrac{E_0 R_L}{R_s + R_L}$ and $m = \dfrac{E_m}{E_0}$.

This indicates that carrier ω_c has been amplitude modulated by the modulating signal at ω_m. The undesired modulation products generated as harmonics are removed by filters that follow the modulator.

Schematic arrangement of a diode balanced modulator is shown in Fig. 7.3(b). Carrier frequency (f_c) at a proper level is fed in pushpull opposite phase to the diodes to switch them on and off, every half cycle. The modulating video frequency (f_m), however, fed in the same phase to the two diodes along with an adjustable dc voltage E_0. The output is the amplitude modulated carrier along with harmonics that are eliminated by filters. These modulators can work equally well as demodulators operable up to frequencies of several MHz.

VSB Filter and Phase Delay Compensator or Equaliser

The VSB filter is designed with lumped components, typically consisting of four sections of low pass ladder network attenuating frequencies beyond 1.25 MHz above the carrier by more than 20 dB. A second frequency conversion in the upconverter makes this attenuated sideband the lower sideband. The phase compensator is built-in with the VSB filter to compensate phase distortion. The maximum delay and the frequency of maximum delay of each of the all-pass sections can be varied continuously.

Frequency Conversion

A second crystal oscillator drive generates a frequency equal to the picture carrier f_0 plus the IF of 38.9 MHz. This is heterodyned with the VSB modulated IF carrier in the upconverter which also employs a diode bridge modulator. A band-pass filter at the output of the upconverter is tuned to f_0 so that only the VSB modulated picture carrier passes through, eliminating the oscillator drive and the sum frequencies and harmonics generated in the mixing process. In the mixing process, upper and lower sidebands are interchanged.

Power Amplifiers

The upconverted frequencies are amplified in linear amplifiers employing ceramic tetrodes in grounded grid configuration using quarter wave mode coaxial re-entrant cavities. The input circuit of the output power amplifier stage is typically single tuned with a very large bandwidth, while the output circuits is double tuned with a variable top capacity coupling to be able to adjust the valley of the double humped bandpass response.[16]

7.7 AURAL EXCITER

As shown in Fig. 7.2, the audio signal is given a pre-emphasis by a high pass network of time-constant of 50 μs, and is amplified in the audio processing unit. It is then fed to a crystal controlled oscillator-modulator for frequency modulation. The audio signal voltage is applied to a varactor diode in parallel with the tuned circuit of an LC oscillator that generates a centre frequency equal to the sound IF carrier. The junction capacitance of the varactor diode varies with the audio bias voltage applied to it, causing corresponding changes in the frequency of the oscillator to produce frequency modulation of the 33.4 MHz sound IF carrier.

The centre frequency of the LC oscillator is stabilized by an automatic frequency control (AFC) circuit that compares this frequency with a crystal drived stable frequency. For this purpose, another varactor diode is connected in parallel with the oscillator tuned circuit. The 33.4 MHz oscillator frequency and the crystal controlled 33.4 MHz frequency are compared in a phase detector that produces an output voltage proportional to the difference between the two frequencies. When the two frequencies are identical and 90° out of phase, the detector produces a zero output voltage. If the FM oscillator drifts to a higher frequency, the phase detector produces a proportional positive voltage, while if it drifts to a lower value, the phase detector produces a proportional negative voltage. The output of the phase detector is directly coupled to the second varactor diode through a low pass filter so that the consequent variation in the varactor diode capacitance alters the frequency to reduce the frequency error to zero.

The frequency modulated IF of 33.4 MHz is then fed to a mixer where it is heterodyned with a frequency equal to (PC + 38.9) MHz to obtain the difference frequency output at the desired sound carrier frequency of the channel. The mixer output is amplified in linear class A, class AB or class B stages that drive the final output stage. The final stage may employ ceramic tetrodes with quarter wave re-entrant cavities as narrow band class C amplifiers.[16]

7.8 DIPLEXER

For reasons of economy, it is a standard practice in television transmission to combine the outputs of both the visual and aural transmitters by means of a diplexer unit for feeding a common broadband transmitting antenna system. The block diagram of a commonly used constant impedance notch (CIN) diplexer unit is shown in Fig. 7.4.

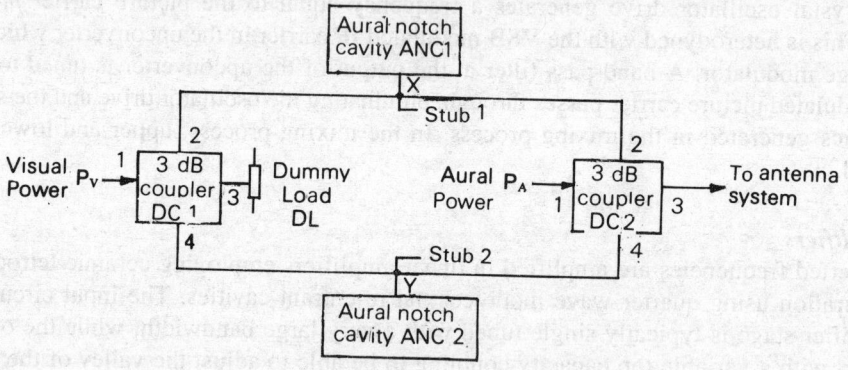

Fig. 7.4 Block diagram of CIN diplexer (*Courtesy*: BEL)

The CIN diplexer combines the visual and aural power to feed a common antenna feeder while maintaining sufficient isolation between the visual and aural inputs. It provides a constant input impedance at the input ports. The diplexer consists of two units of 3 dB directional coupler, two units of aural notch cavities suitably connected by coaxial transmission lines and a swamping dummy load-matched termination as shown in the diagram.

3 dB Couplers

The 3 dB coupler is a four-port device consisting of a pair of parallel plates approximately quarter wavelength long supported symmetrically in an outer conductor which is fitted with end caps. The width of the plates, the distance between them and the position of the end caps are precisely adjusted to achieve the desired properties of a 3 dB coupler, which are as follows:

1. When power P_V is fed to its port (1), half the power appears at port (2) and the other half appears at port (4) while no power appears at port (3). The power at port (4) lags the power at port (2) by 90°, as shown in Fig. 7.5.

2. Conversely, if equal powers at the same frequency are applied at ports (2) and (4), with a phase relation of 90° lagging at port (4) as shown, then the combined output appears at port (3) with no power output at port (1).

3. When port (3) is terminated with the characteristic impedance Z_0, and ports (2) and (4) by any other impedance (both of same magnitude and phase) the impedance at port (1) equals Z_0.

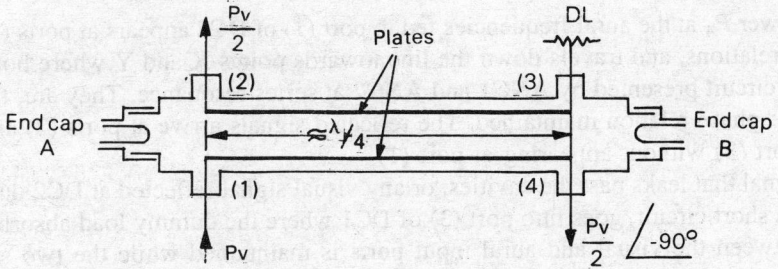

Fig. 7.5 Power flow in a 3 dB coupler (*Courtesy*: BEL)

4. When both ports (2) and (4) are terminated with Z_0, the impedance at port (1) equals Z_0, even if port (3) is terminated with any other impedance.

Aural Notch Cavities (ANC)

An aural notch cavity is a coaxial resonant cavity tuned precisely to present a short circuit at the aural carrier frequency. As shown in Fig. 7.6(a) the cavity consists of an inner conductor approximately quarter-wave long, an outer conductor, a coupling loop and a tuning plunger for fine frequency adjustments. The cavity has a very high Q and very low thermal coefficient because of the Invar material used for the inner conductor, which minimizes the effect of ambient temperature changes on the cavity tuning. In the case of Band III CIN diplexer, the cavities are cooled by forced air from a blower.

As shown in Fig. 7.6(b), a coaxial stub is connected across the notch cavity. The cavity is adjusted for series resonance at aural frequency f_A. The length of the short circuited stub is chosen such that it forms an anti-resonant (parallel resonance) at vision frequency f_v. Thus the cavity presents a very high impedance in the visual band and does not effect visual signals on the line, while the aural band of frequencies are presented a short circuit at the series resonance.

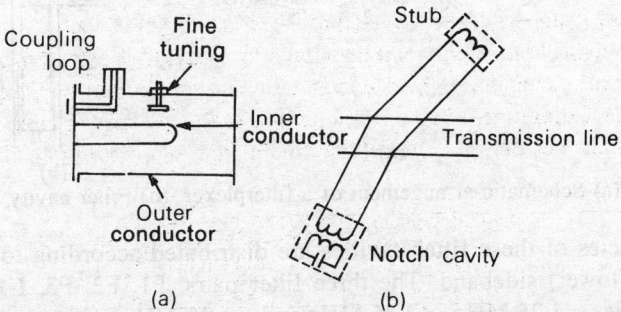

(a)   (b)

Fig. 7.6 Aural notch cavity: (a) Construction, (b) Cavity and stub equivalent circuit (*Courtesy*: BEL)

Working of the Diplexer As shown in Fig. 7.4, two 3 dB couplers DC 1 and DC2 are connected together using coaxial transmission lines of equal lengths. The two notch cavities ANC1 and ANC2 are connected in shunt to the two transmission lines at X and Y, some distance from the coupler DC2. Dummy load DL is connected to port (3) of DC1. Visual power P_V fed to port (1) of DC1 is split into ports (2) and (4), $P_V/2$ appearing at both the ports, with the power at (4) lagging the input power and the power at port (2) by 90°. The divided power travels along the two identical lengths of he line up to ports (2) and (4) of directional coupler DC2, unaffected by the notch filters at X and Y, which behave as parallel resonant circuits over the visual band. The two signals are combined at port (3) and practically no signal appears a port (1) of DC2.

Similarly, power P_A at the aural frequencies fed to port (1) of DC2 appears at ports (2) and (4) of DC2, with 90° phase relations, and travels down the line towards points X and Y where both the components encounter short circuit presented by ANC1 and ANC2 at series resonance. They are, therefore, reflected entirely with 90° phase relation maintained. The reflected signals arrive at ports (2) and (4) of DC2 and recombine at port (3) without appearing at port (1).

Any aural signal that leaks past the cavities, or any visual signal reflected at DC2 due to imperfections of matching and short circuit, goes into port (3) of DC1 where the dummy load absorbs the power. Thus the isolation between the visual and aural input ports is maintained while the two signals combine at output port (3) of DC2 feeding the antenna system.

A meter is provided on the diplexer which monitors the power flowing into the dummy load which is a measure of the performance of the CIN diplexer.

Filterplexer

In high level modulated transmitters, the VSB filter is located at the output so that the filter and the diplexer can be integrated into a single unit called the 'filterplexer'. The schematic arrangement shown in Fig. 7.7(a) consists, as in a diplexer, of two 3 dB couplers and aural notch cavities, with the addition of six more filter cavities serving as wavetraps for suppressing the unwanted part of the lower sideband.

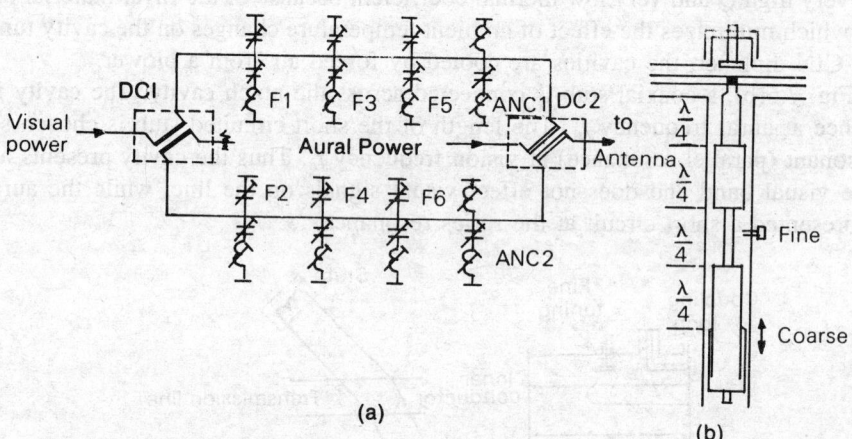

(a)

(b)

Fig. 7.7   (a) Schematic arrangement of a filterplexer, (b) Filter cavity (SIEMENS)

The resonant frequencies of these filter cavities are distributed according to the desired attenuation curve for the vestigial (lower) sideband. The three filter pairs: F1, F2, F3, F4; and F5, F6 are tuned successively to frequencies – 4.25 MHz, – 2.25 MHz and – 1.35 MHz, with respect to the picture carrier. The filters are arranged at equal intervals of approximately one quarter wavelength along the coaxial leads between the two directional couplers. They block the flow of power from the part of the lower sideband to be suppressed, by providing short circuit at the resonant frequencies, in the direction of the antenna, and thus reflecting them back to diplexer DC1 for the picture transmitter. The reflected components maintaining the quadrature phase relation enter ports (2) and (4) and appear together in phase at port (3) where they are absorbed by the dummy load, without any part going to port (1). The frequencies in the stopband of the signal are thus reflected but do not flow back to the picture transmitter. The filter, as a result, has a virtually constant input impedance throughout the large frequency range.

In this filterplexer for band III, the inner conductor of the filter cavity is composed of four sections each of one quarter wavelength, and having different diameters referred to the shorting slide, as shown in Fig. 7.7(b). It is simple to obtain the required slope with this arrangement, making use of the changes in image impedance caused.

7.9 TRANSMITTING ANTENNAS

Generally a common antenna system is used for the aural and visual transmitters. Hence the operating bandwidth of the antenna must be large and the voltage standing wave ratio (VSWR) must be less than 1.1 over the channel bandwidth of 7 MHz. The antenna system should have an omnidirectional pattern with concentration of the radiated power in the horizontal plane as required for a broadcast service.

Television transmitting antennas are basically dipoles arranged at the top of a suitable antenna tower. Wide bandwidth of the dipoles is obtained by employing large diameter elements rather elliptical in section for low wind resistance. Vertical stacking is used to increase directivity in the vertical plane. Circular pattern in the horizontal plane is obtained by employing a turnstile, i.e., crossed dipole arrangement, or by mounting stacked dipole panels with reflectors on the four sides at the top of the antenna tower.

Turnstile Antenna In this antenna, two crossed dipoles are used in a turnstile arrangement as shown in Fig. 7.8(a), and are fed in quadrature, i.e., the currents fed to them are 90° out of phase by means of an extra $\lambda/4$ length in the feeder of one. Each dipole has a figure-of-eight pattern in the horizontal plane, but crossed with each other.

The resultant energy of the two dipoles is a vector sum of the two fields from each dipole, 90° apart, and gives a constant vector sum in all directions. As shown in Fig. 7.8(b), the turnstiles are stacked one above the other for vertical directivity.

Superturnstile or batwing type antenna is composed of a number of crossed dipoles that are modified to give a broadband characteristic. The dipole elements form current sheets, with currents supplied in quadrature as shown in Fig. 7.8(c).

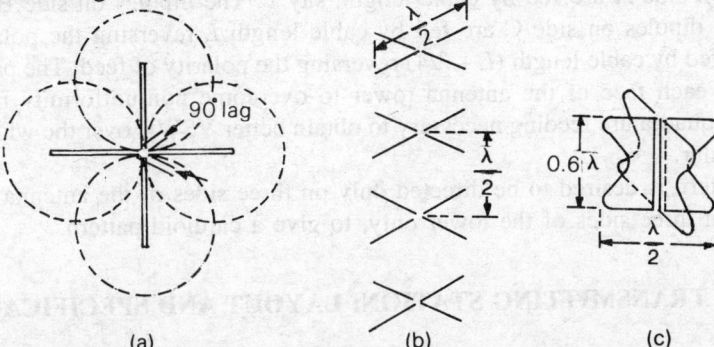

(a)   (b)   (c)

Fig. 7.8 Turnstile antennas: (a) Turnstile arrangement, (b) Stacked turnstile array, (c) Superturnstile

Dipole Panel Antenna System

The antenna system for band I or band III transmitters very commonly consists of dipole panel antennas mounted on the four sides at the top of the antenna tower as indicated in Fig. 7.9(a). Each panel consists of an array of full wave dipoles mounted in front of reflectors. The array is formed by four or eight dipoles that have elliptical flat cross-section to reduce wind resistance and losses as shown in Fig. 7.9(b).

The steel portions of the antenna are hot-dip galvanised and further coated with rust-preventing paint to improve weathering resistance.

The transmitter output coming through the main feeder is split in a power divider and applied to individual branching feeders. Styroflex cables with a spiral spacer insulation of styroflex are used for low

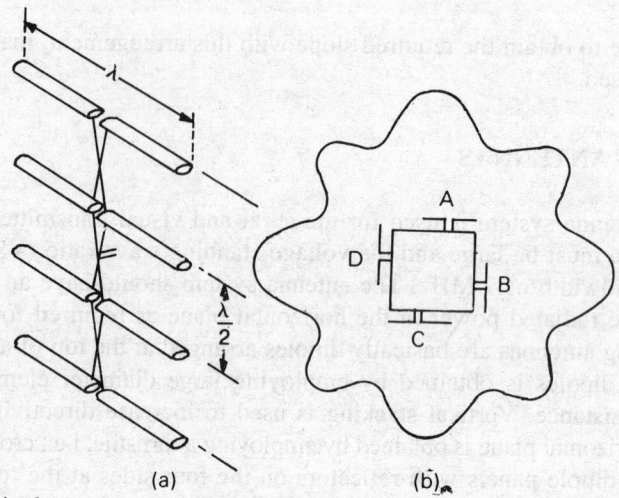

Fig. 7.9　(a) Dipole antenna panel, (b) Polar pattern of four tower mounted antenna panels

losses. Each branching feeder is connected to the balun at the centre of the respective panel. The balun transforms the unbalanced coaxial cable impedance to the balanced feeder line that supplies power to each set of the dipoles, so that the dipole sets secure equiphase feeds with equal amplitudes. The wide band frequency characteristics of the antenna are obtained by utilization of the mutual impedance existing between two adjoining elements. Directivity in the vertical plane depends upon the stacking of the dipoles vertically.

For obtaining omnidirectional pattern with the help of four panels mounted on the four sides A, B. C and D of the antenna tower as shown, the feed of each side is arranged to lag behind the previous side by 90°. The dipoles on side A are fed by cable length, say L. The dipoles on side B are fed by cable length $(L + \lambda/4)$. The dipoles on side C are fed by cable length L reversing the polarity of feed. The dipoles on side D are fed by cable length $(L + \lambda/4)$ reversing the polarity of feed. The panels are mounted slightly off-centre on each face of the antenna tower to overcome non-uniformity in omnidirectional pattern resulting from quadrature feeding necessary to obtain better VSWR over the wide band, and make the pattern more circular.

If the radiation pattern is desired to be directed only on three sides of the antenna tower, the dipole panels are mounted on three sides of the tower only, to give a cardioid pattern.

7.10　TELEVISION TRANSMITTING STATION: LAYOUT AND SPECIFICATIONS

Selection of site for a TV transmitting station, catering to a particular service area, is made on the basis of a number of factors that include clear line-of-sight and required path clearance from obstacles in the service area, propagation phenomena in the region as affected by meteorological and topographical factor, man-made interference, geological factors, etc. Besides these, ancillary facilities like availability of water, reliable power supply, all-weather road communication, telephone lines and security aspects have to be taken into consideration.

To ensure good radio line-of-sight and allow mean free space propagation conditions, the site is chosen such that the first Fresnel zone clearance is obtained over all the obstacles in the service area of the station. As discussed later, in Sec. 7.11, the first Fresnel zone is the zone bounded by points for which the transmission paths from the transmitting antenna and the receiving antenna is greater by one half wavelength. Paths over large bodies of water surface are, as far as possible, avoided as reflections and bending of propagating waves, due to increase in water vapour content in the atmosphere create problems in reception. It is desirable to locate the TV station away from industrial installations, heavy traffic with automobile spark ignition noise and high voltage power lines generating corona arc discharge over its jumpers and resulting RF interference noise. Geologically, the actual site should be on a firm and stable ground foundation, free from seismic activity or floods. Proper mains and RF earthing has to be planned, taking into account the electrical conductivity of the soil. A standby power supply from a diesel-generator set is usually provided for, to take over the operation in the event of power supply failure.

A typical layout of a TV transmitting station is shown in Fig. 7.10.

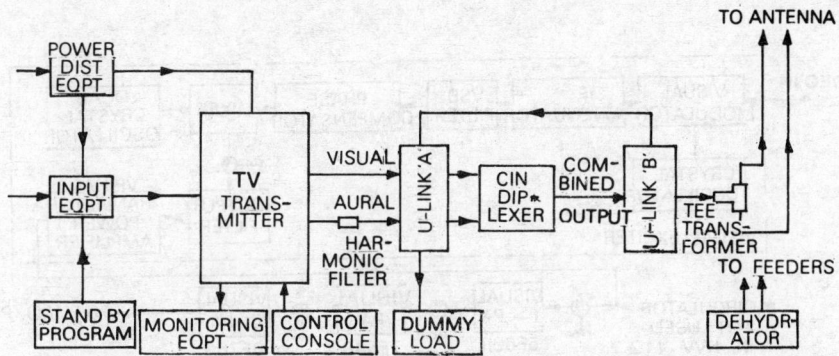

Fig. 7.10 Typical layout of a TV station (*Courtesy*: BEL)

The station consists of the following:

1. Visual and aural transmitters.
2. Input and monitoring equipments for video and audio signals.
3. Control console accommodating the picture and waveform monitors, the program input and monitoring controls.
4. Indoor coaxial equipment consisting of:
   (a) Harmonic filter
   (b) U-link panel
   (c) Dummy load
   (d) CIN diplexer
5. Antenna system with a feeder cable and dehydrator.
6. Automatic voltage regulator (AVR) with power distribution panel.

The U-link panel routes the outputs of the vision and sound transmitters either to the diplexer or the dummy load, and the output of the diplexer to the antenna or alternatively to the dummy load. The antenna system employs dipole panels which are fed from a junction box through branch feeder cables. The feeder line consists of an air dielectric (with styroflex spacers) cable with a dehydrator to maintain dry pressurized air inside the cable. A $1\frac{5}{8}$ in cable is used for a low power (1 kW) transmitter, while a larger size, $3\frac{1}{8}$ in cable is used for a higher power (10 kW) transmitter.

The power distribution equipment incorporates an automatic voltage regulator to overcome mains supply fluctuations and power distribution cubicle.

Standby program facilities, consisting of caption cabinet (CCTV camera) and video tape playback console (VTR) are provided.

Figure 7.11 is a photograph of a typical 10 kW TV transmitter (BEL HVV 11 and 112) for bands III and I. The left side compartment houses the exciter and power amplifiers. The right side compartment houses the visual final stage power amplifier. The plate voltage transformers for aural and visual power amplifiers, silicon rectifiers and blowers are installed outside the transmitter cabinet.

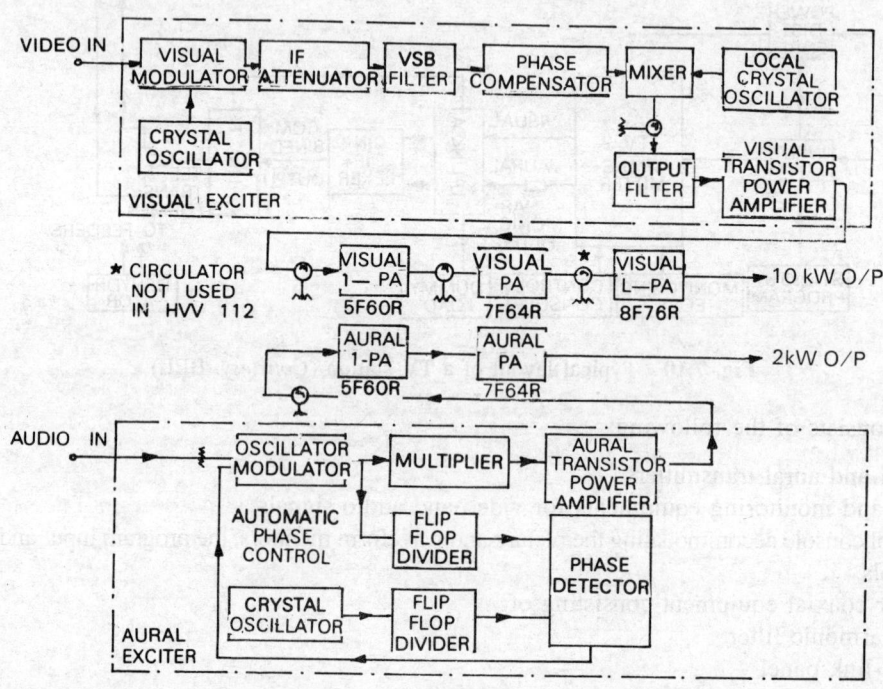

Fig. 7.11 10 kW VHF TV transmitter (*Courtesy*: BEL)

The block diagram of the 10 kW band I/III transmitter is given in Fig. 7.12.
The salient features of the transmitter are as follows:

(i) Solid state devices are used in the visual modultor and exciter as well as in the aural modulator and exciter. Ceramic tetrode tubes are used in the final stages of the transmitter.

(ii) The video carrier is modulated by a diode balanced modulator at an intermediate frequency. The low level modulation minimizes nonlinear distortion.

(iii) The vestigial sideband characteristics are obtained by a built-in VSB filter in the exciter. The VSB filter is of miniaturized lumped circuit type, requiring no periodic maintenance and characteristic checks.

(iv) The VSB filter has a built-in RF phase compensator which fully compensates phase distortion in the VSB filter.

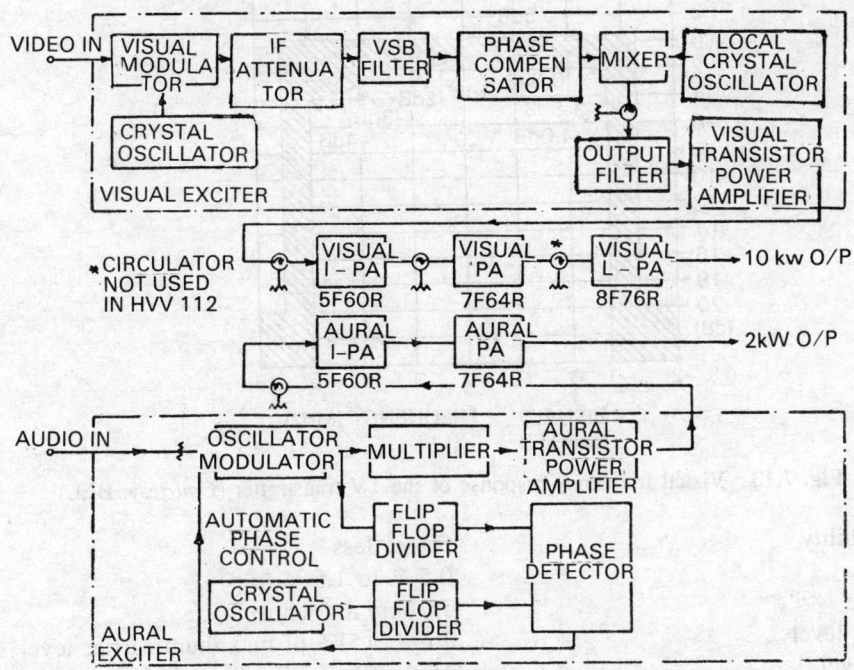

Fig. 7.12 Block diagram of a 10 kW band I/III IF modulated TV transmitter (*Courtesy*: BEL)

(v) The visual and aural excitation outputs from the exciters are amplified by the visual and aural power amplifiers in the succeeding stages. To enable impedance matching between cascade-connected stages of each power amplifier, circulators are used between stages of the visual power amplifier. Therefore, tuning can be carried out easily.

(vi) Same type of vacuum tubes are used for both the visual and aural transmitters, thereby reducing the cost of spare tubes. High efficiency silicon rectifiers for plate supplies reduce operational costs.

Typical technical specifications are as follows:

Visual Transmitter

Type of emission	A5C
Output power	10 kW (sync peak)
	6 kW (mean power in black picture)
Frequency range	Any one CCIR TV channel in Band I or band III
Frequency stability	± 500 Hz or less
RF output connection	$3\frac{1}{8}$ in (77 mm) 50 ohm feeder
Input polarity	Sync negative
Frequency response	As shown in Fig. 7.13
Differential gain	± 5% or less with pre-correction
Differential phase	± 5° or less with pre-correction
Output impedance	50 ohm
Visual blanking level stability	2% or less

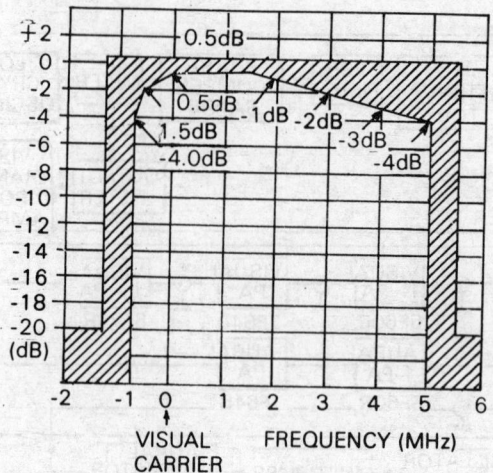

Fig. 7.13 Visual frequency response of the TV transmitter (*Courtesy*: BEL)

Peak power stability	5% or less
Input level	0.5 V to 1.5 V (p-p)
Input impedance	75 ohm ± 5 ohm
Carrier pedestal level	75 (± 1.5)% of maximum carrier level
Carrier white level	15 (+ 10, − 5)% of maximura carrier level
Incidental phase modulation	Better than − 46 dB (50 Hz to 15 kHz)
Noise	Better than − 50 dB
Envelope delay	± 50 ns up to 4.5 MHz
	± 100 ns from 4.5 MHz to 4.8 MHz
Spurious radiation	− 60 dB or less
Wave form distortion	Distortion of sine square wave ($2T$) and that of window wave: K factor within 2%

Aural transmitter

Type of emission	F3
RF output	2 kW
Frequency range	Any one CCIR TV channel in Band I or Band III
Frequency stability	Better than ± 500 Hz
RF output impedance	50 ohm
Output connection	$1\frac{5}{8}$ in coxial feeder
Frequency deviation	± 50 kHz at 100% modulation (modulation capability ± 80 kHz)
Frequency response	Within ± 1 dB at 50 to 15,000 Hz
Input level	+ 16 dBm for 100% modulation (frequency deviation ± 50 kHz)
Input impedance	600 ohm
AM noise level	− 50 dB or less at 100% modulation (50 to 15,000 Hz)
FM noise level	− 60 dB or less at 100% modulation (50 to 15,000 Hz and 50 μs pre-emphasis)

Frequency pre-emphasis	50 μs, 400 Hz standard
Distortion factor	1.5% or less at 50 to 15,000 Hz
	1% or less at 100 to 7,500 Hz

Power requirements

Power supply	415 V ± 10% 3 phase, 3-wire, 50 Hz
Power consumption	50 KVA (max.)

Environmental conditions

Ambient temperature	– 10 to + 45 °C
Relative humidity	95% max.
Maximum altitude	6000 ft

7.11   MICROWAVE TV RELAY SYSTEMS

Microwave communication link is necessary when relaying live programs from places remote from the studio and transmitter, e.g., outside broadcast (O.B.) pick-ups from sports stadium, special functions, etc. They are also employed as studio-to-transmitter links (STL) when the studio is not very close to the transmitter installation. The systems commonly operate in the 2 GHz, 7 GHz and 13 GHz bands. The 7 GHz band is most common for remote pick-ups and STLs.

At microwave frequencies, it is possible to obtain very high power gain by concentrating the radiated power from the antenna into a narrow beam by employing parabolic reflectors for dipole antennas. The power gain (G) of an antenna using a parabolic reflector of an effective area A_e is given by

$$G = \frac{4\pi A_e}{\lambda^2} \qquad (7.3)$$

where λ is the wavelength of radiation, and

$$A_e = \rho_a A_p \qquad (7.4)$$

where A_p is the projected area of the reflector, and ρ_a is the aperture efficiency.

Aperture efficiency reduces the effective area typically by a factor of 0.66 due to non-uniform aperture distribution and blockage due to feed, etc.

For the 7 GHz band, a 120 cm dish typically gives a gain of around 5000 (= 37 dB) and a narrow beam of half power beam width angle of around 3°.

On a smooth surface, the maximum line-of-sight distance between the transmitter and receiver antennas is given by the relation

$$d = 1.22 \left(\sqrt{h_t} + \sqrt{h_r} \right) \text{ in miles, if } h_t \text{ and } h_r \text{ are in feet}$$

and
$$d = 3.7 \left(\sqrt{h_t} + \sqrt{h_r} \right) \text{ in km, if } h_t \text{ and } h_r \text{ are in metres (vide Sec. 8.4)}$$

In addition to the line-of-sight path between the transmitting T_α and receiving R_α parabolic antennas, the radiated energy takes many curved paths depending upon the terrain and the atmospheric conditions as shown in Fig. 7.14.

These are due to upward and downward refraction of the propagating energy. The energy arriving at the receiver over the various paths is categorized into Fresnel zones which are numbered depending upon the phase relationship of the signals arriving at the receiver. The first Fresnel zone is the zone with the smallest radius within which the phase of the received energy has phase difference angle within 180° of

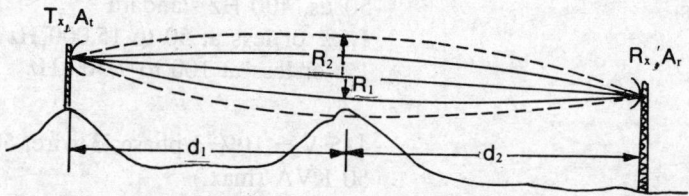

Fig. 7.14 Fresnel zone clearance

the line-of-sight path. A major portion of the energy is received *via* the first Fresnel zone. Energy in the second and other even-numbered Fresnel zones is 180° phased with respect to the energy in the first and odd-numbered zones and hence tends to cancel the latter. It is, therefore, desirable to obstruct the energy from all the zones except the first. However, a clearance of greater than 0.6 times the radius of the first zone is provided for the first Fresnel zone. The radius of the first Fresnel zone is given by the relation:

$$R_1 = \left(\frac{\lambda d_1 d_2}{d} \right)^{1/2} \tag{7.5}$$

where d_1, d_2 are the distances of the obstacle from the transmitter and receiver, respectively and d is the path length, all quantities in the same units.

Performance Factors

For a microwave TV link, the signal-to-noise ratio should be greater than 24 dB if the noise is not to be objectionable. Adequate 'fade margin' to overcome fading due to tropospheric bending and scattering must be provided in addition to the free space path attenuation given by:

$$\text{Attenuation} = 10 \log (P_t/P_\tau)$$

where P_t is the power radiated from the transmitting antenna and P_r, the power available at the output terminals of the receiving antenna.

If A_r is the effective area of the receiving antenna, A_t, the effective area of the transmitting antenna, and d, the distance between antennas and λ the wavelength, we have

$$P_r = A_r P_0, \text{ where } P_0 \text{ is the power flow per unit area}$$
$$= A_r P_t G_t /4\pi d^2$$

Substituting for the antenna gain $G_t = \dfrac{4\pi A_t}{\lambda^2}$,

$$P_r/P_t = A_r A_t /d^2 \lambda^2 = \frac{\lambda^2 G_t G_r}{(4\pi)^2 d^2} \tag{7.6}$$

Path attenuation α between isotropic antennas is, therefore, given by

$$\alpha = P_t/P_r = \left(\frac{4\pi d}{\lambda} \right)^2 = \left(\frac{4\pi d f}{0.3} \right)^2 = \left(\frac{4\pi}{0.3} \right)^2 d^2 f^2$$

where f is in MHz, and d in km.

The path attenuation α in dB is then given by:

$$\alpha = 32.45 + 20 \log f + 20 \log d$$

$$\alpha \approx 33 + 20 \log f + 20 \log d \qquad (7.7)$$

Fade margin for a microwave link depends upon the distance and the reliability objective. For transmission path lengths of around 50 km, it is good engineering practice to provide for possible increase in the signal strength of + 10 dB with respect to free space propagation. A fade margin is allowed depending on the degree of reliability, for example, 10 dB for 90%, 20 dB for 99%, 30 dB for 99.9%, and so on.

For establishing a microwave link for mobile O.B. relay or STL, the path profile should be charted from the topographical data of the region to determine the major obstructions, and then the minimum clearance over the major obstruction should be determined. The heights required for the transmitting and receiving dishes will depend upon the height of the obstruction, the bulge of earth and the difference between the levels of the two points. When the receiving dish can be mounted on the high transmitting tower, the sending dish can be considerably lower depending upon the distance of the obstruction.

When direct line-of-sight path between the sending and receiving points is not available, a common point that affords a line-of-sight path between the two points is found out. A flat reflector of good efficiency and surface of 2 to 3 m^2 area can be used with proper alignment to establish the link. Such a reflecting surface may be formed by a tall steel construction building or an overhead tank, etc.

Block diagram of a typical microwave relay system is shown in Fig. 7.15.

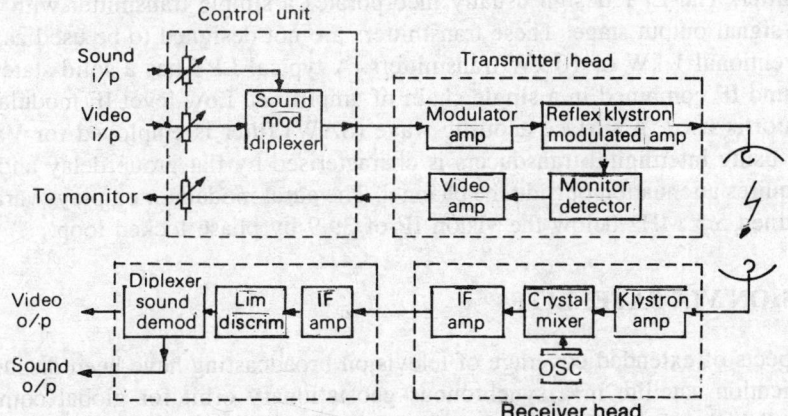

Fig. 7.15 Block diagram of a microwave TV relay system

A microwave relay system employs a microwave tube like a klystron or microwave diodes at the transmitter and receiver heads, with waveguide and button hook couplings to the parabolic dishes. The klystron oscillator at the transmitter is frequency modulated by varying the reflector voltage with the video signal, along with the sound signal. At the receiving head, the microwave signal received is amplified in a klystron amplifier converted to IF in a diode mixer, amplified and demodulated in a limiter discriminator, to get back the modulating video and audio signals.

Terrestrial TV Relay Stations

For interconnection between TV stations, microwave links are commonly employed by terrestrial relay stations spaced about 50–70 km from each other depending upon the terrain conditions. The very high gain parabolic antennas, have high EIRP between them to the extent that transmitted powers of the order of fraction of a watt is more than sufficient. The system may be broadband to handle a number of TV channels. In a relay station, one receiving and one transmitting antenna faces one direction and another similar pair faces the other side. The receiving antenna feeds the weak microwave signal into an amplifier that compensates for the transmission path loss. The amplified output is fed to the transmitting antenna towards the next relay station.

·*Low Power TV Stations (LPT)*

Rebroadcasting at low power is convenient for providing a limited service area over a range of 10–15 km. This also helps reduce interference elsewhere due to use of high power. Translater stations are used to convert the channel frequency of one station to another suitable channel. The program is then rebroadcast at low power in ranges of 10 W, 100 W or 1 kW.

In order to accommodate a larger number of channel assignments, low power TV transmitter stations are being allowed to transmit to provide for reception in small local areas. Subject to transmitter power restriction of 100 W to 1000 W depending on the frequencies and the terrain served, the LPTs can operate on any of the available TV channels. Care has to be taken in choosing the LPT channel such that it does not interfere as image, IF beat, etc. with the regular full-service stations in the area. With the variety of program sources like low cost local studios with professional VCRs, satellite TV, microwave links, cable TV, etc. LPTs are increasingly employed for programme diffusion.

The LPTs (typically 100 W) are simpler in design as compared to the conventional HPTs (1 kW or 10 kW typically), whch have a separate audio transmitter (1 kW), the output of which is diplexed with the video transmitter. The LPT design usually incorporates a simple transmitter with audio as well as video composite signal output stage. These transmitters are not designed to be used as preamplifiers or exciters for conventional 1 kW or 10 kW transmitters. A typical LPT has a solid state exciter with the vision IF and sound IF combined in a single chain of amplifiers. Low level IF modulation with diode-balanced modulator is used. A surface acoustic wave (SAW) filter is employed for VSB shaping. The VSB SAW filter using interdigital transducers is characterised by flat group delay and response in the pass band and requires attenuation outside the channel. The aural modulator employs varactor modulation at the IF, maintained 5.5 MHz below the vision IF of 39.9 by phase locked loop.

7.12   TELEVISION VIA SATELLITE

Some general aspects of extended coverage of television broadcasting have been discussed in Sec. 1.3. Use of communication satellite in a synchronous geostationary orbit for global communication was conceived by the British science fiction writer Arthur Clarke, in 1945. While the launching of the first satellite *Sputnik* by USSR in 1957 heralded the era of satellite communication, the first active communication satellite *Telstar I* was launched in 1962, when TV relay transmissions across the Atlantic were realized between Europe and America, for short durations as the satellite was altitude elliptical orbit visible for communication for hardly 20 minutes.

Satellites orbiting at a height of about 36,000 km from the earth at an orbital speed of about 11,000 km can act as *geostationary satellites*, when the centrifugal force acting on the satellite just balances the gravitational pull of the earth. The orbit has to be equatorial to prevent apparent North-South oscillation due to the tilt in the axis of the earth with respect to the solar orbit of the earth, and small drifts in its position and altitude have to be corrected by small on-board jet motors. The *Syncom* satellites launched in 1963–64 by the USA were the first to prove the viability of geostationary satellites for world communication.

The success of Syncom satellites led to the formation of the *Intelsat organization* with over a hundred member nations investing in the Intelsat satellites system, in proportion to their use of the satellite time. The Insat managed by the American Comsat Corporation has launched a series of Intelsat satellites with increased size and facilities, providing international and domestic lease services over various regions in the world. Currently there are number of Intelsat IV and V satellites operating over the Atlantic, Pacific and the Indian ocean regions to serve 160 countries, through over 700 earth stations to provide international telephone and television services.

Modern communication satellites have powerful transponders that can beam concentrated signals to desired areas and act as direct broadcasting satellites (DBS). Many regional satellites like *Eutelsat* in Europe, *Intersputnik* of USSR, or *Aussat* of Australia, *Insat* in India have been introduced. Major breakthroughs in the sensitive receiving antennas and low noise front-end converters have enabled direct reception of international television channels in homes, the world over.

Frequency Allocations

Frequencies in the upper UHF band and the lower part of the SHF band are least affected by the atmosphere. At lower frequencies, the ionospheric absorption and scintillation degrade the performance of the system. At higher frequencies the attenuation due to moisture, rain or fog becomes significant and Faraday rotation introduces fading. This is avoided by circular polarization of the radiated signals.

Most of the earlier satellites to date, have been using a 2.5/4 GHz frequency for the down-link the first DBS satellite ATS-6 used 860 MHz suitable for low cost receiving equipment in the 70s) and a 6 GHz frequency for up-links. Satellite channels allotted as per CCIR recommendations are Ku band 12–17 GHz, 17.3 to 17.8 GHz for uplinks and 12.2 to 12.7 GHz, for down link. Higher microwave frequency bands, 22.5–23 GHz, 41–43 GHz and 84–86 GHz are also reserved for future worldwide satellite broadcasting systems.

Modulation

The frequency modulation technique requires a larger bandwidth but offers good noise immunity from interference, and requires the least amount of power. The receiving system is also simple and inexpensive. Since the technology of FM transmitters has been fully developed frequency modulation is commonly used although digital modulation can provide greater immunity from interference. Digital techniques have also been developed and are slated for use in the new direct broadcasting satellites.

In a communication link, the carrier to noise C/N ratio determines its performance. In DBS, this depends on the EIRP, rain attenuation and the figure of merit G/T of the earth receiving terminal, where G is the antenna gain and T the total noise temperature (vide Sec. 10.2). A better figure of merit implies increase in EIRP, reduced service coverage area by increase in transmitting antenna gain, and higher performance of the low noise front-end for the TV receiver.

The SITE Experiment

The Government of India, in cooperation with NASA of USA, conducted a Satellite Instructional Television Experiment (SITE) in 1975–76, using the higher power satellite ATS-6 (Application Technology Satellite), positioned at a height of 36,000 km, in a geostationary synchronous orbit with sub-satellite longitude of 33° East. The experiment for the first time demonstrated the potential use of satellite television in providing practical instructions to villagers in remote areas, and in stimulating national development in India. It had important managerial, economic, technological and social implications. The experiment provided experience in development, testing and management of a satellite-based instructional television system, particularly in rural areas, to determine optimum system parameters. Technical details of the system are briefly indicated here.

ATS-6 Satellite Transponder   TV programs from the earth station at Ahmedabad were transmitted to ATS-6, at 6 GHz FM carrier with the help of a 14 m parabolic dish antenna. The FM carrier had a bandwidth of 40 MHz. The satellite employed a C-band horn for the receiving antenna, having a peak antenna gain of 16.6 dB. The signal was transmitted back to earth at 860 MHz by an FM transmitter providing 80 W to a 9 m parabolic antenna that radiated EIRP of 51 dBW peak, with the peak antenna gain of 33 dB, 3 dB beam width of 2.8° and right circular polarization (RCP) by helical feed. The transmitter bandwidth was 40 MHz.

The transmitted signal consisted of a video band of 5 MHz and two audio signals frequency modulated on two audio subcarriers at 5.5 MHz and 6 MHz. The audio signals are thus frequency division multiplexed with the video signal forming a composite base band signal which is then frequency modulated on the 860 MHz carrier. Peak deviation of the carrier by the video signal was up to 6 MHz while the peak deviation of the carrier by the audio subcarriers was up to 1 MHz. The modulation for video is thus FM, while for the audio it is FM/FM.

Direct Reception System (DRS) The receiving terminal for direct reception of the broadcast signals consisted of a 3 m parabolic antenna made out of chicken mesh, and a front end converter. The pick-off of the RCP 860 MHz signals was with the help of a helical antenna at the focal point of the antenna, directed towards the satellite. (Received carrier power at the ground was around – 146 dBW.) The front end converter consisted of two parts. One part consisting of the RF amplifier and converter was mounted at the antenna itself to minimize the signal loss over the antenna feeder. This antenna-mounted unit called head-end, provided the 860 MHz signal a low noise amplification of about 10 dB at a noise figure of 5 dB. The transistorized amplifier used microstrip line distributed network elements for matching. The 860 MHz signal was converted into 70 MHz IF signal by mixing it with a crystal controlled local oscillator that generated 790 MHz signal through a chain of multipliers from a 98.75 MHz crystal oscillator. High stability of the local oscillator was essential because of the frequency modulated signal. A schematic of the direct reception system under the SITE program is shown in Fig. 7.16.

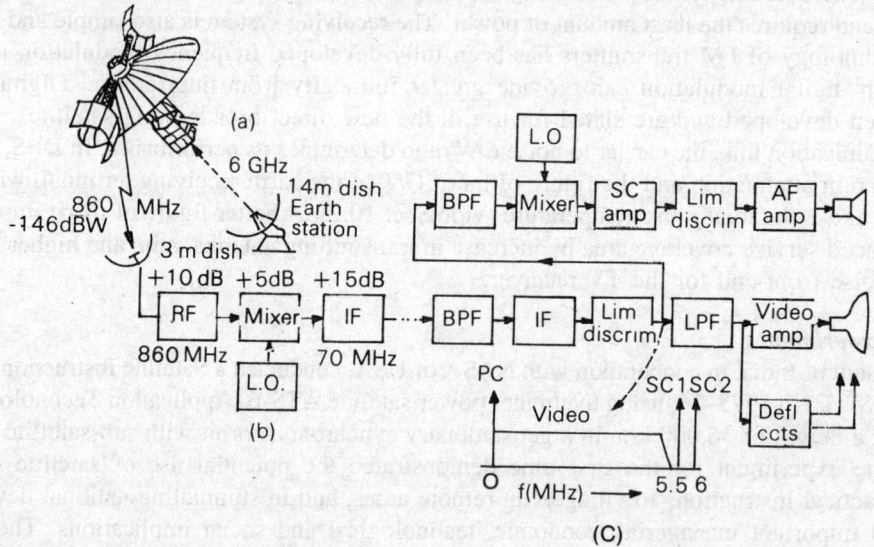

Fig. 7.16 Direct reception system for SITE program: (a) ATS-6 satellite, (b) Front end converter and receiver block diagram, (c) Base band frequencies

The 70 MHz IF signal was provided with a pre-amplification of 15 dB before supplying it to the indoor receiver unit *via* antenna feeder. The indoor or tail-end unit consisted of a 50 dB gain IF amplifier at a bandwidth of 30 MHz followed by a limiter-discriminator that separated the composite base band.

The video signal was separated from this by a low pass filter. The audio subcarriers at 5.5 MHz and 6 MHz are passed through a bandpass filter, mixed with the subcarrier-select local oscillator to obtain the wanted audio subcarrier only. The audio IF output is amplitude limited and demodulated in a discriminator to derive the audio signal. The video signal is fed to the display monitor unit while the audio is supplied to the audio amplifier and loudspeaker.

Limited Rebroadcast Systems

In rural areas where clusters of villages and towns exist, and costly direct reception sets may be not affordable, low power terrestrial transmitters (LPTs) can be used to cover the limited areas. The power of the transmitter can vary from 100 to 1000 W, depending on the area to be served by the rediffussion system. For band III, a field strength of 55 dBu (the minimum specified by CCIR) may be aimed at for the remotest receiver. A 3.5 m or 6.2 m television receive-only (TVRO) terminal giving a noise figure better than 4 dB is used to *receive and relay* the television program through typically 1 kW VHF/UHF TV transmitters (LPTs). Details of the TVRO terminal manufactured by Bharat Electronics Limited for S-band (2.5-3 GHz) reception, are given in the following.

TV Receive only Terminal   The TV receive-only terminal consists of a *satellite relay receiver* (SRR) along with a suitable parabolic antenna for reception of satellite signals. The terminal is capable of processing the received signals and broadcast quality colour video and audio signals for rediffusion. The TVRO typically consists of an adjustable parabolic dish antenna, a front end converter (FEC) and a tail end receiver (TEC).

Antenna   The antenna consists of a 6.2 m parabolic aluminium alloy reflector with back structure and is segmented into eight parts for easy transportation and assembly at site. A helical feed at the prime focus provides for reception of left hand circular (LHC) polarized waves with an axial ratio better than 3 dB. The rough pointing of the antenna in the azimuth is taken care of by suitable orientation of the concrete foundation at site, while fine adjustments in both axes are made possible on the antenna system by suitable hardware and mounting arrangement. The antenna system provides a gain of 41.8 dB with a maximum VSWR of 1.25.

Front End Converter   The front end converter (FEC) is a weatherproof unit mounted on the back structure of the antenna, connected to the antenna feed through a low loss cable. The block diagram of the FEC is shown in Fig. 7.17(a).

The FEC amplifies the 2.575 GHz/2.615 GHz RF signals in a low noise amplifier (LNA), filters the carriers in the respective channel filters, and down converts the RF signal to 70 MHz by mixing with local oscillator signals of LO-1 or LO-2. It also amplifies the 79 MHz IF signal to a level of approximately – 5 dBm for further processing in two separate channels of the TER through shielded coaxial cables.

Tail End Receiver   The Tail End Receiver (TER) is housed in the broadcasting station along with other studio equipment. As shown in the block diagram of Fig. 7.17(b), the TEC receives from the FEC the IF signal at 70 MHz, which is amplified, filtered and amplitude stabilized in the IF AMP/FIL/AGC module. The AGC level is used for monitoring the RF carrier level on CARRIER/VOLTMETER on the front panel of the TER. The output of the module is passed through a series of limiters and Delay Equalisers in the LIM/D.Eq. module.

The amplitude limited 70 MHz signal from the TER is processed by the DISC/BB AMP module to demodulate the FM IF and provide the base band signal. This contains the video as well as the 5.5 MHz FM audio subcarrier. This signal is then fed to both LPF/VID AMP and the BPF DEMOD modules.

The low pass filter of LPF/VID AMP module separates the video signal from the frequency modulated audio, and is further amplified and clamped to obtain a colour composite video signal (CCVS) with a standard 1 Vpp video output. The BPF/SOUND DEMOD module filters a 5.5 MHz sound subcarrier

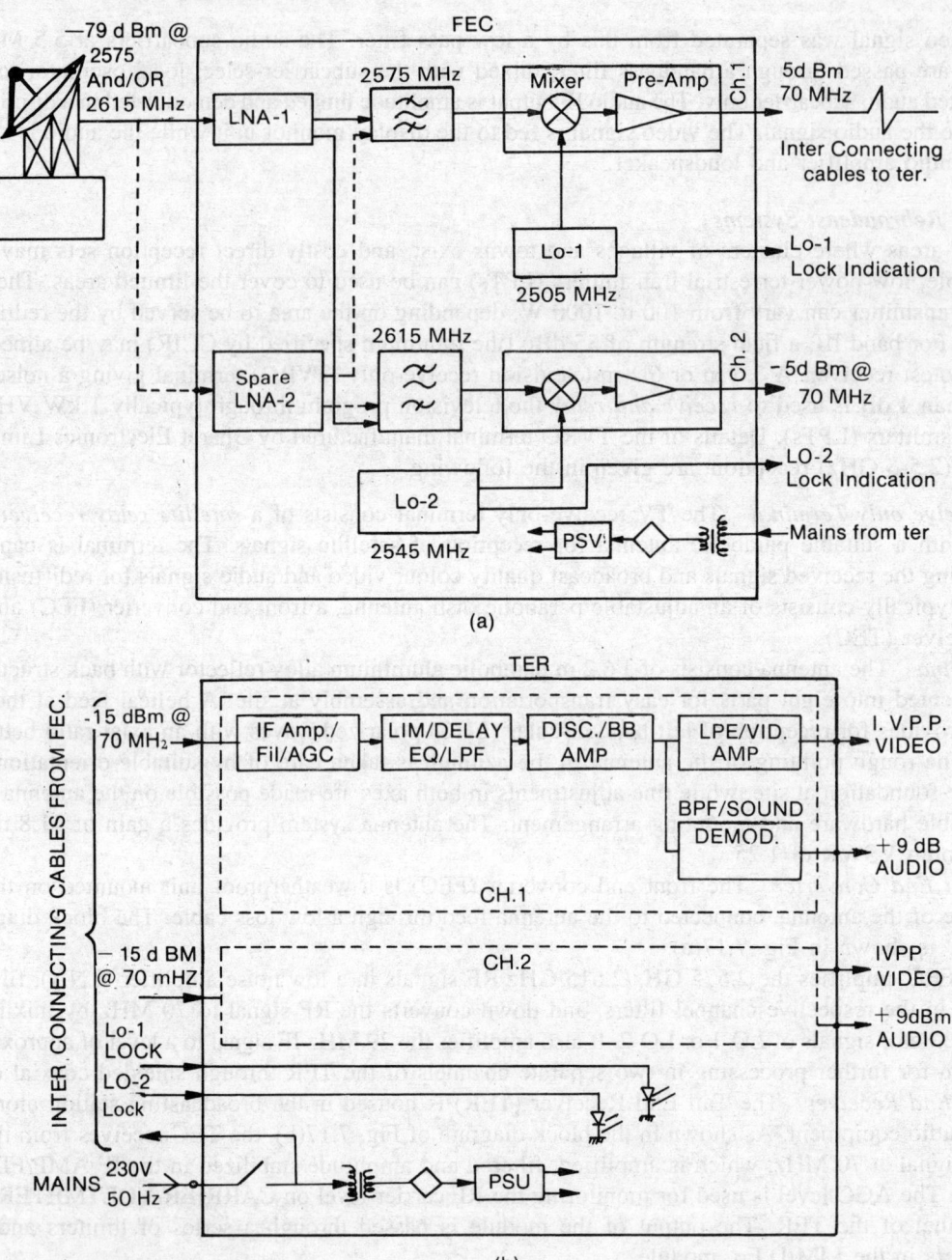

Fig. 7.17 (a) Block diagram of the front end converter (b) Block diagram of the tail end converter
(*Courtesy*: BEL)

from the base band signal and demodulates the FM signal to obtain an audio signal of 9 dBm level at peak deviation.

These professional quality video and audio output signals are available for broadcast with standard

VHF/UHF TV transmitters. The regulated power supplies for the various modules and monitoring circuits are housed in the TER. The monitoring circuits indicate on the front panel the status of various power supplies, lock-in indication of the local oscillators of the AEC and relative RF carrier levels.

REVIEW QUESTIONS

7.1 Explain the television signal requirements in the service area of a television station. Why does the minimum signal strength requirement vary with the frequency band?

7.2 Compare the design features of television transmitters employing high level modulation and low level modulation.

7.3 What are the advantages of IF modulation over direct modulation in a television transmitter?

7.4 Describe, giving block diagrams, the operation of a television transmitter using: (i) high level modulation, (ii) IF modulation.

7.5 Explain the working of the following:
(a) Diode bridge modulator.
(b) Diplexer.

7.6 How is an omnidirectional radiation pattern obtained from dipole antennas or dipole panel antennas?

7.7 How is a satellite used for television broadcasting to cover a wide area? Explain the parameters that affect planning satellite cum terrestrial system. Name the factors that determine the cost-effectiveness of a television system.

7.8 Compare direct reception system with a rediffusion type system in satellite television broadcasting. Discuss the block schematic of a direct reception terminal to receive the SITE programs.

7.9 Give reasons for the following:
(a) Modern transmitters most commonly employ low level IF modulation.
(b) Video modulation is never less than 10%.
(c) It is not possible to obtain sound in a TV receiver even if only the vision transmitter fails and the sound transmitter is on.
(d) Satellite broadcasting links employ circular polarization.

MULTIPLE-CHOICE QUESTIONS

7.1 The minimum TV signal strength per metre, for satisfactory reception in band III, is of the order of:
(a) 40 dBμ, (b) 55 dBμ, (c) 70 dBμ, (d) 80 dBμ.

7.2 The net field strength at moderate transmitting and receiving antenna heights from the ground is: (a) inversely proportional to the distance, (b) inversely proportional to the square of the distance, (c) proportional to the frequency, (d) (b) and (c).

7.3 IF modulation requires: (a) the VSB filter at high power, (b) the filter and diplexer together, (c) linear RF power amplifiers, (d) high power modulating video amplifiers.

7.4 In turnstile antenna, the crossed dipoles are fed: (a) in phase, (b) in quadrature phase, (c) out of phase, (d) one with the video and the other with the audio transmitter power.

7.5 For microwave TV relay links video signals are: (a) amplitude modulated, (b) frequency modulated, (c) pulse code modulated, (d) pulse amplitude modulated.

7.6 Over a smooth earth surface the line-of-sight range between transmitter antenna at a height of 400 m and receiving antenna at 25 m is about: (a) 40 km, (b) 60 km, (c) 90 km, (d) 120 km.

8

Propagation of Television Signals

The antenna for reception of television signals requires considerable attention if the television picture received is to be clear and free from interference. It must pick up the desired channel, developing maximum signal voltage and rejecting unwanted interference including reflected television signals that produce double images or ghosts in the reproduced picture due to different propagation delays. For this purpose, the antenna should have good directional characteristics and impedance match, and its placement must be chosen to optimize a clear ghost- and interference-free picture. A good understanding of the propagation characteristics of television signals at the VHF and UHF frequencies is necessary for planning and installation of the television antennas. Radio wave characteristics and their propagation is, therefore, reviewed in this chapter.

8.1 RADIO WAVE CHARACTERISTICS

The radio wave is an electrical power escaping into free space from the flow of high frequency currents in a suitable aerial circuit. The radiated energy escaping into space is in the form of electromagnetic waves. These waves are characterized by electric field E and magnetic field H at right angles to it, and propagate at right angles to both, as shown in Fig. 8.1.

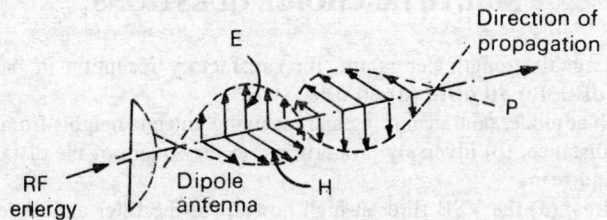

Fig. 8.1 Electromagnetic waves in free space

The E and H fields both go on alternating sinusoidally with respect to time and space, travelling at the speed of 3×10^{10} cm/s in free space. As the radiated power spreads out into space in all directions, radiated power density P is inversely proportional to the square of the distance from the isotropic radiating source radiating equally in all directions according to the relation:

$$P = \frac{P_t}{4\pi d^2} \tag{8.1}$$

where P_t is the transmitted power and d the distance.

The electric field intensity at a point is measured in volts per metre. The relation between electric field intensity E and power density P is analogous to that between voltage and power in an electric circuit so that:

$$P = \frac{E^2}{Z} \qquad (8.2)$$

where Z is the characteristic space impedance.

The characteristic impedance of space is given by $\sqrt{\mu/\varepsilon}$, where μ is the magnetic permeability and ε the electric permitivity of the medium.

For free space, $\quad \mu = 1.26 \times 10^{-9}$ H/m
and $\qquad\qquad\quad \varepsilon = 8.85 \times 10^{-12}$ F/m

Substituting these values, the space impedance is given by

$$Z = \sqrt{\mu/\varepsilon} = 377 \,(= 120\,\pi)\; \Omega$$

Combining Eqs (8.1) and (8.2), and substituting for Z we have

$$\frac{E^2}{Z} = \frac{P_t}{4\pi d^2} \; ; \text{ and hence } E^2 = \frac{ZP_t}{4.\pi d^2} = 30\, \frac{P_t}{d^2}$$

$\therefore$ electric field intensity E is given by

$$E = \sqrt{30\, P_t}/d \qquad (8.3)$$

This indicates that electric field E at distance d from an isotropic source radiating power P_t equally in all directions is proportional to the square root of the power, and is inversely proportional to the distance from the source.

Polarization

Polarization of the electromagnetic waves refers to the physical orientation of the electric field vector which is radiated long the same direction as the electric currents in the antenna wires or rods. Thus a horizontal antenna radiates electromagnetic waves with its electric field vector E horizontal, while the magnetic field surrounds the antenna wire and is perpendicular to it. The waves are then said to be horizontally polarized. A vertical antenna perpendicular to the earth's surface produces vertically polarized radio waves.

Apart from this linear polarization, the electric vector may rotate at the angular frequency of the wave about the line of propagation, the tip of the vector tracing an ellipse, in which case it is referred to as *elliptical polarization*. A special case of the elliptical polarization is circular polarization when the electric vector traces a circle, swinging equally in all directions at right angles to the direction of propagation. If the rotation is clockwise, looking in the direction of propagation, it is *right hand circular* (RHC) polarization, while if it is rotating anticlockwise, it is *left hand circular* (LHC) polarization.

Reception

A wire placed in an electromagnetic field will have currents induced in it proportional to the electric field intensity of the wave in the direction of the wire, acting as a receiving antenna to absorb some of the radiated power. In order to receive a maximum signal, the polarization of the receiving antenna wire or rod must be the same as that of the transmitting antenna generating the transmitted wave. A vertical linear wire dipole antenna will radiate a vertically linear polarized wave. If the antenna wire is at an angle θ the induced field is reduced by polarization loss factor cos θ. The polarization of a wave, depending on its frequency, can get twisted during propagation through the ionosphere or rainy atmosphere, causing *depolarization*, and loss of signal, unless the receiving antenna is rotated to the changed direction of polarization. These problems are encountered at microwaves in satellite antennas. It is advantageous here to use circular polarization, in which depolarization has little effect, once the transmitting and receiving antennas have the same direction, RHC or LHC. The orientation of the transmitting and the receiving antennas towards each other is not so critical. The choice for RHC or LHC polarization provides a means of reducing interference between adjacent main beams of two transmitters on a common satellite serving adjacent countries.

The process of reception is exactly the reverse of the transmission. Transmitting and receiving antennas are freely interchangeable, except for the power handling capability which is larger for the transmitting antennas with a consequent stress on efficiency in their design.

8.2 PROPAGATION PHENOMENA

As the radio waves propagate in the earth's environments, they may undergo changes in their paths due to earthly structures: (i) they may be reflected by ground, mountains, buildings, etc.; (ii) they may be refracted and change their direction as they pass through the layers of the atmosphere which have different densities due to heat, moisture content, etc. or different degrees of ionization; (iii) they may be diffracted around tall massive objects to bend around in their shadows; (iv) the waves at microwave frequencies may be absorbed and attenuated as they pass through media-containing water vapour, oxygen, etc.

(i) Reflection
At low, medium and high frequencies, the radio waves pass through large solid object of dimensions smaller than heir wavelengths, through the attenuation factor rises with increase in frequencies. The VHF and UHF frequencies find it progressively difficult to pass through such objects. Ultimately the object reflects the signal waves back, with the efficiency of reflection increasing with frequency. It follows that reflections at UHF will be more numerous than that at VHF frequencies, but the problem can more readily be overcome by directional aerials at UHF that can be oriented to favour the direct signal and reject the reflected ones. The directional aerials at UHF can be fairly compact because of smaller wavelengths.

Generally speaking, any object of a size larger than the wavelength of the radio waves can act as a reflector, the reflection coefficient or efficiency being greater with metal objects or surfaces. Stones, bricks, wood and like objects reflect signals as well as natural objects like hills and mountains.

At the point of reflection, phase angle changes up to $180°$ can occur depending on the plane of polarization and angle of incidence, besides some absorption of the signal. When the angle of incidence of the ray with the reflecting surface is small, the reflection can be assumed to take place without any change in amplitude and with a reversal of phase irrespective of polarization.

(ii) Refraction
This takes place when waves pass from one medium into another in which the wave velocity is different. If this change of medium density, and hence the velocity change, is gradual refraction takes place

gradually and the waves bend towards the denser medium in which the wave velocity is lower. Just above the earth, atmospheric density changes vary slightly but linearly with height, the refraction bends the waves downwards, thus extending the radio horizon.

(iii) Diffraction

When radio waves encounter an edge of an obstacle, the dimensions of which are of the order of the wavelength, there is reception of the waves behind the obstacle. This is due to the spherical wavefronts from points in the unobstructed side. They produce a succession of interference fringes that become weaker as one moves away from the edge of the obstacle. The VHF and UHF space waves may be received in this manner behind tall buildings and hills.

8.3 SPACE WAVE PROPAGATION

Television signals because of their large bandwidth have been allotted frequencies in the VHF and UHF ranges. At these frequencies, the radio waves follow almost straight, and direct paths into space if the transmitting antenna is mounted on a tall building or tower so that it is many wavelengths above the ground. These waves propagating in straight paths without interruption are called *space waves*.

Because of the curvature of earth's surface, the maximum range of reception due to space waves is naturally limited to the line-of-sight distance from the transmitting antenna. The limiting distance for reception of space waves is apparently the *Radio Horizon* on the earth's surface, as shown in Fig. 8.2.

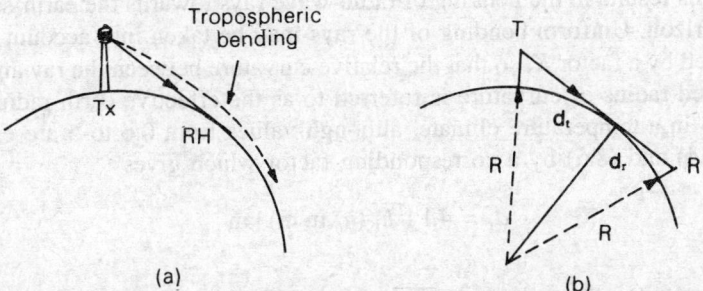

Fig. 8.2   (a) Radio horizon and extended range due to tropospheric bending, (b) Line-of-sight range

The VHF radio waves, however, are refracted and bend towards the earth due to reduction in refractive index with increase in height in the lower region of atmosphere-troposphere, due to temperature and pressure changes. There is also some bending due to diffraction of the waves at the earth's surface. The service area of a station is thus a little beyond the radio horizon distance between the transmitting and receiving antenna.

8.4 LINE-OF-SIGHT RANGE

The maximum line-of-sight distance over which space wave TV signals can be received will be the sum of the radio horizon distance of the transmitting and the receiving antennas. For an antenna at height h_t, the radio horizon distance can be obtained from the geometry of the path and the earth's surface.

With the approximation $h_t \ll R$, the radius of the earth, radio horizon distance d_t is nearly equal to the pathlength. Then,

$$d_t^2 + R^2 = (R + h_t)^2 = R^2 + 2Rh_t + h_t^2,$$

$$= R^2 + 2Rh_t \text{ (neglecting } h_t^2)$$

Hence
$$d_t = \sqrt{2R}\sqrt{h_t}$$

$$= \sqrt{2}\sqrt{6400}\sqrt{h_t} \text{ as } R = 6400 \text{ km}$$

$$= 112\sqrt{h_t} \ (h_t \text{ in km}) \text{ km}$$

$$= 3.7\sqrt{h_t} \ (h_t \text{ in m}) \text{ km} \tag{8.4}$$

$$= 1.22\sqrt{h_t} \ (h_t \text{ in ft}) \text{ miles} \tag{8.5}$$

For a transmitting antenna of height $h_t = 300$ m,

$$d_t = 3.7\sqrt{300} \approx 64 \text{ km}$$

For a receiving antenna of height $h_r = 10$ m,

$$d_r = 3.7\sqrt{10} \approx 12 \text{ km}$$

Under normal atmospheric conditions the refractive index of atmosphere decreases with height above the earth's surface. This results in the bending of radio-wave rays towards the earth's surface, increasing the effective radio horizon. Uniform bending of the rays may be taken into account by considering the radius of earth modified by a factor K, so that the relative curvature between the ray and the earth remains the same. The modified radius of curvature is referred to as the effective earth radius. Factor K has an average value of 1.33 in a temperature climate, although values from 0.6 to 5 are expected in general. This modifies Eqs (8.4) and (8.5) by a corresponding factor which gives

$$d_t = 4.1\sqrt{h_t} \ (h_t \text{ in m}) \text{ km} \tag{8.4a}$$

or

$$d_t = \sqrt{2h_t} \ (h_t \text{ in ft}) \text{ miles} \tag{8.5a}$$

Thus the line-of-sight range between the aerials is typically about 75 km, which is extended by tropospheric bending and diffraction effects up to 80 to 100 km, depending on the terrain conditions.

8.5 SPACE WAVE RECEPTION OVER SMOOTH TERRAIN

The electric field strength at the antenna receiving space waves is due to two rays: (i) direct line-of-sight wave having path length r_0, and (ii) reflected ray r_1r_2 from a point on the earth's surface, with equal angles of incidence and reflection, the curvatures of the earth being neglected, as shown in Fig. 8.3.

The field arriving at the receiver *via* direct ray differs from the field arriving *via* the reflected ray, by a phase angle which is a function of the path length difference.

When $d \gg h_t$ and h_r, angle ϕ is very small, and hence little change of magnitude and phase angle takes place at the point of reflection. By geometry,

The path difference $(r_1 + r_2) - r_0 \approx \dfrac{2h_t h_r}{d}$.

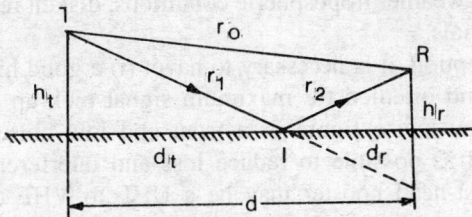

Fig. 8.3 Space wave reception by direct and reflected waves

Hence the phase difference $$\phi = \frac{2\pi h_t h_r}{\lambda d}$$

and $$E = \frac{2 E_0}{d} \sin \phi$$

where E_0 is the field strength of the direct ray at unit distance

$$\therefore \qquad E = \frac{2 E_0}{d} \sin \frac{2\pi}{\lambda} \frac{h_t h_r}{d} \qquad (8.6)$$

E_0/d corresponds to the variation of field strength with distance of a free space wave for a conducting earth surface, the direct and the ground reflected rays add and cancel in their phase providing a maxima and a minima with respect to distance, as the phase angle alternates through zero and π.

For horizontal polarization, the field strength E is practically zero, close to ground, but rises almost linearly with increase in h_r, the height of receiving antenna above the earth's surface.

8.6 DX (DISTANT TRANSMISSION) RECEPTION

Distant (DX) reception of TV signals, far beyond the normal line-of-sight range is sometimes possible due to bending or diffraction of radio waves. The VHF frequencies normally pass through all the ionized layers including the F layer of the ionosphere. At frequencies in the 30 to 100 MHz range, regular but weak propagation by ionospheric bending and scattering can sometimes be obtained. During abnormal conditions of ionosphere, such as periods of high sunspot activity, the lower VHF band I signals may be influenced by the high ionization of sporadic E layer and the refracted back to earth providing fairly strong intermittent propagation at surprisingly great distances, hundreds of miles away. This is often referred to as freak reception as it is rather unpredictable. This mode of propagation is prevalent particularly on the lower frequency channels of band I and can give reception over 1000–2000 km.

The abnormal ionospheric conditions have little influence on the upper VHF band III channels or the UHF channels. These are more affected by the tropospheric conditions. The line-of-sight distance propagation is extended by *tropospheric bending by refraction* through atmospheric layers and *diffraction* effects on earth's curvature and terrain conditions. Under long periods of fine weather conditions, the earth's rising heat creates temperature gradient in air with respect to height, and modifies the refractive index of the troposphere which can trap and guide the waves in a duct along the earth's surface resulting in *duct* propagation far beyond ordinary bending of waves.

Tropospheric propagation occurs mostly on band I and the lower channels of band III, whereby reception up to 500–1000 km may be obtained. The UHF channels are much less affected by tropospheric conditions, but with good clear weather tropospheric conditions, distant reception up to 100–150 km may be possible with high gain aerials.

For a successful distant reception, it is necessary to have: (i) a good high gain directive aerial system tuned to the distant channel and oriented for maximum signal pick-up at a good height, (ii) a highly sensitive receiver of a very low noise figure or a front end low noise booster amplifier, preferably mounted as close to the aerial as possible to reduce loss and interference pick-up of the feeder. For receiving UHF, the aerial mast-head booster may be a UHF to VHF converter to reduce cable loss problems.

8.7   SHADOW ZONES

The space wave propagation of TV signals at VHF and UHF frequencies creates shadow zones behind hills and large buildings which obstruct the waves. Although diffraction effects and tropospheric bending enable reception to some extent in the shadow zones (as shown in Fig. 8.4), the bending effect decreases as higher and higher frequencies are used. The path of UHF signals are thus less curved, and distant reception of UHF signals *via* tropospheric bending is much less common. The diffraction effects are also reduced at UHF signals, and shadow zones due to hills are more pronounced and numerous with UHF signals. Reception in such areas becomes a problem for which the only solution is to mount the aerial on top of the hill or the tall building obstructing the waves, and to use a Community Antenna TV (CA TV) system, i.e., to provide a common antenna signal to a number of community members by a cable distributor. Large buildings and trees can also attenuate signals on UHF channels, and the signal strength may vary between seasons due to the growth of foliage on trees near the signal path to the receiving aerial.

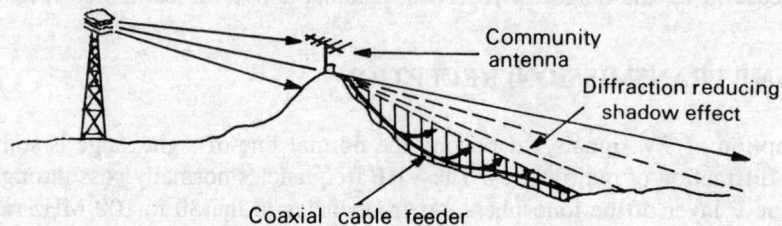

Fig. 8.4   Reception in shadow zone

8.8   CO-CHANNEL INTERFERENCE

Since there are only a limited number of VHF channels available for TV broadcasting, and propagation is over a limited line-of-sight range normally, these need to be shared for a countrywide coverage. The channels are usually shared by stations for apart from each other. The reception of long distance stations on a local channel can be very much annoying in terms of interference, particularly when the same is relayed for extended range by direct reception. This interference may take the form of regular horizontal bars moving up and down the picture (called 'venetian blind interference'), or may take the form of Moire patterns. Under strong signal interference, a ghost picture, plastic in appearance, may be seen drifting and rolling with visible blanking bars with a loud buzz on sound.

Sharing of channels is usually planned carefully so that within the service area of any station, signals from a co-channel station are normally too weak to be perceptible. During abnormal conditions of reception, particularly in summer, signals from distant VHF stations on the lower VHF channels are received very strongly over hundreds of miles, causing co-channel interference in local reception in the grade B service and in fringe areas where local signal strength is rather low. These co-channel interference problems due to DX reception have been causing rethinking in allotment of television channel frequencies. Channel 1 in band I has been removed from television use for the same reasons. There is an increasing trend to use the UHF channels where these effects are considerably less.

Off-set Operation

In order that co-channel stations interfere less with each other, picture carrier frequencies are allowed to have a carrier off-set of + 10.5 kHz or − 10.5 kHz with respect to each other. It is also possible then to use in the aerial down-lead feeder a highly selective filter, such as the high Q bridged-T tuned filter shown in Fig. 8.5, to notch out the unwanted signal carrier. The unwanted carrier may be attenuated by several dB while the wanted signal carrier suffers only a few dB attenuation.

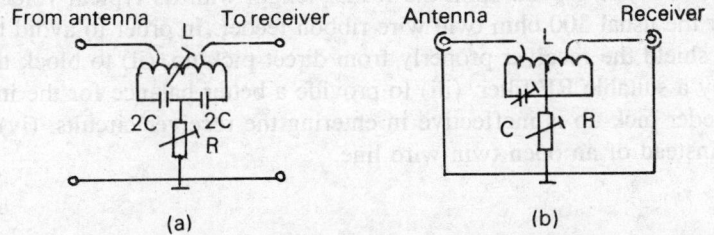

Fig. 8.5 Bridged-T type notch filters: (a) Split capacitor, (b) Split inductance

This can give considerable improvement in reception of the wanted signal by reducing the interference effects.

8.9 GHOST IMAGES

In addition to the direct line-of-sight path, television signals are also received *via* longer paths due to reflection from large objects like big buildings, tanks, hills, etc. The reflected signal arriving a little later with respect to the direct ray, due to the path difference, produces an additional image displayed horizontally by a distance proportional to the time delay between the two paths. As television signals travel at 3×10^5 km/s, each kilometre of path difference involves a time delay equal to $\left(\dfrac{10^6}{3 \times 10^5} = \right) 3.3\ \mu s$.

As the active line scan in the picture takes about 52 μs, the extent to which a ghost image due to the reflected ray will be shifted can be readily found out for a given path difference. Displacement per kilometre of path difference,

$$d = \frac{3.3}{52} \times \text{width of picture tube}$$

$$\approx \frac{3.3}{52} \times 40 \text{ cm for 50 cm tube}$$

$$\approx 3 \text{ cm/km} \qquad (8.7)$$

Usually the ghost picture is of much less contrast than the main picture as the reflected signal is generally weaker. The ghost picture may be negative due to the reversal of phase of the reflected signal. If the time delay is small, the picture may appear fuzzy. A path difference of less than about 15 m will produce a displacement hardly noticeable, falling as it will within almost the resolution limit, 0.5 mm. In strong signal areas with many high rise buildings, the multiple reflections of TV signals from them produces a series of ghosts successively displaced from each other, if a simple aerial like an indoor dipole is used. A good directive and properly located antenna which can reject the reflected signal in preference to the direct ray is essential in such cases.

Leading Ghosts

When a TV receiver is located in an area close to the transmitter, the direct pick-up by the receiver input circuit and the antenna feeder, if the input is not perfectly balanced, will produce a leading ghost. It will be on the left hand side of the main picture produced by the signal received by the antenna which arrives later. The leading distance will depend upon the feeder length with its typical velocity of propagation equal to 252 m/μs for the usual 300 ohm twin wire ribbon feeder. In order to avoid the leading ghosts, it is necessary: (i) to shield the receiver properly from direct pick-up, (ii) to block the pick-up *via* the power supply leads by a suitable RF filter, (iii) to provide a better balance for the input circuit so that the input from the feeder pick-up is ineffective in entering the receiver circuits, (iv) to use a shielded coaxial cable feeder instead of an open twin wire line.

Airplane Flutter

This is caused by interference between the direct signal and the reflected signal from aeroplanes. The reflected signal continuously changes its phase due to changing path difference so that the reflected signal may add and cancel the main direct signal, producing a fluttering picture and sound with a violent beat frequency disturbance as the beat goes through zero frequency. This can be minimized by low angle transmission beams and reception beams by vertically stacked antennas.

8.10 INTERFERENCE PROBLEMS

Unwanted signals or noise which creep into the desired channel frequency range produce disturbance in the picture and sound reproduction. These are of the following types.

(i) Man-made Impulsive Interference

This is caused by electrical discharges or sparks produced from man-made appliances, vehicles, etc., for example, sparking due to ignition systems of motor cars, internal combustion engines, sparking across contacts of electric switches and at the brush contacts of dc motors or the commonly used universal motor in home and workshop appliances, arc welding equipment, and so on.

The interference appears in the form of dark and white spots or dashes on the TV picture with a tendency to disturb the line and field synchronisation, and if severe, causes production of crackles in the sound.

(ii) RF Interference

Regular CW radio frequency signals (often harmonic radiations from various RF equipment like diathermy, other TV sets, etc.) may pass receiver circuits to produce pattern effects on the TV picture.

(iii) Co-channel Interference

Shared channel working can cause interference in the form of the venetian blind effect when unusual ionospheric or tropospheric conditions cause freak reception of long distance TV signals in weak signal areas of a local station, producing rolling horizontal bars described earlier.

(iv) Atmospheric Noise Interference

This is the RF noise originating from thunderstorms, outer space and other natural sources appearing as dots on the screen.

Interference Suppression

The ratio of the wanted television signal from the aerial to the unwanted interference must be large. The maximum permissible levels of interference from various electrical equipment and installations, receiver radiations, etc. are usually governed by regulations in various countries. It is also compulsory for motorists and manufacturers of electrical equipments to keep down the man-made impulsive interference by RF suppressors at the source. Suppressors take the form of a small shunt capacitor or resistance and capacitor combination across the sparking terminals, or the device across which sudden switching occurs. To prevent the interference from going on to the mains·or coming into the receiver from the mains, a low pass RF filter in the form of delta connected capacitors or *LC* filters shown in Fig. 8.6 may be employed.

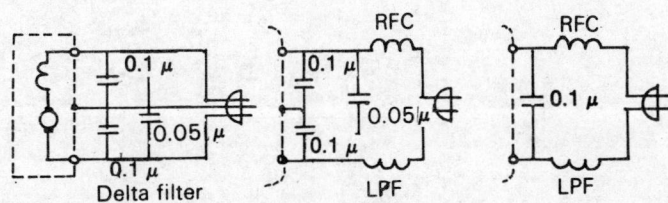

Fig. 8.6   RF interference filters

Capacitors must be of good quality RF type and inductances must be low self-capacitance honeycomb-wound type chokes, for good results. RF chokes of about 100 μH with adequate current carrying capacity are usually employed.

REVIEW QUESTIONS

8.1 Discuss why TV signals have a limited range of propagation. What are the parameters that affect the range?

8.2 Sketch the variation of field strength at the receiving antenna at a height of 10 m with respect to distance from a transmitting antenna at a height of 100 m, assuming smooth terrain.

8.3 Show that when h_t and $h_r \ll d$, the field strength E due to EIRP of P Watts, is given by $E \approx \dfrac{70\, h_t\, h_r\, \sqrt{P}}{\lambda\, d^2}$ V/m.

8.4 How is the space wave propagation at VHF and UHF affected by ionospheric and tropospheric conditions? Explain the possible modes of distant reception.

8.5 What are the causes and effects of co-channel interference? How are they minimized?

8.6 Explain the following:
(a) Ghost images, (b) Leading ghost, (c) Airplane flutter, (d) Off-set operation.

8.7 What are the sources of interference in TV reception? How is interference from electrical equipment to TV receivers minimized?

8.8 A ghost is produced 2 cm to the right of the main image in a 40 cm wide picture tube. What is the difference in the path length of the direct and reflected rays?

MULTIPLE-CHOICE QUESTIONS

8.1 Circular polarization is: (a) involves critical alignment of receiving and transmitting antennas, (b) is useful in reducing depolarization effects on received wave, (c) is useful in discriminating between reception of adjacent beams, (d) (b) and (c).

8.2 Diffraction of radio waves: (a) enables reception in shadow zones, (b) increases the radio horizon, (b) changes polarization, (c) (a) and (b).

8.3 Off-set operation of transmitters is for: (a) operating adjacent TV stations at different channel frequencies, (b) operating the carrier frequencies of the same channels at slightly different frequencies, (c) operating the adjacent transmitters with different polarization, (d) (b) and (c).

8.4 A ghost is produced at 1.5 cm to the right of the main image, on a picture tube of 24 cm width. The path difference between the interfering wave and the direct wave is: (a) 0.5 km, (b) 2 km, (c) 3 km, (d) none.

8.5 DX reception is caused in band III due to: (a) ionospheric scattering, (b) tropospheric bending, (c) duct propagation, (d) (b) and (c).

8.6 RF choke in the interference suppressor units is usually: (a) a multilayer coil, (b) honey-comb wound coil, (c) ferrite core coil, (d) (a) and (b).

Television Antenna Systems

9.1 ANTENNA REQUIREMENTS

A television receiver antenna has the following functions to perform:

(i) It must pick-up the desired signal and develop a maximum signal voltage from the available field strength.

(ii) It must discriminate against unwanted signals like: (a) man-made interference from cars, machines, etc., (b) reflections from buildings that produce ghost signals, and also (c) co-channel interference, if any.

(iii) If it is to receive more channels or bands, it must be capable of wide-band operation and must be rotatable if the stations are located in different directions.

In order to see how these requirements are met, the characteristics of antennas should be known properly.

9.2 RESONANT ANTENNAS AND THEIR CHARACTERISTICS

While in a radio receiver for receiving of signals medium and short waves any length of wire can act as an aerial, this cannot be done for television signals as the length of wire used becomes comparable to the wavelength of the very high frequency used. This drastically affects its reception characteristics, viz. directional response, bandwidth, impedance, etc. It is essential, therefore, to use resonant or tuned antennas that have well designed receiving characteristics for TV reception.

Resonant antennas are just appropriate lengths of wires or rods in multiples of $\lambda/4$, behaving like tuned circuits because of their distributed inductance and capacitance in the form of opened out transmission line of length equal to multiples of $\lambda/4$ or $\lambda/2$. Quarter wave Marconi antenna requires earth reflected $\lambda/4$ image, the two forming a $\lambda/2$ dipole antenna, as shown in Fig. 9.1.

Formed by two $\lambda/4$ sections, a half-wave antenna is the basic resonant antenna used in television reception, operating independent of ground. It is called a *dipole antenna* or a *Hertzian dipole*. Dipole antenna is a very simple antenna of fairly small size at the VHF/UHF TV signals, and has useful

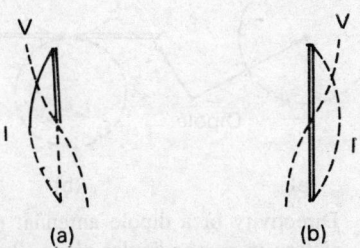

Fig. 9.1 Dipole antennas: (a) $\lambda/4$ Marconi antenna, (b) $\lambda/2$ Hertzian dipole

directional characteristics that can receive the maximum signal when it is at right angles to the dipole, while rejecting signals in the axial direction along its length.

Induced Voltage

In the antenna, induced voltage is proportional to field strength E in volts per metre, in the plane of the antenna. The actual signal V available at the antenna terminals will be the product of the effective length (carrying uniform average voltage or current) l_{eff} and field strength E. Thus.

$$V = l_{eff} \times E = \left(\frac{\lambda}{2} \times \frac{2}{\pi} \right) \times E \text{ for a dipole}$$

$$= \frac{\lambda}{\pi} \times E = \frac{300}{\pi f} \cdot E \approx \frac{96}{f_{min}} \cdot E \tag{9.1}$$

It follows that at lower wavelengths, the signal voltage induced in the antenna is smaller for a particular value of field strength.

The *physical length* of the $\lambda/2$ dipole will be to some extent less than electrical length $\lambda/2$, which is given by

$$\frac{\lambda}{2} = \frac{c}{2f} = \frac{300}{2f \text{ (in MHz)}} \text{ m}$$

This is because of the finite rod thickness which has the current distribution modified at the ends as shown in Fig. 9.2.

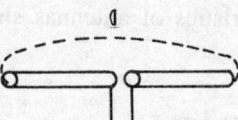

Fig. 9.2 Effect of thickness of rod on current distribution

Reduction factor k varies with the rod diameter, being a function of the d/r ratio. Typically it has values from 0.92 to 0.98 for rod sizes of 12 mm to 6 mm diameter in the VHF range.

Directivity

The directivity of an antenna is the relative ability of an antenna to receive in different directions in space. It is usually plotted in the form of a polar diagram or polar pattern in the plane of interest. Directivity is often expressed by parameters like beamwidth, front-to-back ratio and auxiliary lobes in various planes.

A dipole has a circular polar pattern in the plane at right angles to it, as shown in Fig. 9.3(a).

It can receive or transmit equally well in all directions in that plane. Its directivity pattern in the plane passing through the dipole length is a figure of eight pattern, i.e., a cosine function with respect to angle θ with the normal to the dipole, as shown in Fig. 9.3(b). The pattern is thus a doughnut shaped pattern in a three-dimensioned plot. The pattern alters at frequencies off-resonance due to change in the current distribution at these frequencies. The pattern can be made more directive by addition of reflectors and directors at appropriate distances from the dipole.

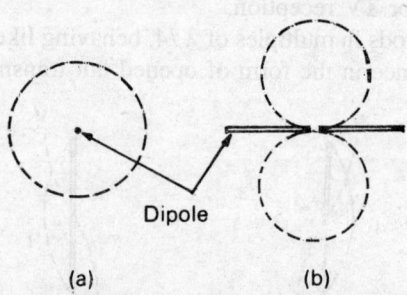

(a)   (b)

Fig. 9.3 Directivity of a dipole antenna: (a) Circular pattern in perpendicular plane, (b) Figure-of-eight pattern in the dipole plane

Beam Width or Lobe Width
The directivity of an antenna is expressed simply by the beam width of the main lobe in the directive pattern as the angle within which the signal received is within 3 dB or half of the maximum power.

Front-to-back Ratio
This is the ratio of the amount of signal received from the front to that received from the back. This indicates the directivity of an antenna.

Directive Gain
This is defined as the ratio of power received by an antenna in a particular direction (usually in the direction of the main lobe) to the power that would be received by an isotropic antenna, i.e., an antenna receiving equally in all directions. Though such an isotropic antenna does not exist, its properties are easy to visualize and calculate.

Another form of gain used in connection with antennas is the *power gain*, defined as the ratio of the power that must be radiated by an isotropic antenna to develop a certain field strength at a distance, to the power that must be fed to the directive antenna to obtain the same field in the direction of its maximum. For example, a half-wave dipole has directive gain of 1.64 over an isotropic radiator.

Antenna Gain
The increase in the signal received by an antenna with respect to a reference antenna, usually the $\lambda/2$ dipole antenna, is termed the antenna gain. The term is generally used to indicate the gain of directional antenna system like a multielement dipole array, and is usually expressed in dB as: $20 \log \dfrac{V_{ant.syst.}}{V_{ref.ant.}}$. An antenna gain of 6 dB of an antenna system has thus a voltage gain of 2 over a simple dipole.

Antenna Impedance
The impedance of a $\lambda/2$ dipole at the resonant frequency is 72 ohm at the centre where the feed current has a maximum and the voltage a minimum as shown in Fig. 9.4. It increases in magnitude as the feed points are widened apart towards the ends and has a value of several thousand ohms at the ends. At frequencies off-resonance, the impedance at the centre is reactive greater than 72 ohm. The impedance of the dipole fed at the centre is usually approximated to 75 ohm at resonance.

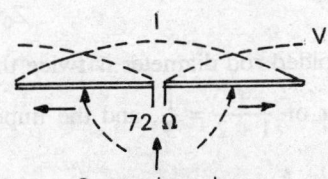

Fig. 9.4   Dipole antenna impedance

Antenna Bandwidth
This refers to the frequency response of the antenna as a resonance circuit. The antenna bandwidth indicates the selectivity of the antenna and is the frequency range over which it provides a *satisfactory* signal within 3 dB or half power ratio. The bandwidth is affected by change in impedance and directivity pattern due to change in the current distribution with respect to the change in frequency, hence the criterion of satisfactory reception is added.

The half-wave antenna is equivalent to a resonant circuit and its Q determines the bandwidth. Large diameter conductors have a lower resistance but much lower reactance (due to equivalent paralleling of inductances or capacitances of smaller conductors connected in parallel to form a larger conductor). Hence fat dipoles employing larger tubes for their conductors or fat dipoles made out of flat strips (as shown in Fig. 9.5) have a lower Q and hence a wider frequency response.

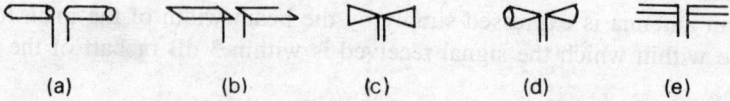

Fig. 9.5 Wide band dipoles: (a) Fat dipole, (b) Flat dipole, (c) Bow-tie dipole, (d) Conical dipole, (e) Parallel wire conductor dipole

The thick diameter with low Q has more uniform impedance over a broader range of frequencies. The thicker antenna length requires more shortening of their physical length due to larger end effects affecting the current distribution. A half-wave dipole with a non-uniform section, tapering flat strips or conical sections, called bow-tie antenna, can operate over a very large bandwidth and is hence suitable for UHF multi-channel reception. They may have thickness around 0.1λ or even greater. The tapering or conical sections maintain more constant impedance over a wide frequency. Besides this, their directional response pattern on the broadside is also maintained over a wide band of frequencies before splitting into multiple lobes.

At lower frequencies, the thickening of the conductors may be conveniently achieved by using conductors separated by 0.1λ or less connected in parallel, in place of strip or conical sections (see Fig. 9.5).

9.3 FOLDED DIPOLE

The impedance of a half-wave dipole is increased to four times by a folded antenna rod construction in which the half-wave dipole is joined at the ends by a continuous rod of the same length near and parallel to it, as shown in Fig. 9.6. For a particular voltage at the antenna feed points, the current in the main dipole is now only a part of the total current.

If the diameter of the two rods are equal, the current in the feed dipole is halved and the impedance at the feed point is increased to four times

$$Z_0 = 72 \times 4 = 288 \approx 300 \ \Omega$$

If the folded rod diameter is twice that of the dipole as shown in Fig. 9.6(b), the current is reduced by a factor of $\dfrac{1}{1+2} = \dfrac{1}{3}$, and the impedance is, therefore, increased to nine times the simple dipole impedance.

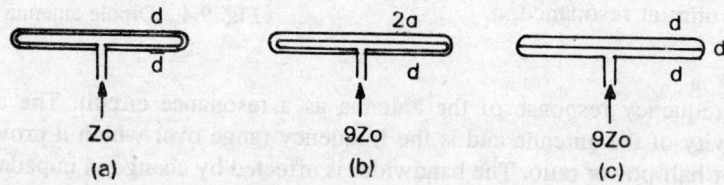

Fig. 9.6 Folded dipole

This is also true if two identical rods are used to fold over, as shown in Fig. 9.6(c). This also multiplies the impedance ninefold.

A folded dipole antenna is conveniently used where higher antenna impedance of 300 ohms or more is desired. It otherwise has similar characteristics in respect of gain and directivity. But because of lowered reactance, implied in the paralleling action of the fold, it has a lowered Q and hence a larger bandwidth.

9.4 PARASITIC ELEMENTS

Additional rod elements placed on either side of a dipole antenna and parallel with it can make the dipole unidirectional in response. These rod elements that have no electrical connection as such are called *parasitic* elements. Depending upon their lengths and how they are placed, these can act as reflectors or directors for the signal being received.

Reflector

A parasitic element usually some 5% longer than the dipole and placed behind it at a distance of about 0.25λ (in practice about 0.2 to 0.23λ), acts like a reflector. Like the dipole itself, the reflector rod absorbs the energy of the incoming signal electromagnetic field E_i and induces in it a standing wave current that lags by 90° because of the extra length which makes it inductive. The standing wave current radiates an electromagnetic field wave E_{rr} lagging 90° in phase with the current. Re-radiated field E_{rr} at the reflector and behind it is thus 180° lagging, or opposite in phase to the original field arriving there [see Fig. 9.7(a)].

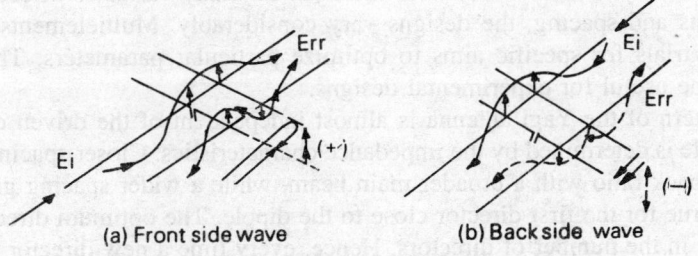

(a) Front side wave   (b) Back side wave

Fig. 9.7   Reflector action of parasitic element

Re-radiated reflector signal E_{rr} acts on the dipole but only after undergoing the delay determined by the distance between the dipole and the reflector. This distance being travelled twice in addition to the 180° phase shift in re-radiation, enables the two waves to add on and strengthen at the dipole up to twice the value.

For a signal arriving from the back side, there is no difference in travelling time between the direct incoming wave and the re-radiated wave from the reflector. However, since the reflector wave is in opposite phase, as seen earlier, the two waves largely cancel each other as shown in Fig. 9.7(b).

Director

A parasitic element shorter than the dipole by about 4 to 5% and placed in front of the dipole at a distance of less than 0.25λ (usually about 0.1 to 0.15λ from the dipole) acts as a director. Because of its shorter length, the induced standing wave current leads the incident incoming electromagnetic field wave. The re-radiated field wave from the director lags this by 90° so that there is no phase difference between the incident field and the field re-radiated by the director. Hence for a signal coming from the front, the direct wave and the re-radiated wave add on since there is no additional phase difference due to identical travelling time. For the signal coming from the back side, the phase difference due to the distance travelled twice by the director wave is 180° if the distance is nearly $\lambda/4$. Hence the two waves tend to cancel each other at the dipole.

It is possible to employ more than one director to increase the directivity further. The addition of parasitic elements as reflector and directors affects the impedance and bandwidth, besides the directivity radiation pattern. There are many combinations of element lengths and spacings that can influence these parameters so that a wide variation of parameters is possible. Addition of parasitic elements to the dipole antenna generally reduces the aerial impedance to some extent. This will introduce a mismatch and a voltage standing wave ratio (VSWR) which may be reduced by proper feed point adjustment and a special folded dipole with a higher impedance step-up ratio. The addition of a reflector affects the dipole impedance less than the addition of a director which needs closer spacing for the same gain.

9.5 YAGI AERIALS

A dipole antenna with a set of usually two or more directors and reflectors is called a Yagi aerial or a Yagi-Uda array after the Japanese engineer who wrote in English about the work originally published in Japanese by Professor Uda, in 1926–28. These Yagi arrays are the most popular of all the aerials used for TV reception in the VHF/UHF range because of their simple construction and low wind resistance.

The performance of Yagi arrays can only be assessed after considering all antenna characteristics, viz. impedance, gain directivity and bandwidth. As there are so many variables affecting each other, with change of dimensions and spacing, the designs vary considerably. Multielements Yagis are designed experimentally with trials for specific aims to optimize particular parameters. The following general characteristics may be useful for experimental designs.

The radiation pattern of the Yagi antenna is almost independent of the driven elements. The length and form of the dipole is determined by the impedance characteristics. Closer spacing of directors results in a higher front-to-back ratio with a broader main beam, while a wider spacing gives a sharper beam. This is particularly true for the first director close to the dipole. The optimum director length decreases slowly with increase in the number of directors. Hence, every time a new director is added, its spacing is adjusted and all the lengths are shortened slightly. As there are too many parameters to be optimized in multielement Yagi arrays, it is simpler to design two or more stacked Yagis with fewer elements than a single one with a larger number of elements.[23, 24]

Yagi Aerial Dimensions for the VHF TV Channels
Typical dimensions of multielement Yagi aerials for the commonly used VHF channels are given in Table 9.1. The aerials are generally constructed out of aluminium rods or pipes of diameter 10 to 12 mm for band I channels and 8 to 10 mm for band III channels. The fold spacing-separation for the dipole should be 60 to 80 mm for band I channels and 50 to 70 mm for band III channels.

Stacked Arrays
Two or more antennas of the same type called bays may be stacked vertically, one above the other, or horizontally, one beside the other. This improves directivity in the respective plane of stacking, providing a power gain of 3 dB with each bay added. The signal from each bay must be phased correctly before combining them together.

Vertical stacking with bays $\lambda/2$ apart provides greater directivity useful in the reduction of noise and interference pick-up from low angle sources like ignition from vehicles. Spacing of $\lambda/2$ is convenient for interconnecting the two bays with two $1/4\lambda$ sections for matching the impedances as shown in Fig. 9.8(a).

Table 9.1 Dimensions (in mm) of Yagi Aerial Elements for VHF Channels*

Channel No.	Refl. R	Refl. spacing r	Folded dipole FD	FD-D_1 spacing d_1	Dir. D_1	D_1–D_2 spacing d_2	Dir. D_2	D_2–D_3 spacing d_3	Dir. D_3
2	2980	1190	2840	900	2690	—	—	—	—
3	2620	1050	2490	790	2360	—	—	—	—
4	2330	930	2220	700	2100	—	—	—	—
2–4	3050	300	2510	600–800	2440	—	—	—	—
5	825	330	785	255	750	305	740	338	730
6	790	325	754	244	720	285	710	325	700
7	765	310	726	235	690	275	680	314	670
8	740	300	723	227	670	265	660	303	650
9	715	290	680	220	650	256	640	293	630
10	690	280	656	212	630	247	620	283	610
11	670	270	636	205	610	240	600	274	590
5–11	760	300	730	160	605	175	590	190	580

Spacings of the further directors may be about the same as D_3 and their lengths made shorter by about 10 mm for band III channels and 30 mm for band I channels.

*Based on the table of dimensions of Yagi antennas for band I to III in 'Practischer Antennenbau' by Herbert G. Mende, *Courtesy*: Franzis-Verlag München, 1977.

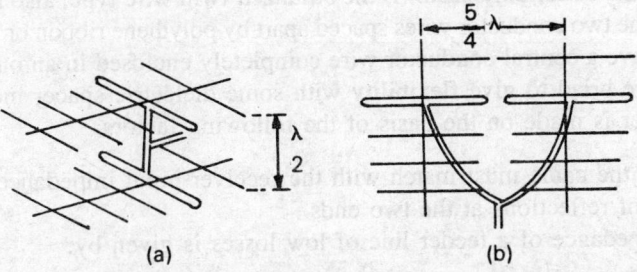

Fig. 9.8 Stacked Yagi arrays: (a) Vertical stacking, (b) Horizontal stacking

Horizontal stacking with the antenna bays placed in the horizontal plane one beside the other provides a sharper beam in the horizontal plane. The minimum spacing between centres will be $\lambda/2$. This provides a gain of 1.8 dB over each. Optimum spacing is about 1 ... $5/4\lambda$ between centres, giving a gain of up to 3 dB. The $3\lambda/4$ length matching section may be used for each bay to combine and match their impedance to the line, as shown in Fig. 9.8(b).

High Gain Yagi Antennas

Figure 9.9 illustrate the construction of a 13 element Yagi antenna for channels 5 to 8 in band III. The antenna provides a gain of over 12 dB, over a 3 dB bandwidth of around 28 MHz, front-to-back ratio of over 20 dB and VSWR less that 2.5, with output impedance of 300 ohm balanced.

9.6 ANTENNA FEEDERS

The antenna feeder is the transmission line which delivers the antenna signal to the receiver with correct

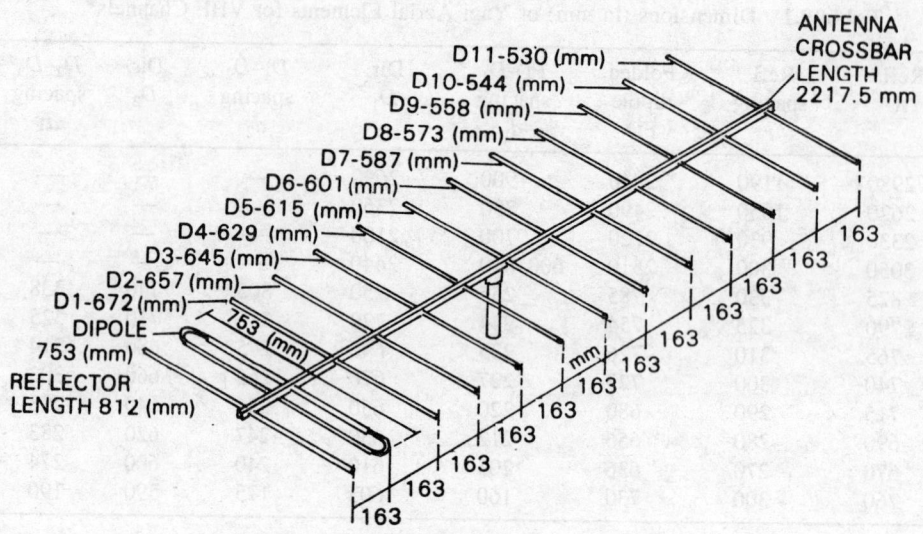

ANTENNA CROSSBAR LENGTH 2217.5 mm

D11-530 (mm)
D10-544 (mm)
D9-558 (mm)
D8-573 (mm)
D7-587 (mm)
D6-601 (mm)
D5-615 (mm)
D4-629 (mm)
D3-645 (mm)
D2-657 (mm)
D1-672 (mm)
DIPOLE 753 (mm)
753 (mm)
REFLECTOR LENGTH 812 (mm)

163 (repeated)

Fig. 9.9 A 13-element Yagi antenna for channels 5 to 8

matching of impedance at the antenna and the receiver input with minimum loss. Losses become important when a long feeder is required and where the signal field strength is low.

Two types of feeders are commonly used: (i) the *balanced* twin wire type, also known as parallel wire ribbon cable, which has the two conductor wires spaced apart by polythene ribbon or tubing; (ii) *unbalanced* shielded cables which have a central conductor wire completely enclosed in an outer metal sheath often made of fine copper wire briad to give flexibility with some dielectric spacer medium.

The choice of a feeder is made on the basis of the following factors:

(i) The impedance of the cable must match with the receiver input impedance for maximum power transfer and prevention of reflections at the two ends.

The characteristic impedance of a feeder line of low losses is given by:

$$Z_0 = \sqrt{\frac{L}{C}} \tag{9.2}$$

where L is inductance per unit length and C the capacitance per unit length

$$Z_0 = \sqrt{\frac{1.8\ \mu H/m}{20\ pF/m}} = 300\ \Omega \text{ for twin wire feeder, typically}$$

(ii) The balanced or unbalanced condition at the antenna and the receiver input must be properly attended to, otherwise there will be a tendency to stray pick-ups by the feeder that result in ghost and interference problems.

(iii) The loss of the feeder cable should be minimum at the frequency of operation. Usually this increases with frequency because of core dielectric or spacer material loss. Dielectric materials are always more lossy than air. Cables with minimum use of dielectric spacing materials (which are commonly polyethelene, teflon, etc.), give lower losses. Weathering conditions like wetness and dirt accumulation on cable increase lossy, if both the conductors and the dielectric between them is exposed in the twin wire ribbon feeder. Special construction like tubular polyethelene spacer or semi-air-spaced *air lead* twin wire feeder reduce these losses considerably. Use of polyethelene foam dielectric for cable gives a lower loss at UHF. The cable loss is usually expressed in dB/100 m, or dB/100 ft, at reference frequencies typically 100 MHz, 200 MHz, and so on.

Velocity Factor

The velocity of electromagnetic waves along the cable using dielectric medium is reduced by the factor $1/\sqrt{\varepsilon}$, compared to the space velocity, where ε is the dielectric constant. This affects the wavelength along the cable, and hence is correspondingly reduced, for example, for the common ribbon feeder the velocity factor is 0.8, and for coaxial cable it is 0.66. Wavelength measurements along these cables correspond to free space wavelengths multiplied by these factors.

Twin Wire Feeders

Out of these parallel wire feeders, the flat ribbon type feeder is the most popular transmission line in use for VHF reception because of its low cost, low loss, convenient balanced 300 ohm impedance and good flexibility. In the simplest form, it has two stranded wire conductors, spaced 6 to 8 mm apart in polyethylene ribbon, as shown in Fig. 9.10.

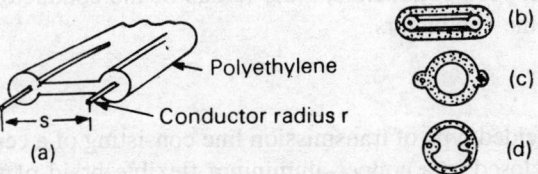

Fig. 9.10   Parallel wire feeder: (a) Flat ribbon, (b) Double clad ribbon, (c) Tubular outer conductor, (d) Tubular inner conductor

The cable is usually standardized at 300 ohm impedance to suit the folded dipole impedance of 300 ohm. Continental designs take the practical folded dipole impedance as $4 \times 60 = 240$ ohm and hence use the 240 ohm standard.

The feeder is essentially balanced and its impedance suits the folded dipole impedance and receiver input impedance of 300 ohm.

The loss of a twin wire ribbon feeder is quite low because of the marginal polyethylene spacing, typically 1.2 dB/100 ft at 100 MHz. This varies with the quality of the polyethylene and the thickness of conductors used. The loss increases up to 7.3 dB/100 ft at 100 MHz when the cable is wet in rainy weather. Old lines which develop minute cracks and accumulate dirt and moisture in them are especially lossy.

Double clad twin wire ribbon with plastic gasket, as shown in Fig. 9.10(b), is less susceptible to weathering and moisture. To make the cable more robust for outdoor use and less susceptible to wetting and weathering, the ribbon cable is provided a plastic jacket or cladding.

Alternative forms suitable for outdoor use employ tubular spacer with the conductor positioned on the inner or outer periphery as shown in Fig. 9.10(c) and (d). These cables have a little higher dry loss than the simple ribbon but the loss in wet condition is lower, about 2.5 dB/100 ft at 100 MHz.

Reduction in the polyethylene spacer between the conductors reduces the line loss as in the airlead feeder or open wire line with spacers as shown in Fig. 9.11.

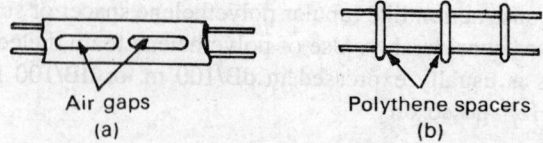

Fig. 9.11 (a) Airlead feeder, (b) Open wire line feeder

These lines are, however, mechanically less robust and liable to shorts or kinks that may disturb signal by mismatch and VSWR. Airlead twin wire line with 80% polythene removed or open wire line with polythene spacers have loss of only 0.2 to 0.35 dB/100 ft at 100 MHz.

Foam dielectric lines which have the two conductors embedded in polyethylene foam are useful particularly at UHF frequencies because of lower loss. The parallel wire line may be of the shielded type encased in a metal foil jacket which can be grounded to reduce interference pick-up. These are, however, costlier.

Characteristic impedance of a parallel wire line is given from its geometry by:

$$Z_0 = \frac{276}{\sqrt{\varepsilon}} \log \frac{s}{r} \tag{9.3}$$

where s is the spacing between conductors, r the radius of the conductors and ε the dielectric constant of the medium between the conductors.

Coaxial Cable Feeder

This is an unbalanced shielded type of transmission line consisting of a central conductor in a polystyrene dielectric completely enclosed in a copper-aluminium flexible braid of wires. A plastic jacket forms a protective coating as shown in Fig. 9.12.

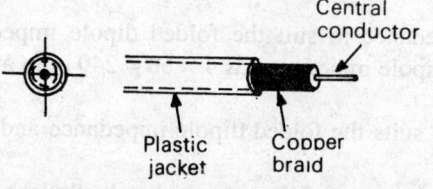

Fig. 9.12 Coaxial cable feeder

Brief cable data

Type	Z_0 Ω	O. diameter, in
RG-59 B/U	75	0.24
RG-58 C/U	50	0.24
RG-11 A/U	75	0.40
RG-8 A/U	50	0.40

The shielded cables have a greater capacitance and the losses due to the dielectric medium are also large, about 3.7 dB/100 ft at 100 MHz, but has the advantage of shielding due to the outer conductor grounding that makes the cable immune to stray interference pick-up and is also unaffected by inclement weather. Losses are reduced in special cables employing low loss foam dielectric, and larger diameters of the conductors.

Characteristic impedance of a coaxial cable is given from its geometry by:

$$Z_0 = \frac{138}{\sqrt{\varepsilon}} \log \frac{D}{d} \tag{9.4}$$

where D is the diameter of the outer conductor, d the diameter of the inner conductor and ε the dielectric constant of the spacer medium.

The cable is useful in aerial installations where stray interference pick-up is high. It is commonly used in the cable distribution systems of closed circuit television and CATV.

9.7 IMPEDANCE MATCHING

It is necessary to match the impedances at the antenna terminals and more importantly at the receiver terminals to ensure maximum power transfer and to reduce the VSWR effects due to reflections at these terminations. Mismatch at either ends of the feeder cable causes reflections and loss of power. Matching at one end, particularly the receiver input, can eliminate reflections at that end and hence the standing waves also. Mismatch is usually expressed in terms of the standing wave ratio (SWR) created on the line

$$SWR = V_{max}/V_{min}$$

where V_{max} and V_{min} are the voltage maximum and minimum on the line. It is given in terms of the impedances by

$$SWR = \frac{1 + |\rho|}{1 - |\rho|} \tag{9.5}$$

where ρ = reflection coefficient $= \dfrac{Z - Z_0}{Z + Z_0}$ (9.5a)

Z_0 is the line impedance and Z, the terminating impedance.

Generally the receiver input impedance is designed to suit the 300 ohm twin wire ribbon feeder, the popularly used low cost cable. This 300 ohm 'balanced' impedance is matched to the 75 ohm 'unbalanced' input impedance of the RF amplifier stage in the TV receiver by a suitable *'balun'* transformer. The 75 ohm input impedance is fairly constant over all the channels, whereas the antenna impedance may vary considerably over a wide range for different channels as its length is no longer $\lambda/2$ and may result in some mismatch loss. This may not matter if the receiver end is properly matched and the signal is large enough to tolerate the loss at the antenna.

Quarter Wave Matching

A quarter wave matching section of a transmission line can be used for matching, as shown in Fig. 9.13.

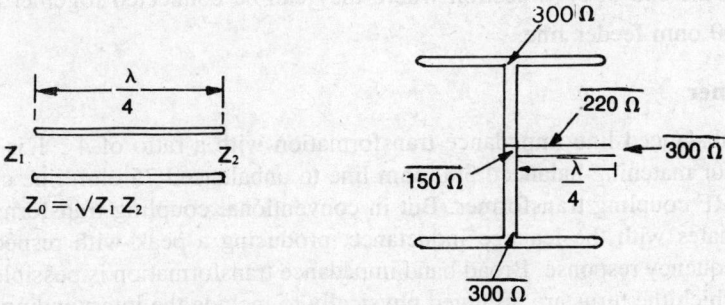

Fig. 9.13 Quarter wave matching

When terminated in Z_2 at one end, it offers impedance Z_1 at the other end given by

$$Z_1 = \frac{Z_0^2}{Z_2}$$

(9.6)

where Z_0 is the characteristic impedance of the line.

Hence for matching two impedances, Z_1 and Z_2, a quarter wave line of characteristic impedance Z_0 equal to $\sqrt{Z_1 Z_2}$ may be constructed to be interposed between the two. Thus a 150 ohm impedance of two parallel connected antennas stacked one above the other, $\lambda/2$ apart, may be matched to the 300 ohm line by a section of $\sqrt{300 \times 150} = 220\,\Omega$ line. The modified diameter of the line conductors with the same spacing, to give this impedance, can be calculated from the impedance expression:

$$Z_0 = \frac{276}{\sqrt{\varepsilon}} \log \frac{s}{r}, \text{ indicated earlier}$$

The matching section length can be $(1/4)$ or $(3/4)$ λ or an odd multiple of quarter wave in general. Longer lengths like $(3/4)\lambda$ may be convenient under some conditions, for example, in connecting together stacked arrays or broadside arrays spaced wide apart. Two Yagi antenna arrays horizontally stacked, can be conveniently connected together by $(3/4)$ λ matching sections, as shown in Fig. 9.14 to match to a similar line of impedance 300 Ω.

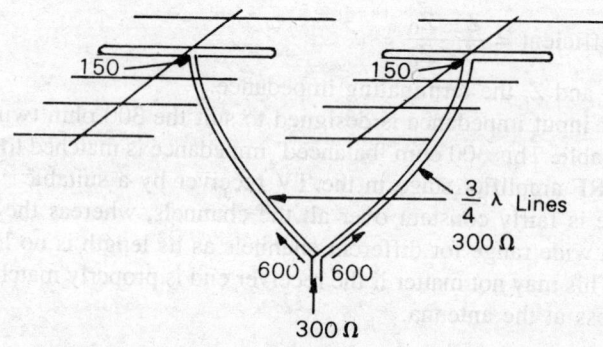

Fig. 9.14 Impedance matching of stacked Yagi aerials

The antenna impedance of about 150 Ω of the multielement Yagi antennas will be stepped up to $\frac{300^2}{150} = 600\,\Omega$ at the end of $3/4\lambda$ section where they can be connected together in parallel to provide matching to a 300 ohm feeder line.

Balun Transformer

A balanced to unbalanced line impedance transformation with a ratio of 4 : 1 is often required in the antenna circuits for matching balanced 300 ohm line to unbalanced 75 ohm line circuits. This could be done by a 2 : 1 RF coupling transformer. But in conventional coupling transformers, the interwinding capacitance resonates with the leakage inductance, producing a peak with respect to frequency. This limits the high frequency response. Broad band impedance transformation is possible by transmission line transformers in which the turns are arranged physically to include the interwinding capacitance as a part of the characteristic impedance of the line. With this technique, bandwidths of the order of hundreds of megahertz can be achieved.

A 4 : 1 impedance transformation and matching enabling a compact balanced to unbalanced connection is possible by combining two quarter wave matching sections as shown in Fig. 9.15(a).

This unit is called a 'balun' allowing balanced to unbalanced impedance matching. The quarter wave sections of characteristic impedance equal to 150 ohm have a pair of terminals in series to provide a match to 300 ohm and the other pair of terminals in parallel to match the 75 ohm. Either of the paralleled ends may then be earthed without affecting the balanced conditions at the other end.

Typically, the lines have the form of wire-pairs bifilar wound on a suitable, core, as shown in Fig. 9.15 (b). The length of the line determines the upper frequency limits. The larger the core permeability, the fewer the number of turns required for a given frequency response. Ferrite core torroids with widely varying core material characteristics, provide a satisfactory response over a wide frequency range. Decreasing permeability with increasing frequency reduces the effective inductance at higher frequencies and produces a wideband flat frequency response. The low frequency limit is set by the primary core inductance.

In practice, the balun consists usually of a two-window ferrite core as shown in the Fig. 9.15 (c), with two windings of four turns each coupled to provide a balanced to unbalanced match and an impedance transformation from 300 to 75 ohm uniform over a wide range of frequencies.

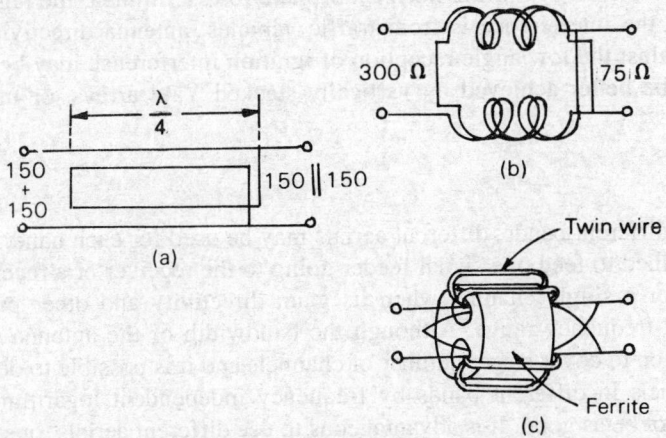

Fig. 9.15   Balun transformer: (a) Equivalent circuit principle, (b) Symbol, (c) Construction

9.8   TELEVISION RECEPTION PROBLEMS

The signal available for the TV receiver depends upon the field strength at the receiving site and the gain of the receiving antenna. The minimum signal strength required for various grades of services, viz. grades A, B and local community have been given in Chapter 7. For grade B service, field strength of the TV signal required is at least 224 μV/m in band I and 631 μV/m for band III. In practice, reception may be affected by local interference from ghost signals, static, ignition systems, etc., depending upon the location, antenna directivity, actual site of antenna installation, and so on. A minimum signal-to-noise ratio of 40 dB is necessary for good picture quality. The type of antenna and feeder are chosen after considering the field strength and interference problems at the location of the receiving site.

In areas near to the transmitter where the signal is quite strong, simple indoor antenna in the form of V-type dipole made of telescopic rods may provide adequate signal to the receiver. The telescopic dipole rods can be adjusted in length, angle and orientation by their swivel base for obtaining best possible reception. It can be a set-top dipole or a separate table top dipole for mounting it at a more favourable signal position through an extention feeder. Such a dipole has 75 Ω impedance and hence includes a matching transformer at its base to step it up to 300 Ω to match it to the standard twin wire feeder.

Strong signal areas are often haunted by ghost signals arriving by reflections from tall buildings and hills nearby. It is difficult then to get a clear picture by use of an indoor dipole or even an outdoor antenna unless a multielement directive Yagi antenna is used, carefully mounted and positioned to minimize the ghost signal pick-up. Ghosts may also be caused by mismatch of the feeder with the antenna and the receiver input, when the feeder is rather long. This can be avoided by proper impedance match at the antenna and the receiver in particular.

These problems become more complex in antenna installations for customers in high-rise buildings with a number of antennas clustering up on the terrace at the top affecting each other by their close proximity. Care should be taken not to mount them directly in front of each other nearer than about three to four times the wavelength. They should be preferably mounted broadside, i.e., sideways or at different heights when in front of each other. A better solution to these problems in high-rise buildings is to employ a shared master antenna or community antenna system where a high gain directive antenna feeds to all the receivers in the building or the locality where the interference is from ghost signals and other man-made sources. Antenna directivity in the horizontal plane to discriminate the signal against these may generally help. When the interference is from traffic vehicles, antenna directivity in the vertical plane that discriminates against the low angle reception of ignition interference may be useful in getting a better picture. This can be better achieved by vertically stacked Yagi arrays, or multielement Yagi aerials.

Signal Combining

When receiving signals on different bands, different aerials may be used for each band, and signals from these may have to be combined to feed one signal feeder going to the receiver or a receiving system. An antenna is best designed for a single channel when its gain, directivity and other parameters can be optimized over the channel-frequency range. Although the bandwidth of the antenna can be increased readily by special construction to cover larger number of channels and it is possible to design a multiband antenna to cover the channels in different bands by frequency independent logarithmic antennas, the performance of these may not be as good. It is advantageous to use different aerials, one for each channel or band and feed their outputs to one common feeder. It is possible to combine the outputs from these different aerials by means of filters in the respective lead, which serve to pass the particular band and isolate it from the other aerials. For example, if channels in the band I and channels in the band III are to be combined, a low pass filter is included in the band I antenna lead and a high pass filter in the band III antenna lead as shown in Fig. 9.16.

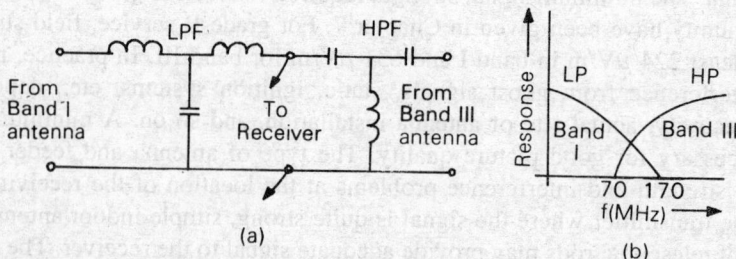

Fig. 9.16 Signal combining with filters: (a) HP-LP filter combination, (b) Response

The filters can be designed with appropriate cut-off frequency as the crossover frequency and terminal impedance. Similar low pass and high pass filters are used to combine band III and band IV-V antenna signals. If the output from FM broadcast band is also to be combined, a bandpass filter is included in the lead for the band II antenna to feed the signal to common feeder. The triple combination forms a triplexer. These combining filters need to be mounted in weatherproof boxes near the antenna masts, in order to avoid separate downlead feeders.

Signal Splitting

If the signal available from the antenna system into the downlead feeder is sufficiently strong, the signal may be split into two or more branches feeding a number of receivers. It is possible to do this simply by resistive star network as shown in Fig. 9.17.

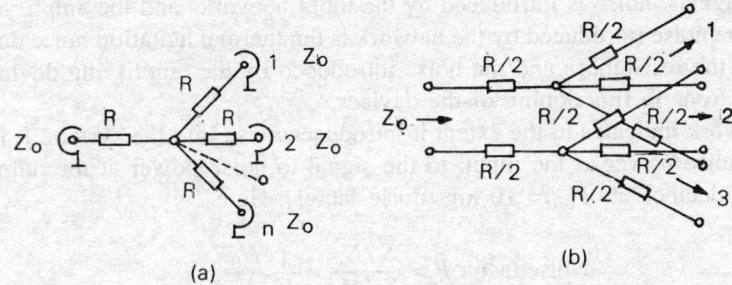

Fig. 9.17 Resistance pad signal splitter: (a) Unbalanced, (b) Balanced arrangement

This maintains the impedance match all the way round the network, when $R = Z_0(n - 1)/(n + 1)$, Z_0 being the common feeder impedance and n the number of outlet connections. For a balanced feeder splitter, a double star balanced network may be used. The signal is of course attenuated. The signal loss to each outlet from the common downlead feeder depends upon the number of outlets. The starpoint splitter can be constructed out of non-inductive carbon composition low wattage resistances in a shielded box. In order to compensate for the losses in the signal splitters and long feeders, amplifiers for the weak signal are necessary. These can be single channel or wideband type to cover the available channels. Low noise transistor amplifiers are popularly used for such shared aerial systems as booster amplifiers.

9.9 TELEVISION SIGNAL BOOSTER AMPLIFIERS

Booster amplifier is a high gain RF amplifier for one or more TV channel frequencies providing a good signal-to-noise ratio to minimize noise spots in the picture and ensure good synchronisation of weak signals. A good booster amplifier can be helpful in the following situations:

(i) Booster amplifier is useful in increasing the sensitivity of a receiver when its gain is moderate. For example, if a receiver has a sensitivity of, say, 100 μV, a booster amplifier which steps up the signal by 10 times increases the basic sensitivity of the amplifier and receiver to 10 μV. Along with the signal, the noise voltages at the input of the amplifier are also amplified.

(ii) The booster amplifier can effect improvement in the signal-to-noise ratio if the noise factor of the amplifier is less than that of the receiver tuner input circuit.

(iii) If the antenna downlead feeder has a tendency to pick-up extraneous noise and interference, a booster amplifier mounted near the antenna mast head may be useful in raising the signal level higher up so that the signal-to-noise ratio is maintained to the level available at the antenna.

(iv) TV signal booster is particularly useful in reception of weak signals in the *fringe areas*. Here the signal may be very weak even though a high gain directive antenna is used to pick-up maximum signal.

(v) The booster amplifier is also useful in lifting the wanted signal, when the wanted signal falls to a low level in a bid to minimize ghosting and other interference, by orientation of the receiving antenna.

The noise introduced by a booster amplifier or a two-port network in general is specified in terms of noise figure, noise temperature, and so on. Some basic terms and concepts used in noise evaluation of a receiving system are discussed below.

Noise Figure

In reception of weak signals, noise is introduced by the input networks and the amplifying devices into the amplifier signal. The noise introduced by the network is the thermal agitation noise due to the random motion of electrons in the resistances and the noise introduced by the amplifying devices is due to the random motion of electrons in functioning of the devices.

Noise figure of network indicates to the extent it introduces noise into the system. It is defined as the ratio of the signal-to-noise power at the input, to the signal-to-noise power at the output. The ratio is generally expressed in decibels as $F_{dB} = 10 \log$ (noise factor). Here

$$\text{Noise factor } F = \frac{S_i/N_i}{S_0/N_0} = \frac{1}{G}\frac{N_0}{N_i} \tag{9.1}$$

where S_i is the signal power at the input, S_0 the signal power at the output, N_i is the noise power at the input, and N_0 the noise power at the output. As ratios, the terms noise figure and noise factor are equivalent.

Noise voltage is commonly generated in electronic circuits in the form of thermal agitation noise due to the random motion of electrons. The noise voltage in a resistance R is given by:

$$E_n^2 = 4kTBR \tag{9.2}$$

where k is the Boltzmann's constant = 1.38×10^{-23} J/°K, T the absolute ambient temperature in °K, B the bandwidth in Hz, and R the resistance in ohms.

For a *matched* termination at the input, the available noise input power is given by:

$$N_i = \left(\frac{E_n}{2R}\right)^2 \times R = \frac{4kTBR}{4R^2} \times R = kTB \tag{9.3}$$

Thus $N_i = kT_0B$ is the available noise power due to the source impedance under the assumption that the temperature equals the standard noise temperature T_0 which is generally 290 °K. Hence

$$F = \frac{1}{G}\frac{N_0}{N_i} = \frac{1}{G}\frac{N_0}{kT_0B} \tag{9.4}$$

where N_0 is the output noise power in watts, k the Boltzmann's constant in Wsec/°K, T_0 the absolute ambient temperature in °K, B the effective noise bandwidth in Hz, and $kT_0 = 4 \times 10^{-21}$ Wsec.

The total noise power at the output N_0 is referred to the amplified reference noise power kT_0B.

Noise Temperature

Noise output power N_0 is composed of the amplifier reference noise power $N_i = kT_0B$, and the noise power N_N introduced by the network. The noise figure may thus be split as:

$$F = \frac{1}{G}\frac{N_0}{N_i} = \frac{GN_i + N_N}{GN_i} = 1 + \frac{N_N}{GN_i} = 1 + F_N \tag{9.5}$$

where F_N represents the contribution of the noisy network. The excess noise power $N_N = GN_{N_i}$, (N_{N_i} is the excess noise power referred to the input). This may itself be expressed as kT_EB, considered as being due to an additional temperature T_E above the reference T_0. Then

$$F = 1 + \frac{N_N}{G_N} = 1 + \frac{G(kT_E B)}{G(kT_0 B)} = 1 + \frac{T_E}{T_0}$$

Here T_E is referred to as the noise temperature of the system

$$T_E = T_0(F - 1) \tag{9.6}$$

Although the concept of noise figure is frequently used for specifying noise properties of a system, it is not particularly convenient in dealing with noise in antennas, low noise amplifiers and devices. Noise temperature concept derived from the early work in radio astronomy, is convenient in the sense that it is additive like the noise power from the various components in the system.

A network that does not add on any noise to the signal just amplifies the input noise. The noise factor in this case is unity and the noise figure is zero dB. An active amplifier network, in addition to amplifying the input contributes its own noise to the system so that the output noise is greater than the input noise power, and the noise factor is greater than 1.

When a number of networks or stages having noise factors $F_1, F_2, ..., F_n$ and gains $G_1, G_2, ..., G_n$ respectively, are cascaded together, the combined single network will have a noise factor given by:

$$F = F_1 + \frac{F_2 - 1}{G_1} + \frac{F_3 - 1}{G_1G_2} + ... + \frac{F_n - 1}{G_1G_2G_3 ... G_{n-1}} \tag{9.7}$$

If the gain of the first stage is large, F_1 the noise factor of the first stage is predominant and is of far greater importance than that of the subsequent stages. The noise figure of the input stage of the RF tuner or the booster amplifier must be as low as possible. Modern transistor booster amplifiers or receiver input circuits in the RF tuner provide noise figures as low as 4 dB in band I and band III.

The noise figure of a receiver can be looked upon as introducing a proportional noise voltage at the aerial input along with the wanted signal. For example, consider the input stage of a tuner RF amplifier using an equivalent noise resistance of 300 Ω and a termination of 300 Ω at the input. The noise voltage at the input, at 17 °C and for 7 MHz bandwidth is given by:

$$E_n = \sqrt{4kTBR}$$

$$= \sqrt{4 \times 1.38 \times 10^{-23} \times 290 \times 7\,M \times (300 + 300)}$$

$$= 6.75\ \mu V$$

For a constant temperature and bandwidth, the network and amplifying device at the input introduce a fixed amount of noise power. For comparison and measurement of noise, the term kT_0 number or count is sometimes used. The reference noise power per hertz of bandwidth is $1\ kT_0$. It is then also the minimum signal power required for signal-to-noise ratio of 1. The receiver sensitivity is expressed in units of kT_0, the number of units representing the noise factor F, so that the noise figure in dB is equal to $10 \log F$ or $10 \log (kT_0$ number$)$.

The noise voltage at the matched receiver input is given by $\sqrt{kT_0\ BR}$, where $B = 7$ MHz, and $R = 300\ \Omega$ for the common TV standards. Then $1\ kT_0$ corresponds to a noise voltage of

$$E_n = \sqrt{4 \times 10^{-21} \times 7\,\text{M} \times 300} = 2.9\ \mu\text{V, across } 300\ \Omega$$

Higher multiples of kT_0 number corresponds to a higher noise voltage proportional to the square root of the multiple number. Thus a $10\ kT_0$ sensitivity corresponds to $\sqrt{10} \times 2.9 = 10\ \mu\text{V}$, $20\ kT_0$ sensitivity corresponds to $\sqrt{20} \times 2.9 = 13\ \mu\text{V}$, and so on. These voltages correspond to signal voltages for signal-to-noise ratio of 1. The signal-to-noise ratios (SNR) required for different picture qualities are generally taken as follows:

Picture quality	SNR-dB	SNR
Barely satisfactory	30	36
Good, noise-free	40	100
Excellent	46	200

In addition to the noise produced at the input stage of a receiver, noise from the antenna picked-up from the extraneous sources also contributes to the total input noise of the receiver and affects the quality of the picture.

Signal-to-Noise Ratio and Picture Quality

Assessment of picture quality in presence of noise interference is a matter of subjective judgement that may vary from viewer to viewer. The assessment may also be affected by the subject matter in the picture. The performance objectives of television receivers with respect to signal-to-noise ratio are generally defined on the basis of tests on viewers conducted by the Television Allocations Study Organisation (TASO) of the USA. The essential results of these tests are indicted in Fig. 9.18 which gives the signal-to-noise ratio of the TV signal corresponding to the rated quality and percentage to viewers.

The TASO signal-to-interference ratio is the ratio of the rms voltage of the TV signal during the synchronisation pulses to the rms noise voltage in the 6 MHz signal bandwidth. The evaluation of picture was based on still images since these are more critically observed than moving images. The scale of picture quality and the corresponding signal-to-noise ratios as stated by 90% of the viewers to be satisfactory (obtained from Fig. 10.3) are given in Table 9.1. The signal-to-noise ratio indicated here is the ratio of the picture signal peak-to-peak voltage to weighted rms noise. The signal-to-noise ratio value is obtained by adding 0.6 to the TASO signal-to-interference ratio.

Antenna Noise Temperature

A receiving antenna feeding into a matched transmission line or a booster amplifier has its impedance equal to the radiation resistance of the antenna at its feed points. The antenna delivers half the power to the receiver and loses the remaining half in its radiation resistance as re-radiated power. Noise generated in the antenna is not due to thermal agitation noise in the radiation resistance. Noise is received by the

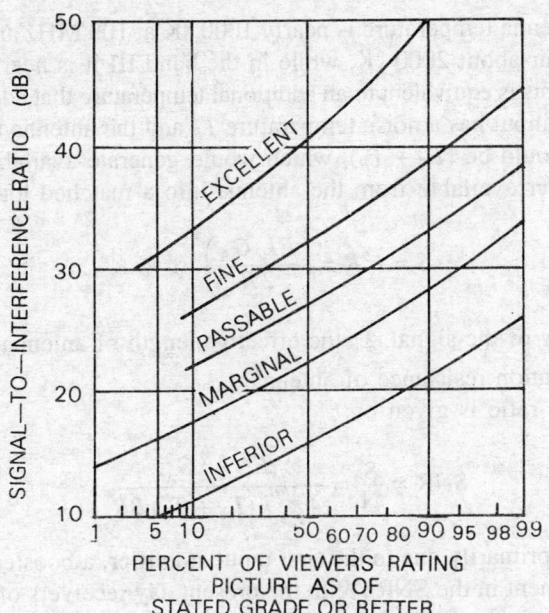

Fig. 9.18 Signal-to-noise ratio and picture quality rating by viewers (TASO), reproduced from 'Investigation of Surface Educational Television Distribution' by Johnson & Enriquez, *IEEE Trans.* BC-22, No. 1, pp. 45–52. (*Courtesy*: IEEE, N Y)

Table 9.1 Picture Quality Scale and Corresponding SNR for the Picture Quality Rating by 90% of the Viewers*

Grade	Quality rating	Description	SNR (dB)
1	Excellent	Picture of extremely high quality as good as could be desired	50.6
2	Fine good	Picture of high quality providing enjoyable viewing	39.4
3	Passable	Picture of acceptable quality, interference not objectionable	32.8
4	Marginal	Picture poor in quality, interference somewhat objectionable	28.6
5	Inferior	Picture very poor but watchable interference definitely objectionable	22.6
6	Unusable	Picture too bad for watching	< 22

*Reproduced from 'Investigation of Surface Educational Television Distribution's by Johnson & Enriquez, *IEEE Trans.,* BC-22, No. 1 (1976), pp. 45–52 (*Courtesy*: IEEE, N Y)

antenna from different sources that include the atmospheric radio noise or static due to lightening discharges, extra-terrestrial solar and cosmic noise, man-made noise from electrical appliances and ignition systems, etc. The actual noise picked-up by an antenna depends upon the frequency range of operation and the directivity. The noise is subjected to diurnal and seasonal variations as affected by the solar activity. The total noise picked-up by an antenna can be accounted for by assigning the antenna noise temperature T_A to its radiation resistance. The thermal agitation voltage in the radiation resistance at that temperature T_A corresponds to the noise pick-up of the antenna.

It is observed that the antenna temperature is nearly 1000 °K at 100 MHz and rises very rapidly below 100 MHz. In band I it is near about 2000 °K, while in the band III it is nearly 290 °K. The man-made noise in industrial environments is equivalent to an additional temperature that adds on to these temperatures.

If the receiver or booster input has a noise temperature T_R and the antenna temperature is T_A, the total system noise temperature would be $(T_A + T_R)$, which would generate available noise power of $k (T_A + T_R) B$, while the signal power available from the antenna into a matched load is:

$$S = i_s^2 R = \left(\frac{EL_e G_A}{2R} \right)^2 \times R$$

where E is the field intensity of the signal, L_e the effective length of antenna $= \dfrac{\lambda}{\pi}$ for a dipole, G_A the antenna gain, the R the radiation resistance of antenna.

Hence the signal-to-noise ratio is given by:

$$\text{SNR} = \frac{S}{N} = \frac{E^2 \lambda^2 G_A^2}{4 \pi^2 k (T_A + T_R) BR} \tag{9.8}$$

Since the system noise is primarily due to the first input amplifier, a booster amplifier with a superior design can provide improvement in the SNR. With the present day receivers of high sensitivity and good noise figure, an additional booster amplifier may hardly be necessary to increase the gain limited sensitivity. In the band I channels the noise temperature of the antenna and the man-made noise is high and predominant. Hence the signal-to-noise ratio cannot be further improved. In the UHF channels, the antenna noise temperature is much lower and hence improvements with low noise boosters in the UHF bands can be considerable.

Booster Amplifier Circuits

Booster circuits are basically RF amplifier circuits designed with special care to minimize the noise figure of the circuit by use of low noise transistors and optimum operating conditions. Grounded emitter circuit gives a higher gain, but grounded base configuration is often used although it gives a lower gain. This is because, this configuration is more stable at high frequencies and at the widely varying temperatures when it is used near the antenna mast-head. The circuit may be a single channel tuned type or untuned resistive load type to operate over a wide range of frequency covering band I through band III. The power required for operating the boosters may be supplied from batteries also mounted on the masts or it may be supplied through the feeder transmission line *via* decoupling RF filters, as shown in Fig. 9.19.

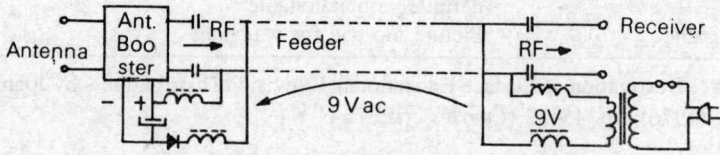

Fig. 9.19   Power supply to antenna booster at the antenna head

The circuit of an antenna booster is shown in Fig. 9.20, providing a gain of 20 dB over the frequency range 175–195 MHz at a noise figure of 6 dB. The circuit uses a high pass filter at the input circuit of the common base input amplifier to allow the band I and band III signals to pass. The interstage coupling and the output load is formed by the bandpass networks.

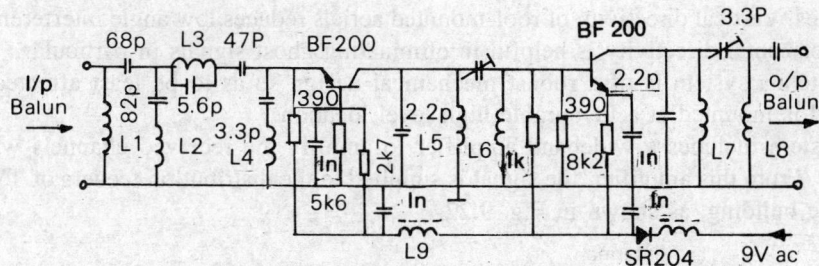

Fig. 9.20 A Band III antenna booster

The power is supplied *via* the feeder at 9 V ac with decoupling RF filters at the booster and the TV receiver ends. Balun transformers are provided at both the input and output of the booster to match them to the balanced feeder.

A 3-stage wideband amplifier in the hybrid IC technique designed for use in mast-head booster amplifiers or as pre-amplifier in maser antenna TV systems is shown in Fig. 9.21.

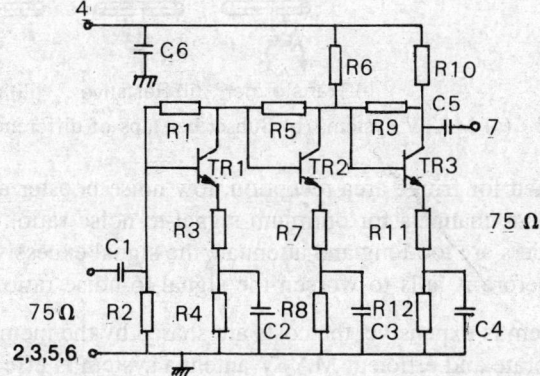

Fig. 9.21 Wide band booster amplifier OM 335 (*Courtesy*: ELCOMA PHILIPS)

The amplifier OM 335 provides a transducer gain of 27 dB over 40–860 MHz at a noise figure of 5.5 dB, and output voltage greater than 98 dBμ at – 60 dB intermodulation distortion. A lower 15.5 dB gain is available with a 2-stage amplifier OM 321.

9.10 SHARED ANTENNA SYSTEMS

These systems employ very efficient high gain antenna systems of robust mechanical design to be shared by a number of receivers in high rise apartment buildings, hotels, or groups of houses. They are called master antenna television (MATV) systems. The MATV system features are as follows:

(i) Since the choice of each receiver can be different, its antenna system includes antennas for receiving all available program channels in the area, oriented in suitable directions and combining the outputs from all of them into a common feeder system. The system may, if required, include radio broadcast receiving aerials particularly for the FM broadcast band.

(ii) The system employs well designed high gain antennas with good directivity in the vertical and horizontal planes. Vertical directivity of roof-mounted aerials reduces low angle interference from vehicle ignition and horizontal directivity is helpful in eliminating ghost signals in particular.

(iii) The antenna system is of a robust mechanical design so as to be least affected by winds and weathering and is mounted at a favourable high level location.

(iv) The system includes a wideband amplifier to amplify the received channels which may be in different bands. From this amplifier, the signal is supplied to the distribution feeders of TV sets at various locations in the building, as shown in Fig. 9.22.

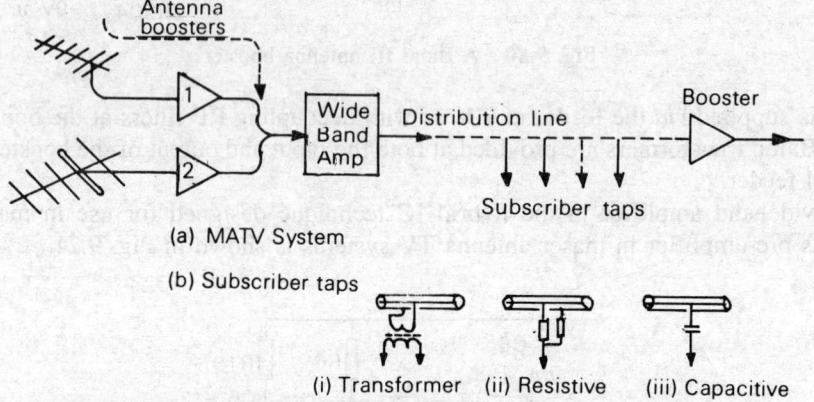

Fig. 9.22   (a) MATV system, (b) Subscriber taps of different types

(v) If the system is designed for fringe area reception, low noise booster amplifiers may be mounted at the antenna of the respective channels for optimum signal-to-noise ratio.

(vi) If the distribution feeders are too long and attenuate the signal excessively, booster amplifiers are necessary to lift the signal before it falls to worsen the signal-to-noise ratio.

Although the antenna system is expensive, the costs are shared by the members. Since the costs of the antennas are shared, an elaborate and efficient MATV antenna system is effectively not very expensive. MATV systems are sometimes referred to as community antenna television (CATV) systems; but the latter term generally implies an antenna system for a wider community group.

A CATV system is a closed circuit television, whereby a particular community—a group of houses or estate in weak signal areas—is supplied television signals from a specially designed antenna system, usually mounted on a high level site like a hill or a high tower. The signals from each antenna for the available program channels in the area, are amplified first by low noise boosters at the antenna mast. These signals are combined and further amplified by wideband amplifiers that feed to the member subscribers *via* subscriber taps. The taps are in the form of balun transformer coupling or capacitive coupling or resistive pads between the cable and the standard receivers at the subscribers' homes as shown in Fig. 9.7(b).

Modern cable television systems having such program options are quite elaborate. In addition to the basic goal of providing high quality signals to the viewers, interactive communication has been developed to allow the subscribers to interact with the program source to request and supply various types of information that can be transmitted over a wideband system. This has a wide potential range of applications.

A CATV system typically consists of a *head end* that receives the signals from antennas, satellites or local studio origin, and processes them; a *trunk amplifier system* that carries a multiplicity of programs through a coaxial cable with minimum distortion; a *distribution system* which carries signals to the subscriber areas from bridging amplifiers tapped on the trunk system, and finally the *subscriber drops* that are fed from taps on the distribution system. The cables are either aerial construction on telephone poles or underground.

A three-stage wideband amplifier in the hybrid technique OM 337 designed for use in MATV systems, and as a general purpose amplifier for VHF and UHF range applications requiring high output level as in distribution amplifiers, is shown in Fig. 9.23.

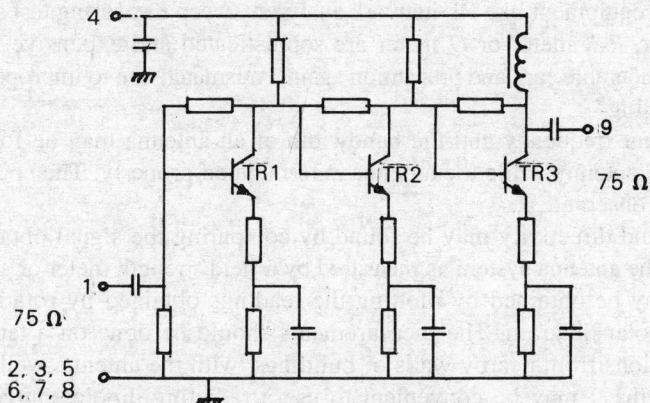

Fig. 9.23 Circuit diagram of hybrid VHF-UHF wide amplifier OM 337 (*Courtesy*: ELCOMA PHILIPS)

It provides a gain of 26 dB over a frequency range of 40 to 860 MHz flat within 1 dB. The output voltage can be up to 112 dBμ at −60 dB intermodulation distortion with a noise figure of 9.8 dB.

A simple distribution amplifier for channel 4, using a common emitter-tuned amplifier is shown in Fig. 9.24.

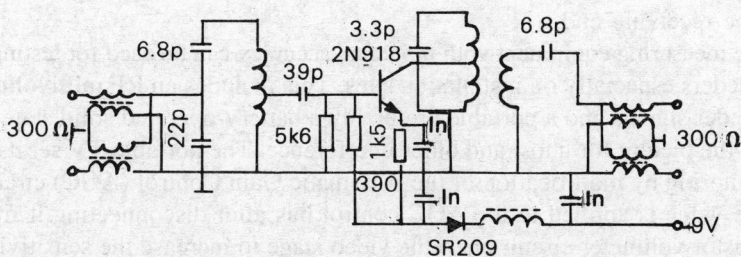

Fig. 9.24 Channel 4 distribution amplifier

The amplifier can distribute the amplifier antenna signal to a number of TV receivers, through resistive signal splitters. The loss is considerable but it is useful in isolating the mutual interference between the receivers. A better signal splitting and isolation between the outlets may be obtained by an emitter follower for each outlet.

9.11 ANTENNA AND FEEDER MEASUREMENTS

Testing and measurements on TV antenna systems and feeders are essential whenever a good and efficient antenna system is being installed to provide a high gain and directivity, and a proper impedance match is to be ensured between the antenna, feeder ends and the receiver input to prevent a high SWR on the feeder that may cause signal loss and ghosts. The measurements include measurement of antenna gain, resonant frequency and band-width, impedance, directivity; feeder impedance and its loss, impedance matching, SWR, and so on. A master antenna system requires a check-up of decoupling between the receivers, isolation and matching of the filters combining the outputs of the different antennas into a common feeder, a check on the gain and possible overdrive in the booster and wideband amplifier used.

Standard laboratory equipment like RF network analyser, sweep measuring test set, tunable RF voltmeter or field strength meter, R-X meter or Q meter are sophisticated and expensive. The RF measurements, in general, require considerable care and precaution against mismatch due to improper connectors, connecting cables and stray coupling.

The antenna resonant frequency and the bandwidth of an antenna may be found by coupling tightly a grid-dip meter to the antenna while the antenna is terminated properly. The coupling is reduced in case multiple resonance is observed.

The antenna gain and directivity may be found by comparing the signal obtained by a simple dipole with that obtained by the antenna system as measured by a field strength meter or a tunable microvoltmeter. Directivity pattern may be obtained by plotting the readings obtained by rotating the antenna through known angles, on a polar diagram. The measurements should be done on a terrace or ground clear of obstructions or reflections from nearby walls or buildings, with the antenna well above the ground by at least 2 to 3 wavelengths. I may be convenient to use a radiating dipole source at a sufficiently large distance fed by a power RF signal generator.

Feeder impedance and losses can be calculated by measurement of L, C and Q of a short length of the feeder sample, at the frequency of interest on a Q meter or an R-X meter. The loss may be estimated by measurement of the sending end and receiving end voltages of the feeder on a selective voltmeter or a field strength meter. A selective or tunable voltmeter is essential as a wideband RF millivoltmeter may give stray RF pick-up readings although this can be confirmed by switching the RF source off. The feeder losses can be directly found by replacement of the length of the feeder by an RF attenuator to obtain the same reading at the receiving end.

Simpler portable measuring equipment with moderate accuracy can be used for testing and measurements on antennas and feeders especially on installation sites. This includes an RF millivoltmeter for measuring the antenna and feeder signals and a portable preferably a battery-operated solid state TV receiver set for visual check-up of the picture for ghost and other interference. The portable TV set itself can be modified for TV signal monitoring by modification of the Automatic Gain Control (AGC) circuit such that a fixed bias of appropriate value is applied to the AGC control bus after disconnecting it from the AGC stage, and adding a transistor voltmeter circuit on to the video stage to increase the sensitivity for measurement of the detected dc voltage which is proportional to the antenna signal. The bias may have to be altered to accommodate the wide variation of the signal in more ranges. This requires tampering with the receiver which may not be quite so simple. Moreover, the readings will be affected by fine tuning which affect the picture as well as sound carrier considerably, and may be useful in relative terms only.

It is, therefore, worthwhile instead, to construct a simple antenna signal strength meter from a good RF tuner by following it up with a high gain video IF stage and simple detector meter circuit or by a transistor voltmeter circuit to increase the measuring sensitivity. The sketch of such signal strength meter is shown in Fig. 9.25(a), while 9.25(b) shows a circuit arrangement of the VIE amplifier and metering circuit. The instrument serves for convenient monitoring and relative evaluation of the antenna signal, and with calibration can be used for measuring purpose also.

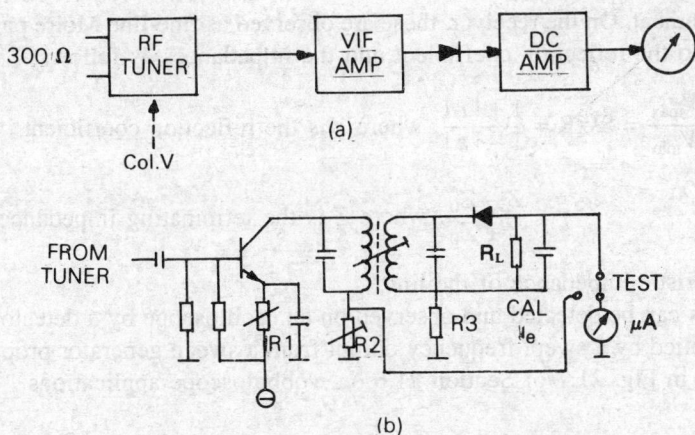

Fig. 9.25   A simple antenna signal strength meter

Detection of Standing Waves   Impedance mismatch between the antenna, feeder, balun transformers, filters and the receiver input is indicated in standing waves on the feeder line, due to power reflected at the mismatch points. A number of measurement set-ups are possible to detect or visualize the standing waves.

When adequate power is available (over 0.5 W), a neon glow lamp can be used to detect the current maxima and minima on the line fed by the RF power at one end and terminated by the usual component at the other. The lamp indicates the current maxima by brightest glow and minima by weakest glow. A uniform glow over the length of the line indicates proper matched conditions.

The standing waves on a twin wire ribbon feeder may be detected by means of a small rigid loop probe connected to the input coaxial cable of a sensitive antenna test instrument or FSM, by moving it along the feeder as shown in Fig. 9.26(a). Figure 9.26(b) shows the resonant loop probe tuned to the received frequency for greater sensitivity.

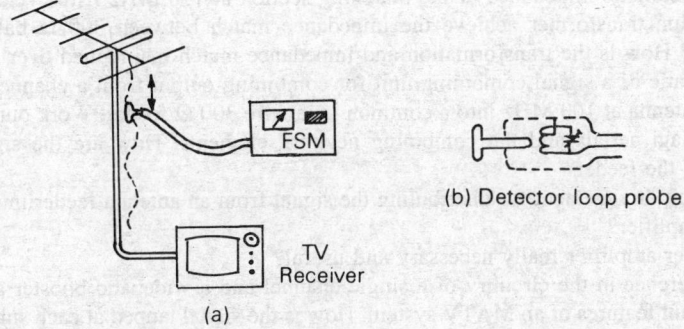

Fig. 9.26   Standing wave ratio detection by loop probe and field strength meter

If a good resonant coupling is desired for higher sensitivity, a tuned circuit may be added in the coupling between the loop and the cable. The measurement is done at a strong signal frequency, or at the local oscillator frequency of the monitoring receiver if the test instrument has the corresponding tuning range. The effects of external oscillator pick-ups should be guarded against, lest they disturb the standing wave measurement. On the receiver, these are observed as fine-line Moire patterns. The standing wave ratio is related to the reflection coefficient and the impedances as follows:

$$\frac{V_{max}}{V_{min}} = SWR = \frac{1 + |r|}{1 - |r|}, \text{ where } r \text{ is the reflection coefficient}$$

and
$$r = \frac{Z - Z_0}{Z + Z_0}, \text{ where } Z \text{ is the terminating impedance}$$

and Z_0 is the characteristic impedance of the line.

The standing waves can be detected and observed on an oscilloscope by a detector probe at the input of the feeder line supplied by a swept frequency output from a sweep generator properly matched to the feeder. This is shown in Fig. 21.7 of Section 21.6 on wobbuloscope applications.

REVIEW QUESTIONS

9.1 Explain the following terms with respect to antennas: directivity, beam width, front-to-back ratio, directive gain, antenna gain and velocity factor.

9.2 Explain how a parasitic element beside a dipole acts as a director or a reflector.

9.3 What is the effect of the addition of reflector and directors to a dipole on the dipole impedance, directivity and bandwidth?

9.4 State the advantages of Yagi aerials that are popularly used for TV reception. When is the vertical stacking and horizontal stacking useful in TV reception?

9.5 Sketch the dimensions of a 3-element Yagi aerial for channel 4, and a 5-element Yagi aerial for channel 5.

9.6 Draw the sketch of a folded dipole that can provide an impedance of 1200 Ω at the centre.

9.7 Compare the properties of a twin wire and a coaxial TV feeder. State where the coaxial feeder is essential.

9.8 A twin wire feeder is to be used as an indoor dipole. If the velocity factor of the feeder is 0.8, find the dimensions for channel 5 reception.

9.9 A quarter wave matching section is used to match 75 Ω dipole impedance to 300 Ω feeder. Calculate the length and characteristic impedance of the matching section at 178 MHz if the velocity factor is 0.8.

9.10 How does a balun transformer achieve the impedance match between 300 Ω balanced line and 75 Ω unbalanced line? How is the transformation and impedance match maintained over a wide band?

9.11 Draw the schematic of a signal combining unit for combining output from a channel 4 Yagi aerial and an FM broadcast antenna at 100 MHz into a common twin wire 300 Ω feeder. Work out the dimensions of the three element Yagi aerials and the combining network elements. How are the signals separated at the receiving end of the feeder?

9.12 Explain the circuit arrangement for distributing the signal from an antenna feeder into four receivers, using a distribution amplifier?

9.13 When is a booster amplifier really necessary and useful?

9.14 What is the difference in the circuitry of a single channel and a wideband booster amplifier?

9.15 State the important features of an MATV system. How is the signal tapped at each subscriber from a coaxial or a twin wire feeder?

9.16 A receiving system has noise figures of the components as follows:
Booster noise figure   = 8 dB in band III
RF amplifier noise figure = 10 dB
Mixer noise figure   = 14 dB
Feeder noise figure   = 2 dB for 30 m of length
If the antenna temperature in band III is 290 °K, calculate the noise figure of the receiving system (a) without booster, (b) with booster mounted near the masthead, and (c) with booster mounted near the receiver.

9.17 Explain how the impedance of an antenna feeder can be measured with the help of a standard signal generator, RF attenuator and an RF voltmeter or a field strength meter.

9.18 Explain how multiple ghost images may be caused by long antenna feeders, as well as multipath reception. Give the methods of reducing the ghost problems.

9.19 What are the sources of impulse noise interference on TV receivers? Discuss the methods to reduce the interference.

9.20 The following measurements are made on a 10 cm length of an antenna feeder, with a Q meter at 100 MHz. The open-circuited line connected across the Q-meter capacitance resonates with a suitable inductance, at $C_2 = 8$ pF indicating $Q_2 = 75$. Without the line, the inductance resonates at the same frequency, with $C_1 = 14$ pF, indicating $Q_1 = 200$. The short-circuited line resonates at 100 MHz with $C = 15$ pF, indicating $Q = 100$. Find the characteristic impedance and the line loss at 100 MHz per 100 m.

Hints: Line capacitance $C_d = C_1 - C_2$

$$\text{Shunt conductance of the line } G_d = \frac{Q_1 - Q_2}{Q_1 Q_2} \cdot \omega C_1$$

$$\text{Series inductance of the line } L_s = \frac{1}{\omega^2 C}$$

$$\text{Series resistance of the line } R_s = \frac{\omega L_s}{Q}$$

Attenuation of a low loss line is given by:

$$A = \frac{1}{2S} \left\{ \frac{G_d Z_0 + \dfrac{R_s}{Z_0}}{1 + \omega^2 L_s C_d} \right\} \text{ Nepers/m}$$

$$= \frac{4.34}{S} \left\{ \frac{G_d Z_0 + \dfrac{R_s}{Z_0}}{1 + \omega^2 L_s C_d} \right\} \text{ dB/m}$$

$$\approx \frac{4.34}{S} \left\{ G_d Z_0 + \frac{R_s}{Z_0} \right\} \text{ dB/m, when } S < \frac{\lambda}{60}$$

where $\quad Z_0 = \sqrt{\dfrac{R_s + j\omega L_s}{G_d + j\omega C_d}} \approx \sqrt{\dfrac{L_s}{C_d}}$

and $S = $ the length of the cable.

10

Broadcast Television Receivers

10.1 BLOCK SCHEMATIC AND FUNCTIONAL REQUIREMENTS

A broadcast television receiver gets from the antenna, a weak VHF or UHF signal containing the audio and video information in the form of amplitude modulation of the picture carrier and frequency modulation of the sound carrier. It amplifies this week signal and processes it so as to reproduce the sound signal through the loudspeaker and the picture signal on the picture tube screen. The TV receiver uses a superheterodyne technique, as does a radio receiver, to obtain a high gain and good selectivity at the fixed IF obtained after frequency conversion. The signal is first amplified at RF and then at the IF until it is large enough to detect it. The detected signal contains: (i) the video signal which is given to the picture tube *via* a video amplifier stage; (ii) the sync signals which are given to the vertical and horizontal deflection system that locks the raster scanning to that at the transmitter camera; and (iii) the sound IF difference carrier which after amplification is demodulated in an IF detector, and the audio signals thus separated drive the audio amplifier and the loudspeaker.

Figure 10.1 is a block diagram showing the signal flow in a typical monochrome television receiver.

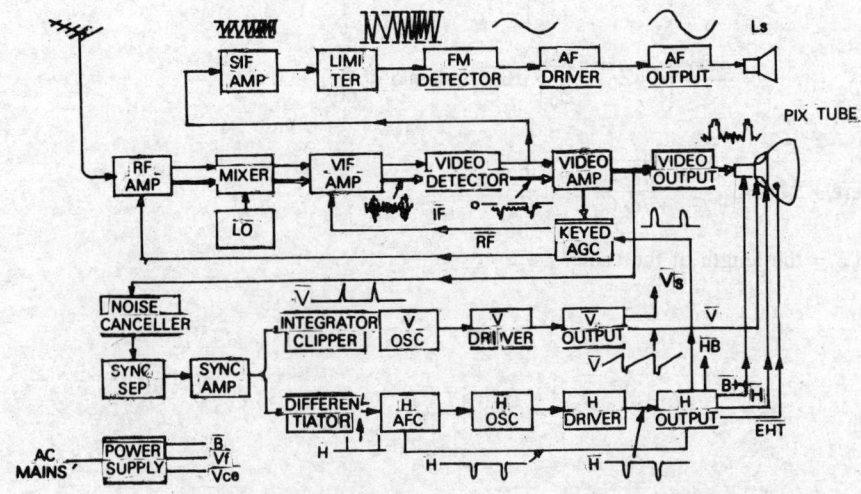

Fig. 10.1   Block diagram of a television receiver

RF Tuner

The signal from the antenna is available to the antenna input terminals of the receiver *via* a 300 Ω twin wire feeder, or *via* a coaxial 75 Ω cable when interference pick-up is to be minimized. The cable should be correctly matched to the antenna input terminal impedance, if necessary, by a 75 Ω to 300 Ω balun.

The RF tuner is a compact well shielded subassembly that pre-amplifies the signal at the RF frequency, and converts it to a fixed intermediate frequencies with the help of a variable frequency local oscillator. The tuner thus consists of three parts: the RF amplifier, mixer and local oscillator. The first RF amplifier stage gets the 300 Ω balanced feeder signal through a 300 : 75 Ω balun at its unbalanced input terminals with one terminal at ground—earth. The RF amplifier has a sufficiently large bandwidth to amplify the full channel frequency range, viz. 7 MHz. It has to amplify the smallest available RF signals and hence provides low noise amplification and also a gain that is controlled by AGC voltage from the detector, to suit the signal strength received.

The second stage in the tuner is the mixer which receives the amplified picture and sound carriers with the respective modulation signal and the local oscillator frequency which beats or heterodynes with the two carriers to produce two IFs, viz. the picture IF, and the sound IF. These are standardized at 38.9 and 33.4 MHz respectively. As the f_{ocs} is generally higher than the picture carrier and the sound carrier, the IF carriers reverse their relative positions. That is, while the picture carrier is lower than the sound carrier, the PIF is higher than the SIF.

The tuning of the RF stage and the oscillator is preset for each channel particularly in the tuners for the VHF channels called VHF tuners. A fine tuning control is provided by the adjustment of the oscillator frequency done manually. This is made automatic in some tuners with the help of AFC circuitry that senses the correctness of the IF and applies a correcting bias to a voltage-controlled oscillator used for tuning.

The RF tuners are generally of two types: The VHF tuners catering to bands I and III channels in the VHF range and (ii) UHF tuners catering to bands IV and V in the UHF range. The output of the tuner is small, from a fraction of mV to several mV. As the tuner is usually a separate unit completely shielded and nearer to the panel controls, the output is coupled to the video IF amplifier through a coaxial cable coupling link.

Video IF Amplifier

The video IF amplifier provides a high gain of the order of 60 to 80 dB and selectivity to provide the IF frequency response [see Fig. 3.5(c)] that compensates for the vestigial sideband reception. It provides the required bandwidth to amplify video signals picture IF and sound IF to the required extent. Picture IF is at 50% (6 dB below) of the peak amplification for frequencies between 35 to 38 MHz containing the video band. The sound carrier is kept at 5 to 10% of (20 dB below) the picture IF so that at the detector they produce a beat which carries the sound modulation. This beat frequency is the intercarrier sound IF that remains constant in spite of any drifts in the oscillator frequency. The adjacent channel sound carrier at 31.9 MHz and the picture carrier at 40.4 MHz are kept 40–45 dB below picture IF to prevent interference from them.

Video Detector

This demodulates the amplitude modulated picture IF to filter the IF and the harmonics so that the output is video signal only without attenuating the high frequencies in it. High frequency compensation techniques are necessary and useful in maintaining even video frequency response and good efficiency of detection. The 5.5 MHz intercarrier sound IF generated by the beating of sound IF with video IF at the detector, is taken off at the output of the detector or from a 5.5 MHz resonant trap circuit that blocks the intercarrier from reaching the picture tube frequency modulated intercarrier.

Sound IF Amplifier

Sound IF of 5.5 MHz with maximum frequency deviation of ± 50 kHz, taken off from the detector, is amplified by one or two stages with adequate bandwidth and limited in amplitude to eliminate any stray frequencies due to incidental amplitude modulation that it may carry.

FM Detector

Amplified IF is given to an FM demodulator to separate the audio frequency signals. The AF output is provided a de-emphasis at high frequencies by a 50 μs low pass filter and then is amplified by the AF amplifier to drive the loudspeaker.

AF Amplifier

This consists of a driver and output power amplifier to amplify the AF signal. Volume and tone controls are provided in this amplifier.

Video Amplifier

The few volts of video signal available at the output of the video detector is amplified by one or two stages of video amplifiers to around 40 to 80 V to drive the picture tube grid cathode with proper polarity and amplitude required for obtaining a good contrast. The polarity of the grid cathode voltage should be such that the black levels are positive going at the cathode to reduce the picture tube beam current to cut-off and produce no brightness on the screen.

Contrast control is usually provided in this stage by the video gain adjustment of the video amplifier or drive voltage adjustment. The signal drives the cathode-grid circuit of the picture tube whereas the dc grid bias adjustment serves the brightness control. Blanking pulses during the flyback are also applied to this grid to drive it to cut-off during the flyback to make the retrace scanning invisible. This is particularly necessary in a video amplifier with ac coupling to the picture tube, when the dc component of the video signal is not restored.

The video amplifier also feeds into (i) sound IF, (ii) sync separator, and (iii) the keyed AGC, besides driving the cathode of the picture tube for brightness modulation. Care has to be taken to minimize the distributed capacitive loading due to these feeds by suitable isolating cathode/emitter follower stages or by series resistors.

The video amplifier response is extended to around 5 MHz by shunt or series HF compensating inductance or both in *RC* coupled amplifiers. The sound IF intercarrier of 5.5 MHz is blocked by a series trap circuit to prevent it from reaching the picture tube and producing Moire interference line patterns. The video circuits after the detector are usually dc coupled as the video signal has dc component also. When ac coupling is used, dc restoration is desirable at the output to get the full contrast range. In tube sets, only one video output stage is adequate for picture tube drive, while in transistor design, a driver video amplifier is usually necessary.

Deflection Circuits

The sync pulses are derived from the video signal in the *sync separator circuit* by clipping out the 70% picture signal portion so that only the synchronising pulses are present in the signal at the output. The sync pulses are given to an *integrator low pass* circuit that produces as output pulse from the relatively wide vertical sync pulse only. This pulse is used for triggering the *vertical oscillator* running at a frequency slightly below 50 Hz, so that the oscillator can synchronise with the sync pulse frequency. The oscillator generates sawtooth drive voltages for the *vertical output amplifier* that sends sawtooth currents through the vertical deflection coils to deflect the beam vertically. The stage also supplies the blanking pulse to the picture tube to blank the vertical retrace positively.

The synchronising pulses separated from the video signal are also given to a *differentiating high pass circuit* of small time-constant that produces sharp output pulses for all sudden changes corresponding to the rise and fall times of the sync pulses. The sharp needle pulses corresponding to the rising edges of the sync pulses are used to control the frequency of the oscillator to synchronise it with the transmitter frequency of line scanning. It is useful to include *noise canceller circuits* in the signal path to the sync separator, so that the noise pulses do not produce any output lest they trigger the horizontal oscillator and disturb the horizontal synchronisation. This possibility is considerably reduced by employing an automatic frequency control (AFC) system for the horizontal oscillator rather than direct triggering. In the AFC system, the sync pulses and the output from the horizontal deflection drive are compared in a *phase discriminator* that produces dc output that is proportional to the difference between the two frequencies. This error voltage corrects the frequency of the horizontal oscillator to reduce the difference in the two frequencies and lock the horizontal oscillator phase to the sync pulses.

The *Horizontal oscillator* generates a trapezoidal-shaped drive voltage required for the horizontal *line output stage* to feed sawtooth currents into the horizontal deflection coils *via* the horizontal output transformer. The retrace being at a high speed induces large flyback pulses in the output circuit. These can be stepped up by the extra high tension (EHT) winding of the horizontal output transformer and rectified by a high voltage rectifier to obtain the EHT supply for the picture tube. The high focusing anode voltage for the picture tube is also obtained from this stage by a *booster diode circuit* that conducts just after the flyback to recover the energy in the inductive load of the deflection coil. The transformer also supplies blanking pulses to the picture tube for blanking of the horizontal retrace. In tube sets, the horizontal and vertical oscillator drive the output stages directly. In transistorized designs, additional driver stages are usually required.

Keyed AGC
The AGC in a TV receiver automatically adjusts the gain of the RF-IF stages to suit the signal strength at the antenna to maintain a relatively constant video signal output. The AGC in a TV receiver is of keyed or gated type in which the AGC voltage is obtained during the sync-blanking interval when the carrier level truely represents the signal carrier level received. The flyback pulses from the horizontal output are used to gate in transistor or a triode rectifier which can conduct only during the flyback interval. The current flow in the circuit is proportional to the video voltage at the base or grid and hence produces a corresponding dc voltage. The filtering time-constant of the output circuit can be small, as the pulse current flow is at a H rate of 15625 Hz, thus making it possible to have a fast acting AGC to overcome flutter fading due to aeroplane reflections, etc.

Power Supply
The power supply circuits provide for the heating of filaments of the tubes used, $B +$ and V_{cc} voltage supplies required for the tubes, transistors and the ICs used in the receiver circuitry. The dc supplies may be obtained from the main rectifier and dc voltage regulators that maintain constant output despite mains variations.

10.2 TRENDS IN CIRCUIT DESIGN

TV receiver circuit design, like other electronic equipment design, has been affected by the available devices and components. Over the last 50 years, the manufacture of systems has changed from the use of tubes to transistor circuits, and now to integrated circuits in reduced numbers but of greater complexity. As a mass-produced consumer item, the TV receiver considerably influenced the design and manufacture of devices and component.

Vacuum Tube and Hybrid Receivers

Early receivers employed vacuum tubes. Special series of tubes designed for optimum operation of the various stages in the TV receiver were available. These included ac mains type 6/12 V E-series and the series filament higher voltage P-series. Some of these tubes were integrated as multifunction tubes to reduce the tube count. The power consumption was as high as 130 to 200 W depending on tube size and design. In the '60s tubes were gradually replaced by the newer solid state diodes and transistors, at first only in small signal circuits of tuner, IF, sync, deflection drive and chroma processing circuits. The power output stages continued to use tube circuits of proven reliability and ease of servicing for the technicians, who had often to merely replace the faulty tube, the most common failure item!

Solid State Receivers

In the early '70s the power output and the high voltage stages also became transistorized and the power consumption dropped to below 60 W for 24-inch, and 10 W for 9-inch sets. Availability of high power high voltage transistors of sufficient or switching response, for use in the video and deflection output stages was the critical factor. The line output stage requires a switching transistor that has a VCEO of 1500 to 2500 V to withstand the flyback voltage encountered at the collector. Its switching speed must be high with a turn-off delay of below 10 μs, and the ON-resistance should be low so that the VCEsat is small compared to the supply voltage. With the availability of such devices with good reliability, TV receivers went in for fully solid-state designs.

To attain a high breakdown voltage in transistors, an intrinsic layer of high resistivity is required near the collector junction, employing *npin* structure. An excessively thin layer of i-layer makes the transistor susceptible to secondary breakdown, while a thick layer would impair the turn-off characteristics, increase saturation resistance and reduce current gain. In transistors with a breakdown voltage in excess of 1000 V surface breakdown is more dominant than the internal breakdown of the collector junction. High surface breakdown voltage has been realized by a suitable choice of the angle of the surface of the junction, clean surface and proper surface treatment.

Integration of the circuitry had progressively reduced the number of devices and components required, as large scale integrated circuits perform TV signal processing functions better, in a more complex manner. Most ICs and components including the tuner can be mounted on a single modest size PCB, while the video drive circuit is mounted close to the colour picture tube socket on a small PCB.

Very large scale integration employing manufacturing processes at the micron level has enabled the functioning of thousands of transistors on a single chip. With digital TV concepts, all TV functions relating to video, deflection and sound and all operating functions and sequences, are controlled by real time signal processing by means of VLSI chips, each containing about 50,000 transistors. A single chip microcomputer controls, measures, coordinates, stores and processes the various signals. Computer controlled TV makes optimal use of the microcontroller during normal operation, service and manufacture of the TV set. Besides such normal functions as tuning, channel display, remote control, analog picture and sound controls, the microcontroller can also control the various alignment functions.

10.3 COLOUR TELEVISION RECEIVER

The block diagram of a colour TV receiver is shown in Fig. 10.2.

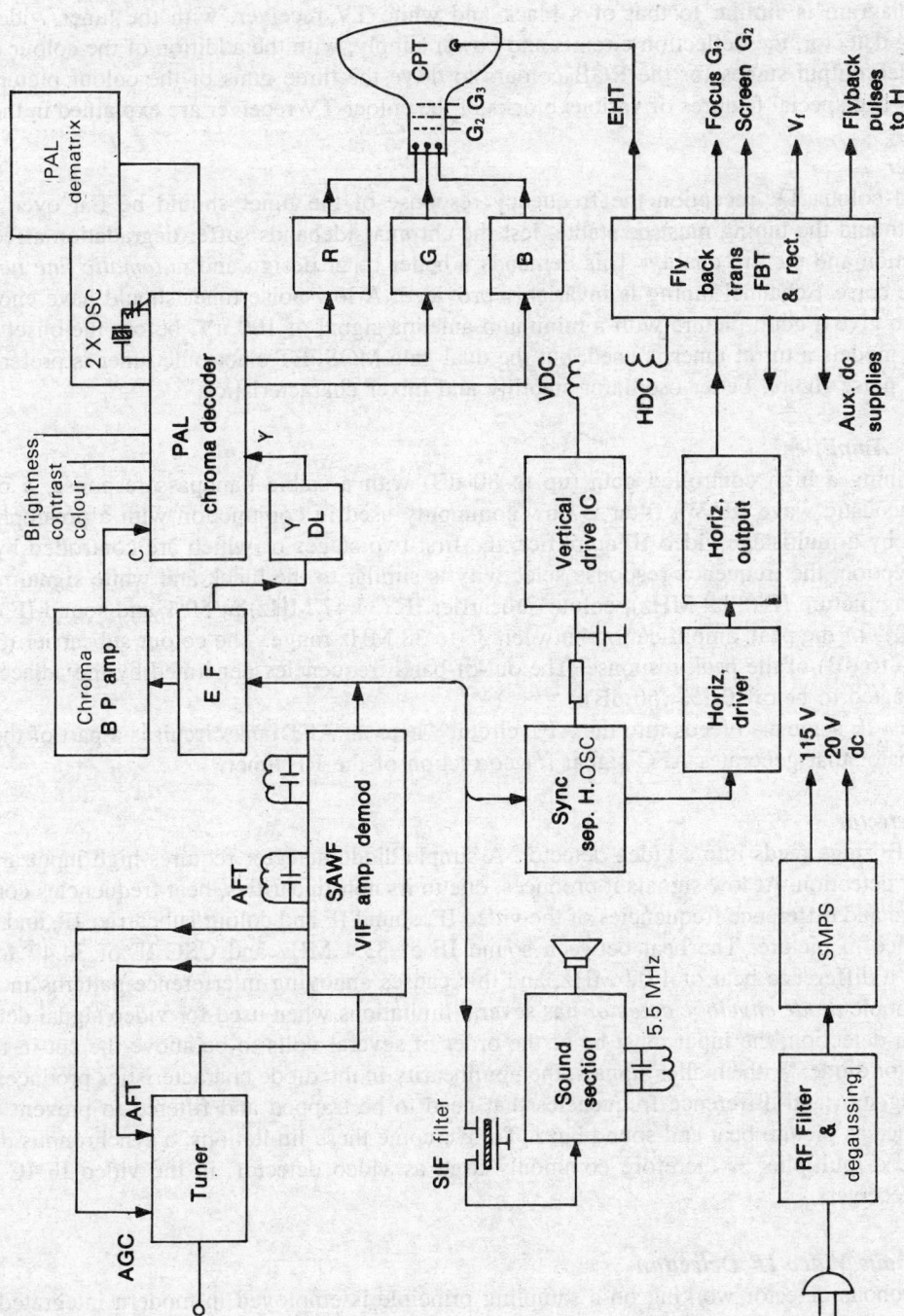

Fig. 10.2 Block diagram of a colour TV receiver

The diagram is similar to that of a black and white TV receiver, with the tuner, video IF stages including detector, the deflection circuits and power supply, with the addition of the colour decoder and three video output stages for the RGB colours to drive the three guns of the colour picture tube. The functions and special features of various blocks of the colour TV receiver are explained in the following.

RF Tuner

For good colour TV reception the frequency response of the tuner should be flat over the channel bandwidth and the tuning must be stable, lest the chroma sidebands suffer degradation affecting colour reproduction and picture quality. This demands a better tuner design and *automatic fine tuning* (AFT), to ensure correct channel tuning is invariably provided. A low noise tuner should have enough gain to be able to give a good picture with a minimum antenna signal of 100 uV, before the onset of AGC. In low cost models a turret tuner is used, but the dual gate MOSFET electronic tuner is preferred because of lower noise figure, better oscillator stability and mixer characteristics.

Video IF Amplifier

For obtaining a high controlled gain (up to 80 dB) with a stable bandpass response, a piezoelectric surface acoustic wave (SAW) filter is now commonly used in conjunction with a preamplifier, and is followed by a multistage video IF amplifier, the first two stages of which are controlled by AGC. For VSB reception, the frequency response selectivity is similar to the black and white signal requirement, amplifying picture IF (38.9 MHz), colour subcarrier IF (34.47 MHz) at 50% and sound IF (33.4 MHz) at 5 to 10% of the peak amplification between 35 to 38 MHz range. The colour subcarrier (34.47 MHz) is at 50% (6 dB) of the peak response. The out-of-band frequencies generated by the adjacent channels are attenuated to below 0.1% (60 dB).

The last IF stage also feeds into the AFT circuit where an AFT tank circuit is a part of the frequency discriminator that generates AFC signal for correction of the RF tuner.

Video Detector

The last IF stage feeds into a video detector. A simple diode detector requires high input around 10 V, for linear detection. At low signals it produces, due to its non-linearities, beat frequencies corresponding to the sum and difference frequencies of the video IF, sound IF and colour subcarrier IF, and these cause interference in picture. The beat between sound IF of 33.4 MHz and CSC IF of 34.47 for example, produces a difference beat of 1.07 MHz, and this causes annoying interference patterns in the picture.

The simple *diode envelope detector* has several limitations when used for video signal demodulation. For linear detection, the input must be of the order of several volts to be above the cut-in threshold of the detector diode. With smaller signals, the nonlinearity in the diode characteristics produces distortion, generating sum and difference frequencies that need to be trapped and filtered to prevent undesirable interference as picture beat and sound buzz. To overcome these limitations, a synchronous demodulator or balanced multiplier is therefore commonly used as video detector, in the video IF IC systems of modern receivers.

Synchronous Video IF Detection

A synchronous detector working on a sampling principle is employed in modern integrated circuits to overcome these problems providing linear detection even with low input signals. The amplitude modulation is detected at the sampling intervals, provided by the modulation-free carrier obtained from the signal itself, by limiting it and selecting the pure carrier by a sharply tuned resonant circuit. The principle of synchronous detection is illustrated in the schematic diagram of Fig. 10.3.

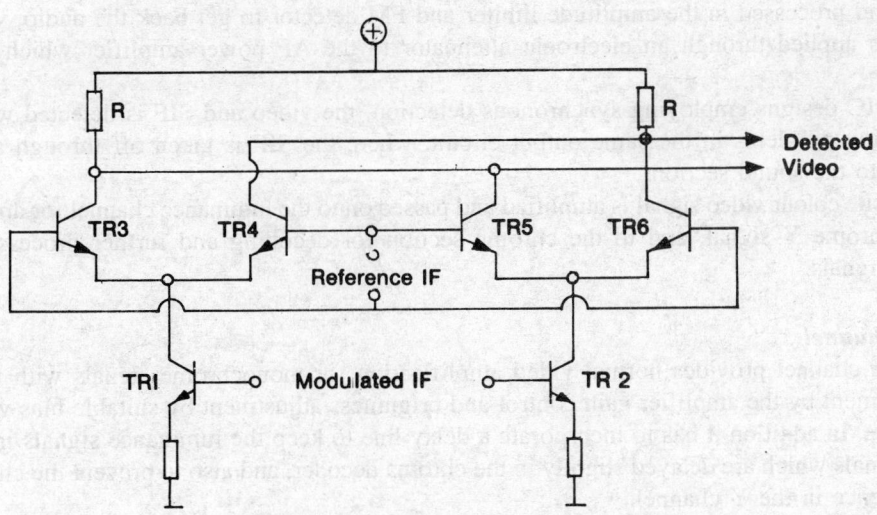

Fig. 10.3 Synchronous detector principle

In addition to the video modulated signal, the synchronous detector requires a reference video IF carrier signal which in effect multiplies with it. The modulated carrier is *sampled* by a pure unmodulated carrier of exactly the same frequency. The phase relation between the reference carrier and the modulated carrier signal are vital in the process of detection. Refer to Sec. 5.7 for mathematical explanation.

The reference signal is fed to the two differential amplifiers comprising TR3-TR4 and TR5-TR6, having their outputs connected in parallel. The modulated video signal is fed to the long tail pair transistors TR1 and TR2 in differential fashion. The output currents in the differential amplifiers vary so that the differential output follows the video amplitude modulation. The video carrier cancels itself in the output of the differential amplifiers because of balanced demodulator configuration. The demodulated output corresponds only to the video signal carrier in synchronism with the reference carrier.

Reference carrier In the video IF IC system, the reference carrier is derived from the video IF signal itself by amplifying and limiting it through clamping diodes in a separate differential amplifier. This *removes the amplitude modulation* and leaves in the output a square wave signal at IF of 38.9 MHz. From this signal the fundamental video IF sine wave can be selected by a resonant tank circuit as the differential load. The harmonics are rejected by the *high-Q* circuit tuned to the video IF, while the IF sine wave reference signal multiplies with the video IF modulated signal as discussed in the preceding, providing synchronous detection.

Synchronous detection, thus, has advantages in reducing IF harmonics, getting rid of harmonic beats, thus eliminating the need for trap filters and elaborate shielding. It provides linear operation even at low signals and produces little or no quadrature distortion, depending on the purity of the reference carrier. The balanced topology of the circuit is easily implemented in IC format.

Sound Separation and Processing

Unlike the monochrome sets, the sound IF signal at 33.4 MHz is often separately detected in a colour TV receiver, taken off at the output of the second video IF stage by means of a tuned circuit which feeds into a diode detector. The video IF and sound IF signals are heterodyned in this detector to produce difference frequency as the intercarrier sound IF at 5.5 MHz. As in the monochrome receiver, the SIF is amplified and processed in the amplitude limiter and FM detector to get back the audio, which after de-emphasis is applied through an electronic attenuator to the AF power amplifier which drives the loudspeaker.

In the VIF IC designs employing synchronous detection, the video and SIF is detected without any cross modulation problems in the same output circuit, where the SIF is taken off through a 5.5 MHz ceramic filter to the sound section.

The composite colour video signal is amplified and passed on to the luminance channel for amplification of the monochrome Y signal, and to the chroma section for decoding and further processing of the chrominance signals.

Luminance Channel

The luminance channel provides normal video amplification for monochrome signals with facility for contrast adjustment by the amplifier gain control and brightness adjustment by suitable bias variation of the picture tube. In addition it has to incorporate a delay line to keep the luminance signals in step with the chroma signals which are delayed slightly in the chroma decoder, and also to prevent the chrominance signal interference in the Y channel.

A sharply resonant *notch filter* can be incorported in this channel to block the colour subcarrier of 4.43 MHz (3.58 MHz in NTSC), which would otherwise appear on the screen as interference dot pattern. As this presents problems in maintaining the frequency and phase response of the higher video frequencies, it has been also a popular design practice in low cost receivers to limit the frequency response to 4 MHz (3.2 MHz in NTSC) and avoid the use of sharply tuned notch filters.

Comb Filter

In better-quality receivers, the comb filter is used to pass only the clusters of monochrome signal energy while *blocking the chroma* signals both of which occur in packets of energy at f_h intervals but interleaved by $f_h/2$, as explained in Ch. 5. The comb filter passes only the specific monochrome chroma frequencies occurring in multiples of $f_h/2$. It is in effect a delay line providing a delay time of one line period H by electromechanical means using a piezoelectric glass substrate and two transducers for input and output.

Chroma Decoder Section

The chroma section amplifies and decodes the chroma signals. Details of the standard PAL-D system are shown in Fig. 10.4. It essentially consists of a two/three stage chroma bandpass amplifier with automatic colour control (ACC) and colour killer connection, a delay line driver, PAL delay line and the PAL matrix. The U and V chroma signals are obtained by synchronous detection using the colour burst as the reference. This requires a local crystal oscillator, burst gate, phase detector and reactance control circuit, and PAL switch for the V phase inverter, as auxiliary stages.

Automatic Colour Control

The composite colour signal from the video detector is passed through the chroma amplifiers which provide a bandpass characteristics of about 2 MHz centered around 4.43 MHz, with a fairly sharp roll-off of the skirts on either side. Manual control of the chroma signal level provides colour saturation control, while the automatic colour control (ACC) circuit controls the gain of the first stage to maintain a nearly constant output. The ACC uses received colour burst amplitude as reference to derive the dc control voltage for controlling the gain.

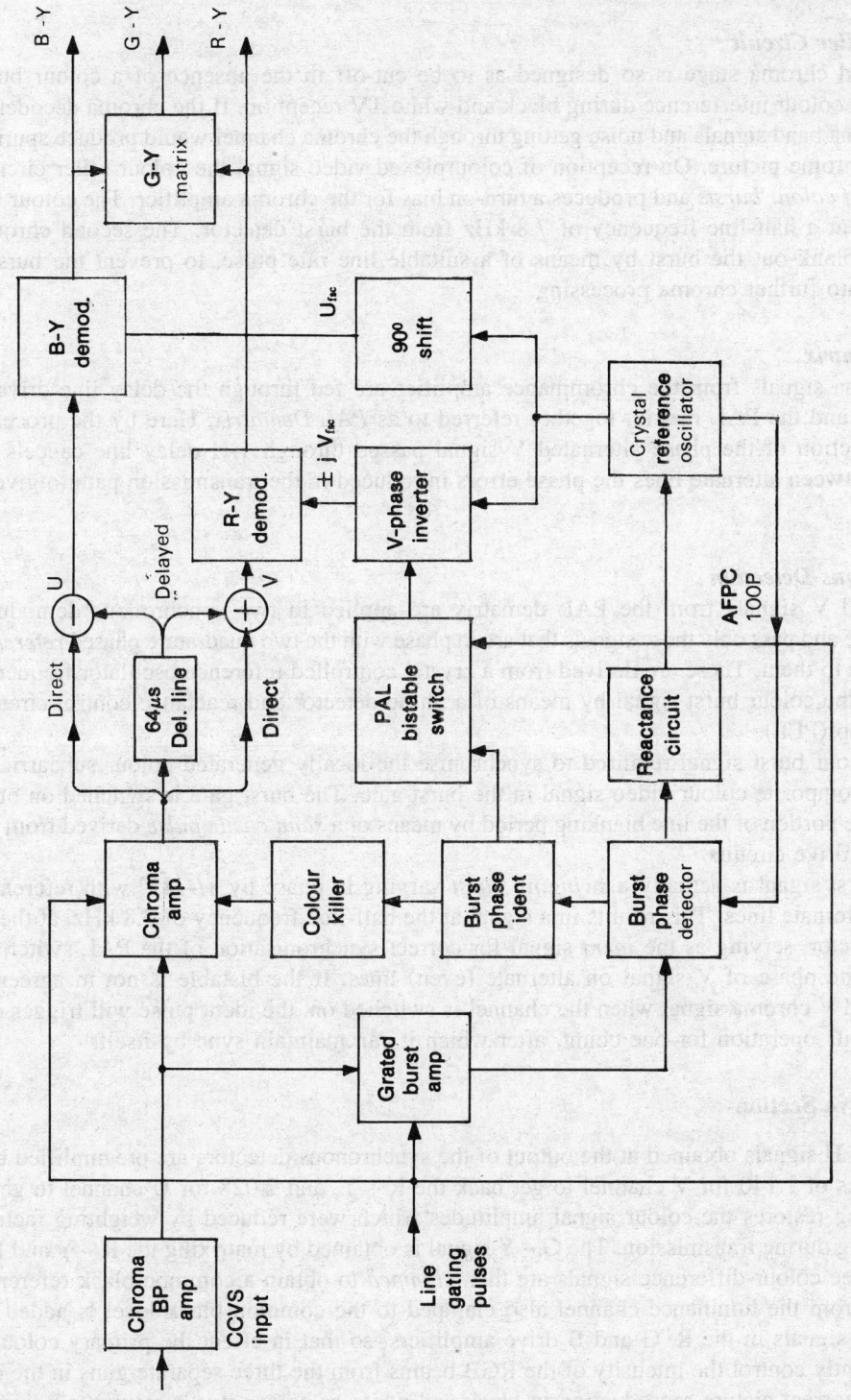

Fig. 10.4 Details of PAL-D decoder

Colour Killer Circuit

The second chroma stage is so designed as to be cut-off in the absence of a colour burst signal for preventing colour interference during black and white TV reception. If the chroma decoder is active, the stray chroma band signals and noise getting through the chroma channel would produce spurious colouring on monochrome picture. On reception of colourplexed video signal the colour killer circuit *detects the presence of colour bursts* and produces a turn-on bias for the chroma amplifier. The colour killer receives the bursts at a half-line frequency of 7.8 kHz from the burst detector. The second chroma stage also serves to blank out the burst by means of a suitable line rate pulse, to prevent the burst signal from entering into further chroma processing.

PAL Dematrix

The chroma signals from the chrominance amplifier are fed through the delay line driver to the PAL delay line and the PAL matrix, together referred to as *PAL Dematrix*. Here by the process of addition and subtraction of the phase alternated V signal passed through 1-H delay line cancels by averaging process between alternate lines the phase errors introduced in the transmission path to give the U and V signals.

Synchronous Detection

The U and V signals from the PAL dematrix are applied to two synchronous demodulators, which demodulate and pass only those signals that are in phase with the two quadrature phased *reference subcarrier* signals fed to them. These are derived from a crystal controlled reference oscillator frequency and phase locked to the colour burst signal by means of a phase detector and reactance control circuit, or a phase locked loop (PLL).

The colour burst signal required to synchronise the locally generated colour subcarrier is separated from the composite colour video signal in the burst gate. The burst gate is switched on only during the appropriate portion of the line blanking period by means of a *sand castle pulse* derived from the horizontal deflection drive circuits.

The burst signal is actually a *swinging burst* varying in phase by +/– 45° with reference to the – U axis, on alternate lines. This results in a signal at the half-line frequency of 7.8 kHz, at the output of the phase detector, serving as the *ident* signal for correct synchronisation of the PAL switch bistable used to invert the phase of V signal on alternate (even) lines. If the bistable is not in agreement with the transmitted V chroma signal when the channel is switched on, the ident pulse will trigger it correctly by inhibiting its operation for one count, after which it can maintain sync by itself.

Video Drive Section

The V and U signals obtained at the output of the synchronous detectors are preamplified by the relative gain factors of 1.140 for V channel to get back the R – Y, and 2.028 for U channel to get B – Y. This *deweighting* restores the colour signal amplitudes which were reduced by weighting factors to prevent overloading during transmission. The G – Y signal is obtained by matrixing the R – Y and B – Y signals.

The three colour-difference signals are then *clamped* to obtain a common black reference level. The Y signal from the luminance channel also clamped to the common black level is added to the colour difference signals in the R, G and B drive amplifiers, so that in effect the primary colour components independently control the intensity of the RGB beams from the three separate guns in the colour picture tube. For correct picture reproduction in black and white or colour it is essential to adjust the guns for identical cut-off points and slope of their transfer characteristics, by what is called *gray scale tracking*.

Gray Scale Tracking

The cut-off points are set by adjusting the bias voltages at the emitters of the three drive transistors. These *black level adjustments*, also called half-tone adjustments, are aimed to produce just visible raster for colour balance on low lights. The slight differences in the slopes of transfer characteristics and also different phosphor efficiencies are adjusted by presetting the potentiometers controlling the drives to the three drive amplifiers on high lights. These are called *white balance* or *monochrome adjustments*, ensuring a picture free of colour stains and aprons.

To achieve time coincidence between the narrowband chroma signals and the wideband luminance signals when they are added after chroma decoding, it is necessary to insert a delay line providing a delay time of the order of 300–800 ns in the Y-channel, depending on the actual circuitry used.

Deflection Section

Functions

Sync separation, synchronisation of the horizontal and vertical oscillators that drive the horizontal and vertical coils, generation of the EHT, the focus anode and other auxiliary voltages, and blanking during flyback are all provided by the deflection section, as discussed in the case of a monochrome TV receiver. Many modern IC chips incorporate the sync separator, line and field oscillators and several other associated functions in a single monolithic IC package.

The deflection system in a colour TV receiver has to be more precise and linear, to ensure correct colour registration by the three beams, and has also to provide a *higher EHT* potential (25 kV) and higher focus anode voltage (7 kV) required by the colour picture tube. The high EHT potential is generated by stepping up the flyback pulses and rectification by voltage multiplier.

In modern receivers a *diode-split transformer* sections arrangement using three or four sections is commonly used. In this arrangement, separate EHT windings are used for each section forming a cell in association with a rectifier diode and a reservoir formed by interwinding layer capacitance. This has better regulation and reliability and is less prone to flash overs. The focus anode voltage is provided by tapping off the output of the first or second EHT cell through a preset focus potentiometer.

The horizontal deflection section also supplies a *sandcastle pulse* to the chroma decoder for blanking and gating the colour burst riding on the back porch of the blanking pulse. The deflection yoke mounted on the neck of the colour picture tube has necessary magnetic adjustments for aligning the three beams to proper convergence and purity of colours. *Degaussing coils* are provided on the colour picture tube for removing stray colouring effects on the screen caused by stray magnetic fields.

Power Supply

In modern colour TV receivers, the switching mode type power supply (SMPS) is used to obtain efficient stabilized output dc voltages needed by the circuitry, even with widely varying ac power supply. In this type of power supply, the unregulated dc voltage is chopped at usually line frequency (15,625 Hz) by a transistor or a thyristor used as an electronic switch. The ON/OFF ratio is varied by a suitable feedback control loop to maintain constant output over a wide range of input and load variations.

10.4 DIGITAL COLOUR TV RECEIVER

The use of digital techniques in television receivers offers several technical and economic advantages, especially when used for modern broadcast information services like teletext or videotext, facilities like PIP, and automatic tuning of any keyed channel. Here, the ready adaptability of a digital microprocessor controlled receiver to teletext decoder, elimination of ghost images and component aging effects are important, and ease of automated production cum testing help reduce costs. The details of digital TV technology are discussed later in chapter 25. A functional block schematic of digital techniques used in a TV receiver is indicated in Fig. 10.5.

The receiver gets the standard video signal in analog form only, as the digitization of video means very large RF bandwidth, and is as such impractical in the VHF-UHF band allocations. The demodulated signal is digitized and presented to *digital signal processing* circuits with the aid of a microcomputer. Before digitization, the video and audio signals are separated to cater to different bandwidths and processing requirements. The clock frequency for the digital circuits is derived from the colour subcarrier burst present in the video sync.

The video signal is digitized at three or four times the f_{sc} with an effective resolution of eight bits per sample. The video processor carries out the operations necessary for filtering, ghost and noise reduction, decoding and so on. The audio signal is digitized at about 32 kHz in 14 bit resolution as the audio processor controls the stereo balance, high fidelity bass/treble manipulations. The deflection processor maintains a highly stable picture providing dynamic corrections for linear deflection and purity of colour reproduction.

The heart of the system is the central control unit (CCU), a microcomputer based device which controls and coordinates all the circuits and signal processing functions, and provides a user interface. This provides a total flexibility to the user who can control the receiver functions, display teletext, multiple *picture-in-picture displays* for simultaneous viewing of different programs, or zooming of the picture into different viewing angles, to cite a few examples.

10.5 DESIGN SPECIFICATION FOR MONOCHROME TV RECEIVERS IN INDIA

The CCIR system-B standards have been adopted in India for television broadcasting on channels 3 and 4 in band I, and channels 5, 6, 7, 8, 9 and 10 in band III in the VHF range, for the present. In view of this and the available technology, the design specifications and performance requirements of monochrome television receivers, based on IS: 4547–1978, may be taken as follows:

Design Specifications

Frequency Range
One or more channels out of the eight channels in the frequency range 54–68 MHz and 174–216 MHz, viz. channels 3 and 4 in band I and 5, 6, 7, 8, 9 and 10 in band III.

Power Supply
Mains supply, ac: 240 V (200 to 250 V), 50 Hz ± 2%. Battery supply: 12 V or 24 V.

Power Consumption
Not to exceed 200 W for tube/hybrid types, in mains operation; current consumption for solid-state type should be less than 2.5 A for battery operation.

Aerial Input Impedance
300 Ω balanced; 75 Ω unbalanced for shielded antenna feeders, if required.

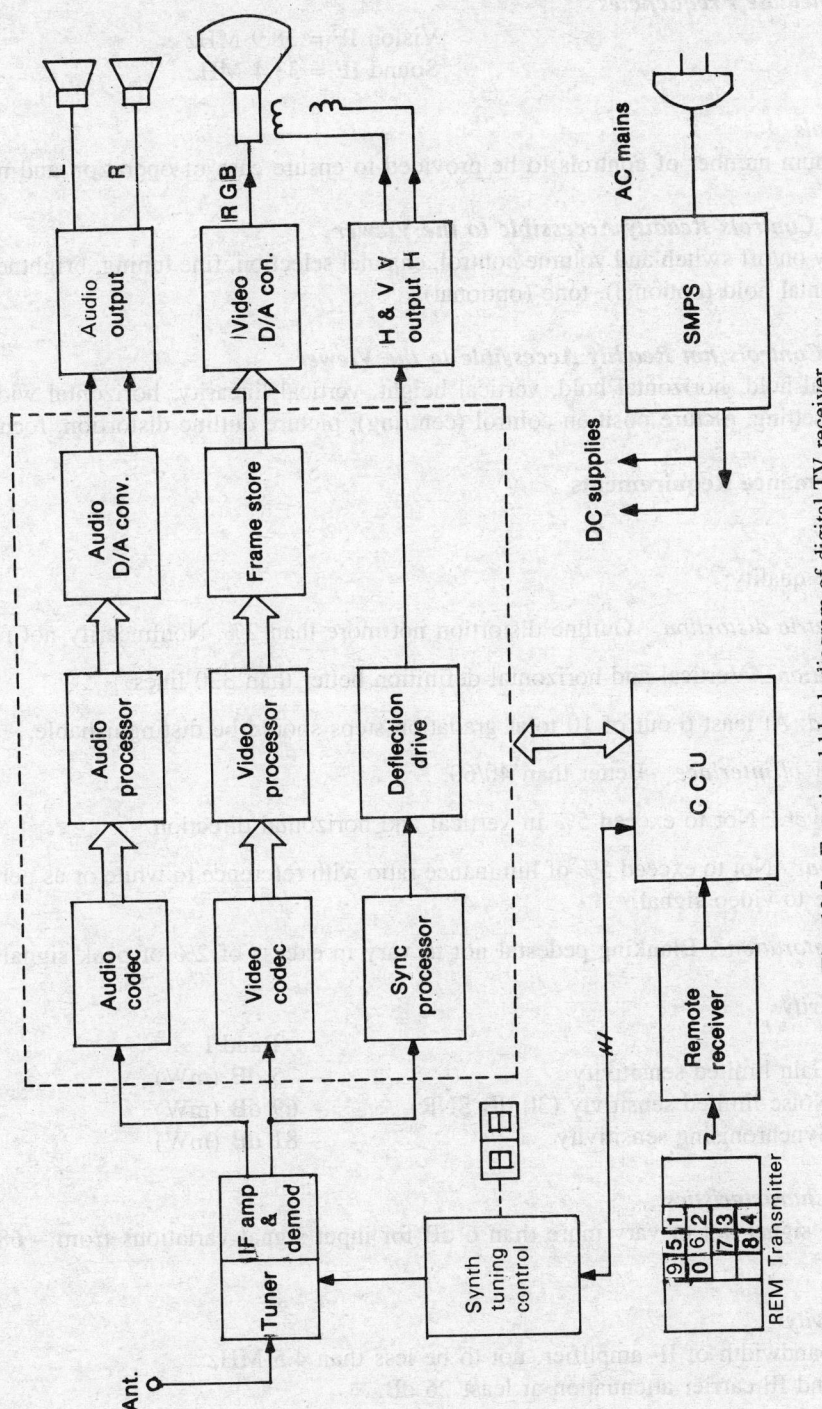

Fig. 10.5 Functional block diagram of digital TV receiver

Intermediate Frequencies

$$\text{Vision IF} = 38.9 \text{ MHz}$$
$$\text{Sound IF} = 33.4 \text{ MHz}$$

Controls
Minimum number of controls to be provided to ensure ease of operation and maintenance.

Panel Controls Readily Accessible to the Viewer
Supply on/off switch and volume control, channel selection, fine tuning, brightness, contrast vertical and horizontal hold (optional), tone (optional).

Prset Controls not Readily Accessible to the Viewer
Vertical hold, horizontal hold, vertical height, vertical linearity, horizontal width, horizontal linearity, AGC setting, picture position control (centring), picture outline distortion, focus.

Performance Requirements

Vision
Picture quality:

Geometric distortion   Outline distortion not more than 2%. Nonlinearity not more than 10%.

Resolution   Vertical and horizontal definition better than 320 lines.

Gamma: At least 6 out of 10 total gradation steps should be distinguishable.

Quality of interlace   Better than 40/60.

Fold-over   Not to exceed 5% in vertical and horizontal direction.

Hum-bar   Not to exceed 5% of luminance ratio with reference to white or as percentage of ratio of hum voltage to video signal.

DC restoration   Blanking pedestal not to vary in excess of 2% of peak signal level.

Sensitivity

	Band I	Band III
Gain limited sensitivity	– 75 dB (mW)	– 69 dB (mW)
Noise-limited sensitiviy (30 dB SNR)	– 69 dB (mW)	– 63 dB (mW)
Synchronising sensitivity	– 81 dB (mW)	– 75 dB (mW)

AGC Characteristics
Output signal not to vary more than 6 dB for input signal variations from – 68 dB (mW) to – 16 dB (mW).

Selectivity
6 dB bandwidth of IF amplifier, not to be less than 4.5 MHz.
   Sound IF carrier attenuation at least 26 dB.
   Adjacent picture carrier suppression at least 26 dB.
   Adjacent sound carrier suppression at least 30 dB.

IF interference suppression better than 40 dB.
Image interference ratio better than 55 dB.

Video Frequency Response
Within 0 ± 1 dB at least up to 3 MHz
Within 0 ± 2 dB at 4.5 MHz
Less than − 40 dB at 5.5 MHz

Step response to pulse modulated carrier:
Rise time better than 0.15 μs; overshoot not to exceed 10%.

Tuning Stability
Local oscillator drift not to exceed 200 kHz.

Synchronising Stability
Drift of free running time base not to exceed ± 1%. Lock-in-range: better than ± 2%.

Sound Section

Audio output power   Not less than 0.5 W for 310 mm or smaller sets.
Not less than 1.0 W for greater than 310 mm.

De-emphasis: 50 μs.

Noise limited sensitivity (30 dB) SNR):
   Better than − 87 dB (mW) for Band I, and
   Better than − 81 dB (mW) for Band III.

Frequency response   Within ± 2 dB over 100–7500 Hz for greater than 310 mm sets.

Hum level   At least 40 dB below the maximum audio output.

Environmental tests   The receiver should withstand the following tests.

Bump test   In accordance with IS: 2106 (Part VII)-1964, the number of bumps being 1000 and height of drop 25 mm (40 g).

Dry heat test   Of severity 55 °C, in accordance with IS: 2106 (Part IV)-1963; duration of recovery-2 h.

Damp heat cycling test   Two cycles of damp heat test in accordance with IS: 2106 (Part II)-1962. Duration of recovery 24 h.
   After recovery, the receiver should conform back to the requirements of synchronising sensitivity, noise-limited video sensitivity, tuning stability of local oscillator and audio output.

10.6   SPECIFICATIONS FOR A COLOUR TV RECEIVER

The technical specifications of a colour TV receiver for the reception of PAL CCIR-B television signals are listed in the following to give an idea of typical parameters and performance.

1. *General*
   (a) Frequency range

VHF Band I	channels 2–4
VHF Band III	channels 5–12
UHF Band IV–V (U)	channels 21–69

   (b) Intermediate frequency

Vision	38.9 MHz
Sound	33.4 MHz
Chroma	34.47 MHz

   (c) Power supply — 140 to 250 V, 50 Hz
   (d) Antenna input impedance — 75 Ω, unbalanced

2. *Vision*
   (a) Picture quality

Picture size	510 mm
Aspect ratio	4 : 3
Geometric distortion	5%
Brightness	200 Nits
Contrast max.	120 : 1
min.	1 : 1
Definition and focus > 300 lines	
Brightness transfer > 7 out of 10	
standard test chart gradations	

   (b) Sensitivity

	Band I	Band III
Sync limited	– 93 dBm	– 84 dBm
Gain limited	– 82 dBm	– 76 dBm
Noise limited	– 77 dBm	– 63 dBm
Colour sensitivity	– 89 dBm	– 79 dBm
Colour killer sensitivity	68 mV	
AGC range for 6 dB variation	93 dB	63 dB
Max input signal	0 dBm	
Video response	colour	3.8 MHz
	B/W	4.2 MHz

   (c) Interference

Self sound carrier rejection		20 dB
Adj. channel pict. carrier rejection		40 dB
Adj. channel sound carrier rejection		60 dB
IF rejection	Band I	30 dB
	Band III	70 dB
Image rejection		40 dB
Mains RFI suppression		53 dB

   (d) Fidelity

Monochrome video response (6 dB)		3.5 MHz
Chrominance video response		0.6 MHz
Step response		
	Overshoot	10%
	Rise time	200 ns
DC component distortion		5%
Primary signal matrixing error		8%

(e) Stability

Tuning frequency		200 kHz
Synchronising range		
	Vertical lock-in	20%
	Horizontal lock-in	+ /– 3.5%

3. *Sound*
   - (a) IF intercarrier   5.5 MHz
   - (b) Max. useful power output   2 W
   - (c) De-emphasis   50 μs
   - (d) Fidelity 100 Hz to 7.5 kHz   2 dB
   - (e) Audio noise limited sensitivity

	Band I	– 84 dB
	Band III	– 77 dB

   - (f) Interference

	AM suppression	40 dB
	Hum	38 dB

4. *Power consumption*   60 W

5. *Environmental*
   - (a) Temperature

	Operating	0 to + 55 °C
	Storage	– 10 to 70 °C

   - (b) Damp heat   .95% RH at 40 °C
   - (c) Bump and vibration   Type tested to qualify for normal handling and transportation

REVIEW QUESTIONS

10.1 Draw the block diagram of a television receiver. Sketch the signal waveforms at various points and indicate the function of each stage briefly. Identify the positions of function controls.

10.2 Discuss whether it is possible to use a straight TRF-tuned radio frequency type design for single channel reception in a strong signal area, with some economy.

10.3 Compare the design features and performance of tube and transistor TV receivers.

10.4 Draw and explain briefly the functional block diagram of a colour TV receiver.

10.5 In what respects do the requirements of the antenna, the tuner and the deflection system in a colour TV receiver differ from those of the monochrome?

10.6 Give the schematic of a digital TV receiver, indicating the merits of the digital techniques.

MULTIPLE-CHOICE QUESTIONS

10.1 Which of the following stages incorporates contrast control in a colour receiver?
   - (a) chroma amplifier
   - (b) Luminance amplifier
   - (c) video detector
   - (d) colour killer circuit

10.2 Which of the following stages incorporates colour saturation control?
   - (a) chroma amplifier

    (b) chroma demodulator

    (c) colour subcarrier generator

    (d) colour killer circuit

10.3 PAL dematrix consists of:

    (a) a chroma demodulator and matrixing arrangement to separate the RGB colours

    (b) a delay line and matrixing arrangement to separate U and V components

    (c) a delay line and matrixing arrangement to separate the three colour difference components

    (d) delay line to compensate for delay in luma path.

10.4 A comb filter is introduced in high-quality TV receivers to:

    (a) pass only the chroma frequencies through the colour demodulator

    (b) block the monochrome signals through the chrominance channel

    (c) block chroma frequencies through the luminace channel

    (d) none of these.

RF Tuner and AGC Circuits

11.1 FUNCTIONAL REQUIREMENTS

The RF tuner often called the front end of a TV receiver has the following functions to perform.

(i) It selects the channel to be received by switching pre-tuned circuits in the RF stage and the oscillator.

(ii) It matches the impedance of the line at its input and amplifies it to maintain a good signal-to-noise ratio. The gain is controlled by AGC voltage, to suit the input signal strength.

(iii) It converts the RF into IF by mixing it with the local oscillator frequency, to feed into the video IF amplifier.

(iv) It blocks the interfering antenna pick-up signals in the IF range and prevents them from entering the receiver and mix up with the local oscillator and thence to the IF amplifier.

(v) It isolates the local oscillator signals from the antenna, due to the RF amplifier acting as a buffer, preventing radiation and interference to other receivers.

(vi) I rejects the image frequencies by means of the RF selective circuits.

Depending upon the tuning range required, the tuner is designed as a single channel, multichannel VHF or UHF tuner. The VHF tuner may be a step-switch type tuner, using preset tuned circuits on a turret or a rotary switch. The UHF tuner uses continuous tuning usually employing transmission line or strip-line tuned circuits.

11.2 BLOCK DIAGRAM OF A VHF TUNER

The block schematic of a tuner is shown in Fig. 11.1.

Balun

At the antenna input terminals, the tuner must have input impedance equal to the characteristic impedance of the aerial feeder so that the matching provides a maximum power transfer and avoids reflections on the line. The standard impedance is 300 Ω, corresponding to the twin wire ribbon feeder commonly used. The balun matches this impedance to the 75 Ω input impedance of the RF amplifier. It consists of a ferrite core upon which are four tightly coupled and evenly spaced bifilar windings of a couple of turns each, in the form of two quarter wave lines each of 150 Ω, that provide by series connection a 300 Ω balanced impedance on one side and a parallel connected 75 Ω impedance unbalanced, on the other side, by grounding of one terminal.

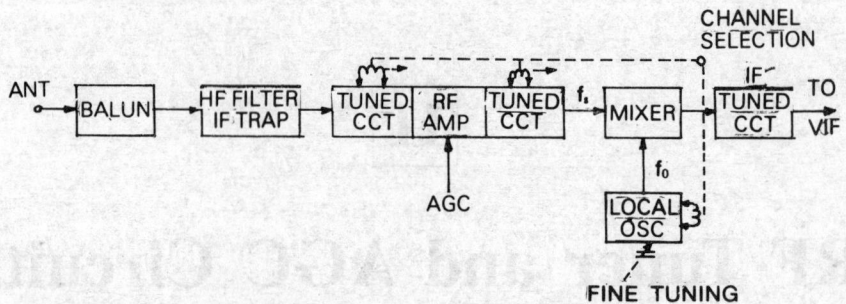

Fig. 11.1   Block schematic of an RF tuner

A pair of two small capacitors 470 pF each are included in each lead to prevent or block the dc path from chassis to antenna and prevent damage due to lightning. The 2 Meg shunting resistors discharge any static accumulated charge on these capacitors.

IF Trap and HP Filter

Unwanted spurious signals in the IF range of 33 to 40 MHz are blocked by the IF trap and high pass (HP) filter as the rejection of these by the low Q RF circuits, may not be adequate due to broad bandwidth, particularly on the lowest channel number 2, viz. 47–54 MHz. It is very difficult to reject them once they reach the mixer and IF stages. The trap usually consists of an HP filter with pass band beyond 40 MHz, in the form shown in Fig. 11.2.

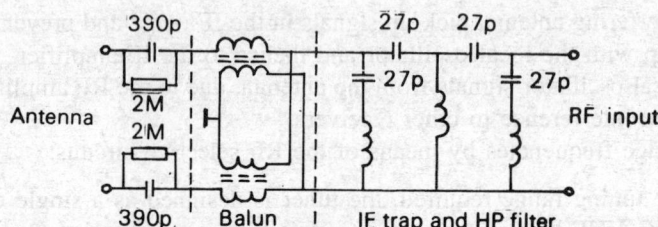

Fig. 11.2   Tuner input balun and filters

RF Amplifier

This provides a gain to the input signal ensuring better signal-to-noise ratio, isolation to the local oscillator radiation and better image selectivity. Its primary function is to provide adequate gain (20 dB) to weak signals maintaining a good signal-to-noise (S/N) ratio at the mixer. The mixer generates more noise because of its heterodyne function.

The equivalent noise voltage at the input of the RF amplifier sets a limit to the minimum signal that can be received. The noise voltage is visible on the screen of the picture as a snowy background of black and white randomly moving spots. The noise voltage is typically around 10 μV. The RF stage is best suited for AGC because the signal level is small and the gain control is most effective here producing minimum distortion. Application of AGC to RF stage is usually delayed suitably in order to maintain good S/N ratio at weak signals.

The stage provides isolation by acting as a buffer between the local oscillator and antenna terminals to minimize radiation from the local oscillator. These radiations can be a source of considerable interference to neighbouring receivers showing as diagonal line patterns on the screen. The field strength should be less than 100 μV/m at 100 ft in the VHF band or as per regulations prevailing. A judicious placement of the coupling coils, windows, use of RF chokes, feed-through capacitors in the tuner supply lines, viz. V_{cc}, B +, filament supply wires and proper tuner case shielding, are all important to bring down the radiation below specified limits. Separate chassis ground return is generally often employed for tuner circuits. Feed-through capacitor construction shown in Fig. 11.3, provides very efficient bypassing and decoupling because of its coaxial construction.

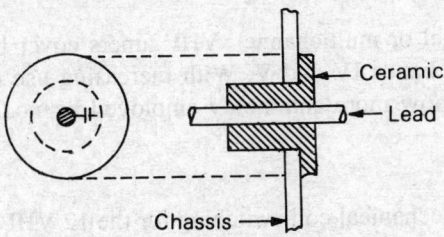

Fig. 11.3   Feed-through capacitor

Reception of image frequency corresponding to (f_s + 2 IF) which also produces the IF at the mixer, should also be adequately suppressed by the RF tuned circuits.

RF Response
The RF amplifier has a passband broad enough to pass the channel selected and also allow for the variation in the local oscillator fine tuning and variation due to the AGC voltage that should not affect the RF gain. Double tuned filters with a suitable dip between peaks can provide broad bandwidth of over 11 MHz and also good transient response. Figure 11.4 shows a typical response with a dip of 1 dB between the peaks.

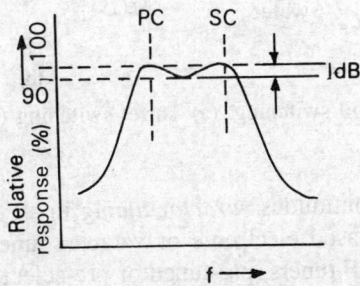

Fig. 11.4   RF amplifier response

Mixer
The mixer produces the IF signal by heterodyning the RF signal with the local oscillator frequency. As there are two carriers in the RF signal, viz. picture carrier and sound carrier, two IFs are produced, the picture IF equal to 38.9 MHz and the sound IF equal to 33.4 MHz. The local oscillator frequency is higher than the RF carriers so that for channel 4 for example, the LO frequency is (62.25 + 38.9 =) 101.15 MHz; and for channel 5, it is (176.25 + 38.9 =) 215.15 MHz.

Local Oscillator
This provides the local oscillator frequency which should be essentially stable, free from drifts due to

temperature aging of components or small changes in the supply voltages, etc. It should have minimum harmonic content. The oscillator frequency control is basically the fine tuning control of the receiver. This may be manual varactor tuned or automatic fine tuning using varactor diode bias control from the frequency discriminator at the IF.

11.3   TYPES OF RF TUNERS

RF tuners may be single channel or multichannel. VHF tuners cover bands I and III only, while UHF tuners cover the UHF range of bands IV and V. With increasing use of UHF bands, electronic tuners covering VHF/UHF bands are now more commonly employed in modern receivers.

VHF Tuners

VHF tuners generally employ mechanical coil switching for the 12 VHF bands in low cost versions. This can be done by turret type or wafer type switch arrangement for coil changing.

In *coil turret type switching*, the relevant coils for each channel are mounted on the periphery of the turret drum. The rotation of the turret changes the coils in contact with the spring clips for coil strips for each channel, as shown in Fig. 11.5(a).

In *wafer type switching*, the coils in incremental arrangement or in printed circuit form are mounted on wafers switched to the contacters, as shown in Fig. 11.5(b). The rotary wafer switch having series connected coils, progressively shorts out parts of coils as channel frequency is increased.

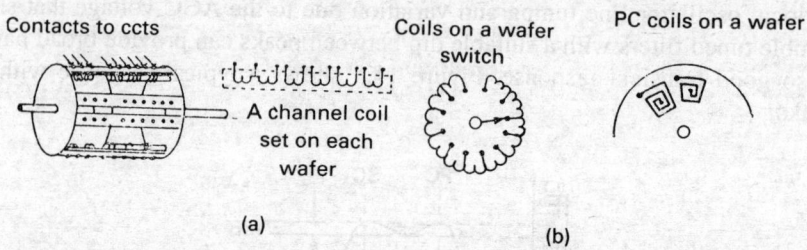

Fig. 11.5   Tuner coil switching: (a) Turret switching (b) Wafer switching

Electronic Tuners

The modern practice is to employ continuous *varactor tuning* in all the bands. With no moving parts or mechanisms, these varactor diode based electronic or varactor tuners are very compact but relatively expensive. With electronic VHF/UHF tuners, pre-tuned or pre-select positions are often provided in the receiver, as generally a limited number of stations can be received at a place. One can then tune in to any desired channel at any of the preset positions, with their individual tuning controls and the band switches (I/III/U). The band switching can also effected by biased diodes, switching tuning coil taps. Varactor tuning can be easily adapted to automatic fine tuning (AFT) and remote selection of channels by keypads.

Automatic Fine Tuning

The fine tuning in the tuner of a colour TV receiver is critical because any small shift in the local oscillator and hence the video IF can result in severe colour distortion or loss of colour. All colour TV receivers, therefore, necessarily employ automatic fine tuning (AFT) to check the video IF generated at the tuner by means of a frequency discriminator circuit in the video IF amplifier and apply a correction to the AFT varactor diode, if it deviates from the correct value of 38.9 MHz. An AFT ON/OFF switch is often provided on the panel to disconnect the AFT for optional manual fine tuning.

UHF Tuners

Tuners meant for receiving band IV and band V may not employ on RF amplifier, to reduce cost and complexity. The UHF antenna signal, usually via a coaxial cable, is coupled to the input tuned circuit and diode mixer that gives out the IF. This is fed via the unused channel 1 cum UHF position of the VHF tuner to its mixer stage, which acts as IF amplifier to make up for the lower gain of the UHF tuner, and thus supplies equal signals to the video IF amplifier on UHF and VHF reception. Input traps are not necessary at the UHF frequencies, and the oscillator radiations limits at the UHF are also less stringent for UHF. The RF amplifier is incorporated in some UHF tuners. Mechanical UHF tuners employ rotating capacitors to fine tune stripline inductors. The current trend in electronic tuners is to design VHF and UHF tuner circuits on the same PCB, in one well-shielded enclosure.

11.4 TRANSISTOR TUNERS

With their low power drain, miniature size and reliability, transistors are universally used in the tuner stages of a modern TV receiver. The transistors used must have the required high frequency performance providing adequate gain and low noise figure in order to achieve good overall sensitivity of the receiver.

Configurations

Grounded emitter or grounded base configuration may be used for the RF amplifier circuit. Grounded emitter configuration can provide a higher power gain and may require neutralization unless designed with proper transistors allowing for mismatch loss for stability. Grounded base configuration gives a lower gain but does not require neutralization. Its input impedance is quite low, but is constant over the band III range, along with other parameters. The common base configuration hence requires a large transformation ratio which reduces the available gain. The signal handling capacity is low. The grounded emitter configuration has a larger signal handling capacity, but the parameter variation due to AGC is also greater, introducing higher cross-modulation. This can be reduced by a small unbypassed emitter resistance while terminating the input circuit so that the resistive part of the input impedance is 75 Ω, to match it to the feeder impedance through the balun.

The dual gate MOSFET transistor like BEL 3N200 can also be used for RF amplification with the advantages of good power gain, high input impedance, low noise factor and especially good cross-modulation performance without overload distortion. The MOSFET has a wide dynamic range, and its low feedback capacitance provides a stable performance without neutralization over a wide band. The dual gate feature permits the design of simple AGC circuitry requiring very low control power. A FET RF amplifier of this type useful as front end in the TV tuner, is shown in Fig. 11.6. Gate G1 has input signal as a grounded source stage, while the AGC is applied to gate G2 bypassed to ground by a capacitor. Bypassing the gate G2 shields the gate G1 from the output at the drain. Neutralization is therefore, not needed. The AGC may be applied to the gate G1 also partially by a potential divider to ground as shown in Fig. 11.6(b).

The design of transistor RF stage and mixer can be done on the basis of stability criterion derived by Stern to introduce a minimum mismatch loss to ensure stability in the stage. The method has been outlined later in Sec. 13.5 on design of IF amplifiers. The circuit of a typical transistor VHF channel tuner is shown in Fig. 11.7.

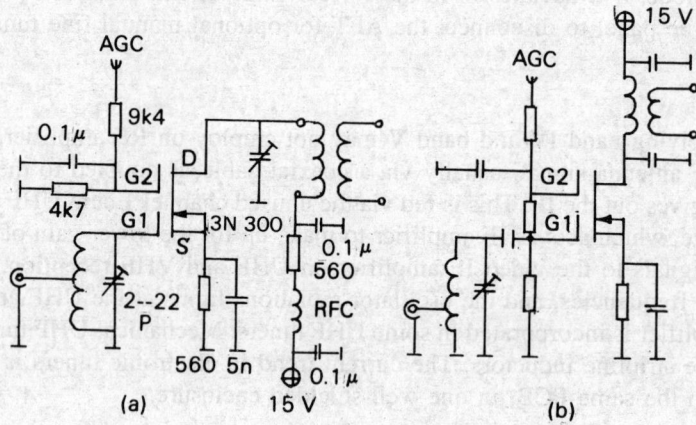

Fig. 11.6 (a) FET RF amplifier (BEL), (b) FET RF amplifier with AGC to both gates

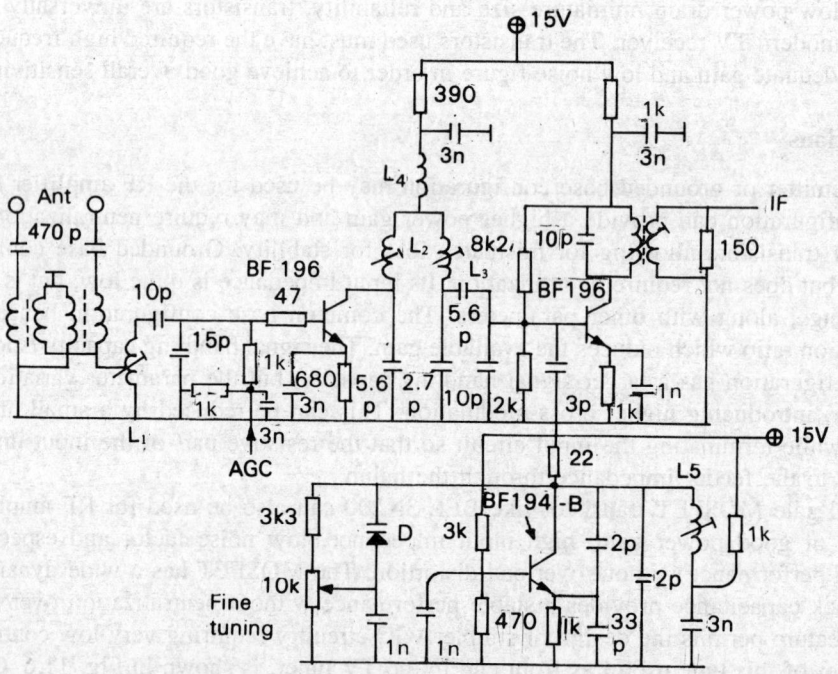

Fig. 11.7 A Transistor VHF tuner

The tuner employs BF 196 for the RF amplifier and mixer stages, and BF 194B for the local oscillator. The input tuning circuit formed by L1 and the 10 pF and 15 pF capacitors, is matched to the 75 Ω impedance of the balun on the antenna side by a tap on the coil L1 and to the input impedance of the transistor by the capacitive tap of the two capacitors. The RF transistor is provided a forward AGC *via* the 1k base resistances decoupled to ground by the 3 nF capacitor. The double tuned interstage coupling with bottom coupling inductance L4 is used between the RF stage and the mixer input to achieve required RF wideband response with a small dip between the two peaks at the sound and picture carriers.

The mixer stage gets the local oscillator signal over the 2 pF capacitor and 22 Ω resistor, at the base along with the RF signal, to heterodyne the two frequencies and select the difference IF signal over the collector output IF transformer. The If is coupled from the secondary of the transformer providing 75 Ω impedance, to the video IF amplifier over a coaxial link coupling.

The local oscillator uses BF 194B in a Colpitt circuit with the capacitance tap between the 2 pF and 33 pF capacitors. Fine tuning of the oscillator is done by the varactor diode D. The reverse bias of the diode and hence its junction capacitance is varied by the 10 k potentiometer. The biasing tap is bypassed to ground so that the capacitances of the leads of the panel mounted potentiometer do not affect tuning.

VHF/UHF Tuner with Diode Tuning

With the use of UHF channels becoming more and more common, combined VHF/UHF tuner with electronic tuning and band switching to cover all the TV broadcast bands is often employed in modern TV receivers. Typically the tuner is built on a single low loss printed wiring board carrying all components, in a metal housing made of a rectangular frame and front and rear covers. The supply voltages, AGC voltage, tuning and switching voltages are made *via* feed-through capacitors in the underside of the frame which can be mounted on to a printed wiring board. The tuner consists of a VHF and an UHF part, with the two aerial connections on the opposite sides of the frame.

The VHF part of the tuner is shown in Fig. 11.8. The VHF aerial signal is fed *via* an IF trap combined with a high-pass filter, to a tuned input circuit which is connected to the emitter of the input transistor BF 200. The collector load of this transistor is formed by a double tuned circuit, transferring the signal to the base of the mixer transistor BF 182. The oscillator is equipped with a transistor BF 194. The four RF circuits are tuned by four capacitance diodes D7-D10 (BB 106). Switching between VHF band I and band III is achieved by four switching diodes D2-D5 (BA 243/244).

The collector circuit of the mixer transistor is a single tuned IF resonant circuit, at the low end of which the IF signal is capacitively coupled out of the tuner. An IF injection point is provided at the collector of the mixer, for aligning this circuit together with the IF amplifier of the TV receiver.

The tuner requires transistor supply voltages of + 12 V, a switching voltage of + 12 V, AGC voltages variable from + 2.4 V (normal operating point) to about + 7.5 V (maximum AGC) and a tuning voltage variable from + 0.5 V to + 28 V.

The tuner typically provides a gain of 28 dB with an AGC range of over 40 dB, and a noise figure of 6.5 dB.

Program Preselection

While tuning in to a particular channel is relatively easy in a turret or switch type of VHF tuner, it becomes a tedious process to select quickly a desired channel in UHF or combined VHF/UHF electronic tuners, which have continuous tuning control. Since the number of channels that can be received at a location are few, it is a common feature in such tuners to incorporate some 8 to 12 preselect switch positions that provide for pretuning of any desired channel in the band I/III/UHF. Any of these push button or touch sensitive switches can be used to select any desired channel in the band I/III/U, by means of a corresponding band-switch and continuous electronic tuning potentiometer. When selection is done via infrared remote controls, the keyed signals are digitally processed and interpreted to synthesize the required tuning voltage.

Figure 11.9 shows a circuit for preset channel selection to provide required switching voltage to the I/III/U diode switches inside the electronic tuner and the fine tuning voltage VT. Each of the three-way preselector switches routes the DC supply voltage and the potentiometer selected varactor tuning voltage

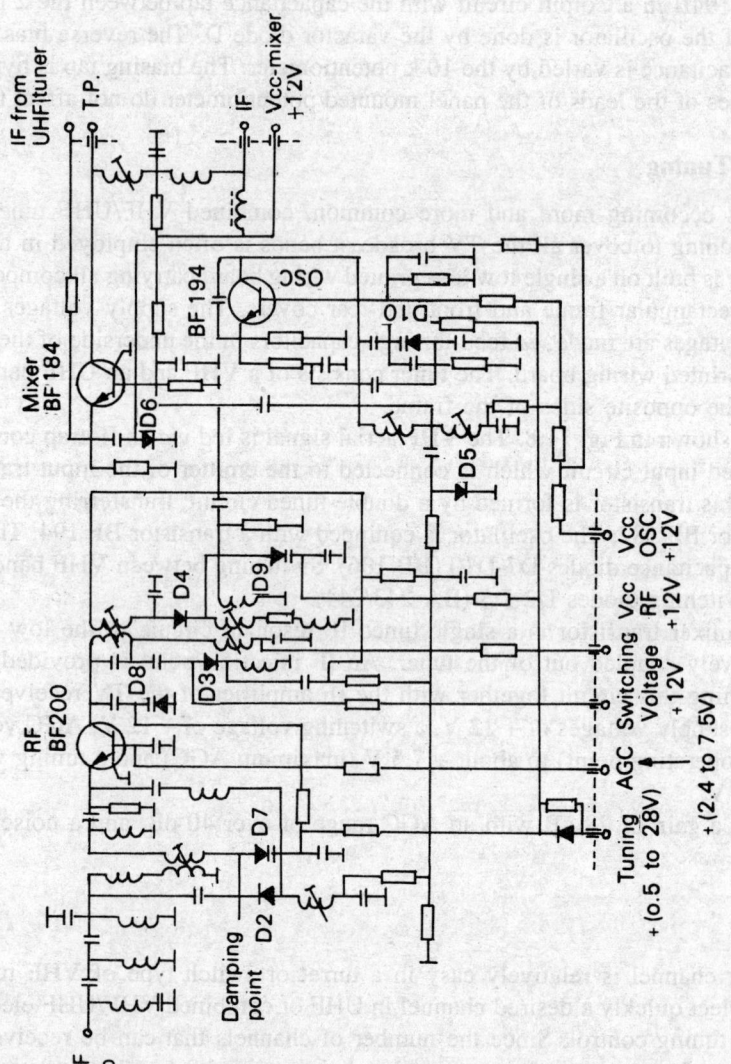

Fig. 11.8 A transistor tuner with diode tuning and switching (*Courtesy:* ELCOMA PHILIPS)

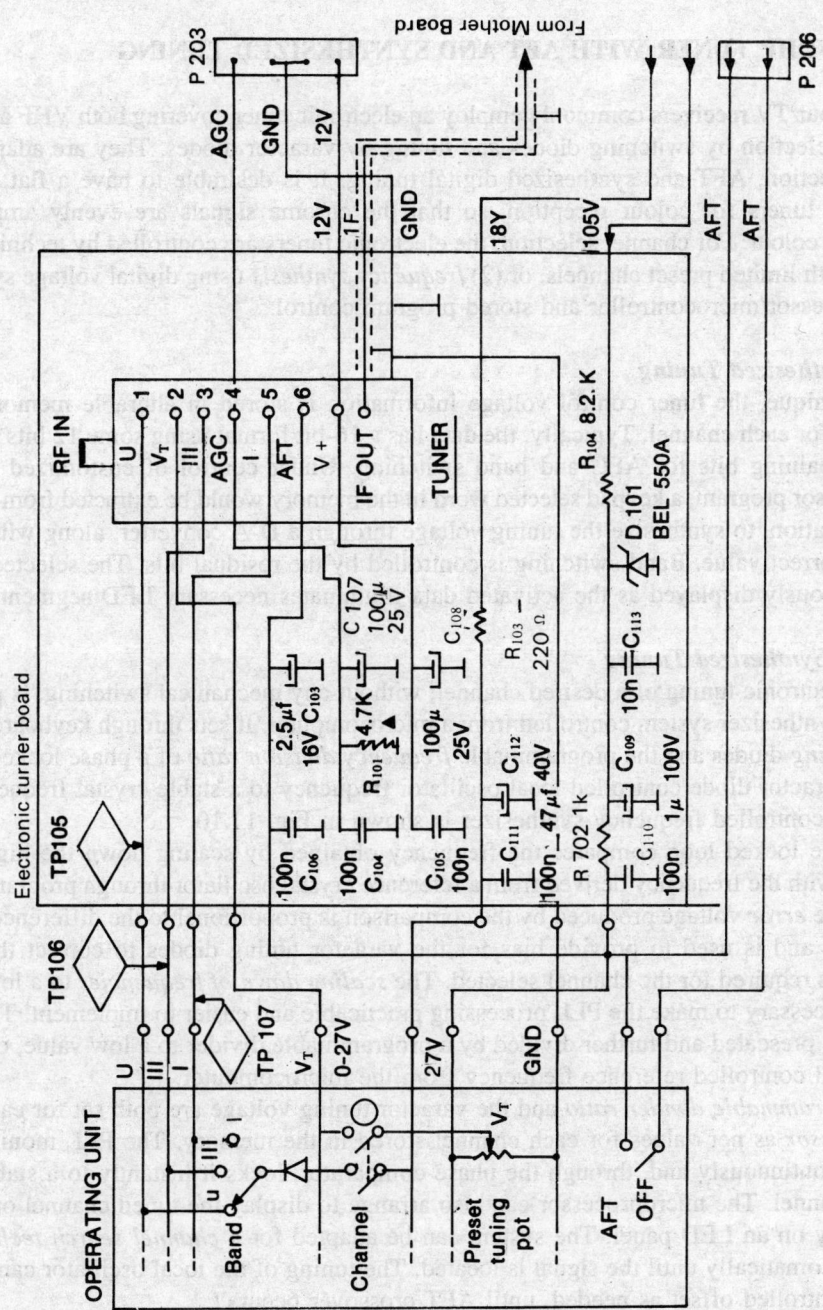

Fig. 11.9 Circuit for preset channel selection with electronic tuner board (BEL)

for the wanted channel, to the tuner. For the tuning to be stable, a separate zener stabilized voltage supply is necessary, and is supplied in some circuits from a separate IC regulator.

11.5 VHF/UHF TUNER WITH AFT AND SYNTHESIZED TUNING

Modern colour TV receivers commonly employ an electronic tuner covering both VHF and UHF channels, with band selection by switching diodes and tuning by varactor diodes. They are adaptable to pretuned channel selection, AFT and synthesized digital tuning. It is desirable to have a flat channel bandpass response of tuners for colour reception, so that the chroma signals are evenly amplified to prevent distortion in colour. For channel selection, the electronic tuners are controlled by technique of (1) *voltage synthesis* with limited preset channels, or (2) *frequency synthesis* using digital voltage synthesis involving a microprocessor/microcontroller and stored program control.

Voltage Synthesized Tuning
In this technique, the tuner control voltage information is stored in alterable memory (EAPROM) as digital data for each channel. Typically, the data has a 16-bit format using some 12 bits for tuning voltage and the remaining bits for AFC and band switching. Under control of customized logic circuitry or microprocessor program, a keypad selected word in the memory would be extracted from the corresponding memory location, to synthesize the tuning voltage through a D-A converter, along with an AFT voltage to set the correct value. Band switching is controlled by the residual bits. The selected channel number is simultaneously displayed as the activated data illuminates necessary LED segments.

Frequency Synthesized Tuning
Accurate electronic tuning of a desired channel, without any mechanical switching, is possible by use of frequency synthesizer system controlled from a microcomputer. It sets through keyboard input the correct *band switching* diodes and the programmable *frequency division ratio* of a phase locked loop (PLL), and locks the varactor diode controlled local oscillator frequency to a stable crystal frequency. A schematic of the PLL controlled frequency synthesizer is shown in Fig. 11.10.

The phase locked loop compares the frequency obtained by scaling down the high local oscillator frequency, with the frequency derived from a reference crystal oscillator through programmable frequency dividers. The *error* voltage produced by the comparison is proportional to the difference between the two frequencies, and is used to provide bias for the varactor tuning diodes to correct the local oscillator frequency as required for the channel selected. The *scaling down of frequencies* to a low value, typically 1 kHz, is necessary to make the PLL processing practicable and easier to implement. The local oscillator frequency is prescaled and further divided by a programmable divider to a low value, equal to the scaled down crystal controlled reference frequency from the microcomputer.

The *programmable divider ratio* and the varactor tuning voltage are both set for each channel by the *microprocessor* as per values for each channel stored in the memory. The PLL monitors the oscillator frequency continuously and, through the phase comparator, locks it instantly to a stable value required for each channel. The microprocessor can also arrange to display the tuned channel on the screen itself or separately on an LED panel. The system can be adapted for a *channel search technique* to scan the channels automatically until the signal is located. The tuning of the local oscillator can be made through software controlled offset as needed, until AFT crossover occurs.

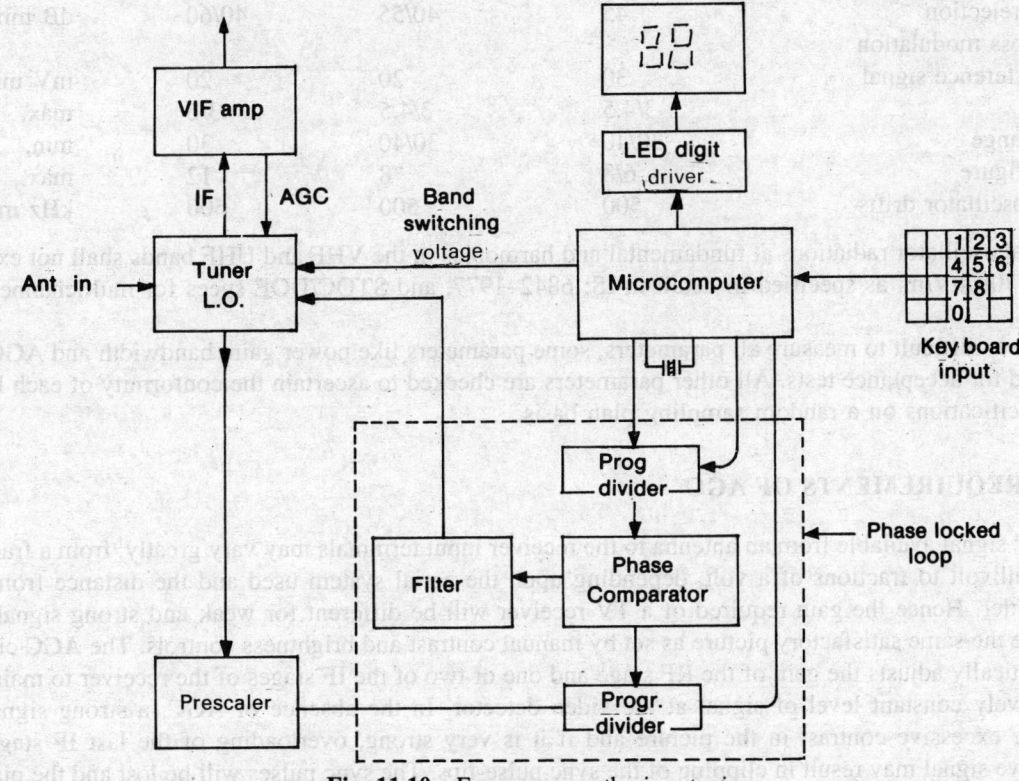

Fig. 11.10 Frequency synthesizer tuning

11.6 TV TUNER PERFORMANCE SPECIFICATIONS

Important performance specifications of typical TV tuners for systems B and G are given in the following:

Channels	VHF band I	channels	2 – 4		47 to 68 MHz
	III	channels	5 – 12		174 to 230 MHz
	UHF U	channels	21 – 69		470 to 862 MHz

RF bandwidth	11 MHz min.	
IF bandwidth	4 MHz min.	
Fine tuning	+/ – 1 to 4 MHz	
AFT range for CTV	+ /– 100/250/500 kHz in band I/III/U	
Input impedance:	75 Ω unbalanced or 300 Ω balanced	
Output impedance:	75 Ω unbalanced	
Supply voltage:	12 V +/– 10%	for Mechanical tuner
	12 V +/– 5%, or 24 V +/– 2%	for Electronic tuner

Band	I	II	III	
Mech (M)/Elec (E) tuner	M/E	M/E	M/E	
Power gain	26	26	20	dB min.
Power gain variation for channels in same band	3	3	3	dB max.
IF rejection	40–50	50	50	dB min

Image rejection	45	40/55	40/60	dB min.
1% Cross modulation				
interference signal	30	20	20	mV min.
VSWR	3/4.5	3/4.5	3/6	max.
AGC range	30/40	30/40	30	min.
Noise figure	6/8	8	12	max.
Local oscillator drifts	500	500	500	kHz max.

Local oscillator radiations at fundamental and harmonics in the VHF and UHF bands shall not exceed 400 to 700 μV/m, as specified in detail in IS: 6842–1977, and STQC/DOE specs for multichannel TV tuners.

As it is difficult to measure all parameters, some parameters like power gain, bandwidth and AGC are checked for acceptance tests. All other parameters are checked to ascertain the conformity of each lot to the specifications on a random sampling plan basis.

11.7 REQUIREMENTS OF AGC

The TV signal available from an antenna to the receiver input terminals may vary greatly, from a fraction of a millivolt to fractions of a volt, depending upon the aerial system used and the distance from the transmitter. Hence the gain required of a TV receiver will be different for weak and strong signals, to produce the same satisfactory picture as set by manual contrast and brightness controls. The AGC circuit automatically adjusts the gain of the RF stage and one or two of the IF stages of the receiver to maintain a relatively constant level of signal at the video detector. In the absence of AGC, a strong signal ill produce excessive contrast in the picture and if it is very strong, overloading of the last IF stage by excessive signal may result in clipping of the sync pulse-tips. The sync pulses will be lost and the picture may fall out of synchronisation, appearing as a torn-out picture that may roll vertically also. Overloading of video stage may result in a negative picture as the stage may no longer function as a normal amplifier with phase inversion of signal but may simply pass the signal without phase reversal.

The AGC circuit must be able to act fast to control the gain of the RF and IF stages if the signal fades or fluctuates rapidly due to interference between the direct ray and reflected signals from moving objects like aeroplanes. This necessitates a fast acting AGC. The AGC filter time-constant should not be too long. In view of this and also the fact that the average dc level at the output may vary greatly with the mean brightness level, as shown in Fig. 11.11, it is not possible to use the detector output voltage for obtaining the AGC bias as is done in radio receivers.

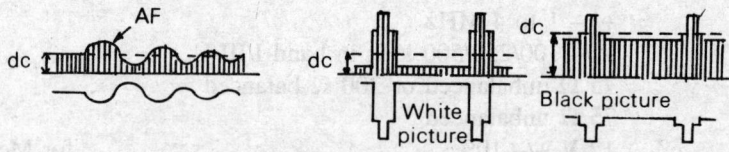

Fig. 11.11   dc output levels in detected signal

11.8 SIMPLE AGC LIMITATIONS

At simple AGC system as used in a radio receiver is shown in 11.12. Here the control voltage proportional to the signal is obtained by rectifying the signal from the IF amplifier and passing it through a low-pass filter to remove the variations at the modulating signal frequencies.

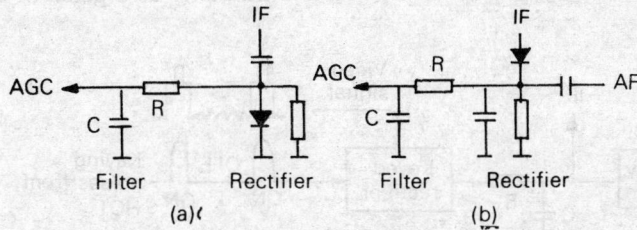

Fig. 11.12 Simple AGC systems: (a) Shunt rectifier, (b) Series rectifier circuits

The filtered AGC voltage is applied as a bias voltage that reduces gain of HF stages with increase in signal amplitude received. The filter time-constant R-C is kept large enough, viz. around 0.2 sec so that it filters the lowest frequency in the signal, the 50 Hz vertical sync pulse rate with the period of 20 ms. If the filter time-constant is smaller, the vertical sync voltage can appear on the AGC line and the reduction in gain due to it will reduce vertical sync amplitude, that may cause bending in picture by loss of vertical sync. On the other hand, a long time-constant will not allow the AGC voltage to alter fast enough if the signal fluctuates rapidly. This does happen in flutter fading due to interference between direct ray and reflected signal from aeroplanes and other moving objects.

Another difficulty in the use of the simple AGC system in the TV receiver is due to the dc component in the modulating signal. While in a radio signal the mean carrier level is independent of modulation, in a TV signal the mean carrier amplitude or the mean dc level in the video signal may vary greatly depending on the mean brightness level in the picture, as indicated in Fig. 11.11.

This implies that the rectified mean carrier level of the TV signal is not a very convenient parameter for AGC bias. The peak levels of the TV signal corresponding to the 100% modulation sync pulses are levels independent of picture modulation and represent the received signal level amplitude. A peak detector circuit can be used to obtain corresponding AGC voltage. Such a detector has the disadvantage of responding to noise pulses and the AGC may produce a large control voltage when strong noise pulses interfere. This may cause unwanted reduction in gain. In the fringe area where the signal is weak and noise strong, the reduction in gain may cause an unnecessary flat picture and the weakening of the IF output may lead to failure in synchronisation. This problem is solved by employing an AGC system that responds to the amplitude of H sync and blanking pulses only, by a suitable gating technique.

11.9 KEYED AGC SYSTEM

The keyed AGC or gated AGC system is receptive to the incoming video signal during the flyback duration only when the sync pulses occur. During the remaining scan period, the AGC rectifier is cut-off and hence is immune to noise pulses during that long period. Another important feature of keyed AGC is that the keying pulses are at the rate of the horizontal frequency of 15,625 Hz. The AGC rectifier current pulses conduct at the same rate. Hence the ripple in the AGC voltage is at 15,625 Hz which can be filtered by a low time-constant AGC filter. Consequently the keyed AGC can respond to rapid fluctuations of the signal due to aeroplane flutter to counteract the effect.

Keyed AGC thus offers a more efficient solution to aeroplane flutter and makes the AGC line less susceptible to noise in the fringe areas where the signal is weak and impulsive noise interference may be quite strong.

The keyed AGC system employs a pentode, triode or a transistor as a gated or coincidence rectifier indicated in Fig. 11.13.

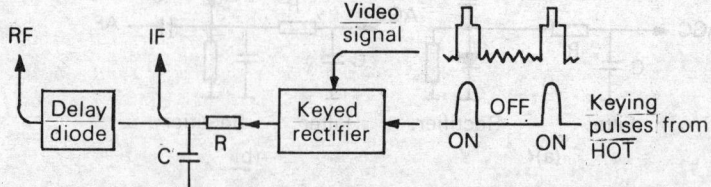

Fig. 11.13 Keyed AGC system

The devices are biased to cut-off by a dc bias and conduct only during the period of the keying pulses from the horizontal output stage, obtained during flyback of the horizontal scanning. During this period, the current flow in the gated stage is made proportional to the amplitude of the sync pulse in the direct coupled video signal so that a proportional AGC voltage is developed across an *R-C* circuit in the path. This is further filtered by low-pass *R-C* section, and then supplied to the HF stages for gain control.

Delayed AGC

It is desirable to operate the RF amplifier at its maximum gain for weak signals so that the mixer gets maximum signal possible to keep the signal-to-noise ratio at the mixer high. The RF AGC bias is, therefore, delayed by a suitable pre-bias circuit so that until the antenna signal rises above 500 to 1000 μV, there is no AGC bias for the RF stage. Stronger signal produces a higher AGC voltage that overrides the pre-bias of opposite polarity and controls the RF stage gain. Stronger signals reduce the RF gain to minimum preventing cross-modulation in the mixer. The RF AGC is thus usually separate from the IF AGC line.

Amplified AGC

A dc amplifier stage is often used for the AGC bias developed in the AGC rectifier so that small changes in the bias due to small changes in the signal can produce larger changes in the AGC voltage applied to the controlled stages. This improves the control and maintains a more constant video output, independent of the signal fluctuations.

Transistor Keyed AGC Circuit

A transistor AGC using a transistor keyer and an AGC amplifier is shown in Fig. 11.14.

Transistor TR2 functions as the stage gated-in by the positive going keying pulses from the horizontal output transformer, while normally a cut-off bias at its emitter is provided from the adjustable potential divider formed by the 33 kΩ, 330 kΩ and 1 kΩ potentiometer. The direct-coupled video signal applied at the base of TR1 puts it on during the positive going sync tips. This develops a voltage across the 2.5 μF capacitor in the collector circuit, that can serve as AGC bias. The polarity of the voltage for the *npn* transistor circuit is negative going. This voltage is amplified and also inverted in polarity to obtain positive going AGC bias required for the commonly used forward AGC of *npn* transistors in the RF and video IF amplifiers.

This is done by the dc amplifier stage of TR1, from the collector of which the AGC voltage is fed to the RF and IF stages *via R-C* filters. The transistor TR1 is fully on in the absence of the video signal and voltage across the 2.5 μF condenser is + 0.7 V. As the signal amplitude increases, the negative voltage across the condenser increases. This reduces the forward bias of TR1 to reduce its collector

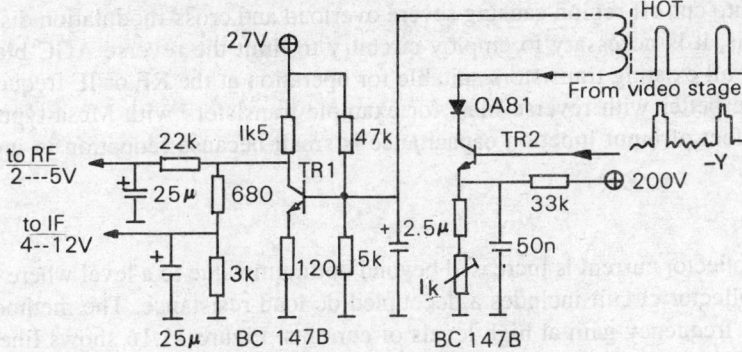

Fig. 11.14 Transistor keyed AGC circuit

current and raises the collector voltage to a more positive value. The diode OA 81 is included in the collector circuit of TR2 to block the AGC voltage developed from the collector. The polarity of the voltage is such that in the absence of the diode, it may provide forward bias to the collector junction and discharge the condenser through the transistor.

11.10 AGC OF TRANSISTOR RF/IF AMPLIFIERS

The gain of the transistor tuned amplifiers (RF or IF) can be controlled by two methods: (1) Reverse gain control in which the forward biasing voltage to the base emitter junction is reduced to decrease the collector current and hence the gain, and (2) Forward gain control in which the forward bias of the base emitter junction is increased to increase the collector current, and reduce the V_{CE} with the help of resistance in the collector circuit. It is thus possible to decrease the gain by decreasing or increasing the forward bias if the operating point of the transistor is properly set and a suitable transistor is selected.

(i) Reverse AGC

In this method, the transistor collector current is reduced by the reverse biasing AGC voltage superposed on the existing base-emitter forward bias. The dc load in the collector circuit is quite small so that the V_{CE} remains fairly constant, or decreases only slightly, as shown in Fig. 11.15.

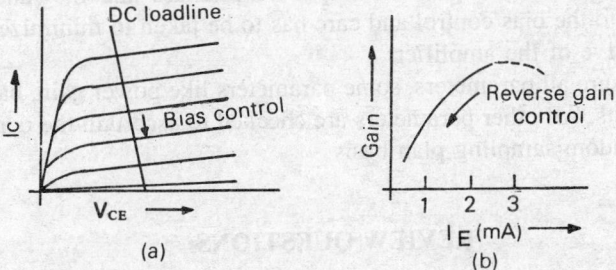

Fig. 11.15 Reverse AGC operation: (a) Operating point bias variation, (b) Gain control characteristics

With the reduction in the collector current, the forward transfer admittance and hence, the gain of the stage is reduced. The reduction in gain is quite step with respect to current and is considerably nonlinear at high currents because of changes in the transistor parameters with current.

The AGC control range is then rather limited, and very strong signals can produce AGC bias that will drive the amplifier into cut-off region causing severe overload and cross-modulation distortions in strong signals. To avoid this, it is necessary to employ circuitry to limit the reverse AGC bias. Reverse AGC can be employed for all existing transistors suitable for operation at the RF or IF frequency. Some types of transistors operate better with reverse bias, for example transistors with Mesa-type construction. In reverse AGC, the effect of input junction capacitance is small because reduction in input bias causes a smaller change in it.

(ii) Forward AGC

In this method, the collector current is increased beyond optimum value to a level where the amplification starts falling. The collector circuit includes a decoupled dc load resistance. The method is based on the deterioration of high frequency gain at high levels of currents. Figure 11.16 shows lines of constant y_{fe}, the forward transconductance in the $I_c \times V_{CE}$ plane.

Continuous lines are for transistors, such as BF 167, specially designed for forward gain control, while the broken lines are for others not suitable for this type of control. The high frequency knee is made rather large for forward gain control transistors. As the operating pint moves up the dc load line, y_{fe} decreases, reducing the gain. Since forward gain control makes use of deterioration of the high frequency properties of the transistors, this method of gain control is possible at relatively high frequencies. A typical gain control characteristic is shown in Fig. 11.16.

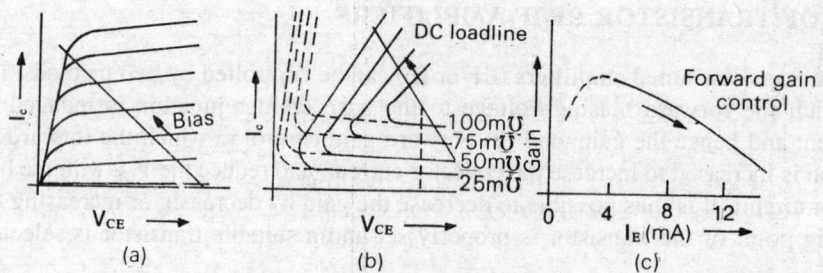

Fig. 11.16   Forward AGC operation: (a) Operating point bias variation, (b) Y_{fe} variation, (c) Gain control characteristics

Forward AGC has a greater gradual control of gain, being proportional to the AGC bias. With a strong signal, the transistor will, at the most, tend to saturate, V_c approaching a low saturating voltage but not cut-off as in the reverse AGC. The input and output admittances and the junction capacitances are considerably affected due to the bias control and care has to be taken to minimize the effect of these on the amplitude response curve of the amplifier.

As it is difficult to measure all parameters, some parameters like power gain, bandwidth and AGC are checked for acceptance tests. All other parameters are checked to ascertain the conformity of each lot to the specifications on a random sampling plan basis.

REVIEW QUESTIONS

11.1   State the functional requirements of the following in the RF tuner: RF amplifier, high pass filter and balun at the input, feed-through capacitors, and varactor diodes.

11.2   What are the methods of fine tuning in RF tuners? Draw and explain the basic circuit of electronic tuning. What is automatic fine tuning?

11.3 Discuss the technique of synthesized tuning used in TV receivers.

11.4 What are the important performance specifications of a TV tuner?

11.5 State the types of tuners and explain the electrical design techniques used for VHF/UHF tuners. Discuss their merits and demerits.

11.6 What are the advantages of transistors and MOSFETs in particular in the tuner circuit?

11.7 Explain giving reasons:
   (a) The RF stage is most suited for AGC control.
   (b) The tuner is shielded extremely well in a completely closed box.
   (c) Feed-through capacitors are essential in the supply leads of the tuner.
   (d) Electronic tuning is invariably used in UHF tuners.
   (e) Preset tuning is essential with combined VHF/UHF tuner.
   (f) Highly regulated supply is used for an electronic tuner.

11.8 In what way do the AGC requirements of a television receiver differ from those of a radio receiver?

11.9 What are the advantages of keyed AGC over the other forms?

11.10 Give reasons for the following:
   (a) A keyed AGC can have a filter with a smaller time-constant.
   (b) An RF amplifier is given a delayed AGC voltage.
   (c) A high AGC bias can cause a weak picture, and a low AGC bias can cause disturbed synchronisation and a negative picture.
   (d) Peak detector AGC cannot be used in a TV receiver.
   (e) A keyed AGC can overcome aeroplane flutter.

11.11 What is the difference in the principle of operation and performance of forward and reverse AGC in transistor amplifiers?

MULTIPLE-CHOICE QUESTIONS

11.1 Pick out the wrong statement:
   (a) RF tuner includes an IF trap filter at the input to prevent radiation to the aerial
   (b) Bandwidth of the RF tuner is 7 MHz, (b) Fine tuning adjusts the tuning in the RF amplifier,
   (c) The local oscillator frequency is higher than the signal frequency,
   (d) AGC to the RF stage is different from that for video IF stages.

11.2 RF amplifier bandwidth is kept a little larger than the channel width,
   (a) to take care of the local oscillator drifts
   (b) to allow fine tuning of the channels,
   (c) to improve noise factor of the stage,
   (d) a and b.

11.3 Pick out the wrong statement:
   (a) RF stage improves the noise figure
   (b) mixer stage reduces the noise figure,
   (c) RF stage contributes to RF radiation from the receiver,
   (d) Mixer stage generates RF interference signals from the tuner.

11.4 Pick out the incorrect statement:
   (a) UHF tuner feeds its output into VHF tuner,
   (b) Some UHF tuner employ a diode mixer,
   (c) VHF/UHF tuner employs diode switching,
   (d) IF radiation can be a serious problem in UHF tuners.

12

Video IF Amplifier

12.1 REQUIREMENT OF GAIN AND BAND PASS RESPONSE

In a TV receiver, the signal frequencies are converted into fixed IF by superheterodyne technique in the mixer stage of the tuner where these beat with the local oscillator frequency. The sensitivity and selectivity of a TV receiver is decided largely by the video IF amplifier where the fixed IF frequencies are amplified.

(i) The aim in a video IF amplifier is to obtain a high and stable gain so that the low fraction of millivolt or greater amplitude of the IF signal from the mixer stage is raised to a level of a few volts to obtain linear video detection without distortion. The gain should be also controllable, AGC controlled to suit the requirements of the signal which may vary within wide limits. The video IF amplifier must thus have a gain of about 60–80 dB from the tuner output to the detector load.

(ii) The video IF amplifier must have adequate bandwidth of about 5 MHz to pass all the video frequencies in the signal and have the bandpass response curve to suit the reception of the vestigial sideband TV signal to give equal detector output for all video frequencies, and the sound.

(iii) The amplifier must have good adjacent channel selectivity, i.e., it must reject the frequencies of the adjacent channels particularly the picture carrier and sound carrier on either side.

(iv) The shape of the bandpass response curve should be sufficiently independent of AGC variations in bias to provide a gain variation of about 60 dB by control of one or two stages.

Choice of IF

The value of IF is a compromise between several conflicting factors. It should be as low as possible from the considerations of high and stable gain which is easier to obtain at lower frequencies but it should be high enough so that it can have the required bandwidth and selectivity.

The IF chosen should be such that interference due to spurious reception of image frequencies (f_{osc} + 2 IF), direct IF range signals, difference frequencies between frequencies of stations, which may mix up and produce IF, etc., is minimum. To a certain limit, higher IF is thus preferable from consideration of spurious response, good selectivity and easier filtering at the video detector. But it is not easy to build high gain stable circuits at higher frequencies. Increased losses in circuit components, tendency to radiate to cause interference to other receivers nearby, and instability due to feedback, bring in considerable design difficulties. Because of high IF, a high oscillator frequency is required, which is usually less stable and creates tuning problems.

With these considerations, the CCIR-B standard receivers employ 38.9 MHz as the picture IF and 33.4 MHz as the sound IF (The American CCIR-M standard receivers employ 45.75 MHz for video IF and 41.25 MHz for sound IF). The requirements of IF response curves are specified as shown in Fig. 12.1.

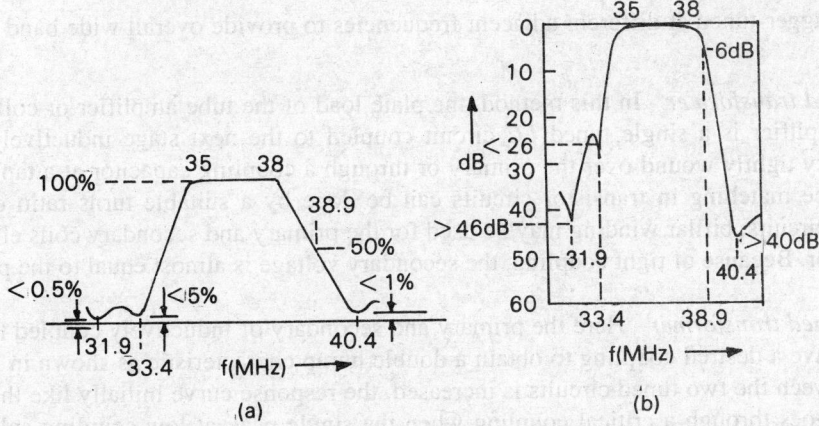

Fig. 12.1 Video IF response requirement: (a) Linear scale (b) Log scale

(i) The picture IF of 38.9 MHz lies on the so-called Nyquist flank, 6 dB below (at 50% of) the maximum response.

(ii) The bandwidth of the response curve between the points at which the response is 6 dB down, should be 5 MHz.

(iii) The sound IF of 33.4 MHz should be 20 dB below (at 10% of) the level of picture IF.

(iv) The adjacent channel sound carrier at 40.4 MHz should be suppressed below 40 dB of the maximum response.

(v) The adjacent channel picture IF at 31.9 MHz on the other side should be at least 46 dB below the maximum response.

(vi) The response should be flat within 2 dB between the range 35 ... 38 MHz.

Phase Response

Besides gain amplitude response with respect to frequency, the phase response can significantly affect the picture quality. Phase variation and envelope delay is more pronounced where amplitude response changes sharply. In double sideband (DSB) systems, the steep flanks of the amplifier band-pass response are situated at the extreme high frequency ends of the sidebands and hence their effects are less important; but in vestigial sideband (VSB) systems, the carrier and the major frequencies lie on the flanks and hence envelope delay distortion occurs more in the main video frequency range. This distorts the overall video response and results in so called relief effects in reproduced picture. Some transmitters compensate for this by suitably modifying the transient response.

12.2 INTERSTAGE COUPLING METHODS

For obtaining good sensitivity of the TV receiver as demanded in fringe area reception, the video IF amplifier must have a gain of about 60 ... 80 dB from the tuner output to the detector load and require the use of cascoded multistage tuned band-pass amplifiers. Generally three stages are adequate to provide this gain, with the first stage or sometimes the first two stages having automatic gain control facility.

Various methods can be used for the coupling between the stages of the multistage tuned band-pass amplifier, along with single tuned or double tuned circuits to obtain the required response curve of the amplifier as a whole. The tuning of the stages may be synchronous, i.e., at the same frequency or the stages may be stagger-tuned at different adjacent frequencies to provide overall wide band response and good selectivity.

(i) Single tuned transformer In this method, the plate load of the tube amplifier or collector load of the transistor amplifier is a single tuned *LC* circuit coupled to the next stage inductively through an untuned secondary tightly wound over the primary or through a coupling capacitor at a tap on the tuned circuit. Impedance matching in transistor circuits can be done by a suitable turns ratio or tap on the circuit. In tuned circuits, bifilar winding may be used for the primary and secondary coils eliminating the coupling capacitor. Because of tight coupling, the secondary voltage is almost equal to the primary tuned circuit voltage.

(ii) Double tuned transformer Here the primary and secondary of inductively coupled tuned circuits are adjusted to have a desired coupling to obtain a double hump characteristic as shown in Fig. 12.2. As the coupling between the two tuned circuits is increased, the response curve initially like that of a single tuned response, goes through a critical coupling when the single peak at low coupling splits apart into a double hump characteristic.

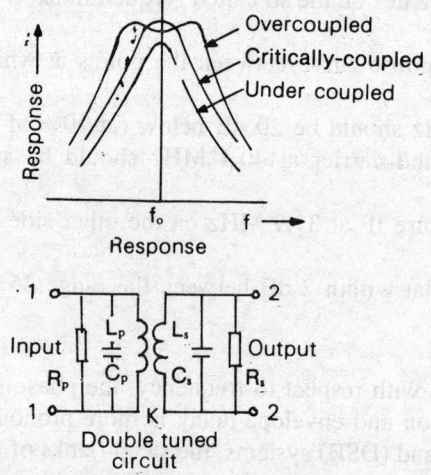

Fig. 12.2 Double tuned transformer coupling

The humps can be reduced by damping or loading resistors reducing *Q* and gain but giving larger bandwidth with more flat top.

(iii) Impedance coupling This utilizes a common impedance between the output circuit of one stage and the input circuit of the next stage. A mutually common impedance between the current loops of the two circuits on the earth side as in a T-type circuit provides a bottom coupling; while a mutually common impedance across the nodes of the two circuits as in a π type circuit is said to provide a top coupling. Combinations of the two can also be used. Impedance coupling circuits along with appropriate tuning elements at the output and the input can form a band-pass filter to provide desired IF response curve. Some of the commonly used impedance coupling filters are shown in Fig. 12.3.

Figure 12.3(a) shows a filter formed by C_1, C_2 and *L*. This splits the tuned circuit into two parts and is advantageously used when a shielded cable is required for coupling between two stages, e.g. tuner output of the IF input. Cable capacitance can form a part of the filter capacitance thus avoiding its otherwise shunting effect. Double tuned circuits with variations in bottom couplings are used as shown

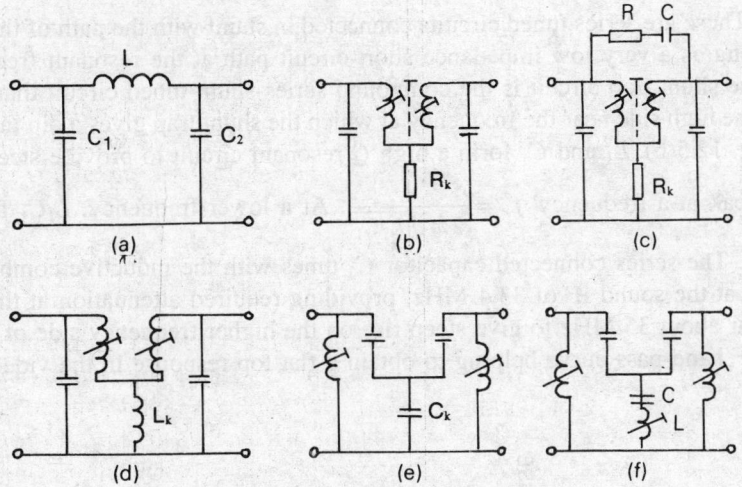

Fig. 12.3 Impedance coupling filters of different types

in Fig. 12.3(b)-(e). Bottom resistance coupling shown in Fig. 12.3(c) also serves for necessary damping for broad bandpass curve. Additional *R-C* top coupling provides a more favourable phase delay response. Inductive bottom coupling is shown in Fig. 12.3(d), and capacitive bottom coupling in Fig. 12.3(e). Adjustment of the coupling element can provide a variable coupling. The series tuned circuit in the bottom coupling paths of the two circuits as shown in Fig. 12.3(f), provides a third tuned circuit adjustment.

Coaxial Link Coupling

This coupling as shown in Fig. 12.4, is used to couple two stages rather apart as in the case of the mixer stage in the tuner and the first IF input stage. A coaxial cable at low impedance level prevents radiation from the lead as well as pick-up from outside interference signal. The cable is often a mini-size 75 Ω type with plug and socket at one end to be able to remove the connection readily.

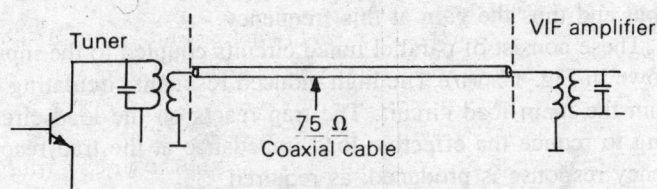

Fig. 12.4 Coaxial link coupling

12.3 TRAP CIRCUITS

In the video IF amplifier, it is necessary to suppress the unwanted frequencies like the nearby adjacent channel carriers and the sound carrier of the tuned channel to an adequate level by means of traps that absorb energy at these frequencies. Basically these are high Q LC-resonant circuits tuned to the frequencies and coupled to the amplifier circuits or placed in the IF signal path. Various types of these traps are used in practice.

(i) Shunt traps   These are series tuned circuits connected in shunt with the path of the signal as shown in Fig. 12.5(a), acting as a very low impedance short-circuit path at the resonant frequency.

A variation of the shunt trap circuit is the compound series-shunt tuned circuit that provides a steep rising response on the high side near the frequency at which the shunt trap gives a dip in the characteristic as shown in the Fig. 12.5(b). L_1 and C_1 form a high Q resonant circuit to provide steep rising response with a resonance peak at a frequency $f_r = \dfrac{1}{2\pi\sqrt{L_1 C_1}}$. At a lower frequency, $L_1 C_1$ form an inductive circuit combination. The series connected capacitor C_2 tunes with the inductive combination to form a shunt trap typically at the sound IF of 33.4 MHz, providing required attenuation at the sound IF while $L_1 C_1$ is tuned to near about 35 MHz to give steep rise on the higher frequency side of the sound carrier as required in the IF band-pass curve helping to obtain a flat top response in the video frequency band of 35 to 38 MHz.

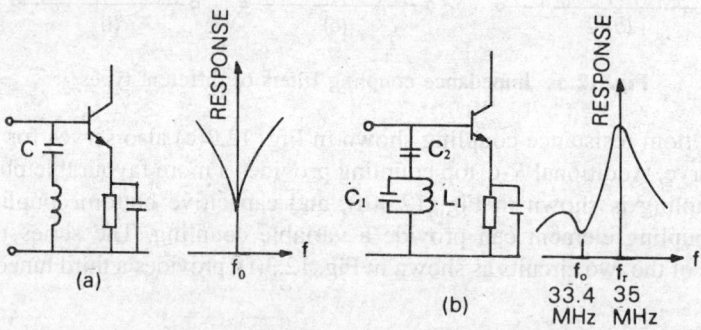

Fig. 12.5   (a) Shunt trap, (b) Series-shunt trap

(ii) Series traps   These consist of parallel tuned circuits connected in series with the input lead as shown in Fig. 12.6(a). The tuned circuit impedance and the input impedance of the following circuit act as a voltage divider. At the resonant frequency, the tuned circuit impedance is very high equal to $\omega_0 L Q$, reducing the effective input and thus the gain at this frequency.

(iii) Absorption traps   These consist of parallel tuned circuits coupled to the input or output coils of the tuned amplifier as shown in Fig. 12.6(b). The high induced resonant circulating current in the tuned circuit absorbs power from the main load circuit. The trap reacts on the load circuit of the amplifier through inductive coupling to reduce the effective load impedance at the trap resonant frequency and hence a dip in the frequency response is produced, as required.

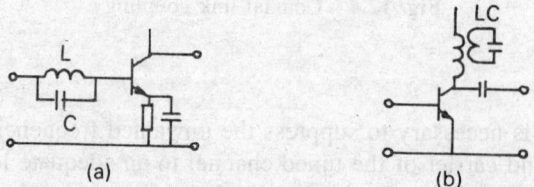

Fig. 12.6   (a) Series trap, (b) Absorption trap

(iv) Degenerative traps   A parallel resonant circuit introduced in the unbypassed emitter or cathode circuit of the amplifier introduces large negative feedback at the resonant frequency at which it offers

a high impedance. The ac voltage drop across the trap circuit opposes the input signal voltage reducing the effective input to the amplifier device, thus reducing the gain at the trap frequency. The trap circuit may be directly in the cathode or emitter lead or the same may be coupled inductively to a coil in the lead, as an absorption trap, as shown in Fig. 12.7(a).

(v) Bridged-T trap This consists of a modified series trap in which a resistance is bridged across the centre tap of the coil or split capacitor and the common ground as shown in Fig. 12.7(b). Balance is obtained at resonance when reactance of the capacitance is equal to the reactance of the inductance and resistance R in the bridge arm is 1/4 of the parallel impedance of the resonant circuit at the resonant frequency.

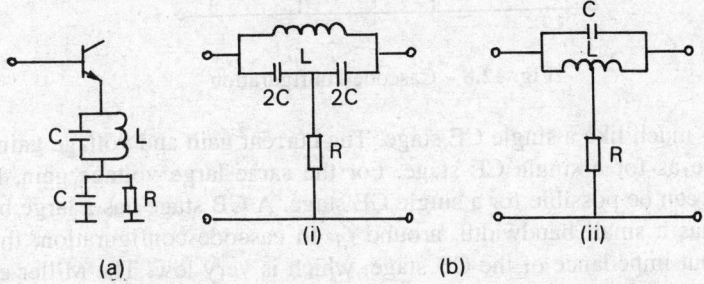

Fig. 12.7 (a) Degenerative trap, (b) Bridged-T traps

Much sharper null and greater rejection can be obtained by these bridged-T traps than the other types or traps referred to above, as the shunt resistance R of the bridge arm can be adjusted to cancel the losses in the tuned circuit formed by L and C.

The wave traps are generally included at the input of the video IF amplifier to keep down the undesired signals at the base of the input transistor as low as possible, thus reducing the possibility of cross-modulation.

There are considerable variations in video IF amplifier designs depending on the choice of load, coupling circuitry and traps. Stages using transistors and integrated circuits are now most commonly used in place of vacuum tube amplifier.

12.4 TRANSISTOR IF AMPLIFIER DESIGN

Transistors are very commonly used in IF amplifiers in TV receivers to obtain a compact and reliable circuit design with low power consumption. At each interstage coupling between the stages, impedance matching is necessary between the high impedance (several kilo-ohms) collector circuit and the low impedance (less than a kilo-ohm) base input circuit of the next stage. Transformer coupling with proper step-down ratio or suitable impedance coupling can be used.

High frequency transistors like BF 167, BF 173, BF 196, BF 197, with high f_T of around 400 MHz and high forward transfer conductance y_{fe}, of around 100 mA/V are now available for video IF amplifiers, that can be designed to have adequate stability by sacrifice in gain by deliberate mismatch loss, avoiding need for neutralisation as such.

Common emitter stages are widely used because of their higher power gain in preference to common base circuit, though the latter can provide better feedback isolation and stability and can operate at higher frequencies. Cascode configuration is sometimes used combining the high power gain and better feedback isolation features of the two configurations.

Cascode Configuration   The circuit configuration is a cascode of common emitter (CE) and common base (CB) connection, as shown in Fig. 12.8.

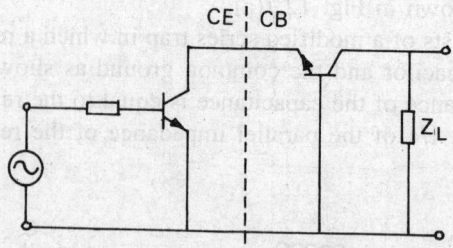

Fig. 12.8   Cascode configuration

The circuit acts very much like a single CE stage. The current gain and voltage gain of the circuit are approximately the same as for a single CE stage. For the same large voltage gain, however, it has a greater bandwidth than can be possible for a single CE stage. A CB stage has a large bandwidth, around f_α, while a CE stage has a small bandwidth, around f_β. In cascode configuration, the load of the CE configuration is the input impedance of the CB stage, which is very low. The Miller effect feedback is, therefore, considerably reduced. The overall bandwidth of the cascode circuit is close to f_β even if load R_L used for the circuit is large, as against in a single CE stage where the bandwidth is considerably reduced at large values of R_L.

The circuit has an another desirable feature in that its reverse transmission is extremely low. The internal feedback is almost completely negligible. Hence interaction of succeeding stages can be ignored. The CE-CB cascode stage is thus less prone to instability problems. Cascode tuned amplifier can hence employ a synchronously tuned design, or a stagger tuned design which otherwise presents instability problems.

Stability in Multistage Transistor Bandpass Amplifiers   A transistor will give maximum unilateralized gain, if optimally terminated at the input and output and is perfectly unilateralized. Perfect neutralization is practically impossible, however, because of parameter variation due to bias changes caused by AGC requirement, drifts, etc. particularly in a bandpass amplifier having wide bandwidth. For ensuring stability in absence of neutralization, an approach based on 'stability factor' defined by Stern is useful to establish a stability criterion. The minimum power gain that should be sacrificed from the maximum unilateralized power can be found out by using Stern's stability equation given by:

$$(g_{11} + G_g)(g_{23} + G_L) = \frac{K}{2}(L + M) \tag{12.1}$$

where $L = |y_{12}y_{21}|$ = product of transadmittances, M = real part of $(y_{21} y_{21})$, G_g = source conductance, G_L = load conductance reflected at the collector.

Terms g_{11} and g_{22} are the real parts of transistor input and output admittances.

For stability, K must be greater than 1. If it is assumed that the output is terminated optimally, then for a given stability factor and y parameters, G_g can be calculated:

$$G_g = \frac{K(g + M)}{4g_{22}} - g_{11}$$

The loss occurring due to mismatch at the input for this value of G_g is given by

$$\text{Mismatch loss} = 10 \log \frac{(g_{11} + G_g)^2}{4g_{11}G_g} \qquad (12.2)$$

This gives the minimum mismatch loss required for stability of G_g is chosen for $K = 1$.

The loss required for stability may be introduced at the input or output loading.

A transistor interstage equivalent circuit is shown in Fig. 12.9, where g_t is the tuned circuit load conductance and g'_{11} is the input conductance of the next stage, reflected through the transformer turns ratio.

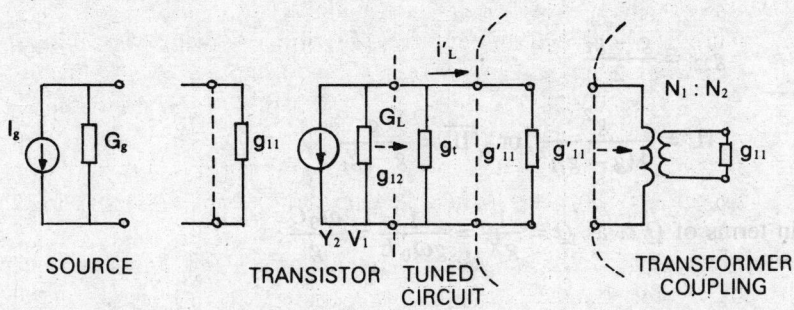

Fig. 12.9 Equivalent circuit of a transistor tuned interstage

In the tuned amplifier, the collector load conductance G_L is formed by the reflected load in parallel with the tuned circuit load conductance.

$$G_L = g'_{11} + g_t$$

where

$$g'_{11} = \left(\frac{N_2}{N_1}\right)^2 g_{11}, \text{ and } g_t = \frac{\omega_0 C}{Q_0}$$

Hence it is possible to introduce mismatch of g'_{11}, the actual load or the increased loading of the tuned circuit g_t.

If the loss is taken due to increased loading and insertion g_t, the unloaded Q_u, L, C and damping resistance for the tuned circuit may be calculated.

$$\text{Insertion loss IL} = \frac{\text{Max. available power from generator}}{\text{Actual power into load}}$$

$$= \frac{\text{MAP}}{P_L} \qquad (12.3)$$

Now $\text{MAP} = \dfrac{I_g^2}{4g_{22}}$, when $g_{22} = g'_{11}$ (matched load),

and

$$P_L = \frac{(i'_L)^2}{g'_{11}} = \left(I_g \frac{g'_{11}}{g}\right)^2 \frac{1}{g'_{11}}$$

$$= \frac{I_g^2 g_{11}'}{g^2}, \text{ where } g = g_{22} + g_t + g_{11}'$$

$$\text{IL} = \frac{\text{MAP}}{P_L} = \frac{I_g^2}{4g_{22}} \cdot \frac{g^2}{I_g^2 g_{11}'} = \frac{g^2}{4g_{11}' g_{22}} \quad (12.4)$$

$$= \frac{g^2}{4g_{22}^2} \text{ when } g_{22} = g_{11}' \text{ and the loss is introduced by the coil.}$$

Now, $$g = g_{22} + g_t + g_{11} = 2g_{22} + g_t$$

i.e. $$g_{22} = \frac{g - g_t}{2}$$

Hence $$\text{IL} = \frac{g^2}{4(g - g_t)^2}, \text{ or } \sqrt{\text{IL}} = \frac{g}{g - g_t}$$

Substituting for g in terms of Q's, as $Q = \dfrac{1}{gX} = \dfrac{1}{g\omega_0 L} = \dfrac{\omega_0 C}{g}$,
we have

$$\sqrt{\text{IL}} = \frac{Q_u}{Q_u - Q_L}, \text{ where } Q_u = \frac{1}{g_t X}, \text{ and } Q_L = \frac{1}{gX} \quad (12.5)$$

Q_L can be found from the bandwidth requirements and Q_u may be realized with the help of a suitable shunt resistance, as the coil inherent unloaded Q will be quite high.

The design procedure is illustrated with the help of an example.

Example: Design a video IF amplifier to provide a gain of about 80 dB using a transistor with the following data at $I_c = 1$ mA and $V_c = 10$ V, at 35 MHz.

$g_{11} = 0.85$ mΩ,	$C_{11} = 19$ pF,
$y_{12} = 125$ $\mu\Omega$,	$\phi_{12} = 270°$,
$y_{21} = 35$ mΩ,	$\phi_{21} = 345°$,
$g_{22} = 6$ $\mu\Omega$,	$C_{22} = 1.6$ pF

In the Stern's equation

$$L = | y_{12} y_{21} | = 125\mu \times 35 \text{ m} = 4.38\mu$$

$$M = \text{Re} \, (y_{12} y_{21}) = \text{Re} \, (125 \angle 270° \times 35 \angle 345°)$$

$$= -1.02\mu$$

$$\therefore \quad (L + M) = (4.38 - 1.02)\mu = 3.36\mu$$

For stability, minimum loss may be found out for $K = 1$. Assuming output to be matched, $G_L = g_{22}$. Hence the Stern's equation is given by:

$$2g_{22}(g_{11} + G_g) = \frac{L + M}{2}$$

Substituting for g_{11}, g_{22} and $(L + M)$

$$2 \times 6\mu \,(0.85 \text{ m} + G_g) = \frac{3.36}{2}\,\mu$$

$$G_g = 0.139\Omega = 139 \text{ m}\Omega$$

Hence the minimum loss required for stability is given by:

$$\text{Mismatch loss} = 10 \log \frac{(g_{11} + G_g)^2}{4 g_{11} G_g} \text{ dB}$$

If $G_g \gg g_{11}$, this simplifies to

$$\text{Mismatch loss} = 10 \log \frac{G_g}{4 g_{11}}$$

$$= 10 \log \frac{139 \text{ m}}{4 \times 0.85 \text{ m}}$$

$$= 16.1 \text{ dB}$$

Maximum unilateralized gain is given by:

$$\text{MUG} = \frac{y_{21}^2}{4 g_{11} g_{22}} = \frac{(35)^2\,\mu}{4 \times 0.85 \text{ m} \times 6\mu} = 58400$$

$$= 10 \log 58400 \text{ dB} = 47.7 \text{ dB}$$

Hence maximum available gain per stage

$$\text{MAG} = \text{MUG} - \text{Minimum loss for stability}$$

$$= 47.7 - 16.1 \text{ dB} = 31.6 \text{ dB}$$

Since overall gain required being 80 dB, three stages of amplifiers are enough. The loaded Q of each stage for obtaining the required bandwidth of 5 MHz at the IF centre frequency of 35 MHz, in the 3 stages of amplification is given by:

$$Q_L = \frac{f_0}{\text{BW}} \sqrt{2^{1/n} - 1} \qquad\qquad (12.6)$$

$$= \frac{35}{5} \sqrt{2^{1/3} - 1}$$

$$= 3.7$$

In Eq. (12.6) n is the number of stages and hence $n = 3$.

If a loss of 20 dB per stage is introduced instead of the minimum loss of 16.1 dB required for stability, the corresponding unloaded Q may be found out by equating this to the insertion loss of the tuned circuit load.

$$\text{IL} = 20 \log \frac{Q_u L}{Q_u - Q_L} = 20 \text{ dB}$$

$$\therefore \qquad \frac{Q_u}{Q_u - Q_L} = 10,$$

$$\therefore \qquad Q_u = \frac{10\,Q_L}{9} = \frac{3.57}{0.9} \doteq 4$$

The tuned circuit capacitance C and the damping resistance R_d to realize the unloaded Q can be found from the relations:

$$Q_L = \frac{\omega_0 C}{g} = \frac{\omega_0 C}{2g_{22} + g_t}, \text{ and } Q_u = \frac{\omega_0 C}{g_t}.$$

$$\therefore \qquad \omega_0 C\left(\frac{1}{Q_L} - \frac{1}{Q_u}\right) = 2g_{22},$$

$$\therefore \qquad C = \frac{2g_{22}Q_u Q_L}{\omega_0(Q_u - Q_L)} \qquad\qquad (12.7)$$

Substituting the values,

$$C = \frac{2 \times 6\mu \times 4 \times 3.57}{2\pi \times 35\,M \times (4 - 3.57)} = 4.7 \text{ pF}$$

$$L = \frac{1}{\omega_0^2 C} = \frac{1}{(2\pi\,35\,M)^2 \times 4.7\,p}\,H$$

$$= 4.4 \ \mu H$$

and

$$R_d = \frac{1}{g_t} = \frac{Q_u}{\omega_0 C} = \frac{4}{2\pi\,35\,M \times 4.7\,p}$$

$$= 3.85 \text{ k}\Omega$$

$$g_t = \frac{\omega_0 C}{Q_u} = \frac{1}{3.85\,k} = 260\,\mu\Omega$$

$$g = \frac{\omega_0 C}{Q_L} = \frac{2\pi\,35\,M \times 4.7\,p}{3.57} = 288\,\mu\Omega$$

$$g' = g - g_t - g_{22} = 288 - 260 - 6 = 22\,\mu\Omega$$

Transformer turns ratio $\left(\dfrac{N_1}{N_2}\right)^2 = \dfrac{g_{11}}{g_L'} = \dfrac{850\mu}{22\mu} = 37$

$$\therefore \qquad \frac{N_1}{N_2} \approx 6$$

The turns ratio is thus chosen to give the required mismatch at the input and required bandwidth. The biasing circuit may be designed for the required dc stability.

12.5 TRANSISTOR VIDEO IF AMPLIFIER CIRCUITS

Transistor Amplifier with Cascode Interstage

Figure 12.10(a) shows a typical 3-stage transistor video IF amplifier circuit employing BF 196 and BF 197. The amplifier provides a high gain with a good stability by use of a cascode configuration interstage that ensures better isolation between the stages. The frequency response and the gains of the stages are indicated in Fig. 12.10(b) as the detector output with respect to the signal at the base of the first, second and the third IF stages in sequence.

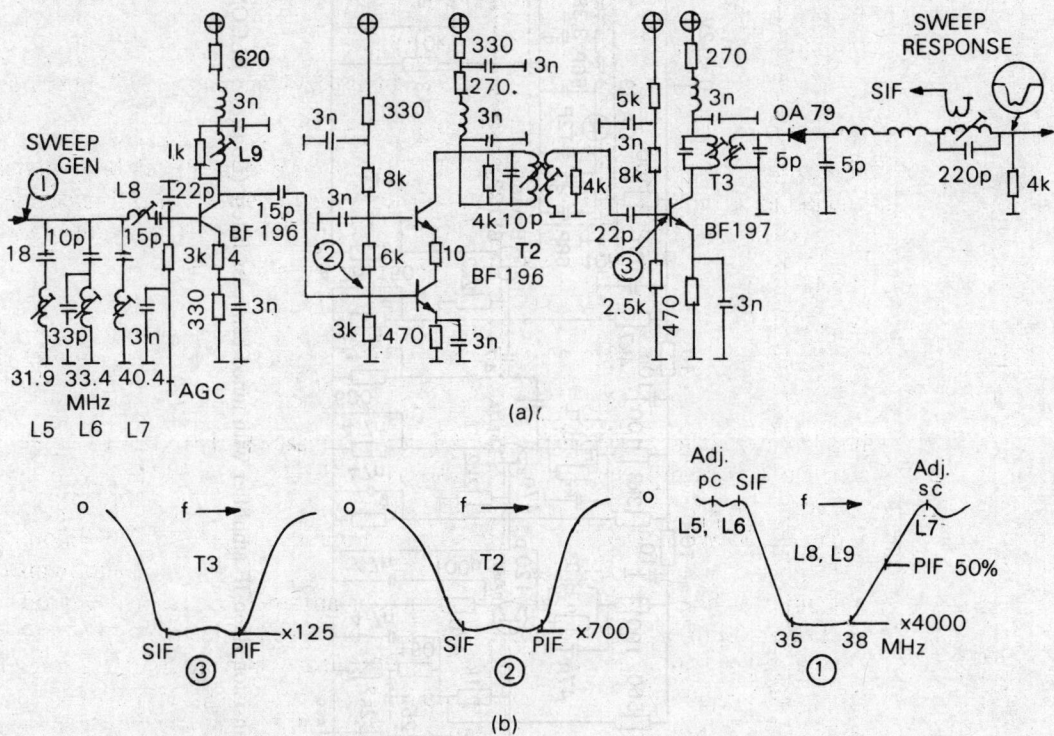

Fig. 12.10 (a) Transistor video IF amplifier with cascode interstage, (b) Sweep response curves for stagewise feed points

All the traps are included at the input of the amplifier as shunt traps. The 33.4 MHz IF sound trap is a shunt-series trap to provide steep rising response on the higher frequency side.

The decoupling *RC* filters in the collector supply leads include RF chokes for more effective decoupling at high frequencies. A zener regulated 27 V supply is used for collector supply.

Figure 12.11 shows another transistor video IF amplifier circuit having three stages and four tuned bandpass filters, providing a gain of over 90 dB. The AGC is applied to the input stage employing BF 167. The forward AGC voltage reduces the collector emitter voltage from 9 V to 1V by increase of the emitter current from 4 mA to 6 mA. This gives a gain control range exceeding 60 dB. The wave traps are mounted at the input of the amplifier with a view to keep down the undesired signals at the base of the input transistor at minimum and reduce the risk of cross-modulation.

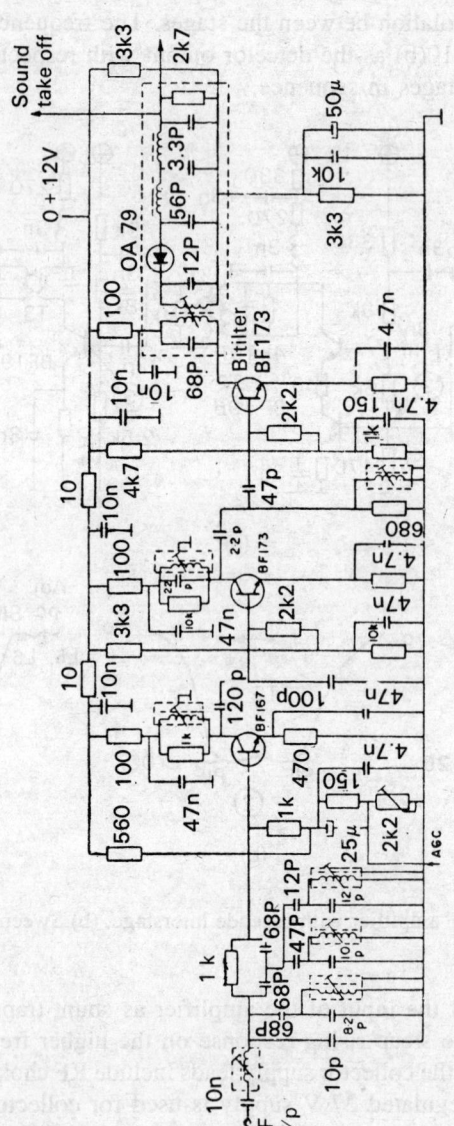

Fig. 12.11 A 3-stage transistor video IF amplifier with tuned bandpass filters (*Courtesy:* ELCOMA PHILIPS)

The interstage employs a BF 173, biased at 7 mA, 10 V. Double tuned circuit with a capacitive top coupling is used in its output circuit, with proper damping resistances for ensuring a high stability factor.

The output stage also employs a BF 173 with a double tuned bandpass filter with bifilar-wound inductive coupling to the video detector diode OA 79. Radiation of harmonics of the picture IF carrier generated at the detector is prevented by enclosing the complete BP filter in a screening can and blocking them by the low-pass filter following the diode. The trans-impedance of the complete amplifier including the video detector is 18.5 MΩ. The power gain from the base of the input transistor to the output of the detector is 92 dB. The 3 dB bandwidth of the amplifier is 4.4 MHz.

Another video IF amplifier employing BF 167 and BF 173, followed by the synchronous demodulator IC TCA 540 is shown in Fig. 12.12.

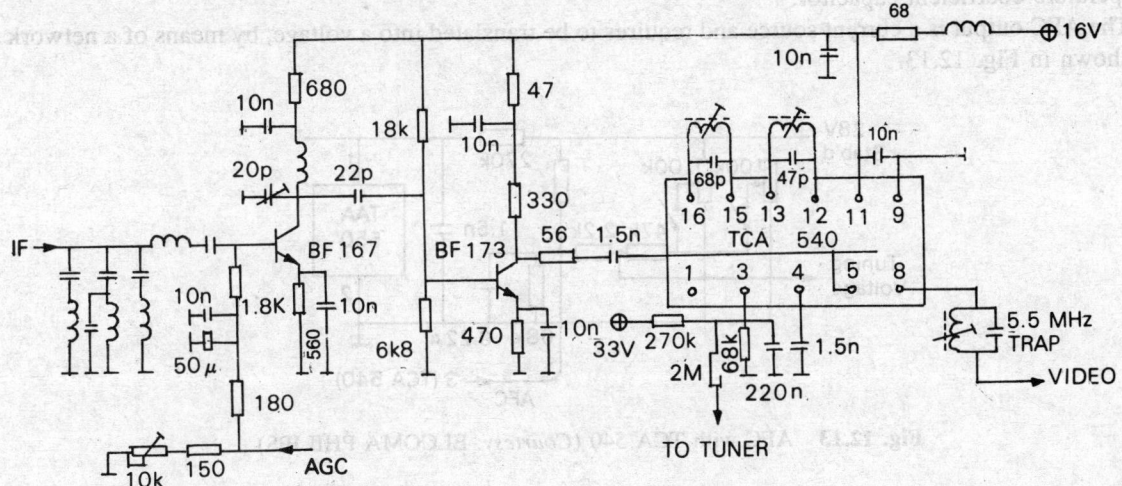

Fig. 12.12 Video IF amplifier with demodulator IC TCA 540 (*Courtesy*: ELCOMA PHILIPS)

The amplifier provides a power gain of 85 dB with an AGC range of 60 dB, by forward AGC control of the first stage using BF 167. The second stage is *RC* coupled, feeding to the synchronous demodulator IC TCA 540. The IC combines the following functions:

   (i) Synchronous regeneration of the reference carrier.
   (ii) White spot inventor.
   (iii) Video pre-amplifier.
   (iv) AFC circuit.

A balanced input for the IC has been provided at pins 4 and 5. Since the dc level is set internally, the signal should be applied from a floating transformer winding or through coupling capacitors. Unbalanced input is applied to either pin by decoupling the other to earth by about 1.5 nF capacitor. The input impedance of the IC is typically 6 kΩ ‖ 4.7 pF. About 70 mV are required at the input of the IC to give a peak output signal of 3V pp.

The video output signal at pin 8 is with negative going sync (+ Y) with TBA 890 signal processing IC. The zero signal dc output level is 6V at pin 8 and 5.7V at pin 7.

A 5.5 MHz trap is connected in series with the video output to block the sound carrier from reaching the video drive.

A synchronous demodulator causes white peaks in the video output signal when interferences are received. For this reason, a white spot inventor is used in the IC which detects the white peaks, reverses them to black level.

A reference tuned circuit is connected between pins 12 and 13 to provide the carrier filtering. The damping impedance between the pins is about 6 kΩ. The choice of L/C ratio of the tuned circuit is a compromise between good figure for differential gain and intermodulation products. A practical value of 47 pF gives good results. The unloaded Q of this circuit has to be greater than 50.

The AFC tuned circuit is loosely coupled to the reference tuned circuit of the demodulator by means of the collector-base capacitance of transistors in the IC. The unloaded Q of this circuit has to be greater than 70. A good temperature stability of the AFC requires use of a miniature coil former and NP_0 temperature coefficient capacitor.

The AFC output is a current source and requires to be translated into a voltage, by means of a network as shown in Fig. 12.13.

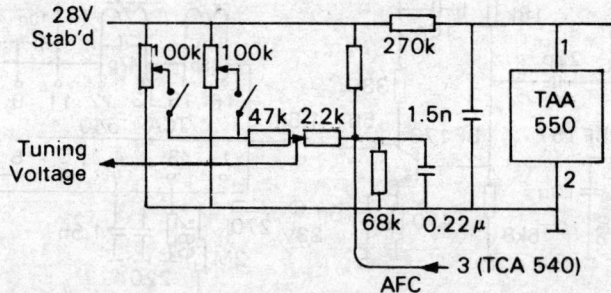

Fig. 12.13   AFC with TCA 540 (*Courtesy*: ELCOMA PHILIPS)

The network loading is about 50 kΩ. With this load, a sensitivity of about 40 mV/kHz is obtained for the AFC.

12.6 SAW FILTERS

With the development of ICs, which could supply all the required gain of IF stages and carry out demodulation in a single block, in the late 60's, the need to replace the LC filter by a *block filter* matching the IC technology was obvious. The hunt for such devices led to the development of surface acoustic wave (SAW) filters. Surfce acoustic waves are waves propagating along the surface or boundary of the media, similar to the elastic waves caused by earthquakes. If these mechanical-acoustic waves are created on a piezoelectric surface by suitable electrodes, frequency selective filtering action can be arranged by interference effects produced by waves generated by different sets of electrodes. First commercialized for FM tuners in 1976, these became soon available for video IF selections of TV receivers on a mass production scale, to replace the LC filters by solid state block filters.

Construction

The basic configuration of a surface acoustic wave (SAW) filter comprises piezoelectric substrate measuring 4 to 8 mm on a side and 0.4 mm thick, upon which are deposited two interdigital transducers (IDTs) of comb shaped electrodes, 10 to 20 μm in width, with their teeth interleaved, as shown in Fig. 12.14.

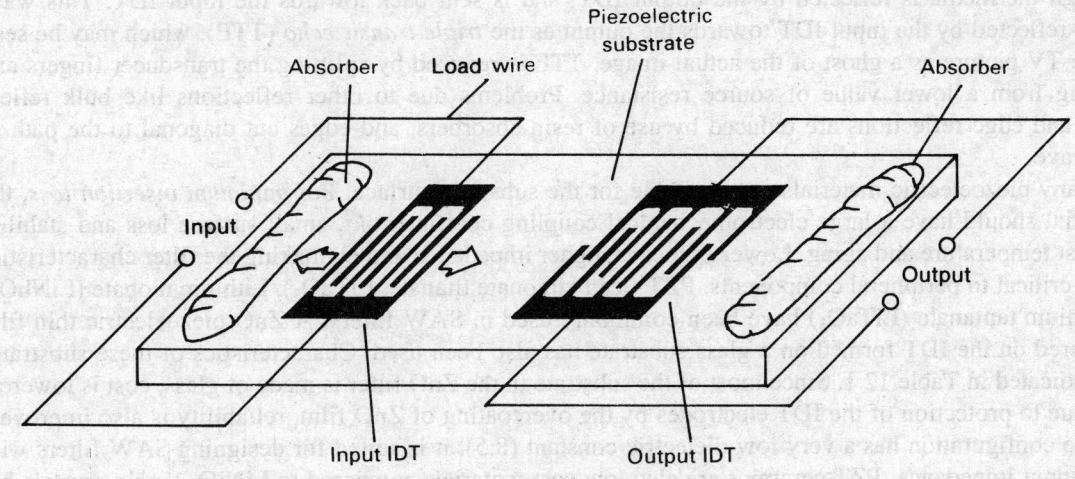

Fig. 12.14 Basic structure of the SAW filter (*Courtesy*: Murata)

The surface acoustic wave is generated by the piezoelectric property when an electric signal applied to the input IDT, and then propagates along the surface of the substrate. Those travelling to the right towards the output IDT are converted back into electric signals by the output transducer, while those propagating to the left are absorbed by the absorber layer. When the pitch of the comb shaped electrodes of the output transducer is constant, the output is maximum for surface acoustic wave having half wavelength equal to the pitch. At other wavelengths, cancellation occurs between the signals induced on each comb shaped electrode. The centre frequency of the SAW filter is determined by the pitch of the comb shaped electrodes.

When a large number of the same electrodes are arranged at the same pitch to form an IDT, with the spaces between the electrodes constant, the filter has a frequency characteristic expressed by the (sin x)/x function. The synchronous frequency f_0 at which the maximum energy is generated is given by

$$f_0 = \text{surface wave acoustic velocity}/\lambda_0$$

where λ_0 is the uniform pitch distance

However the frequency characteristics can be modified by changing the shape of the envelope of the overlapping comb electrodes as in the output IDT shown in the figure. In actual practice, when desired frequency characteristics are to be obtained, the shape of the electrodes is designed by performing an inverse Fourier transformation by means of computer aided design. By changing the length of fingers and the overlap or *apodising*, optimum VIF band pass and trap response with optimum amplitude and phase characteristics can be obtained. Apodising is equivalent to connecting several transducers having different resonant frequencies and band pass response in parallel. Variation in the aperture spacings, distance between transducers, and use of passive *multistrip coupler* line patterns in between the transducers, enable adjustment of the response and reduction in distortions due to reflections of propagating energy from the bottom of the substrate.

The SAW filter is a *travelling wave device*, and hence the reflections on the surface at any obstructions, at the edges, and within the bulk of the substrate, cause unwanted interference with the desired waves, resulting in ripples and echos in the response. A part of the surface acoustic wave propagating through the media is reflected by the output IDT, and is sent back towards the input IDT. This wave again reflected by the input IDT towards the output as the *triple transit echo* (TTE), which may be seen in the TV picture as a ghost of the actual image. TTE is reduced by splitting the transducer fingers and driving from a lower value of source resistance. Problems due to other reflections like bulk reflections and edge reflections are reduced by use of resin absorbers, and edges cut diagonal to the path of the wave.

Many piezoelectric materials are available for the substrate surface. For *minimum insertion loss*, the material should have a large electromechanical coupling coefficient k^2, small surface loss and stability against temperature and aging. Lower k^2 means higher impedance and Q, making the filter characteristics more critical to peripheral components. PZT (lead zirconate titanate, $LiTaO_3$), Lithium niobate ($LiNbO_3$) or lithium tantanate ($LiTaO_3$) have been commonly used in SAW filters. A ZnO piezoelectric thin film sputtered on the IDT formed on a glass substrate has also been used. Characteristics of these substrates are indicated in Table 12.1. Since most of the substrate in the ZnO filter is made of glass, cost is lowered, and due to protection of the IDT electrodes by the overcoating of ZnO film, reliability is also improved. As the configuration has a very low dielectric constant (8.5), it is suited for designing SAW filters with high input impedance. PZT ceramics are also low cost materials compared to $LiNiO_3$ single crystals but have relatively larger dielectric constant (400), and hence are considered suitable for low impedance filters. Characteristics of various substrate materials are given in Table 12.1.

Table 12.1 Characteristics of substrate materials (Murata)

Material	Cut	Propagation plane	Wave velocity	Coupling $(k^2/2)$	Temp coef. ppm/°C	Dielectric const.
$LiNbO_3$	y	z	3488	0.0241	− 94	50.2
$LiTaO_3$	y	z	3230	0.0033	− 35	47.0
$Bi_{12}SiO_2$	100	110	1675	0.0062	− 118	38.3
ZnO film	—	—	2560	0.0072	− 28	8.5
PZT	—	—	2420	0.0122	− 38	400.0

Electrical Characteristics

Typical amplitude characteristics and group delay characteristics of a SAW filter are shown in Fig. 12.15.

The filter has, typically, an insertion loss of 15 to 20 dB and therefore requires a preamplifier to maintain a satisfactory signal-to-noise (S/N) ratio in the receiver. The SAW filters are non-minimum phase networks; the designer, therefore, has the independence to tailor the amplitude and phase response. *Group delay* is large, some 1.2 μs for the distance of 3 mm between the IDTs on a ZnO substrate having a propagation velocity of 2600 m/s. This is held flat within 50 ns in most designs, a value as good as the best discrete stage designs. The independence of phase response has made it possible to produce SAW filters with greatly differing group delay for the luminance sidebands compared with the chrominance sidebands, and thus helps eliminate the need for a separate delay line in the luminance channel of the colour TV receiver.

Application

The insertion loss of a SAW filter is compensated for commonly by a discrete transistor preamplifier aiming to reduce its cost. In the case of a grounded emitter the amplifier has a low input impedance of

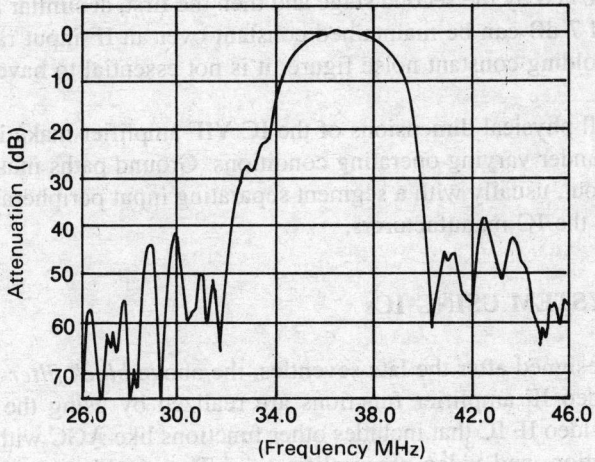

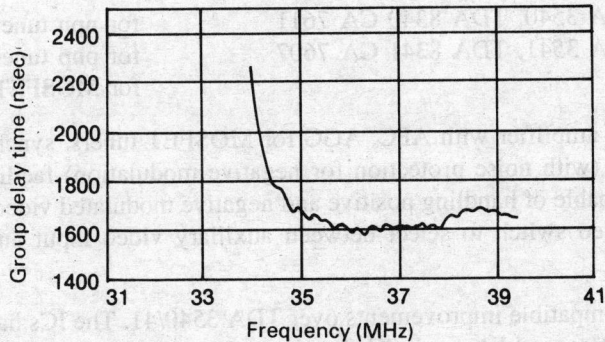

Fig. 12.15   (a) Amplitude characteristics of a SAW 38.9 MB filter (b) Group delay characteristics of SAW 38.9
filter (*Courtesy*: Murata)

the order of 50 to 100 Ω, while the output impedance depends upon the collector load resistance and the
output impedance of the transistor in parallel, in actual case in the range of 100 to 1000 Ω. The input
impedance of one-chip IC for VIF is of the order of 1k ohm to 3k ohm. The insertion loss of the filter
depends upon the *impedance matching* condition. In general, when the external terminal impedance is
made equal to the impedance of the SAW filter (including the tuning coil if used), the insertion loss
becomes minimum, but when the level of TTE is limited to one which does not lead to the formation
of ghost images. With no tuning coil or with parallel tuning, the filter must be used under mismatched
termination to the lower impedance side; the ripple on the group delay time characteristics due to the TTE
is then reduced.

The *IC gain block* has the basic requirements of high gain, low noise, low distortion, and large linear
gain control range with high stability under operating conditions. The basic differential amplifier providing
a gain of around 20 dB and gain control range of about 24 dB, common in many linear ICs, can meet
these requirements in a cascade of three stages. This yields an overall gain of more than 60 dB and gain
control range of over 64 dB. The gain control system internal to this IC begins to reduce the third stage
gain at an IC input level of 100 μV, VIF carrier. With increasing input signal level, the gain of the third

stage reduces to 0 dB, followed by the second stage and then the first, at similar levels. By this technique, the noise figure of around 7 dB can be maintained constant over an IF input range of 40 dB. In such a design of VIF amplifier holding constant noise figure, it is not essential to have a preamplifier ahead of the SAW filter.

The high gain and small physical dimensions of the IC VIF amplifier make it obligatory for the PCB layout to ensure stability under varying operating conditions. Ground paths must be designed for lowest impedance and proper layout, usually with a segment separating input peripheral components and output circuitry, as suggested by the IC manufacturers.

12.7 VIDEO IF SUBSYSTEM USING ICs

In modern TV receivers designed after the late seventies, the above *block filter-block gain* approach has become common. The video IF amplifier functions are realized by using the SAW filter to give the required selectivity and a video IF IC that includes other functions like AGC with noise gating, AFC with on/off switch, video detection, and video preamplification. The combination of the SAW filter with a transistor preamplifier to compensate the SAW filter, the VIF IC with the AFT and detector coils forms the video IF subsystem. A number of monolithic VIF ICs are now available in two groups:

TDA 2540, TDA 3540, TDA 8340 CA 7611	for npn tuners
TDA 2541, TDA 3541, TDA 8341 CA 7607	for pnp tuners
TDA 2549	for MOSFET tuners

IC TDA 2549 is a video IF amplifier with AFC, AGC for MOSFET tuners, synchronous demodulation and video preamplification (with noise protection for negative modulation) facilities for multistandard television receivers. It is capable of handling positive and negative modulated video signals in colour and B/W receivers and has video switch to select between auxiliary video input and demodulated video signal.

ICs TDA 8340/41 are pin compatible improvements over TDA 3540/41. The ICs have additional features like integrated filter to limit second harmonic IF signals, and a wide supply variation range. The block diagram of the IC along with a preamplifier and SAW filter is shown in Fig. 12.16.

The IC has a input sensitivity of 40 μV and a gain control range of 67 dB. The blocks internal to the IC function are as follows:

IF amplifier This is a three-stage, gain-controlled IF amplifier with a wide dynamic range. On-chip capacitors in the dc feedback loop of the amplifier maintain stability at maximum gain. Internal stabilization of the supply voltage ensures the desired sensitivity and gain control range over the whole supply voltage range and also gives very good power supply ripple rejection in this part of the circuit.

Demodulator The redesigned IF demodulator is a quasi-synchronous circuit that employs passive carrier regeneration and logarithmic clamping to give improved signal handling. The demodulator input is as coupled to the IF amplifier to reduce dc offsets and thus minimizes residual IF carrier in the output signal.

Video amplifier The linearity and bandwidth of the video amplifier are sufficient to meet all wide band requirements, e.g. for teletext transmissions. Second harmonics of the IF carrier are effectively reduced by a Sallen-key low pass interstage filter between the demodulator output and the video amplifier input. An integrated filter in the noise inverter reduces the sensitivity of the video amplifier for high sound

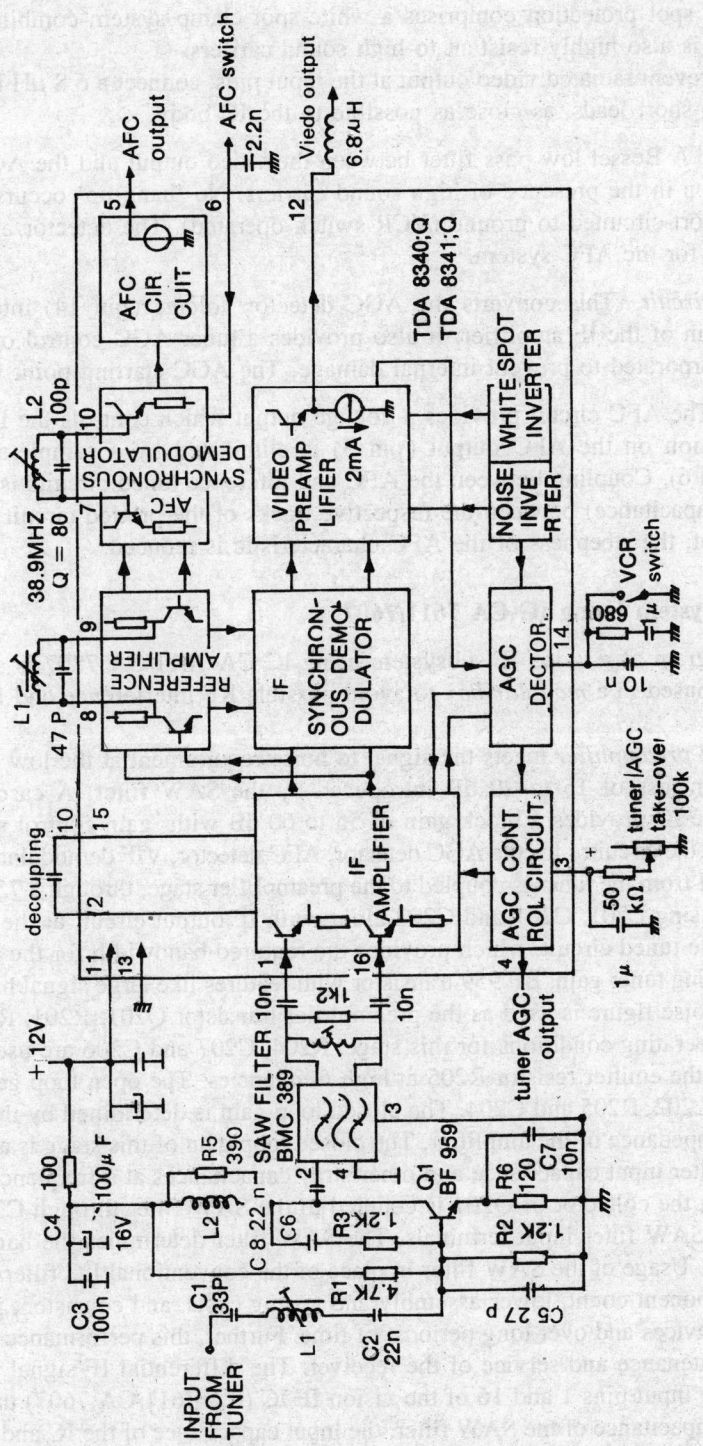

Fig. 12.16 Block diagram of Video IF IC TDA 8340/41 (*Courtesy:* PHILIPS)

carriers. White spot protection comprises a white spot clamp system combined with a delayed-action inverter which is also highly resistant to high sound carriers.

Note To prevent radiated video output at the input pins, connect a 6.8 μH inductor in series with pin 12 and fit with short leads, as close as possible to the IC body.

AGC detector A Bessel low pass filter between the video output and the AGC detector improves the detector function in the presence of high sound carriers. No 'hang up' occurs in the detector after pin 14 has been short-circuited to ground (VCR switch operated). The detector also generates the sample-and-hold pulse for the AFC system.

AGC control circuit This converts the AGC detector voltage (pin 14) into a current signal which controls the gain of the IF amplifier. It also provides a tuner AGC control output from pin 4, current limiting is incorporated to prevent internal damage. The AGC starting point is adjusted via pin 3.

AFC circuit The AFC circuit provides a voltage output which controls the IF frequency of the tuner. Video information on the AFC output (pin 5) is eliminated by a sample-and-hold circuit (external capacitor at pin 6). Coupling between the AFC and reference tuned circuits is via two small capacitors (for parasitic capacitance) between the respective tracks of the printed circuit board. If the capacitance is less than 1 pf, the steepness of the AFC characteristic is reduced.

Video IF Subsystem Using IC CA 7611/7607

The circuit diagram of a video IF subsystem using IC CA 7611/CA 7807 is shown in Fig. 12.17. The subsystem is housed in a *modular box* to avoid possible RF interference and is soldered onto the main PCB.

A *wide band preamplifier* meets the signal to noise requirement at the low signals, and compensates for the insertion loss of 15 to 20 dB introduced by the SAW filter. A cascode of three differential amplifiers in the IC provides a block gain of 55 to 60 dB with gain control range of over 60 dB. The IC also contains the circuitry for the AGC detector, AFC detector, VIF demodulator and video preamplifier.

The IF signal from the tuner is coupled to the preamplifier stage, through a 75 Ω RF cable. A *matching network* comprising L201, C201 and C202 along with IF output circuit at the mixer stage of the tuner forms the double tuned circuit, which provides the required bandwidth for the sound and vision carriers without sacrificing tuner gain. BF 959 transistor with features like large signal handling capability, higher fT and lower noise figure is used as the preamplifier transistor Q201, R201, R202 and R205 determine the quiescent operating conditions for this stage. R204, C207 and C206 are used for supply decoupling. C204 bypasses the emitter resistor R205 at high frequencies. The open loop gain of this preamplifier is determined by L202, R205 and C204. The closed loop gain is determined by the feedback resistor R203 and the input impedance of the amplifier. The closed loop gain of this stage is around 26 dB. L202 tunes out the SAW filter input capacitance and other stray capacitances at a frequency f_0 equal to 36.15 MHz. The IF signal at the collector of Q201 is coupled to the SAW filter through C205, thus isolating the dc voltages at the SAW filter input terminals. The SAW filter determines the band pass characteristics of the TV receiver. Usage of the SAW filter in place of the conventional LC filters offers many advantages like lower component count, lower assembly and testing costs, and consistent performance over a large population of devices and over long periods of time. Further, this performance cannot be tampered with during the maintenance and service of the receiver. The differential IF signal at the SAW filter output is applied to the input pins 1 and 16 of the vision IF IC (CA 7611/CA 7607) through C208. L203 tunes out the output capacitance of the SAW filter, the input capacitance of the IC and other stray capacitances. Better video frequency response is obtained by having higher loaded Q for this coil and tuning the output

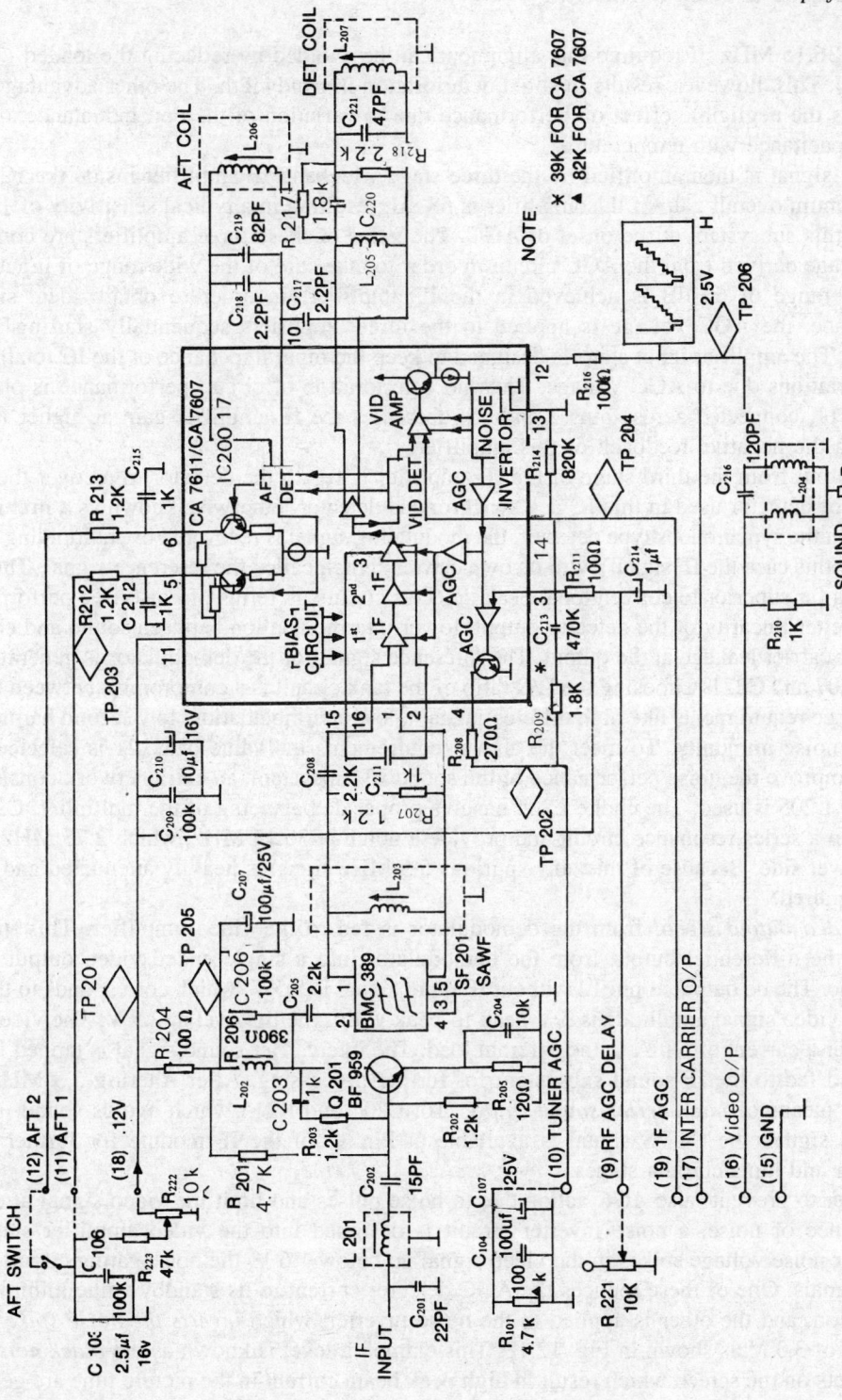

Fig. 12.17 A video IF subsystem (*Courtesy:* BEL)

circuit to 36.15 MHz. If required this alignment can be avoided by reducing the loaded Q (by damping it heavily). This, however, results in slight deterioration in bandwidth. The other advantage of low Q for this coil is the negligible effect on performance due to variation of the coil inductance and SAW filter output capacitance with temperature.

The IF signal is then amplified by the three-stage wideband IF amplifier inside the IC CA 7611/07. The maximum overall gain of this amplifier is 58 dB, resulting in a typical sensitivity of 100 μV at 38.9 MHz for this subsystem at the onset of AGC. The gains of these three amplifiers are controlled by the AGC voltage derived from the AGC circuit in order to take care of the wide range of input signal levels. An AGC range of 60 dB is achieved in the IF amplifier. In order to obtain ideal signal to noise performance, the AGC voltage is applied to the three amplifiers sequentially starting with the third amplifier. The amplifier input stage is designed to keep the input impedance of the IC totally independent of the variations due to AGC voltage. Thus no deterioration of circuit performance is observed due to AGC. C211, connected across pins 2 and 15, decreases the IF amplifier gain at higher frequencies by increasing the negative feedback of this amplifier.

The output from the third stage of the IF amplifier is fed to the detector to recover the video signal. The type of detector used in this IC is a synchronous detector, otherwise known as a multiplication type detector. In the synchronous type detector, the modulating signal is recovered by multiplying the modulated carrier (in this case the IF signal) with its own carrier signal, called the reference signal. The synchronous detector is far superior to conventional peak detector circuits in terms of noise-free performance at weak signals, better linearity of the detector output, lower intermodulation between sound and chroma carriers and lower carrier leakage at the output. The reference signal for the demodulator is generated by the tank circuit L207 and C221. Choosing the L/C ratio of the tank circuit is a compromise between the conflicting performance requirements like high differential gain, low intermodulation, low second harmonic distortion and high noise immunity. To meet the above requirements the value of C221 is selected as 47 pF. In order to improve the noise performance of the sound IF subsystem, an extra network consisting of R217, C220 and L205 is used. The choke L205 ensures proper dc balancing of the multiplier. C220, C221 and L207 from a series resonance circuit and provide a notch at 36.15 MHz, which 2.75 MHz less than VIF on the lower side. Because of this, any spurious 5.5 MHz signal is heavily attenuated and sound quality is not impaired.

The *video output signal* from the demodulator is fed to the video amplifier. This video amplifier converts the differential output from the demodulator into a single ended video output with negative going sync. The dc output at pin 12 without IF input signal is 5.5 V, which corresponds to the white level. The total video signal amplitude is 3 V peak-to-peak with sync tip level at 2.5 V. The video output stage can deliver a current of 6 mA to the external load. The intercarrier sound signal is tapped from the video output and fed to the IF sound subsystem for further processing. After filtering 5.5 MHz in the video signal by means of *intercarrier sound trap* C216, L204 and R215 which avoids sound interference on the video signal, the CCVS signal is available at Pin 16 of the IF module for further processing in deflection and luma/chroma stages.

In order to prevent false *AGC* action due to noise pulses and limit the video signal amplitude during the presence of noise, a noise inverter circuit is designed into the video amplifier stage of this IC. Whenever noise voltage spikes in the video signal go below 1.6 V, the noise gating circuit generates two output signals. One of these reduces the AGC detector current to its standby value minimizing spurious AGC action, and the other is applied to the noise inverter, which *inverts the noise spike and clamps* it at a level of 3.3 V as shown in Fig. 12.18. This clamping level is known as the *black noise clamp level*. White spots on the screen which result in high peak beam current in the picture tube are generally caused by over modulating noise signals which are faithfully detected by the synchronous demodulator. These

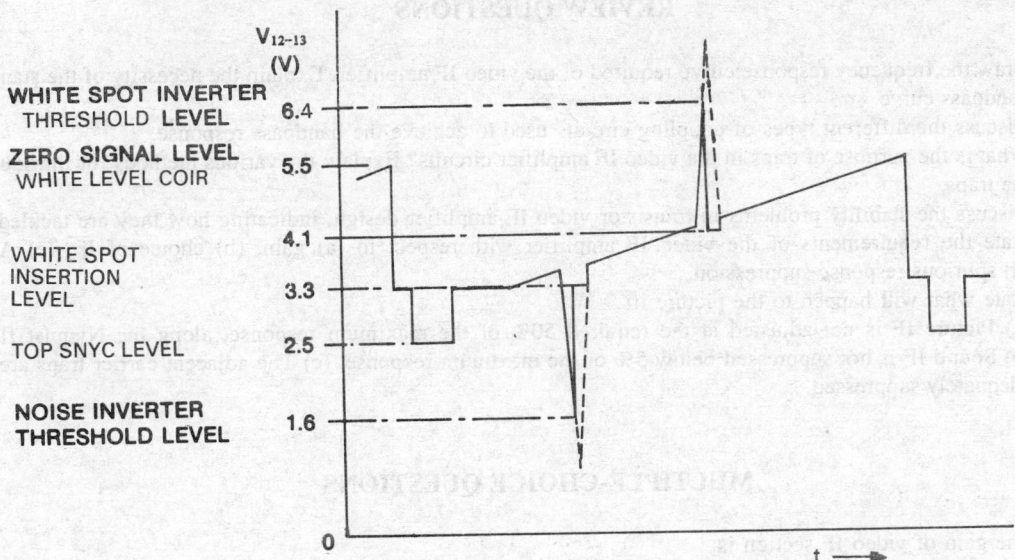

V$_{12-13}$ (V)

WHITE SPOT INVERTER THRESHOLD LEVEL — 6.4

ZERO SIGNAL LEVEL WHITE LEVEL COIR — 5.5

WHITE SPOT INSERTION LEVEL — 4.1 ... 3.3

TOP SNYC LEVEL — 2.5

NOISE INVERTER THRESHOLD LEVEL — 1.6

0

t

Fig. 12.18 Noise inversion in the video output of CA 7611/07 (*Courtesy:* BEL)

white spots are eliminated by the *white spot inverter* within the IC which detects white noise spikes exceeding 6.4 V and clamps them to 4.1 V.

After noise invertion, the video signal is applied to the *AGC amplifier* to derive the required AGC voltages. The voltage across capacitor C213 at pin 14 controls the current through IF amplifiers and thus controls the IF stage gain. Normally under maximum gain condition the voltage at pin 14 is 10 V. When the sync tip levels of the video signal exceeds the reference level set in the AGC comparator, the capacitor discharges, thereby reducing the AGC voltage at pin 14. This in turn reduces the current through the IF amplifiers, thus reducing the gain. The AGC controlled voltage at pin 14 is compared with the voltage at pin 3 to generate the *delayed AGC* voltage required for the tuner at pin 4. The amount of delay required can be adjusted by varying the voltage at pin 3 by R221. For *VCR operation*, pin 14 is grounded there by applying full IF and tuner gain controls and taking out of circuit the video output emitter follower. This presents a high impedance at pin 12 and avoids overloading of the external video signal.

To obtain optimum performance from the synchronous demodulator, it is essential that the reference tuned circuit have a high Q to obtain low harmonic distortion and better intermodulation performance. However, high selectivity makes the tuning critical. An AFT circuit is thus required to correct the effect of the slight mistuning and drift in the tuner/detector tuned circuit to a point where they have no adverse effect on the performance of the demodulator. The *AFT detector* generates current which varies in proportion to any deviation from the nominal frequency of the IF input to the IC. This current is converted into an AFT control voltage at Pins 5 and 6 and is fed to the AFT pin of the tuner.

REVIEW QUESTIONS

12.1 Draw the frequency response curve required of the video IF amplifier. Explain the necessity of the standard bandpass curve.

12.2 Discuss the different types of coupling circuits used to achieve the bandpass response.

12.3 What is the purpose of traps in the video IF amplifier circuits? Explain the various methods for introducing the traps.

12.4 Discuss the stability problems in transistor video IF amplifier design, indicating how they are tackled.

12.5 State the requirements of the video IF amplifier with respect to (a) gain, (b) choice of IF, (c) AGC, (d) spurious response suppression.

12.6 State what will happen to the picture if:
(a) Picture IF is not adjusted at the required 50% of the maximum response, along the Nyquist flank.
(b) Sound IF is not suppressed below 5% of the maximum response. (c) The adjacent carrier traps are not adequately suppressed.

MULTIPLE-CHOICE QUESTIONS

12.1 The gain of video IF section is:
   (a) controlled by delayed AGC
   (b) set by manual AGC
   (c) controlled by AGC bias
   (d) fixed

12.2 The response of video IF amplifier is:
   (a) flat within the video bandwidth
   (b) flat over one sideband only
   (c) complementary to the transmitted bandwidth
   (d) peaked at the middle of the band

12.3 The bandpass characteristics are obtained by a SAW filter by:
   (a) resonance effects in the piezoelectric material
   (b) interference of surface waves generated on a piezoelectric substrate
   (c) absorption of surface waves by suitable layers
   (d) electrical interaction between interdigital transducers.

12.4 The material more suitable for construction of low impedance SAW filters is:
   (a) lithium neobate
   (b) lithium tantalate
   (c) zinc oxide
   (d) PZT ceramics

12.5 Pick the incorrect statement:
   (a) The centre frequency of a SAW filter is determined by the pitch of the IDT comb shaped electrodes.
   (b) When a number of identical electrodes are arranged at the same pitch, the frequency characteristics of the filter is expressed by $(\sin x)/x$.
   (c) The response is maximum when the acoustic wave has its wavelength equal to the pitch.
   (d) The response can be modified by shaping the overlap of the comb electrodes.

12.6 In a video IF subsystem, the AFT voltage is obtained:
   (a) at the video detector
   (b) at a separate AFC detector
   (c) at a separate synchronous detector
   (d) at the sound IF detector.

13

Video Circuits

In a TV receiver the video signal is extracted from the modulated IF signal at the video detector and is amplified in the video amplifier sufficiently to provide the picture tube with the required drive voltage to vary its beam current from cut-off to maximum value for full brightness condition. The detector and the video amplifier must have adequate bandwidth and phase response without introducing distortion in the video signal. The video signal must also be supplied to the sync separator and keyed AGC circuits as shown in Fig. 13.1. The sound IF intercarrier frequency is also taken off at the detector or from the sound trap circuit in the video amplifier.

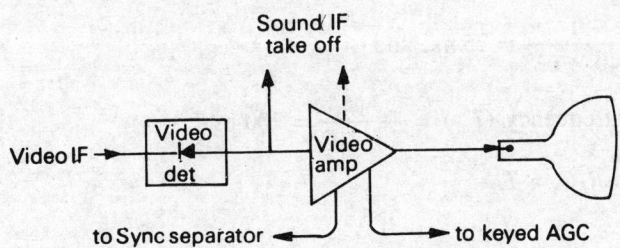

Fig. 13.1   Video circuits

13.1   VIDEO DETECTOR

A diode detector is commonly used to demodulate the amplitude modulated video IF signal by the rectification process. The IF component in the rectified output signal is eliminated by a low-pass filter or by a simple bypass reservoir filter capacitor, as indicated in Fig. 14.2.

General principles of detection require that the detection efficiency should be maximum and the distortion due to clipping be avoided. For maximum detector efficiency, which is the ratio of peak-to-peak detector output to the peak-to-peak signal in the modulated envelope, detector load R_L must be much greater than diode forward resistance r_f, and filter capacitor C_s must be much greater than diode capacitance C_d; that is,

$$R_L \gg r_f \text{ and } C_s \gg C_d$$

However, C_s cannot to too large as time constant $R_L C_s$ must be low enough to prevent diagonal clipping at high modulating frequencies due to the slow discharge of C_s as shown in Fig. 13.2.

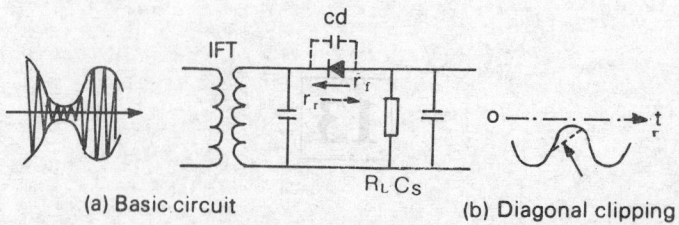

Fig. 13.2 Diode detector: (a) Basic circuit, (b) Diagonal clipping

In a television receiver, the IF is 33 MHz to 40 MHz and the bandwidth of the modulating video signal is large too, viz. 5 MHz. The ratio of IF to the highest modulating frequency, f_m is 8 : 1, low compared to the ratio in a broadcast receiver. This renders the filtering problems more acute.

(i) Choice of Diode
Diode forward resistance r_f and diode capacitance C_d should be low. Point contact germanium diodes hence are quite suitable; OA 70, OA 79 or OA 80, OA 90 with $C_d \approx 1$ pF and $R_f \approx 400\ \Omega$ are commonly used. They have adequate reverse blocking voltage rating of more than 30 V, and low forward cut-in voltage of about 0.2 V. Their small size makes them suitable to be mounted conveniently in the shielded IF can.

(ii) Choice of R_L and C_s

Period of IF $(T_{IF}) = \dfrac{1}{40 \times 10^6} = 25$ ns, and

Period of max. video frequency $(T_m) = \dfrac{1}{5 \times 10^6} = 200$ ns

It is desirable to have $R_L C_s \gg T_{IF}$

but $\ll T_m$,

i.e., 200 ns $\gg R_L C_s \gg 25$ ns.

A time-constant of $R_L C_s$ around 100 ns may be aimed at.

For maximizing the detector efficiency, the load resistance $R_L \gg r_f$, the forward resistance of the diode. C_s should be chosen to be much greater than C_d, the diode capacitance. In practice, C_s is formed by the stray capacitance of wiring in addition to the externally added capacitance of a few picofarads, usually 4–5 pF. The total C_s may be of the order of 20 pF, so that for a time-constant of 100 ns,

$$R_L = 5\ k\Omega$$

(iii) Frequency Response
The load impedance of the detector should not fall off due to shunting by C_s. This shunting effect is reduced by the use of a lower R_L, or by using a small peaking inductance in series with R_L that increases load impedance at high frequencies and gives some high frequency compensation.

(iv) IF Filter
The IF carrier component in the detected output signal must be completely removed. The detected output

contains fairly strong harmonics and beat frequencies that may radiate back to the tuner and cause interference in the signal in the form of annoying patterns in the picture. A harmonic of the IF falling near the received signal carrier frequency may give rise to an output component from the frequency changer which is not very far from the video IF. The beat frequency produced from these two at the detector may fall in the video frequency range and pass through to the picture tube to produce pattern lines in the picture. To minimize this, the detector diode and the IF filter may be conveniently enclosed in the last video IF transformer shielding can.

Because of low R_L, RC filters are not practicable as they would attenuate the video signal. For this reason LC filters are used. The filter L_f and C_f shown in Fig. 13.3 has various possibilities.

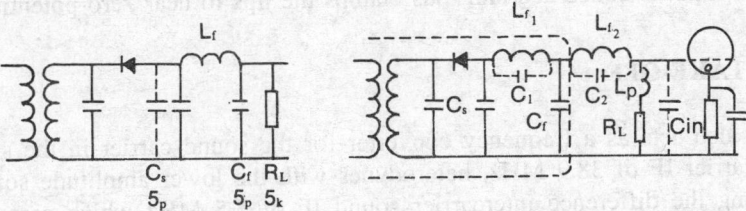

Fig. 13.3   Video detector with filters

Inductance L_f may resonate with its self-capacitance at the IF to block it. It may resonate with C_f to give a 'series' peaking at the high end of the video frequencies. Filter capacitor C_f may be formed by C_s or C_{in} of the video amplifier.

More than one series L_f may be used to give more effective filtering. The self-resonant frequencies of the series inductances may be at the IF and its harmonics that may be troublesome, providing satisfactory attenuation at the unwanted frequencies but not appreciable loss at the video frequencies.

Detector Polarity

The polarity of the detected output signal can be positive going or negative going, depending on the diode connection. The actual polarity is decided by the number of stages because of the reversal of phase at each stage, and the type of picture drive, whether the cathode is supplied the video signal or the grid. The ultimate aim is to cause proper intensity modulation of the picture tube beam so that white levels cause increase in the beam current to produce a positive picture. The sync and blanking tips of the video signal must cut off the beam. A wrong polarity would produce reversal of black and white in the picture, displaying a negative-picture.

13.2   DC COMPONENT RESTORATION

The detected signal contains a dc component corresponding to the mean brightness level in the scene which must be preserved by using direct coupling in the succeeding video circuits feeding to the picture tube grid-cathode. If ac coupling is used, the dc component should be restored by a dc restorer circuit using a clamping diode as shown in Fig. 13.4.

Capacitor C charges during the positive or negative going part of the signal, depending on the polarity of the diode. Time constant RC is very long compared to the period of the waveform so that v_C, the voltage on capacitor v_C remains more or less constant, to be replenished by small currents through the diode in

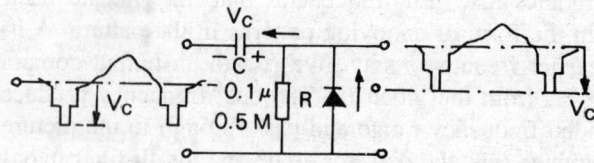

Fig. 13.4 Basic dc restorer circuit

case it falls or increases due to signal amplitude changes. This voltage v_C comes in series with the ac signal as far as the output is concerned and thus clamps the tips to near zero potential.

13.3 SOUND IF TAKE-OFF

The video detector also acts as a frequency converter for the sound carrier in the signal. The larger amplitude picture carrier IF of 38.9 MHz, heterodynes with the lower amplitude sound carrier IF of 33.4 MHz, producing the difference intercarrier sound IF of 5.5 MHz which carries the frequency modulation of the sound signal. This can be coupled to the sound IF circuit for amplification, detection and further AF amplification.

The sound IF lies just beyond the video band and if allowed to reach the picture tube, can produce picture modulation in the form fine diagonal interference line pattern called Moire pattern that may vary slightly with sound. The pattern, also called Herringbone weave, gives a wormy appearance to the picture as these lines alter with the frequency modulation due to sound.

The Moire pattern interference is prevented by introducing a parallel resonant circuit in the series path, tuned to the intercarrier sound frequency of 5.5 MHz. The trap may be provided in the detector load series path as shown in Fig. 13.5, or in the video amplifier coupling to the picture tube. The sound IF may be taken at the detector terminal prior to the trap or through a coupling coil on the resonant trap circuit itself. The trap with the coupling may be included in the video output circuit in order to obtain somewhat higher sound IF signal.

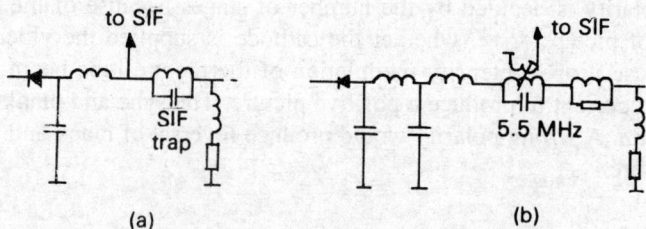

Fig. 13.5 Sound take-off circuits

13.4 VIDEO AMPLIFIER REQUIREMENTS

The contrast in a reproduced picture depends upon the video amplifier gain, while the sharpness and resolution depends upon bandwidth and phase response.

Gain

The contrast, i.e., the black to white saturation in the reproduced picture depends upon the picture tube drive from the video amplifier. The detected video signal is around 2 to 5V. The picture tube requires a video voltage drive of about $80V_{p-p}$ for full contrast. Thus the minimum gain required is of the order of 16 to 40. The gain is usually made variable by a negative feedback in the cathode circuit, or by a screen voltage control to serve as the contrast control.

Bandwidth Frequency and Phase Response

For faithful reproduction of the picture, the shape and form of the video waveform must be preserved during amplification. The shape of the complex waveform depends not only on the frequencies contained in the signal but also upon the relative phases. It is, therefore, necessary that:

(i) all the frequencies must be amplified equally to maintain the same relative amplitudes, i.e., the frequency response must be flat over the video bandwidth of 0 to 5 MHz; and

(ii) the relative phases of all the frequency components in the output must be the same as at the input; otherwise distortion results in the shape of the waveform.

Phase shift implies relative time shift of various frequency components in the signal waveform. It must be noted that the phase shift introduced by a system should either be zero or proportional to the frequency ($\phi = 2\pi ft$, i.e., $\phi \propto f$); if all the frequency components are to arrive at the output of the amplifier with the same relative phases that they have at the input, all the component frequencies should undergo the same time delay. This is illustrated in Fig. 13.6, where the effect of a constant phase shift of 90° in the frequencies f and $2f$ is shown in (a) and that of 180° is shown in (b). For example, at $f = 50$ Hz, a phase shift = 90° implies time delay $t_d = 5$ ms; at $f = 100$ Hz, a phase shift = 90° implies time delay $t_d = 2.5$ ms, while a phase shift = 180° implies time delay $t_d = 5$ ms.

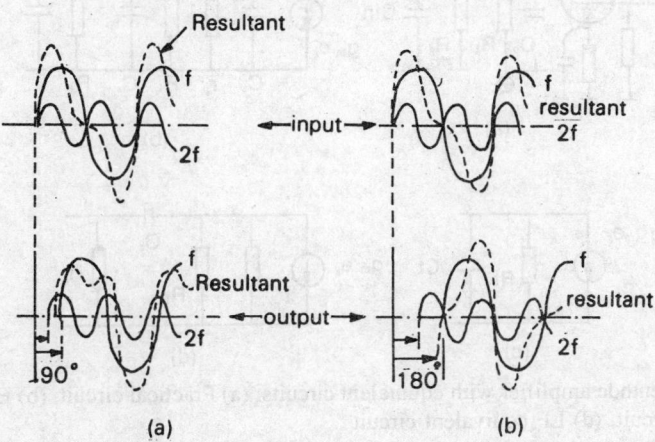

Fig. 13.6 Effect of phase shifts on waveforms containing harmonics

The combined waveform constituting the two frequencies is seen to change shape unless the phase shift is made proportional to frequency.

In an amplifier, the phase shift delay will be zero only if the load is purely resistive. It should be noted that the so-called 180° phase shift in a single stage amplifier is not a phase shift or delay in the real sense. It is just an inversion of signal polarity which is equivalent to a 180° phase shift for a waveform symmetrical about the time axis, but not for a complex video waveform.

In practice, the amplifier has reactive components in the load impedance and reactive components are invariably present. These include the coupling capacitor which affects the gain at low frequencies and the shunting stray capacitances of the wiring and the amplifying devices, which affect the gain at high frequencies. In a single stage or two-stage video amplifier with adequate bandwidth, the phase shift difficulties are not significant. But in multistage amplifiers for video frequency working, it is necessary to employ phase equalising circuits to give the required proportionality of phase shift and frequency.

13.5 VIDEO AMPLIFIER DESIGN

The video amplifier required to handle video frequencies is essentially a wideband amplifier with bandwidth from dc or low frequency of 25 Hz to high frequency of 5 MHz typically. The simplest wideband amplifier is the conventional *R-C* coupled amplifier. The frequency response of the amplifier is limited by the coupling capacitor on the low frequency side and by the stray shunt capacitance on the high frequency side. It may usually have to be extended by suitable compensation methods. On the low frequency side, the problem may be eliminated by direct coupling which is readily possible for one or two limited number of stages. In circuits with capacitor coupling, suitable compensation for low frequency response is possible, the dc being restored by dc restorer diode clamping circuits. High frequency compensation is done by shunt-series peaking coils in the load circuits. These coils help increase the effective load impedance at high frequencies and increase the gain.

Consider a pentode wideband *R-C* coupled amplifier without compensation, as shown in Fig. 13.7(a).

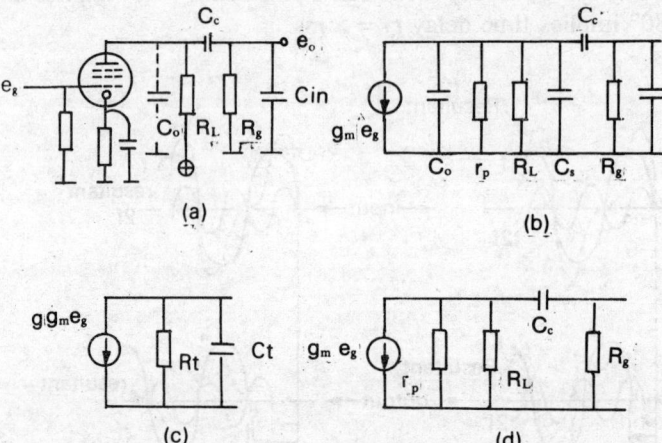

Fig. 13.7 *RC* coupled pentode amplifier with equivalent circuits: (a) Practical circuit, (b) Equivalent circuit, (c) HF equivalent circuit, (d) LF equivalent circuit

The equivalent circuit of the stage is shown in Fig. 13.9(b), along with the output capacitance C_o of this stage, the input capacitance C_i of the next stage, and the stray shunt capacitance of wiring. If the stage has

$$g_m = 10 \text{ m}\Omega, \; r_p = 1\text{M}, \; R_L = 2.5\text{k}, \; R_g = 100 \text{ k},$$

$$C_t = 20 \text{ pF and } C_o = 0.1 \; \mu\text{F},$$

the mid-band gain $A_m = g_m R_L$ $\hspace{4cm}$ (13.1)

$$= 10 \text{ m} \times 2.5 \text{ k} = 25$$

From the high frequency equivalent circuit shown in Fig. 13.7(c), the high frequency gain is given by

$$|A_h| = \frac{g_m R_L}{\sqrt{1 + (\omega C_t R_L)^2}} \tag{13.2}$$

$$= \frac{1}{\sqrt{2}} \cdot |A_m|, \text{ when } \omega C_t R_L = 1;$$

i.e.,
$$f = \frac{1}{2\pi C_t R_L} \triangleq f_2 \tag{13.3}$$

This is the upper 3 dB frequency at which voltage gain falls to $1/\sqrt{2}$ times the mid-band gain.

$$f_2 = \frac{1}{2\pi C_t R_L} = \frac{1}{2\pi \, 20 \, \text{p} \times 2.5 \, \text{k}} = 3.2 \, \text{MHz}$$

The value of f_2 can be increased by minimizing C_t, the total shunt capacitance, and reducing load resistance R_L.

From the low frequency equivalent circuit shown in Fig. 13.7(d) the low frequency gain is given by

$$|A_t| = \frac{g_m R_L}{\sqrt{1 + \dfrac{1}{\omega C_c R'}}} \text{ where } R' = R_g + (r_p \parallel r_L) \tag{13.4}$$

$$= \frac{1}{\sqrt{2}} |A_m| \text{ when } \omega C_c R' = 1,$$

i.e.,
$$f = \frac{1}{2\pi C_c R'} \triangleq f_1 \tag{13.5}$$

This is the lower 3 dB frequency at which voltage gain falls to $1/\sqrt{2}$ times the mid-band gain. For the given stage, this frequency is given by

$$f_1 = \frac{1}{2\pi C_c R'} = \frac{1}{2\pi \times 0.1\mu \times 100 \, \text{k}} = 15 \, \text{Hz}$$

At the output of the video amplifier, a large video voltage swing is required to drive the picture tube. Because of low R_L needed to obtain large bandwidth, large output current swing is required in the output circuit.

A 50 V peak-to-peak swing across the 2.5 kΩ load resistance implies a current swing of (50/2.5 =) 20 mA. Hence even if just a voltage drive is required at the output, it becomes necessary to use a power tube for the video stage.

RF pentodes and circuits with inherent low C_o and C_s allow the use of a larger R_L for the same bandwidth, i.e., f_2, and reduce the peak current swing and hence the power dissipation requirements in the tube as well as the load resistance.

13.6 HF COMPENSATION

The aim of high frequency compensation techniques is to increase the upper half-power (3 dB) frequency maintaining the same mid-band gain. Various methods are possible, some of these are discussed below.

(i) Shunt Peaking

In this method, a small inductance L_p is added in 'shunt' with total capacitance (C_t) of the stage. This is done by placing it in series with load resistance (R_L) as shown in Fig. 13.8.

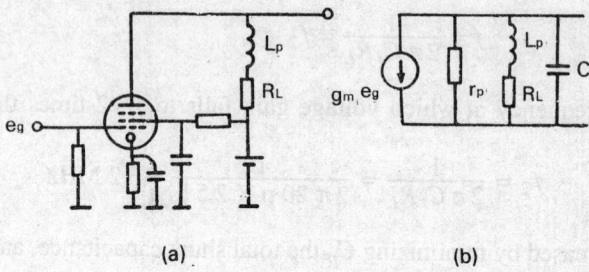

(a)   (b)

Fig. 13.8   Shunt peaking circuit: (a) Amplifier, (b) Equivalent circuit

The inductance raises the impedance of the load circuit at high frequencies and hence partly compensates for the decrease in reactance of C_t. This is effectively an *LCR* circuit, but practically non-resonant because of high R_L and small L_p. The relative gain at high frequencies can be expressed in the form:

$$\left|\frac{A_h}{A_m}\right| = \sqrt{\frac{1 + n^2\left(\dfrac{f}{f_2}\right)^2}{1 + (1 - 2n)\left(\dfrac{f}{f_2}\right)^2 + n^2\left(\dfrac{f}{f_2}\right)^4}}$$

where $n = \dfrac{L}{CR_L^2}$

$$\tag{13.6}$$

and the phase shift angle

$$\phi = \tan^{-1}\frac{f}{f_2}\left\{1 - n + \left(\frac{f}{f^2}\right)^2 n^2\right\} \tag{13.7}$$

The curves for the relative gain and phase response can be plotted at different values of n as the parameter. These are of the nature shown in Fig. 13.9.

L_p can be designed by the formula

$$L_p = nR_L^2 C_t, \tag{13.8}$$

where n is a numerical factor that can be chosen suitably from the response curves for different values of n, usually between 0.25 to 0.7, the commonly used value is 0.5. The numerical factor n is effectively the damping constant of the *LCR* resonant circuit, being equivalent to Q^2 at the resonant frequency, where Q is the quality factor of the circuit at the resonant frequency f_0.

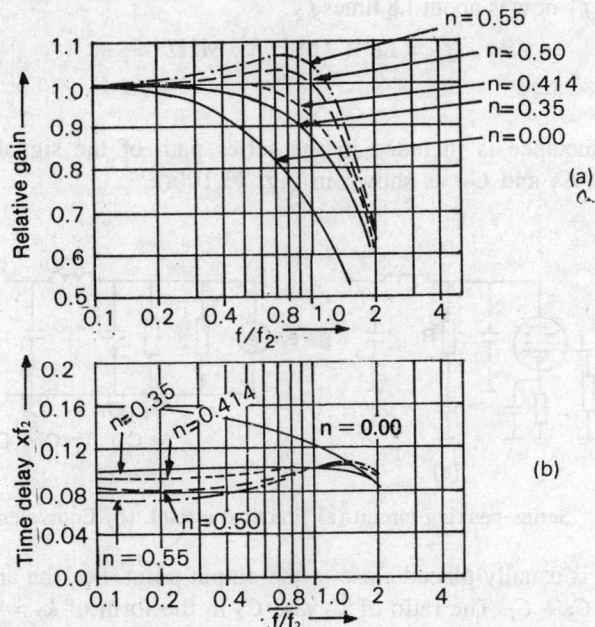

Fig. 13.9 Shunt peaked amplifier: (a) Gain-frequency response, (b) Time delay characteristic, relative to uncompensated amplifier in the mid-frequency range (After curves from *Fundamental of Television Engg.* by Glasford G.M., *Courtesy*: McGraw-Hill Book Co. Inc., New York)

$$n = \frac{L_p}{C_t R_L^2} = Q^2 \tag{13.9}$$

Typical values of n and the corresponding response characteristics are as follows:

1. $n = 0.25$ gives critical damping (without overshoot in the transient response).
2. $n = 0.322$ gives optimum linear phase response (providing phase shift proportional to frequency).
3. $n = 0.414$ gives optimum maximally flat frequency response.
4. $n = 0.5$ gives complete equalisation of the gain at the 3 dB frequency f_2 to the mid-band gain value. There is a small hump of about 1.2 dB at a frequency just below f_2 and a slight but tolerable, perhaps beneficial, tendency towards overshoot in suddenly changing transient step response.

$$\text{Modified 3 dB frequency } f_2' \approx 1.8 f_2 \tag{13.10}$$

In the pentode amplifier having the tube $g_m = 10$ mA/V, $R_L = 4$ kΩ, and $C_t = 20$ pF.

$$A_m = g_m R_L = 4k \times 10m = 40, \text{ and } f_2 = 2 \text{ MHz}$$

with shunt peaking, choosing $n = 0.5$ lifts f_2 by 3 dB.
The peaking inductance $L_p = nR^2 C_t$

$$= 0.5 \times 4^2 k^2 \times 20 p$$
$$= 160 \ \mu H$$

The new 3 dB frequency f_2' now is about 1.8 times f_2.

$$f_2' = 1.8 \times 2\,M = 3.6\,MHz$$

(ii) Series Peaking

Here a small peaking inductance is included in the series path of the signal, splitting the shunting capacitance into two parts, C_1 and C_2, as shown in Fig. 13.10(a).

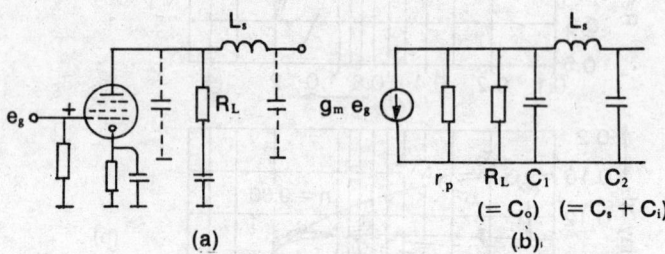

(a) (b)

Fig. 13.10 Series peaking circuit (a) Practical circuit, (b) Equivalent circuit

The series peaking coil is usually placed close to the output point, i.e., the anode pin or the collector so that $C_1 = C_o$ and $C_2 = C_s + C_i$. The ratio of C_1 and C_2 in the form of $k_2 = C_2/(C_1 + C_2)$ is another design parameter, besides the numerical factor n which can be chosen for a desired response. If properly optimized, the circuit performance is better than shunt compensation. It gives better rise time in transient response and the frequency response roll-off is steeper as often desired. The circuit is, however, more critical to adjust.

The derivations of the HF gain and phase response are more complex in this case, and so are the results too. The basic design formula for the series peaking inductance L_s, is again:

$$L_s = nR_L^2 C_t, \text{ where } n \text{ is a numerical factor usually between 0.5 to 1.0.} \tag{13.11}$$

For maximally flat frequency response,

$$k_2 = \frac{C_2}{C_1 + C_2} = 0.75 \tag{13.12}$$

i.e.,

$$\frac{C_1}{C_2} = \frac{1}{3}, \text{ and } n = 0.67$$

The new 3 dB frequency f_2', in this series compensated case is about 2 times f_2. Commonly used values for series peaking are $C_2/C_1 = 2$, and $n = 0.67$.

A variation of this series peaking compensation is by connecting L_s in series with R_L but taking the output at the junction of the two, as shown in Fig. 13.10(b).

The circuit is similar to the previous one but R_L appears at the output end, and the results hold with

the ratio $k_2 = \dfrac{C_1}{C_1 + C_2}$.

(iii) Series-shunt Peaking

The shunt and series peaking techniques may both be used to obtain a still better optimum performance if values of the components are properly adjusted. This is shown in Fig. 13.11(b). The values of L_p, the

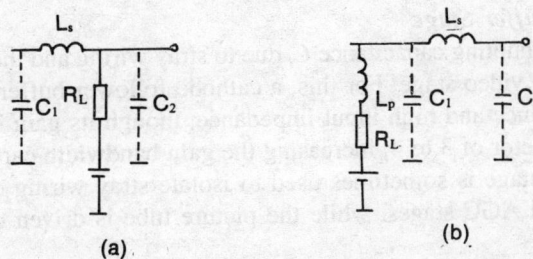

Fig. 13.11 Series peaking modified circuit, (b) Series-shunt peaking circuit

parallel or shunt peaking inductance and L_s, the series peaking inductance will be affected by capacitance distribution of C_1 and C_2. The values may be calculated by the following formulae. In practice, they will have to be adjusted empirically for the desired response.

$$L_p = n_p R_L^2 C_t, \; L_s = n_s R_L^2 C_t, \; k_2 = \frac{C_2}{C_t} = \frac{C_2}{C_1 + C_2} \qquad (13.13)$$

For given values of R_L and C_t, n_p and n_s and k_2 are adjusted for obtaining the required response. The approximate values required for different types of responses are as follows[1].

	n_p	n_s	k_2
Optimum frequency response	0.14	0.58	0.6
$f_2' \approx 2.6 \, f_2$			
Optimum phase response	0.1	0.46	0.72
$f_2' \approx 2.2 \, f_2$			
Critical damping response	0.068	0.39	0.8
$f_2' \approx 1.8 \, f_2$			

Commonly used values for combined series-shunt peaking are:

$$\frac{C_2}{C_1} = 2, \; n_p = 0.12, \text{ and } n_s = 0.52$$

The peaking coils may introduce peaks in the response curve or oscillatory transient response if they resonate with their self-capacitance. The coils are designed to have low self-capacitance so that their natural resonant frequency is high. These may be damped by shunting resistances across them to swamp the ringing oscillations in transient response. A convenient practice is to wind the coil on the damping resistance and seal it.

(iv) Cathode or Emitter Compensation

In this methods negative feedback is introduced at low and middle frequencies relative to high frequencies by low value cathode or emitter bypass condenser. This boosts the gain at high frequencies relative to middle frequencies. The time-constant of cathode circuits $R_k C_k$ should be of the same order (0.5 to 2 times) as time-constant $R_L C_t$ of the plate circuit.

---

[1]*Television Engineering Handbook*, D.G. Fink (McGraw-Hill), 1957.

(v) Cathode Follower Buffer Stage

A substantial part of the shunting capacitance C_t due to stray wiring and the picture tube input capacitance may be isolated from the video stage. For this, a cathode follower buffer stage is used because this has a very low input capacitance and high input impedance, though its gain is near unity only. Capacitance C_t can be reduced by a factor of 3 to 4, increasing the gain bandwidth capability by the same factor. The cathode follower buffer stage is sometimes used to isolate stray wiring capacitance of the feeds to the sync separator and keyed AGC stages, while the picture tube is driven directly.

13.7 LOW FREQUENCY COMPENSATION

If capacitive coupling is used in video amplifier, the low frequency performance is affected; lower half-power frequency f_1 is given by

$$f_1 = \frac{1}{2\pi C_c R_g} \tag{13.14}$$

Value f_1 may be made low by increasing C_c and R_g but the large value of C_c may increase stray shunt capacitance to worsen the high frequency response. Even if dc coupling is used, inefficient cathode and screen bypass condensers lower the LF response by introducing negative feedback at low frequencies.

The LF response may be compensated by decoupling components R_d, C_d inserted in the plate supply circuit. The effective load at low frequencies is increased by increasing reactance of C_d, giving a higher gain.

Partial dc Coupling

The low frequency response of video amplifiers in TV receivers is often deliberately rolled down by a series RC filter to reduce the annoying flutter or pumping effect caused by the interference of direct signal and reflected signal from aeroplanes. The resulting low frequency beat changes its frequency through zero due to the constantly changing phase of the signal from passing aeroplane. Although fast acting AGC circuits are designed to reduce the flutter effect on the screen, it can be made less annoying by further including a long time-constant RC filter circuit in series with the coupling path, eliminating very low frequency variations.

The low frequency filter formed by $R_c C_c$, as shown in Fig. 13.12, reduces the dc component of the signal going to the picture tube by factor $R_c/(R_c + R_k)$ while the middle and high frequencies are coupled through by C_c. This is known as partial dc coupling.

Besides reducing the effect of aeroplane flutter, partial dc coupling also reduces *blooming* of the picture, an effect due to change in picture size caused by change in brightness levels. This is due to poor regulation of the EHT supply which affects the deflection sensitivity, giving larger deflections and larger picture size with fall in the EHT. Removal

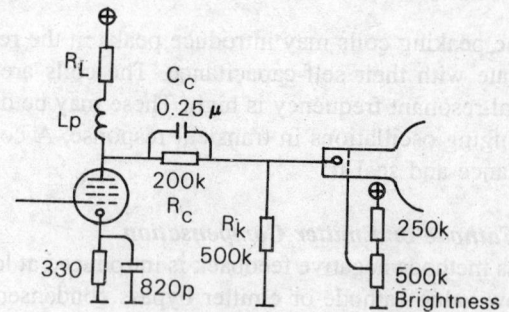

Fig. 13.12 Partial dc coupling

of the dc component in the picture restricts the overall contrast range handled by the picture tube as the mean brightness level that changes from scene to scene is removed. Hence partial dc coupling reduces the demands on the EHT supply regulation.

DC Restoration

In receivers that employ ac coupling in the video stages, it is desirable to restore the dc component in the signal by use of dc restorer circuit of the type shown in Fig. 13.13(a) at the picture tube drive terminal.

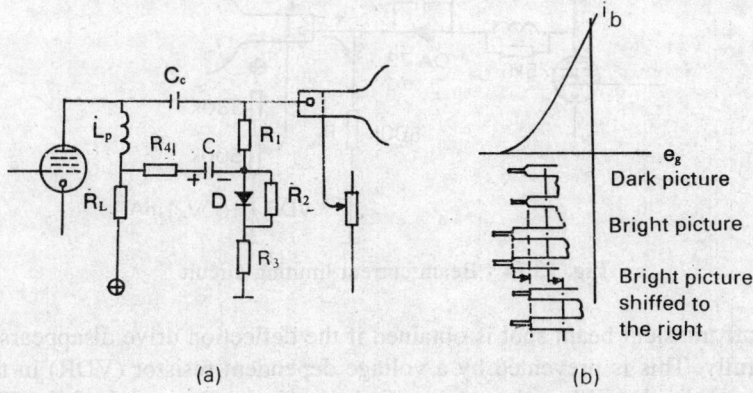

(a)   (b)

Fig. 13.13   (a) DC restorer circuit and (b) Its effect on drive to picture tube

Clamping diode D conducts during the positive going half cycle of the video signal across load resistance R_L to charge capacitor C to the dc voltage proportional to the brightness level. The discharge time-constant of the circuit formed by C and resistance R_L, R_4, R_2 and R_3 is long compared to the line period, but not too long with respect to the frame period so that the capacitor voltage can change frame to frame if the mean brightness level changes. Resistance R_s can affect the clamping voltage as the voltage drop across it pre-biases the diode D. Peaking inductance L_p and resistance R_1 isolate the circuit from the HF signals which may be otherwise affected by added circuit wiring capacitance, etc.

For economy of design, the dc restorer circuit is often omitted in commercial TV sets. Loss of the dc component tends to make the overall picture darker as the contrast range shifts to blacke₁ portions in the absence of proper black level clamping as indicated in Fig. 13.13(b). The brightness has, therefore, to be increased. This may make the retrace lines visible in the picture. They are further blanked in such sets by applying negative going flyback pulses from the deflection stages to control grid G1 and screen grid G2 of the picture tube.

13.8   BEAM CURRENT LIMITING

It is desirable to prevent excessive beam current which may overload the EHT supply and also damage the tube screen. The beam current can be limited by means of a diode capacitor coupling between the video output and the picture tube cathode, and cathode resistance R_k in the picture tube beam circuit, as shown in Fig. 13.14.

In normal operation, resistances R_L and R_k along with diode D form a potential divider between B+ and ground so that the video signal passes through the conducting diode to the picture tube cathode.

If the brightness is increased excessively, the beam current flowing through the picture tube and resistance R_k raises the cathode potential above the video output anode terminal. This reverse-biases the diode, to block the dc and prevent flow of excess current *via* the video stage. A capacitor in shunt with the diode couples the ac picture signal to the picture tube although loss of dc shifts the mean brightness level to a somewhat lower level.

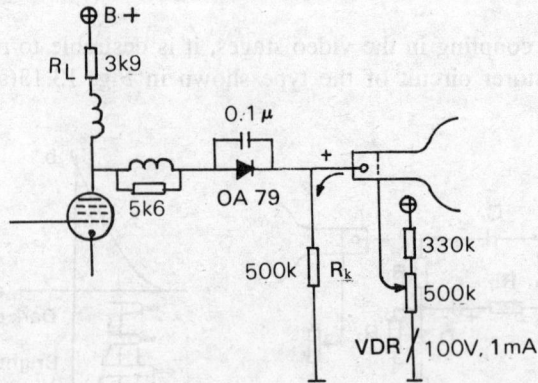

Fig. 13.14 Beam current limiting circuit

A bright switch-off transient beam spot is obtained if the deflection drive disappears before the EHT voltage discharges fully. This is prevented by a voltage dependent resistor (VDR) in the lower arm of the brightness potential divider. When the set is switched-off, the $B+$ supply falls. This increases the resistance of the VDR to slow down the discharge of the positive potential at the grid, and the resulting high beam current discharges the EHT speedily before the deflection disappears.

13.9   VIDEO RESPONSE TESTING

In a picture signal, the brightness changes along the scanning lines are often sudden due to borders between black and white areas. The corresponding video signal voltage has more or less sharp transient step changes. The frequency response and phase response characteristics of video amplifiers can be an indication of its performance. Actual characteristics may differ substantially from the optimum shapes causing changes in the relative phase of the various frequency components and it may be extremely difficult to visualize the effects of these deviations on the picture.

Thus, it is not easy to evaluate the performance of a video amplifier system or equipment by the gain and phase response plot with respect to frequency as it does not give any direct indication of the ability of the system to reproduce the picture details faithfully. Although the gain response of a system can be readily observed by sweep frequency techniques, it is extremely difficult to plot the phase response in practice.

It is more convenient to use transient signal waveforms for the testing and evaluation of video systems. A simple square wave signal can be a rough guide for this, and is often used in the simple vertical bar or cross-hatch pattern generator for the testing of TV sets.

Square Wave Testing

A square wave signal is a typical transient signal containing a wide range of harmonics of the fundamental repetition frequency of the square wave. Because of this, the square waveform can be used for checking the frequency response over a wide range by appropriate choice of the fundamental frequency. The effect of deficiencies in a video amplifier on the output square waveform are shown in Fig. 13.15.

A fall in the low frequency response produces a sag in the output waveform, appearing as a smear and streaking in the corresponding bars in the picture. A poor HF response increases the rise-time of the output waveform, reducing sharpness of the borders implying loss of resolution and smear in the picture.

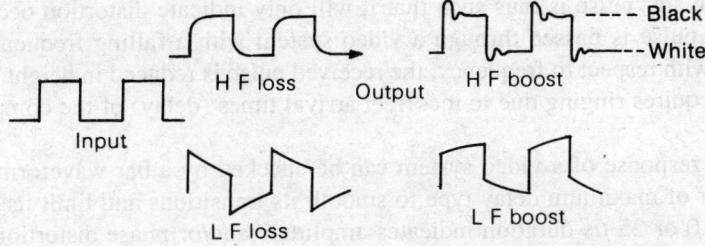

Fig. 13.15 Square wave response of a video amplifier

An underdamped HF compensation by peaking coils appears as ringing bars near the vertical borders due to oscillatory output.

Sin² Pulse and Bar Testing

The waveform used for testing video systems must be typical of video signal, of a shape that would convey a direct impression of distortion and an immediate evaluation of the resolution capability of the system. The waveform chosen for this purpose is based on sine-squared pulse defined by:

$$G(t) = \sin^2 \frac{\pi}{2} \frac{t}{\tau}, \text{ when } (0 \le t \le 2\tau) \tag{13.15}$$

$$= 0, \text{ when } (t < 0, \ t > 2\tau)$$

where τ is the duration of half-amplitude level of the pulse. The sin² pulse and its spectrum against normalized frequency $f\tau$ is shown in Fig. 13.16. It will be observed that the amplitude of the pulse is half of the maximum at $f\tau = 0.5$, and the amplitude is zero at $f\tau = 1$. The width at the base is 2τ, and a very small portion of the energy of the pulse falls above the frequency $f_0 = 1/\tau$, at $f\tau = 1$. The shape of the sin² pulse is very close to the characteristic of the scanning beam in TV pick-up tubes.

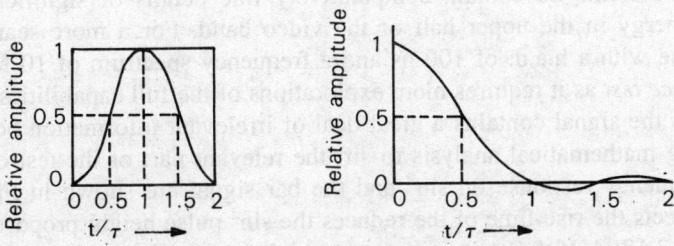

Fig. 13.16 Sin² pulse waveform and its spectrum

A sin² pulse can be produced by applying a narrow pulse (impulse of some 20 ns duration) to a filter of maximally flat delay type. The filter is designed to produce a sin² pulse whose half amplitude duration (h.a.d.) is equal to one half-period of the nominal cut-off frequency of the video system. This is known as a *T* pulse.

$$T = \frac{1}{2f_U} = \frac{1}{2 \times 5} \mu s = 100 \text{ ns for the 5 MHz 625 line system, and}$$

$$= \frac{1}{2 \times 4} \mu s = 125 \text{ ns for the 4 MHz line system.}$$

The spectrum of the $\sin^2$ pulse is thus such that it will only indicate distortion occurring in the band of interest. If the $\sin^2$ pulse is passed through a video system with a falling frequency response, but a linear phase response with respect to frequency, the received pulse is reduced in height. If phase distortion is present, the pulse acquires ringing due to incorrect arrival times (delay) of the component frequencies of the pulse.

The low frequency response of a video system can be checked by a bar waveform which is also fed through the same filter of maximum delay type to smooth its transitions and limit its HF spectrum. The bar of the nominally 10 or 25 μs duration indicates amplitude and/or phase distortion in the lower part of the frequency spectrum.

$\sin^2$ pulse and bar signal are used for testing video systems. These consist of a $\sin^2$ pulse of half amplitude duration switchable to T or $2T$, a bar of 10 to 40 μs duration, and a staircase or sawtooth waveform. For the 625 line 5 MHz system, the $\sin^2$ pulse has a h.a.d. of 100 ns or 200 ns and the bar has a duration of 10 μs, as shown in Fig. 13.17.

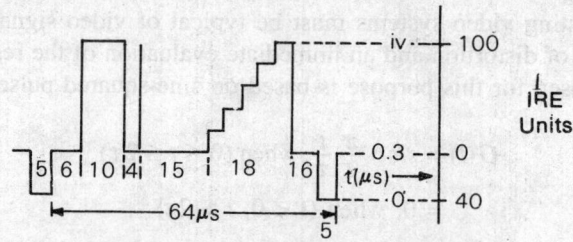

Fig. 13.17 $\sin^2$ pulse and bar signal

The bar is most sensitive to frequencies up to about 0.5 MHz while the T pulse is sensitive to higher frequencies. The $2T$ pulse is indicative of the performance from 0.5 MHz to 5 MHz and is used for *routine or maintenance test*, since it is suitable for ensuring that the performance of the system is up to standard.

Some monochrome systems do contain comparatively fine details of significance and substantial amounts of spectral energy in the upper half of the video band. For a more searching test at higher frequencies, the T pulse with a h.a.d. of 100 ns and a frequency spectrum of 10 MHz is used. This is termed as the *acceptance test* as it requires more explorations of the full capabilities of the system under test. In many instances the signal contains a great deal of irrelevant information because of the out-of-bound effects requiring mathematical analysis to sift the relevant part of the test output.

Effects of poor frequency response on $\sin^2$ and the bar signal are shown in Fig. 13.18. Poor high frequency response affects the rise-time of the reduces the $\sin^2$ pulse height proportionally, as shown in Fig. and changes step 13.18(a). If the phase response at high frequencies is not satisfactory, unsymmetry is introduced in the $\sin^2$ pulse as shown in Fig. 13.18(b). Poor low frequency response causes a sag in the bar as shown in Fig. 13.18(c), while unsatisfactory phase response at low frequencies can cause a bar lift as shown in Fig. 13.18(d).

Measurement of Linearity

The staircase waveform adjustable from 5 to 10 steps or optionally a sawtooth waveform is provided for linearity measurement. The staircase signal reveals quickly nonlinearity in the amplifier response, in individual portions of the amplifier range. The staircase waveform, when differentiated and filtered, produces spikes from the risers in the staircase which show the degree of nonlinearity by their departure from constant height.

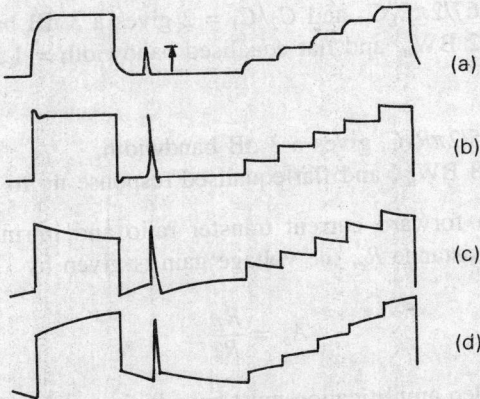

Fig. 13.18 $\sin^2$ pulse and bar responses of video amplifiers: (a) Poor HF response, pulse height reduced proportionately, (b) Phase response not satisfactory at HF unsymmetrical $\sin^2$ pulse, (c) Poor LF response, bar sag, (d) Phase response at LF not satisfactory, bar lift

A better way to measure nonlinearity of a video system is to superpose a high frequency sinewave on the staircase or sawtooth. The output waveform from the network under test is then passed through a high-pass filter or a narrow band filter tuned to the frequency of the sine-wave signal. This eliminates the sawtooth or the staircase component, and the amplitude variations in the sinewave signal reveal the nonlinearity present, if any, on an oscilloscope triggered by the line sync in the signal. This method is known as the *differential gain method* and is most convenient for nonlinearity tests on television transmission systems.

Differential phase may be measured by an extention of the above method using additional equipment which compares the phase of the sinewave on each step of a staircase with the phase at the black level.

13.10   TRANSISTOR VIDEO CIRCUITS

In order to obtain the required gain and bandwidth without loading the detector stage, the video amplifier in a TV receiver using transistors generally incorporates two or three stages. Usually two stages are adequate, the first one having a rather high input impedance serving as the driver to the final video output stage. The driver stage is often an emitter follower that provides impedance matching, prevents loading of the detector by the low input impedance of the transistor and helps to obtain the required gain and bandwidth in the output stage.

AC coupling is sometimes used to simplify the design. The output stage may employ the series-shunt combined peaking which gives maximally flat frequency response with the series peaking inductance L_{se} = 0.52 $C_t R_L^2$, and the shunt peaking inductance L_{sh} = 0.12 $C_t R_L^2$ with the C_2/C_1 ratio equal to 2. The 3 dB bandwidth with this compensation is improved by a factor of over 2.6, compared to the uncompensated bandwidth (BW_{uc}) of the amplifier. The bandwidth (BW) is thus given by:

$$BW = 2.6/2\pi R_L C_t = 2.6 \; BW_{uc}.$$

The bandwidth for flat equalised response is up to 1.8 BW_{uc}, that is, $1.8/2\pi R_L C_t$.

The combined peaking requires critical adjustment to obtain both the gain and phase response satisfactorily. Series or shunt compensation only may often be sufficient to achieve a satisfactory gain and bandwidth response. Series compensation with

$$L_{se} = 0.67/2\pi R_L C_t, \text{ and } C_2/C_1 = 2 \text{ gives a 3 dB bandwidth,}$$
$$\text{BW}_{se} = 2.2 \text{ BW}_{uc} \text{ and flat equalised bandwidth} = 1.5 \text{ BW}_{uc}.$$

Shunt compensation with

$$L_{sh} = 0.5/2\pi R_L C_t \text{ gives a 3 dB bandwidth,}$$
$$\text{BW}_{sh} = 1.8 \text{ BW}_{uc}, \text{ and flat equalised response up to } f_2$$

For a transistor with a high forward current transfer ratio and intrinsic emitter resistance, small compared to external emitter resistance R_e, the voltage gain is given by

$$A_v = \frac{R_L}{R_e}$$

The transistor selected for video amplification must have high gain-bandwidth product, low collector base capacitance and output capacitance. In order to amplify video output voltages upto 80–90V without distortion, the supply voltage must be greater than 100V, and the BV_{CEO} of the transistor must be higher than this supply voltage. The transistor must be capable of dissipating the resultant collector dissipation which is of the order of 5 W for loads in the range of 5 kΩ. Suitable video output transistors are BF 457–8–9 with total power dissipation up to 6 W, $h_{FE} \geq 26$, $f_T = 90$ MHz and rated $V_{CE\text{max}}$ of 100–250–300 V, respectively. Other transistors with similar ratings are 2N 3501, BD 115, BF 174, BF 178, etc.

The load line and the large signal operating conditions in an ac-coupled video-output transistor amplifier are shown in Fig. 13.19.

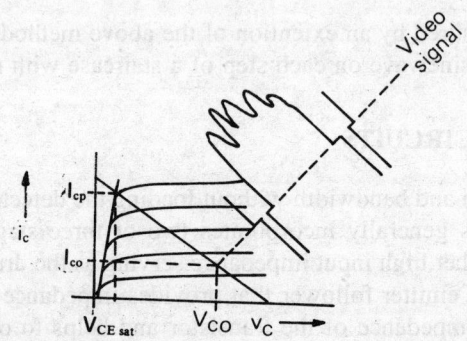

Fig. 13.19 Operating conditions in an ac-coupled video output transistor

The maximum transistor output voltage is given by

$$V_{o\text{max}} = (V_{cc} - V_{CE\text{sat}}) \frac{R_L}{R_e + R_L}$$

If the operating point is set at $I_{co} = \dfrac{I_{cp}}{3}$, corresponding to the black level,

$$I_{co} = \frac{V_{ce} - V_{CE\text{sat}}}{3(R_L + R_e)}, \text{ and } V_{co} = V_{ce} - I_{co}(R_L + R_e)$$

This can be set by a suitable potential divider for biasing.

The bandwidth of the video driver stage is limited by R_o the output resistance of the stage in combination with C_{in}, the input capacitance of the video output stage, given by

$$BW = \frac{1}{2\pi R_o C_{in}},$$

where
$$R_o = \frac{R_e\left(\dfrac{R_g}{h_{fe}} + 1\right)}{R_e + \left(\dfrac{R_g}{h_{fe}} + 1\right)}, \; R_g \text{ being the output impedance of the detector.}$$

The input capacitance C_{in} of the video output stage is primarily the Miller capacitance $C_c(A + 1)$, where C_c is the base-collector capacitance.

For driver stage of the emitter follower type, the bandwidth is quite high and hence the overall characteristics of the video amplifier are generally limited by the output stage. A typical two stage ac-coupled amplifier of this type is shown in Fig. 13.20.

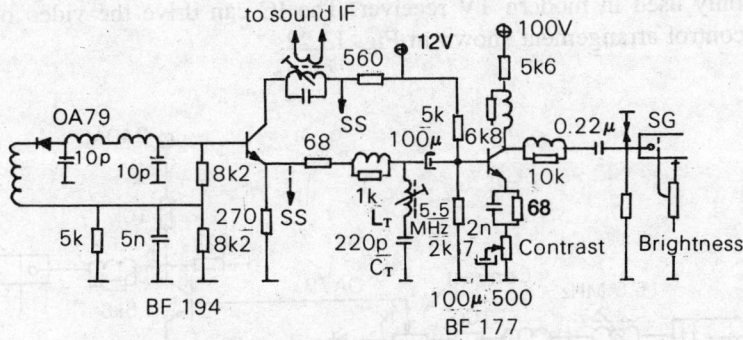

Fig. 13.20   ac-coupled 2-stage transistor video amplifier

The stages are ac-coupled with series peaking in the coupling from the emitter follower to the output stage which employs combined series-shunt coupling. The sound take-off is arranged from a resonant circuit in the collector circuit of the driver while the sound trap is a shunt trap formed by L_T and C_T, across the emitter load. The sync separator is fed from the emitter of the driver. An additional resistive load in the collector circuit may be used alternatively to tap-off the feed to the sync separator to isolate the shunt capacitance due to this connection from the video path. A variable bypass for the emitter resistance of the output stage serves for the contrast control in the form of variable negative feedback in the emitter circuit. A low time-constant R-C circuit is in addition used in the emitter circuit to boost the HF gain relative to the midband gain due to better bypass action and reduced emitter degeneration at high frequencies.

In direct coupled circuits, a bridge arrangement is used at the input of the video output stage when the contrast control is incorporated in the coupling network between the driver and the video output stage. This prevents variation of the operating point of the output stage with the variation of contrast, so that the black level is kept constant over the control range. An arrangement of this type is shown in Fig. 13.21.

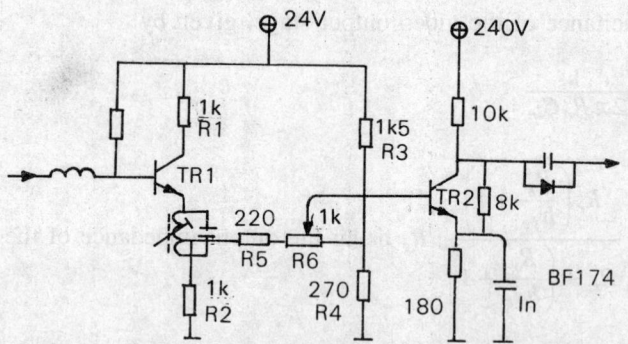

Fig. 13.21   dc-coupled transistor video amplifier

In this circuit, resistances R1, R2, R3 and R4 form the bridge with no dc cross resistances R5 and R6. Variation of contrast control R6 does not affect the dc operating point of output transistor TR2.

The video pre-amplifier with emitter follower output is incorporated in the video processing ICs like TBA 890, commonly used in modern TV receiver. The IC can drive the video output stage directly through contrast control arrangement shown in Fig. 13.22.

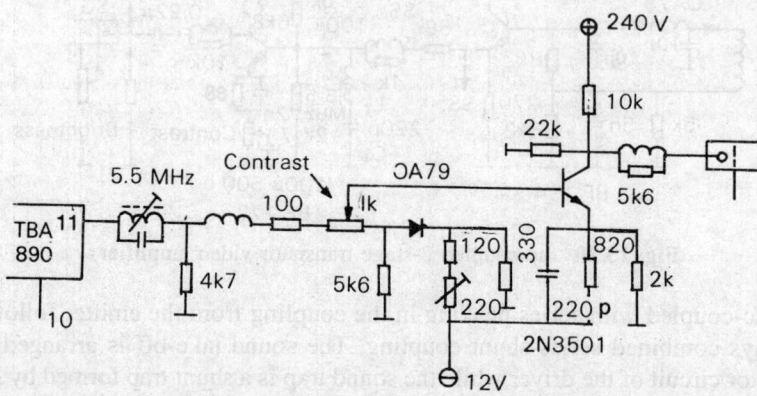

Fig. 13.22   dc-coupled video output amplifier driven by the IC TBA 890 (*Courtesy*: TELERAD)

The circuit features a diode clamp at the low end of the contrast control to avoid the possibility of retrace lines becoming visible at low contrast and high brightness. The video output at pin 11 of TBA 890 is very stable in amplitude against signal variations and is also internally blanked, so that the black level is always at – 7V and the blanking pulse is of 5 V, going to – 12V. On full modulation, the black to white video signal amplitude is about 3.5V as illustrated in the circuit diagram of Fig. 22.20.

To take full advantage of the video signal and to have a true dc-coupling in the video circuit, the contrast control is wired in a balanced circuit, wherein the cold end of the contrast control is connected to a low impedance divider formed by 330 Ω + 120 Ω + 220 Ω preset potentiometer, which is adjusted to keep the cold end exactly at the black level potential.

However, with this arrangement, the blanking pulse of 5V also gets attenuated by the contrast control, especially on low contrast settings, and hence, the blanking amplitude at the picture tube cathode also gets reduced. Consequently, on low contrasts, there is a danger of the retrace lines becoming visible if the brightness control is set a little high. In order to avoid this, the diode clamp using OA 79 is added in series with the contrast control. This remains conducting during scan but opens during flyback. Hence the blanking pulse is always maintained at the full 5 V for all positions of the contrast control, ensuring complete blanking of retrace lines.

Resistance 5k6 provides a small forward bias to the diode so as to avoid gray scale distortion near the black level.

REVIEW QUESTIONS

13.1 What are the causes of spurious responses in the video detector? How are these reduced?

13.2 Explain why phase distortion is important in video circuits particularly at low frequencies.

13.3 Explain with diagrams the effects of poor HF response and poor LF response on the picture quality.

13.4 What is the effect of ac-coupling in video amplifiers on the picture?
Explain the working of a dc restorer used in the video amplifier.

13.5 What is partial dc-coupling? What are the advantages and disadvantages of the same in the video output amplifier?

13.6 Explain with the help of circuits, the methods for (i) beam current limiting (ii) suppression of switch-off transient spot in a picture tube.

13.7 Give the various methods for contrast control, brightness control and stabilization in the video output circuits.

13.8 Discuss the methods for testing the response of a video amplifier by (a) square wave input, (b) $\sin^2$ and bar signal input. Bring out the ease with which the $\sin^2$ pulse and bar signal can be used in evaluating the video response.

13.9 Discuss the methods of high frequency compensation indicating the improvements achieved by each.

13.10 Draw the circuit of a transistor video output amplifier. How is the effect of contrast variation on brightness minimized in direct coupled circuits?

13.11 Derive the expressions of Eq. (14.6) for relative gain and Eq. (13.7) for phase shift angle, for a shunt peaked amplifier in the normalized forms.

MULTIPLE-CHOICE QUESTIONS

13.1 Pick the wrong statement:
  (a) The average dc level for a dark picture is closer to the blanking level than for a light picture.
  (b) Absence of dc coupling to the picture tube results in low brightness in the picture
  (c) The gamma of a TV picture tube is greater than 1
  (d) Phase shift in the video amplifier must be proportional to frequency.

13.2 Pick the incorrect statement:
  (a) A video diode detector uses point contact germanium diode for better linearity.
  (b) The time constant of a detector load circuit should be smaller than the period of the video IF.
  (c) The polarity of output at the video detector is related to the picture tube drive.
  (d) LC type filters are used to filter out the video IF and harmonics.

13.3 Shunt peaking circuit for optimum video response is:
  (a) critically damped
  (b) slightly underdamped
  (c) slightly overdamped
  (d) none of these

13.4 An excessive gain at 3.2 MHz will cause:
   (a) smear in picture
   (b) ringing in picture
   (c) Moire patterns in picture
   (d) none of these
13.5 A sine-squared pulse is useful in evaluating:
   (a) HF response
   (b) HF phase response
   (c) LF response
   (d) (a) and (b)
13.6 Pick the wrong statement:
   (a) AT pulse has an HAD of 100 ns.
   (b) A 2T pulse indicates the HF performance of a video system up to 5 MHz
   (c) A bar signal indicates the amplitude and phase response at low frequencies
   (d) A lift in the bar indicates gain boost at low frequencies
13.7 A lagging tilt in the 2T pulse indicates:
   (a) lagging phase at HF
   (b) lagging phase at LF
   (c) leading phase at LF
   (d) leading phase at HF

14

Sound Section

The sound section of a TV receiver contains the sound IF amplifier, the amplitude limiter, FM demodulator, AF pre-amplifier and AF power amplifier that ultimately feeds the audio power to the loudspeaker. The sound IF is the intercarrier frequency of 5.5 MHz generated as the difference between the sound carrier and the picture carrier intermediate frequencies at the video detector. The intercarrier sound IF of 5.5 MHz with its frequency modulation side bands of $\pm$ 50 kHz, is taken-off at the detector or from the coupling on the sound IF trap coil in the video amplifier. The amplitude is of the order of a few to several millivolts.

14.1 SOUND IF AMPLIFIER AND LIMITER

The small (mV) sound IF signal from the video detector is amplified in the sound amplifier to a level of a few to several volts level required for the FM demodulator or detector. This requires one or two amplifying stages. The second stage is often used as a limiter to eliminate any amplitude variations in the FM signal, due to noise interference or unequal amplification of the frequencies in the sidebands. The IF amplifier is designed for adequate bandwidth to pass $\pm$ 50 kHz at 5.5 MHz, using signal tuned or double tuned, inductively or capacitively coupled stages with damping resistors across the tuned circuits to obtain the wide bandwidth.

The limiter stage is also an IF tuned amplifier but operates at much reduced operating dc voltages so that the stage operates as a saturated amplifier. Up to a certain low signal input, it gives an output proportionally increasing; but as the output swing approaches the supply voltage magnitude, the output gets clipped and amplitude limited. The tuned circuit load allows only the sine wave IF signals to be present in the output to be fed to the FM detector. Limiter stage is particularly necessary when FM discriminator is used as the FM detector. Ratio detector does not require an AM limiter as it has AM rejection properties.

14.2 BALANCED SLOPE DETECTOR-FM DISCRIMINATOR

The aim of an FM detector is to obtain voltage proportional to the frequency deviation in the frequency modulated signal, so that the original modulating signal can be extracted. This can be readily available if the FM signal frequencies are aligned along the sloping side of the IF response curve as shown in Fig. 14.1, where f_0 is the resonant frequency of the LC circuit and f_c the FM carrier.

The output voltage has an amplitude that varies with the frequency of the sidebands in the FM signal.

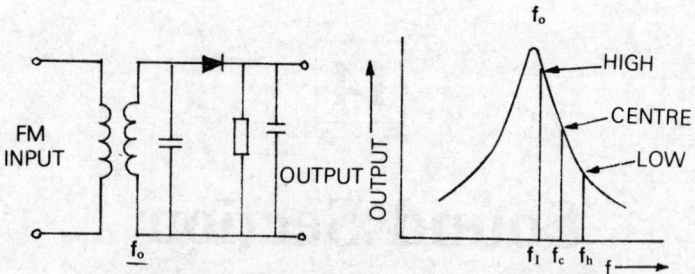

Fig. 14.1 Slope detection of FM signal

The amplitude modulated signal thus obtained may be given to a diode detector to obtain the original modulating signal.

Two such resonant circuits tuned to slightly different frequencies f_1 and f_2, with their slope-detected outputs connected back-to-back form the basis of a *balanced slope detector or FM discriminator*. The basic arrangement is shown in Fig. 14.2(a) along with output response in Fig. 14.2.

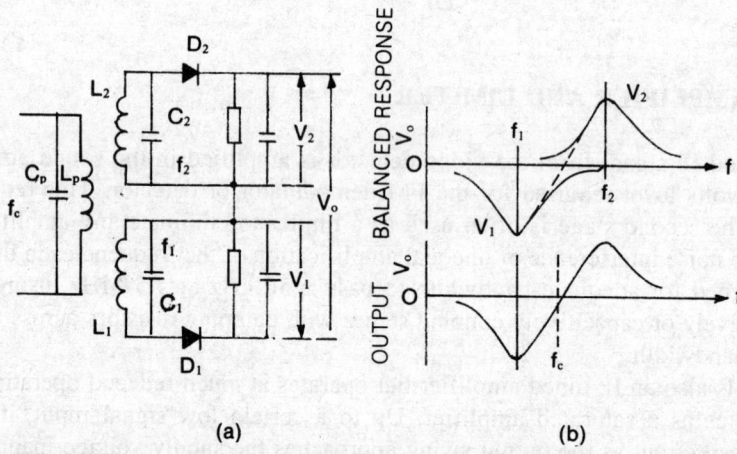

Fig. 14.2 (a) Simple FM discriminator—balanced slope detector (b) output response

In this simple circuit, the primary is tuned to f_c the carrier and the two secondary resonant circuits are tuned to a little higher and lower frequencies say, 5.3 and 5.7 MHz. Secondary L_1C_1 produces an output when the carrier deviates to lower frequencies, while L_2C_2 produces output when it deviates to higher frequencies. The output voltage V_o which is the difference between the two, varies along a resultant S-shaped curve. If the response curves are properly adjusted, this gives a linear output in the intermediate frequency range between f_1 and f_2, crossing through zero at the carrier frequency f_c, to which the primary resonant circuit L_pC_p is tuned.

This balanced slope detector involves three tuned circuits and becomes difficult to align. The output is affected by the amplitude modulation also.

14.3 PHASE SHIFT DISCRIMINATOR

It is possible to obtain the S-shaped detector response from a double tuned circuit with both the primary and secondary tuned to the same centre carrier frequency of 5.5 MHz, as shown in Fig. 14.3. This greatly simplifies alignment of the circuit and yields better linearity than that in slope detection. The transformer distributes the IF signal to both the diodes proportional to the phase angle between the primary and the secondary voltages.

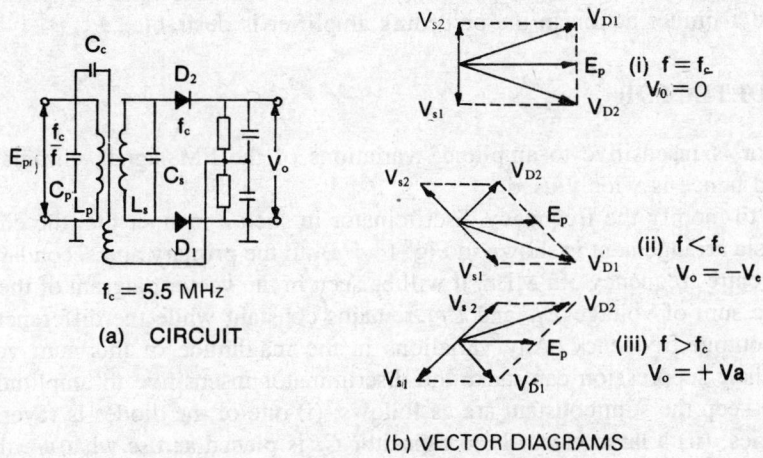

Fig. 14.3 Foster-Seely discriminator

The IF signal voltage from the primary circuit is coupled to the centre tap of the secondary coil through a coupling and dc blocking capacitor C, so that the voltage applied to each diode detector is the vector sum of this primary voltage and the half secondary voltages V_{s1} and V_{s2}, which are 90° out of phase with it, when the input frequency is exactly equal to the carrier frequency to which the circuits are tuned. This can be seen as follows:

The induced voltage in the secondary $E_s = \pm j\omega M I_p$, where M is the mutual inductance between L_p and L_s, and I_p is the current in L_p.

$$E_s = j\omega M I_p = j\omega M \frac{E_p}{j\omega L_p} = \frac{M}{L_p} E_p$$

Hence the secondary voltage drop across half the coils given by:

$$V_{s1} = \frac{j\omega L_s}{2} \cdot I_s = \frac{j\omega L_s}{2} \cdot \frac{E_s}{Z_s}$$

$$= \frac{j\omega L_s}{2} \cdot \frac{1}{R_s + j\left(\omega L_s - \dfrac{1}{\omega C_s}\right)} \cdot \frac{M}{L_p} \cdot E_p$$

$$= \frac{j\omega L_s}{2} \cdot \frac{M}{L_p R_s} \cdot E_p, \text{ at resonance}$$

Voltage drop V_{s2} is equal and opposite in phase to V_{s1}, both being at right angles to the primary voltage E_p, at resonance. Hence at the carrier frequency, the two diodes get equal voltages V_{D_1} and V_{D_2} that are vector sums of E_p and V_{s1}, and E_p and V_{s2} respectively as shown in Fig. 14.3(b), to give a zero difference output signal. As the carrier deviates to a higher or lower frequency, the secondary voltages differ from 90° phase with the primary voltage and the diodes get unequal voltages for detection. There is thus a difference output from the two detectors, positive going or negative going, depending on the deviation to a higher or lower frequency with respect to the centre frequency. The output voltage thus follows the frequency deviation of the input signal. The output is also sensitive to the amplitude of the incoming signal and hence a limiter action in the preceding amplifier is desirable.

14.4 RATIO DETECTOR

This FM detector is insensitive to amplitude variations of the FM signal, eliminating necessity of a limiter stage and hence is widely used.

It is possible to modify the frequency discriminator in such a manner that the circuit itself provides limiting. The basic arrangement is shown in Fig. 14.4. Both the primary and secondary tuned circuits are tuned at the IF centre frequency 5.5 MHz. It will be seen in the vector diagram of the discriminator, that by and large, the sum of voltage E_{D_1} and E_{D_2} remains constant while the difference varies because of the changes in output frequency. Any variations in the magnitude of this sum voltage are spurious unwanted, and their suppression can make the discriminator insensitive to amplitude modulation. The modifications to keep the sum constant are as follows: (i) one of the diodes is reversed so that the two now work in series, (ii) a large electrolytic capacitor C_L is placed across what are the output terminals in the discriminator, (iii) output is taken at the detector load common point with respect to the mean voltage across the electrolytic capacitor.

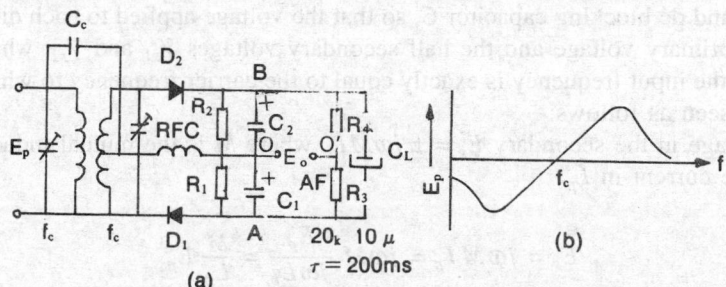

Fig. 14.4 Balanced ratio detector (a) Circuit, (b) Response

As the two diodes are now series aiding, the voltages of the two detectors give a sum voltage across the terminals AB and the large capacitor C_L charges to keep the sum voltage constant. The output is taken across OO′, one of these points being grounded. The output voltage E_o is obtained only when the ratio between the two E_{AO} and E_{BO} changes. Hence the circuit has the name *ratio detector*.

$$E_{o'o} = E_{O'B} - E_{OB} = \frac{E_{AB}}{2} - E_{OB} = \frac{E_{AO} + E_{OB}}{2} - E_{OB}$$

$$= \frac{E_{AO} - E_{BO}}{2}$$

The ratio detector output is thus equal to *half* the difference between the output voltages from the individual diodes, which as in the phase discriminator, varies with the frequency deviation. The ratio detector thus gives S-shaped output curve with respect to frequency, near the resonant frequency of the tuned circuits.

Amplitude Limiting Action

The ratio detector provides a diode variable damping, to the tuned amplifier, varying the gain of the amplifier by changing the damping of its tuned circuits. This maintains a constant output voltage despite changes in the amplitude of the input signal.

When the input signal voltage E_p is constant, C_L charges up and remains at the sum voltage E_{AB} and there is no current flow to charge or discharge the capacitor. The capacitor impedance is thus infinite. The total impedance for the two diodes D_1 and D_2 is R_1 plus R_2, neglecting R_3 and R_4 which are much larger. If E_p increases, C_L in the first place prevents rise in output by its fairly long time-constant and does not allow the voltage across the capacitor to change rapidly. Thus as the input voltage rises, extra diode current flows into the capacitor C_L without much change in its voltage. The current in the diode load is equivalent to a decrease in load impedance and the resultant increase in damping on the tuned circuit reduces the stage gain to counteract the initial rise in the input voltage. If the input voltage falls, the diode current falls, the load voltage remaining more or less constant because of the large time-constant of C_L. This amounts to a reduced damping that increases the gain of the driving amplifier to counteract the fall in amplitude of the input voltage.

In practice, a number of variations of the circuit are used. A practical economy circuit is shown in Fig. 14.5. This is an unbalanced or single ended version because the diodes are not equally balanced against the ground.

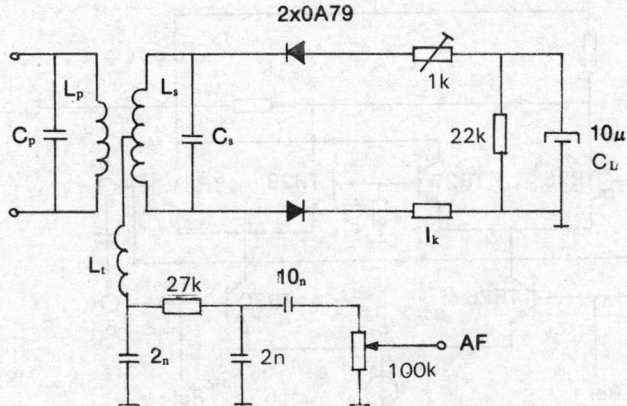

Fig. 14.5   Unbalanced ratio detector

The tertiary winding L_t of the ratio detector transformer is used to couple the primary voltage to the centre tap of the secondary coil L_s. L_t can be adjusted to match the low impedance of the secondary to the high impedance of the primary circuit, since the operation of the primary is improved if its dynamic impedance is kept high. L_t gives a voltage step down to reduce damping of the primary by the ratio detector action.

A resistance in series with the tertiary or 1 kΩ series resistances in each diode lead limit the peak diode currents to improve the balance of the dynamic input capacitance of the diodes at higher signal conditions. One of the two 1 kΩ series resistance may be adjustable for better balance.

The time-constant of the electrolytic capacitor C_L in parallel with the diode load resistances is so large that the circuit does respond neither to fast amplitude changes due to noise nor to the slower changes in amplitude due to spurious amplitude modulation. With $C_L = 10 \ \mu F$ and $R_1 + R_2 = 20 \ k\Omega$, the time-constant is 200 ms. A larger time-constant makes the alignment difficult.

The 27 $k\Omega$ resistance and the 2 nF low-pass circuit in the AF output lead serves as de-emphasis circuit with the time-constant of 50 μs required in the FM system.

14.5 SOUND IF INTEGRATED CIRCUITS

Monolythic integrated circuits are now available from different manufacturers, to perform the functions of sound IF limited-amplifier, FM detector, and AF pre-amplifier. These include Philips TBA 750 A, TAA 570, TAA 661; BEL CA 3065, which are commonly used in modern circuits. Some of these are discussed below.

The monolythic ICs employ more complex and versatile circuitry minimizing use of inductors and transformer couplings. The detectors use new techniques like quadrature detection, differential peak detection or phase locked loops.

The Philips TBA 750 A has a limiter amplifier that consists of a four-stage differential amplifier that gives a very good noise interference suppression. All the four-stages receive their collector voltage *via* a multiemitter transistor that reduces the stage supply voltages and provides a self-limiting action on the voltage swings at the collectors. Its FM detector is of the balanced quadrature type. A simplified quadrature detector equivalent circuit is shown in Fig. 14.6(a).

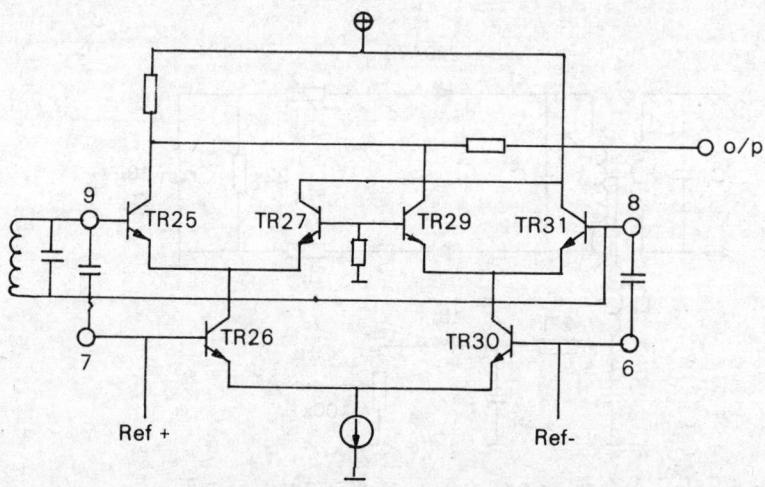

Fig. 14.6 (a) Simplified equivalent circuit of a quadrature detector

The balanced phase detector requires only one tuned circuit for obtaining phase shift proportional to the frequency modulation. The detector operates in a switching mode, switching on and off alternate pairs (TR 25–27, TR 29–31) of transistor differential amplifiers, of which one transistor in each pair has a common load resistor. The input signals to these pairs, at pins 8 and 9 are amplitude limited FM signals appearing out of phase on the alternate sides of each differential amplifier. At the centre frequency, these signals are 90° phase shifted with respect to the reference signal derived directly from the limiter output and applied differentially to the transistors TR 26–30 in long tail pair connection, at the pins 6 and 7. At the centre frequency, the two input signals are in exact quadrature with respect to the reference. As the frequency goes off-resonance, the phase shifts away from 90° and switches the amplifiers on-off for different durations to make available at the common load, an average output level proportional to the frequency variations of the input signal.

The dc volume control stage provided after the detector has a control range of more than 80 dB. The AF pre-amplifier that follows, can drive a triode-pentode output stage or a class A transistor driver stage. The circuit application information of the IC is indicated in Fig. 14.7.

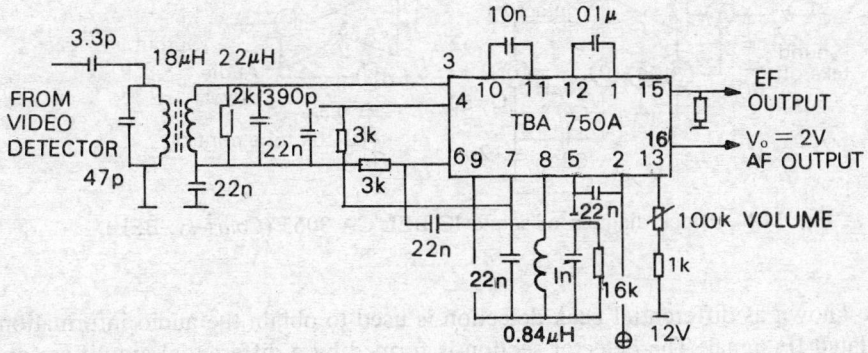

Fig. 14.7 (b) Circuit application of the sound IC TBA 750 A (*Courtesy*: ELCOMA PHILIPS)

Quick Reference Data (TBA 750A)
Supply voltage = 12 V, Total current drain = 30 mA,
f_0 = 5.5 MHz, Input limiting voltage 200 μV
AM rejection = 45 dB, at V_i = 1 mV,
AF pre-amplifier voltage gain (pin 1 to 16) = 10,
Input resistance (R_i) = 35 kΩ,
AF output voltage at pin 16 : 2 V for Δf = ± 15 kHz,
R_o at pin 16 = 4.7 kΩ (collector load).

BEL CA 3065
This is a monolythic TV sound system combining the functions of a multistage IF amplifier, a limiter, an FM detector, an electronic attenuator, a regulated supply and an audio amplifier-driver. The block diagram of the IC is shown in Fig. 14.8.

Regulated Power Supply
The built-in regulated power supply provides a constant voltage to the different stages of CA 3065. The main supply line to the IC is regulated through a zener diode combination. Four other regulated supply lines are derived from this main supply to improve decoupling between stages and hence stability.

IF Limiter Amplifier
This consists of three stages of amplifier-limiter section of CA 3065, each in a single ended configuration. The quiescent currents for the different transistors are determined so as to obtain the best possible limiting characteristic for good AM rejection and capture ratio. The limiting amplifier is followed by an active low-pass filter which suppresses the higher order harmonics from the signal. The output of this filter is then applied to the detector.

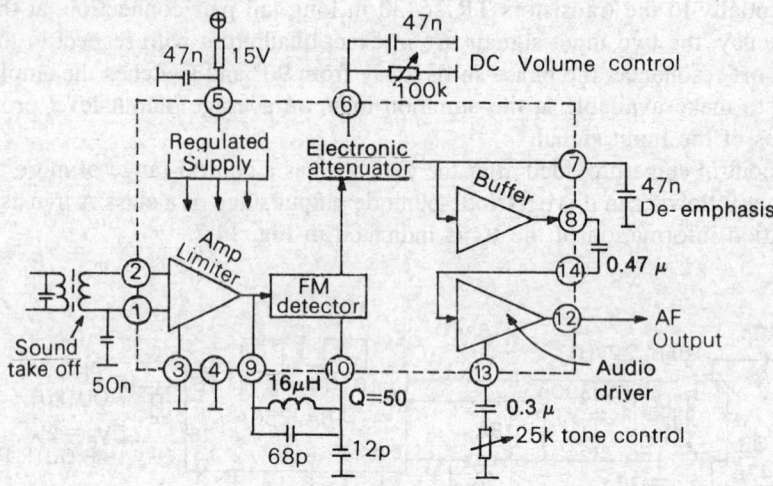

Fig. 14.8 Block diagram of sound IC BEL CA 3065 (*Courtesy*: BEL)

FM Detector

A new technique known as differential peak detection is used to obtain the audio information from the frequency modulated IF signal. The detector section is formed by a differential amplifier configuration with emitter followers at the inputs to provide high impedance at each input pin number 9–10 of the detector and thus isolate the detector from the IF amplifier. The differential amplifier transistors operating at a low current of approximately 10 μA, together with associated capacitors perform peak detection of signals fed to the differential input to the pins 9–10. The output signal appears at the output of the electronic attenuator as the difference between the two peak detected signals. Hence the detector is called differential peak detector. This type of FM detector requires only a single tuned coil as shown in Fig. 14.9.

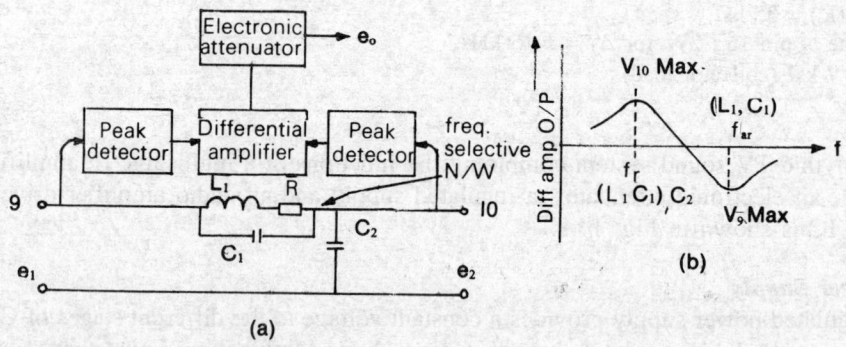

Fig. 14.9 Differential peak detector of BEL CA 3065: (a) Equivalent circuit arrangement, (b) Detection curve (*Courtesy*: BEL)

The frequency sensitive network comprises a parallel *LC* network and a series capacitor. The voltage e_1 across the entire network is applied at the terminal 9. The voltage e_2 across the capacitor C_2 is applied to the terminal 10. The analysis of the equivalent circuit shown in Fig. 16.9 indicates that $e_1 - e_2$ follows the normal S-curve. The linearity of this detector is good and the total harmonic distortion is less than 1%. The tuning coil can be externally loaded to reduce the distortion considerably; at the cost of the output audio signal, which also decreases.

At low frequencies, the voltages e_1 and e_2 are nearly equal and the differential output current of the difference amplifier is nearly zero. As f increases, L_1C_1 parallel circuit series-resonates with C_2, so that the pin 9 experiences a very low impedance and the voltages at pin 10 (V_{10}) is maximum at the series resonant frequency (f_r).

At a higher frequency, L_1C_1 provides parallel resonance so that V_{10} is very low and the voltage at pin 9(V_9) reaches a maximum at the anti-resonant frequency (f_{ar}). This variation of the voltages across V_9 and V_{10}, along with differential operation provides an S-shaped detection curve required for FM detectors. This is shown in Fig. 14.9(b).

Electronic Attenuator and Buffer Amplifier

The output of the differential peak detector is fed to the electronic attenuator, which performs the function of a volume control. This is achieved by a single ended differential amplifier that forms a cascade active load for the detector. The gain through this differential amplifier is varied by varying the quiescent current in the transistors by varying the dc potential of one of the inputs while he other input potential is kept constant. This is done by an external variable resistance connected between the terminal 6 and the ground. Since there is no audio signal present across the variable resistance forming the volume control, this type of volume control is called a *dc volume control*. In this method, there is no possibility of any hum or noise pick-up in the audio signal. In most cases, only a single unshielded wire is required between the IF board and the volume control.

The output of the attenuator is fed to the buffer amplifier which is an emitter follower isolating the attenuator from the external load. The detected variable output is available at the terminal 8 of the IC and the de-emphasis capacitor can be connected to terminal 7.

An *audio driver* is provided to give a gain of about 20 dB. It is sandwiched between two emitter followers for input and output isolation. Tone control can be incorporated at the terminal 13.

The sound IF system can drive a PCL 84/PCL 86 audio stage in hybrid sets or a solid state complementary transistor output stage. A complete sound IF circuit along with a complementary transistor output amplifier capable of delivering 2.5 W, is shown in Fig. 14.10. Each of the output transistors have to be mounted on a 3 mm thick aluminium plate heat sink of at least 12.5 sq. cm area. The circuit draws a no-signal current of 55 mA and a maximum signal current of 300 mA at 15 V, while delivering 2.5 W at less than 3% total harmonic distortion. The RF sensitivity for full output with frequency deviation of 15 kHz at 1 kHz, is 30 μV.

PLL FM Detectors

Although most sound ICs use differential peak detectors, phase locked loop (PLL) can also be employed to obtain FM detection without the need for a coil and resonant circuit. A PLL consists essentially of a voltage controlled oscillator locked to the incoming frequency through a phase detector, that compares the frequency and phase of the VCO with the incoming frequency and provides the error correction voltage to the VCO, through a lowpass filter and dc amplifier. When the input to the PLL is frequency modulated, the error voltage has obviously to follow the modulating signal in order to keep the VCO locked to the input. The VCO control voltage is thus the FM detected output if the input frequency is in the locking range of the PLL.

14.6 SOUND SECTION SUBSYSTEMS

As the trend towards integration stretched into power amplifiers, combinations of power amplifiers ICs,

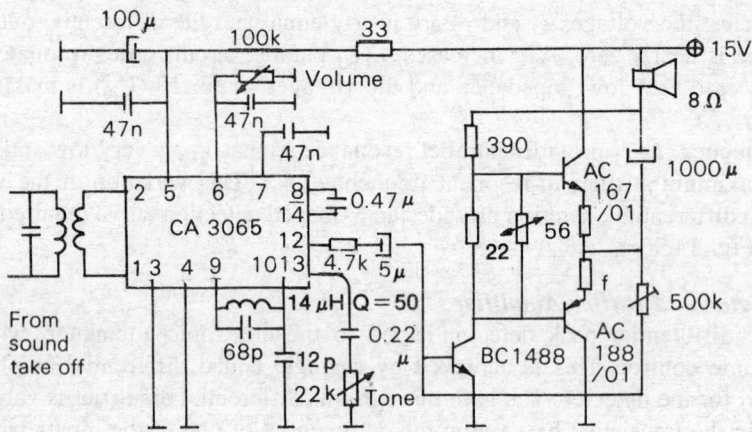

Fig. 14.10 Sound IF amplifier with audio output amplifier (*Courtesy*: BEL)

such as CA 810, TDA 2610A, TDA 2611A, TDA 1013A with the IF ICs, were used to form the sound section. Combination ICs were soon available to integrate the complete sound section, incorporating the sound IF signal and audio signal processing required to drive the loudspeaker. Some of these ICs are TDA 1190Z, CA 1190, TDA·3190, and TDA 4190. Functional details provided in CA 1190 are given below. For stereo TV reception, there are stereo/dual TV sound processor/decoder ICs, such as TDA 2795, TDA 3800G/GS, and TDA 3803A.

Sound Subsystem CA 1190

The circuit diagram of the sound IF subsystem using BEL IC CA 1190 is shown in Fig. 14.11.

The ICs provide on-chip centralized power supply, stabilized with a reference Zener diode. From this main reference voltage, separate series pass control transistors supply currents to the FM detector and IF amplifier limiter stages, reducing unwanted coupling between various stages and improving overall stability. The IF amplifier limiter comprises six stages of cascaded differential amplifiers, and the bias voltage of the differential amplifier transistors is kept low to provide good limiting action.

The amplified IF signal is passed through an active low pass filter, which shapes the response around 5.5 MHz and eliminates the unwanted HF content in the signal produced by limiting action. The filtered output is given to the FM detector which provides *differential peak detection* with the help of the SIF resonant circuit formed by the inductance $L = 16~\mu H$, and the 68 and 10 pF capacitors at pins 6 and 7. The detected output drives the AF power amplifier through the dc volume control which provides a dynamic range of about 90 dB. The audio amplifier consists of the preamplifier and the output stage. The overall gain of the audio amplifier is determined by the external negative feedback network connected between pins 9, 11 and 15 as shown. Frequency compensation for the amplifier and biasing for class AB operation of the output transistors is provided internally in the IC.

14.7 STEREO TELEVISION SYSTEMS

With the popularity of high-fidelity stereo systems, stereo sound has been introduced in several countries for television broadcasting also. Inputs for TV receiver available from Hi-Fi VCR, video disc, computer and satellite receiving systems have increased the need for improved vision as well as sound quality. The receiver has to satisfy the stringent demands of external professional equipment and adapt to recent

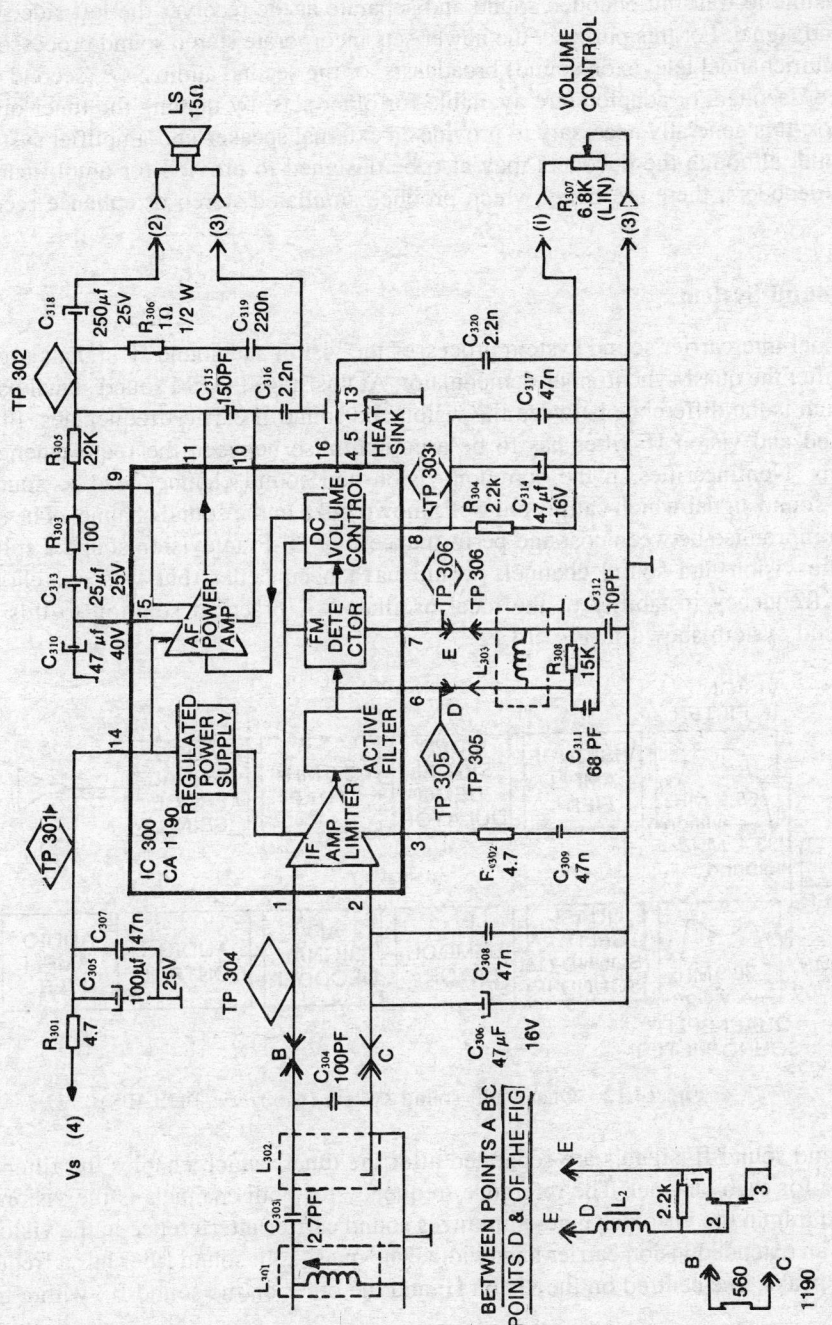

Fig. 14.11 Functional block diagram of CA 1190 (*Courtesy:* BEL)

innovations like stereo or dual language transmissions, teletext data, and so on. In a stereo system, it should be possible to transmit encoded sound and separate at the receiver the left side sound signal and right side sound signal. For this purpose, the newer sets incorporate stereo sound processors for the *stereo TV* or *MTS* (multichannel television sound) broadcasts, or the second audio *SAP* (second audio program). External stereo decoders or adaptors are available for older sets, by tapping the tuner output through an MPX connector. It is generally necessary to provide an external speaker-and-amplifier system to reproduce the stereo sound, although the decoders may also be designed to provide for amplification. In addition to true stereo decoders, there are many which produce simulated stereo or enhance reception of stereo TV.

Quasi-split Sound System

The conventional intercarrier sound system processes the vision and sound IF signals together until they are separated after the quasi-synchronous demodulator. At this stage the FM sound signal has an intercarrier frequency which is the difference between the vision and sound If carrier frequencies. In this system the design of sound and vision IF filter has to be a compromise between the requirements of vision and sound channels. Nonlinearities in the common vision and sound channel lead to spurious amplitude modulation of sound signal which causes the well-known buzz in the sound channel. The system however represents a compromise between cost and performance. For Hi-Fi television sound a split sound system having separate vision and sound channels would have been better, but was not chosen because of problems like frequency instability of the local oscillator. A practical solution to this problem is the quasi-split sound system shown in Fig. 14.12.

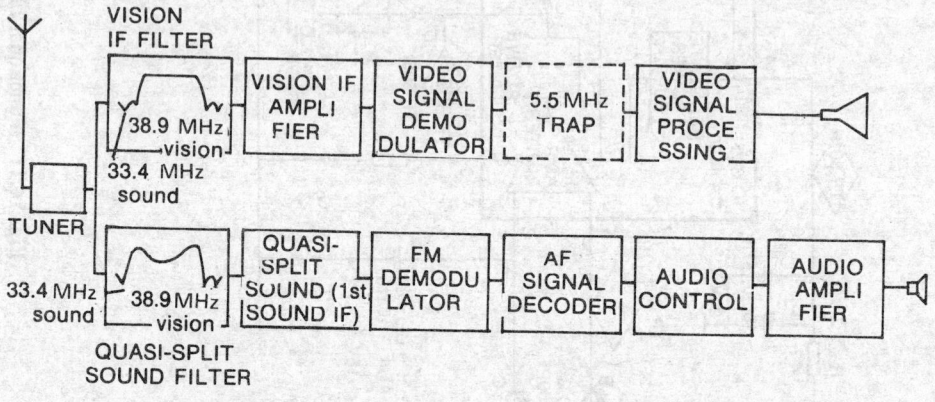

Fig. 14.12 Quasi-split sound system (*Courtesy*: PHILIPS)

The vision and sound IF signals are separated after the tuner, which enables the filter characteristics to be optimized for each channel. The reference frequency for both channels is the vision IF 38.9 MHz. The deep sound trap in the vision channel minimizes sound carrier interference in the vision channel, and thereby allows an extended vision carrier bandwidth. The quasi-split sound filter has a frequency response consisting two peaks, one centred on the vision IF and the other on the sound IF, with a trough between them.

The filter attenuates the unwanted IF frequencies that cause intermodulation products at the intercarrier frequencies. Since there is no attenuation at the sound IF carrier, the SNR of the quasi-split system is higher than that of the conventional system. The group delay characteristics of the split-sound system should be identical in the vision and sound passbands. Because Nyquist slope is not necessary, amplitude and phase characteristics are symmetrical around the vision carrier frequency. This is important for linear phase regeneration of the vision carrier for the sound IF carrier demodulator. A linear multiplier is used as a sound demodulator as it generates few intermodulation products which affect the SNR. Quasi-split sound provides improvement in SNR regardless of whether the noise is random or picture-related.

Stereo/Dual-channel Sound

Typical stereo/dual channel sound B/G transmission standards created in Germany and widely adopted in many other countries have been given in Ch. 3.

The IF spectrum consists of vision carrier, a first sound IF carrier at 5.5 MHz and a second sound IF carrier at 5.7421875 MHz from the vision carrier. To minimize interference, the two carriers are separated by $15.5 \times f_H$. The first sound IF is modulated with $(L + R)$ information so that mono TV receivers can still reproduce mono sound when stereo is being transmitted. The second sound IF carrier is frequency modulated with 2R information. To identify whether the transmitted signal is mono, stereo or dual sound, the second sound IF carrier is also frequency modulated with a +/- 2.5 kHz identification signal. The identification signal is a 54.6875-kHz pilot carrier signal which is 50% amplitude modulated with 117.5 Hz for stereo or 274.1 Hz for dual transmission. The frequency spectrum of the intercarrier IF signal is shown in Fig. 14.13(a), and the audio spectra after filtering and demodulation in Fig. 14.13(b).

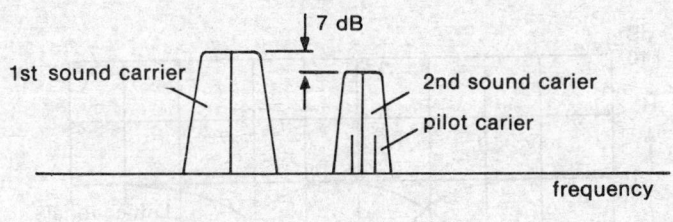

Intercarrier signal frequency spectrum

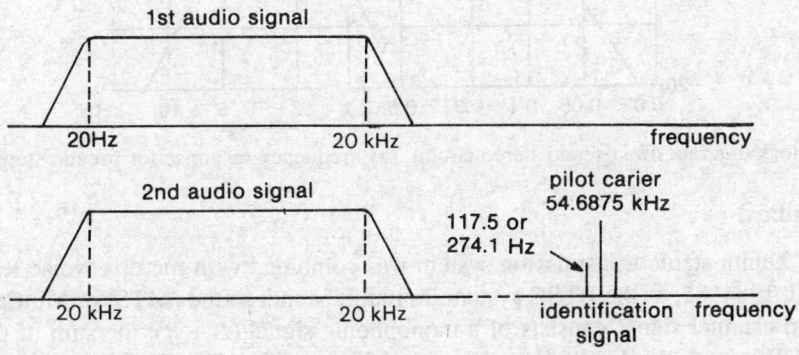

Audio signal frequency spectra including indentification signal

Fig. 14.13 (a) Intercarrier IF signal frequency spectrum, (b) Audio signal frequency spectra including identification. (*Courtesy*: PHILIPS)

The stereo information can therefore be simply dematrixed by doubling the amplitude of the first audio signal and subtracting the demodulated second audio signal from it, i.e. $2(L + R) - 2R = 2L$. After the identification signal is demodulated, it is used to activate logic circuits to route the required information to the sound outputs, under control of the viewer.

Pseudo Stereo Effect

In order to obtain a kind of stereo effect from received monophonic audio signals, the signal is usually split into several frequency ranges and the two left and right channels are supplied with different frequency bands as shown in Fig. 14.14(a).

The left channel is fed a mono signal filtered by a bandpass filter. The right channel gets the original mono signal *minus* the filtered signal of the left channel, thus decreasing the amplitude in the frequency range of the (filtered) left channel signal. In this way a stereo impression is produced. The frequency determining coefficients are contained in the coefficient ROM, C-ROM, of the audio processor. A typical frequency response for pseudo stereo is shown in Fig. 14.14(b).

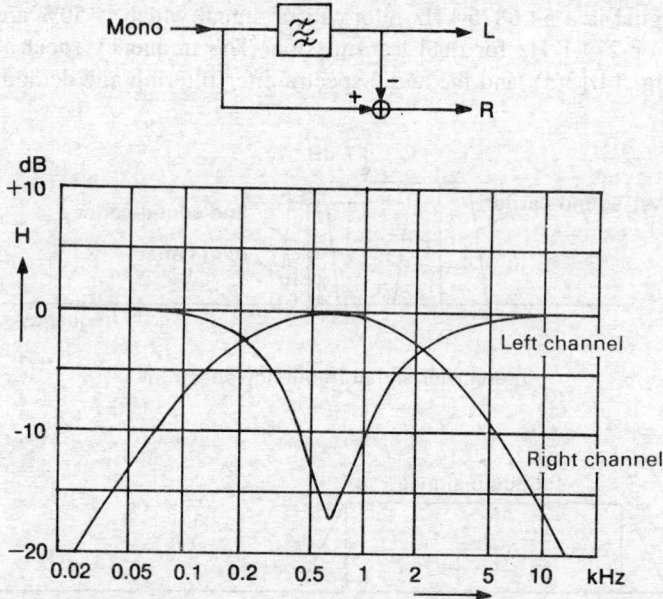

Fig. 14.14   (a) Block diagram of a pseudo stereo circuit, (b) Frequency response for pseudo stereo (*Courtesy*: ITT)

US Stereo Standard

In the USA, the Zenith stereo transmission system was combined with the dbx Noise Reduction System and introduced in 1984–85, as the BTSC system. In multichannel sound (MTS) or Multiplex audio signal (MPX), the main channel signal consists of a monophonic signal $(L + R)$, the sum of the left and right signals, L and R. This is identical with the conventional TV sound, and can be received by any conventional receiver. $(L - R)$, the difference between the left and right channels is sent as an additional stereo signal, so that the L and R signals are separated. The $(L - R)$ stereo signal is sent on a subcarrier $2 \times f_H$, as a double sideband suppressed carrier (DSB-SC), requiring twice the bandwidth. A *pilot signal* is inserted between the main $(L + R)$ channel and the stereo $(L - R)$ channel, as shown in the base band spectrum of Fig. 14.15(a). This serves as reference for the subcarrier and also indicates the presence of stereo signal. The signal processing block diagram for the MPX/SAP system is shown in Fig. 14.15(b).

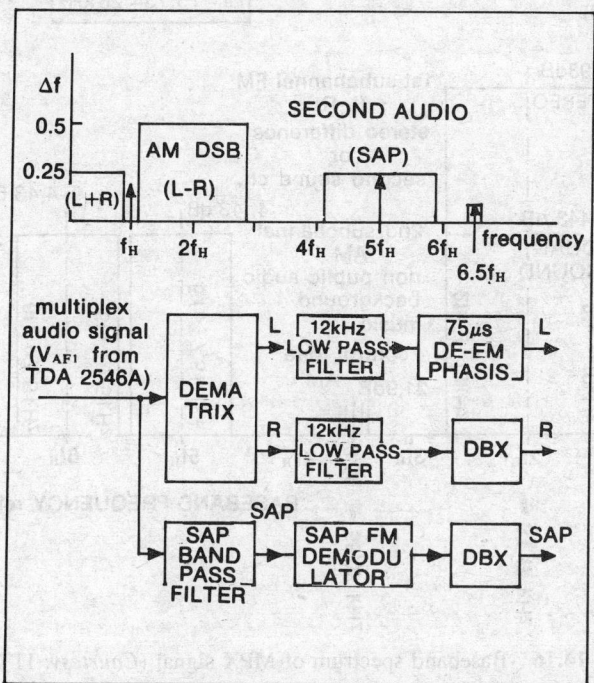

Fig. 14.15   (a) MTS-MPX/SAP base band spectrum, (b) Block diagram of Signal processing Lenk 1.5

The second audio program (SAP) signal is sent on frequency modulated on another subcarrier locked to $5f_H$. Another professional channel for voice or data frequency modulating a frequency of $6.5\,f_H$, is meant for business use information services. The monophonic $L + R$ signal of 50 Hz to 15 kHz is given a pre-emphasis of 75 μs, peak FM deviation of $+/- 25$ kHz, while the stereophonic (L-R) and SAP are encoded by a noise reduction system—the dbx companding system. The professional subchannel sends FM voice or FSK data with pre-emphasis of 150 μs, and peak deviation of 3 kHz.

Japanese FM-FM Multiplex Standard

This FM-FM multiplex system is designed to provide dual/stereophonic sound transmission, while maintaining compatibility with the existing TV receivers and networks. The second audio channel of about 15 kHz is multiplexed with the first (main) audio signal by a frequency modulated subcarrier and then transmitted over single aural carrier. A control signal (pilot) transmitted along with the second audio channel enables the receiver to identify the three programme transmission modes: monophonic, stereophonic and dual sound (bilingual). Supplementary voice and data channels can also be multiplexed, and this MPX signal is used to frequency modulate the aural carrier. The baseband spectrum of the MPX signal is shown in Fig. 14.16.

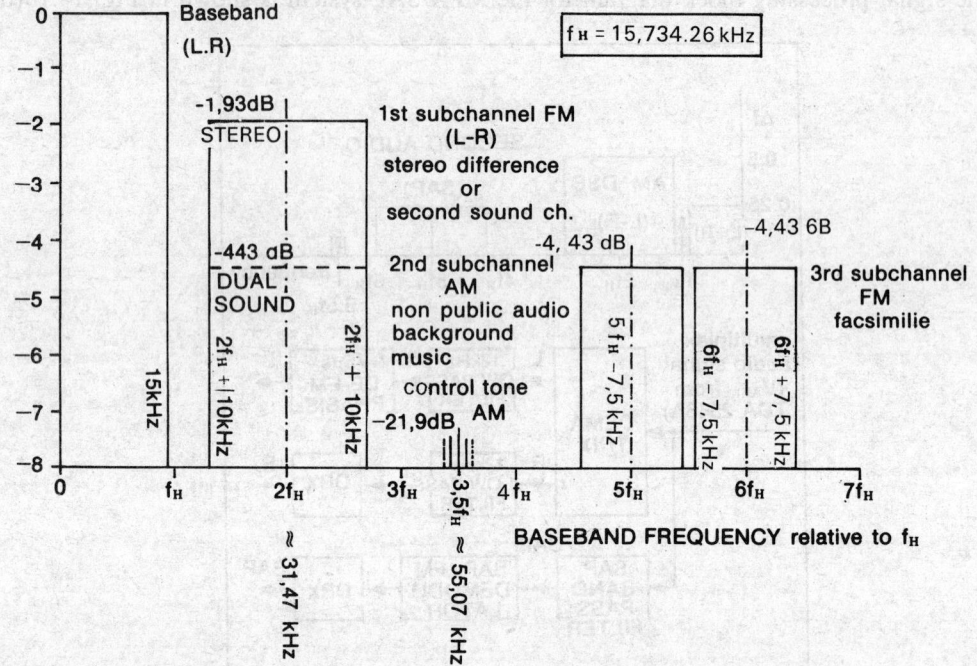

Fig. 14.16 Baseband spectrum of MPX signal (*Courtesy*: ITT)

Dual Sound Transmission in MPX/SAP System

Main/Monophonic Channel

Frequency deviation	+/– 25 kHz
AF range	50–15,000 Hz
Pre-emphasis	75 µs

Sub-channel

Multiplex system	FM-FM
Subcarrier centre frequency	$2 \times f_H$
Modulated bandwidth	16–47 kHz
Frequency deviation of	+/– 15 kHz for bilingual
main carrier by subcarrier	+/– 20 kHz for stereo
AF range	50–15,000 Hz
Pre-emphasis	75 µs

Stereophonic Transmission

Compatible signal $M = (L + R)/2$	Same as main channel
Stereo signal $S = (L – R)/2$	Same as sub-channel except freq. deviation main carrier subcarrier
Polarity of freq. deviation	Left signal produces deviations of subcarrier and main carrier in the same direction

Modulation by left and right signals	Same modulation percentage for sub and main. Max. 50% for each
Frequency deviation of main carrier by subcarrier	+/– 20 kHz
Compensation for delay time of stereo signal S	20 μs

Control Signal

Subcarrier frequency	3.5 fH or 55.07 kHz
Modulation frequency	
for dual sound transmission	922.5 kHz
for stereo transmission	982.5 kHz
Modulation	AM to 60%
Max. freq. deviation of main carrier by control signal SC	+/– 2 kHz

Digital TV sets generally provide stereo facility.

Quasi-split Sound/FM Demodulator Stages

A block diagram of the IF/sound carrier is shown in Fig. 14.17. The quasi-split sound filter is a *double bandpass filter* with symmetrical amplitude characteristics in the vision carrier double sideband frequency range. Asymmetry around the vision carrier frequency causes increased demodulation products at the quasi-split sound stage output which leads to reduction in video related S/N ratio.

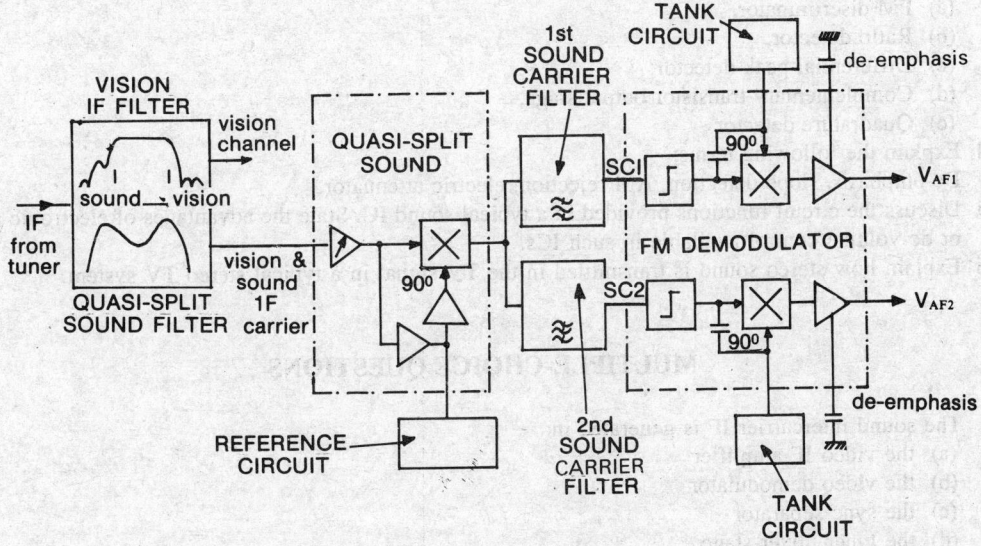

Fig. 14.17 IF/sound carrier signal processing (*Courtesy:* PHILIPS)

The quasi split sound stage must have a *gain control* range adapted to the vision IF channel and sufficient bandwidth to amplify vision and sound carriers linearly. The AGC is generated internally and is not affected by the video modulation content or the ratio of the vision-to-sound IF carrier levels. Generation of the intercarrier signals using a quadrature demodulator results in high suppression of the vision content in sound signals. Using a *linear multiplier* for quadrature demodulation results in a further reduction in the video related components to achieve a high S/N ratio.

The FM demodulator stage is adapted for simple ceramic sound carrier filters and improved quality filters for dual sound applications. The total harmonic distortion is primarily determined by the tank circuits and sound carrier filters. Using two tank circuits in parallel considerably improves the total harmonic distortion. The FM demodulator stage incorporates output audio signal de-emphasis to suppress any high spurious signals from the quadrature demodulator, and to compensate for transmitter pre-emphasis.

ICs for Stereo/dual Sound Systems

IC TDA 2545A is a quasi-split sound circuit incorporating a three-stage AGC controlled IF amplifier and a linear multiplier for quadrature demodulation of the vision and sound IF carriers. TDA 2546A consists of TDA 2545A plus a single FM demodulator. It is suitable for MPX stereo systems in use in the USA or Japan. TDA 2555 and TDA 2557 are dual TV sound demodulator circuits containing twin FM demodulators to demodulate the sound carrier signals from TDA 2545A quasi-split sound circuit.

REVIEW QUESTIONS

14.1 What are the advantages of frequency modulation over amplitude modulation?

14.2 Discuss the operating conditions of a limiter IF amplifier.

14.3 Explain giving circuit diagrams the operation of the following:
  (a) FM discriminator,
  (b) Ratio detector,
  (c) Differential peak detector,
  (d) Complementary transistor output stage,
  (e) Quadrature detector

14.4 Explain the following terms:
  De-emphasis, slope detection, AM rejection, electric attenuator.

14.5 Discuss the circuit functions provided in a typical sound IC. State the advantages of electronic attenuator or dc volume control available in such ICs.

14.6 Explain how stereo sound is transmitted in the TV signal in a typical stereo TV system.

MULTIPLE-CHOICE QUESTIONS

14.1 The sound intercarrier IF is generated in:
  (a) the video IF amplifier
  (b) the video demodulator
  (c) the sync separator
  (d) the tuner mixer stage.

14.2 The FM detector used in integrated TV receivers is:
  (a) phase locked loop
  (b) a differential peak detector
  (c) a ratio detector
  (d) (a) and (b)

14.3 A differential peak detector employs:
  (a) slope detection in a single resonant circuit
  (b) slope detection with two resonant circuits
  (c) slope detection with single tuned circuit having dual resonance
  (d) quadrature detection

14.4 IF the sound trap is not properly tuned in the video IF amplifier
  (a) the sound output will be too low
  (b) the sound will contain a picture buzz
  (c) the picture will contain sound bars
  (d) (b) and (c)

14.5 Pseudo stereo effect can be produced by:
  (a) feeding different ranges of frequencies fed to the two channels
  (b) decreasing amplitude of a limited band of frequencies from one channel
  (c) feeding a limited band of frequencies from one channel
  (d) (b) and (c)

14.6 Quasi split sound separates the sound IF:
  (a) in the video IF stage
  (b) in the tuner-mixer stage
  (c) in the video demodulator
  (d) in a separate filter after the tuner

14.7 A pilot carrier (CW) is sent with the stereo sound transmission:
  (a) to regenerate the suppressed sound carriers
  (b) to identify the stereo/dual sound transmission
  (c) to identify the mono/stereo sound
  (d) (b) and (c)

15

Monochrome Picture Tube and Deflection Evils

The TV picture tube is a cathode ray tube with a large-sized screen. For compactness, it is designed with a narrow neck and a large deflection angle of the beam. It consists of a sturdy glass bulb with a cylindrical narrow neck, a fluorescent screen on the inside of the face of the tube and an electron gun situated in the neck of the tube. The tubes are usually designated by the maximum deflection angle and the screen size along the diagonal. In this chapter, the construction and characteristics of monochrome picture tubes are described. Some basic relations in electron optics are discussed initially to lead to the construction of electron gun and focusing methods used in cathode ray tubes.

15.1 ELECTRON OPTICS

Production and focusing of an electron beam in a cathode ray tube to obtain a thin scanning beam is referred to as 'electron optics' because it is achieved by means of electric and magnetic fields that serve as lens systems for electrons. The electric beam obtained from the tube cathode is refracted and focused by electronic and magnetic fields, just as light rays are refracted and focused by an optical lens system. The study of motion of electrons in electric and magnetic fields in general is termed as 'electron ballistics'. Some basic relations in electron ballistics are summarized below:

Motion of an electron in an electric field   Force **F** experienced by an electron in an electric field E is given by:

$$\mathbf{F} = -e\mathbf{E}$$

where e is the charge on an electron = 1.6×10^{-19} coulombs, and **E** is the electric field in V/m.

An electron accelerated through the electric potential E_α acquires a velocity v given by

$$eE_\alpha = \frac{1}{2}mv^2, \text{ so that}$$

with $E_1 < E_2$. The equipotential lines are shown in between the two electrodes. Since the beam is accelerated in the process of refraction and bending the lens is often called an *accelerating lens*. The lower potential left hand side of the lens produces a converging action while the higher potential right hand side of the lens produces a diverging action. But since the velocity of electrons is proportional to $\sqrt{E}$ at respective points in the path, the beam has a lower velocity at the low potential field and higher velocity at the high potential field. Hence the converging action of the low potential field is more than the diverging action of the high potential field, with the result of a net converging lens action.

A modified electrostatic lens arrangement of a tetrode type gun, called an *equipotential lens* is more commonly used in most cathode ray tubes. It comprises three coaxial cylindrical electrodes as shown in Fig. 15.1(c). The inner electrode is at a much lower potential E_1 than the outer ones which are usually at the same higher potential E_2. The lens becomes actually unipotential when $E_1 = 0$. The lens system is equivalent to two accelerating lenses formed by (i) E_2/E_1 followed by (ii) E_1/E_2. The lens formed by E_2/E_1 is actually of decelerating type. Hence the electrons pass through the converging lens portion for a longer duration, than through the diverging portion and the net effect is a convergent action as shown by the electron beam paths across

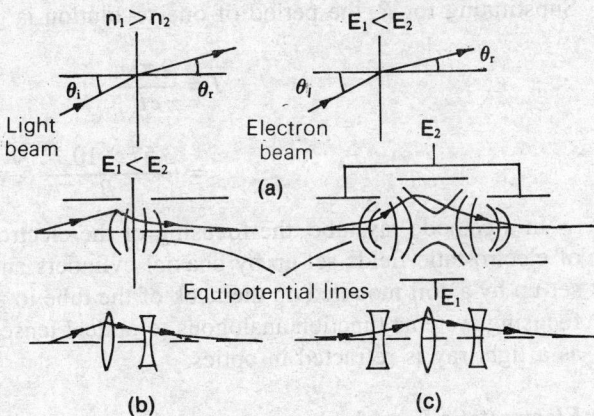

Fig. 15.1 (a) Refraction of light beam at an optical boundary and that of an electron beam at an electrostatic boundary, (b) Accelerating electrostatic lens, (c) Equipotential electrostatic lens

the equipotential lines in the lens system. It may be observed that the presence of a divergent lens in the system does not have any undesirable effect on the focusing, in fact, this reduces spherical aberrations to obtain a sharper focusing as in a compound lens. It is also worth noting that the refraction, the degree of bending and hence the focal

$$v = \sqrt{\frac{2e}{m} E_\alpha} \qquad (15.1)$$

Motion of an electron in a magnetic field An electron moving at a velocity v across a magnetic field B experiences a force F given by:

$$\mathbf{F} = e(-\mathbf{v}) \times \mathbf{B} = e\mathbf{B} \times \mathbf{v} \qquad (15.2)$$

where **F**, **B** and **v** are vectors.

When the velocity v and the magnetic field B are mutually at right angles to each other, the force F is at right angles to both and follows the direction of a right-hand screw whose tip moves in the direction of B. The force $F = Bev$, causes of electron to follow a circular motion with a radius R such that the radial centrifugal force $\dfrac{mv^2}{R} = Bev$.

Hence $$R = \frac{mv}{eB} \qquad (15.3)$$

The angular velocity $\omega = \dfrac{v}{R}$, and the period of a circular revolution

$$T = \frac{2\pi}{\omega} = \frac{2\pi R}{v}.$$

Substituting for R, the period of one revolution is given by

$$T = \frac{2\pi m}{eB} \tag{15.4}$$

$$= \frac{3.57 \times 10^{-11}}{B} \text{ s, where } B \text{ is in webers/} m^2.$$

In a cathode ray tube, the focusing of the electron beam produced at the cathode is done by means of electrostatic fields set-up by coaxial cylinders and diaphrams or by means of electromagnetic fields set-up by a coil mounted on the neck of the tube to produce an axial magnetic field inside the tube. The focusing systems function analogous to optical lenses, bending and refracting the beam in the same way as a light ray is refracted in optics.

Electrostatic Focusing

A pencil beam of electrons crossing a boundary between two spaces having different potentials E_1 and E_2, as shown in Fig. 15.1(a), is refracted in much the same way as refraction of a light ray at the boundary of two media of different refractive indices n_1 and n_2, according to Snell's law: $n_1 \sin \theta_i = n_2 \sin \theta_r$, where θ_i is the angle of incidence and θ_r is the angle of refraction.

In the case of electrons having velocities proportional to the $\sqrt{E}$, as given by Eq. (15.1).

$$\sqrt{E_1} \sin \theta_i = \sqrt{E_2} \sin \theta_r \tag{15.5}$$

Figure 16.1(b) shows a simple electrostatic lens formed by two coaxial cylindrical electrodes having potential E_1 and E_2 with respect to the cathode, length of the system is independent of the charge and mass of the electrons or other charge carriers, and depends only on the ratio of the potentials.

Electromagnetic Focusing

Focusing of an electron beam in a cathode ray tube can also be achieved by an axial magnetic field established parallel to the electron gun axis by means of a coil surrounding the tube. This method was used in earlier TV picture tubes and is commonly used in TV camera tubes.

Electrons are released from the gun cathode with some initial velocities within a small angle of less than 5° with the axis and thus have some axial and some transverse components of velocities, as they are accelerated by the gun anode structure. The transverse component of the electron reacts with the axial magnetic field B perpendicular to it, to exert force at right angles to both the directions, imparting a circular motion to the electron and sends it gyrating or spiralling along the axis because of the simultaneous axial velocity. The gyrating electrons leaving from one point near the cathode all converge at particular points along the path, the number of these convergent points or nodes depend upon the magnitude of the magnetic flux density, which can be varied by adjusting the current in the focusing coil.

In a camera tube, the axial magnetic field strength and the tube wall anode voltages are chosen so that an integral number of gyrating cycloidal loops are produced between the gun aperture and the target surface being scanned.

15.2 ELECTRON GUN

The electron gun produces a thin beam of electrons which strikes the fluorescent screen to cause a luminous spot at the point of impact. The construction of an electron gun used in a 110° electrostatically focused picture tube is shown in Fig. 15.2, with typical electrode potentials.

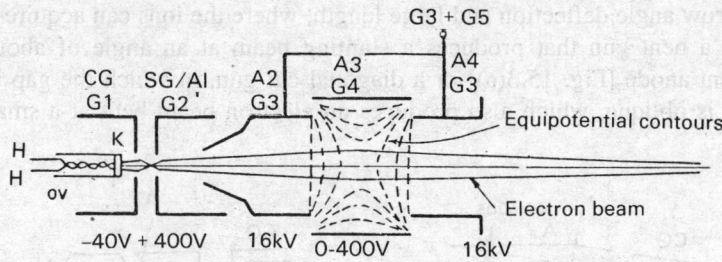

Fig. 15.2   Construction of an electron gun

The electrodes are all cylindrical to produce radially symmetrical fields, and are arranged in line to produce a pencil beam of electron along the axis. The first grid G1 is the control grid CG. Potential difference between control grid CG and cathode K controls the beam current. The video signal with a proper polarity is fed to the cathode while the control grid G1 is given a variable dc potential that serves as the brightness control. The grid G2 serves as the screen grid and is often called the first anode A1.

Grids G1 and G2 have only small apertures on their axis, and with the difference in their potentials they form a strongly convergent electrostatic lens, which is also affected by the cathode potential. The electrons emerging through the hole in G1, are accelerated by + 400 V at G2. They are forced in paths at right angles to the equipotential lines between the electrodes with the result that they converge towards the axis crossing over at a point somewhere between G1 and G2, called the cross-over point, that gives a pre-focusing for the beam. After this, the beam diverges again to be focused on the screen by the main lens formed by G3, G4 and G5. Of these G3 and G5 are at the EHT potential of 16 kV, while G4 is set at a lower potential of 0 to 400 V to give a better fine focus. Moreover, because of symmetrical arrangement, a certain amount of independence of focus from voltage fluctuation and deflection angle is achieved.

The potential on the accelerating grid G3 (also called the accelerating anode) is very high, typically 16 kV or greater so that only a small beam current can produce a given light intensity. If this was to be done by a larger beam current, the potential on G2 will have to be increased. This would increase the grid bias drive requirement and, in addition, because of larger surface area of emissive cathode and increased mutual repulsion amongst the electrons, also increase the beam spot size. The accelerating grid G3 is connected internally to the aquadag coating of the bulb with the result that all the space in the tube beyond this grid is free from electric fields.

In magnetic focusing used in earlier picture tubes, the axial magnetic field produced by the neck mounted coil, deflected the diverging electrons in screw paths with their axes along magnetic flux lines. The electrons thus cut the axis, in nodes along the axis at regular intervals, the position of which can be adjusted to be at the screen.

Ion Trap
Some ions are always present in the electron beam, no matter how carefully the tube is degassed or how well the cathode coating is applied. Negative ions are formed by either gas molecules or molecules of cathode coating acquiring electrons. The negative ions are also accelerated towards the screen. Because they are many times heavier than electrons, they are hardly deflected by the magnetic fields. They hit the screen, therefore, mostly at the centre and can damage the screen phosphors causing a permanent black patch called an *ion burn*.

Ion trapping arrangement is provided in the electrons guns of some picture tubes, particularly in the older tubes of narrow angle deflection and large length, where the ions can acquire large velocities. This consists of either a bent gun that produces a slanting beam at an angle of about 11° to the axis of the tube into a bent anode [Fig. 15.3(a)] or a diagonal cut gun, in which the gap between the first and the second anode is oblique, which also produces an electron beam bent at a small angle as shown in Fig. 15.3(b).

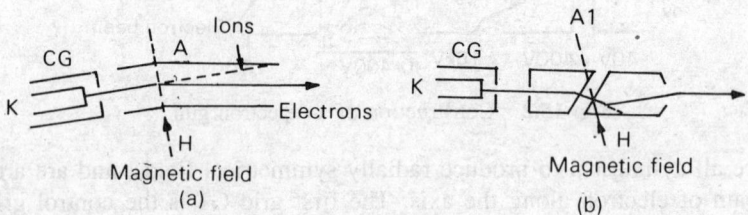

Fig. 15.3 Ion traps: (a) Bent gun, (b) Diagonal cut tilted lens

An external permanent magnet deflects the electron beam along the axial paths while the heavier ions are not much deflected because of their larger mass and are caught by the anode structure.

15.3 LUMINESCENT SCREEN AND FILTERS

The face plate of a picture tube is coated from the inside by a thin layer of phosphors, materials that emit light by excitation of energy from the impinging beam of electrons. This effect called 'luminescence' occurring during excitation, is referred to as 'fluorescence', and that immediately after the excitation is removed, as 'phosphorescence'. The phosphorescence is usually of the same colour as the fluorescence except in a few cases like P7 (Blue/Yellow), P14 (Blue/Yellow-orange) long persistance phosphors, where the luminous radiation occurs at a lower frequency during phosphorescence.

The rate at which the phosphorescent output decreases after removal of excitation is the phosphor delay characteristic or persistence expressed usually as a decay time to 10% level. Phosphors having persistence greater than 100 ms are known as long persistance phosphors, while those with lower values are called medium-short (1 ms to 100 ms) and short (less than 1 ms) phosphors.

The phosphors are composed of inorganic base materials in crystalline forms as the sulphides and oxides of cadmium, zinc, manganese, etc. activated by metals like silver, manganese, copper, etc. They can be formed to give different light emission and persistance characteristics to suit various applications, and are accordingly designated by their type numbers.

In a monochrome or black and white picture tube, P4 phosphor is commonly used. It is formed out of mixtures of blue phosphorescent cadmium tungstate ($CdWO_4$) and yellow phosphorescent zinc sulphide (ZnS) or similar complementary colour phosphors that give out an additive white light with some shade-blue or yellow, if desired. Persistance of phosphor is also a consideration in choice of phosphor for the picture tube. It should be less than the scan time of 1/50 second. Longer persistence will give blurring of fast movements in picture while too short a time will exhibit flicker. Typical value used is 5 ms, a medium-short persistance.

Contrast Problems
The design of the picture tube screen must be such that a good brightness and high contrast image is reproduced. The light emitted by the glowing spot distributes itself in various directions. About 30% only

goes out towards the observer and more than 50% emits back into the picture tube, while the rest is lost in the glass by total internal reflections.

The light internally reflected from the glass-air border falls back on the phosphor to produce a halo ring illumination around the bright spots as shown in Fig. 15.4.

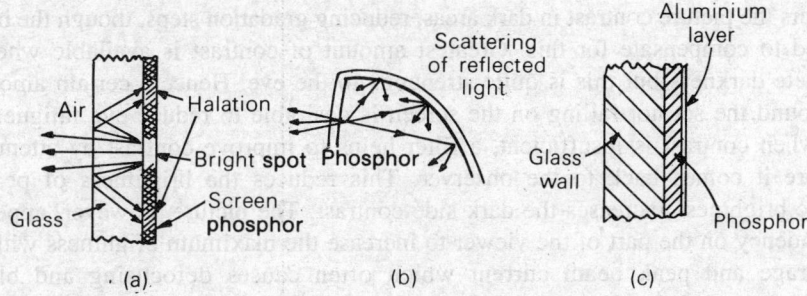

Fig. 15.4 (a) Halation, (b) Internal scattering of light, (c) Metallised screen

This halation effect reduces the detail-contrast ratio by reducing the darkness around the fine bright points or borders. External ambient light falling on the screen also reduces field contrast of the overall picture by illuminating the screen, even at places of total darkness. These effects can be reduced and contrast improved, by use of a filter glass with light gray tinge which attenuates the internally reflected rays and the ambient light more than the direct outcoming glow light as the former rays have to pass through the glass twice while the direct ray passes out only once.

Light emitted back into the tube scatters from the tube interior by reflections from the glass surfaces and also illuminates the screen from the back side. This can be considerably reduced by a metal film backing deposited on the scan side of the phosphor.

Aluminized Screen

A thin (0.1 to 0.14 μm) aluminium layer backing increases the light output from the phosphor by a factor of about 2, by reflecting forward the back-emitted component of light. At high accelerating anode voltages of more than 15 kV, very little of the energy of electrons is wasted in the thin aluminium coating. At lower voltages, however, this loss is predominant, and light gain is less. The aluminium film also protects the phosphor coating from damage due to ions by absorbing their energy. The ions have much lower velocities because of their large mass and hence cannot penetrate through the metal film readily, if the film is adequately thick.

The aluminium backing is conducting and is merged at borders with the internal graphite aquadag coating on the walls of the bulb extending into the reck and connected to the accelerating grid of the electron gun. The charge on the screen due to the impinging electrons is thus conveyed back to the anode EHT circuit. The aluminium coating also improves contrast in the picture by preventing the back-scattered light; but the ambient light front is reflected back towards the front to a greater extent. This must be reduced by use of a neutral filter glass to improve contrast.

Screen Burn

A screen burn at the centre is not necessarily due to the ions. The dark spot burn can also be caused by intense beam of electrons immediately after the TV set is switched off, if the deflection drives disappear earlier than the discharge of the EHT capacitor which has a large time-constant because of the small beam current.

Time-constants of the G1, G2 and the time base supply line circuits must be large enough so that sufficient beam current is maintained to discharge the EHT capacitor before deflection collapses after the set is switched off.

Filters

Ambient light blurs the picture contrast in dark areas, reducing gradation steps, though the basic brightness may be increased to compensate for this. Greatest amount of contrast is available when viewing the picture in complete darkness but this is quite strenuous to the eye. Hence a certain amount of indirect ambient light around the set not falling on the screen is desirable to reduce eye fatigue.

In daylight, when contrast is insufficient, a filter helps to improve contrast by attenuating ambient light twice before it comes back to the observer. This reduces the brightness of peak whites, but lowering of basic brightness increases the dark side contrast. The picture, however, appears less sunny and there is a tendency on the part of the viewer to increase the maximum brightness with a consequent increase of average and peak beam current which often causes defocusing and blurred picture. Hence absorption factor of the filter must not be too great. Actually the requirement of a filter for each ambient light condition is different, for obtaining optimum results. Selective filters with blue or golden yellow shade are sometimes used instead of neutral gray filters. They should be adapted to suit the emission characteristic shade of the picture tube and that of the artificial ambient light at night.

The face plate screen of a picture tube is rectangular in shape in the ratio around 4 : 5 or 1 : 27 rather than the aspect ratio of 3 : 4. About 6% of the picture information in the horizontal direction is thus lost in overscan. This period can be useful in blanking the ringing in the horizontal time base that produces a few black and white vertical bars on the left hand side of the picture. Ringing is the parasitic oscillatory condition in the deflection currents.

Anode Connection

The anode in the tube is also formed by the conductive aquadag coating on the inside wall of the glass bulb extending from the face plate to the accelerator anode spring contacts in the neck. A separate anode connection is provided on the side of the glass bulb as a cavity to accommodate a spring clip of the EHT supply wire. External surface of the bulb is also provided graphite coating for connection to ground with the aid of a wire spring brushing against it. The two coatings, viz. the external grounded coating and the internal anode coating, along with the glass bulb wall as dielectric, form a filter capacitor for the EHT rectifier. The capacitance is typically about 1500–2000 pF with a very high leakage resistance.

15.4   TUBE SIZES AND BULB CONSTRUCTION

The television picture tubes are designated by diagonal size of the screen and the maximum angle of deflection measured across the diagonal. With the advance of tube technology, the tube deflection angles and the screen sizes have increased from the earlier deflection angles of 55°, ..., 70°, 90 to 110°, 114° of the present day tubes, with screen sizes from 28, 31 cm for portable TV sets to 47, 50, 56, ... 61, 66 cm for home sets. The large deflection angles employed, besides reduction in the length of the neck, thanks to modifications in the electron gun, have enabled construction of tubes of smaller length as indicated in Fig. 15.5. This has enabled a compact receiver cabinet design. Reduction of depth of cabinet due to smaller length of tubes have contributed much to the portability of television receiver sets.

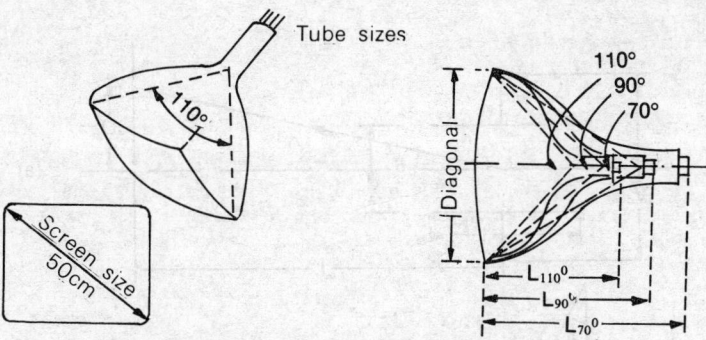

Fig. 15.5 Picture tube sizes and deflection angles

Implosion Protection

A modern picture tube is banded with mounting lugs or is clamped on a metal frame to protect it and reduce the danger of 'implosion', as the television picture tube is completely evacuated of air and hence must withstand the high atmospheric pressure of about 1 kg/cm^2. For safety, the face plate is about 8 mm thick. In most picture tubes, a protective glass panel filter usually made of a gray tint glass is sealed to the face plate. The joining layer between the screen and the tube bulb being most sensitive to breakage, a modern picture tube is banded along this layer with a metal rim band to reduce the danger of 'implosion'. In the case of monopanel TV picture tube, a protecting screen or filter of transparent material should be placed, as such tubes do not have built-in integral protection.

Precautions must be taken by handling the tube with care especially at the neck to prevent the risk of implosion inwards before the tube explodes outwards in splinters. Stresses on the neck and bumps or scratches on the tube must be avoided.

15.5 DEFLECTION OF ELECTRON BEAM

Deflection of electron beam in cathode ray tubes is generally obtained by electrostatic fields produced by horizontal and vertical pairs of deflecting plates in the tube itself. For a television tube, electrostatic deflection is not suitable because of the large angle of deflection necessary for large screen display. For large screen deflection, the deflection voltage requirement for the high accelerating anode potential used, is very high. Magnetic deflection requires sawtooth currents, which are easier to produce and a considerable portion of the energy stored in the field established can be recovered at the end of the scan. In electrostatic deflection, the beam spot suffers greatly from defocusing and astigmatism during large deflection angles. This can be kept small in magnetic deflection. Electrostatic deflection draws no current from the driving circuits and deflection is solely dependent on the applied voltage. Magnetic deflection is very much dependent on current and frequency but in the TV applications the form and frequency of scanning are constant so that they can be taken into account in the circuit design and the energy spent can be partly recovered. Let us analyse the mechanism deflection of an electron beam in electrostatic and magnetic fields as shown in Fig. 15.6.

Electrostatic Deflection

An electron of mass m, charge e and velocity v, entering an electric field produced between two parallel deflecting plates of length l, and spaced distance d apart will be deflected by the field V_d in the y-direction, by a force given by

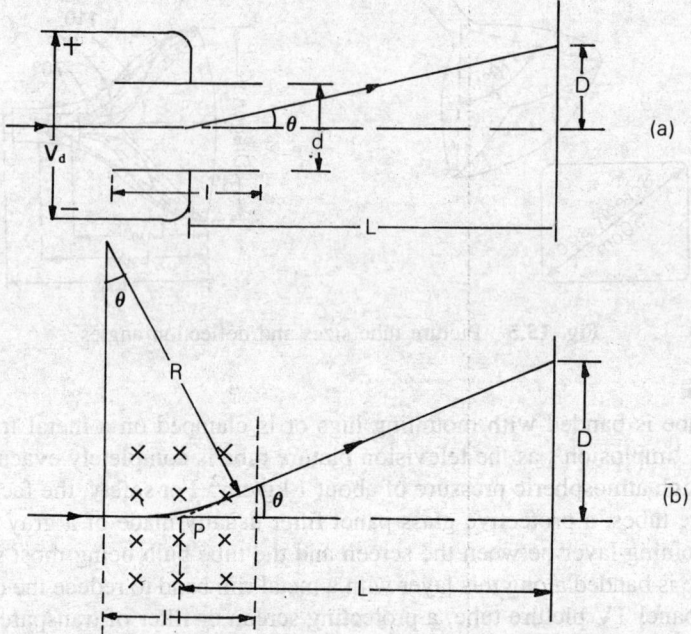

Fig. 15.6 (a) Electrostatic deflection, (b) Magnetic deflection

$$F = e \cdot \frac{V_d}{d} = m\dot{y} \tag{15.6}$$

where $\dot{y}$ is the acceleration it acquires along the y-direction.

The deflection y_l as it comes out of the field will be

$$y_l = \frac{1}{2} \frac{F}{m} t^2, \text{ where } t \text{ is the time taken in the field and,}$$

$$= \frac{1}{2} \frac{eV_d}{md} t^2 \text{ if the initial values of } \dot{y} \text{ and } y \text{ are zero}$$

The value of t depends upon the velocity along the axis and

$$t = \frac{l}{v}, \text{ where the velocity } v \text{ is given by the accelerating potential } E_a \text{ in the relation:}$$

$$\frac{1}{2} mv^2 = eE_a, \text{ i.e. } v^2 = \frac{2eE_a}{m}, \text{ so that } t^2 = \frac{l^2}{v^2} = l^2 \frac{m}{2eE_a}$$

Substituting for t^2,

$$y_l = \frac{1}{2} \frac{eV_d}{md} \frac{l^2}{2} \frac{m}{eE_a} = \frac{1}{2} \frac{V_d}{E_a} l \frac{l/2}{d} \tag{15.7}$$

From Fig. 15.6(a) illustrating the electrostatic deflection,

$$\frac{D}{L} = \tan \theta = \frac{y_l}{l/2} = \frac{1}{2} \frac{V_d}{E_a} \frac{l}{d}$$

Hence the deflection sensitivity is given by:

$$\frac{D}{V_d} = \frac{L}{2E_a} \frac{l}{d} \qquad (15.8)$$

The deflection in electrostatic field is thus independent of the mass of the charge carriers and is inversely proportional to the accelerating EHT potential.

Magnetic Deflection

An electron with velocity v, entering a magnetic field of flux density B at right angles to its path bends around at right angles to both in a circular path of radius R, as shown in Fig. 15.6(b). The radius R is given by:

$$v = \omega R$$

The radial force $\frac{mv^2}{R} = Bev$, hence $R = \frac{mv}{eB}$

For a small deflection angle θ,

$$\frac{D}{L} = \frac{l}{R}, \text{ hence } D = \frac{lL}{R} = lL \cdot \frac{eB}{mv}$$

Now, the velocity of an electron accelerated through a voltage E_a is given by:

$$eE_a = \frac{1}{2} mv^2, \text{ i.e. } v = \sqrt{\frac{2eE_a}{m}}$$

Therefore,

$$D = lLB \frac{e}{m} \sqrt{\frac{m}{2eE_a}}$$

$$= \frac{lLB}{\sqrt{E_a}} \sqrt{\frac{e}{2m}} \qquad (15.9)$$

The deflection in magnetic field thus depends directly upon the length of the magnetic field, the screen distance and varies inversely as the square root of accelerating potential and the charge mass.

15.6 GEOMETRICAL DISTORTIONS IN RASTER

For faithful reproduction of the picture, the raster produced by the deflected beam must be rectangular. The long sides of the rectangle must be horizontal and the picture must possess the same linear relationship as in the original transmitted picture. Various types of distortions do occur, however, in practice. These are: (i) Nonlinear distortion, (ii) Pincushion distortion, (iii) Barrel distortion, and (iv) Trapezoidal or Keystone distortion, as shown in Fig. 15.7.

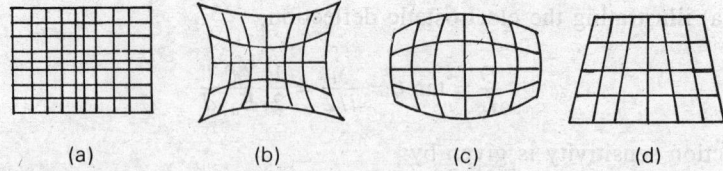

Fig. 15.7 Geometric distortions in the raster: (a) Nonlinear, (b) Pincushion (c) Barrel, and (d) Trapezoidal

(i) Nonlinear distortion This is caused by flatness of the screen. It has been seen that in a uniform magnetic field that begins and ends abruptly over a distance l, the electron beam is deflected along a circular path of radius $R = \dfrac{mv}{eB}$. From Fig. 15.8(a).

$$\sin \theta = \frac{l}{R} = \frac{leB}{mv} = \frac{lB}{E_a} \sqrt{\frac{e}{2m}} \qquad (15.10)$$

For a *spherical* screen with the deflection point P as centre, the displacement of the spot on screen D' is given by $D' = L \sin \theta$, as shown in Fig. 15.8.

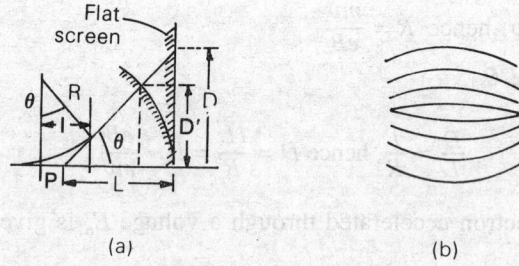

Fig. 15.8 Nonlinear distortion in a picture tube: (a) Effect of flat screen, (b) Barrel shaped field

Deflection D' is thus proportional to the flux density B. If B varies uniformly with time, the spot moves over the screen linearly at a constant velocity.

For a *flat* screen, $D = L \tan \theta$, and the deflection increases more rapidly with B. A cross-hatch equidistant line pattern is then compressed at the centre and spreads out at the edges. This can be prevented by making the deflecting field non-uniform or by increasing it less rapidly with time. The field is made maximum on the axis decreasing towards the edges. This type of barrel-shaped field (Fig.15.8(b)] causes pincushion distortion. Hence it is preferred to vary the field nonlinearly less rapidly with time by passing currents through the deflection coils, which are S-shaped rather than pure sawtooth.

(ii) Pincushion distortion This is also caused by the flatness of screen and its rectangular shape. The distance between the gun and the screen goes on increasing as the beam moves towards the corners of the screen. The deflection is no more proportional to $\sin \theta$ in both horizontal and vertical directions, but to $\tan \theta$ and hence when both deflections are present the corner deflections are greater giving a pincushion appearance to the rectangular raster.

Pincushion distortion can be countered by making the deflection field itself pincushion-shaped. The fields increase towards the edges in the direction of deflection, as shown in Fig. 15.9.

Because the two deflection fields are cushion-shaped making acute angles with each other towards the corner edges, the forces acting on the electron beam at right angles to the fields, make obtuse angle with

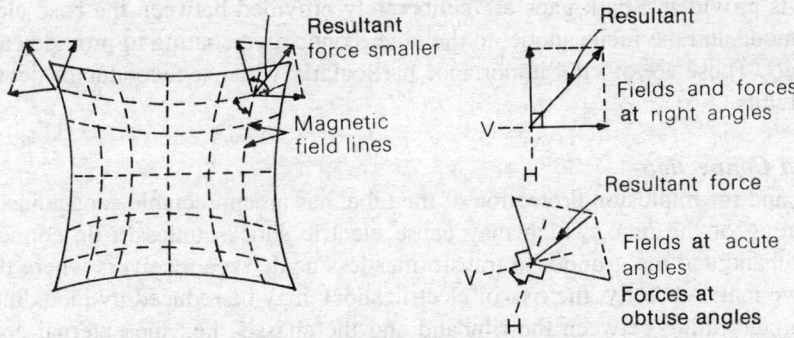

Fig. 15.9 Pincushion shaped fields and their effects on deflection

each other, so that the resultant force is smaller. The electrons are then less deflected in the diagonal direction than they would be in uniform rectangular fields. Towards the edges, the horizontal or vertical deflection fields are also increased in the direction of deflection, so that the deflection increases nearer to the edges. Both these effects overcome the pincushion distortion. The increase of fields towards the edges increases the nonlinear distortion but this can be countered by S-shaped current drive as seen earlier.

(iii) Barrel distortion This is opposite in nature to the pincushion distortion and may be caused by nonuniform fields in both directions.

(iv) Keystone distortion This can occur when both the fields are present and is due to the gradual increase and then decrease of the magnetic field along the axis of symmetry. The assymmetry can arise out of assymmetry in the coil construction, when the inductance and resistance of the coil halves are not exactly equal. The coil halves assembled must be of equal parameters, their windings being kept within strict tolerances.

Centring
In practice, the electron beam in the tube may not enter the deflection field along the tube axis, but at a small angle and the deflection may produce picture raster off the centre and introduce assymmetrical nonlinearity, pincushioning and keystone distortions. This can be due to a number of causes such as eccentricity in construction of the tube or the deflection coils, and external fields such as earth's magnetic fields or fields originating from transformers or loudspeakers in the TV set.

The deflection coils carry two centring magnet-rings that produce a resultant adjustable magnetic field at right angles to the tube axis to deflect the beam and hence the raster in a desired direction.

15.7 PROTECTIVE CIRCUITS

Flash-over protection Because of the very high voltage at the accelerating anode and the internal coating, TV picture tubes are prone to flash-over or arcing, a rather unpredictable phenomenon caused by build-up of high voltages between electrode connected to the EHT capacitor and electrodes terminated in pins on the base of the tube, for example the focus anode. High potentials reaching several kilovolts can appear at the tube base terminals and the resulting transient currents and voltages may be of sufficient magnitude to cause damage to the tube itself and to other components in the drive circuitry, unless spark gap protection is provided. Spark gaps are deliberately provided between the base electrodes, typically from the first anode and the focus anode, to the outer conductive coating to provide easier routes for the transient currents. These are of vital importance particularly when semiconductor devices are employed in the drive circuits.

Metal Rimband Connection

The metal rimband for implosion protection of the tube, has an appreciable capacitance with the internal conductive coating of the tube, which may cause electric shocks unless a dc connection is provided between the rimband and the ground. In transformer less ac/dc type receivers where the chassis can get connected to live mains directly, the risk of electric shock may be reduced by including a 2 MΩ resistor of adequate voltage rating between the rimband and the chassis, i.e., the external coating of the tube. During flash-over, high voltages may, however, be induced on the rimband. To reduce this danger, the 2 MΩ resistor may be by-passed by a 5 nF capacitor rated at a peak voltage determined by the voltage divider formed by this capacitor and the capacitance of the metal rimband to the internal conductive coating of the tube. The capacitor also serves to reduce the radiation from the band.

Prevention of Screen Burns

Care should be taken to prevent permanent damage to the screen phosphors by not operating the tube with stationary pictures at high beam currents or with intense beam spot without deflection drives. Since the time-constant of the EHT circuit is quite high, it is necessary to ensure a rapid discharge of the EHT capacitor by choosing higher discharge time-constants of the $B+$ supplies for the control grid G1 and the first anode G2 of the tube to maintain sufficiently high beam currents, and also for the time base supply lines circuits to maintain deflection for some time after the TV set has been switched-off.

An alternative method is to suppress the beam current on switching-off by providing a negative bias to the control grid from a line output stage rectifier and allowing the EHT capacitor to discharge slowly, while normally the control grid has a superposed positive bias. As the cathode cools down very slowly, a spot may, however, appear on the screen shortly after the switch-off.

15.8 TRANSFER CHARACTERISTICS

The picture tube converts the electrical video signal into light information. The transfer characteristics plotted as light output against electrical signal on log-log scale is not linear but follows a square law. Gamma, i.e., the slope of this characteristic plotted on a log-log scale is around 2.2. The gamma of the TV camera and studio chain is adjusted at 0.45 to 0.5 to make the overall system linear.

The mutual transfer characteristics of a picture tube is plotted as the beam current i_b in μA against the control grid voltage V_{gk}, as shown in Fig. 15.10.

Grid G2 (also called screen grid or first anode) has considerable influence on the beam current. The smaller the V_{G2}, the smaller the beam current, and, consequently, the brightness. The beam current can thus be blanked out by feeding negative blanking pulses of adequate amplitude to G2 also.

The picture tube can be driven by the video signal

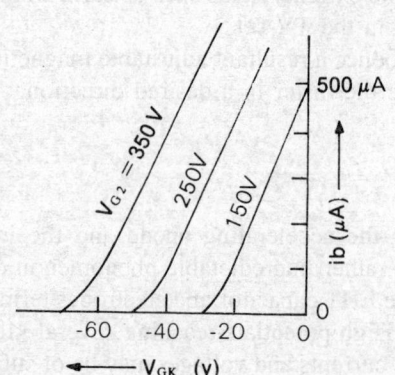

Fig. 15.10 Picture tube transfer characteristics

at the cathode or the control grid. Cathode drive gives a better transfer slope because of the effect of G2 on the beam current. Cathode drive helps to increase the effective cathode to G2 voltage. This means a greater beam current change for a certain video signal drive from the video amplifier.

The cathode drive is also safer from the point of view of the open-circuit failure of the amplifier device using positive supply, as this raises the picture tube cathode to the positive supply voltage to cut-off the beam current. Cathode drive requires a negative modulated (– Y) video signal. Generally, the sync separator and keyed AGC circuits also require negatively modulated (– Y) video signal which can be fed from the video amplifier output.

15.9 TYPICAL MONOCHROME PICTURE TUBE DATA

The following picture tubes are commonly used in modern present day TV receivers: 310 C1 P4 (12″), 470 C1 P4 (19″), 500 C1 P4 (20″), 590 C1 P4 (23″), 610 C1 P4 (24″).

The 19″ and 20″ tubes have the same electrical characteristics and the 23″ and 24″ tubes have the same electrical characteristics. The 20″ and 24″ tubes have a more rectangular screen format and hence are more commonly used.

All these tubes use electrostatic focusing and electromagnetic deflection. The screen employs P4 phosphor and is aluminized. The other salient data for the 310 mm (12″), 500 mm (20″) and 610 mm (24″) tubes are given in Table 15.1.

Table 15.1 Typical Monochrome Picture Tube Data (*Courtesy*: BEL)

Type	310 C1 P4 (12″)	500 C1 P4 (20″)	610 C1 P4 (24″)
Deflection angle	110°	114°	110°
Filament voltage	12 V	6.3 V	6.3 V
Heater current	75 mA	300 mA	300 mA
Acc. anode voltage (V_a)	10 kV	16 kV	18 kV
Focusing anode voltage *(V_{G4})	0–400 V	0–400 V	0–400 V
Cut-off voltage[†] (V_{G1})	– 33 to – 77 V	– 41 to – 99 V	– 41 to – 99 V
Dimensions length	240 mm	311 mm	362 mm
Diagonal	312 mm	505 mm	613 mm
Width	276 mm	428 mm	496 mm
Height	224 mm	344 mm	392 mm

Notes: [†]Visual extinction of undeflected focused spot.

*Individual tubes have satisfactory focus at some value between 0 to 400 V.

In a picture tube, deflection of electron beam for scanning purposes is achieved by two sets of deflection coils. These form an external accessary to the picture tube, called 'deflection unit' or 'yoke' which is mounted on its neck to produce vertical and horizontal magnetic fields changing with respect to time to deflect the electron beam in the horizontal and vertical direction at the required rate. Although the electrons in the beam do not take up any energy from the magnetic field, the repetitive build-up of magnetic fields in the volume of the coils requires energy. This can be recovered, but not without unavoidable losses. The energy required depends upon the coil dimensions and the picture tube voltage and its maximum deflection angle. The design of coils, the dimensions and winding distribution must also minimize geometrical distortions in the raster and beam with proper field distribution.

15.10 DEFLECTION ENERGY REQUIREMENT

Consider a uniform deflection field of strength H_0 perpendicular to the tube axis and extending over a length l_0 along the tube axis z, as shown in Fig. 15.11.

The peak deflection energy required to deflect the electron beam through an angle θ is given by:

$$W_m = \frac{a^2 \sin^2 \theta \times E_a}{0.7200 \, l_0} \text{ ergs}$$

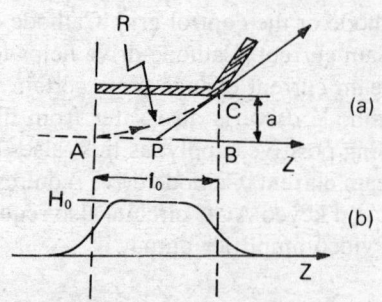

Fig. 15.11 Magnetic deflection of electron beam

This can be derived as follows: In practice, the distribution of the field along the axis will be gradual rather than abrupt, as shown in Fig. 15.11. The 'effective length' of the coil will then be given by:

$$l_0 = \left\{ \int_{-\infty}^{+\infty} H \, dz \right\} \Big/ H_0 \qquad (15.10)$$

over which H_0 can be assumed constant. The maximum permissible effective length of the coil can be found by geometry under the condition of a deflected beam avoiding the neck shadow due to corner cutting. From Fig. 15.11 we have:

$$l_0 = AP + PB = PC + PB = \frac{a}{\sin \theta} + \frac{a}{\tan \theta} = \frac{a}{\tan (\theta/2)}$$

An electron beam of velocity v passing through the field H_0 is deflected along a circular path of radius $R = \dfrac{mv}{eH_0}$; also, $\dfrac{l_0}{R} = \sin \theta$, so that $\sin \theta = \dfrac{l_0 e H_0}{mv}$. Substituting for v in terms of E_a, we have

$$\sin \theta = \frac{l_0 e H_0}{m} \sqrt{\frac{m}{2eE_a}}, \quad Q \quad \left(\frac{1}{2} mv^2 = eE_a \right)$$

$$= \sqrt{\frac{e}{2m}} \frac{l_0 H_0}{\sqrt{E_a}}$$

$$= \frac{0.3 \, l_0 \, H_0}{\sqrt{E_a}},$$

where l_0 is in cm, H_0 in oersteds and E_a in volts.

$$\therefore \qquad H_0 = \frac{\sin \theta}{0.3 \, l_0} \sqrt{E_a} \qquad (15.11)$$

Now the peak magnetic energy of field H_0 per unit volume is given by:

$$W_m = \frac{\mu H_0^2}{2 \times 4\pi} = \frac{H_0^2}{2 \times 4\pi}, \text{ for air, and } H_0 \text{ in Oersteads.}$$

The total energy in the magnetic field W is given by:

$$W = \frac{H_0^2}{2} \times \frac{\text{Volume}}{4\pi}, \text{ where volume} = \pi a^2 l_0$$

$$W = \frac{\sin^2 \theta \times E_a \times \pi a^2 \times l_0}{0.09 \, l_0^2 \times 8\pi} = \frac{a^2 l_0 \sin^2 \theta}{0.09 \, l_0^2 \times 8}$$

$$= \frac{a^2 \sin^2 \theta \times E_a}{0.72 \, l_0} \text{ ergs}$$

$$= \frac{a^2 \sin^2 \theta \times E_a}{7200 \, l_0} \text{ mJ} \tag{15.12}$$

Energy W per kV of accelerating potential is given by

$$\frac{W}{E_{a \, (\text{in kV})}} = \frac{a \sin^2 \theta \times \tan (\theta/2)}{7.2} \text{ mJ, because,} \left[\frac{a}{l_0} = \tan \left(\frac{\theta}{2} \right) \right] \tag{15.13}$$

The equation for energy indicates that the energy requirement for deflection is larger, for larger deflection angles. Modern 110–114° tubes require much larger energy than the earlier 70° or 90° tubes. For a particular tube, the energy depends upon the EHT used.

The energy requirement is actually greater than this value because of the finite size of the electron beam of radius of about 1 to 2 mm, at the deflection point. To allow for this, the useful deflection radius of the neck is reduced by the beam radius r, i.e., $(a - r)$, in place of a, in the denominator of the expression for W. The energy of the field is not limited to the cylindrical space of diameter $2a$, but to a space with diameter $2(a + b)$, where $2b$ is the increase in the diameter of the coil field due to the thickness of the tube glass walls and the thickness of the coil. The expression then modifies to

$$W \text{ per kV} = \frac{(a + b)^2 \sin^2 \theta \cdot \tan (\theta/2)}{7.2 \, (a - r)} \text{ mJ} \tag{15.14}$$

The thickness b has considerable influence on the energy. A close tolerance on its variations can save a considerable amount of energy.

For large deflection angle tubes, a good deal of saving of energy is possible by use of flared deflection coils, with their flaps lying against the conical part of the bulb instead of mounting them up to the neck only. By doing so, the effective length of the coil can be kept larger while keeping the centre of deflection at the same place, thus avoiding neck shadow.

15.11 AMPERE-TURNS REQUIREMENT

The magnetic field strength requirement for obtaining deflection of the beam through an angle θ, by a coil of effective length l_0, has been seen in the last section as:

$$H_0 = \frac{\sin \theta}{0.3 \, l_0} \sqrt{E_a} \text{ Oe}$$

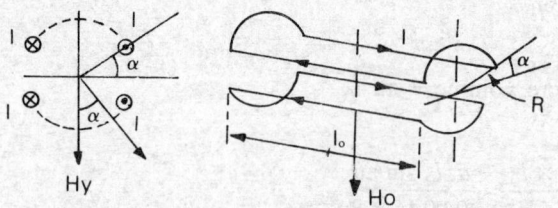

Fig. 15.12 Two saddle coils producing vertical magnetic field

The number of ampere-turns required to produce this field strength can be calculated, for two simple air-cored rectangular coils bent around the neck along the cylindrical plane of radius R as shown in Fig. 15.12.

The contribution made by the curved sides of the coils is small compared to that by the straight sides, which may be the contribution of each straight conductor to the field strength at the axis. This is obtained from the relation:

$\int H \cdot dl = 0.4\,\pi I$. Integrating along a circle whose centre is on the conductor, $H. 2\pi R = 0.4\pi I$.

$\therefore\ H = 0.2\ I/R$. Hence the field strength along the vertical axis due to each of the four conductors in a single turn is given by: $H_v = \dfrac{0.2\,I}{R}\cos\alpha$, where α is the angle of the conductor with the horizontal axis.

The total vertical magnetic field due to n number of turns is:

$$H_0 = \frac{4 \times 0.2\,nI}{R}\cos\alpha \tag{15.15}$$

For a more or less uniform field, α has to be around 30°. A value of α greater than 30° gives a barrel-shaped field, while a value of α less than 30° gives a pincushion-shaped field. For $\alpha = 30°$,

$$\frac{0.8\cos 30° \times nI}{R} = H_0 = \frac{\sin\theta \times \sqrt{E_a}}{0.3\,l_0}$$

$$nI = 4.8\,\frac{R}{l_0}\sin\theta\,\sqrt{E_a} \tag{15.16}$$

This applies to air-cored coils. For coils with ferrite core, approximately half this number of ampere-turns is sufficient. The required number of ampere-turns is larger than this because of finite l_0 and the effect of curved sides and some reserve against corner cutting that may arise out of tube and coil manufacturing tolerances. This makes it possible to move the nominal coil and the deflection point a few mm back along the nominal tube axis, before corner cutting takes place. This means some extra ampere-turns and somewhat shorter l_0. The expressions provide a starting point from which the actual required number of ampere-turns can be determined empirically. The number of turns will depend upon the current the driving circuit can deliver safely. The gauge of wire to be used is a compromise for low resistance and space availability.

A deflection coil has a self-inductance L and winding resistance r. If the coil requires peak currents of $+ I_m$ and $- I_m$ to deflect it to the maximum deflection angle, the maximum magnetic energy stored in the coil is:

$$W = \frac{1}{2} L I_m{}^2 \tag{15.17}$$

The deflecting currents rises in a sawtooth form from $- I_m$ to $+ I_m$ collapsing back to $- I_m$, at the scan rate frequency f. The power required will, therefore, be equal to $\frac{1}{2} LI_m^2 f$. Out of this, there will be a resistive loss of $\frac{1}{12} r I_m^2$ due to sawtooth currents flowing in the resistance.

15.12 DEFLECTION DRIVE VOLTAGES

The power requirement of horizontal deflection coils is considerable, of the order of about 15 to 40 VA; e.g., for a coil of 2 mH requiring 2 A for peak deflection, the power required is:

$$\frac{1}{2} LI_{m^2} f = \frac{1}{2} \times 2\,m \times 2^2 \times 15{,}625 = 62\text{ VA/s.}$$

The deflection circuit, therefore, must be so designed that the energy stored at the end of each scan is recovered. The resistance loss here should be minimum. The resistance of coils can also cause nonlinearity unless the voltage drive to the coil is corrected. For linear deflection, the coils must carry linear sawtooth current as shown in Fig. 15.13(a), as represented by the equation:

$$t = \frac{I_{pp}}{2}\left(\frac{2t}{t_s} - 1\right) \tag{15.18}$$

If the inductance only is considered, the coil requires a voltage given by:

$$v_L = L\frac{di}{dt} = L\frac{2I_m}{t_s} = L\frac{I_{pp}}{t_s}$$

Because of the sawtooth current in the coil resistance r, a corresponding voltage drop $i \cdot r$ must be provided for. Hence the total voltage requirement is given by:

$$v = v_L + v_r = \frac{LI_{pp}}{t_s} + r\frac{I_{pp}}{2}\left(\frac{2t}{t_s} - 1\right) \tag{15.19}$$

as shown in Fig. 15.13(b). The voltage is thus a sum of a step and a ramp voltage having a trapezoidal waveform.

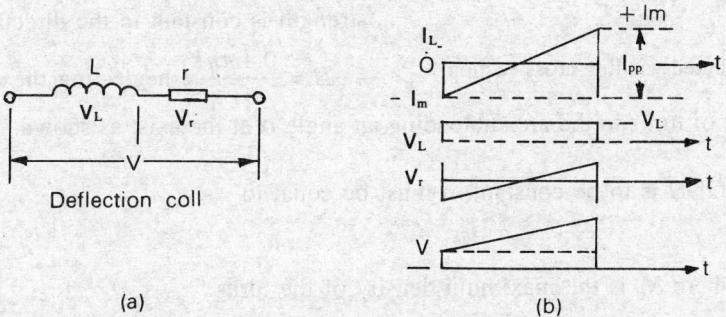

Fig. 15.13 Deflection drive voltage across a coil: (a) Deflection coil, (b) Drive waveforms

In practice, a more or less constant voltage, a little larger than that at the end of the scan, is applied. The difference between this voltage and the required voltage v is taken up by a nonlinear inductance called the linearity coil that shows high impedance at the beginning of the scan and low impedance at the end of the scan.

Vertical Deflection

As the vertical scanning frequency is low (50 Hz only), the reactive power in the vertical deflection coils is negligible compared to the resistive losses which are then significant though low, a fraction of a watt in practice.

The sawtooth current through the coil must be:

$$i_v = \frac{I_{pp}}{2} \left(\frac{2t}{t_v} - 1 \right)$$

where t_v is the vertical scan period and I_{pp} is the peak-to-peak deflection current.

The drive voltage for the coil is:

$$V_r = r_v \cdot i_v = r_v \frac{I_{pp}}{2} \left(\frac{2t}{t_v} - 1 \right)$$

If the coil inductance of the coil is taken into account, a constant voltage $v_L = \dfrac{L_v \cdot I_{pp}}{t_v}$ must be added to V_r.

15.13 WINDING DISTRIBUTION OF COILS

The present day wide angle deflection tubes require coils that are flared over the conical portion of the tube and have to produce field to overcome the geometrical distortions in the raster, due to the flat face of the tube. The windings of the coils are for this purpose distributed along the periphery of the core in a plane perpendicular to the axis as shown in Fig. 15.14, where the coil's cross-section is shown shaded.

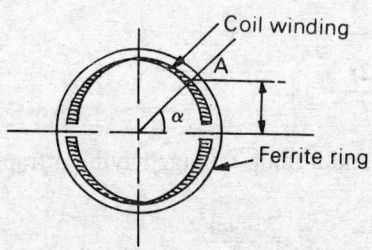

If the distribution of turns is such that the field strength is constant in the direction of deflection,

$H = \dfrac{0.4 \, \pi n I}{l}$, neglecting the core susceptance.

Fig. 15.14   Distributed winding cross-section

Here n is the number of turns in the arc subtending an angle α at the axis, as shown. Since $l = R \sin \alpha$,

$H = \dfrac{0.4 \, \pi n I}{R \sin \alpha}$, and if H is to be constant, n must be equal to

$$\int N_0 \cos \alpha \, d\alpha,$$ where N_0 is the maximum density of the turns.

A cosine distribution of the turns along the periphery of the core thus gives a uniform field. In general, n can be $N_0 \cos^p \alpha$. If $p > 1$, the density of winding decreases more rapidly than with $\cos \alpha$, and the resultant field is pincushion shaped, while if $p < 1$ a barrel-shaped field is produced.

15.14 CONSTRUCTION OF DEFLECTION COIL UNIT

The deflection coils are constructed in two forms: (i) saddle coils (ii) toroidal coils. Saddle coils are used mainly for the horizontal deflection while the toroidal construction coils are restricted for use in vertical deflection. The coils are mounted together in a deflection unit assembly that carries other accessories like pincushion magnets, centring rings, etc.

(i) Saddle coils These coils have the shape of a saddle as shown in Fig. 15.15(a). One end of the coils is flared, i.e., bent conically so that it can lie against the conical neck portion of the tube. This reduces the physical coil length minimizing deflection energy. The coils are completely enclosed by a ferrite core of negligible susceptance so that the ampere-turns required are minimum. As the ends of the coils are rather extended to bend round, the windings have rather a long average turn and hence a higher *R/L* ratio. Because of the ferrite core enclosing the coils the stray fields are, however, minimum and the coils are, therefore, suitable for horizontal deflection where the magnetic energy is a more important consideration.

The coils are wound on a metallic mandrel with a special self-bonding copper wire coated with thermoplastic material. A heavy current of short duration is passed through the coil so that the heat melts the plastic coating which bonds the coil turns together into a saddle-shaped coil after cooling.

(ii) Toroidal coils These are wound on the opposite sides of the ferrite core toroidal ring in a distributed manner as shown in Fig. 15.15.

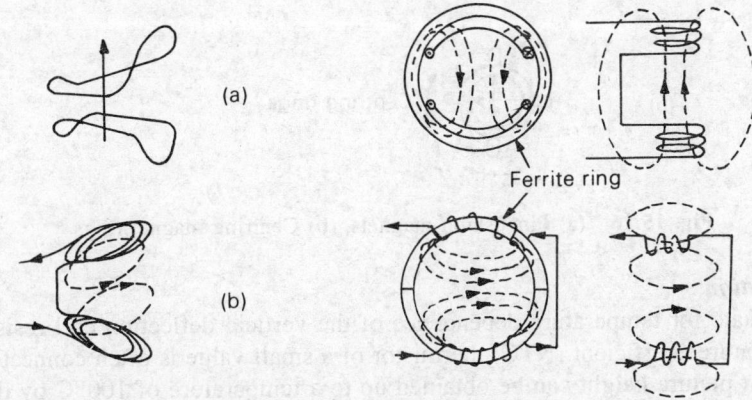

Ferrite ring

Fig. 15.15 (a) Saddle coils, (b) Toroidal coils

These coils produce a horizontal magnetic flux inside the ring because of the opposing directions of flux produced by the coils in the ferrite core. A lot of the flux leaks around the outside of the deflection unit and hence the construction is not suited for horizontal deflection requirements.

The average length of the turn is, however, smaller because of the direct winding on the core and hence have a low *R/L* ratio. This makes them suitable for vertical deflection where the coil impedance is primarily resistive, and the resistive losses are of primary importance.

Pincushion Correction Magnets

In the design of the coils, the pincushion distortion of the raster is prevented by making the fields themselves pincushion shaped. But if the pincushioning of the field is too strong, the beam spots suffer astigmatic distortion. Hence the pincushion character of the fields is not made too strong so that some pincushion distortion remains. This residual distortion is prevented by means of four permanent magnet rods or toroidal rings, attached on the sides of the deflection unit. These magnets give the electron beam an extra horizontal or vertical deflection field that the curved sides of the raster are pulled straight, as shown in Fig. 15.16(a). Often only two magnets fitted on the short sides of the cone of the tube are used, which correct the vertical pincushion distortion.

Centring Magnets

The eccentricity of the picture tube and the deflection unit can be corrected by means of two separate adjustable centring magnets of plastic-bonded ferrite magnets of ring shape, magnetized diametrically. By turning the rings, the resultant magnetic field at right angles to the beam, can be adjusted to remove the eccentricity. These are shown in Fig. 15.16(b).

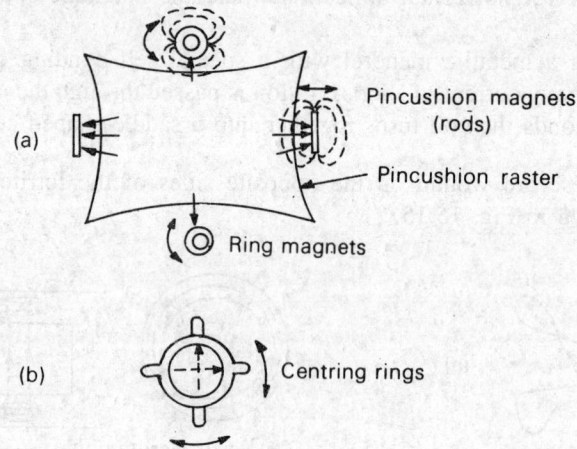

Fig. 15.16 (a) Pincushion magnets, (b) Centring magnet rings

Thermal Compensation

In order to compensate for temperature dependence of the vertical deflection coil resistance, a built-in the negative temperature coefficient (NTC) thermistor of a small value is often connected in series with the coils. A constant picture height can be obtained up to a temperature of 100°C by this arrangement, with voltage feedback in the drive circuit to linearize the sawtooth voltage.

15.15 TYPICAL DEFLECTION COIL DATA

Electrical data of some commonly used deflection coil units are given in Table 15.2.

Table 15.2 Deflection Coil Unit Data (Philips, BEL)

Type No.	Line coils			Frame coils		
	I_{pp} (A)	L_H (mH)	R_H (ohm)	I_{pp} (A)	L_V (mH)	R_V (ohm)
AT 1030 (Tube-circuits)	2.29	2.9	4.6	0.44	82	38 + 18 (NTC)
AT 1040 (Tube/transistor)	2.82	2.1	3.9	0.545	66	29 + 33 (NTC) 30 + 10 (NTC)
AT 1040/05 (Tube/transistor)	2.82	2.1	3.9	0.545	10.2	4.5
BEL DU 43 (Tube circuits)	2.8	2.1	3.9	0.55	66	30 + 10 (NTC)
AT 1040/15 (Transistor circuits)	2.3	3.3	5.7	1.1	16.5	7.5
AT 1072 (Transistor, 31 cm pic. tube)	8.2	0.107	0.2	—	29.9	6.4

REVIEW QUESTIONS

15.1 Explain the construction of a black and white picture tube, indicating the function of each electrode.

15.2 Explain in the following terms with respect to the picture tube: Screen size, Deflection angle, Corner cutting, Screen burn.

15.3 Write notes on the following:
(a) Ion trap, (b) Screen filters, and (c) Implosion protection.

15.4 Name the various geometrical distortions in the picture tube deflection and explain the methods to eliminate these.

15.5 How does the aluminized screen help improve the performance of the tube and increase its life?

15.6 Explain the methods of protection of the picture against screen burn and flash-over.

15.7 If the accelerating anode potential in a cathode ray tube falls by 20%, other things remaining constant, what will be the effect on the deflection of the electron beam in the case of (a) electrostatic deflection, (b) magnetic deflection.

15.8 A cathode ray tube with an accelerating potential of 2 kV has deflection plates of length 2 cm, with a distance of separation between them equal to 0.5 cm. Find the deflection of the electron beam on the screen at a distance of 20 cm from the deflection plates.

15.9 In a picture tube the magnetic deflection unit creates a uniform magnetic field extending over a distance $l = 5$ cm to deflect the electron beam accelerated through a potential of 10 kV, by 10 cm along a flat screen at a distance of 20 cm from the deflection centre. Find the flux density of the magnetic field required. What will be the changes in flux density if the deflection is on a curved screen with a radius of 70 cm?

15.10 Deduce the expression for energy requirement for magnetic deflection of an electron beam accelerated by potential E_a in a TV picture tube and deflected through an angle θ by a magnetic field extending over length l_0 in the neck of radius a.

15.11 Find the ampere-turns required for obtaining the deflection through the angle θ, by a saddle coil of length l_0 and saddle radius R.

15.12 Discuss the effect of distribution of winding turns of the deflection coils on the magnetic field and geometric distortions.

15.13 Give reasons for the following:
   (a) Saddle coil construction is used for horizontal deflection while toroidal construction is used for vertical deflection coils.
   (b) A thermistor is connected in series with some vertical deflection coils.
   (c) Cosine distribution of the winding turns is used along the periphery of the deflection coils of the saddle type.
15.14 What are the adjustments provided on a deflection coil unit?

MULTIPLE-CHOICE QUESTIONS

15.1 Pick the wrong statement:
   (a) Modern picture tubes use the accelerating lens system
   (b) Instant-on tubes have specially constructed cathodes
   (c) Aluminized screen does not need an iron trap magnet
   (d) Filters are necessary to increase contrast in the day time.
15.2 In a picture tube with electromagnetic deflection, the deflection is proportional to:
   (a) accelerating anode potential E_a
   (b) inversely proportional to E_a
   (c) inversely proportional to the square of E_a
   (d) inversely proportional to the square root of E_a
15.3 A Pincushion raster is produced due to:
   (a) flatness of the screen
   (b) barrel shaped field
   (c) pincushion shaped field
   (d) (a) and (b)
15.4 Screen burn is prevented by:
   (a) making the deflection supply rectifier time constant large
   (b) making the picture tube grid and screen supply time constants large
   (c) increasing the EHT rectifier time constant
   (d) none of these.
15.5 Weak emission from the picture tube cathode causes:
   (a) dark picture
   (b) low contrast
   (c) silvery gray picture
   (d) (b) and (c)
15.6 Deflection energy required for a picture tube is proportional to:
   (a) neck diameter
   (b) square of the neck diameter
   (c) length of the tube
   (d) none of these.
15.7 Deflection coils are flared in order to:
   (a) reduce the length of the tube
   (b) reduce the deflection energy required
   (c) avoid neck shadow
   (d) (b) and (c)
15.8 Under uniform magnetic field, the ampere-turns required are proportional to:
   (a) deflection angle θ
   (b) $\sin \theta$
   (c) $\tan \theta$
   (d) none of these.

15.9 Pick the correct statement:
    (a) Saddle coils have higher R/L ratio than toroidal coils
    (b) Toroidal coils minimize leakage flux outside the deflection unit
    (c) Toroidal coils are more suitable for vertical deflection
    (d) The average length per turn of toroidal coils is larger.

15.10 Centring magnets consist of:
    (a) two permanent magnet rings magnetized diametrically
    (b) two rings magnetized radially
    (c) two permanent magnetic rods on the side of the yoke
    (d) four magnets on the four sides of the raster.

16

Colour Picture Display Devices

Colour picture display devices are based on the principle of *additive impression of colours* on the eye. This is achieved either by light from Red-Green-Blue dot trios on a screen, or by projecting three red-green-blue pictures on a white screen which reflects all the three colour images superimposed, to give combined impression. Colour display devices include colour CRTs, flat panel displays like LCD, plasma LED and projection systems using three CRTs or three LCD panels. Large flat panel displays are not expected to seriously rival the CRT until the mid '90s. With its low cost, multicolour, brightness, high resolution, and gray scale capability, the colour picture tube has a dominant position in television. It has no serious contender yet, except where its depth, weight or power limitations preclude its use. Pocket LCD TV, laptop computers and portable medical instruments now use LCD and plasma displays, as their capabilities have immensely improved.

A colour picture tube has minute RGB phosphor dots or stripes coated on the face of the picture tube. The phosphor dots are excited by three electron beams that are intensity-modulated with corresponding RGB signals. The three beams simultaneously scan the phosphor screen aligning themselves on the respective colour phosphors during the entire scanning process. This is achieved with the help of a shadow mask of a thin perforated steel plate, mounted a small distance behind the screen. The colour picture tube based on this principle was first developed by Dr Goldsmith and his team in 1950, at the RCA laboratories in the USA. The tube was called *shadow mask tube*. The tube embodied all the basic features of a modern colour picture tube. The shadow mask tube evolved into many derivatives that improved upon its low beam efficiency and critical convergence adjustments. These include the single electron gun Trinitron (1968) with inline cathodes by Sony Corporation and the Precision In-line (PIL) tubes with the three separate electron guns precisely aligned horizontally to illuminate vertical strips of phosphor through slotted mask. These include 20 AX (1974), 30 AX (1979) and full square hi-bri tubes by Philips, and other PIL tubes by ITT and other manufacturers which are now in common use.

16.1 SHADOW MASK TUBE

In the shadow mask tube widely used in earlier colour receivers, the screen has over 300,000 closely spaced *triads* of red, green and blue phosphor dots of about 0.42 mm diameter each spaced some 0.72 mm apart triangularly. A delta-gun assembly consisting of three separate electron guns positioned 120° apart in the rear neck of the tube, generates three high velocity (25 kV acceleration) electron beams. The guns are slightly tilted towards the centre line of the tube to make the beam converge near the face plate at the apertures in the mask, and fall on the respective colour phosphor dots on the face plate screen, some 9 mm away from the aperture mask as shown in Fig. 16.1.

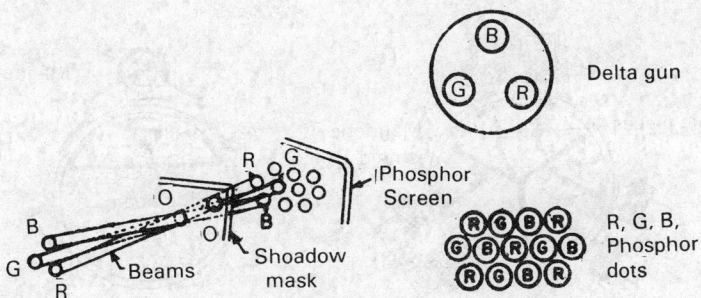

Fig. 16.1   Shadow mask tube principle

There is one aperture hole for each phosphor triad and the converged beams move over the mask, momentarily illuminating a phosphor trio as they pass through each aperture hole. The beams are thus blanked or masked out while moving between the apertures. This is an essential function of the shadow mask, in the absence of which the beams would excite all the dots in sequence, mixing up the colour reproduction. The shadow mask thus blocks an appreciable portion of the beam current dissipating several watts of power. The phosphor screen gets less than 20% of the beam current reducing the brightness capabilities of the colour tube. This is why colour picture tubes use higher anode voltage and need larger beam and cathode currents than monochrome tubes.

The three electron guns have separate cathodes and control grids so that individual colour signals can be coupled to the grid cathode circuit to modulate beam intensity. Each gun has a separate screen grid, the voltage at which is adjusted for the desired cut-off characteristics for each gun. The accelerating anodes are supplied EHT of about 25 kV, while the focusing grids are provided an adjustable potential of about 5 to 7 kV for optimum focus.

In front of the delta gun assembly structure is located the convergence electrode assembly for the three guns. This includes a magnetic shield that allows each beam to be shifted in position without affecting others, with the help of the internal pole pieces as shown in Fig. 16.2(a). They are so shaped that they concentrate the magnetic field produced by the external convergence yoke coils, across the beams. This enables to shift them radially, in order to make them converge through the apertures in the mask. For each beam there is a small rotatable permanent magnet and a coil on a U-shaped ferrite core.

The permanent ring magnets are adjusted for static convergence in the central area of the screen. The radial adjustment is not enough to ensure convergence of all the three beams at a common point; it is necessary to be able to control at least one beam in a second direction as well.

For this purpose, a fourth magnet called the blue lateral magnet is used as shown in Fig. 16.2(b). The blue beam is made to pass across the radial flux produced by the blue lateral magnet and its pole pieces to move it laterally left or right, when the blue beam is at the top of the tube.

Apart from the static convergence at the centre, it is necessary to ensure dynamic convergence when the beams are deflected to form a raster. As the three guns are tilted slightly towards the tube axis and the practical TV screens are made almost flat, each beam produces its own slightly displaced patterns of pincushion distortion. Dynamic convergence at the other edges of the screen during the scanning process can be provided by correction currents in the convergence coils. For this purpose two types of *dynamic magnetic magnetic fields* are used. One is produced by additional winding on the yoke ring of the deflection yoke, driven by adjustable sawtooth currents. The other is generated by sawtooth and parabolic currents synchronised with the scanning currents. Purity ring magnets, which are essentially centering magnets, enable adjustment of purity of single colours, red, green and blue separately, by cutting out two of them at a time, starting usually with red purity first.

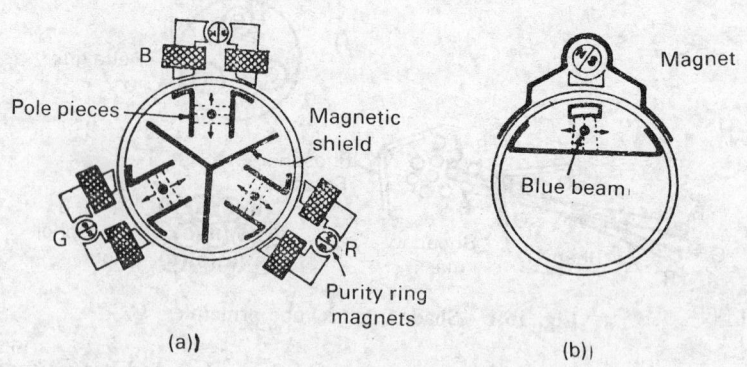

Fig. 16.2   (a) Convergence assembly, (b) Blue lateral magnet

The conventional shadow mask tube with the three guns arranged in delta fashion had the best possible ratio of gun-to-neck diameter, allowing for maximum diameter of the focus anodes of the three guns within the neck for better focusing. However the tube suffered from a number of disadvantages, as discussed in the following.

The shadow mask absorbed over 80% of the beam currents amounting to heat dissipation of over 20 W. The beam convergence was a complex process requiring skill and experience to adjust over a dozen controls. The focus and convergence were again compromise adjustments, if one were to try to achieve them over the entire screen area. These *disadvantages* are overcome to a large extent in in-line beam construction employing vertical stripes of phosphors.

16.2   COLOUR PHOSPHOR SCREEN MANUFACTURE

For correct registration of colours the alignment of phosphor dots or strips with the apertures in the mask is very critical, it being essential that errors in the shadow mask pattern not affect colour reproduction in the tube. This is ensured by forming the *phosphors in alignment with the apertures* in the mask with the help of the ultraviolet source used for fixing them. The face plate of the screen is initially coated with green phosphor and photoresist on top of it. The shadow mask is accurately positioned relative to the faceplate and an ultraviolet light originating from the deflection centre position of green gun beam is shone through the mask on the coating. This hardens the dot or strip areas exposed to ultraviolet light. The remaining phosphor is then etched away and a coating of blue phosphor with photo resist on top of it is applied to the screen. The ultraviolet light source is then moved to the blue gun deflection centre position and shone through the mask on the screen to harden the areas of blue phosphor dots or strips. The red dots or strips are similarly hardened in accurate positions. The aluminizing thin metallic layer is then applied and the faceplate along with the mask is sealed to the glass cone.

For increasing brightness and contrast ratio, various refinements have been carried out by manufacturers. Improved gun design, higher EHT and screen grid voltage reduce the beam spot size increasing the striking electron density. Very sharp focus can be obtained by modern high bi-potential or tri-potential lens tubes. With EHT of 25 kV, this calls for a focus anode voltage of 6 to 8 kV, a value at 28 to 31% of the EHT voltage.

X-ray radiation emitted from the colour picture tube should not exceed 0.5 mR/h. This is ensured by the manufacturers for operation below design maximum ratings. The maximum anode voltage at which the x-ray radiation does not exceed 0.5 mR/h at an anode current of 0.1 mA is of the order of 35 kV. In large screen tubes having EHT of this order, a safety circuit to disable the line oscillator in case the limit of 30 kV is exceeded is essentially incorporated in the receiver.

Colour Phosphors

Generally used phosphors for the three colours are zinc sulphide for red, zinc silicate for green and europium and yttrium for red. Their compositions are chosen according to colour temperature desired and may show a little variation in colour shades. The three phosphors are together designed as P22 and have medium short persistance, about 1 ms. The phosphor materials used are pigmented and dark coloured to absorb ambient light falling on the screen and reduce reflection. The space left between the phosphor dots or strips is also filled with a light absorbing black pigment serving as black mask. For striped phosphors these form black stripes in between the vertical phosphor. This leaves some guard band for beam landing errors, besides improving contrast.

16.3   TRINITRON PICTURE TUBE

The Sony Corporation of Japan developed the trinitron tube in 1968, employing, for the first time, in-line beam construction, as shown in Fig. 16.3. The tube has just a single gun structure with three separate in-line cathodes for each of the red, green and blue beams. The shadow mask consists of a vertically slotted aperture grille etched out of a metal sheet and the phosphor screen is in the form of vertical RGB stripes corresponding to the each of the slots in the aperture grille. Four thin horizontally stretched platinum wires are microwelded to the mask to ensure stability against mechanical vibrations.

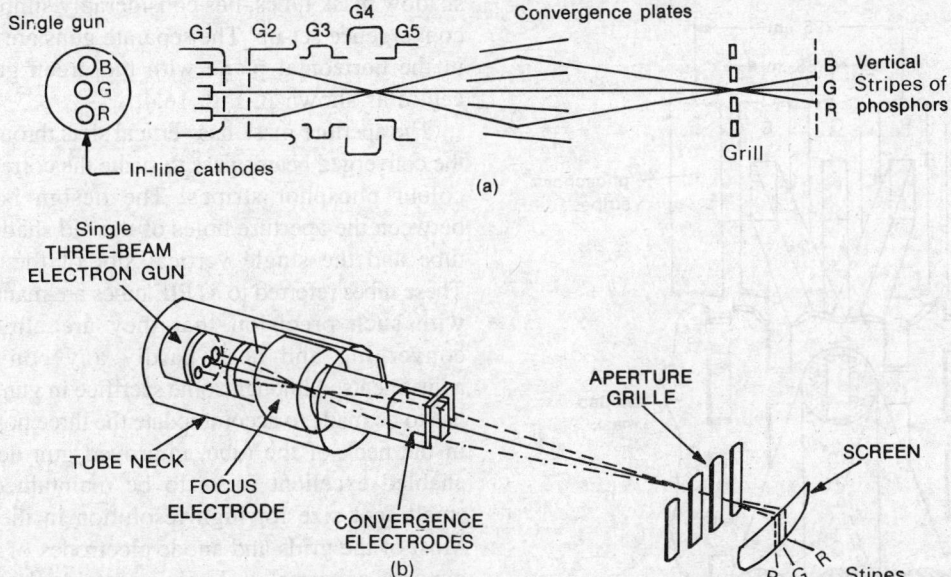

Fig. 16.3   Trinitron colour picture tube with in-line beams: (a) Electrode structure; (b) Gun and aperture grille construction

The control grid G1 is a single-cup electrode with three apertures for the three beams. The screen and the focus grids are common for all the beams. The beams cross over to converge within the focus field of the single-gun lens before leaving the gun to cross again at the slots in the shadow mask. This gives the advantage of passing all the three beams through the central portion of the relatively large common electrostatic lens, minimizing spherical aberrations in the outer beams. The three beams are in focus together at the centre of the main focus electrode.

Convergence of beams at the slots in the aperture grille is achieved electrostatically by four convergence plates, with the green beam passing through two central plates, both of which are connected to the anode. The red and blue outer beams passing between the outer plates can be moved horizontally to converge with the central green beam by adjusting the voltage applied to the outermost plates. This voltage is a little lower than the anode voltage and forms the static convergence control. It is associated with the EHT voltage source and is supplied to the convergence plates via the EHT cable.

Misconvergence of the electron beams in the horizontal plane is symmetrical and very small because of the symmetrical construction of the gun and the convergence electrode assembly. Hence only static and parabolic line convergence correction is required, involving only six adjustments.

On very close viewing, the picture formed by the trinitron tube appears vertically striped whereas these formed by other tubes appear dotted. Improved vertical resolution is noticeable in trinitron as it is limited only by the number of scanning lines. Larger aperture of the grille and common gun assembly structure minimizes restrictions on the beam currents passing through, and gives a greater transparency for the electron beams, almost doubling the brightness obtained.

16.4 PRECISION-IN-LINE PICTURE TUBES

Mounting the three electron guns horizontally in line rather than on the delta structure of the earlier shadow mask tubes, has considerably simplified the convergence set-up. The separate guns are mounted in the horizontal plane with the green gun at the centre as shown in Fig. 16.4.

The aperture mask has vertical slots through which the converged beams pass to strike the corresponding colour phosphor stripes. The design is a cross between the aperture holes of the old shadow mask tube and the single vertical slot of the trinitron. These tubes referred to as PIL tubes are manufactured with such precision that they are almost self-converging and need hardly any convergence adjustments. Although some sacrifice in gun diameter has to be made to accommodate the three horizontally in the neck of the tube, improved gun design has enabled excellent focus to be maintained with a small spot size for high resolution in the picture. Most of the grids and anode electrodes of the three guns are common, thus helping avoid relative drifts.

The PIL tube carries with it the deflection yoke and magnet assembly for purity and convergence, factory adjusted and bonded on to the tube neck. The deflection yoke employs a *torroidal yoke* having precision wound coils and is less bulky. Very precise

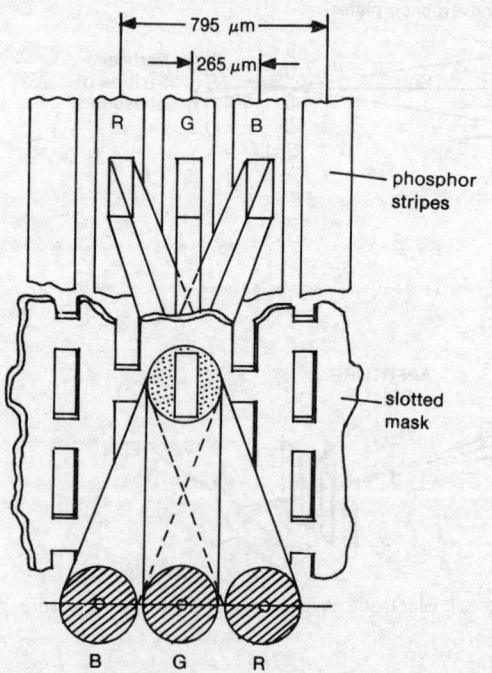

Fig. 16.4 Precision-in-line gun and slotted aperture mask arrangement (Philips)

field shipping is achieved by positioning and distribution the individual turns of wire in the notches cast in the supporting frame. The fields can be modified by magnetic shunts or enhancers fitted inside the tube or the scan yoke which affect the local magnetic flux exactly as per requirements of self convergence.

Geometric Pincushion Correction

Modern colour tubes having wider angle of deflection and flat square screen entail pincushion distortion which is corrected dynamically by amplitude modulating the line scanning current with field rate parabolic waveform, and superimposing the line rate sawtooth waveform on the field scanning current. This increases the deflection near the central portions of deflection to counter E-W axis pincushioning. The amplitude modulation can be obtained by varying the line output stage supply parabolically, at the vertical rate by an RC circuit in the supply lead. Similarly parabolic waveform at line rate is superposed on the vertical scanning waveform to counter N-S axis pin cushioning. With improvements in the yoke design, recent PIL tubes of 20 AX and 30 AX systems do not require any such vertical pincushion correction.

16.5 PURITY AND CONVERGENCE

Purity

For exact reproduction of colour, the effective source of the beams viz. the deflection centres, must coincide exactly with the position of the light source used for dot fixing during the manufacture of the tube. Otherwise one or more of the beams will spill over onto the adjacent phosphors producing a wrong colour for that beam. This is prevented by a pair of two pole magnetic rings provided with the deflection yoke on the gun side. They are rotated with their tabs to shift all the three beams relative to the axis of the tube, in the same way as the centering magnets in a monochrome receiver. The field direction and orientation can be adjusted by relative movement of the two rings. The three beams can be moved in any direction together relative to the axis of the tube to give necessary purity at the screen centre. The entire deflection yoke is then moved and set axially to establish coincidence between the deflection centre and the original dot-fixing source position. With this setting, purity is established over the entire screen.

Convergence

The three electron beams, controlled independently by the RGB signals, are scanned by a common deflection fields and draw together the full coloured picture. It is necessary to register the three colour images precisely, so that they overlay at all points on the phosphor screen. Convergence adjustments aim at this by positioning adjustments for each of the three beams, preventing colour fringes on borders and outlines in the reproduced picture. This has to be done for *static convergence* when the beams are in the small central area of the screen under no deflection conditions, and for dynamic convergence over the full screen with beams deflected over the entire screen by the deflecting magnetic fields.

For static convergence two sets of multipole ring magnets are arranged on the yoke towards the neck of the picture tube. They produce well defined and adjustable magnetic fields in the region of the paths of individual electron beams within the electron guns. A combination of four-pole and six-pole ring magnets is used to set-up such fields. A pair is used in each case as usual so that by rotating them relative to each other and together, the strength and direction of the resulting magnetic field can be adjusted to set the beam positions exactly as required. This is illustrated in Fig. 16.5.

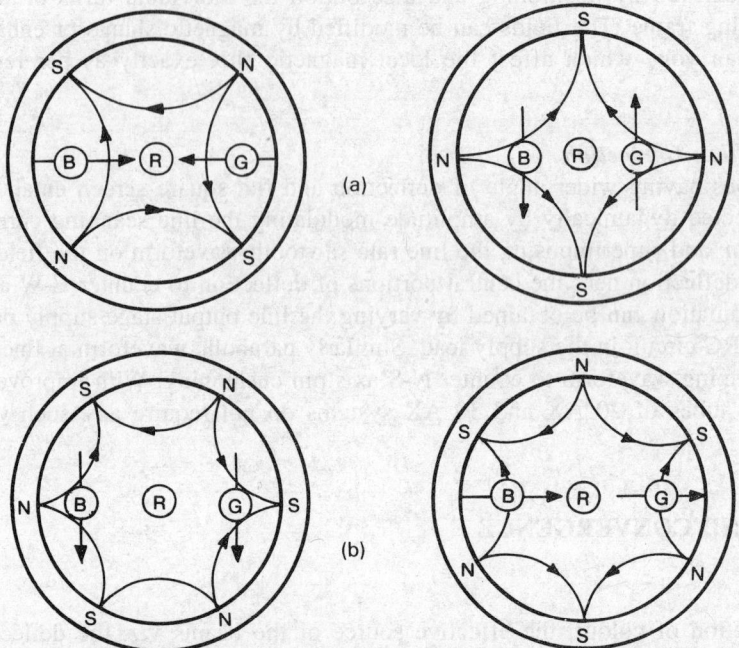

Fig. 16.5 (a) Four-pole and (b) six-pole ring magnets

In the four-pole ring arrangement shown Fig. 16.5(a), the central green beam is not affected. Depending on the orientation of the poles, the outer red and blue beams can be *moved differentially* in the horizontal or vertical direction, depending on the resultant field and the amount of shift controlled by the field strength.

In the six-pole field, shown in Fig. 16.5(b), the outer beams can be moved together in the same direction depending on the orientation of the field, which is controllable by the relative movement or combined movement of the two rings. Misregistration is easier adjusted on crosshatch or dot matrix pattern where it is most visible.

Dynamic convergence is necessary because of the flat screen of the picture tubes. Since the outer beams are off the tube axis horizontally, the point of convergence falls progressively short of the flat screen as the deflection is increased. This *beam parallax*, and beam tilt caused by the radial velocity component in the beams due to predeflection convergence, causes misregistration. Dynamic convergence field correction may be applied by generating parabolic waveforms with line rate deflection flux variation by means of a strap-on convergence yoke fed from the line time base. The need for separate dynamic convergence circuits has essentially been eliminated in the in-line gun constructions, with the use of predistorted yokes whose magnetic field is tailored to give exact dynamic correction.

16.6 AUTOMATIC DEGAUSSING

Any extraneous magnetic field can deviate the path of electron beams causing errors in beam landing and hence in purity of colours. This can be due to magnetic objects like toys, loudspeakers, etc. Even the earth's magnetic field can upset the beams to affect colour purity. Modern picture tubes have internal magnetic shields to prevent the external magnetic field from entering inside. The mask and mounting frame may still have a tendency to get magnetized by external sources. The residual magnetism in the mask structure and the magnetic shield can change in different situations and time, affecting purity. To overcome this problem, the shield and the shadow mask may be provided with an automatic degaussing system, consisting usually of two coils covering the top or bottom cone parts, or one large coil. These are energized by the 50 Hz supply at switch-on by a *decaying burst* of 50 Hz magnetizing current with the help of a pair of thermisters.

For proper degaussing an initial magnetomotive force of over 300 ampere-turns is required in each of the coils. This has to be gradually decreased by appropriate circuitry, such that in steady state no significant mmf should be residual in the coils, less than 0.3 ampere-turns in practice. Degaussing circuits now most commonly employ a dual PTC thermistor in series with the degaussing coils as shown in Fig. 16.6. Examples of double-coil and single coil are shown in (a) and degaussing circuit using dual thermistor is shown in (b).

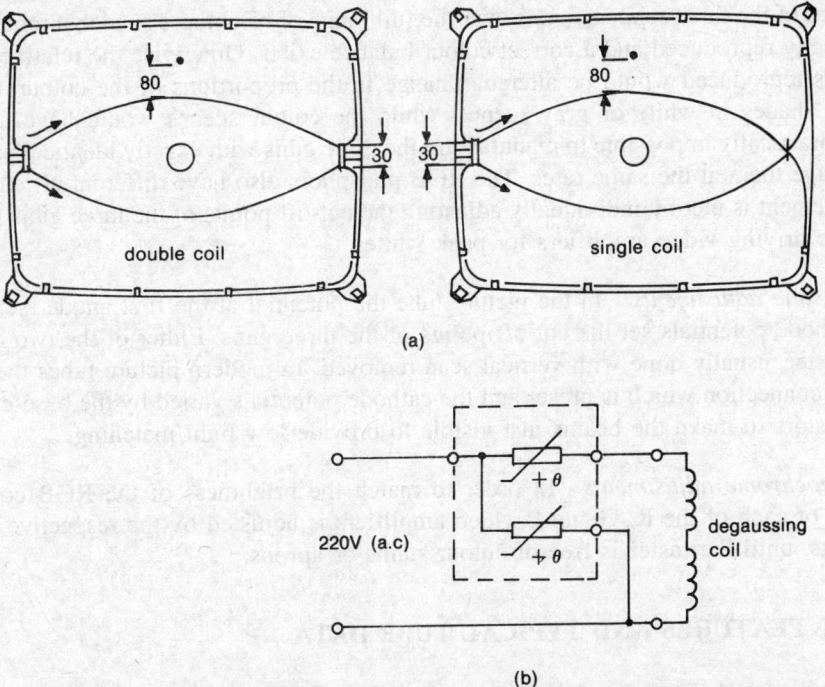

Fig. 16.6   (a) Position of degaussing coil on the picture tube. Coils: 60 turns, 0.35 mm Cu-wire, 12.5/25.1 Ω
Circumference: single 117 cm, double 237 cm; (b) Automatic degaussing circuit using dual PTC thermister

When the receiver is switched on, the cold resistance of the thermistor is very small, causing a surge of 50 Hz degaussing current to flow through the coil. As the coil series thermistor heats up due to the high current, its resistance rises to a high value gradually, reducing the current flowing into the coil slowly to a small negligible value of a few milliamperes. The shunt thermistor keeps the dual unit warm enough to keep the coil current low after the degaussing surge. The degaussing coil operates for about a second after switching-on of the receiver.

Whenever the receiver is moved to a different position, it may be necessary to switch off the receiver for a few minutes before switching it on again, to allow the thermistor to cool down and allow the full degaussing current to flow. Loudspeakers, motorized toys, and other devices containing strong magnets should not be brought close to the TV receiver. They may strongly magnetize the tube mountings, to such an extent that powerful external degaussing coil may be necessary to remove magnetisation and the colour splashes due to it.

The external degaussing coil with current on, is brought close to the screen and the sides of the tube a few number of times and then withdrawn to a distance of over two metres, before being disconnected from the mains. The receiver may be on or off during the degaussing process and shows colour splashes as the coil is moved. Should colour shading persist on the screen, it is necessary to adjust colour purity and convergence.

16.7 GRAY SCALE TRACKING

Since there are three separate electron guns controlled by the three colour drives, it is essential to *match* the characteristics of the three *separate guns* over the full range of brightness such that monochrome gray shades are correctly reproduced and a correct colour balance exists. Otherwise the relative intensities of the three colours reproduced would be altered. Change in the proportions of the colours mixed would result in colour shades in white or gray scenes, while the colour scenes would have altered hue or saturation. It is practically impossible to manufacture the three guns with exactly identical parameters and expect them to age them at the same rates. The RGB phosphors also have different efficiencies. Hence the above requirement is met by individually adjusting the cut-off points of the three guns for black, and the slopes of the driving video amplifiers for peak whites.

Black level half tone adjustments   In the picture tube the potential at the first anode (screen grid G2) and the grid-cathode potentials set the cut-off points of the three guns. Either of the two can be set for cut-off of the raster, usually done with vertical scan removed. In modern picture tubes the screen grids have a common connection which is preset and the cathode potentials varied by the base-emitter bias of the driver transistors to make the beams just visible to provide low light matching.

White level monochrome adjustments   In order to match the brightness of the RGB colours at high lights, the slope of each of the R, G and B video amplifiers is adjusted by the respective monochrome drive adjustments, until the raster is free of colour stains or aprons.

16.8 SPECIAL FEATURES AND TYPICAL TUBE DATA

Modern colour picture tubes employ full square and flat screen with sharply cornered edges that give visibly enlarged TV picture, approaching the aspect ratio of 4 : 3. The cathode is a quick-heating type, and it requires less than 5 seconds for a clearly visible picture to emerge. This is achieved by employing smaller cathode structure and concentrating the heater element just behind the front surface of the cathode. Pigmented phosphors and high light transmission efficiency of face plate glass (over 68%) help reduce cathode beam requirements. Black stripes between the colour phosphors provide guard bands against cross colour due to beam landing errors. Pigmented phosphors with variable pitch give improved colour saturation of the TV picture. Modern colour picture tubes employ yoke winding techniques that hardly require any raster correction along the E–W and N–W directions.

Typical CPT-Yoke Data

Type	Defln (deg)	Diag (mm)	Light (%)	Pitch mm	V_f/I_f (V/mA)	EHT (kV)	V_{g3} (kV)	V_{g2} (V)	L_H (mH)	I_h (A)	L_v (mH)	I_v (A)
A51–231X	90	480	68	0.82	6.3/700	25	7	600	1.93 2 Ω	3	28.5 14 Ω	0.8
A51–590X AT1480 FST-A59–	90	480	64	0.8	6.3/685	25	7	600	1.9 2.2 Ω	3.1	29 13.6 Ω	0.86
EAK-00X FST-A66–	110	590	67	0.8	6.3/310	25	8	700	1.85 1.85 Ω	4.1	11 6.5 Ω	1.7
EAF00X1	110	534.5	68	0.82	6.3/700	25	7	620	1.5	4.65	24	1.40
M37–120X	90	277.3	86	0.31	6.3/630	25	7	380	1.2	3.70	24	0.90

16.9   FLAT PANEL DISPLAYS

Flat panel matrix-addressed displays suitable for wall mounting have been a long-cherished dream of television engineers. Several technologies have been pursued towards this end with only limited success in achieving good resolution, contrast and brightness. Most of these technologies are based on electroluminescence (EL) as in LED, gas discharge plasma (PL), liquid crystals (LC) and vacuum fluorescence (VF) using phosphor coated anodes. The pixels are arranged in a matrix at the intersections of several hundred rows and columns. LEDs serve as good pixels with different colours but dissipate large amounts of power, besides being costly. LCDs are cheaper are need little power but are slow in response as passive displays and smear is noticeable in moving objects. Improved LCDs are under development with improved technologies. Gas discharge and thin film electroluminescent displays hold some promise. The addressability of the pixels totalling up to a million, and getting uniform light output from each of them, have been the major technological problems. Three popular standards for screen resolution derived from video display units (VDU) for personal computers define the 'number of rows × columns' matrix for video graphic displays as follows:

— CGA: 640 × 200 (2 colours) or 320 × 200 (4 colours)
— EGA: 640 × 350 (16 colours)
— VGA: 640 × 480 (16 colours) or 320 × 200 (256 colours)

Display Technologies

Flat panel displays generally have a front glass plate. In a 640 × 480 VGA display, for example, 640 vertically oriented transparent conducting column stripes (of tin oxide) are placed on the inside of the glass plate. The next layer of the panel, placed between the front glass and the back panel, contains 480 horizontally oriented conducting stripes (of aluminium) forming rows. In between these stripes is placed the luminous medium comprising a solid (EL), a liquid (LC), or a gas (PL). In dc display the emitting solid or gas makes contact with the pixel stripes and the load is resistive, while in ac displays the stripes have insulating layer between stripes making the load capacitive. In vacuum fluorescent (VF) displays, the load is capacitive until emission occurs, after which the load turns resistive. The ac plasma display, which is capacitive, requires an ac waveform drive. LCDs also require ac waveforms to help prevent degradation of some type of liquid crystal material.

The transfer characteristics of light output against the driving voltage is important. Light emission begins at a certain threshold V_{th}, above which light increases proportional to the drive voltage until it reaches saturation to level off at V_{max}. The V_{th} for different displays varies greatly, from about a volt for LCDs, to several hundred volts for EL and PL displays. For *gray scale capability*, the slope between the threshold voltage V_{th} and V_{max}, the voltage for maximum light output, has to be wide and must be stable with temperature lest it should cause brightness changes. Generating a gray scale is not possible directly, if the slope is very steep offering almost on/off light output from the pixels. It is nevertheless possible to generate a limited gray scale, if the pixel display technology is fast-acting. By exciting the pixels from a number of variable short bursts, an integrated impression of 8 to 16 gray tones can be created on the eye.

Drive Methods

Drivers for the pixels arranged in a set of rows and columns are usually located on all sides of the panel, as shown in the typical arrangement of Fig. 16.7.

The actual drive schemes are complicated because the mechanical constraints of the packaging leads which are wider than the pixel spacings. For overcoming this to some extent, half the pixels are driven from the left and top while the other half are driven from the right and bottom. The drivers comprise serial-to-parallel converters with high voltage outputs. The data are fed to the chain of column drivers until all the columns for a row are filled. The data are then latched and the row drivers are turned on to excite the pixels in the row. Each row is driven in turn, as fresh column data enter the driver to proceed to the next row.

Some EL panels tend to degrade and output dimmer light if driven by repeated pulses of the same polarity. This has led to drive schemes that alternate polarity of drive pulses. In the asymmetric drive, all the rows are written into the display in the same direction, and at the end of the frame, a pulse of opposite polarity is sent by all the rows to refresh the display, before the scanning of the next frame. As the number of ON pixels is indeterminate, refresh pulse may not exactly cancel the energy delivered by the drive, and some pixels may be still on if quasistatic images are displayed for long periods. This results in a sort of burned-in image called latent image.

16.10 PLASMA DISPLAYS

Plasma (PL) displays are currently available as large luminescent flat panels for display of information or images. Plasma cells are arranged in 2-D matrix addressed by application of proper row and column voltages that create a continuous gas discharge while enough voltage is maintained. The plasma cells depend on their operation on the excitation of phosphors overlaying a discrete matrix of display cells, by means of the ultra-violet rays radiated through gas discharge created in each cell. The discharge in each cell is controlled by providing driving voltage to the correct row and column by bus lines in the 2-D matrix of cells required to depict the image, as shown in Fig. 16.8.

Each display cell embodies a controlling switch for activation of the pixels. The plasma is formed in a low pressure inert gas like xenon or mercury vapour which is excited by either dc or ac. fields. For colour TV displays each pixel is formed by a RGB trio of phosphors arranged in alternate sequence that are excited proportionately as in a colour picture tube. Panels of 1.5 m diagonal size (1000×1000) pixels) have been made.

Recently ac plasma displays have been exhibited (P33 ac plasma display system from Plasmaco, Inc. Highland, NY) that mount address driver chips directly on the glass panel rather than on electronic printed circuit boards. The single board with surface mounted components, replaces up to four PCBs,

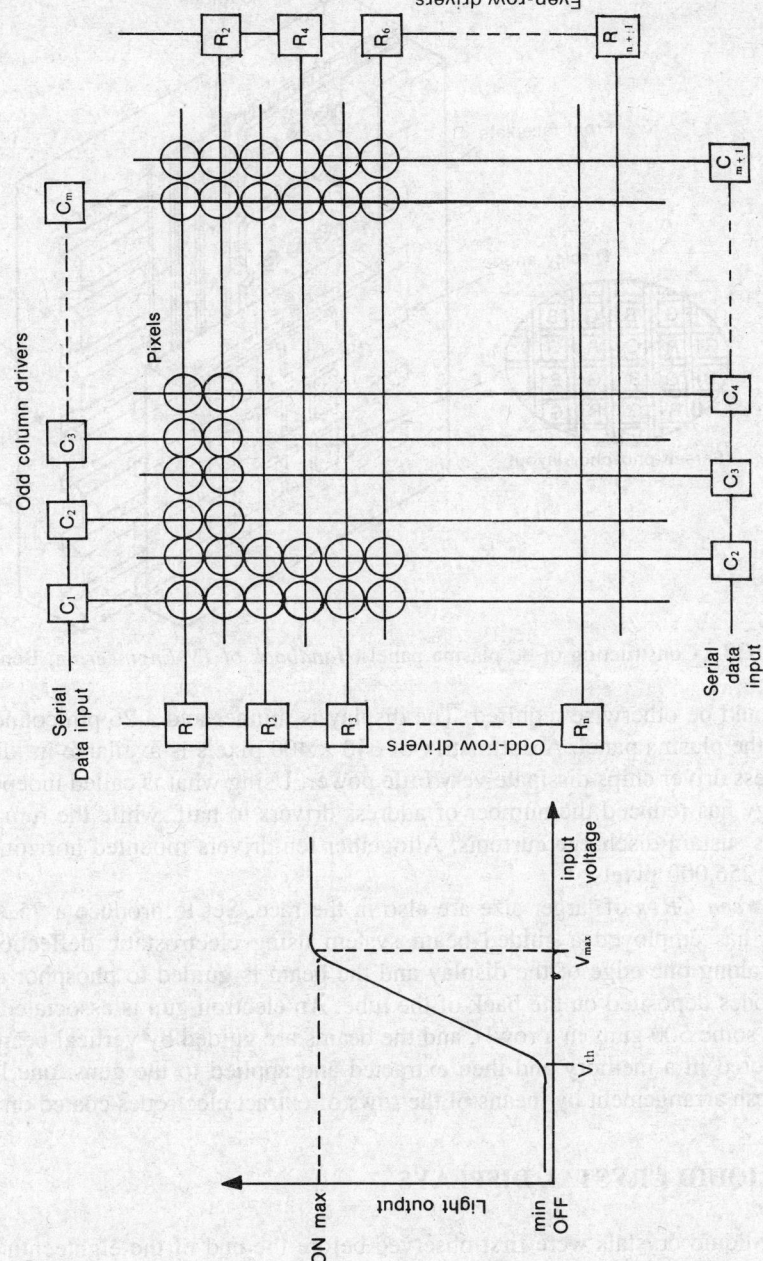

Fig. 16.7 (a) Transfer characteristics of pixel; (b) Interlaced display drivers for flat panel displays

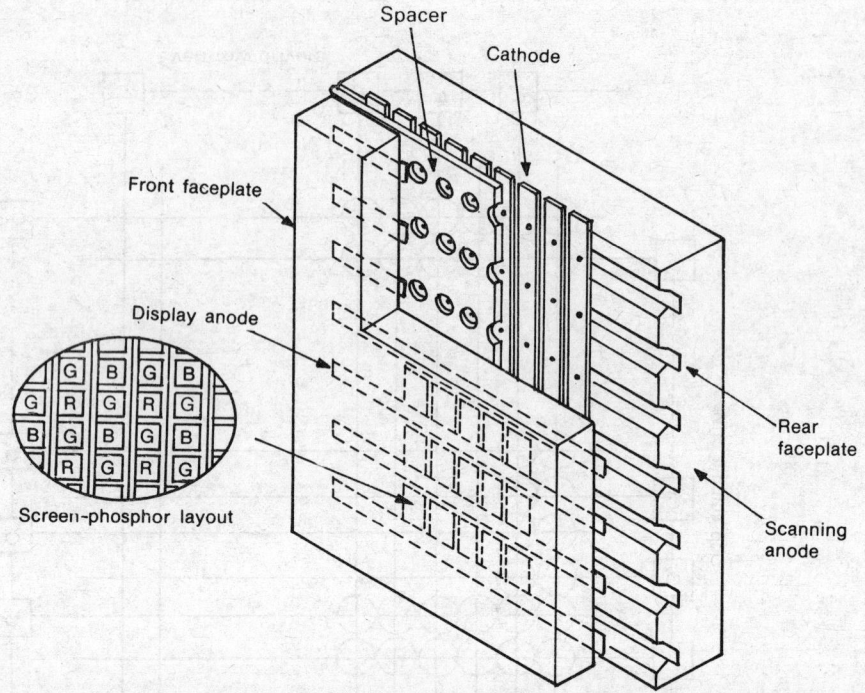

Fig. 16.8 Construction of dc plasma panel (*Handbook of TV Engineering*, Benson, McGraw-Hill, NY)

which would be otherwise required. The display is thinner and a 26 pin connector is enough to link the board to the plasma panel. A dot matrix of 640 × 400 pixels is available in display area of 21 × 13 cm. The address driver chips dissipate very little power. Using what is called independent sustain and address technology has reduced the number of address drivers to half, while the remaining half need to supply much less sustain discharge currents. Altogether ten drivers mounted horizontally, and eight vertically, drive the 256,000 pixels.

Flat screen CRTs of larger size are also in the race. Set to produce a 75 × 100 cm flat screen tube, the RCA has employed a guided beam system using electrostatic deflection. The electron guns are arranged along one edge of the display and the beam is guided to phosphor dots by means of voltages on electrodes deposited on the back of the tube. An electron gun is associated with each of the columns (totalling some 500 guns in a row!), and the beams are guided by vertical beam guides. The video signal is first stored in a memory and then extracted and applied to the guns, one line at a time, through the ladder mesh arrangement by means of the rows of extract electrodes coated on the back plate of the tube.

16.11 LIQUID CRYSTAL DISPLAYS

Although liquid crystals were first observed before the end of the eighteenth century, they came to the fore only at the beginning of the '70s with large scale applications in electro-optic displays. Liquid crystals are organic compounds, whose macroscopic behaviour resemble that of a liquid, but which show physical properties only found in crystals. The molecules of liquid crystals have an elongated and somewhat rigid structure. Their orientation affects all their physical parameters, such as viscosity, elastic and dielectric behaviour, magnetic susceptibility, electric conductivity and refractive index, which are dependent on direction i.e. anisotropic.

Liquid crystal molecules, usually elongated rod-shaped, have strong electric dipole moments and are slightly polarizable. They have definite diffraction patterns similar to solid crystalline substances. Liquid crystals are divided into three groups according to the type of special arrangement between the molecules: (i) nematic, (ii) cholesteric and (iii) smectic crystals, as shown in Fig. 16.9. Nematic liquid crystals have only one type of alignment, where the long axes of the cigar shaped molecules lie parallel to each other and the molecules can glide past each other freely, in a rather fluid state. Cholesteric liquid crystals are similar to nematic liquid crystals, with the long axes lying parallel to each other in a suitable plane. In any given plane, this privileged direction is slightly *twisted* relative to an immediately adjacent parallel plane. Cholesteric crystals can be said to have a twisted nematic structure, with a *helical structure* with a definite pitch between 200 nm to 20,000 nm. The smectic crystal structure is closest to that of normal solid crystals, with the difference that the molecules are not assigned fixed position, but are merely bound to certain planes in layers which can be shifted against each other as a whole.

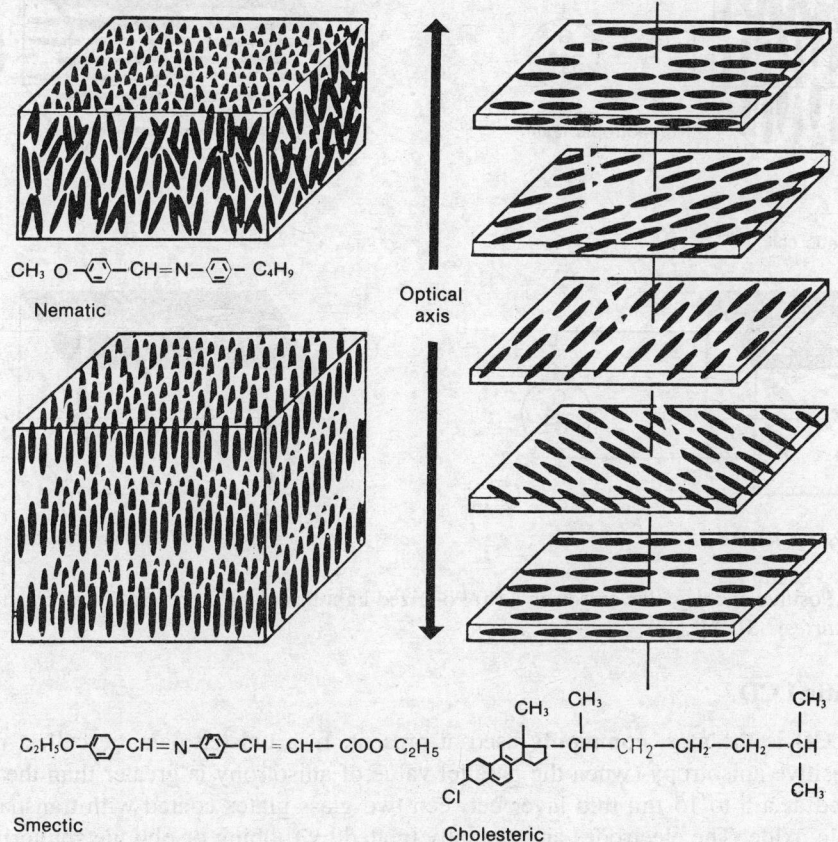

$CH_3 O - \langle \rangle - CH = N - \langle \rangle - C_4H_9$

Nematic

Optical axis

$C_2H_5O - \langle \rangle - CH = N - \langle \rangle - CH = CH - COO \ C_2H_5$

Smectic

Cholesteric

Fig. 16.9 Types of liquid crystals and compounds (*Courtesy*: SIEMENS)

Nematic Crystals

Of the three basic groups of liquid crystals, nematic crystals are the most commonly used. Liquid crystals generally have a high anisotropic dielectric constant. When an electric field is applied, a torque is exerted on each molecule of the material which tries to orient the liquid crystals along the field. Dielectric anisotropy $E = E_\parallel - E_\perp$ in a nematic crystal can be positive or negative, depending on whether the parallel value $E_\parallel$ is greater or the perpendicular value $E_\perp$ is greater. Accordingly, the nematic liquid crystals adjust themselves parallel or perpendicular to the electric field, as shown in Fig. 16.9(a). The resultant deformation of the structure causes a change in the *anisotropic optical properties* of the crystal. In a non-driven crystal, linear polarized light is rotated as it passes through, while in an electrically driven crystal linear polarized light is not rotated, as shown in Fig. 16.10(b). This can be used to pass or block polarized light propagating through.

Liquid crystal with dielectric anisotropy in the electric field

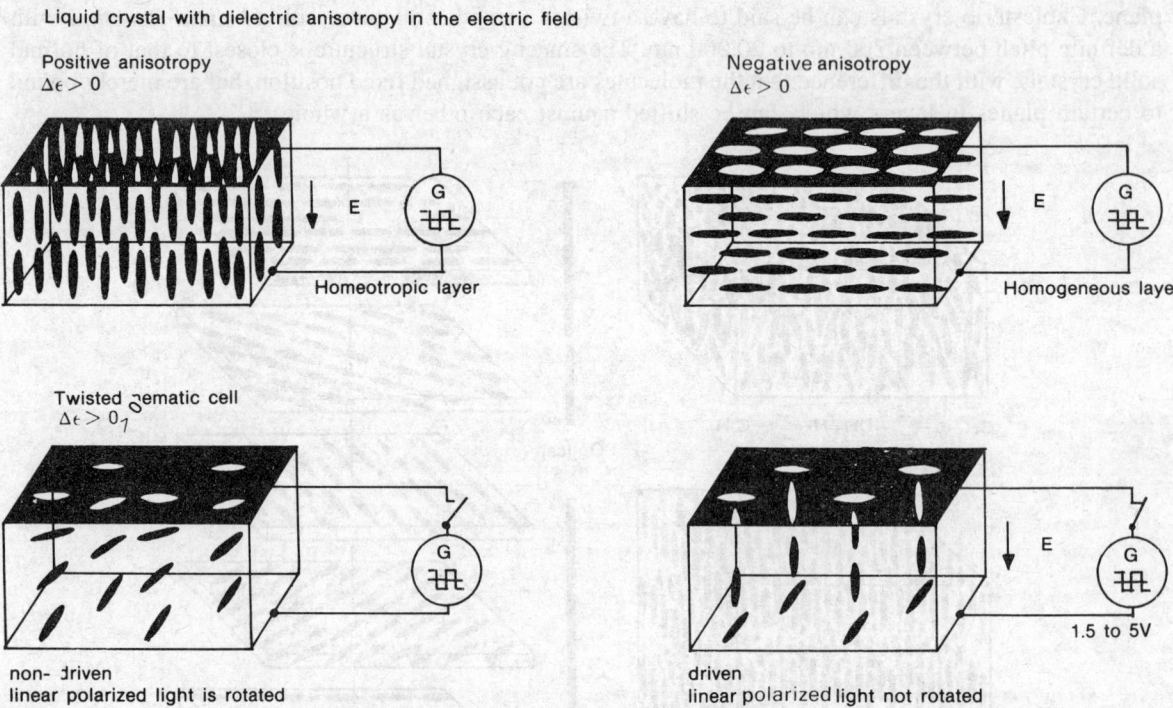

Positive anisotropy
$\Delta\epsilon > 0$

Negative anisotropy
$\Delta\epsilon > 0$

Homeotropic layer

Homogeneous layer

Twisted nematic cell
$\Delta\epsilon > 0$

non- driven
linear polarized light is rotated

driven
linear polarized light not rotated

1.5 to 5V

Fig. 16.10   (a) Positive and negative anisotropy; (b) Polarized light through a driven and non-driven nematic crystal (*Courtesy*: SIEMENS)

Twisted-nematic LCD

This type of LCD is the most commonly used at present. In a twisted nematic cell, a nematic liquid crystal with positive anisotropy (when the parallel value of anisotropy is greater than the perpendicular value) is located as a 5 to 15 μm thin layer between two glass plates coated with transparent electrode layers of stannic oxide. The electrodes are specially treated by rubbing or oblique sputtering, so that the molecules are forced to align in the privileged direction on the surface of the electrode. The orientation of the two electrode layers are perpendicular to each other, so that in the absence of an electric field the liquid crystals in the space in between them align themselves in a helical pattern. This twists polarized light propagating through the cell by 90°, and channels it along its new axis through an exit polarizer to be seen as a lighted pixel. If a voltage is applied to the electrodes the crystal molecules are aligned parallel to the electric field, by the elasto-electrical deformation of the crystal structure. Linear polarized light is no longer twisted, and is blocked by an exit polarizer to be seen as a black pixel. This can be reversed also, when the voltage off-state can provide a dark pixel while voltage on-state gives a bright pixel. Colour filters can be added to produce R, G or B light pixels.

A non-driven twisted nematic cell between two polarization filters (polarizer and analyzer) forms a *transmissive display*. The cell becomes opaque when a voltage of 1.5 to 5 V is applied to it, acting as light switches and are suitable for projection, as shown in Fig. 16.11(a). For normal use, back illumination is necessary and is provided by a fluorescent tube Construction of *reflective displays* using a twisted nematic cell is shown in Fig. 16.11(b). A polarizer with reflecting layer is placed at the back of the cell. A diffused reflector is preferred to avoid undesirable reflections.

LCD Problems
Individual pixels that transmit or block light are formed at the points defined by row and column lines that pass at right angles and crystals act as dielectric between the voltage carrying axes of the matrix. As the row and column axes of a pixel are activated, adjacent pixels are partly affected due to 'crosstalk', producing smear and reduced contrast. The effect is reduced if the transition between on/off voltages is reduced. However, this conflicts against higher resolution which demands sharper transition between the on/off voltages.

Supertwisted-nematic and *active matrix twisted-nematic* liquid crystal technologies have been introduced recently, to overcome these problems to some extent. These types of LCDs are expensive but have been used in recent video displays.

Double Layer Supertwisted-nematic LCD
This is a variation in the twisted-nematic structure that allows the liquid crystal molecules to twist beyond

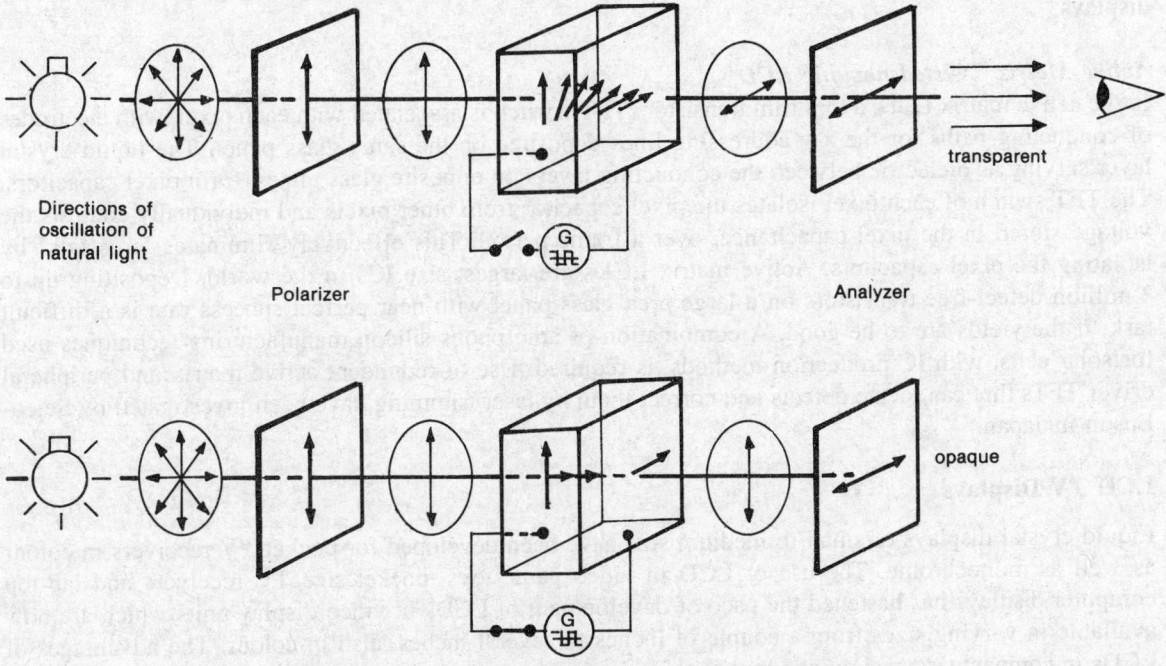

Fig. 16.11   (a) Twisted nematic cell transmissive display

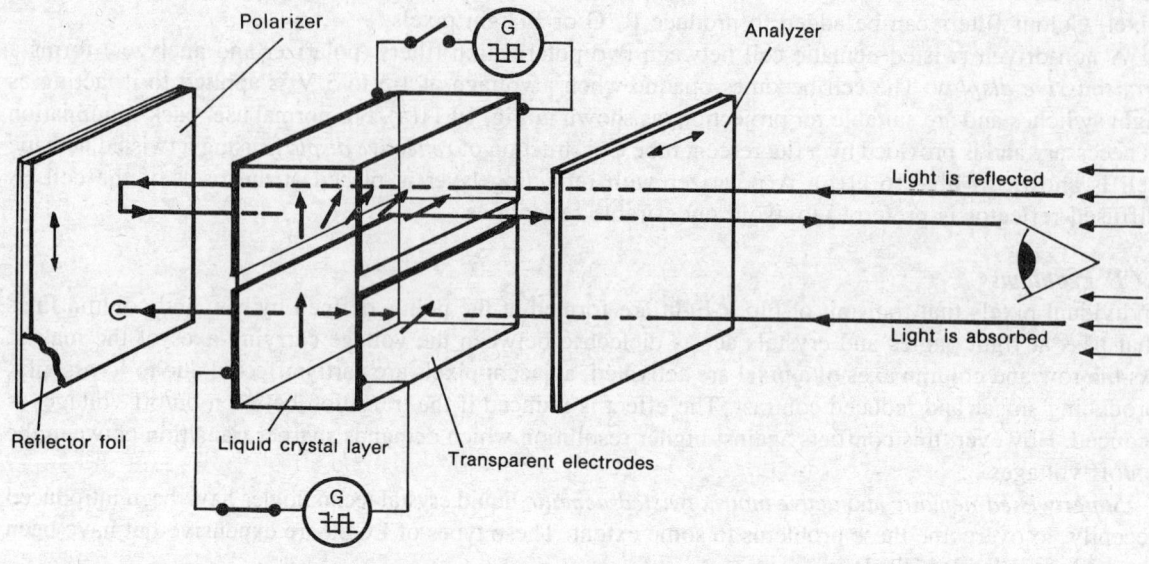

Fig. 16.11 (b) Twisted nematic cell reflective display (*Courtesy*: SIEMENS)

90 degrees up to as much as 270 degrees (supertwist). The greater twist requires a more linear voltage and more lines can be displayed. The sharper voltage transition required for increasing the resolution can work against producing a more gradual voltage change desirable for producing gray scale in colour displays.

Active Matrix Twisted-nematic LCD

In the active matrix LCD, a thin film transistor (TFT) switch is associated with each pixel, with electrodes of conducting paths for the x-y addressing lines deposited on the same glass panel. The liquid crystal layer serving as dielectric between the conducting layers on opposite glass plates, form pixel capacitors. The TFT switch of each pixel isolates the pixel capacitor from other pixels and individually controls the voltage stored in the pixel capacitance, over a frame period. This effectively eliminates 'crosstalk' by isolating the pixel capacitors. Active matrix LCDs are largest size ICs in the world. Depositing up to 3 million defect-free transistors on a large area glass panel with near perfect success rate is a difficult task, if the yields are to be good. A combination of amorphous silicon manufacturing techniques used for solar cells, with IC production methods, is required. Use of redundant active matrix and peripheral driver TFTs that can locate defects and correct them by laser trimming have been investigated by Seico-Epson in Japan.

LCD TV Displays

Liquid crystal displays of small to medium size have been developed for pocket TV receivers in colour as well as monochrome. The use of LCD in video game toys, pocket size TV receivers and lap-top computer displays has hastened the pace of development of LCDs in video display units which are now available in varying sizes from a couple of inches to several inches, also in colour. The advantages of LCDs in compactness, weight and low power consumption are obvious, while the limitations of contrast and brightness are noticeable.

A 2-inch LCD display consists of a P-channel MOS switching matrix of 240×220 pixels. The gates of all transistors in a horizontal row are connected to each horizontal common bus, while the drains of all the transistors in vertical column are connected to common vertical buses as shown in Fig. 16.12.

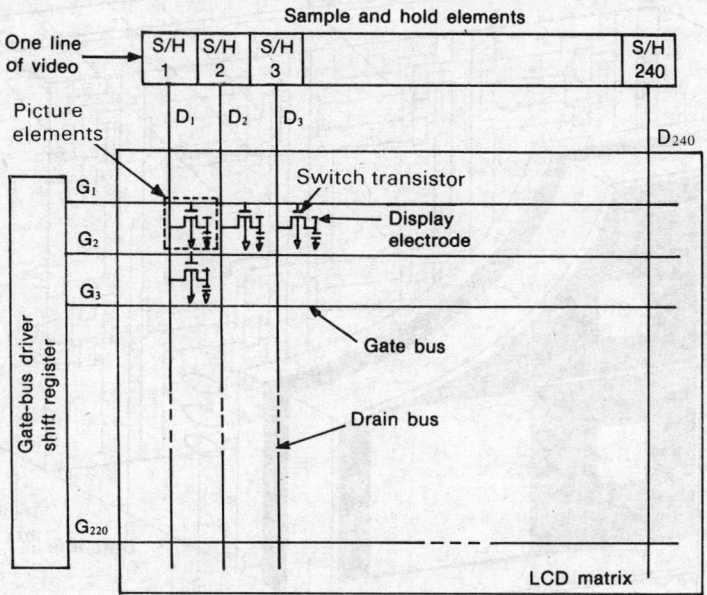

Fig. 16.12   LCD TV display (Reproduced from *Handbook of TV Engineering*, Benson) (*Courtesy*: McGraw-Hill, NY)

The horizontal gate buses providing row addressing are driven sequentially from a 220-stage shift register. The video information is placed on each column bus of drain connections, during the horizontal scan via a set of 240 sample and hold (S/H) stages, which hold and store the respective line pixel information. This creates a line sequential display, and the information on each drain line is updated only once every horizontal period of 64 μs through the S/H stages. The stray capacitance and the cross under-drain resistances associated with the drain buses present loading problems for the S/H stages. As only 220 lines are available for display instead of 625, both the interlaced fields are scanned on the same rows, while the extra lines of each field are eliminated during the top and bottom overscan.

Colour LCD Panels

An *active matrix TFT* approach to colour is taken by a thin film transistor (TFT) LCD display developed in a joint effort by IBM and Toshiba. The schematic arrangement used in the 14-in prototype display is shown in Fig. 16.13.

A diffusion plate spreads the fluorescent backlighting uniformly over the polarizing filter, as the light channels through an array of amorphous silicon TFTs, addressed by row (address) and column (data) lines. The TFTs can apply a voltage to the liquid crystal material, which along with the display and common electrodes, forms a capacitor for storing the voltage for the duration of the frame. The pixels are here in *quadruplet subpixels of red, green, blue and white* translucent dots (equipped with filters). When no voltage is applied, the liquid crystals twist the polarized light to make it either parallel or

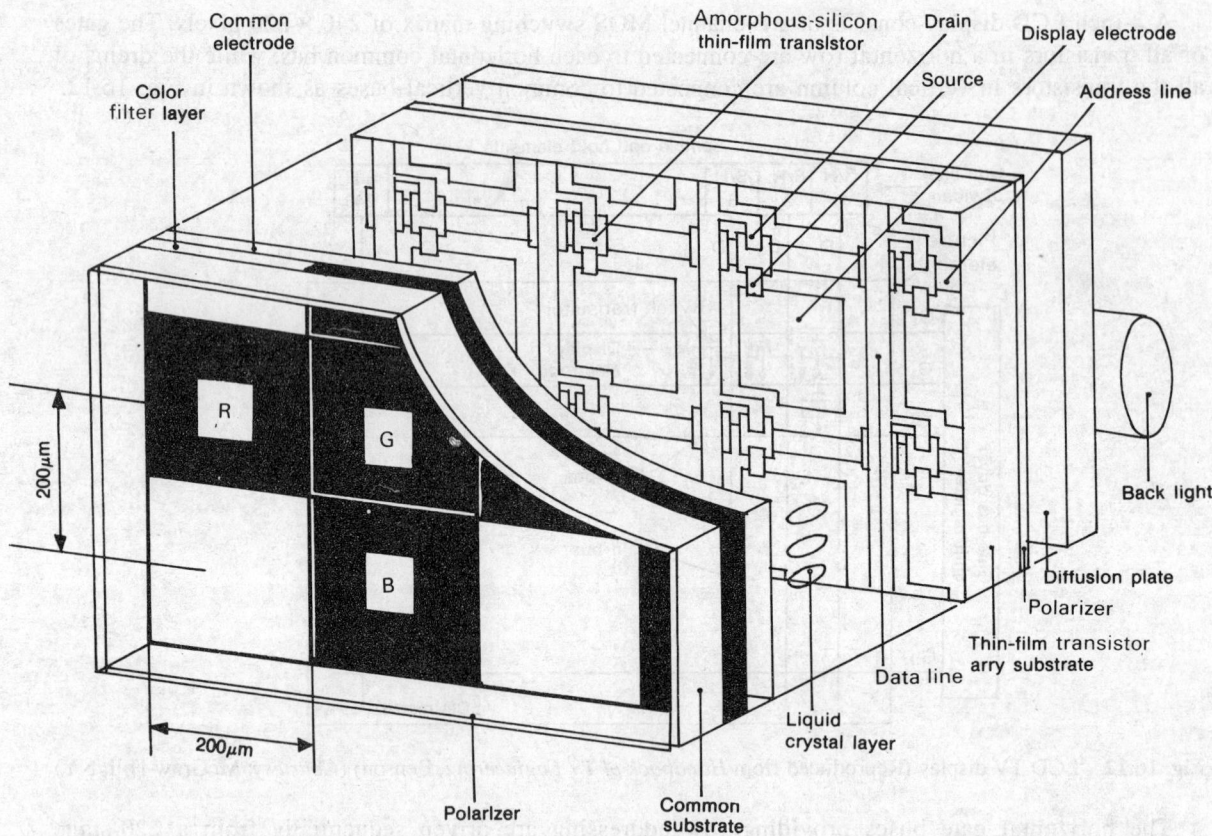

Color filter layer

Common electrode

Amorphous-silicon thin-film transistor

Drain

Display electrode

Source

Address line

200μm

R

G

B

200μm

Polarizer

Common substrate

Liquid crystal layer

Data line

Thin-film transistor arry substrate

Polarizer

Diffusion plate

Back light

Fig. 16.13   TFT LCD 14-in colour display developed by IBM-Toshiba (*Courtesy*: IEEE Spectrum)

perpendicular to a second polarizer, forming a lighted or dark pixel. If a voltage is applied by means the TFT assigned to the dot, the crystal orientation is disrupted, leaving the polarized light unaffected, and the light shines through the pixels. Depending on the combination of primary colours expressed or suppressed, the pixels assume one of 16 colours. On a 36 cm diagonal panel, 720×550 colour pixels, or 1440×1100 monochrome pixels are available.

Power consumed by the fluorescent tube for backlighting the LCD is 6–10 W for 11-in. display, an overhead that increases with colour filters that affect transmission characteristics and brightness, and erodes the power advantage over CRTs.

CRT vs Flat Panels

Despite impressive technological progress, flat panel displays remain inferior to the formidable 100-year-old CRT except when its large size, weight and power preclude its use in portable computers and some mobile applications in aircrafts. The colour resolution and addressability of the high energy electron beam CRT coupled with its low cost are hard to beat. Unwilling to rest on their laurels, CRT designers continue to push the tube performance to its limits. There is now a 38 cm and 48 cm monochrome monitor tube (Flanders Research Inc., N.J.) with a pixel resolution of 3300×2560 suitable for demanding applications in CAD, DTP, and imaging applications in medicine and defence.

Even with vast funding for HDTV display research by agencies like Defence Advanced Research Project Agency (DARPA), no significant flat panel technologies are on the horizon yet to compete with the CRT which may dominate the consumer TV displays until the late '90s. The flat panel display systems would however continue to evolve fast, with increasing applications in portable personal computers. No single technology may dominate this VDU field, because the display to suit customer needs, and his subjective perception, are often decisive factors in favour of laptop computer manufacturers.

16.12 LARGE SCREEN DISPLAYS

Construction of large size CRTs have been limited to less than 1 metre diagonal, for reasons of glass strength and weight factors. Projection systems employing high intensity CRTs can provide large optical magnification to expand images to full life size or greater. Here, the high intensity projection CRTs, the optical assembly and the viewing screen are the important constituents to determine the performance levels. Two basic methods are employed for projection:

(i) *Front projection*, in which the image is projected on the screen from the side it is viewed. This needs a screen with good reflectivity. Ambient light tends to dilute image contrast and colour purity, although a screen with highly directional reflectance can reduce this effect. A front projector uses available space more economically, as the audience area provides path for light; but then a passer-by may block the image. A ceiling mounted projector can be used to eliminate this problem.

(ii) *Rear projection*, in which the image is projected on the screen from side opposite to the viewing side. The light path is confined inside the cabinet in an area behind the screen and is abridged by use of folding mirrors and wide angle lenses. The screen used in this method should have high transmission of brightness and low reflectance so that contrast dilution by reflected ambient light is reduced. A technique similar to that employed in black-stripe colour picture tubes can be used here to improve contrast. With nonreflective black stripes coated on the viewing side and lenticular surface (lens-shaped double convex segments behind the gaps intervening the stripes) on the rear side of the screen, it is possible to have high transmittance from the rear projection side and negligible reflectance. A fresnel sheet lens behind the screen collimates the diverging light from the image towards the viewer, preventing the formation of a bright area at the centre. The vertical lenticular lens controls the horizontal viewing angle while diffusers embedded in the lens scatter the light to provide a wide viewing angle. Colour correction is provided by the lenticular surface moulded on the front surface, while the vertical junction portions carry black stripes. Screens of this type permit viewing in well-lit rooms with a fair amount of ambient light, and have been used with LCD plate projectors as well.

Optical assembly requires good light collection efficiency of the lens with required magnification and image geometry for the throw distance, optical path transmission, weighed against factors of weight, cost, etc. Geometric distortions which are often severe, are generally corrected in the deflection circuits electronically. Lens astigmatism, coma, and spherical and chromatic aberration affect resolution. Additional display elements, such as dichroics and optical path folding mirrors, also affect final display. For high resolution images, an innovative dual focus design with a four-element glass-and-polymer hybrid optic has been devised with separate focus adjustment for edges and the full screen. A low cost *fresnel lens* in front of the CRT is the simplest means of limited optical magnification in home TV sets.

Projection CRTs Sources of high intensity images with high resolution require CRTs of special design. The three-CRT system is now common, in which three mono-colour picture tubes produce three rasters in red, green and blue colours, each being driven by corresponding R, G and B drive signals. These three rasters are projected through wide aperture lenses on to a highly reflective screen, with orientation of the tubes adjusted accurately for good colour registration. In order to overcome the low efficiency of the lens arrangement, the *Schmidt objective* system, employing concave spherical mirror optics with a corrective lens of suitable shape, is often used for projection. Besides increasing the efficiency about tenfold, the system considerably reduces spherical aberrations to give a clearer and better contrasted picture. For screen sizes in excess of 10–15 m^2 the EHT required in the picture tube is very high, with the attendant problems of shielding of x-ray radiations, limited life of phosphors, and so on.

High luminance pictures are obtained from special high voltage (over 30 kV) projection CRTs, for screen sizes of 2 to 3 m. High brightness liquid cooled CRTs typically 7-in are used. The front of the CRT holds an additional glass plate that encloses a liquid coolant. This prevents overheating that could otherwise be caused by the high power quadra-potential-focus (QPF) electron gun, which includes an extra electrode to improve focus and sharpness. Luminous output of 650 lumens at peak white, and RGB resolution of 1000 lines and video resolution of 650 lines is possible, and suitable for computer displays.

Flood Beam CRT Displays

Arrays of flooded screen CRTs have been used in obtaining very large displays of up to 20 m diagonal for audiences in some large sport stadiums. The display element consists of a trio of flood beam tubes (giving 2.5 to 15-cm diameter bright pixel dots!) one each for R G and B. Requiring over 140,000 tubes for each colour individually addressable, the connections and drive electronics is complex and costly. This has been reduced in practice by a factor of 10 by special matrix addressing techniques using multiple gating signals.

LCD Projector

LCD panels are currently available for display through an overhead projector. These have been useful for projecting compute displays for large screen viewing as well as video images from S-VHS VCRs with a good resolution of around 350×240 lines. Prototypes of projectors for display of an image from three small RGB LCD panels inside a unit onto a wall or screen with a diagonal size from 20 to 100 inches, have been recently exhibited by Sharp, Toshiba and Sanyo. Weighing less than 14 kg, the projection systems can zoom the image size without loss of picture clarity up to 300 to 350 lines. The systems are based on three active matrix LCD panels that employ twin thin film transistors. Sharp system model XV-100 employs three 7.5 cm diagonal thin film transistor—LCD panels separate the source beam into R, G and B components to provide a picture over 268,000 pixels. The projection schematic of the system is shown in Fig. 16.14.

Light from a 100 W RGB lamp, bouncing off the regular mirror M, passes through the ultra-violet (blocking protection) filter, and is split into red, green and blue components by the two dichroic mirrors as shown, before each component hits the corresponding LCD. The three LCD plates controlled by the respective video sources to activate required image pixels, are illuminated by corresponding collimated colour beams. The illuminated LCD panels are projected onto the screen by he variable focus zoom lens to recombine the colour image.

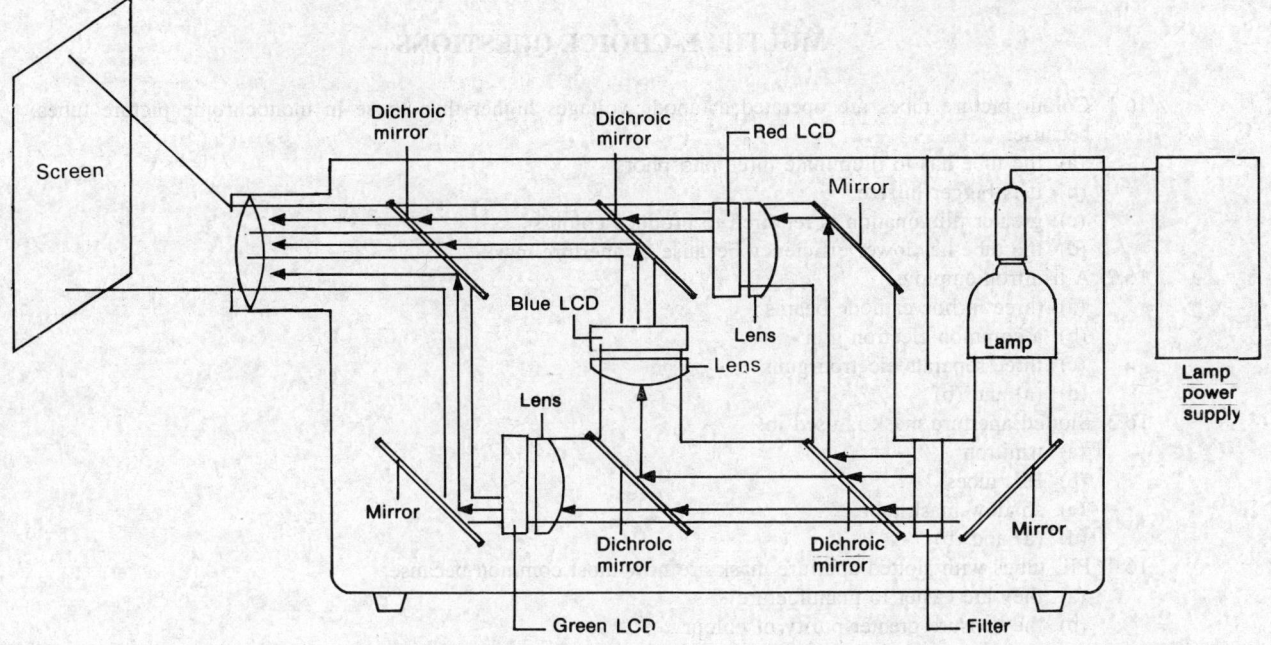

Fig. 16.14 Projection schematic of LCD Projector (Sharp Corp) (*Courtesy*: IEEE Spectrum, Aug. 1989)

REVIEW QUESTIONS

16.1 Give the construction of the shadow mask tube. Explain the operation and give reasons for its low efficiency. Describe the convergence technique employed.

16.2 Explain the terms purity and convergence, and the need for static and dynamic convergence. How are these arranged in modern colour picture tubes?

16.3 What are the advantages of in-line cathode construction compared to the delta gun structure? What are the special features of trinitron that make it more attractive?

16.4 Explain how the tricolour phosphor screen is manufactured to ensure correct registration of colours?

16.5 Discuss with the help of diagrams the use of multipole ring magnets along with the yoke to adjust purity and convergence.

16.6 What is degaussing? Why is it necessary in a TV receiver? Describe the automatic degaussing technique used in modern receivers. When is external degaussing necessary?

16.7 Explain the need for gray scale tracking. How is it achieved by the adjustments of the drive signals to the CPT?

16.8 What is the principle of LCDs. Discuss the limitations and technological solutions to improve the performance of LCDs for use in television displays.

16.9 Describe with the help of a schematic diagram the working of an LCD TV display. How are the colours displayed?

16.10 Describe the construction and working of one of the following: (a) Trinitron, (b) Precision-in-line colour tubes (c) LCD projector, (d) TFT LCD colour flat panel display.

16.11 Discuss the relative performance of the different colour picture tubes. What does the term 'black stripe picture tubes'?

16.12 Compare the technologies suitable for flat panel displays. Discuss recent trends in plasma type displays.

MULTIPLE-CHOICE QUESTIONS

16.1 Colour picture tubes are operated at anode voltages higher than those in monochrome picture tubes, because:
    (a) the tube has to illuminate three phosphors
    (b) it is bigger in size
    (c) greater illumination is required to produce colours
    (d) the tube has lower efficiency because of aperture mask

16.2 A trinitron employs:
    (a) three in-line cathode beams
    (b) a common electron gun
    (c) three separate electron guns
    (d) (a) and (b)

16.3 Slotted aperture mask is used in:
    (a) trinitron
    (b) PIL tubes
    (c) shadow mask tube
    (d) (a) and (b)

16.4 PIL tubes with slotted aperture mask are now most common because:
    (a) they are easier to manufacture
    (b) they ensure greater purity of colour
    (c) they are self-convergent

16.5 Automatic degaussing employs two thermistors,
    (a) one each for sensing the temperature of the two deflection coils
    (b) for generating a decaying current in the deflection yoke
    (c) for generating decaying ac current in a separate set of magnetizing coils
    (d) (a) and (b)

16.6 Complexity of drive electronics in flat panel displays for large screen displays is reduced by:
    (a) multiplexing
    (b) gate signal coding
    (c) matrixing
    (d) (b) and (c)

16.7 Liquid crystal displays more commonly employ:
    (a) cholesteric crystals
    (b) nematic crystals
    (c) twisted nematic crystals
    (d) sematic crystals

16.8 In twisted-nematic cells polarized light is rotated:
    (a) when electric field is applied across the crystal layer
    (b) when electric field is absent
    (c) when the applied electric field changes
    (d) none of these

16.9 Pick the wrong statement:
    (a) Liquid crystal displays are realized by field effect cells
    (b) Supertwisted-nematic LCDs allows more lines to be displayed than active matrix TFT LCDs.

16.10 Black level half tone adjustments are needed:
    (a) to adjust the gamma characteristics of a colour picture tube
    (b) to match the colours at low lighting levels
    (c) to adjust the cut-off points of a colour picture tube
    (d) to match the colour mixing at half-way light levels.

17

Horizontal Deflection Circuits and EHT Generation

17.1 DEFLECTION CURRENT REQUIREMENTS

Because of the flat face of the picture tube, variation of current through the deflection coils, required to obtain linear deflection of the spot on the screen, is not purely linear. The actual variation of the deflection currents required for a given picture tube and deflection coils can be determined experimentally by passing a direct current through the coils to obtain displacements of the spot, along vertical and horizontal scales on the tube face. In order to prevent screen burn due to intense focused spot, the unused coil may be supplied an alternating current to obtain a line deflection in the other direction.

It will be observed that for linear deflection or displacement of the spot, the required current variation with respect to time has a maximum near the centre of the screen, particularly in large deflection angle tubes with flat screen, as shown in Fig. 17.1.

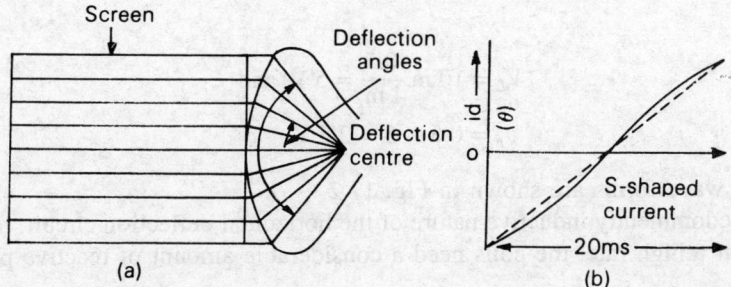

Fig. 17.1 Deflection current requirement of a flat screen large deflection angle picture tube: (a) Deflection angles, (b) Current requirement

The current drive required will be somewhat S-shaped rather than linear sawtooth. The deviation from linearity is not excessive, and the circuit requirement may be calculated on the basis of linear currents.

The voltage drive across the deflection coil consists of an inductive component $V_L = L \dfrac{di}{dt}$, and a resistive component of $v_r = ir$. As already seen in the last chapter, in horizontal deflection, the inductive component is much greater than the resistive drop, as the horizontal deflection rate is quite high. During the forward scan,

$$V_L = L_H \frac{di}{dt} = L_H \frac{I_{pp}}{t_s}$$

where I_{pp} is the peak-to-peak deflection current required in the coil L_H, and t_s is the forward scan time. For a typical horizontal coil $L_H = 2$ mH, and $r = 4\ \Omega$

$$I_{pp} = 3\text{A}\ ;\ t_s = 54\ \mu\text{s};\ t_f = 10\ \mu\text{s}$$

∴
$$V_L = 2\,\text{m} \times \frac{3}{54\,\mu} = 111\,\text{V, and}$$

$$v_r = i \times r = 3 \times 4 = 12 V_{pp}$$

During flyback, assuming linear current fall,

$$V_L = L_H \frac{di}{dt} = 2\,\text{m} \times \frac{3}{10\,\mu} = 600\,\text{V, and}$$

$$v_r = i \times r = 3 \times 4 = 12\ V_{pp}$$

Thus $V_L \gg v_r$, for the vertical deflection coil.
For a typical vertical coil $L_v = 10$ mH, and $r_v = 5\ \Omega I_{pp} = 0.5$ A.
Taking scan time $t_s = 19$ ms and flyback time $t_f = 1$ ms

$$V_L = L_v \cdot \frac{di}{dt} = L_v \cdot \frac{I_{pp}}{t_s} = 10\,\text{m} \times \frac{0.5}{19\,\text{m}} = 0.26\,\text{V, and}$$

$$v_r = i \times r = 0.5 \times 5 = 2.5\,\text{V}$$

Thus $v_r \gg V_L$, for the vertical deflection coil.
During flyback,

$$V_L = 10\,\text{m}\ \frac{0.5}{1\,\text{m}} = 5\,\text{V, and}$$

$$v_r = 0.5 \times 5 = 2.5\,\text{V}$$

These drive voltage wave forms are shown in Fig. 17.2.

Because of the predominently inductive nature of the horizontal deflection circuit, requiring magnetic energy to alternate at a high rate, the coils need a considerable amount of reactive power given by:

$$\frac{1}{2} L_H \left(\frac{I_{pp}}{2}\right)^2 f_H = \frac{1}{8} L_H I_{pp}^2 f_H = \frac{1}{8} \times 2\,\text{m} \times 3^2 \times 15625$$

$$= 35\,\text{W}$$

Resistive power in the coil due to sawtooth currents flowing in the resistance r_H is given by

$$\frac{1}{12} I_{pp}^2 \cdot r_H = \frac{1}{12} \times 3^2 \times 4 = 3\,\text{W}$$

The time-constant $L_H/r_H = \frac{2\,\text{m}}{4} = 500\ \mu\text{s}$, which is quite high compared to the horizontal scan period.

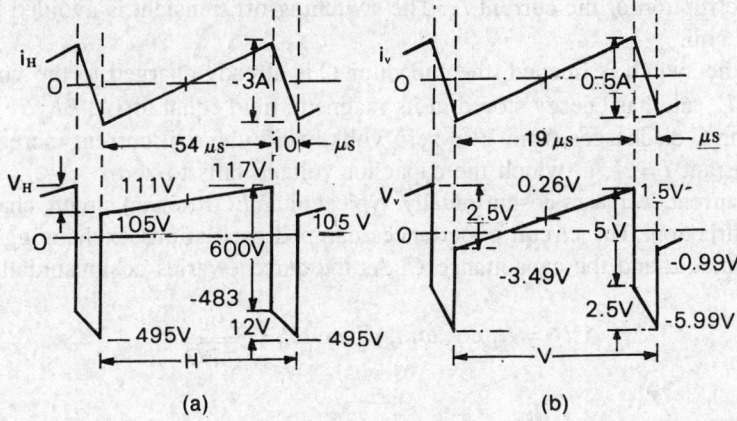

Fig. 17.2 Voltage drive for: (a) horizontal, and (b) vertical deflection

This implies that the horizontal coil is essentially inductive over the scan period. The reactive power in the coil is quite high, a major percentage of the total power requirement of the receiver. It is, therefore, very desirable to recover the energy from the magnetic field and use it again during the next scan period. The resistive loss of power only, may then be drawn from the supply.

17.2 BASIC HORIZONTAL DEFLECTION CIRCUIT

If the resistive drop of the coil is neglected, a constant voltage applied across the deflection coil gives a linear sweep current through the coil. Hence the basic circuit for horizontal deflection consists of a direct voltage V connected *via* an active device having characteristics of a switch, to the coil in parallel with a capacitance C normally consisting of the parasitic self-capacitance of the coil, as shown in Fig. 17.3.

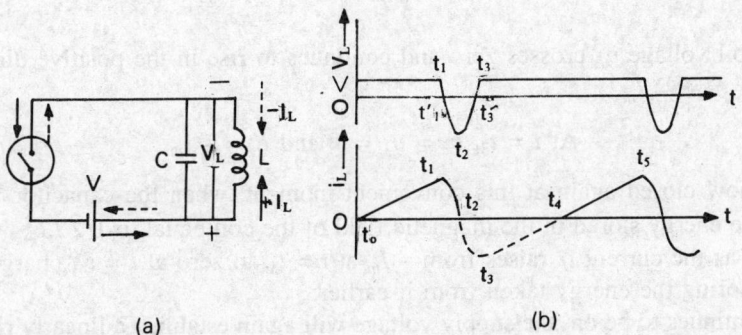

Fig. 17.3 (a) Basic horizontal deflection circuit. (b) drive waveforms

On closing of the switch at $t = t_0$, a current linearly increasing at the rate $\dfrac{di}{dt} = \dfrac{V}{L}$, flows through the coil, while the capacitor also takes a sudden charging current to charge to the voltage V, until the switch is opened at $t = t_1$. By this time, the linearly rising current has attained the value $I_m = \dfrac{V}{L}$.

The circuit being highly inductive, very high arcing voltage will be induced across the switch depending upon the rate of interruption of the current I_m. The switching-off transient is avoided by the capacitor C in parallel with the coil.

At $t = t_1$, when the switch is opened, the capacitor C is already charged to the voltage V, while the coil carries current I_m, and has energy stored in its magnetic field equal to $1/2 \, LI_m^2$ J. On opening of the switch, the capacitor C discharges from V to zero volts, while the coil current increases a little, not as linearly until the instant $t = t_1'$, at which the capacitor voltage falls to zero.

After $t = t_1'$, the current decreases cosinusoidally, typical of an LC resonant circuit, charging the capacitor C in the opposite direction. The circuit experiences damped oscillations exchanging the energy stored between the inductance L and the capacitance C. As the current varies cosinusoidally, the coil current

$$i_L = I_m \cos \omega_0 t, \text{ where } \omega_0 = \frac{1}{\sqrt{LC}} \qquad (17.1)$$

Voltage across the coil given by $L \dfrac{di}{dt}$ increases sinusoidally,

$$v_L = V_m \sin \omega_0 t, \text{ where } V_m = \omega_0 L I_m \qquad (17.2)$$

At $t = t_2$, i.e., after a quarter period of the frequency $f_0 = \dfrac{\omega_0}{2\pi}$, the current through the coil reaches zero and v_L reaches the peak value V_m. The capacitor now carries this voltage, storing energy in its electrostatic field, given by $1/2 \, CV_m^2$.

Now, Energy in C (at $t = t_2$) = Energy in L + Energy in C, (both at $t = t_1$)

Hence
$$1/2 \, CV_m^2 = 1/2 \, LI_m^2 + \frac{1}{2} CV^2 \qquad (17.3)$$

This gives
$$V_m = \sqrt{\frac{L}{C} I_m^2 + V^2} \qquad (17.4)$$

At $t = t_3'$, the coil voltage v_L crosses zero and continues to rise in the positive direction, as the coil current crosses the peak $-I_m$.

At $t = t_3$, $v_c = v_L = V$ and $i_L = -I_m$

The switch is now closed again at this convenient moment, when the capacitor voltage equals the supply voltage. The energy stored in the magnetic field of the coil equal to $1/2 \, LI_m^2$ it is returned to the supply by the coil as the current i_L raises from $-I_m$ at $t = t_3$, to zero at $t = t_4$, charging the dc voltage source V, thus restoring the energy taken from it earlier.

If the switch continues to be on, the supply voltage will again establish a linearly rising current at the rate V/L, to reach $+I_m$ at $t = t_5$ when the switch is opened again.

The interval t_3 to t_5 is the forward scan when i_L rises from $-I_m$ to $+I_m$ linearly at the rate $\dfrac{V}{L}$ A/s, while the interval t_1 to t_3 is the flyback period when the current swings over to the opposite value with the aid of the capacitance C.

The switch conducts in both directions. The replacement of the switch in practice, therefore, cannot be by just one electron device-pentode, transistor, or thyristor which pass current in one direction only and the recovery of energy is not possible, without additional diode for conducting current in the opposite direction. The controlled electron device as well as the diode should be good switches, offering low internal impedance to be able to carry large currents with reduced dissipation.

For proper operation, the duration of the flyback must be determined exclusively by the natural period of the oscillation of the LC combination and must not be affected by the switch opening time if it differs from this. The flyback period is somewhat longer than half the period of oscillation. The interval from t_1 to t_1' is usually included in the flyback as the current increase during that interval is only slight and nonlinear.

17.3   PEAK FLYBACK VOLTAGE

The relation between the peak voltage V_m, which appears across the coil during the flyback and the voltage V, supplied to the coil can be found out in terms of the flyback ratio $p = \dfrac{t_f}{t_s + t_f}$, by relating it with the oscillation period. We have the energy equation:

$$\frac{1}{2} CV_m^2 = \frac{1}{2} LI_m^2 + \frac{1}{2} CV^2$$

where,

$$I_m = \frac{V t_s}{2L}$$

This gives

$$V^2 = V_m^2 \frac{4LC}{t_s^2 + 4LC} \tag{17.5}$$

Now $LC = \dfrac{1}{\omega_0^2}$, and $V = V_m \cos\left(\omega_0 \dfrac{t_f}{2}\right)$, because $v_L = V_m \sin \omega_0 t$

At $\quad \omega_0 t = \dfrac{\pi - \omega_0 t_f}{2}, v_L = V = V_m \sin\left(\dfrac{\pi - \omega_0 t_f}{2}\right) = V_m \cos\left(\dfrac{\omega_0 t_f}{2}\right)$

Therefore,

$$V_m^2 \cos^2\left(\frac{\omega_0 t_f}{2}\right) = V^2 = \frac{4/\omega_0^2}{t_s^2 + (4/\omega_0^2)} V_m^2$$

$$t_s^2 = \frac{4}{\omega_0^2 \cos^2\left(\dfrac{\omega_0 t_f}{2}\right)} - \frac{4}{\omega_0^2}$$

$$= \frac{4}{\omega_0^2} \frac{1 - \cos^2\left(\dfrac{\omega_0 t_f}{2}\right)}{\cos^2\left(\dfrac{\omega_0 t_f}{2}\right)}$$

$\therefore \qquad t_s = \dfrac{2}{\omega_0} \dfrac{\sin\left(\dfrac{\omega_0 t_f}{2}\right)}{\cos\left(\dfrac{\omega_0 t_f}{2}\right)}$

Now

$$\frac{\omega_0 t_f}{2} \approx 90°, \text{ therefore, } \sin\left(\frac{\omega_0 t_f}{2}\right) \approx 1$$

and

$$\cos\left(\frac{\omega_0 t_f}{2}\right) \approx \sin\left(\frac{\omega_0 t_f}{2} - \frac{\pi}{2}\right) \approx \left(\frac{\omega_0 t_f}{2} - \frac{\pi}{2}\right)$$

$$\therefore \qquad t_s = \frac{2}{\omega_0} \frac{1}{\dfrac{\omega_0 t_f}{2} - \dfrac{\pi}{2}} = \frac{4}{\omega_0^2 t_f - \pi\omega_0}$$

$$\therefore \qquad \frac{4}{t_s} = \omega_0^2 t_f - \pi\omega_0. \text{ Multiplying by } \frac{t_f}{\pi^2}$$

$$\frac{4}{t_s} \cdot \frac{t_f}{\pi^2} = \frac{\omega_0^2 t_f^2}{\pi^2} - \frac{\omega_0 t_f}{\pi}$$

$$\therefore \qquad \frac{4}{\pi^2} \times \frac{t_f}{t_s} = \frac{\omega_0 t_f}{\pi}\left(\frac{\omega_0 t_f}{\pi} - 1\right) \qquad (17.6)$$

Now

$$\frac{t_f}{t_s + t_f} = p, \text{ hence } \frac{t_f}{t_s} = \frac{p}{1-p}$$

Substituting in Eq. (17.6), we have,

$$\frac{\omega_0 t_f}{\pi}\left(\frac{\omega_0 t_f}{\pi} - 1\right) = \frac{4}{\pi^2}\frac{p}{1-p} \qquad (17.7)$$

Here $\dfrac{\omega_0 t_f}{\pi} = \dfrac{2\pi f_0}{\pi} t_f = \dfrac{t_f}{T_0/2} (\approx 1)$, T_0 being the period of oscillation.

We have,

$$\frac{V_m}{V} = \frac{1}{\cos\dfrac{\omega_0 t_f}{2}} = \frac{1}{\sin\left(\dfrac{\omega_0 t_f}{2} - \dfrac{\pi}{2}\right)} \approx \frac{1}{\dfrac{\omega_0 t_f}{2} - \dfrac{\pi}{2}}$$

$$= \frac{2/\pi}{\left(\dfrac{\omega_0 t_f}{\pi} - 1\right)}$$

Substituting for the denominator from Eq. (17.7)

$$\frac{V_m}{V} = \frac{2/\pi}{\dfrac{4}{\pi^2}\dfrac{p}{1-p}\cdot\dfrac{\pi}{\omega_0 t_f}} = \frac{\pi}{2}\frac{1-p}{p}\left(\frac{\omega_0 t_f}{\pi}\right) \tag{17.8}$$

As $\dfrac{\omega_0 t_f}{\pi} \approx 1$, we have from Eq. (17.7)

$$\left(\frac{\omega_0 t_f}{\pi} - 1\right) \approx \frac{4}{\pi^2}\frac{p}{1-p}$$

$$\therefore \qquad \frac{\omega_0 t_f}{\pi} \approx 1 + \frac{4p}{\pi^2(1-p)}$$

Substituting this in Eq. (17.8)

$$\frac{V_m}{V} = \frac{\pi}{2}\frac{(1-p)}{p}\left(1 + \frac{4p}{\pi^2(1-p)}\right)$$

$$= \frac{\pi}{2}\frac{1-p}{p} + \frac{2}{\pi}$$

$\therefore$ The flyback-to-scan voltage ratio:

$$F_p \overset{\Delta}{=} \frac{V_m}{V} = \frac{\pi}{2}\frac{1-p}{p} + \frac{2}{\pi} \tag{17.9}$$

This is plotted as a function of p as shown in Fig. 17.4.

The above relations assume a sinusoidal flyback. In practice, a higher frequency third harmonic voltage is superposed on this flyback voltage which reduces the V_m on the switching device, to an extent of 20%. The device must be able to withstand the peak voltage with adequate safety margin. For the normal values of p between 14 to 18%, the ratio $\dfrac{V_m}{V}$ is about 10 to 8.

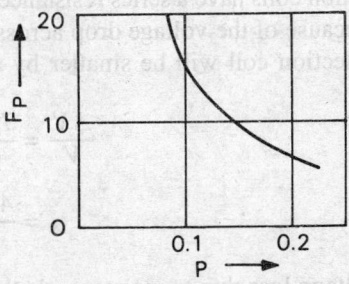

Fig. 17.4 Flyback-to-scan voltage ratio F_p as a function of flyback ratio p

17.4 ENERGY RECOVERY SYSTEM

In the basic deflection circuit, it has been observed that the energy stored in the coil during the interval t_0 to t_1 is returned to the battery dc source, *via* the switch. In practice, the switch function during this period is performed by a separate diode.

The energy returned can be utilized in two ways, to form either a parallel efficiency configuration as in Fig. 17.5(a) or a series efficiency configuration as shown in Fig. 17.5(b).

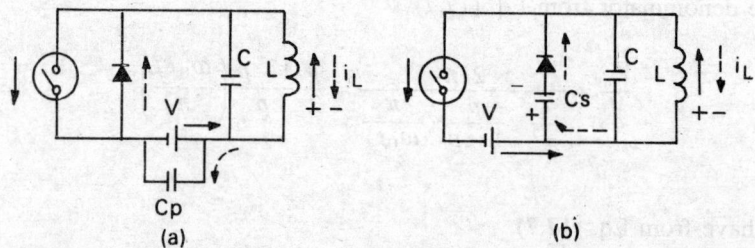

Fig. 17.5 Energy recovery circuits: (a) Parallel efficiency circuit, (b) Series efficiency circuit

In the parallel efficiency circuit, the diode conducts to charge the supply reservoir condenser so that the average current from the supply is much less compared to the peak current of the device. This configuration is more commonly prevalent in transistorized stages in which the current can be larger but the voltage ratings are limited.

In Fig. 17.5(b), the series efficiency diode charges capacitor C_s called the booster capacitor. The additional voltage available from the recovered energy, can be much higher than the dc source battery voltage V. This higher voltage can be used to feed stages operating at high voltages, and with some more modifications, the output stage itself which can then operate at a higher 'boosted' voltage. This configuration is more commonly used in vacuum tube stages, because with a higher supply voltage, the currents in the circuit are reduced to suit valves, and the plate dissipations in the valves are also reduced.

The diode and also the coil are more conveniently connected to the output pentode or transistor through a transformer which can provide a good matching for the deflection coil, so that the required deflection current can be obtained in the deflection coil with available supply voltage.

17.5 HORIZONTAL LINEARITY CORRECTION

The deflection coils have a series resistance r_H. During the scans, the voltage available to the coil inductance reduces because of the voltage drop across the resistance at increasing currents, and the voltage available to the deflection coil will be smaller by a factor,

$$\frac{\Delta V}{V} = \frac{I_m r_H}{V} = \frac{r_H f_s}{L} \quad \left(\because I_m = \frac{V}{L} t_s \right)$$

$$= \frac{4 \times 54\mu}{2\,\text{m}} \approx 0.1 \text{ typically}$$

The voltage loss due to resistance is thus about 10–15% in practice. The *di/dt* at the end of the scan will then be correspondingly smaller and will cause the picture to stretch on the left and compress on the right side of the screen. A constant voltage across the deflection coil is also not obtainable because of the on-resistance of the switching device which is not zero. The nonlinearity in deflection, therefore, is appreciable and a linearity correction to compensate for it, is essential.

A commonly used method for linearity correction is to use a linearity coil as shown in Fig. 17.6. It consists of a small coil wound on a ferrite core which is pre-magnetized by means of a permanent magnet mounted adjacent to it. The coil is connected in series with the horizontal deflection coil, and behaves as a saturable reactor, the reactance of which can alter with saturation of the magnetic flux in the ferrite core. At the beginning of the scan, the direction of current is such that the induced flux reduces the field

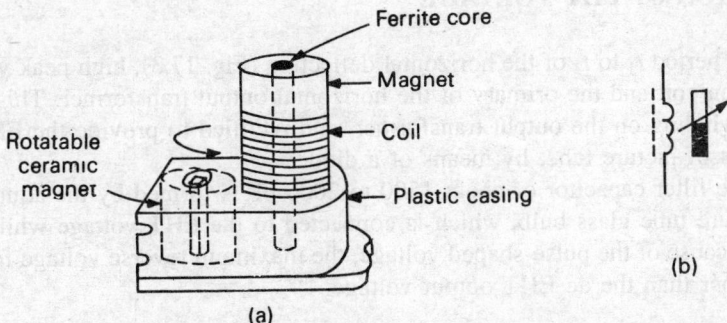

Fig. 17.6 Horizontal linearity coil: (a) Construction, (b) Symbol

of the permanent magnet. The core saturation is then minimum and thus the coil reactance and hence the drop across it is maximum. After half scan, the current changes direction and the ferrite is driven further into saturation. Its self-inductance and hence the voltage drop across it is reduced.

The magnet is adjustable by rotating it, so that the change in voltage drop across the deflection coil is exactly compensated by an equal and opposite change in the voltage drop across the linearity coil. This gives quite a satisfactory result at the price of only a small extra voltage drop lost across the coil. The loss due to pre-magnetization is 2 to 3% while in the absence of it, the loss is around 15%. Inclusion of a linearity coil in series with the deflection coil may give rise to parasitic oscillations immediately after flyback, in liaison with self-capacitance and stray capacitance, causing vertical stripes on the left side of the screen. This can be reduced by adequate damping for the coil by an extra shunt resistance.

Symmetrical Nonlinearity S-Correction

It has been noted at the beginning of the chapter that modern wide angle picture tubes require S-shaped deflection currents because of the almost flat screen. A capacitor in series with the coil as shown in Fig. 17.4(a), provides the required S-correction by reducing the voltage available to the coil at the larger values of deflection current. The value of the capacitor is such that the resonant frequency of the series circuit formed is far below the line frequency. The sawtooth current which flows through the capacitor produces an approximately parabolic voltage across it.

During the first half of the period, the capacitor C is charged by a decreasing current of the deflection coil so that its voltage is rising parabolically. During the second half, the voltage falls back to zero parabolically due to the current that is reversed. The voltage across C then starts increasing in the opposite direction. The voltage across the deflection coil during scan is the sum of the constant voltage across the transformer and the parabolic voltage across the capacitor. The current, therefore, rises at the fastest rate at the centre and becomes S-shaped as seen in Fig. 17.7. If the value of the capacitor (C) is reduced by the series resonant frequency approaches the line frequency, and the parabolic form of current may become more sinusoidal. A lower tapping on the transformer would then be necessary, to achieve proper correction.

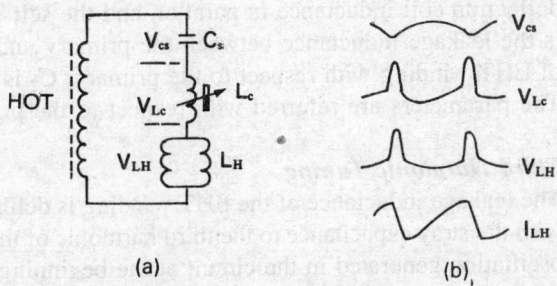

Fig. 17.7 Horizontal nonlinearity correction: (a) Circuit, (b) Waveforms

17.6   GENERATION OF EHT VOLTAGE

During the flyback period t_1 to t_3 of the horizontal deflection (Fig. 17.3), high peak voltage pulses occur across the deflection coil and the primary of the horizontal output transformer. This can be stepped up by an extra EHT winding on the output transformer, and rectified to provide the EHT required for the accelerating anodes of picture tube, by means of a diode.

The high voltage filter capacitor of about 1500 to 2000 pF is formed by the aquadag coating on the outside of the picture tube glass bulb, which is connected to the EHT voltage while the outer coating forms the earth. Because of the pulse-shaped voltage, the maximum reverse voltage for the EHT rectifier is only a little higher than the dc EHT output voltage.

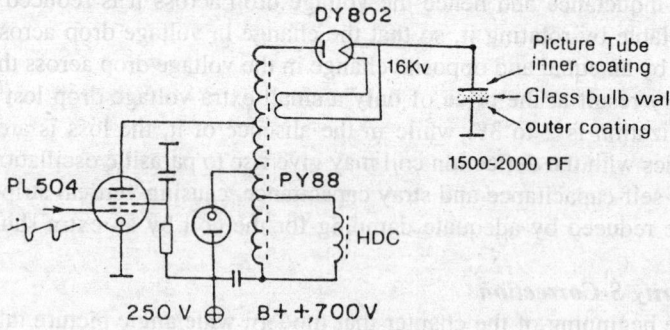

Fig. 17.8   Horizontal output stage with EHT circuit

Care is taken in fitting the EHT winding on the transformer. The EHT rectifier draws pulse current largely from the parasitic capacitance at the anode of the rectifier. For pulse-shaped load current, the unavoidable leakage inductance forms an impedance isolation between the EHT rectifier and the larger primary winding capacitance. The equivalent circuit of EHT portions of the output stage with reference to the primary is shown in Fig. 17.9, where L_1 and C_1 represent respectively the primary inductance, and deflection coil inductance in parallel, and the self-capacitance and the tuning capacitance in parallel. L_k is the leakage inductance between the primary and the secondary EHT windings. C_k is the capacitance of EHT winding with respect to the primary. C_2 is the parasitic capacitance of EHT winding to ground. The parameters are referred with respect to the primary impedance level.

Third Harmonic Tuning

The leakage inductance of the EHT winding is deliberately made greater (L_k is about $L_1/2$) so that it tunes with the stray capacitance to the third harmonic of the flyback primary resonant frequency, f_0. The parasitic oscillation generated in the circuit at the beginning of the flyback is cancelled out again during the end of the flyback, so that there is minimum ringing during the scan time. The peak voltage at $3f_0$ occurring across the leakage inductance at the middle of the flyback period adds on to the input voltage giving a

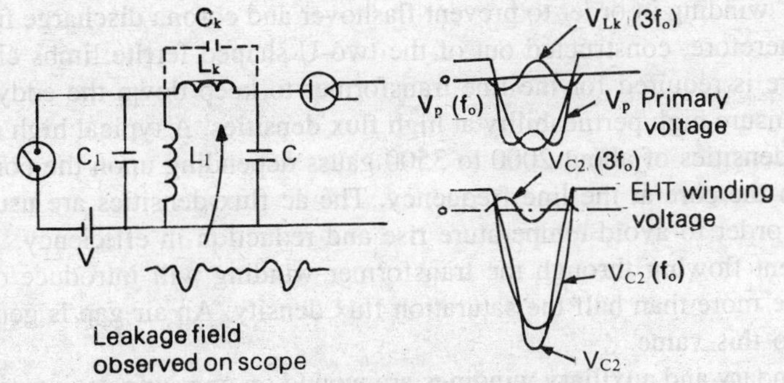

Fig. 17.9 EHT winding equivalent circuit and third harmonic tuning

higher EHT pulse as shown in Fig. 17.9. While due to reaction in the oscillation current on the primary of the transformer, the primary peak voltage is reduced by about 15%. This allows a greater magnetic power to be handled by the switching device.

The correct tuning can be obtained by bringing the leakage inductance to the required value by adjusting the placement of the EHT winding on the core with respect to the primary or by short-circuiting turns around the stray fields by mounting a spacing ring between the two windings. The tuning can be simply observed on the oscilloscope with the help of a single turn coupling loop placed in the stray field near the core of the transformer. The tuning is correct when the parasitic oscillations have greater amplitude during flyback than during the forward scan, as shown in Fig. 17.9.

17.7 EHT REGULATION

The EHT voltage falls with increase in the beam current load due to the internal impedance of the EHT rectifier circuit. Fall in EHT causes widening or 'blooming' of the picture as the deflection sensitivity increases with decreasing EHT, being proportional to $\sqrt{E_a}$. The internal impedance should be as low as possible and should not exceed 5 MΩ. This is kept low by minimizing the internal resistances of the EHT rectifier (adequate heater power) and output valve (low impedance operation) or by a feedback circuit to stabilize the EHT.

The increase in deflection sensitivity caused by decrease in EHT may just be compensated by maintaining the ratio between deflection current and $\sqrt{EHT}$, constant. A low internal impedance is hence not quite essential for good regulation of picture width. However, lower internal impedance maintains more constant EHT which is desirable from the point of greater brightness and better focusing at larger beam current.

As the field sensitivity also increases with fall in EHT, the field-deflection can be linked with EHT by using supply voltage for the vertical deflection oscillator, which can be set-up to produce proportional sawtooth voltage drive to the output stage so that the deflection currents produced, compensate for the change in sensitivity.

17.8 HORIZONTAL OUTPUT TRANSFORMER

The transformer core and size must be of material and dimensions to ensure good efficiency. The losses in the deflection circuit due to parasitic oscillations caused by leakage inductance must be minimized. It is necessary to have a very large window opening for the core to ensure adequate spacing between the primary and the EHT winding in order to prevent flashover and corona discharge from parts carrying the EHT. The core is, therefore, constructed out of the two-U-shaped ferrite limbs clamped together.

Special ferrite core is required for the line transformer to keep down the eddy current losses by its high resistivity and ensure high permeability at high flux densities. A typical high resistivity core allows high saturation flux densities of about 2000 to 3500 gauss depending upon the core temperature, which rises due to losses in the core at the line frequency. The ac flux densities are usually not kept beyond 2000–2500 gauss in order to avoid temperature rise and reduction in efficiency.

The average current flowing through the transformer winding will introduce dc flux density in the core, which cannot be more than half the saturation flux density. An air gap is generally used to reduce the dc flux density to this value.

The primary, secondary and auxiliary windings are wound on one support, impregnated with polyester resin and placed on one limb of the core. The EHT winding, encapsulated suitably in polyester resin, is placed usually on the other limb. All the windings terminals are brought to a terminal plate and the EHT terminal is brought to a well insulated valve cap or EHT solid state rectifier cap. In some line output transformers like AT 2048/12 designed for transistor operation, the primary, auxiliary and EHT windings are all situated on one leg of the core.

17.9   TRANSISTOR LINE OUTPUT STAGE

The switching transistor employed for the horizontal output stage must have a high current rating as well as a high reverse pulse voltage rating; and it should have medium-fast switching speed to operate satisfactorily at the 15,625 Hz switching rate. Suitable transistors with high current switching at fast speeds, along with high reverse breakdown capabilities are now available and solid state line output stages are commonly found in modern TV sets that provide improved efficiency and lower power dissipation.

Transistor line output stages generally employ deflection coils of lower impedance design, as this keeps down the voltage rating required. The deflection coils for transistor stages have a high inductance compared to their resistance so that the drive voltage requirement is almost rectangular. The transistor in such a situation can function as just a switch put on by the base current drive to connect the supply voltage, stepped down through the output transformer ratio, cross the deflection coil. This sends a sawtooth current through the coil until the transistor switch is made off by the base drive, periodically.

As discussed in the Section 17.5, there is a slight reduction in the voltage across the resistance of the coil, as the current builds up. This introduces a compression in the horizontal deflection to the right that can be corrected by nonlinearity coil in series with the deflection coil, functioning as a nonlinear compensating inductance.

The transistor is given a rectangular base drive that generally cuts-off the transistor for a longer period than the retrace time for reducing the current requirement and improving the efficiency. The base drive ON/OFF duty cycle is adjusted to limit the current drawn from the supply, by use of energy recovery in the process of flyback.

The transistor output stage should have a bias system that prevents excessive currents in the event of drive failure. Over current and over voltage protection must be provided to prevent damage to the transistor due to flashover, etc.

A basic transistor line output circuit is shown in Fig. 17.10(a)

The circuit employs a parallel efficiency circuit for recovery of energy *via* the damper diode D1. The linearity coil *Lc* compensates for the drop in the small coil resistance that tends to compress the deflection on the right hand side. The capacitor C_s provides S-correction for the coil current as discussed in Sec. 17.5. The drive and the output voltage and current waveforms are shown in Fig. 17.10(b). The total load

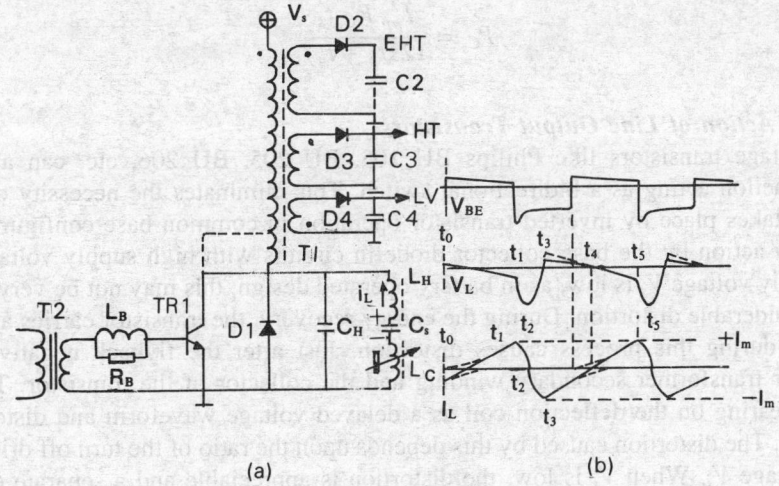

Fig. 17.10 Transistor line output stage: (a) Circuit arrangement, (b) Waveforms

of the output circuit is thus formed by the deflection coil L_H, the S-correction capacitor C_s and the linearity coil Lc, all connected in series.

The capacitor C_s is initially charged to the supply voltage V_s and functions as an auxiliary voltage source. At $t = t_0$, the positive base drive v_{BE} puts the transistor *on* deep into saturation and a linearly increasing current i_L flows through the deflection coil L_H. The voltage drop across the transistor at the end of the scan is $V_{CE_{sat}}$. If the small drop (less than 2%) across the linearity coil Lc is neglected the voltage drop across the deflection coil at the end of the scan period is $V_s - V_{CE_{sat}}$.

At $t = t_1$, the transistor is turned *off* by the negative going base drive voltage and i_l decreases cosinusoidally from its peak value $+ I_m$ at $t = t_1$, to the negative peak value $- I_m$ at $t = t_3$, crossing through 0 at $t = t_2$ during the half cycle of oscillation determined by L_H and C_H. During this flyback period the diode D1 is reverse biased by the positive pulse voltage at the collector with respect to the emitter. As this voltage reverses at the end of the flyback, the diode D1 conducts the reversed current $- i_L$ continues to flow through the deflection coil to charge C_s, restoring the energy taken from it during the period t_0 to t_1. The voltage drop across the coil at the beginning of the scan is thus $V_s + V_{D_1}$ and $V_s - V_{CE_{sat}}$ at the end of the scan.

The nonlinearity in the forward characteristics of the diode D1 may introduce a jump in the voltage cross the coil if V_{D_1} is comparable to V_s, causing distortion. This is reduced by connecting the diode to a tap on the transformer, which reduces its effect on the voltage across the coil.

The line output transformer primary does not actually carry the deflection current. It carries only a small magnetizing current and the average load current required. The flyback tuning capacitor C_H and the S-correction capacitor C_s must have minimum series loss resistance. The power loss in the series resistance R_{CH} of the flyback tuning capacitor C_H carrying sinusoidal current during the flyback time t_f is given by

$$P_{CH} = \frac{I_{pp}^2 \, R_{CH} t_f}{8 \, (t_s + t_f)}$$

The power loss due to the sawtooth current through the capacitor C_s during the scan period t_s is given by

$$P_{C_s} = \frac{I_{pp}^2 R_{C_s} t_s}{12 (t_s + t_f)}$$

Efficiency Diode Action of Line Output Transistors

Modern high voltage transistors like Philips BU 105, BU 205, BU 206, etc. can also perform the efficiency diode action acting as a bidirectional switch. This eliminates the necessity of the diode D1. Energy recovery takes place by inverted transistor operation in common base configuration, providing parallel efficiency action by the base collector diode in circuits with high supply voltages.

When the supply voltage V_s is low, as in battery operated design, this may not be very convenient and may result in considerable distortion. During the energy recovery, the transistor carries a reverse current. The nonlinearity during this process causes distortion. Just after the flyback negative current flows through the driver transformer secondary winding and the collector of the transistor. The v_{BE} turn off pulse will be appearing on the deflection coil as a delayed voltage waveform and distort the left hand side of the picture. The distortion caused by this depends upon the ratio of the turn off drive voltage pulse to the supply voltage V_s. When V_s is low, the distortion is appreciable and a separate efficiency diode is necessary.

In the design of the line output stage, the choice of supply voltage, the transistor, the line output transformer and deflection coil unit are all linked together. The supply voltage is generally around one tenth of the breakdown voltage ratio BV_{CEX} of the transistor used.

The supply voltage requirements can be estimated from the deflection coil inductance the peak current requirements. The voltage drop V_0 across the deflection coil is given by

$V_0 = L_H \dfrac{I_{pp}}{(1 - p) T}$, where I_{pp} is the peak-to-peak current, L_H the inductance of the coil, p is the flyback ratio $= (t_f/T)$ and T the line period.

This may be increased by about 6% overscan to ensure adequate width against tolerances and by $V_{CE_{sat}}$ of around 5 V for the line horizontal voltage output transistor of high voltage type.

A small resistance is generally introduced in series with the transistor supply for flashover protection. This resistance along with the primary winding resistance of the output transformer is allowed typically a drop of around 10% of the supply voltage, increasing the supply voltage requirement correspondingly.

Flashover Protection

Flashover is the EHT arcing of the high voltage generated by the EHT winding on the output transformer and the EHT rectifier. The EHT arcing can take place inside the picture tube for very short durations as flashover across the tube coating EHT capacitor. The arcing can also occur outside the picture tube due to accidental or intentional short on the EHT lead. The EHT arcing causes short-circuit across the EHT and results in an increase of the current by a factor of about 3 and a decrease in the retrace time by a factor of 2. The flyback voltage would increase by a factor of 2.5. For the increase in the collector currents, the base drive may now be not enough to keep the transistor in saturation. If it comes out of saturation in the active region, there will be a large dissipation in the transistor during the scan period increasing the transistor temperature. These factors may result in avalanche breakdown and failure of the transistor. The series resistance limits the collector current to within safe limits.

Additional protection against flashover can be obtained by generating some auxiliary supply by connecting a diode to the collector of the transistor. This limits the magnitude of the peak value of the collector-emitter voltage during arcing.

Generation of EHT and Auxiliary Supplies

As discussed in Sec. 17.6, the large flyback pulse occurring across the deflection coil is stepped up with the help of EHT winding, to a high value and rectified to provide the EHT voltage required for the accelerating and the final anode of the picture tube. Solid state rectifiers like BY 185, TV 18, TV 20 are now commonly employed for EHT rectification in place of the thermionic rectifiers. *Third harmonic tuning* discussed in Sec. 17.6 is particularly desirable in the solid state designs for reducing the peak flyback voltages at the collector, while increasing the EHT pulse on the secondary side.

Fifth Harmonic Tuning

The horizontal oscillator transformer EHT winding is usually tuned to the third harmonic of the fundamental frequency of the flyback pulse (210 kHz). If the secondary winding could be tuned to the fifth harmonic wave, the top of the flyback pulse would become flatter, i.e., the conduction angle of EHT rectifier would become wider than in the case of the third harmonic wave and as a result, output resistance of the power supply circuit for EHT can be reduced to 2 MΩ without stabilization. It is extremely difficult, however, to tune the secondary winding to the fifth harmonic without set-by-set adjustment of coupling of primary and secondary coils, on the production line.

Auxiliary supply voltages required for the first anode of the picture tube are of the order of 400 to 500 V. These can also be obtained by transforming the flyback pulses and rectifying them by peak rectifier diodes of fast recovery type capable of 15,625 Hz pulse operation, like BY 184, BYX 55–77, etc. The actual load current of these anodes is negligible. The supplies may have to be adjusted by employing potentiometers which may, however, themselves draw appreciable load currents.

Low Voltage Supplies

Low voltage auxiliary supplies may be obtained by rectification of the suitably transformed horizontal output during the scan period. This requires the diode to withstand the high reverse flyback pulses during which the diode has to block. The diode conducts during the relatively long scan period. The peak current is, therefore, relatively small and internal impedance of the supply is kept low.

17.10 LINE DRIVER STAGE

Base Drive Requirements

The base drive circuit is usually a transformer coupled to supply the necessary base current for the output transistor. The base drive circuit includes often an inductance that acts similarly of the order of 25 μH in order to slow down the turnoff pulse by a few μs and allow the excess saturation charge to leak off. Alternatively the driver transformer is designed with leakage inductance. The inductance causes ringing during the switching off period of the transistor, which may tend to turn on. The ringing is damped by a shunting resistance across L_B, or a resistance across the base-emitter.

The base drive should be timed so that the forward bias is not applied to the transistor before the end of the flyback period. The turn-off delay time for BU 205 is 10 to 12 μs while the flyback period is 12 μs. The actual reverse bias pulse width should be larger than 22–24 μs. Normally the leading edge of the drive waveform occurs between the end of the flyback and the zero collector point of the forward scan. The ON-OFF duty cycle, therefore, typically corresponds from 0.7 to 1.7. This means that the line output transistor is reverse biased for about 10 to 20 μs after commencement of the scan period. The base series resistance R_B is determined from the recommended value of the end base current required for desired peak collector current, and the available turn-on base drive amplitude. The driver transformer turns ratio is given by:

$n < \dfrac{BV_{\text{CBO}}}{V_{\text{OFF}} + V_{\text{ON}}}$, where BV_{CBO} is the collector breakdown voltage of the driver transistor and $(V_{\text{OFF}} + V_{\text{ON}})$ is the peak-to-peak base drive. A typical down stage is shown in Fig. 17.11.

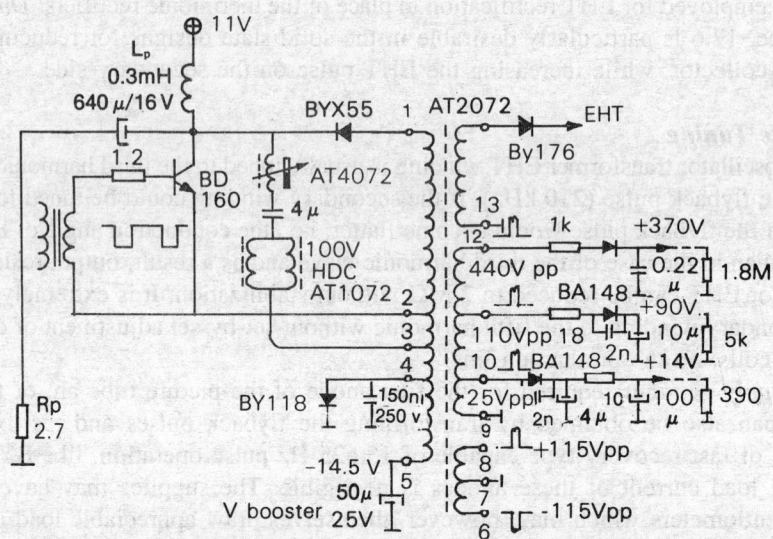

Fig. 17.11 Transistor line output stage for 31 cm tube deflection (*Courtesy:* ELCOMA PHILIPS)

The driver and the line output stage operate in a 'non-simultaneous' mode with respect to each other. The driver transistor is off when forward drive is applied to the output transistor. This forward drive draws energy from that stored in the inductance of the drive transformer during the preceding conduction period of the driver stage. During the period the driver is off, a large voltage pulse appears across the driver transformer primary and at the driver transistor. Ringing may occur due to the leakage inductance of the driver transformer and the inductance L_B in the base circuit of the output transistor. The ringing can be minimized by a suitable *RC* network across the primary. This also prevents the surge voltage at the collector from rising above the transistor voltage rating.

17.11 TYPICAL TRANSISTOR LINE OUTPUT CIRCUITS

A transistor line output circuit for a 12-in. (31 cm) picture tube for a television set working on 11 V, is shown in Fig. 17.11. The design uses deflection unit AT 1072, line output transformer AT 2072 and linearity coil AT 4072.

The line output circuit is designed around the line output transformer BD 160, the parallel efficiency diode BYX 55, series efficiency diode BY 118 and the EHT rectifier diode BY 176. The circuit operates at a stabilized power supply of 11 V that can be fed from a mains rectifier or a 12 V accumulator battery.

The diode BYX 55 and the diode BY 118 are used for recovery of the magnetic energy. The series efficiency diode enables a boost voltage of 14.5 V to be available for the stage. L_s and R_P function as a protection for flashover hazards. L_t serves as a source during the energy recovery period. The 150 nF flyback capacitor and the various parasitic capacitances of the transformer act as a resonant *LC* circuit controlling the flyback. Since during a flashover, the capacitor may have to withstand a high current and a sharp voltge rise, a polystyrene capacitor is recommended with a nominal voltage rating of 250 V and a capacitance tolerance of 5%. To prevent spurious oscillations which would manifest themselves as ringing bars in the picture, the connecting leads should be short. The 150 nF capacitor should be particularly close to the series diode BY 118/BYX 71.

The deflection unit employed is AT 1072 which has parallel connected line deflection coils with $L = 107 \ \mu H$, $R = 0.2$ ohm; and series connected frame deflection coils with $L = 29.9$ mH, $R = 16.4 \ \Omega$.

The line output transformer AT 2072 has its magnetic circuit comprising one U and one I ferroxcube (ferrite) core. The primary winding 1–5 and the secondary windings 6–8, 9–12 are situated on the opposite legs. The EHT winding is polyester, encapsulated with a well-insulated lead for the anode of the rectifier BY 176. The stage draws 0.95 A at 11 V with a boost of 14.5 V. Deflection current is 8.2 A with an overscan of 10%. The EHT voltage is 10.2 kV at zero beam current and 9.7 kV at 100 μA beam current.

Where a dc voltage with low spread in value and low internal resistance is required, it is best obtained by rectification during the scan period. Less critical voltages at low drains can be obtained by pulse rectification during flyback.

A transistor line output stage circuit for a 20″ (50 cm) set working on ac/dc transformerless power supply is shown in Fig. 17.12.

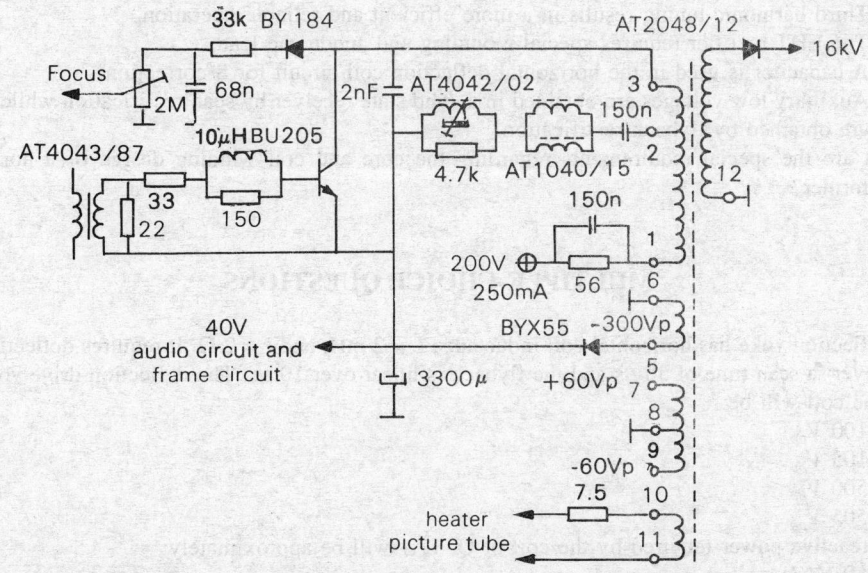

Fig. 17.12 Transistor line output stage for 50 cm tube deflection (*Courtesy*: ELCOMA PHILIPS)

The output stage works on a 200 V collector supply obtained from a regulated power supply. The stage produces an auxiliary low voltage of 40 V required for AF and RF-IF stages, by scan rectification and also supplies the picture tube filaments a part of the output energy through a winding on the horizontal output transformer (HOT).

REVIEW QUESTIONS

17.1 A deflection coil has horizontal coil inductance = 1 mH, resistance = 2 ohm. Determine the drive voltage required for sending a sawtooth current of $2A_{pp}$ through the coil.

17.2 Calculate the amount of inductive power required and the resistive loss in the deflection coils of the above question.

17.3 Discuss the working of a basic drive circuit for a horizontal deflection coil using (i) a pentode, (ii) a transistor.
Indicate how the magnetic energy is recovered every cycle by series or shunt efficiency circuit.

17.4 Deduce the expression for flyback to scan voltage ratio as a function of the flyback ratio, assuming a sinusoidal flyback.

17.5 What is the need for linearity correction in the horizontal deflection circuit? How is it effected by means of a linearity coil?
Give the construction of a typical linearity coil.

17.6 What is blooming? What are the causes and cures for blooming?

17.7 Explain giving reasons:
(a) A modern wide angle deflection tube requires a somewhat S-shaped deflection current instead of exact linear sawtooth.
(b) The horizontal output tube or transistor must be a high speed power device with high voltage rating.
(c) Third harmonic tuning results in a more efficient and reliable operation.
(d) The EHT rectifier requires special mounting and anode cap lead.
(e) A capacitor is used in the horizontal deflection coil circuit for S-correction.
(f) Auxiliary low voltages are obtained in a solid state receiver by scan rectification while high voltages are obtained by flyback rectification.

17.8 What are the special requirements regarding the core and coil winding design of a horizontal output transformer?

MULTIPLE-CHOICE QUESTIONS

17.1 A deflection yoke has horizontal coil inductance $L = 2$ mH, and $r = 2\ \Omega$. It requires deflection current 2.5 A_{pp} over a scan time of 50 μs and the flyback is linear over 10 μs. The deflection drive voltage required by the coil will be:
(a) 100 V_{pp}
(v) 105 V_{pp}
(c) 500 V_{pp}
(d) 505 V_{pp}

17.2 The reactive power required by the coil of Q. 17.1 will be approximately:
(a) 50 VA/s
(b) 100 VA/s
(c) 120 VA/s
(d) none of these

17.3 The resistive power loss in the coil of Q. 17.1 will be approximately:
(a) 1 W
(b) 2 W
(c) 3 W
(d) none of these

17.4 Energy recovery from the horizontal deflection coil takes place during:
(a) the flyback portion
(b) the initial small part of the scan

  (c) the initial half portion of the scan
  (d) the middle portion of the scan

17.5 A linearity coil compensates for the nonlinearity due to:
  (a) the flatness of the picture tube screen
  (b) the resistance of deflection coil
  (c) the ferrite core of the deflection coil
  (d) none of these

17.6 Third harmonic tuning is the tuning of the horizontal deflection circuit to:
  (a) the third harmonic of the horizontal frequency
  (b) third harmonic of the frequency corresponding to the flyback time as its period
  (c) third harmonic of the frequency obtaining half sinusoidal flyback within the flyback duration
  (d) none of these

18

Synchronising and Vertical Deflection Circuits

The synchronising circuits in the deflection system of a television receiver ensure that the scanning at the receiver takes place in exact synchronism with the transmitter. The video signal contains horizontal and vertical synchronising pulses which are used for this purpose. The synchronising pulses from the composite video signal are separated in the sync separator stage. The horizontal sync pulses are then used to control the frequency of the horizontal oscillator by means of an automatic frequency control system to lock the line scanning to the transmitter, while the vertical sync pulses are used to trigger the vertical oscillator so that it holds the picture frame locked to the transmitter. If the horizontal oscillator does not lock to the transmitter, the reproduced picture tears into diagonal bars of dark and gray-white shades. If the vetical oscillator fails to synchronise with the transmitter, the picture rolls vertically. The synchronising circuits in a TV receiver must ensure that impulsive noise interference does not disturb synchronisation as far as possible.

18.1 SYNC PULSE SEPARATOR REQUIREMENTS

The sync separator stage separates the synchronising pulses from the composite video signal available at the video amplifier. Since the sync pulses lie on the side of the video signal, this can be done by clipping out the picture information on the other side to leave only the sync pulses in the output of the amplitude separator circuit. The separated horizontal and vertical sync pulses are in their turn separated from each other by differentiating and integrating circuits. The vertical field pulses produced must be identical in shape and their timing must be such that the interval between the start of the transmitter field pulse and the triggering of the vertical field oscillator is the same, so that a good interlace is ensured.

The sync separator output must not be affected by the variations in amplitude of the video signal. The picture signal must not appear at the output, if the video amplitude falls or rises, in spite of the AGC action to keep it constant.

The output of the sync separator should not contain interference noise pulses as they disturb the horizontal and at times the vertical synchronisation also. It must suppress noise as far as possible.

Basic circuit principle Sync pulses having an amplitude 30% of the maximum composite video amplitude, lie at the top (or bottom) peaks of the composite video waveform. Hence it is easy to separate them by a simple amplitude separator or clipping circuit. A transistor amplifier generates a high cut-off bias for itself in its base circuit, by means of an RC circuit in its base-emitter diode that causes the transistor to operate in class C conditions. The transistor conducts heavily to bottom the collector to saturation level, only during the tips of the sync pulses, and remains biased to cut off for the rest of the signal because of the high discharge time constant of *RC*. A basic sync separator circuit employing an *npn* transistor is shown in Fig. 18.1.

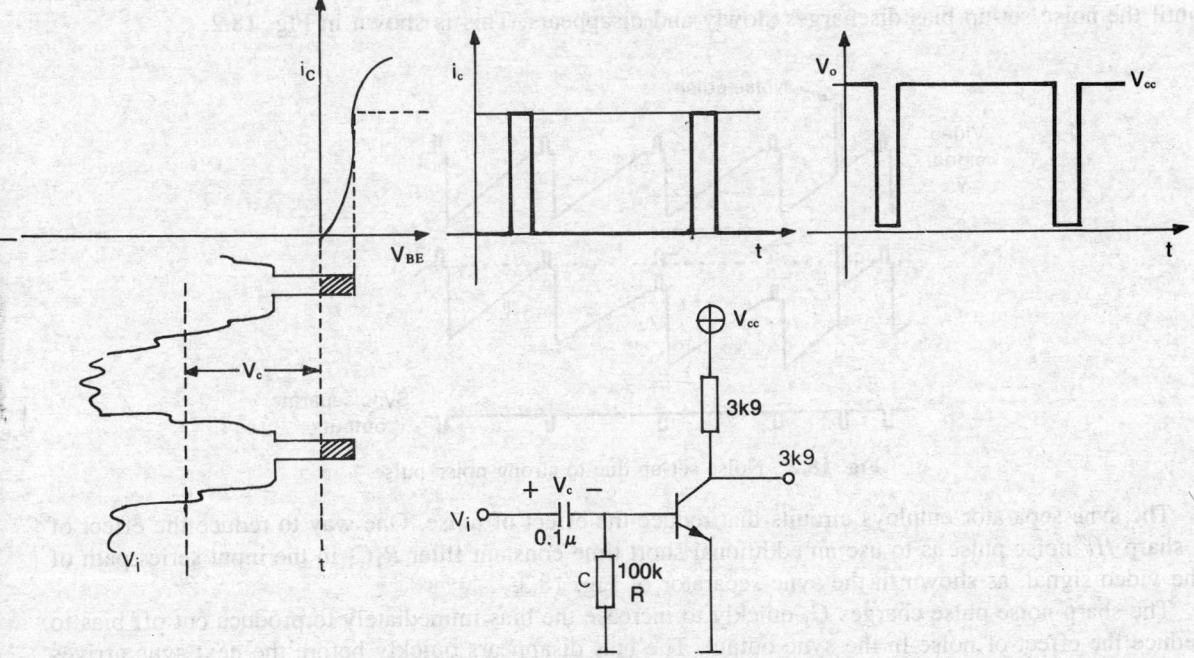

Fig. 18.1 Sync separator circuit principle

The composite video waveform must have sufficiently high amplitude, with the sync tips positive going, so that the *npn* transistor is forced into conduction by the sync pulses (negative going sync tips will be required for the *pnp* transistor stage). With no signal and no bias at the base, the transistor is cut off. When the composite video signal is applied to the base through the coupling capacitor C, the transistor gets a forward bias during the positive going waveform and conducts through the base emitter diode to quickly charge the capacitor C via the source resistance R_s and diode forward resistance r_f, which are both small. The charging time constant is small, with $(R_s + r_f)C \ll t_h$, the horizontal period. The capacitor develops across it a voltage nearly equal to the positive going amplitude of the composite waveform. This voltage acts as a reverse bias for the transistor during the subsequent period.

The discharge time constant when the transistor does not conduct, is large, with $RC \gg t_h$, the horizontal period. Hence the bias leaks off very slowly, and is also replenished during the sync tips to stabilise at a certain cut-off value providing class *C* conditions. In effect the *sync tips are clamped* to the near zero potential, whatever may be the variation in the total composite video due to brightness changes in the picture. As the transistor conducts during the sync tips only, negative going sync pulses are produced at the collector output, where the differentiating and integrating circuits can separate the *H* and *V* pulses.

18.2 NOISE CANCELLOR CIRCUITS

Impulsive noise interference from vehicle ignition systems, arcing brushes, sparking on the mains supply lines as also the atmospheric static noise can arrive at the sync separator along with the video signal and disturb the synchronisation by false early triggering of the horizontal and vertical oscillators. This causes the picture to tear off or roll temporarily until the sync pulses lock it again. This is particularly troublesome with weak signals. A strong noise pulse can produce a very high bias set-up in the base circuit of the sync separator, which cuts off the stage to such an extent that the sync pulses cannot appear at the output until the noise set-up bias discharges slowly and disappears. This is shown in Fig. 18.2.

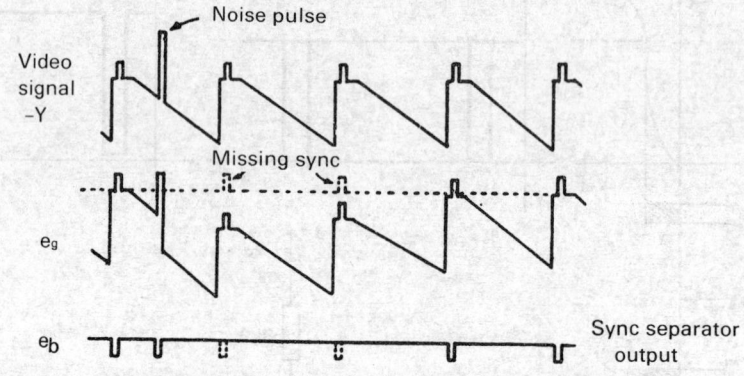

Fig. 18.2   Noise set-up due to strong noise pulse

The sync separator employs circuits that reduce the effect of noise. One way to reduce the effect of a sharp *HF* noise pulse is to use an additional short time constant filter R_1C_1 in the input series path of the video signal, as shown in the sync separator of Fig. 18.3.

The sharp noise pulse charges C_1 quickly to increase the bias immediately to produce cut off bias to reduce the effect of noise in the sync output. The bias disappears quickly before the next sync arrives because of the short time constant of R_1C_1 of about 50 μs. The sharp noise pulse does not produce significant bias on the capacitor C which is much larger. The double time constant thus maintains average bias due to sync pulses only. In case of repetitive or periodic noise the bias on the short time constant filter can develop cumulatively to cause noise set-up problems.

The *noise inverter, TR$_2$* adjusted to cut off by the emitter bias over 1k preset for normal signal, is forced into conduction only by a strong noise pulse input. The noise pulse available at the collector of TR_2 is applied to the base circuit of the sync separator to cancel the noise input to the sync separator.

In the noise gate circuit diagram of Fig. 18.4, the sync separator transistor TR_1 has the noise gate transistor TR_2 in the emitter lead, normally kept ON through the 47 k base resistor. Diode D is biased such that the negative going sync pulses are not adequate to conduct the reverse biased diode and do not reach RT_2. A stronger noise interference pulse larger than the bias provided conducts the diode D and, acting over it through the capacitor C_1, cuts off TR_2 at once. This suppresses the sync separator output for the short duration of the noise pulse, making the noise pulse ineffective.

During the sync pulses of the video signal, transistor TR_2 bottoms or nearly saturates to give out negative-going sync pulses. The transistor must be capable of fast switching to give out steep leading edge of the sync pulses and have adequate current gain. The video input need not be very high as the cut-in sync pulse amplitude is 0.6 to 0.8 *V* only for silicon transistors. A certain amount of fixed bias ensures good bottoming giving out clean output of clipped pulses by *saturation limiting on one side and cut-off bias limiting on the other side.*

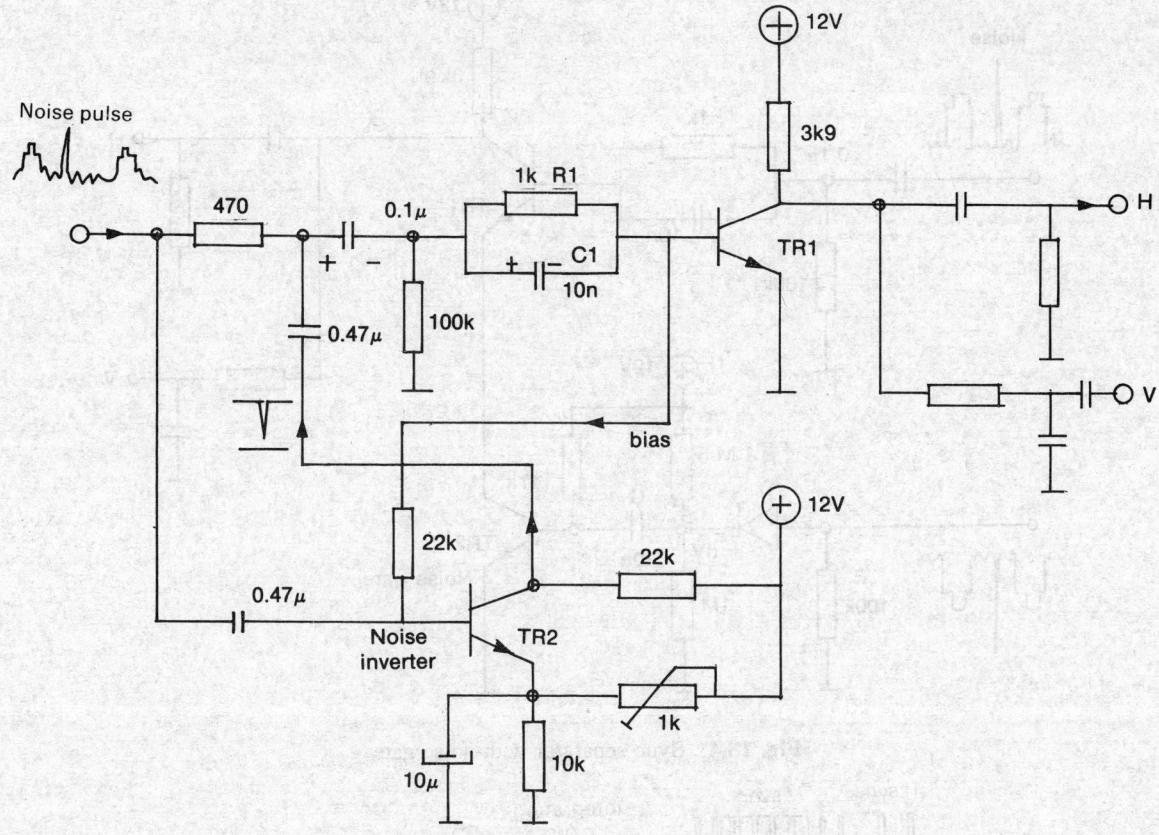

Fig. 18.3 A transistor sync separator with noise inverter

18.3 SEPARATION OF LINE AND FIELD SYNC PULSES

After the sync pulses (both horizontal and vertical) have been separated from the video signal, the pulses are supplied to a differentiating high-pass R-C filter and also to an integrating low-pass R-C filter as shown in Fig. 18.5.

The time-constant R_1C_1 of the HP differentiator circuit is about 0.5 μs, small compared to the horizontal pulse duration of 5 μs. The circuit transfers the full amplitude of the sudden change in voltage through the series capacitor which cannot change its voltage instantaneously. As the capacitor charges exponentially at the rate determined by R_1C_1, the output voltage falls exponentially to zero level. The output is thus in the form of spikes which are given to the phase comparator of the horizontal AFC system for correcting the horizontal oscillator. The spikes are generated at both the rising and falling edges (with opposite polarities) of the sync or equalising pulses. Only the front edge spikes with full line intervals between them are effectively used for the phase comparison.

Time-constant R_2C_2 of the LP integrator circuit is 30-600 μs, large compared to each field sync pulse duration of about 27 μs. The integrator circuit output cannot change instantaneously because of the capacitor at the output which cannot change its voltage suddenly but charges exponentially at a rate

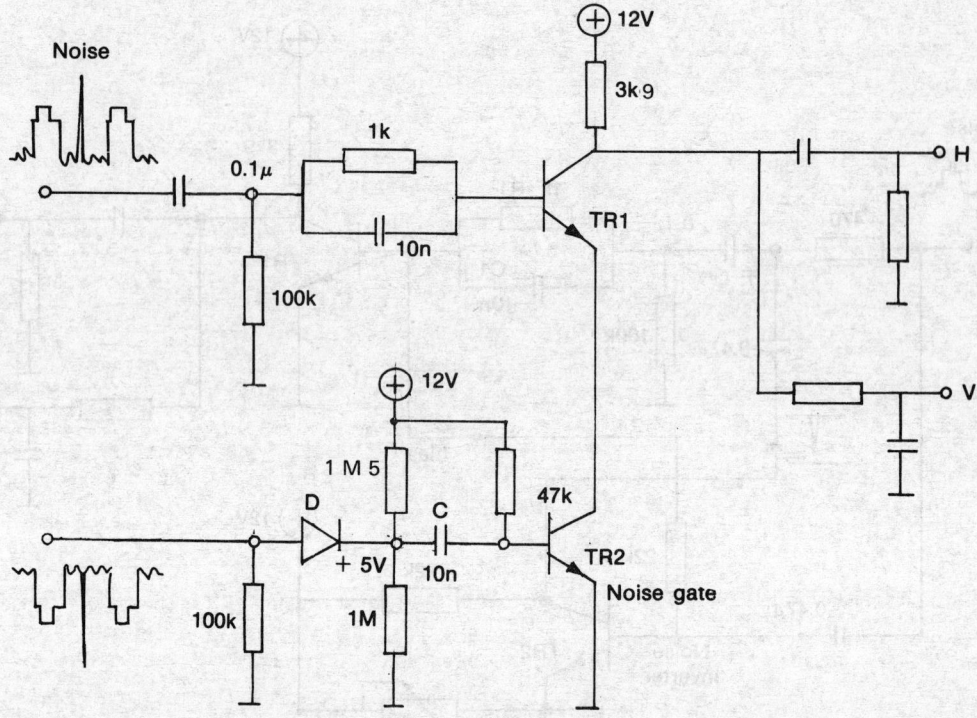

Fig. 18.4 Sync separator with noise gate

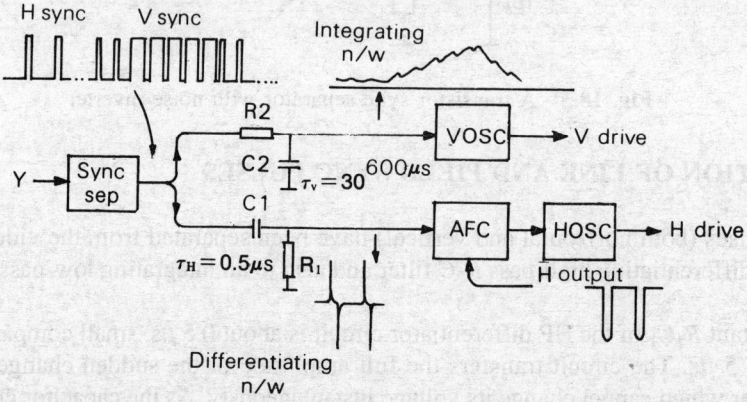

Fig. 18.5 Separation of horizontal and vertical sync pulses

determined by R_2C_2. The exponential change of the output from the integrator continues until there is a change in the input voltage again. For the duration of the positive going sync pulse the output rises exponentially and at the falling edge of the sync pulse the output decays exponentially, at the rate of R_2C_2. For the small H sync pulse duration, very little output builds up until the field sync pulses arrive to produce or develop a vertical sync pulse rising exponentially, as shown in Fig. 18.6. The output tends to fall only slightly during the very small intervals of the serrations. The resulting output pulse is coupled to the vertical oscillator so that it is triggered in step with the vertical sync pulse.

Time-constant of the Integrator

If the time-constant is too long, the out-put voltage rises too slowly. Though this will remove the ripple due to the horizontal sync pulses and serrations more effectively from the output, the integrated output of the VSP will also be small and the pulse will not be sharp enough to trigger the vertical oscillator properly. If the time-constant is not long enough, the output will be higher but there will be a considerable *H* ripple in the output due to quicker charge and discharge. A logical solution to reduce the ripple is to introduce an additional *R-C* low pass filter. A two-section or three-section filter is therefore, often used, as shown in Fig. 18.6.

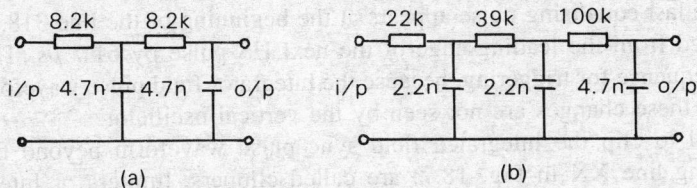

Fig. 18.6   Field pulse integrator using: (a) two-sections, (b) three-sections

The field pulse waveforms obtained through the *R-C* integrator circuit are shown in detail for odd and even fields in Fig. 18.7, to bring out the purpose and effectiveness of the equalising pulses.

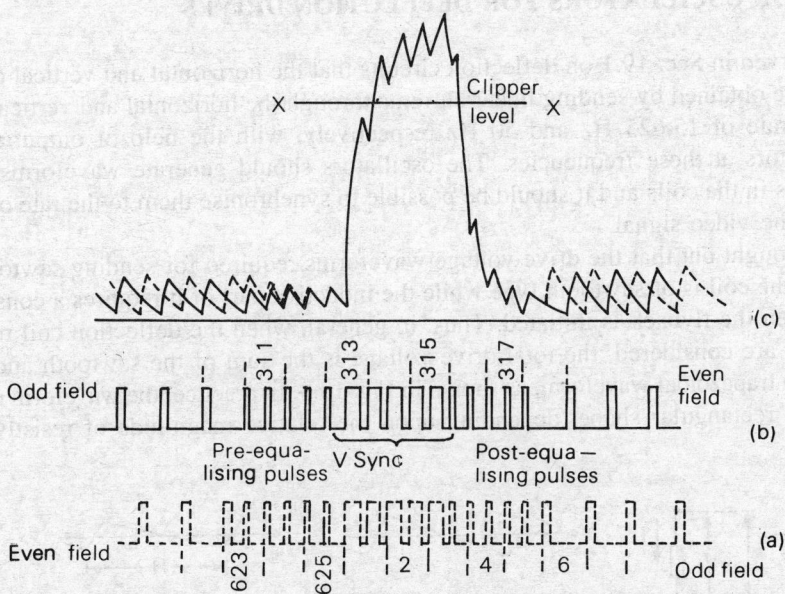

Fig. 18.7   Field pulse integrator waveform: (a) Sync pulse waveforms at the end of odd field, (b) Sync pulse waveform at the end of even field, (c) Integrated sync waveform

The inclusion of equalising pulses before and after the field pulses isolate the integrator from the difference which does occur as a consequence of inserting the field pulse waveform in the *middle* of one line on the field and at the *end* of a line on the alternate field.

At the end of even fields, the first equalising pulse appears in the middle of the line 623. The interval between the leading edge of this pulse and the leading edge of the last HS pulse before it, is 31.7 μs.

The integrator capacitor has not fully recovered from the effect of the line pulse, when this first equalising pulse arrives. At the end of odd fields, the first equalising pulse appears at the end of the line 310, and the capacitor has recovered during the full line period. It will be thus observed that there is a difference in the waveforms at the beginning of the equalising pulses. Thereafter, the applied pulse waveforms are identical. At the end of the two and half line equalising pulse period, the effect of the initial discrepancy on the integrator disappears and the integrated waveforms across the capacitor are coincident until the end of the post equalising pulses. At the end of the equalising pulses also, there is a similar difference. On the beginning of the odd field the last equalising pulse appears in the middle line 5. The trailing edge of this equalising pulse is separated from the leading edge of the next HS pulse by 30 μs. On the beginning of the even field, the last equalising pulse appears at the beginning of the line 318. The trailing edge of this pulse is separated from the leading edge of the next HS pulse by 61.7 μs. These differences are, however, of no consequence for triggering, because the integrator field pulse waveform is usually clipped at a rising level and these changes are not seen by the vertical oscillator.

Circuits employed to clip the integrated field sync pulse waveform beyond the level reached by line pulses (say along line XX in Fig. 18.7) are called clippers, limiters or interlace diode circuits. These may employ biased dc coupled diode clipper or ac coupled diode clipper circuits. The vertical pulse is directly used for synchronisation of the vertical oscillator as this can readily produce an interlaced raster.

18.4 CONTROL OSCILLATORS FOR DEFLECTION DRIVES

It has been observed in Sec. 19.1 on deflection circuits that the horizontal and vertical deflections of the electron beam are obtained by sending linear currents through the horizontal and vertical deflection coils at the required rate of 15,625 Hz and 50 Hz respectively, with the help of output stages driven by sawtooth oscillators at these frequencies. The oscillators should generate waveforms that give linear sawtooth currents in the coils and it should be possible to synchronise them to the rate of the sync pulses separated from the video signal.

It was also brought out that the drive voltage waveforms required for sending sawtooth currents in a resistive part of the coil is of sawtooth type while the inductive part of it requires a constant voltage that must reverse when the flyback is initiated. Thus, in general, when the deflection coil resistance as well as its inductance are considered, the total drive voltage is the sum of the sawtooth and the rectangular waveforms, viz. a trapezoidal waveform, as shown in Fig. 18.8. In practice, the waveform may approximate to a sawtooth or rectangular shape, depending upon the relative magnitude of resistive and inductive voltage drops.

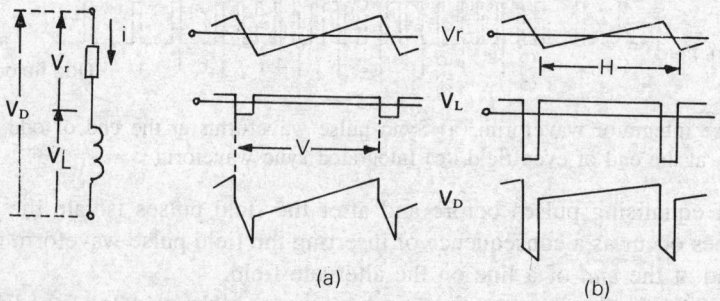

(a)   (b)

Fig. 18.8 (a) Vertical deflection drive, (b) Horizontal deflection drive

In tube vertical deflection circuits, the ratio of the vertical deflection coil reactance to its resistance is appreciable, and in their drive circuits, the drive waveform is trapezoidal. In the horizontal deflection coils, the coil reactance at the high horizontal deflection frequency is much higher compared to the resistance and a rectangular wave is required for the sawtooth current through the coil.

In transistor vertical output amplifiers the input to the transistor vertical output stage is closer to the sawtooth form while the input to the horizontal output stage is closer to the rectangular shape.

Generation of sawtooth waveform is basically done by charging of a capacitor through a high resistance for the forward scan period and discharging the capacitor through a tube or transistor switch forming part of an oscillator, as shown in Fig. 18.9(a).

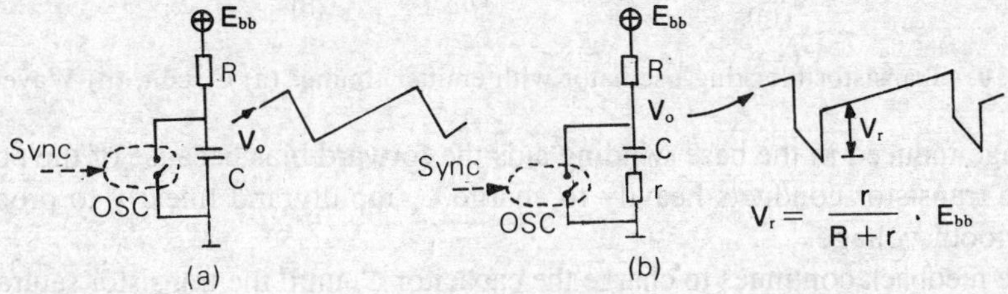

$$V_r = \frac{r}{R+r} \cdot E_{bb}$$

Fig. 18.9   (a) Generation of sawtooth voltage, (b) Generation of trapezoidal voltage

Trapezoidal voltage waveform can also be readily generated by adding a small resistance in series with the capacitor and taking the output voltage across both as shown in Fig. 18.9(b). The voltage drop across the resistance r due to the nearly constant current charging of the capacitor is constant forming the rectangular waveform, while the capacitor voltage rises linearly in a sawtooth form.

For discharging the capacitor, the conducting tube or transistor of a deflection oscillator acts as an on-off switch. Blocking oscillators or multi-vibrators are conveniently used for this purpose as they can be readily synchronised by trigger pulses or can be controlled by a dc AFC voltage, that can automatically control its frequency to make it coincide with the transmitter deflection frequency. Sine wave oscillators are also used along with a reactance tube control to make them voltage controlled oscillators. A regenerative two transistor switch or Miller integrator is also used for sawtooth voltage generation particularly for the vertical deflection oscillator.

Blocking Oscillator

This had been one of the most popular deflection oscillators used to obtain the sawtooth drive waveforms because of its single tube or transistor construction using a transformer coupling for positive feedback that puts the tube or the transistor into oscillations. The hard *on* condition of the tube or transistor is followed by a cut-off condition due to the grid or the base bias developed in the process. This makes the oscillator a very convenient switch for the *R-C* charging circuit that generates the sawtooth waveform.

Transistor Blocking Oscillator

A number of circuit variations are possible with base timing or emitter timing. The circuit of a transistor blocking oscillator with emitter timing is shown in Fig. 18.10.

The transformer with its windings n_1 and n_2 in the base and collector circuits provide a tightly coupled positive feedback from the collector to the base of the transistor. In order that the transistor conducts, the base is provided a bias through the potential divider R_1R_2.

When the collector voltage is applied, the forward bias through the potential divider makes the transistor conducting and the capacitor C_e starts to charge towards positive. As the collector current

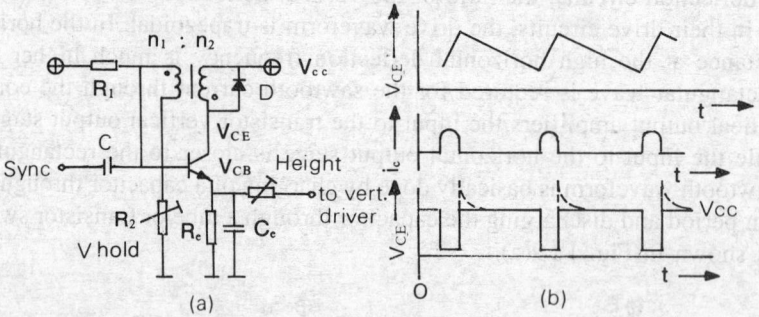

Fig. 18.10 Transistor blocking oscillator with emitter timing: (a) Circuit, (b) Waveform

increases, the voltage induced in the base winding aids the forward bias because of the positive feedback connection and the transistor conducts heavily to charge C_e rapidly and linearly to produce the retrace portion of the sawtooth voltage.

The regenerative feedback continues to charge the capacitor C_e until the transistor saturates. This causes the collector current to arrest at the saturation level and the transformer voltage induced in the secondary base circuit winding drops to zero. The voltage across C_e acts against the base bias to cut-off the transistor. The energy stored in the magnetic field of the transformer now collapses, inducing a voltage pulse in both the windings. The positive collector pulse is clipped due to the diode D that gets forward biased by the pulse. The energy stored in the magnetic field is dissipated over the diode D which conducts.

With the transistor now cut-off, the capacitor C_e discharges through R_e producing the forward scan sawtooth waveform in the negative direction. This continues until the voltage across C_e drops low enough to make the base forward bias effective. The transistor conducts again and the cycle repeats itself.

Synchronisation
Synchronisation of a blocking oscillator, used for the vertical drive oscillator, can be readily accomplished by keeping the natural free running frequency of the oscillator slightly lower than the sync pulse frequency. The sync pulses then hasten the tube or transistor into cut-in or conduction to force the period of oscillations of the blocking oscillator to that of the sync pulses, when the sync pulses trigger the tube or transistor into conduction earlier. In the absence of these, the period is longer as shown by the dotted curves.

dc Voltage Control
The blocking action can be controlled by a dc voltage superposed on its grid or base bias so that its frequency can be varied automatically by an AFC circuit loop for the horizontal oscillator. Increase in the forward bias hastens the conduction of the tube or transistor and reduces the frequency and vice versa.

Because of bulkiness and the cost of the transformer, the blocking oscillator is less commonly used in modern TV receivers that prefer RC oscillators like multivibrators or Miller-integrators.

Multivibrator Circuits
A multivibrator is a pulse generator formed by two R-C coupled amplifier stages cascaded into a feedback loop. Each stage inverts the polarity of the signal so that the output returned to the input of the first stage amounts to a positive feedback resulting in oscillations. The oscillations are in the form of on-off conductions alternating for the two stages, depending upon the coupling capacitor time-constants. These

astable or free-running multivibrators can be constructed using triodes or transistors. They are classed as plate coupled or collector coupled according to whether the output is from the plate or the collector of one stage that drives into the grid or base of the second. The multivibrator is called cathode-coupled when the plate of one stage drives into the grid of the second while the coupling from the second stage back into the first is through a common cathode resistor. Cathode coupled multivibrators conveniently produce an unsymmetrical output required for sawtooth generation. An emitter-coupled multivibrator also has a common emitter resistance coupling while only one stage couples the collector output to the base of the second.

Multivibrators are conveniently used as drive oscillators because they do not require any coils and can be synchronised by pulses or can be voltage controlled by a dc bias, for AFC purposes. Vacuum tube multivibrators, particularly the cathode-coupled type were sometimes used in earlier sets. Transistor multivibrators are now commonly employed in solid state TV sets. A transistor collector-coupled multivibrator suitable for the vertical deflection oscillator is shown in Fig. 18.11.

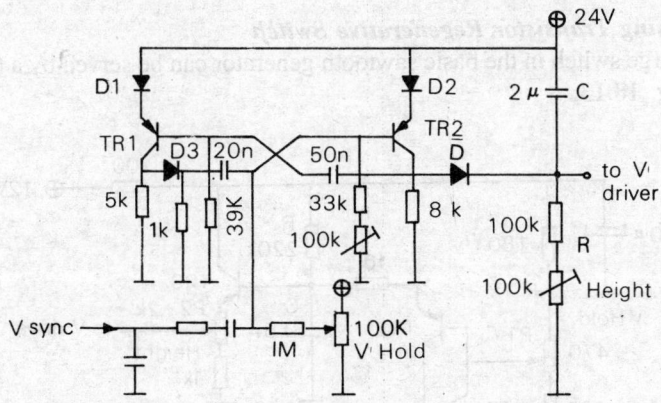

Fig. 18.11 Transistor multivibrator as vertical oscillator

The diodes D1 and D2 in the emitter circuits of the two transistors serve to protect the emitter base junction against reverse voltage transients in the process of switching. The present potential in the base circuit of the transistor TR2 serves as the hold control to adjust the free running frequency of the multivibrator. The vertical sync pulses are coupled to the base of the transistor TR2 over a capacitor. A superposed dc bias of TR2 varied by the potentiometer P2 serves as an additional vertical hold control on the panel of the receiver. The coupling diode D3 along with the resistance 1 kΩ improves the rise-time of the collector voltage waveform, by providing a lower time-constant path for charging the 20 nF capacitor. The collector and emitter of the transistor TR2 are coupled across the sawtooth charging capacitor *C* so that the capacitor changes so long as the transistor TR2 is *off* and discharges rapidly as TR2 becomes *on* to initiate the flyback.

Another collector-coupled multivibrator suitable for horizontal deflection drive is shown in Fig. 18.12.

The multivibrator is dc voltage controlled by the AFC voltage from the phase discriminator over the dc amplifier with 1 k load. A superposed dc bias obtained from a potentiometer provides for the frequency setting of the multivibrator, serving as the horizontal hold control. The coupling diode D and the 1.5 kΩ resistor serve to improve the rise-time of the collector waveform by reducing the charging time-constant. The collector voltage feeds into the base of the driver stage transistor TR4, which drives the horizontal output transistor BU 105 *via* the transformer coupling. The drive voltage requirement, as

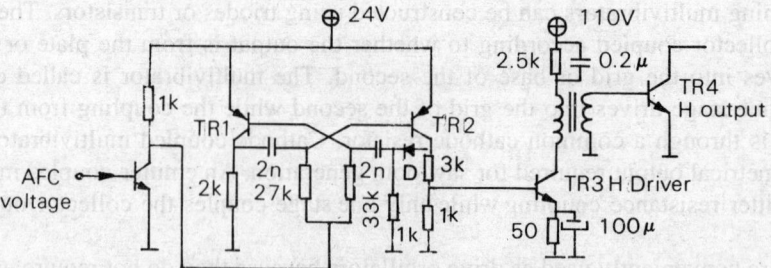

Fig. 18.12 Transistor multivibrator as horizontal oscillator

has been brought out in Sec. 19.9, is a rectangular-shaped wave; hence sawtooth forming circuit is not necessary.

Sawtooth Oscillator Using Transistor Regenerative Switch

The function of a discharge switch in the basic sawtooth generator can be served by a transistor regenerative switch as shown in Fig. 18.13.

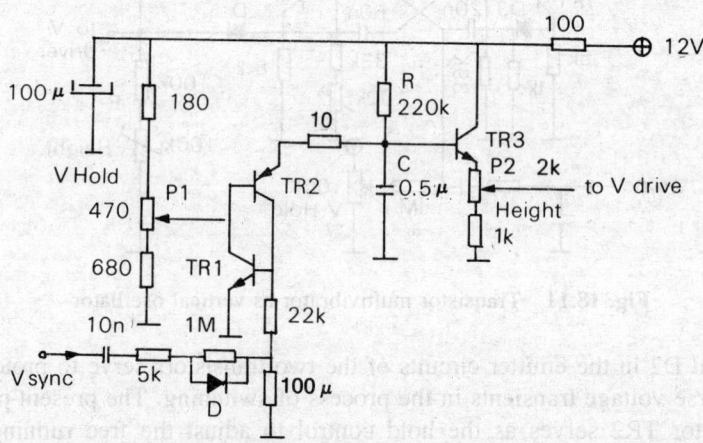

Fig. 18.13 Transistor sawtooth oscillator using regenerative switch

The two transistors TR1 and TR2, a *pnp-npn* pair, are connected with the collector current of TR2 forming the base current of TR1 and the collector current of TR1 forming the base current of TR2, thus forming a regenerative connection, if a trigger pulse of base current is sent in one of the transistors. In this circuit, the base of TR2 is held to an adjustable positive potential so that TR2 cannot conduct until its emitter rises above this by a base emitter cut-in voltage.

The emitter voltage of TR2 rises exponentially towards V_{cc} with the time-constant R-C. As the emitter voltage of TR2 rises, the base-emitter bias of TR2 cuts in to send in a base current. This is amplified h_{fe} times in the collector circuit of TR2. The collector current i_{c2} forms the base current of TR1 which amplifies it further h_{fe} times, in its collector circuit. The collector current i_{c1} forms the base current of the transistor TR2 and the circuit thus establishes large saturation currents in the two transistors. The capacitor C, across which the two transistors are connected, discharges rapidly causing the flyback of the sawtooth waveform. Once the capacitor is discharged below the saturation voltage, the transistor regenerative switch is off again because of the positive bias on the base of TR2. The capacitor then starts charging, to generate sawtooth rising waveform again. The timing of the switching on and hence the frequency of the sawtooth can be varied by the potentiometer P1 that sets the bias for the base of TR2.

The circuit can be triggered into on state by providing a current pulse to the base of TR1, so that sawtooth period synchronises with the period of the triggering pulses. For use as the vertical sawtooth oscillator, the field sync pulses are fed to the base of the transistor TR1 with appropriate polarity and pulse shaping. The free-running frequency must of course be somewhat smaller than the sync pulse frequency for the sync pulses to be effective in earlier triggering of the switch.

Sine Wave Oscillators

For the horizontal or line drive, *LC* oscillators are also frequently used to generate the 15,625 Hz sine wave. This sine wave is then suitably shaped to produce the necessary sawtooth or rectangular drive signal, by arranging the horizontal oscillator to conduct for a short duration. Sine wave oscillators are inherently more stable than the blocking oscillators or multivibrators which are effected by the supply fluctuations and impulsive interference. The *LC* tank circuit in the sine wave oscillators maintains its frequency because of its flywheel-like effect and it is not so easily brought out of synchronisation.

The frequency of the *LC* oscillator can be controlled by means of a varactor diode or a reactance tube or transistor in parallel with the tuned circuit. The bias of the varactor diode or the reactance tube is varied by the AFC voltage that alters the capacitive or inductive reactance of the device to correct frequency until it coincides with the transmitter line frequency. Circuits of this type are given later in Sec. 20.8, discussing the sync processor IC TBA 890.

18.5   HORIZONTAL AFC

The horizontal or line oscillator driving the horizontal output stage must be synchronised by the narrow horizontal sync pulses obtained after differentiation of the sync pulses from the sync separator. These pulses could be used for triggering the horizontal-line oscillator. However, extraneous impulsive noise interference also produces output at the differentiator circuit in spite of the noise canceller and suppressor circuits and these noise pulses may also trigger the horizontal oscillator and put it out of synchronisation. It is, therefore, not desirable to feed these pulses directly to the horizontal oscillator for direct triggering and synchronisation as in the case of vertical synchronisation.

The horizontal oscillator can be more immune to the impulsive interference by employing an automatic frequency control circuit. This circuit controls the frequency of the horizontal oscillator indirectly by a dc error voltage obtained from a phase discriminator circuit, that compares the phase of the horizontal sync pulses with that of the horizontal output waveform from the horizontal output stage. The schematic of such an AFC system is shown in Fig. 18.14.

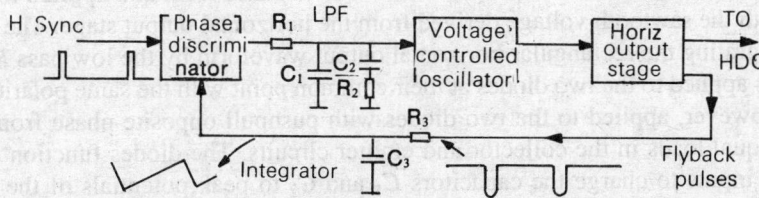

Fig. 18.14 Block schematic of the horizontal AFC system

The sync pulse signal separated in the sync separator stage is compared with the horizontal oscillator output waveform in the phase discriminator where a control voltage proportional to the frequency and phase difference between the two, is produced. This control voltage is filtered by a low-pass filter and is fed to the voltage-controlled oscillator like a blocking oscillator or multivibrator, the frequency and phase of which is then corrected to make it coincide with that of the sync pulses.

In such a configuration, the phase discriminator does not produce any output from the noise pulses. The average frequency of the horizontal oscillator is maintained constant, locked to the transmitter sync frequency. Any deviation in the oscillator frequency produces a correction dc voltage in the discriminator which corrects it in the proper direction. This method is also called flywheel synchronisation since the mean frequency of the horizontal oscillator is maintained by the governor-type action of the phase discriminator.

Sync Phase Discriminators

A simple duo-diode phase discriminator is most commonly used for phase comparison of the sync pulses with the horizontal output voltage. The circuit basically uses two diode detector circuits giving output differentially to a common output point. The two diodes of the detectors are identical, a matched pair of two separate diodes is also available as a single 'duo-diode' package. The two detectors are fed the two signals to be phase compared. One of the two signals has to be fed to the two diode detectors with a 180° out of phase polarity, while the other signal is applied in series common to both the circuits. Depending on the mode of supplying the opposite polarity, the discriminator is called pushpull or single ended.

Duo-diode Pushpull Phase Discriminator

A circuit of the pushpull type phase discriminator is shown in Fig. 18.15.

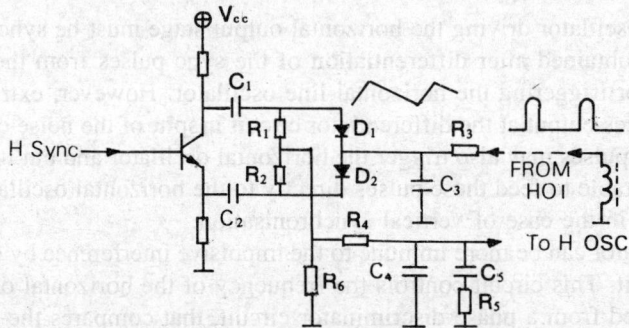

Fig. 18.15 Pushpull sync phase discriminator

Here the sync pulses are obtained in pushpull 180° out of phase form and applied to the two detector diodes in series with the sawtooth voltage derived from the horizontal output stage. The sawtooth voltage is obtained by integrating the rectangular horizontal output waveform by the low pass R-C network. The sawtooth voltage is applied to the two diodes at their common point with the same polarity. The horizontal sync pulses are, however, applied to the two diodes with pushpull opposite phase from a phase splitter amplifier having equal loads in the collector and emitter circuits. The diodes function as peak rectifiers for the sync pulse inputs to charge the capacitors C_1 and C_2 to peak potentials of the resultant applied waveform. The amount of the conduction through the two diodes depends upon how the sync pulses are phased with respect to the flyback slope of the sawtooth voltage. This is indicated in the waveforms of Fig. 18.16.

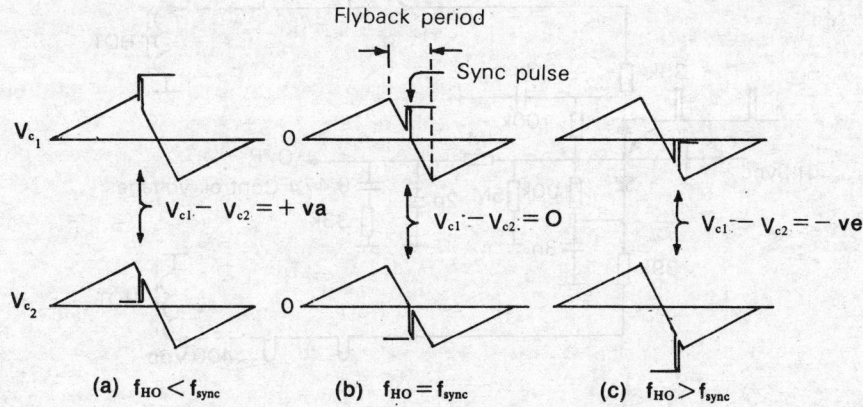

Fig. 18.16 Phase discriminator waveforms

The peak voltages to which the capacitors C_1 and C_2 charge, send currents to the common detector load and filter formed by R_6 and R_4C_4. When the two diodes conduct by the same amount, the net voltage developed across the $R_6R_4C_4$ network is zero. This occurs when the frequencies of the oscillator and the sync pulses are the same and the sync pulses are phased at the centre of the flyback slope of the sawtooth as shown in Fig. 18.16. When the oscillator frequency is above or below the sync frequency, D_1 or D_2 get higher or lower peak voltages that produce positive or negative control voltage to correct the oscillator frequency.

AFC Filter
Current in the detector load R_6 flows in the form of narrow pulses during flyback at the horizontal frequency. This is filtered by the low pass integrating filter R_4C_4. The time-constant of the filter must be much greater than the line period to bypass the horizontal frequency components and noise interference and produce an average dc control voltage. A value of 5 ms is typically used. Too large a time-constant may affect the dc voltage, increasing the pulling in time for the synchronisation during switching of channels. The vertical sync pulse may also affect the control voltage causing bend at the top of the picture. The delay introduced by the low-pass filter in control voltage actually applied to the oscillator may result in overcorrection swinging the frequency above and below the desired frequency, called hunting. Hunting may produce edge ripple or gear-tooth effect along the vertical borders in the picture. This is reduced by R_5C_5 across the output, which reduces the time delay by reducing the capacitive nature of the circuit, a lead compensation to obtain stability in the feedback system.

Alternatively, a pushpull phase discriminator may be constructed by applying the sync pulses to the two diodes in common series lead, while sawtooth voltages are applied to the diode detectors, in pushpull 180° out of phase. These are obtained by integration of the output flyback pulses from two pushpull windings on the horizontal output transformer. An arrangement of this type is shown in Fig. 18.17.

Single Ended Phase Discriminator
The need for a pushpull input for phase comparison could be avoided by coupling the sync pulse input and the sawtooth voltages from the output stage as shown in Fig. 18.18.

Here the horizontal sync pulse input is connected to the common cathode of the two diodes, while the sawtooth voltage formed from the output of the horizontal transformer is applied to the two diodes D1 and D2 in series opposition effectively bringing one half of the total voltage to appear across each diode.

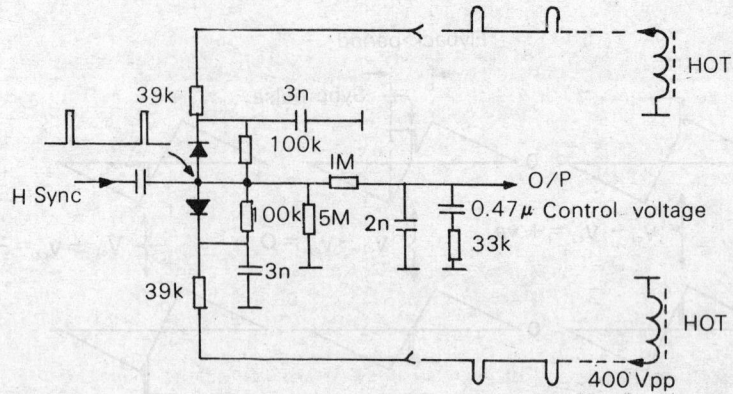

Fig. 18.17 Duo-diode phase discriminator with pushpull sawtooth from HOT

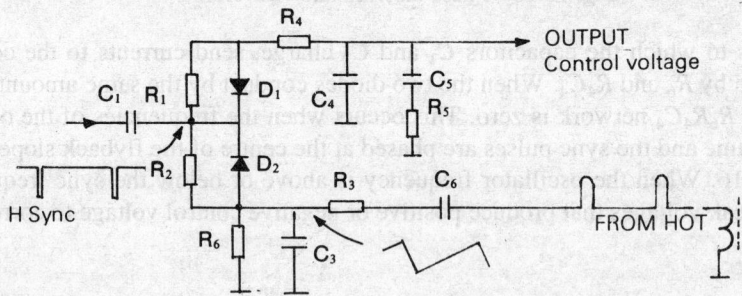

Fig. 18.18 Single ended phase discriminator

With respect to the common cathode, each diode gets an opposite phase sawtooth voltage. Along with the sync pulse input, the two diodes function as differential detectors to produce an error control voltage proportional to the frequency and phase difference between the sync and the oscillator frequency.

18.6 SIGNAL PROCESSING INTEGRATING CIRCUITS

The many complex signal processing functions in a TV receiver are performed by ICs like Philips TBA 890/900 or TBA 920. The IC TBA 890 offer in combination with the line oscillator IC TBA 720 A, possibility of covering the complete low level signal processing circuitry of a monochrome receiver. These circuit functions include:

   (i) video pre-amplifier with emitter follower output and short circuit protection,
  (ii) blanking facility for video circuitry,
 (iii) gated AGC detector supplying AGC voltages to the VIF and tuner circuits,
 (iv) noise cancelling circuit in the AGC and sync separator circuits,
  (v) sync separator,
 (vi) automatic horizontal phase detector,
(vii) vertical sync pulse separator.

Thus a large number of active and passive components can be replaced by these ICs and can offer effective saving in costs, development time and reliability with improved performance.

The TBA 890 gives out AGC suitable for *npn* transistors in the tuner and VIF amplifier, while TBA 900 gives out AGC suitable for *pnp* transistor stages. The input of the video pre-amplifier section of these ICs is designed such that these circuits can be used in sets equipped with normal diode video detector.

The video input characteristics are such that signals can be handled from either normal video detectors or from the IC synchronous demodulators like TCA 540. The video output voltage is designed for use with transistorized video output stage. The sync output can drive both tube or solid state vertical and horizontal oscillators.

Brief Data for TBA 890

Supply voltage	V_{1-16}	typ.	12 V
Video input voltage	$V_{9-16\ (pp)}$	typ.	2.7 V
Voltage gain of the video amplifier	G_v	typ.	7 dB
AGC voltage for video IF	V_{7-16}		1.7 to 12 V
AGC voltage for tuner	V_{6-16}		0.3 to 12 V
Output voltage from phase detector	V_{2-16}		2 to 10 V
Vertical sync output positive pulse	$V_{14-16\ (pp)}$	typ.	11 V
Maximum supply voltage	V_p		20 V
Maximum power dissipation	P_{tot}		700 mW

Application of the IC for various circuit functions is discussed below.[32]

Video Amplifier

The *video input* is given to pin 9 where a prebias voltage of 6 V is required. This is supplied by the resistance divider R_1 and R_2, from the supply as shown in Fig. 18.19(a). The video detector must be able to supply 3 V_{pp} video with negative going sync (+Y), into the pin 9. The peak sync (3 V) level is determined by the AGC detector in the IC.

The video output voltage of 6 V_{pp} is available at the pin 11. The black level is about 5 V. A circuit for driving a video output transistor is indicated in Fig. 18.19(a). A contrast control is applied which keeps the black level constant over the complete contrast control range, by means of the bridge arrangement. If no blanking is applied, the sync pulse is suppressed in the output waveform when the load resistance

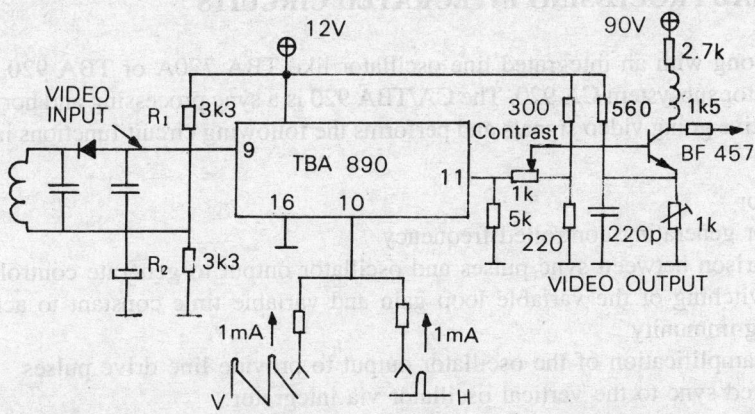

Fig. 18.19(a)   Video input, video output and blanking with TBA 890
(*Courtesy:* ELCOMA PHILIPS)

at pin 11 equals 4.7 kΩ. Full sync pulse is available in the output with a load resistor of 1 kΩ.

Video blanking can be obtained when necessary, by feeding positive horizontal and vertical flyback pulses to pin 10. These pulses have to be more than 1 V above the negative supply line (but not exceeding 5 V). The input impedance of pin 10 is about 1 kΩ. When no video blanking is required, pin 10 should be connected to the negative supply line (pin 16). The output voltage during blanking is clamped to zero so that the blanking is effective when contrast is controlled in the base circuit of the video output stage.

Automatic Gain Control (AGC)

In the TBA 890, the current flowing into pins 6 and 7 decreases for increasing gain control, so counteracting the current variation of the control stages of the IF stages and tuner. Figure 18.19 shows the external circuit required for AGC operation with TBA 890.

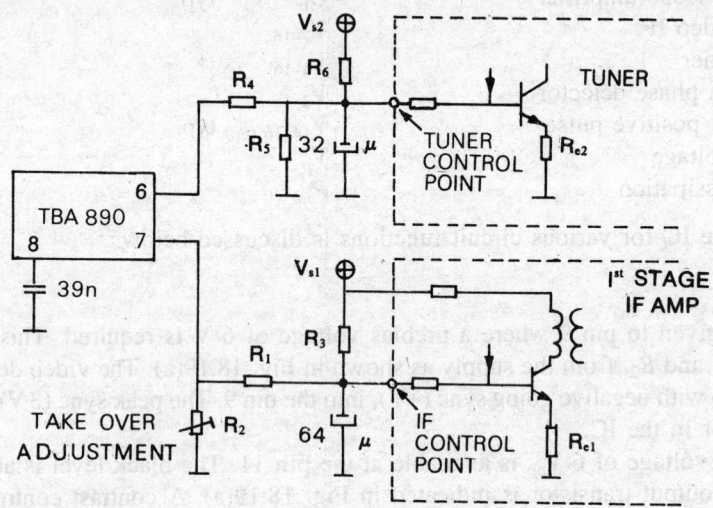

Fig. 18.19(b) AGC bias from TBA 890 (*Courtesy:* ELCOMA PHILIPS)

18.7 SYNC SIGNAL PROCESSING INTEGRATED CIRCUITS

TBA 890 is used along with an integrated line oscillator like TBA 720A or TBA 920 (BEL CA 920).

Horizontal oscillator subsystem CA 920. The CA/TBA 920 is a sync processing and horizontal oscillator IC that accepts negative going video signals and performs the following circuit functions in a monochrome receiver:

— sync separator
— line oscillator generating controlled frequency
— phase comparison between sync pulses and oscillator output to generate control voltage
— automatic switching of the variable loop gain and variable time constant to achieve very high synchronising immunity
— shaping and amplification of the oscillator output to provide line drive pulses
— provides mixed sync to the vertical oscillator via integrator.

The functional diagram of a typical line flywheel sync system using CA 920 is shown in Fig. 18.20.

In the horizontal AFC loop, the phase comparison can take place (i) either between the sync pulse and

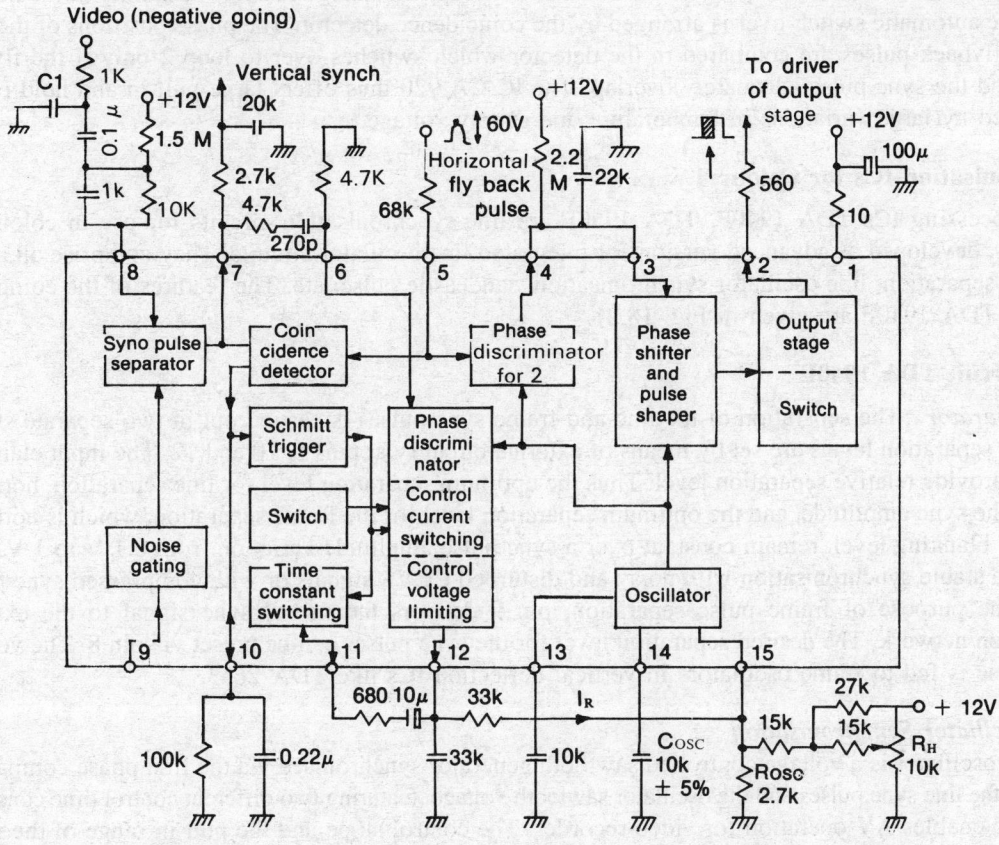

Fig. 18.20 Block diagram of CA 920 with peripheral components (BEL)

the line flyback signal or (ii) between the sync pulse and the oscillator signal. With the phase comparison between the sync pulse and the line flyback signal, the various phase shifts that are associated with the oscillator, the line driver and the line output stages, are automatically compensated for. However, the disadvantage of this method is that the amplitude, shape and duration of the line flyback pulse vary, dependent on picture tube beam current and other dynamic variations. Hence the design of filtering circuits cannot be optimum for good noise immunity. In the second method, where phase comparison with the oscillator is used, the controlled pulse phase position is independent of the amplitude, shape and duration of the line flyback pulse. Therefore, the disadvantage is that the static and dynamic changes in the output stage may affect performance.

The problem is comprehensively solved in CA 920, combining both the methods, by incorporating two control loops. The first control loop consists of the phase discriminator 1, the filter and the oscillator, providing synchronisation of the oscillator to the sync pulses. The second control loop consisting of phase discriminator 2, the filter, the controlled phase shifter and the output stage compensates for the output stage effects. Discriminator 2 for oscillator comparison. For rapid synchronisation in the event of noise interference, the overall loop gain should be high and the LP filtering effect weak. Thus, for rapid pull-in, the comparator slope (loop gain) is high and the filtering time constant short.

Once the oscillator is synchronised, the comparator slope (loop gain) is reduced by switching automatically to a lower value and the filtering time constant is increased achieving good noise immunity in the holding state. The automatic switch-over is arranged by the coincidence detector. The phase positions of the sync and the flyback pulses are compared in the detector which switches over to loop 2 only if the flyback pulses and the sync pulse adequately overlap. The IC CA 920 thus offers large pull-in and hold ranges unaffected by large variations in temperature and supply voltage.

Synchronisation ICs for Colour TV

Sync processing ICs TDA 1940F, TDA 1950F are line synchronisation circuits for use in colour TV receivers, developed as advanced versions of integrated line oscillator circuits. They comprise all stages for sync separation, line oscillator synchronisation, sandcastle pulse, etc. The features of the commonly used IC TDA 1940F are given in Fig. 18.21.

Line Circuit TDA 1940F

Sync separator   The separation of the line and frame sync pulses is carried out in two separate stages. Both the separation levels are set by means of external circuitry at pins 8, 10 and 12. The input clamping circuits provide relative separation level. Thus the optimum separation level for line separation, normally 50% of the sync amplitude, and the optimum separation level for the frame separation, which is normally closer to blanking level, remain constant over a sync pulse amplitude variation from 0.1 V to 1 V. This ensures a stable synchronisation with noisy and disturbed CCVS signals or with compressed sync pulse.

For the purpose of frame pulse separation, pin 9 delivers the mixed sync signal to the external integration network. The desired separation level for the sync pulses can be preset via pin 8. The vertical sync pulse is fed to frame oscillators in vertical deflection ICs like TDA 2653.

Line Oscillator Synchronisation

The line oscillator is a voltage controlled sawtooth generator synchronised via the first phase comparison between the line sync pulses and the oscillator sawtooth voltage, featuring two different control time constants. This also enables AV operation for video recorder. The control slope and the pull-in range of the phase comparator are externally adjustable via RC networks at pins 13 and 16. The pull-in range is independent of the supply voltage. For AV operation, the control time constant of the first phase comparison is set differently by means of a dc voltage applied to pin 5, by connecting this pin to ground or supply voltage.

A second phase comparison gives the phase position of the picture by comparing the line flyback pulses and the oscillator sawtooth voltage. If required, the phase position can be adjusted by varying the dc voltage fed externally to pin 17.

A sandcastle pulse generator delivers a burst gate pulse for the chroma decoder. The pulse is superposed on the line blanking pulse and the frame blanking pulse. The IC provides for muting of the sound channel in the absence of TV signal. The signal at the coincidence detector is used as indicator. The voltage at pin 7 is used through a muting circuit to the sound IC to block its operation.

Combination ICs

The next group of synchronisation ICs are combination ICs that include:

TDA 2577	sync circuit with vertical oscillator and driver stages
TDA 2578A	sync circuit with vertical oscillator and driver stages
TDA 2579	sync circuit with synchronised vertical divider system and output stages
TDA 2593	horizontal combination
TDA 2594	horizontal combination with transmitter identification
TDA 2595	horizontal combination with transmitter identification and protection circuits

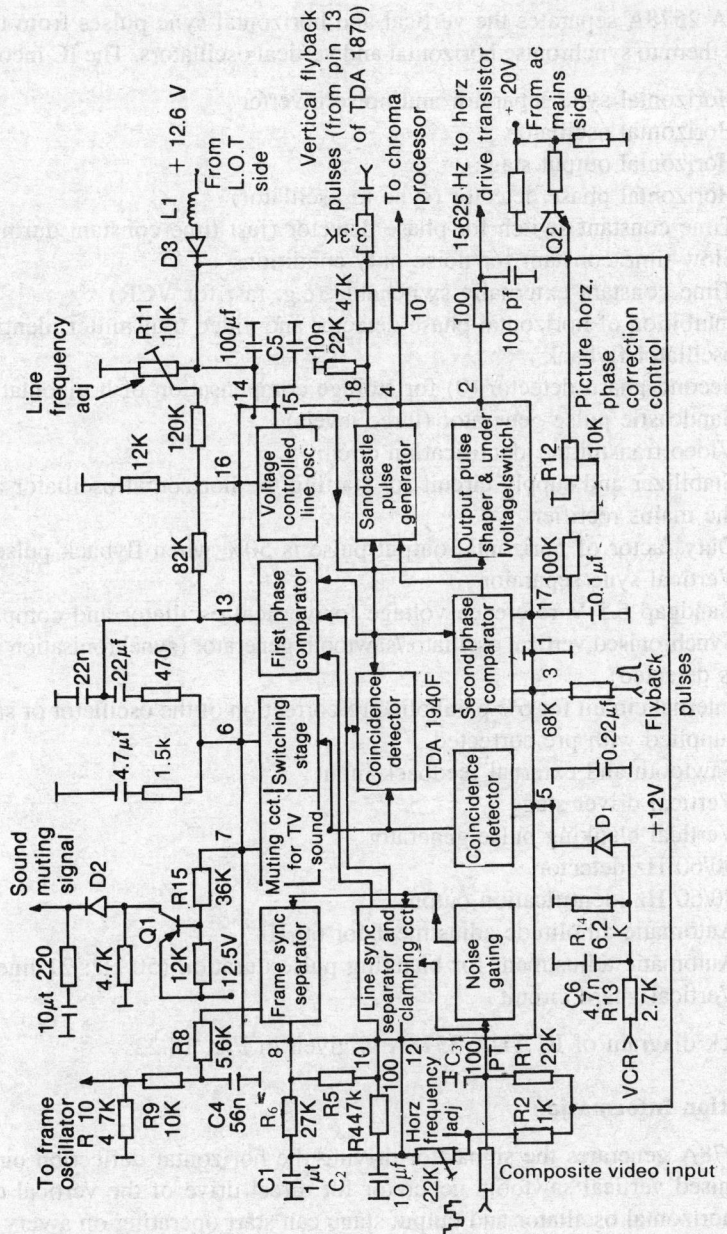

Fig. 18.21 Block schematic and external circuit details of IC TDA 1940F

Sync IC TDA 2578A

The TDA 2578A separates the vertical and horizontal sync pulses from the composite TV video signal and uses them to synchronise horizontal and vertical oscillators. The IC incorporates the following features:

— Horizontal sync separator and noise inverter
— Horizontal oscillator
— Horizontal output stage
— Horizontal phase detector (sync to oscillator)
— Time constant switch for phase detector (fast time constant during catching)
— Slow time constant for noise only conditions
— Time constant externally switchable (e.g. fast for VCR)
— Inhibition of horizontal phase detector and video transmitter identification circuit during vertical oscillator flyback
— Second phase detector (0) for storage compensation of horizontal deflection stage
— Sandcastle pulse generator (three levels)
— Video transmitter identification circuit
— Stabilizer and supply circuit for starting the horizontal oscillator and output stage directly from the mains rectifier
— Duty factor of horizontal output pulse is 50% when flyback pulse is absent
— Vertical sync separator
— Bandgap 6.5 V reference voltage for vertical oscillator and comparator
— Synchronised vertical oscillator/sawtooth generator (synchronisation is inhibited when no transmitter is detected)
— Internal circuit for 6% parabolic pre-correction of the oscillator or sawtooth generator. Comparator supplied with pre-corrected
— Sawtooth and external feedback input
— Vertical driver stage
— Vertical blanking pulse generator
— 50/60 Hz detector
— 50/60 Hz identification output
— Automatic amplitude adjustment for 60 Hz
— Automatic adjustment for blanking pulse duration (50 Hz: 21 lines; 60 Hz: 17 lines)
— Vertical guard circuit

The block diagram of IC TDA 2578A is given in Fig. 18.22.

Application Information

TDA 2578A generates the signal for driving the horizontal deflection output circuit. It also contains a synchronised vertical sawtooth generator for direct drive of the vertical deflection output stage.

The horizontal oscillator and output stage can start operating on a very low supply current ($I_{16} >$ or = 4.5 mA), which can be taken directly from the mains rectifier. Therefore, it is possible to derive the main supply (pin 10) from the horizontal deflection output stage. The duty factor of the horizontal output signal is about 65% during the starting-up procedure. After starting-up, the second phase detector (0) is activated to control the timing of the positive-going edge of the horizontal output signal.

A bandgap reference voltage (6.5 V) is provided for supply and reference of the vertical oscillator and comparator stage.

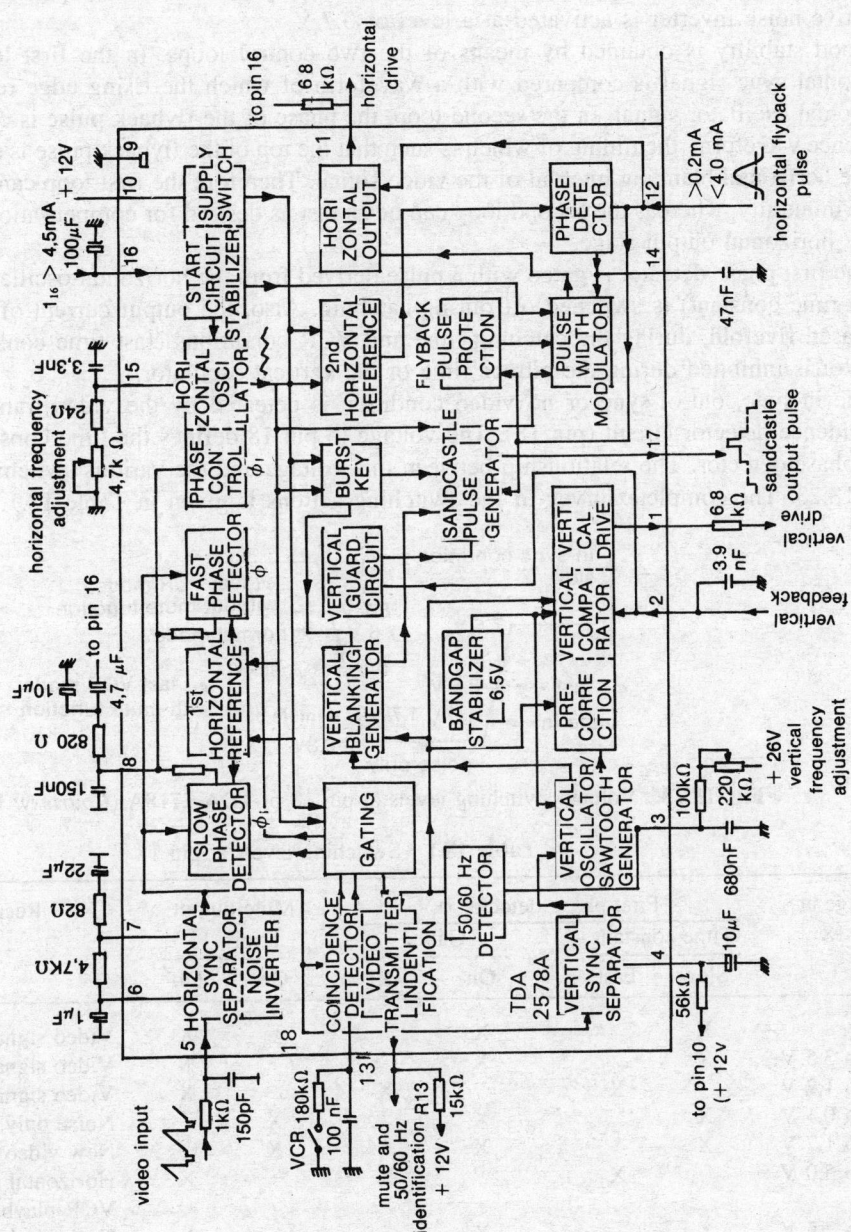

Fig. 18.22 Block diagram of the Sync IC TDA 2578A (*Courtesy: PHILIPS*)

The *slicing level* of the horizontal sync separator is independent of the amplitude of the sync pulse at the input. The resistor between pins 6 and 7 determines its value. A 4.7 kΩ resistor gives a slicing level at the middle of the sync pulse. The nominal top sync level at the input is 3.1 V. The amplitude selective noise inverter is activated at a level of 0.7 V.

Good stability is obtained by means of the two control loops. In the first loop, the phase of the horizontal sync signal is compared with a waveform of which the rising edge refers to the top of the horizontal oscillator signal. In the second loop, the phase of the flyback pulse is compared with another reference waveform, the timing of which is such that the top of the flyback pulse is situated symmetrically on the horizontal blanking interval of the video signal. Therefore the first loop can be designed for good noise immunity, whereas the second loop can be as fast as desired for compensation of switch-off delays in the horizontal output stage.

The first phase detector is gated with a pulse derived from the horizontal oscillator signal. This gating (slow time constant) is switched off during catching. Also, the output current of the phase detector is increased fivefold, during the catching time and VCR conditions (fast time constant). The first phase detector is inhibited during the retrace time of the vertical oscillator.

The in-sync, out-of-sync or no video condition is detected by the video transmitter identification/ coincidence detector circuit (pin 18). The voltage to pin 18 defines the time constant and gating of the first phase detector. The relationship between this voltage and the various switching levels is shown in Fig. 18.23. The complete survey of the switching actions is given in Table 18.1.

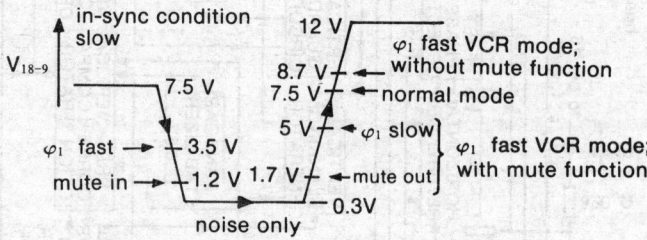

Fig. 18.23 Voltage switching levels at pin 18 of TDA 2718A (*Courtesy:* PHILIPS)

Table 18.1 Switching levels at pin 18

Voltage at pin 18	First phase detector φ_1				Mute output at pin 13		Receiving conditions
	Time constant		Gating				
	Slow	Fast	On	Off	On	Off	
7,5 V	X		X			X	Video signal detected
7,5 to 3,5 V	X		X			X	Video signal detected
3,5 to 1,2 V		X		X		X	Video signal detected
1,2 to 0,1 V	X		X		X		Noise only
0,1 to 1,7 V	X	*	X	*	X		New video signal detected
1,7 to 5,0 V		X		X		X	Horizontal oscillator locked
							VCR playback with mute function
5,0 to 7,5 V	X		X			X	Horizontal oscillator locked
8,7 V		X		X		X	VCR playback with mute function

Where: * = 3 vertical periods.

The stability of displayed video information (e.g. channel number), during noise only conditions, is improved by the first phase detector time constant being set to slow.

The average voltage level of the video input on pin 5 during noise only conditions should not exceed 5.5 V, otherwise the time constant switch may be set to fast due to the average voltage level on pin 18 dropping below 0.1 V. When the voltage on pin 18 drops below 100 mV, a counter is activated which sets the time constant switch to fast, and not gated for three vertical periods. This condition occurs when a new video signal is present at pin 5. When the horizontal oscillator is locked the voltage on pin 18 increases. Nominally a level of 5 V is reached within 15 ms (one vertical period). The mute switching level of 1.2 V is reached within 5 ms ($C_{18} = 47$ nF). If the video transmitter identification circuit is required to operate under VCR playback conditions the first phase detector can be set to fast by connecting a resistor of 180 kΩ between pin 18 and ground.

The supply for the horizontal oscillator (pin 15) and horizontal output stage (pin 11) is derived from the voltage at pin 16 during the start conditions. The horizontal output signal starts at a nominal supply current into pin 16 of 4.2 mA, which will result in a supply voltage of about 5,5 V (for guaranteed operation of all devices, $I_{16} > 4.5$ mA). It is possible that the main supply voltage at pin 10 is 0 V during starting, so the main supply of the IC can be taken from the horizontal deflection output stage. The start of the other IC functions depends on the value of the main supply voltage at pin 10. At 5.5 V all IC functions start operating except the second phase detector (oscillator to flyback pulse). The output voltage of the second phase detector at pin 14 is clamped by means of an internally loaded *npn* emitter follower. This ensures that the duty factor of the horizontal output signal (pin 11) remains at about 65%. The second phase detector will close if the supply voltage at pin 10 reaches 8.8 V. At this value the supply current for the horizontal oscillator and the output stage is delivered by pin 10, which also causes the voltage at pin 16 to change to a stabilized 8.7 V This change switches off the *npn* emitter follower at pin 14 and activates the second phase detector. The supply voltage for the horizontal oscillator will, however, still be referred to the stabilized voltage at pin 16, and the duty factor of the output signal at pin 12 is at the value required by the delay at the horizontal deflection stage. Thus switch-off delays in the horizontal output stage are compensated for. When no horizontal flyback signal is detected the duty factor of the horizontal output signal is 50%.

Horizontal picture shift is possible by externally charging or discharging the 47-nF capacitor connected to pin 14.

The IC also contains a synchronised vertical oscillator/ sawtooth generator. The oscillator signal is connected to the internal comparator (the other side of which is connected to pin 2), via an inverter and amplitude divider stage. The output of the comparator drives an emitter-follower output stage at source of 26 V or higher. The sawtooth amplitude is not influenced by the main supply at pin 10. The feedback signal is applied to pin 2 and compared to the sawtooth signal at pin 3. For an economical feedback circuit with less picture bounce, the sawtooth signal is internally pre-corrected by 6% (convex) referred to pin 2. The linearity of the vertical deflection current depends on the oscillator signal at pin 3 and the feedback signal at pin 2.

Synchronisation of the vertical oscillator is inhibited when the mute output is present at pin 13.

To minimize the influence of the horizontal part on the vertical part, a 6.7 V bandgap reference source is provided for supply and reference of the vertical oscillator and comparator.

The sandcastle pulse, generated at pin 17, has three different voltage levels. The highest level (11 V) can be used for burst gating and black level clamping. The second level (4.6 V) is obtained from the horizontal flyback pulse at pin 12 and used for horizontal blanking. The third level (2.5 V) is used for vertical blanking and is derived by counting the horizontal frequency pulses. For 50 Hz the blanking pulse duration is 21 lines and for 60 Hz it is 17 lines. The blanking pulse duration and sawtooth amplitude is automatically adjusted via the 50/60 Hz detector. The timing diagram of the IC TDA 2578A is shown in Fig. 18.24.

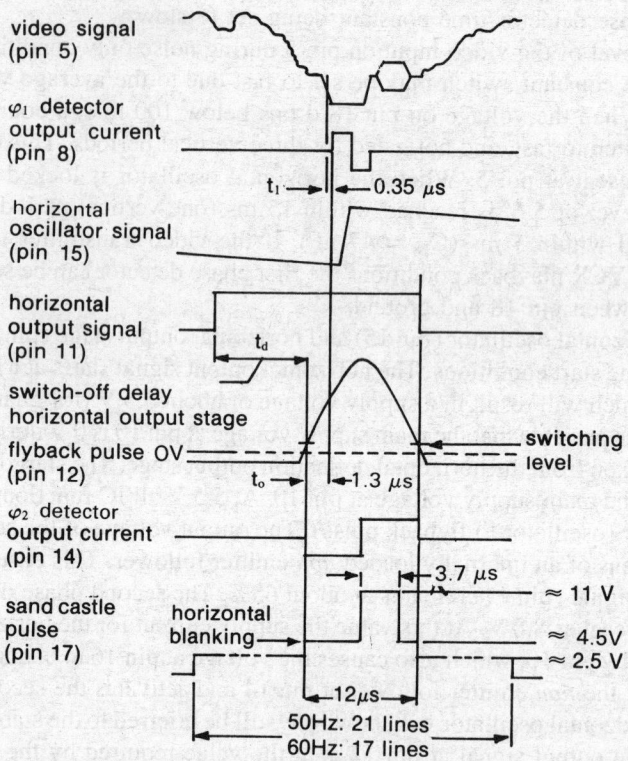

video signal
(pin 5)

φ_1 detector
output current
(pin 8)

$t_1 \rightarrow \ \leftarrow$ 0.35 μs

horizontal
oscillator signal
(pin 15)

horizontal
output signal
(pin 11)

t_d

switch-off delay
horizontal output stage

flyback pulse 0V
(pin 12)

$t_o \rightarrow$ $\leftarrow$1.3 μs

switching
level

φ_2 detector
output current
(pin 14)

$\rightarrow$ $\leftarrow$3.7 μs

sand castle
pulse
(pin 17)

horizontal
blanking

$\approx$ 11 V

$\approx$ 4.5V

$\approx$ 2.5 V

$\leftarrow$12μs$\rightarrow$

50Hz: 21 lines
60Hz: 17 lines

Fig. 18.24 Timing diagram of the IC TDA 2578A (*Courtesy:* PHILIPS)

The IC also incorporates a vertical guard circuit, which monitors the vertical feedback signal at pin 2. If this level is below 3.35 V or higher than 5.15 V, the guard circuit will insert a continuous level of 2.5 V into the sandcastle output signal. This will result in complete blanking of the screen if the sandcastle pulse is used for blanking in the TV set.

18.8 VERTICAL DEFLECTION DRIVE REQUIREMENTS

Vertical deflection circuits are power amplifiers to drive the 50 Hz sawtooth current into the vertical deflection coil of the deflection unit mounted on the neck of the picture tube. The drive waveform required is a sawtooth or rather trapezoidal, because of the inductance of the deflection coil. In practice the flatness of the screen requires less slope near the edges for which an S-correction is required.

In the early designs an output transformer or choke coupled class A output stage was used to drive the coil, to match the load and prevent the dc through the coil. These designs required expensive transformer or choke, and needed linearity correction for the finite inductance, besides increase in the dissipation that required a thermistor to compensate for the increase in the coil resistance due to receiver warm-up. The finite primary inductance of the transformer or choke needed additional parabolic magnetizing current to maintain a sawtooth voltage across it. Hence, class A circuits have yielded to transformerless class AB or class B output stages. The quiescent current of these class B circuits has to be properly set to minimize crossover distortion. These circuits consist of an audio power amplifier with current negative feedback, wherein the current sensing resistance should be temperature stable.

With the need for presetting height, linearity (top, central and bottom), frequency hold, and arrangements for retrace blanking, S-correction, etc. The vertical deflection circuits were rather complex and cumbersome to adjust. Integrated circuits incorporating all these factors have been developed for driving the vertical deflection coils directly, to provide a stable linear deflection.

18.9   TRANSISTOR VERTICAL OUTPUT STAGES

Transistor output stages for vertical deflection may be a single transistor operating in class A amplifier or pushpull class B or AB amplifier employing a similar complementary pair of *pnp-npn* power transistors, similar to the output stages of transformerless output stages in audio Hi-Fi amplifiers. Some circuits are in fact only slight modifications of audio output circuits of this type to include linearity adjustment and flyback spike protection required because of the inductive load.

In a single transistor class A amplifier, the deflection coil usually of a low impedance design, can be choke-capacitance coupled when the supply voltage is low in the range of 24 V, but when the supply voltage is high (more than 100 V), transformer coupling is usually necessary unless a higher impedance coil is chosen. This is necessary for matching the impedance coil to the optimum load requirements of the power stage. The circuit of Fig. 18.25 shows a typical choke-capacitor coupled vertical output stage employing a single transistor in class A operation.

The 2000 μF coupling capacitance blocks the dc from the output stage so that the deflection is symmetrical. The deflection coils have a damping resistance of 680 Ω across them to damp out the ringing oscillations during flyback when the collector current suddenly reverses and the inductive load oscillates with the parasitic capacitance. The NTC thermistor compensates for increase in coil resistance with temperature.

Sawtooth trapezoidal drive is obtained across the two 2.5 μF capacitors C_1C_2 charging over the 27 kΩ resistor and 100 kΩ potentiometer and discharging through the input drive transistor TR4 of the vertical multivibrator oscillator, *via* the diode D and the 1 kΩ resistor. Transistor TR3 amplifies this voltage across C_1C_2 to drive the transistor TR2 in Darlington connection with the output transistor TR1. During retrace, the large flyback voltage spikes at the collector of the output transistor TR1 should not exceed the V_{CEB}, the breakdown voltage of the transistor. The rating must be high enough or else a VDR protection across the choke may be used to limit the spike amplitude to safe limits.

The positive feedback from the emitter of TR2 to the centre of the split capacitors C_1C_2 over the 50 Ω resistor and 1 kΩ potentiometer provides a faster charging, improving linearity by the bootstrap action. The 1 kΩ potentiometer feedback adjustment serves for linearity control.

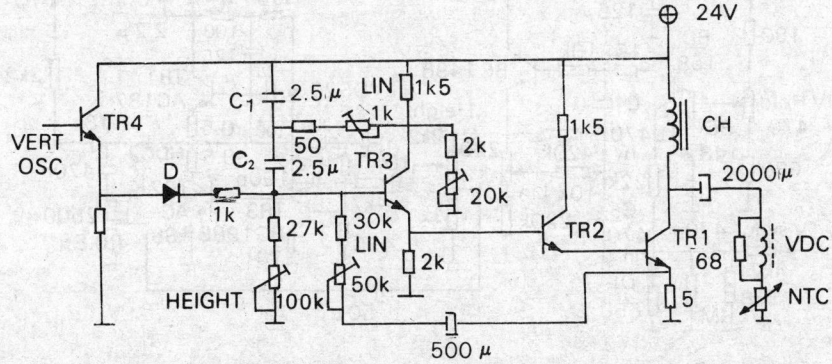

Fig. 18.25   Choke-capacitor coupled transistorized vertical deflection circuit

The negative feedback from the emitter of TR1 over the 500 μF capacitor to the base of TR3 serves to linearize the vertical deflection. The 20 kΩ potentiometer setting the emitter bias of TR3, controls the top linearity of the deflection. The 100 kΩ potentiometer in the charging path of $C_1 C_2$ serves for the height adjustment control.

The circuit of a transformer coupled vertical output stage for high supply voltage operation is shown in Fig. 18.26.

The circuit is basically similar to that of Fig. 18.25. The output deflection coil has a series thermistor with NTC to compensate for the temperature rise, and a series resistance of 2.7 Ω for current negative feedback to the emitter of BC 178 B. The output transformer primary carries a VDR protection to limit the voltage spikes of flyback to 220 V, and R-C combination of 10 kΩ and 22 μF to prevent ringing. The sawtooth discharging circuit employs a transistor regenerative switch triggered by vertical sync pulses from the sync separator.

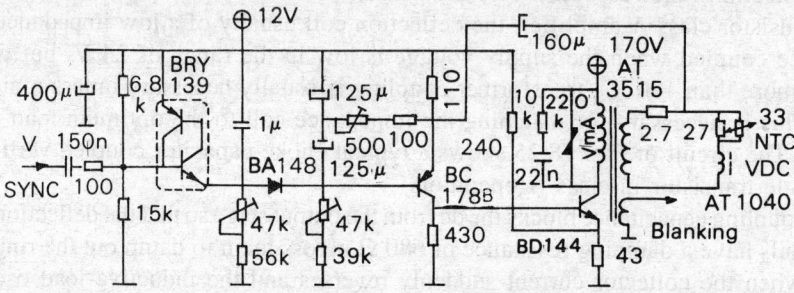

Fig. 18.26 Transformer coupled vertical output stage using transistors (*Courtesy:* ELCOMA PHILIPS)

Complementary symmetry amplifier circuit employing *pnp-npn* pair of transistors is very commonly used for the vertical amplifier stage. The transformerless pushpull configuration can be operated in class B or class AB providing a greater efficiency for the class stage reducing the power requirements to some extent. Each of the two transistors conducts for half of the sawtooth current sweep. As the stage draws fluctuating current, the power supply to the stage has to be well regulated.

A typical circuit for 12 in. (31 cm) tube vertical deflection employing AC 187 and AC 188 complementary pair is shown in Fig. 18.27.

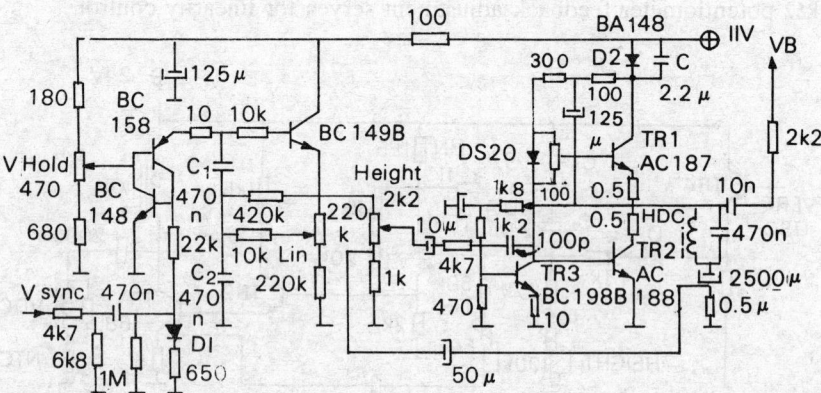

Fig. 18.27 Vertical deflection circuit for 31 cm tube using complementary transistor pair output stage (*Courtesy:* ELCOMA PHILIPS)

The operating point of the two transistors, TR1, TR2 are set by adjusting the base potentials of the two complementary transistors, by the 100 Ω potentiometer across the forward clamped voltage of the diode D1. Here one silicon diode voltage drop is adequate for the prebiasing as the two transistors are germanium types. Two diodes in series are necessary if silicon complementary pair like BD 233 and BD 236 are used for higher power drive circuits for 20 in or 24 in picture tubes. The operating points is adjusted for minimum crossover distortion. Stabilization of the operating points of the direct coupled stages of TR1, TR2 and of TR3 is obtained by the dc negative feedback from the emitters of TR1 and TR2 to the base of TR3 over 1.8 kΩ and 1.2 kΩ resistors ac bypassed by 470 μF capacitor. Negative current feedback from the deflection coil series resistance of 0.5 Ω to the input over the 50 μF blocking capacitor, stabilizes the deflection current amplitude.

Linearization of the sawtooth generated over the capacitors $C_1 C_2$ is obtained by positive feedback by bootstrapping the emitter follower output of TR4 to the lower capacitor *via* a fixed 420 kΩ resistor and the adjustable 220 kΩ potentiometer.

18.10 INTEGRATED CIRCUITS FOR VERTICAL DEFLECTION

BEL 1044

One of the earlier ICs available for monochrome TV receivers was BEL 1044. The IC features:

— Linear sawtooth generator
— Geometric S-correction circuit
— Flyback booster
— Output deflection currents   up to 1.5 A$_{pp}$
— Positive blanking pulse of   20 V$_{max}$
— Wide power supply range   12 to 27 V

Block diagram of the IC BEL 1044 in a typical circuit application is given in Fig. 18.28.

A number of integrated circuits have been designed for vertical deflection drive. One group of ICs included all the functions of the vertical deflection section, from the sync pulse input controlling the vertical oscillator to the output amplifier driving the deflection coils. These vertical deflection ICs include:

TDA 2653 A	26 V	I_{pp} = 2.2 A	50/60 Hz switch, CTV
TDA 2654 S	12 V/25V	I_{pp} = 2 A	Tiny CTV/mono
TDA 2655 B	22 V	I_{pp} = 0.45 A	50/60 Hz switch, CTV 90%

The ICs incorporate the following circuit functions:

Oscillator, 50/60 Hz switch capability, Synchronisation circuit,
On-chip voltage stabilizer, Flyback generator,
Blanking generator with guard circuit, Sawtooth generator-buffer,
Output stage with thermal and short circuit protection,
S-correction and linearity circuit, Frequency detector,
Preamplifier with fed-out inputs.

TDA 2653A

The block diagram of TDA 2653 A is given in Fig. 18.29.

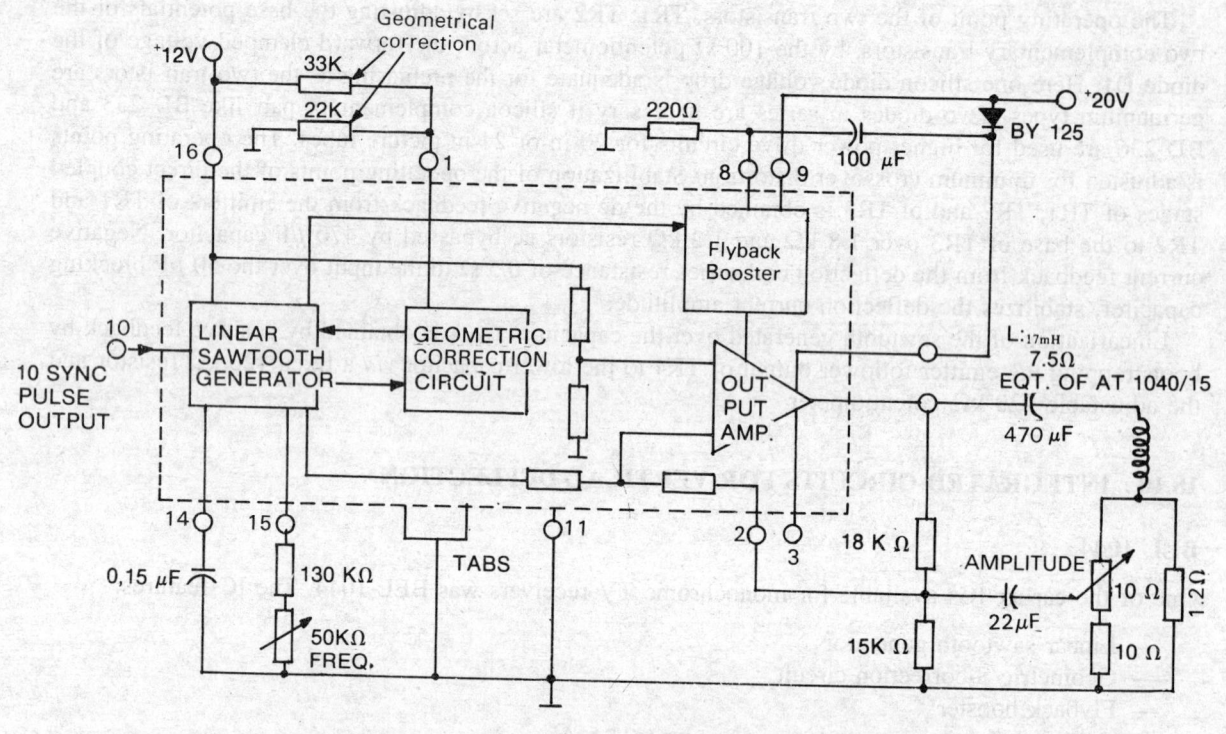

Fig. 18.28 Block diagram of IC BEL 1044 (*Courtesy:* BEL)

The typical application circuit is given in Fig. 18.30. The pinwise functions are as follows:

1, 13 The oscillator frequency is set by the potentiometer at pin 1 and capacitor at pin 13.

2 Combination of sync input and blanking output. The oscillator is synchronised by a positive going pulse between 1 and 12 V. The integrated frequency detector delivers a switching pulse at pin 12. The blanking pulse is 20 V at 1 mA load.

3 Sawtooth generator output. The sawtooth signal is fed via a buffer stage to pin 3. It delivers a signal which is used for linearity control, and drive of the preamplifier. The sawtooth is applied via a shaping network to pin 11 (linearity) and via a resistor to pin 4 (preamplifier).

4 Preamplifier input. The dc voltage is proportional to the output voltage (dc feedback). The ac voltage is proportional to the sum of the buffered sawtooth voltage at pin 3 and the voltage with opposite polarity, at the feedback resistor (ac feedback).

5 Positive supply of the output voltage. This supply is obtained from the flyback generator. For proper operation of the flyback generator, an electrolytic capacitor is connected between pins 7 and 5, and a diode between pins 5 and 9.

6 Output of class B stage. The vertical deflection coil is connected to this pin, via a series connection of a coupling capacitor and a feedback resistor to ground.

7 Flyback generator output. An electrolytic capacitor is connected between pins 7 and 5 to complete the flyback generator.

8 Negative supply (ground).

9 Positive supply. The supply voltage at this pin is used to supply the flyback generator, voltage stabilizer, blanking generator and buffer stage.

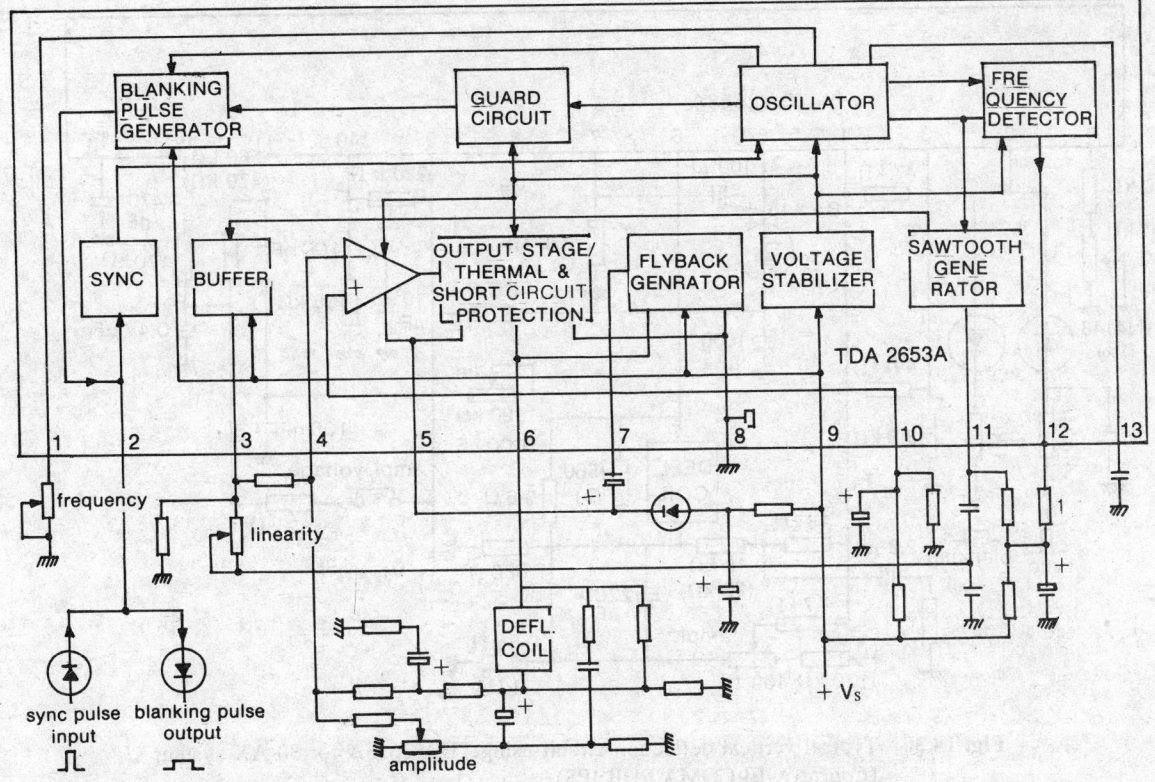

Fig. 18.29 Block diagram of TDA 2653A (*Courtesy:* PHILIPS)

10 External adjustment and decoupling of reference voltage of the preamplifier is provided at this pin.

11 Sawtooth capacitor at this pin is split to realize linearity control.

12 50/60 Hz switching levels. LOW for 50 Hz and HIGH for 60 Hz, are delivered by this pin.

Later, when combination ICs like TDA 1940F, TDA 2577A/2578A/2579 were developed for supplying synchronised vertical and horizontal oscillator drive signals to both horizontal and vertical deflection amplifiers, another group of vertical amplifier ICs that incorporated only the driver and output stage to match to the different deflection coils and picture tube sizes were brought out. These vertical deflection ICs include:

TDA 3651	A, AQ	$V_s = 50$ V	$I_{pp} = 2$ A
TDA 3652	,Q	$V_s = 40$ V	$I_{pp} = 3$ A
TDA 3653	,B/C	$V_s = 40$ V	$I_{pp} = 1.5$ A
TDA 3654	,Q	$V_s = 40$ V	$I_{pp} = 3$ A 90% and 110% deflection

These ICs operating at higher supply voltages up to 40/50 V, feature:

On-chip voltage stabilizer, flyback generator, driver and output stage with thermal SOAR protection circuits against excessive dissipation.

TDA 3653 and 3854 are provided with an internal blanking guard circuit which blanks the raster in case of absence of deflection current.

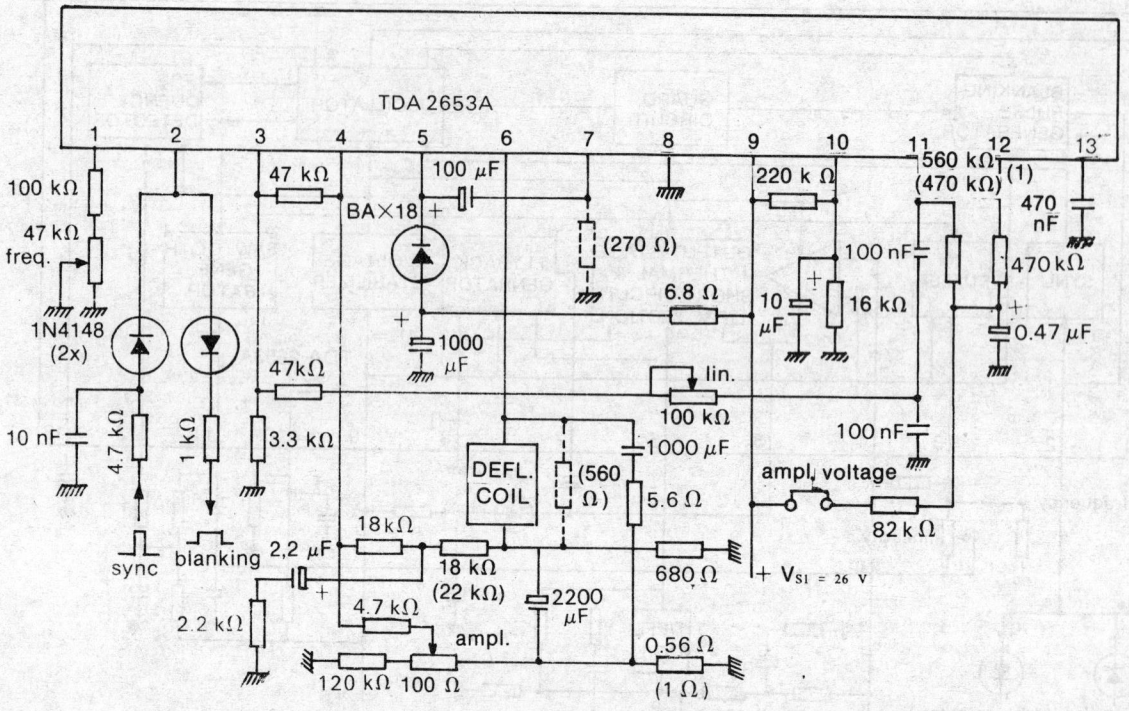

Fig. 18.30   Typical vertical deflection circuit using TDA 2653A for 30 AX system
(*Courtesy:* ELCOMA PHILIPS)

TDA 3653 B/C

The block diagram of TDA 3653A is given in Fig. 18.31.

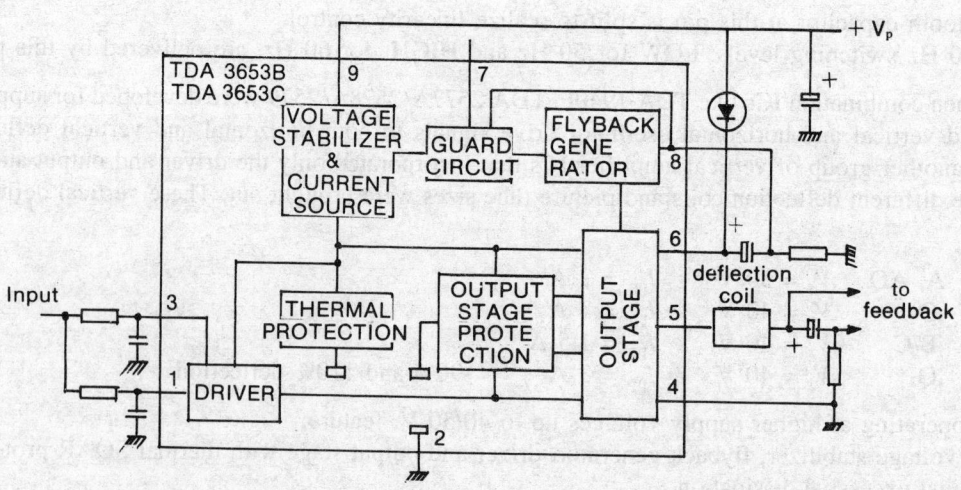

Fig. 18.31   Block diagram of TDA 3653 B/C (*Courtesy:* PHILIPS)

Output Stage and Protection Circuit

The output transistors of the class B output stage can deliver 0.78 A maximum and are protected such that their operation remains within the SOAR area. This is achieved by the co-operation of the thermal protection circuit, the current voltage detector, the short circuit protection and the special measures in the internal circuit layout.

The upper output transistor is protected against short circuit currents to ground. The lower power transistor is connected to the ground during flyback and is thus protected against far too high flyback pulses which may occur during adjustments. The circuit is protected thermally against excessive dissipation by a circuit which operates at temperatures above 175° causing the output current to drop to a such a value that dissipation cannot increase.

Driver and Switching Circuit

The input drive signal is also supplied to the switching circuit. When flyback starts, the switching circuit rapidly turns off the lower output stage and so limits the turn-off dissipation. It also allows a quick start of the flyback generator.

Flyback Generator

During scan, the capacitor between pin 6 and 8 is charged to a level which is dependent on the value of the resistance at pin 8. During normal operation, the voltage at pin 8 may not be lower than 2.2 V. When the flyback starts and the voltage at output pin 5 exceeds the supply voltage at pin 9, the flyback generator is activated. The supply voltage V_p is then connected in series, via pin 8, with the voltage across the capacitor during the flyback period. This implies that during scan the supply voltage can be reduced to the required scan voltage plus the saturation voltage of the output transistors. The amplitude of flyback voltage at the supply pin 6 of the output stage will then be a maximum of 2 V_p. A lower voltage can be obtained by changing the value of the external resistance at pin 8.

Guard Circuit

When there is no deflection current and the flyback generator is not activated, the voltage at pin 8 reduces to less than 2 V. The guard circuit will then produce a dc voltage at pin 7, which can be used to blank the picture tube and thus prevent screen damage.

Voltage Stabilizer

The internal voltage stabilizer provides a stabilized voltage of 6 V to drive the power output stage, which prevents the direct current of the output stage being affected by supply voltage variations.

The typical application circuit diagram of IC TDA 3653 used along with the combination synchronisation IC TDA 2578A is given in Fig. 18.32.

The deflection coils AR 1236/20 ($L = 29$ mH, $R = 13.6$ Ω) require, without overscan, a deflection current of 0.82 A_{pp}. The coils are ac coupled to the IC output at pin 5. Vertical feedback is given to pin 2 of the synchronising IC TDA 2578A. The current negative feedback adjustment at 100 Ω serves for amplitude control.

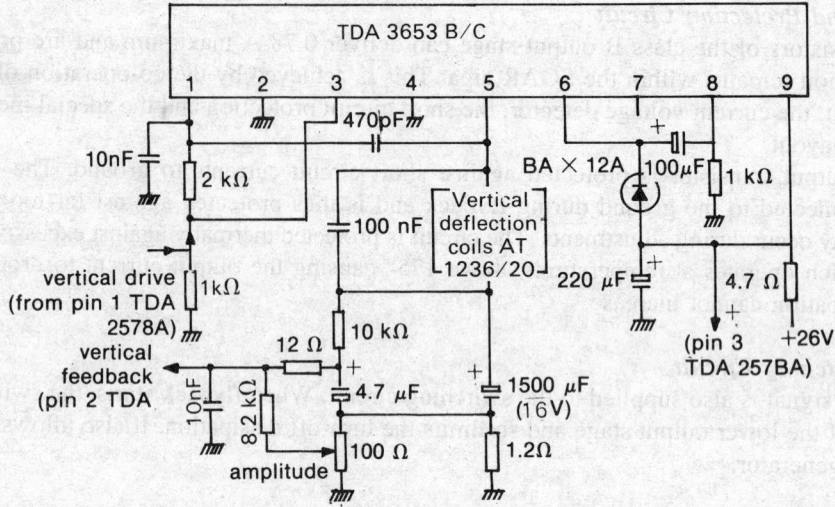

Fig. 18.32 Application circuit of vertical deflection circuit TDA 3653 with synchronisation IC TDA 2578A

REVIEW QUESTIONS

18.1 How are the sync pulses separated from the composite video signal? Explain giving a basic circuit, using a transistor. Design the circuit for a video signal of 5 V_{pp} and typical supply voltages and device data.

18.2 Explain the importance of noise rejection in a sync separator. Discuss the circuits to get rid of noise pulses in the output of the sync separator.

18.3 What are the considerations for choice of the differentiating and the integrating circuit used for separating the H and V sync pulses? Draw the waveforms of vertical sync pulses at the output of the integrator circuit to indicate the effect of equalising pulses during odd and even fields.

18.4 Draw circuit diagrams of the following to explain their use as control oscillators for deflection drive: (a) blocking oscillator, (b) multivibrator, and (c) sawtooth generator using a regenerative switch.

18.5 Compare the use of sine wave oscillators with other types for a horizontal oscillator with AFC. What is meant by the catching range and holding range of the AFC circuit?

18.6 How far is the frequency of a horizontal oscillator drifted away if four diagonal bars are observed in the torn picture? How much is the vertical oscillator frequency if the picture rolls upwards at a rate of three pictures per second?

18.7 What is the effect of a 50/100 Hz ripple on the dc supply for a sync separator?

18.8 Explain, with the help of waveforms, the working of a phase discriminator circuit. Explain the need of a proper filter circuit for the output circuit generating the control voltage.

18.9 An LC resonant circuit is sometimes included in the collector circuit of the multivibrator used as a horizontal oscillator. What is the purpose of the circuit?

18.10 Explain the facilities provided in a typical video signal processing IC.

18.11 What is the effect of an output transformer in the vertical output stage, on the drive voltage requirements? Deduce the current requirement of the stage using a transformer of finite inductance.

18.12 How is the nonlinearity in the vertical deflection compensated for in a transistor driving circuit?

18.13 Explain with circuits how negative current feedback is used to improve vertical deflection.

18.14 Give reasons for the following:
(a) The vertical deflection circuit includes more linearity adjustments than the horizontal deflection circuit.
(b) The voltage waveform to a drive transformer coupled vertical output stage is parabolic rather than linear.
(c) A thermistor is used in some vertical deflection coil circuits.
(d) A resistance is shunted across certain vertical deflection coils.
(f) A VDR is shunted across the primary of the drive transformer in a vertical output stage of a transistor circuit.

18.15 What are the functional requirements of the vertical deflection drive amplifier? What is the necessity for internal vertical blanking?

18.16 Compare the designs of transformer coupled, choke-capacitance coupled and complementary pair transistor output stages for vertical deflection.

18.17 Draw the block diagram of a typical vertical deflection IC. Explain how thermal protection is provided for the output stage. How can the amplitude of flyback voltage be chosen?

MULTIPLE-CHOICE QUESTIONS

18.1 Pick the incorrect statement:
  (a) A sync separator operates in class A condition
  (b) A noise inverter circuit operates in class C condition.
  (c) A strong noise pulse cuts off the sync separator bias for a while
  (d) A sync separator contains all sync pulses.

18.2 Field sync pulses can be separated by an integrator circuit because:
  (a) the field frequency is lower than horizontal frequency
  (b) field sync pulses are in a group of pulses
  (c) field sync pulses are wider than the line sync pulses
  (d) (b) and (c).

18.3 Pick the correct statement:
  (a) Vertical synchronisation is obtained by triggering from the ac mains.
  (b) Low frequency vertical oscillators are more stable than horizontal oscillators.
  (c) Noise does not affect low frequency triggered oscillator circuits.
  (d) Little noise is present in field sync pulses separated from sync pulses.

18.4 Pick the incorrect statement:
  (a) A vertical drive oscillator employs a voltage controlled oscillator.
  (b) A vertical oscillator waveform is shaped to drive the output stage by a sawtooth voltage.
  (c) Vertical flyback is linear.
  (d) A vertical integrator circuit time constant is more than 30 μs.

18.5 Pick the incorrect statement. The Horizontal AFC filter includes:
  (a) low pass filter to filter out noise pulses
  (b) an LP filter to filter out line pulse ripple
  (c) an antihunt RC filter to prevent edge ripple
  (d) (b) and (c).

18.6 Pick the wrong statement:
  (a) No horizontal output results in a bright vertical line.
  (b) A large error in horizontal frequency results in a torn picture.
  (c) A horizontal drive is predominantly a sawtooth waveform.
  (d) A guard circuit is incorporated in the vertical drive integrated circuit to prevent damage to the picture tube.

18.7 A vertical deflection coil has $L = 30$ mH and $R = 10\ \Omega$. If it requires 1 A_{pp}, for full deflection at 50 Hz, and flyback time is 1 ms, the peak-to-peak drive voltage is:
  (a) 10 V
  (b) 30 V
  (c) 40 V
  (d) 31.5 V

19

Miscellaneous Topics in TV Receivers

19.1 POWER SUPPLY CIRCUITS

The power supply requirements of a TV receiver depend upon the electron devices and circuitry used in the receiver, viz. vacuum tubes, transistors, ICs and their combinations. Often the available power has to be kept in mind while designing the circuitry. Generally the ac mains 230 V, 50 Hz supply is available from which the voltage required for the circuitry may be derived. A battery operation may sometimes be desired for portable small size sets. These may be operated from 12 V battery or chargeable accumulator. In battery-operated sets, the circuitry has to be essentially designed for the available low voltage and maximum efficiency.

In the mains operated sets, power supply circuits must convert the available ac mains supply to feed the filaments, $B+$ anode voltages, $\pm V_{cc}$ transistor collector supplies and the IC supplies. Besides the circuits, suitable fuses, switches and RF suppressor circuits are also included in the power supply. The mains operated sets are often designed for transformerless ac/dc operation, in order to reduce the cost and weight. For the television set requiring 60 to 200 W of power, the transformer is not only bulky and costly but also introduces stray magnetic field problems that may affect the sync and deflection circuits to cause a hum bar in the picture; attempts to reduce the stray fields through special cores only adds to the cost. A trend is, therefore, to avoid the transformer not only in the tube sets but also in the solid state receivers by designing them for high voltage operation with the modern high voltage solid state devices now available. In colour TV sets, SMPS are most commonly employed as they provide good regulation against mains supply fluctuation and reduce dissipation.

Hybrid Receivers

The employed mixed circuitry with tubes, transistors and ICs. The sets were designed for transformer or transformerless operation. The sets employ tubes of proven reliability for the output stages where transistors made inroads subsequently.

In transformer type circuits, the E-series tubes for parallel heater operation may be used with a filament winding. The low voltage dc required for transistor and IC stages was obtained from a low voltage rectifier and the tube $B+$ supply, by direct half-wave rectification from the supply (or with an autotransformer arrangement).

The low voltage supply may have a regulator circuit to provide a constant voltage. In transformerless circuits, the P-series tube may be used for 300 mA series filaments operation. As the number of tubes was small, the total filament voltage is for from the supply voltage of 230 V, so that a large series ballast resistor may have to be used implying considerable loss of power. This could be economically overcome by: (i) Half-wave heating through a series diode D or (ii) Reactive voltage drop across a condenser C (iii) Series ballast capacitor for reactive voltage drop.

Solid State Receivers

Solid state TV receiver sets may be designed for working on a low battery voltage or ac mains operation with transformer or transformerless circuitry.

(i) *Battery-cum-mains operated sets*   These are generally designed for working on the common 12 V accumulator or chargeable cells with a facility for ac mains operation through a low voltage regulated supply of the required current rating. A circuit of this type is shown in Fig. 19.1.

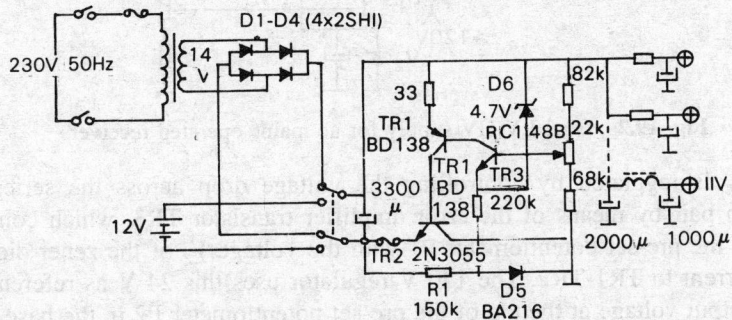

Fig. 19.1   LV supplies for battery-cum-mains operated sets (*Courtesy:* ELCOMA PHILIPS)

For mains operation, the bridge rectifier provides 15 V to the series-pass transistor regulated supply, to obtain a constant 11 V which is supplied to the different stages through decoupling *R-C* circuits.

In the circuit shown, TR2 is the series regulator transistor the voltage drop across which varies with bias from TR1 and TR3. The output voltage tapped at the base of TR3 is compared with the zener voltage of D6 to provide an error signal current to the base of TR3. This is amplified further by TR1, since it is also the base current of TR2, the voltage drop across the same is also changed in such a way that the output voltage remains constant. The diode D5 is normally on, its current flowing *via* TR1 and TR3, but in case of accidental short-circuit conditions at the output, D5 gets a reverse biasing voltage at its anode via R_1, so that the TR3, TR1 and hence TR2 are cut-off, preventing excessive current to flow.

The horizontal output stage in this circuit uses a series efficiency diode to boost the supply for that stage by 14.5 V. The high V_{cc} of 100 V required for the video output stage, and the focusing anode voltage of around 300 V are obtained from the horizontal output stage flyback pulse rectifiers.

(ii) *ac mains transformer circuits*   Solid state circuits normally work on low supply voltages. The video output stage, however, essentially requires a higher supply voltage in order to be able to drive the picture tube about 40 to 80 V_{pp} video signal. The horizontal output circuit also typically employs a high voltage transistor that requires about 120 V to operate efficiently. These types of circuits, therefore, employ a transformer to derive regulated dc supplies: one at about 24 V catering to the low level stages and the other at about 100 to 120 V to feed the video output stage and the horizontal output stage. The picture tube gets its filament supply from the heater winding and the focus anode voltage as well as the EHT supply are obtained from the flyback rectifier circuits. A power supply of this type is shown in Fig. 19.2.

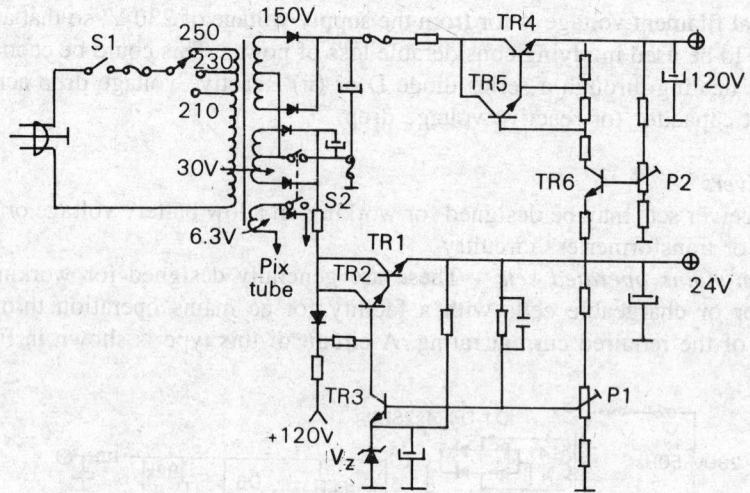

Fig. 19.2 Dual LV-HV supply for ac mains operated receivers

The 24 V supply is regulated by controlling the voltage drop across the series-pass transistors TR1-TR2 Darlington pair by means of the error amplifier transistor TR3, which compares the output voltage at the tap of the pre-set potentiometer P1 with the voltage V_z of the zener diode, and provides a correcting base current to TR1-TR2. The 120 V regulator uses this 24 V as reference and compares it with part of the output voltage at the tap of the pre-set potentiometer P2 in the base emitter circuit of TR6. This transistor amplifies the difference error voltage and controls the voltage drop across the Darlington pair series-pass transistors TR4-TR5 by providing them suitable base currents, so that the output voltage remains constant.

(iii) *Transformerless solid state circuits* With the availability of high voltage power transistors, it has been possible to design transformerless circuits operating on the dc voltage of around 200 V derived from direct rectification of the ac mains voltage, followed by a series-pass transistor regulated supply to eliminate the mains fluctuation problem. The 200 V feeds the video output stage, the picture tube and the horizontal output stage. A typical circuit after Philips-Elcoma design is shown in Fig. 19.3. The picture tube heater supply is derived from the deflection energy. The low voltages required for the audio and vertical deflection are also derived from a HOT winding rectifier. The line oscillator gets its supply directly from the mains rectifier through a Zener regulator. This is essential for the circuit to have a self-starting oscillator for driving the output stage.

The standby power loss is fairly low in solid state and hybrid receivers as the number of tubes is small. The warm-up standby operation, besides enabling instant-on picture and sound, reduces surge current through the filaments in cold condition and extends the life of the tubes. The set interior is kept warmer in the standby mode reducing the moisture content and the failure-liability of the circuits.

19.2 SWITCHED MODE POWER SUPPLIES

Operating Principle and Advantages

Switched mode power supplies (SMPS) have been the more efficient and more versatile solution to power supply requirements of TV sets which have to operate on a wide range of mains supply voltages

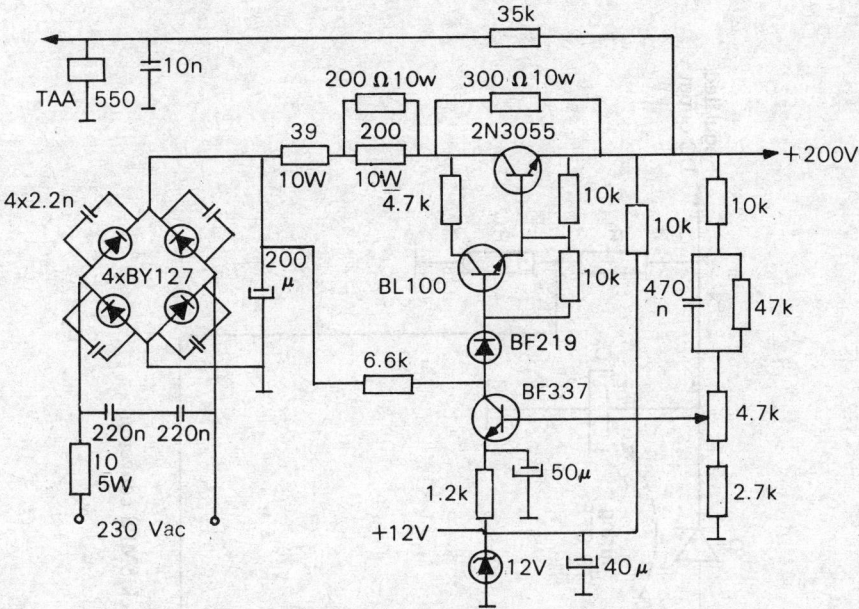

Fig. 19.3   Power supply for transformerless ac/dc solid state receiver circuits (*Courtesy:* ELCOMA PHILIPS)

from 90 to 270 V, this feature being referred to as the *autovolt* feature. This takes care of not only wildly fluctuating power supplies but also takes case of possible burnout situations, as when a 110-V operated set is accidentally connected to 230-V Supplies. Switching mode implies a controlled ON/OFF switching operation of a series device, to obtain a chopped dc output. The chopped dc is filtered and maintained constant by feedback control of the ON/OFF duty cycle. The chopped output can also be applied to the primary of a pulse transformer so that any desired value of dc output can be obtained by rectification of the secondary induced voltages, by fast recovery power diodes and filtering.

As the switching frequency is chosen high in the range of 10 to 100 kHz, the size of the pulse transformer and the filter components is considerably reduced. The ripple content in the output and the harmonic feedback into the mains supply are also low. In TV receivers the switching frequency is kept identical to the horizontal line frequency of 15,625 Hz. This is taken from the horizontal flyback pulses, avoiding possible beat interference if some other frequency were chosen. Interference from harmonics and ringing of the switching pulses can appear as bars or dots on the screen. The SMPS unit is therefore kept shielded away from the tuner and IF stages. Line synchronisation of the interference makes visible interference less annoying. Besides the wide range of fluctuations it can take care of, SMPS makes for very low dissipation in the series transistor, and can provide electrical isolation from the ac mains supply with a more compact transformer operating at higher switching frequencies in the kilo hertz range. Depending upon how the energy from the input is transferred from the input to the output, the SMPS operation can be classified as forward, flyback or pushpull type. It is interesting to note that the horizontal output stage in TV sets, in fact, functions as a switching mode converter to generate the EHT, focus and other supply voltages.

Flyback SMPS

The schematic circuit of a flyback SMPS converter is shown in Fig. 19.4.

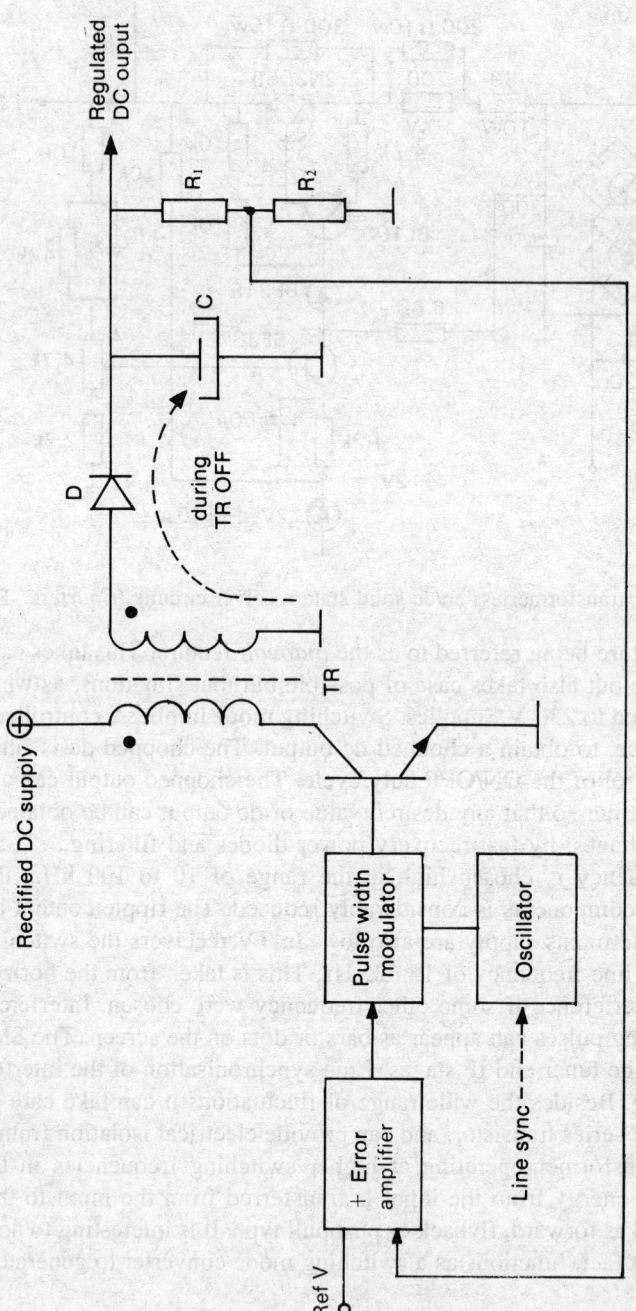

Fig. 19.4 Schematic circuit of Flyback SMPS converter

In the flyback type of SMPS, the fluctuating mains ac voltage is rectified to obtain an unregulated dc voltage. This is applied to the primary of the pulse transformer in series with the switching transistor Tr. The series transistor Tr is switched ON and OFF by pulses from the pulse waveform generator. The ON-OFF (duty cycle) ratio of switching is controlled by the *pulses of variable width* obtained from the pulse width modulator. The width of the pulses obtained from the line synchronised oscillator is varied by the control voltage from the error amplifier which compares the output voltage with desired reference voltage. The polarity of the error correcting control voltage is such that any fall or rise in the output due to supply fluctuation or load variation produces necessary increase or decrease in the pulse width, and the output is stabilized. The PWM generator can be a monostable circuit, triggered from the line sync pulses.

During the time the transistor Tr is switched ON, the transformer primary gets connected across the unregulated dc input voltage. This step input sets up a linearly rising current in the primary winding, the slope being given by the voltage divided by its inductance. This induces in the secondary winding a voltage which, in the flyback converters, has a polarity so connected that the rectifier diode D does not conduct during the ON interval. Here the *energy is stored in the rising magnetic field* due to the current. Before any saturation can occur, the transistor is switched off by the PWM drive, and the primary winding is disconnected from the supply. The energy stored in the magnetic field starts collapsing inducing a reverse polarity voltage in the secondary, so that the diode can now conduct and retrieve the energy to charge the output capacitor and supply the load current, during the OFF period.

As the flyback SMPS *transfers the energy only during the OFF periods,* the regulation and ripple factor performance are poor at higher loads. It is more suitable for higher output DC voltages at modest loads. As a separate filter inductor is not required it is cheaper. It is commonly used in TV receivers, where the output power is below 100 W, and multiple outputs are required. The transformer can have additional secondary windings with appropriate turns ratios and rectifier-filter circuits to produce any specific output voltages. These outputs do not require any filter chokes for each, as in the forward SMPS.

Forward SMPS

A forward SMPS configuration is shown in Fig. 19.5. The basic arrangement for the feedback control of the duty cycle of the ON-OFF duty cycle to regulate the output is similar to that of the flyback converter, apart from the difference in mode of energy transfer. In the forward SMPS converter, the rectifier diode D conducts and *transfers energy* from the input to the output storage capacitor C, *during the ON period* of the transistor switch. In this process magnetic flux built up in the filter choke L and the energy stored in its inductance are released to the output circuit during the OFF period of the transistor switch via and diode D across it, which is often called a 'free-wheeling' diode.

The forward converter requires a filter choke in each of the output circuits if multiple outputs are taken through transformer windings. In the basic schematic, the energy storage function is served by the filter choke. Hence the transformer could be avoided if the output is lower than the minimum unregulated input, and isolation from the input is not mandatory. The series pass transistor switch can be used to provide a chopped input to the choke capacitor filter in the output circuit. Because of the choke input filter the forward SMPS can give a better regulation and is suitable for higher loads.

Push-pull SMPS

In this arrangement two pushpull connected transistors are switched on alternately by opposite polarity pulses from the PWM feedback circuit. The transformer coupled output circuit supplies the DC output in full-wave rectifier configuration through choke-capacitor filter. The conduction periods of the two

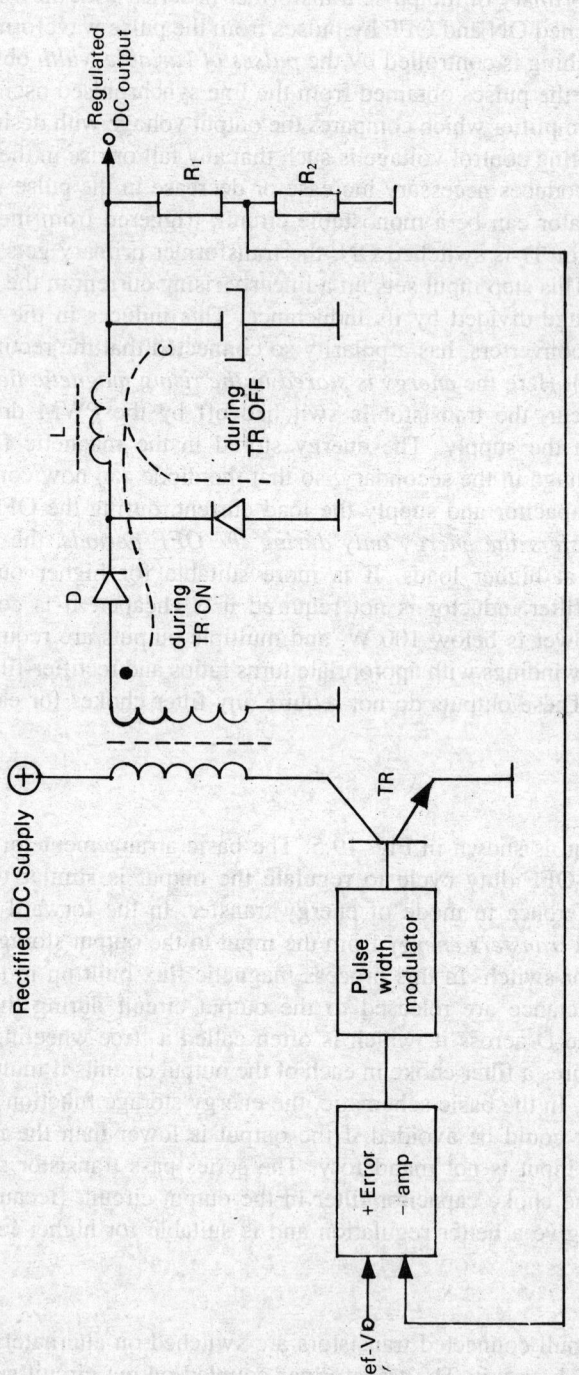

Fig. 19.5   Schematic circuit of Forward SMPS

transistor switches are varied by the feedback control voltage to maintain constant output in spite of load current or input supply variation. This type of SMPS is suitable for higher power requirements with more stringent requirements of regulation and ripple factor.

Protection Circuits

SMPS circuits are prone to failure of switching transistor and allied components due to switching transients at the instants of switching on and due to overload conditions. These problems are taken care of by the protection circuits that control the ON-OFF duty cycle to prevent the damage. The *sudden inrush of current* to charge the large filter capacitor, at the moment of switching on, is reduced by *soft-start circuits* that ensure the duty cycle is initially very low. As the capacitor charges slowly during the short ON periods the duty cycle is slowly increased to settle to the required value. Excessive load currents or short-circuits can be detected by the protection circuits which arrange to switch off the series pass transistor. For transient overloads the protection circuits can operate in a *start-up cycle* to check the overload conditions periodically.

Typical SMPS for a Colour TV Receiver

The circuit diagram of a typical SMPS for a colour TV receiver is given in Fig. 19.6.

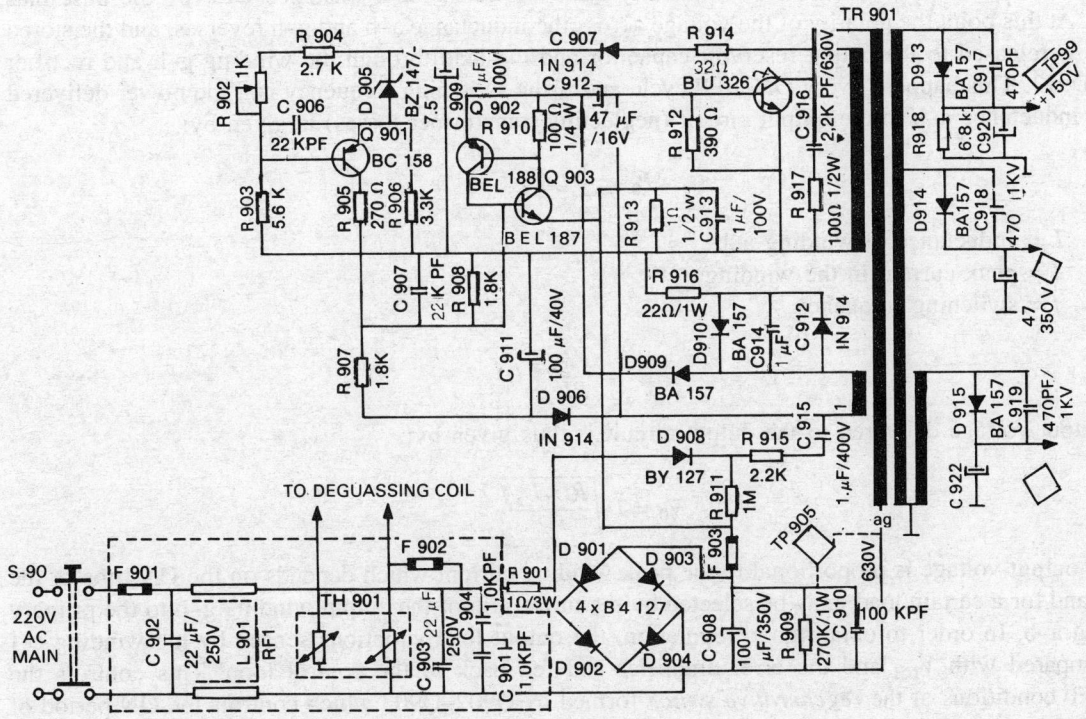

Fig. 19.6 Circuit diagram of an SMPS power supply (*Courtesy:* BEL)

The circuit operates as a *non-synchronised self-oscillating dc-to-dc converter,* providing complete isolation of the circuit from the ac mains supply. This eliminates shock hazards at the antenna terminals, audio/video sockets for VCR, and so on, and offers advantages of high efficiency and compact size. The circuit provides stabilized dc output voltages of 150 V/40 mA, 105 V/400 mA and 18 V/800 mA, over a wide voltage fluctuation of 140 to 250 V, protected against open circuit and short circuit conditions. The line regulation is 0.3% (150–250 V_{ac}), load regulation is 2% (25–50 W), and the ripple content is 400 mV_{pp}.

The mains input circuit includes a pi-filter formed by 0.22-kpF capacitors C902, C903 and L901 to suppress RF interference from the mains entering the receiver and also prevent switching pulses from the SMPS from going to the mains line. The inductance is a *bifilar wound choke* on a special high-frequency ferrite core. The 1-kpF capacitors C901 and C904 are safety capacitors providing earth as per safety regulations. The degaussing coil is connected to the ac mains supply through the thermistor TH 901, for performing the demagnetization of the picture tube, when the receiver is switched on.

The small 1 Ω/3W resistance R901 is use in series for *limiting the inrush of current* at the time of switch-on. The fuse F901 protects against excessive current. The mains voltage is rectified by the diode bridge D901–904 to develop about 330 dc voltage across the filter capacitor C908. For safety a discharging path is provided by the 270-k resistor R909, while the 100-kpF capacitor C910 provides HF bypass for decoupling. A functional block diagram of this circuit is given in Fig. 19.7.

The circuit functions essentially as a *blocking oscillator* which is switched ON/OFF by the heavy positive feedback provided by the winding c–d. The switching transistor Q904 (BU 326) conducts during the ON condition to store energy in the inductance of the primary winding a–b of the switching transformer. The current in the inductance a–b rises linearly until the transistor is switched OFF by the base bias circuit. At this point the polarity of the voltage across the inductance a–b and g–h reverses, and the stored energy is released to the output reservoir capacitor and the load through the winding g–h and rectifier diode D913. This happens every ON-OFF cycle and if the switching frequency is f, the power delivered by the inductance a–b to the output circuit (neglecting transformer losses) is given by:

$$P_0 = \frac{1}{2} L I^2 f$$

where L = inductance of winding a–b
  I = peak current in the winding
  f = switching frequency

$$= \frac{V_0^2}{RL}$$

The output voltage delivered to the output circuit is thus given by:

$$V_0 = I \sqrt{\frac{RL \cdot L \cdot f}{2}}$$

The output voltage is proportional to the peak winding current which depends on the ON time of the cycle, and for a certain load, may be selected by the turns ratio of the output winding g–h to the primary winding a–b. In order to obtain good regulation, the output load variations sensed by the winding e–f are compared with V_{ref} and the error amplifier and fed back to the control loop. This controls the ON-OFF conditions of the *regenerative switch* formed by Q902–Q903 which controls the ON period of the switching transistor by short circuiting the base to the emitter an adjusts the ON-OFF duty cycle accordingly.

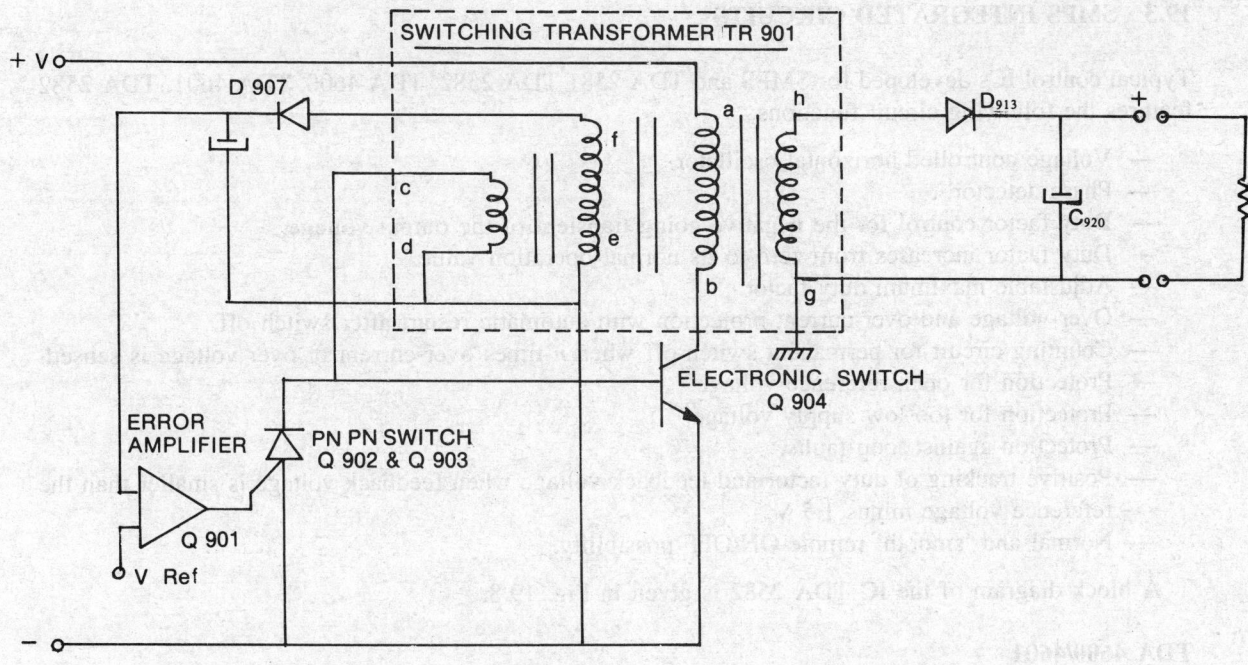

Fig. 19.7 Functional diagram of a self-oscillating SMPS

Circuit Details

The dissipation of the switching transistor Q904 is very low. The switching losses are further reduced by the RC network R917–C916. The base drive from the winding c–d of the transformer is applied across the base emitter junction through D910, D912, C914 and R916. R916 limits the maximum base current during the conduction period.

Normally the gate of the PNPN regenerative switch Q903–Q903 is held at negative potential with respect to the cathode, by a voltage derived from the drive winding c–d through D906, C911, R907 and R908. The PNPN switch is turned ON by the positive pulse supplied by the error amplifier Q901. The error sense winding e–f closely coupled to the secondary load supply winding g–h, provides the sense voltage to the base of the error amplifier Q901, through rectifier D907–C909 and potential divider R902–R903–R904. The reference voltage is derived from the same rectified voltage via Zener diode regulator D901–R906. The potentiometer R902 can set the desired output voltage.

To start the switching mode action of Q904, pulses of duration 5 ms, generated from the ac mains voltage by the diode-RC network D908, R915 and C915 applied to its base. Under open circuit load or low load current conditions, the power supply runs into an intermittent 50 Hz operation. The switching frequency of the oscillator increases and the ON period becomes shorter than the turn-off time of the PNPN switch. The switch then remains turned ON, and the oscillations are interrupted until the next start pulse is provided by the diode-RC network.

The resistance R918 serves as the minimum load, preventing very high output voltage under open circuit conditions. By proper design of the transformer and choice of ratio of R907 and R908, the stored energy in the transformer and the current through output diodes can be limited during short circuit conditions.

19.3 SMPS INTEGRATED CIRCUITS

Typical control ICs developed for SMPS and TDA 2581 TDA 2582, TDA 4600, TDA 4601. TDA 2582 features the following circuit functions:

— Voltage controlled horizontal oscillator.
— Phase detector.
— Duty factor control for the negative going transient of the output voltage.
— Duty factor increases from zero to its normal operation value.
— Adjustable maximum duty factor.
— Over-voltage and over-current protection with automatic restart after switch off.
— Counting circuit for permanent switch-off when *n*-times over-current or over-voltage is sensed.
— Protection for open reference voltage.
— Protection for too low supply voltage.
— Protection against loop faults.
— Positive tracking of duty factor and feedback voltage when feedback voltage is smaller than the reference voltage minus 1.5 V.
— Normal and 'smooth' remote ON/OFF possibility.

A block diagram of the IC TDA 2582 is given in Fig. 19.8.

TDA 4600/4601

Because of wide operational and high-voltage stability even at high load changes, these ICs are used not only in TV receivers and VCRs but also in power supplies of Hi-fi sets and active speakers. Their salient features are:

— direct control of switching transistor
— low start-up current
— reverse-going linear overload characteristic curve
— collector current proportional to the base current input.

The internal block diagram of TDA 4600 is given in Fig. 19.9.

Circuit Description

(a) *Start-up sequence* During the start-up three sequential steps are followed, to ensure correct switching of the transistor. The sequence guarantees the supply to the switching transistor through the coupling electrolytic capacitor. (i) An internal *reference voltage is built up* which feeds to the voltage regulator and enables supply to the coupling capacitor and switching transistor. For a supply voltage of up to 12 V at pin 9, the pin 9 current is below 3.2 mA. (ii) *Release of the internal reference* voltage at pin 1 = 4 V, when V9 is 12 V, enables all the blocks in the IC to be supplied from the control logic, a thermally stable and overload protected current supply. (iii) As soon as the reference voltage is available, the *control logic is switched on* through an additional stabilization circuit.

(d) *Normal operation* Zero crossing of the feedback coil is registered at pin 2, and passed on to the control logic. Rectified amplitude variations of feedback coil are applied to the pin 3, which serves for regulation of input, overload and standby recognition. The *regulating amplifier* works with an input voltage of 2 V, and a current of 1.4 mA. Together with the collector current simulation at pin 4, the overload recognition defines the operating conditions, depending on the internal reference voltage. The simulation of collector current is obtained from an external RC network at pin 4, and an internally set voltage level.

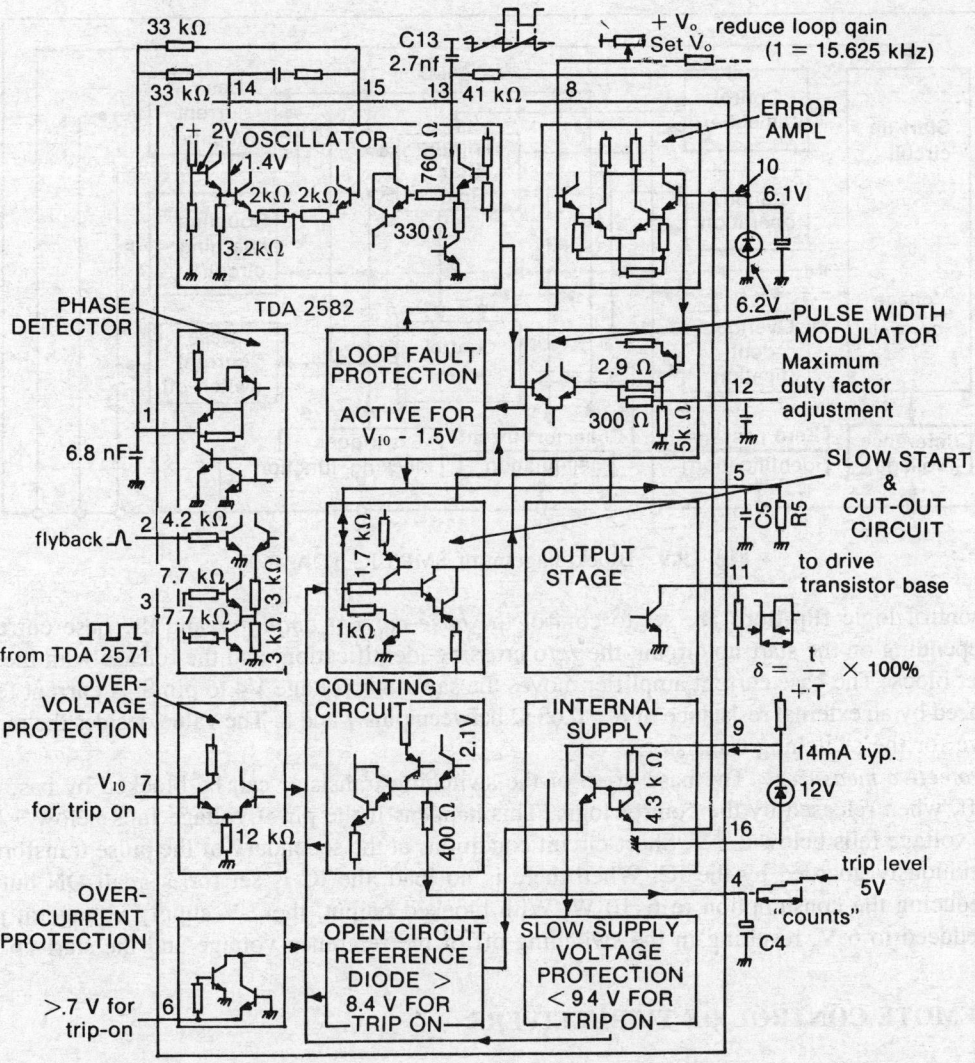

Fig. 19.8 Block diagram of the SMPS IC 2582 (*Courtesy:* ELCOMA PHILIPS)

The regulation lies between 2 V clamped dc voltage and ac sawtooth voltage rising up to 4 V, the reference voltage.

A *reduction of the secondary load* down to 20 W causes the switching frequency to rise to about 50 kHz, at an almost constant ON pulse duty factor of about 30% of the period. A further reduction of the secondary load down to about 1 W results in changing the switching frequency to around 70 kHz, and the drop in the ON duty factor to about 9%. The peak collector current also falls at the same time, below 1 A.

The output level of the control amplifier, the overload recognition and the collector current simulation are compared in the trigger block, and instructions passed on to the control logic for *base current switch-off.* Pin 5 provides for additional triggering and blocking of the output at pin 8, which can be blocked if pin 5 voltage is less than 2.2 V.

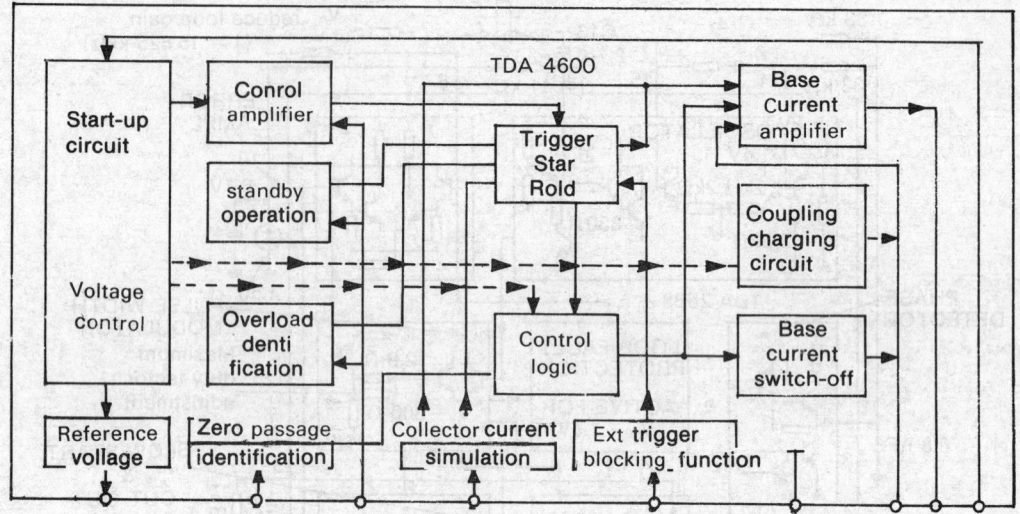

Fig. 19.9 Block diagram of SMPS IC TDA 4600

The control logic flip-flops are set to control the *base current amplifier* and the base current shut down, depending on the start-up circuit, the zero crossing identification, and the release with the help of the trigger block. The base current amplifier moves the sawtooth voltage V4 to pin 8. A current feedback is introduced by an external resistance of $R = 0.68\ \Omega$ between pins 7 and 8. The value decides the maximum base drive for the switching transistor.

(c) *Protective measures*   The base drive of the switching transistor can be blocked by base current switch-off, when released by the control logic. This happens if the pin 9 voltage falls below 7 V, or if the pin 5 voltage falls below 2.2 V. Short-circuit conditions at the secondary of the pulse transformer are also continuously guarded by the IC. When there is no load, the IC is set for a small ON duty pulse factor, reducing the consumption to 6–10 W. With blocked output, the 7-V supply voltage at pin 9 is further reduced to 6 V, resulting in the switching off of the reference voltage and the start-up.

19.4   REMOTE CONTROL OF TV RECEIVERS

As a television receiver is viewed from a considerable distance, it is useful to have a remote control to manipulate its controls. One may often feel inclined to change the volume or mute it at times, or alter the brightness and contrast, and change stations if alternate channels are available, sitting cozily at a place. Switching the receiver set on/off remotely is also a desired function. The remote control was obtained in the early years by *wired* extensions. Later, wireless technique was used employing *ultrasonic* waves in the range of 35 to 45 kHz as the link. Modern receivers commonly use the *infrared* wave as the link with PCM which can encode a large number of control commands.

Wired Control

A step in this direction is to provide a wired extension of the volume, brightness and ON/OFF control to the receiver. For example, an extension volume control at the end of a shielded cable can be provided by including a variable resistance potentiometer in the earthy end of the set mounted volume control. The latter being dc/electronic type, the capacitance of the extension cable does not affect audio frequency response. This wired extension may be introduced into the receiver by a socket and plug arrangement which substitutes the internal potentiometer by an extended one taking its position at the appropriate circuit points.

Ultrasonic Remote Control

A wireless remote control is possible by use of ultrasonic waves of around 40 kHz as link between the receiver and the hand control unit. Ultrasonic waves are suitable for short distance control like this, because they can provide a greater directivity with small size transducers and are confined to the room only, reducing the possibility of interference to similar sets nearby.

The ultrasonic signals can be generated in the control unit either mechanically or electronically. In the early mechanical system, these were generated by banging metal rods with spring loaded hammers. Each rod produced ultrasonic waves characteristic of its dimensions, for a short duration. A separate frequency was used for each function. When the piezoelectric ceramic transducers became available, they were conveniently used to produce the ultrasonic control signals. Electronic transistorized oscillators excited the piezoelectric transducer plate of barium titanate which converted the electric excitation at its faces into mechanical vibrations at the ultrasonic frequencies.

Ultrasonic frequencies were quite suitable for this purpose as they present no electromagnetic radiation problems, as only a moderate distance usually 3 to 10 m, is involved, without any obstructions. Frequencies in the range of 35 to 45 kHz were used. Care had to be taken to make the system selective enough so that it did not readily respond to stray sounds, such as jingles of keys, and so on. Good directivity was readily possible and interference from outside was negligible. Reflections from walls, ceilings and floors enabled operation without critical directional aim except at large distances.

The ultrasonic control receiver for ultrasonic remote control again required a microphone type piezoelectric transducer that converted the ultrasonic waves into electrical signals. The weak microvolts signal developed by the transducer was amplified by three to four transistorized R–C coupled and tuned stages, each tuned to the particular control frequency. The first stage was an emitter-follower type to match the high impedance transducer. This was followed by three stages of amplifiers, one having a tuned load with adequate bandwidth to cover all the control frequencies. The last amplifier stage had a number of tuned loads, and operated at low collector voltage to limit the peak resonant swings. The output selective tuned circuits using ferrite pot core coils, each tuned to the particular control frequency, provided a high output to the associated relay driver stages. These operated a stepper relay and ratchet, or relays to switch dc geared motors which performed the desired control functions mechanically. Later versions of such remote controls employed fully electronic methods for varying the controlled parameters, when dc voltages could be used for channel selection, volume control, etc.

Infrared Remote Control

Infrared Remote Control Transmitter

With the development of infrared LEDs and infrared photodiodes, infrared light has been widely used as the link for remote control of TV receivers. *Pulse code modulated* infrared light serves for the transmission of remote control commands, the information being defined by the varying time intervals between a sequence of very short infrared pulses. This pulsed operation enables the IR LED to the driven with high current pulses of over 1A, achieving a large transmission distance and high interference immunity while ensuring a long battery life for the control unit.

SAA/IRT 1250

Integrated transmitter IC SAA 1250 is a typical IC developed for infrared remote control of colour TV receivers. IRT 1250 is a further development on it, differing mainly in respect of the address selection code and higher output current capability. As a CMOS IC, it consumes minimum current and is highly flexible in application. On account of its large instruction repertoire, up to 1024 commands can be transmitted by one SAA/IRT 1250 transmitter IC. An infrared remote control receiver IC like the SAA 1251/SAA 1290 is meant to be a receiver. A block diagram of an infrared remote control using IRT 1250 at the transmitting end is shown in Fig. 19.10.

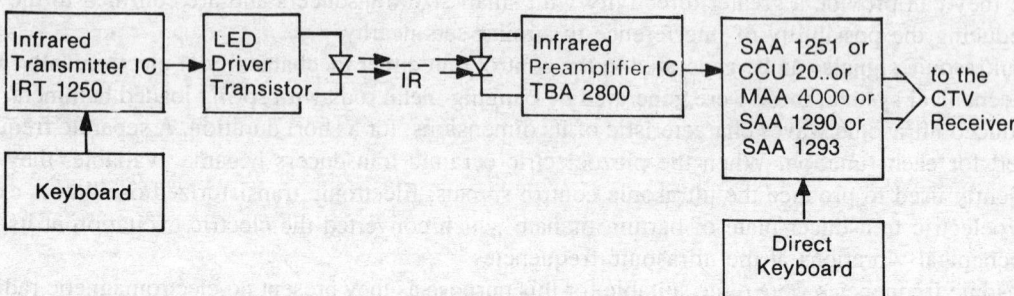

Fig. 19.10 Block diagram of an infrared remote control using IRT 1250 at the transmitter end (*Courtesy:* ITT)

At the receiving end, a photodiode converts the received IR transmissions into electrical signals which are amplified in the preamplifier IC TBA 2800. The PCM signals are *decoded and converted into respective commands,* e.g. ON/OFF switching, selection of the channel to be received, the setting of the analog control values for volume, contrast, colour, etc., control of teletext display, and so on. The TV set ON/OFF is effected by a triac in its mains circuit, switched ON/OFF by the mains flip-flop through an optocoupler for isolation. The circuits for volume, contrast, colour, etc. are provided analog signals from the D/A converters to set the desired values.

Each remote control signal word contains 10 information bits. The transmitter SAA 1250 is therefore capable of delivering 1024 different signals. The word is usually structured into 4 bits and 6 bits offering 16 addresses and 64 commands. However, this separation may not be adhered to. The receiver IC allows an operational mode in which commands and addressable additional ICs become practically unlimited.

Synthesis of Remote Control Signals

The signals are transmitted by means of infrared light in the shape of *packaged pulses.* For transmission of a 10-bit word, 14 pulses are required. The binary information of a bit is contained in the time interval between two pulses. The time interval T (approx. 1000 μs) between any two pulses is defined as the *basis for each code* employed. A short interval of duration T between any two pulses designates bit 0, while a long interval of the duration 2 T signifies bit 1, as seen in Fig. 19.11.

For a ten-bit data word, eleven data pulses are required. In addition, every signal contains a preliminary pulse, a start pulse and a stop pulse. The spacing between the preliminary pulse and start pulse is 3 T. This is followed after 1 T by the eleven pulses and terminated after 3 T by the stop pulse. Consequently, a command in which the bit 0 occurs ten times has a total duration of 17 T. Likewise, a command containing ten 1s has a duration of 27 T.

Interference Immunity

In order to reduce interference from IR sound transmissions, a preamplifier like TBA 2800, equipped

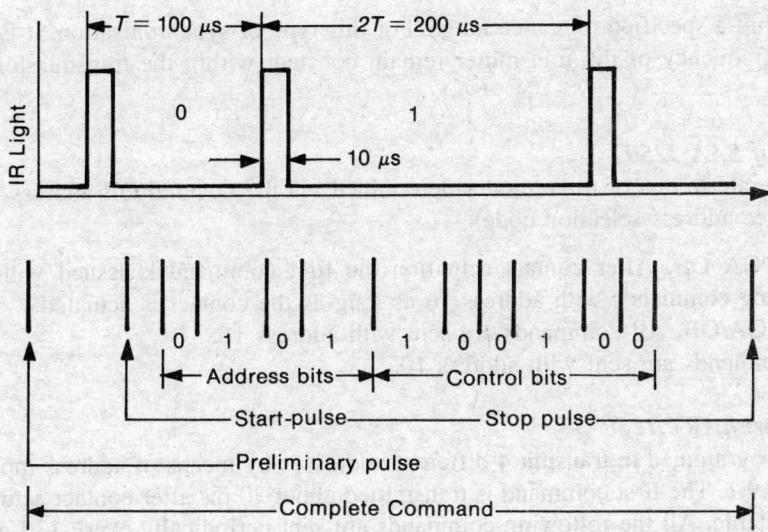

Fig. 19.11   (a) Representing bit 0 and bit 1 by time interval (b) Command word 0101100100, as an example
(*Courtesy:* ITT)

with AGC is employed in the receiver. The gain is set relative to the interference signal by means of the preliminary pulse. Further, the IR input of the receiver IC is blocked after each received pulse, and is only reopened after a time T for a short time slot t_F, as in gated AGC. If a pulse is received during t_F, a 0 signal is indicated. If no pulse is detected, then a time slot is again opened after a further period T. A pulse detected during this time slot signifies a logic 1.

With this mode of transmission, a pulse is bound to be detected *at least every second time slot* while information is being transmitted. If such is not the case, faulty reception is indicated and the receiver evaluation is interrupted and the receiver input reopened after the error has been detected. Exceptions to this rule are the first data pulse and the stop pulse. The start pulse and the subsequent detected data pulses are counted. After the 12th pulse, the checking conditions are changed as follows: No pulse should appear during the following two time slots, and the *stop pulse should be detected during the third time slot*. The interference-prevention system therefore recognizes undesired pulses coinciding with the time slots, and also the suppression of an information pulse because in that case the above mentioned condition for stop pulse is not met.

The preliminary pulse only serves as control signal for the AGC of the preamplifier. The receiver IC treats it like any other interference. Evaluation of the pulse train starts only after the start pulse.

Synchronisation of Receiver

The infrared remote control ICs IRT 1250 and SAA 1251 offer the advantage of requiring a quartz crystal (0.4 to 4.4 MHz) only in the receiver, whereas the transmitter contains merely a *simple RC oscillator*. Satisfactory synchronisation between the transmitter and receiver ICs, necessary to obtain the above mentioned interference immunity, is achieved in the receiver by measuring the interval between the start pulse and the first data pulse. This value is stored and, from this measurement, the value of the base interval T between the time slots is determined.

The duration of the time slot t_F for the first data pulse is therefore extended by comparison with the later time slots, so that the first data pulse is recognized without fail, provided the oscillator frequency falls within a specified tolerance range. For this type of synchronisation, it is only necessary that the oscillator frequency of the transmitter remain constant within the transmission period of one command.

Operating Modes of SAA 1250

SAA 1250 may operate in one of three modes determined via the OA and OB address inputs, HH/HL/ LH, (and LL for free address selection code).

Option I: HH at OA OB, After contact actuation the first command is issued with address 1, all following commands with address 16 as long as the contact is actuated.
Option II: HL at OA/OB, All commands are sent with address 16.
Option III: All commands are sent with address 10.

Address Selection with IRT 1250

IRT 1250 can be programmed to transmit 4 different addresses by means of address inputs A1 and A2, as shown in Table 19.1. The first command is transmitted about 20 ms after contact actuation, using the first address in the table. All the following commands are sent periodically every 130 ms, as shown in the program flowchart of Fig. 19.12.

Table 19.1   Code A1 and A2 address inputs

Option	A1	A2	Addresses
1	H	H	1, 16
2	L	H	3, 14
3	H	L	7, 10
4	L	L	4, 13

In this mode the required address must be initially entered into the address register of the transmitter IC, using one of the commands 17 to 32. Then all the following commands are transmitted together with the stored addresses, including commands 17 to 32, executing the FAS off commands 2, 3, 33 and 39, which clear the address registers and reset the FAS flip-flops. Immediately following this the command is processed in the preselected option.

If the commands are to be transmitted consecutively to various addresses with free address selection, the L signal must be permanently applied to both the address inputs. The FAS off commands 2, 3, 33 and 39 will under these conditions clear only the address register.

Functional Blocks of IRT 1250

A block diagram of the IC IRT 1250 is given in Fig. 19.13. The frequency of the *RC oscillator* is determined by the external RC element R_3, C_2 at pins 2 and 3. The resistor R_4 compensates for the dependence of the oscillator frequency on the supply voltage. If the receiver IC is driven by a 4.4-MHz crystal, then the oscillator frequency of the transmitter must be within 160 to 220 kHz, requiring that the R_1C time constant = 1.8 μs +/– 4.5%. For other crystal frequencies the time constant is altered in proportion.

The clock generator produces a *two-phase clock* for all the circuit elements of IRT 1250, except for actuation monitoring. The latter is static and blocks the oscillator as long as no input is being activated. This ensures negligible current consumption of the transmitter IC in the inactive condition.

Program control regulates the time sequence of all functions.

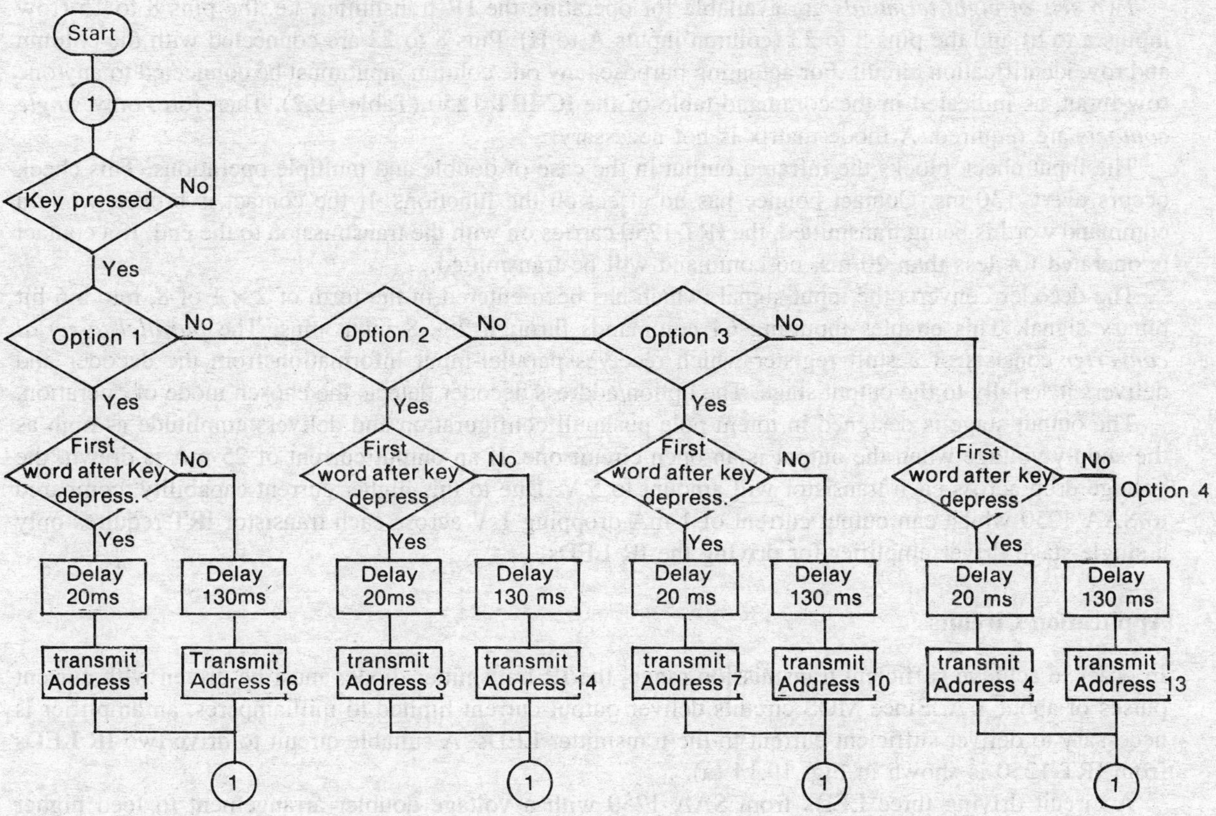

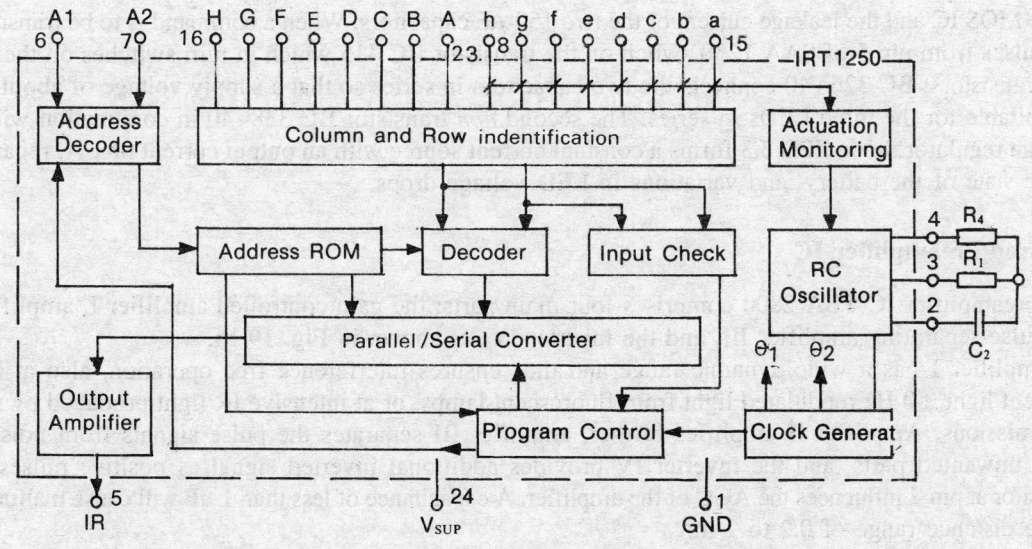

Fig. 19.12 Program flowchart of SAA 1250 (*Courtesy:* ITT)

Fig. 19.13 Block diagram of the IC IRT 1250 (*Courtesy:* ITT)

Two sets of eight terminals are available for operating the IR transmitter, i.e. the pins 8 to 15 (row inputs a to h) and the pins 8 to 23 (column inputs A to H). Pins 8 to 23 are connected with the column and row identification circuit. For actuating purpose, any one column input must be connected to any one row input, as indicated in the command table of the IC IRT 1250 (Table 19.2). Therefore, only *single contacts* are required. A diode matrix is not necessary.

The input check blocks the infrared output in the case of double and multiple operations. This check occurs every 130 ms. Contact bounce has no effect on the functions. If the contact is broken while a command word is being transmitted, the IRT 1250 carries on with the transmission to the end. If a contact is operated for less than 20 ms, no command will be transmitted.

The decoder converts the input signal, which has been entered in the form of 2×1 of 8, into a 6-bit binary signal. This enables inputting 64 commands through 2×8 input pins. The *parallel-to-serial converter* consists of a shift register which receives parallel-input information from the decoder and delivers it serially to the output stage. The option/address decoder detects the chosen mode of operation.

The output stage is designed in totem-pole pushpull configuration and delivers amplitude as high as the supply voltage when the output is an open-circuit one. If an output current of 25 mA is drawn, the voltage drop across each transistor will amount to 5 V. Due to this higher current capability, compared to SAA 1250 which can output current of 1 mA dropping 1 V across each transistor IRT requires only a single-stage driver amplifier for driving the IR LEDs.

Application Circuits

In order to achieve sufficient transmission range, the IR transmitter diodes must be driven with current pulses of about 1 A. Since MOS circuits deliver output current limited to milliamperes, an amplifier is necessary to deliver sufficient current to the transmitter LEDs. A suitable circuit to drive two IR LEDs from IRT 1250 is shown in Fig. 19.14 (a).

A circuit driving three LEDs from SAA 1250 with a voltage doubler arrangement to feed higher voltage to the LEDs is shown in Fig. 19.17(b). In the steady condition and the spaces between pulses all three transistors are blocked and the two 150-μF capacitors charge up to the battery voltage. When this condition is reached the current consumption of the circuit is determined only by the current consumption of the MOS IC and the leakage current of the two 150-μF capacitors. When a command is to be transmitted the pulses from pin 5 of SAA 1250 switch on the transistor BC 415 which in turn switches on the other two transistors. BC 326–40 connects the two capacitors in series so that a supply voltage of about 15 V is available for the three LEDs in series. The second *npn* transistor BC 338–40 in conjunction with the voltage regulator diode ZPD 3.3 forms a constant current source with an output current of 1 A, regardless of the state of the battery, and variations in LED voltage drops.

Infrared Preamplifier IC

The preamplifier IC TBA 2800 comprises four main parts: the gain controlled amplifier I, amplifier II, the pulse-separating amplifier III, and the inverter IV, as shown in Fig. 19.15.

Amplifier I has a wide dynamic range and thus ensures interference free operation, also at bright ambient light, 50 Hz modulated light from fluorescent lamps, or at intensive IR light produced by sound transmissions. Amplifier II amplifies further, amplifier III separates the pulse signals from noise and other unwanted parts, and the inverter IV provides additional inverted signal as positive pulses. The capacitor at pin 2 influences the AGC of the amplifier. A capacitance of less than 1 μF will cause malfunction in the distance range of 0.2 to 2 m.

Table 19.2 Command table of IR transmitter IRT 1250 (ITT)

Command No.	a	b	c	d	e	f	g	h	A	B	C	D	E	F	G	H	Option 1 Addresses 1 and 16	Option 2 Addresses 3 and 14	Option 3 Addresses 7 and 10	Option 4 Addresses 4 and 13
1	×								×											
2	×									×										
3	×										×									
4	×											×								
5	×												×							
6	×													×						
7	×														×					
8	×															×				
9		×							×											
10		×								×										
11		×									×									
12		×										×								
13		×											×							
14		×												×						
15		×													×					
16		×														×				
17			×						×											
18			×							×										
19			×								×									
20			×									×								
21			×										×							
22			×											×						
23			×												×					
24			×													×				
25				×					×											
26				×						×										
27				×							×									
28				×								×								
29				×									×							
30				×										×						
31				×											×					
32				×												×				
33					×				×											
34					×					×										
35					×						×									
36					×							×								
37					×								×							
38					×									×						
39					×										×					
40					×											×				
41						×			×											
42						×				×										
43						×					×									
44						×						×								
45						×							×							
46						×								×						
47						×									×					
48						×										×				
49							×		×											
50							×			×										
51							×				×									
52							×					×								
53							×						×							
54							×							×						
55							×								×					
56							×									×				
57								×	×											
58								×		×										
59								×			×									
60								×				×								
61								×					×							
62								×						×						
63								×							×					
64								×								×				

Option 1: First command is transmitted 20 ms after key actuation with Address 1, further commands periodically in a distance of 130 ms with Address 16.

Option 2: First command is transmitted 20 ms after key actuation with Address 3, further commands periodically in a distance of 130 ms with Address 14.

Option 3: First command is transmitted 20 ms after key actuation with Address 7, further commands periodically in a distance of 130 ms with Address 10.

Option 4: First command is transmitted 20 ms after key actuation with Address 4, further commands periodically in a distance of 130 ms with Address 13.

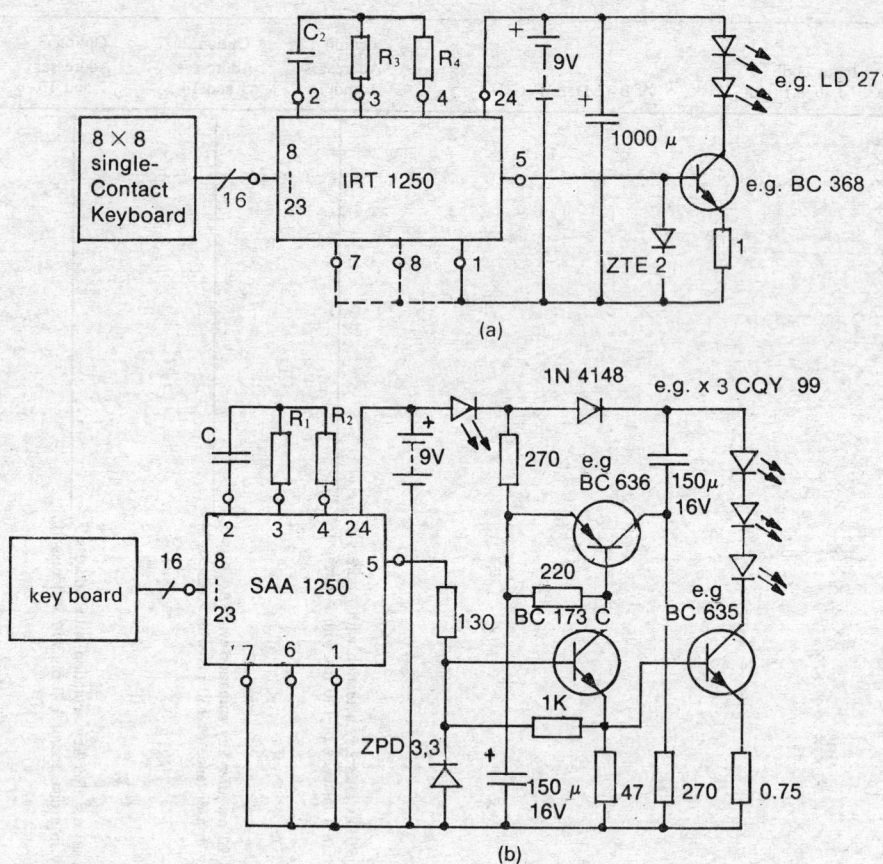

Fig. 19.14   (a) Transmitter amplifier for IRT 1250 (b) Transmitter amplifier using voltage doubler for SAA 1250 (*Courtesy:* ITT)

IR Receiver IC SAA 1251

Operational Modes

The SAA 1251 can be made to function in four different modes, by connecting the option pin 18 to different pins of the IC.

Option I   Connect pin 18 to pin 1. This functional mode is intended for remote control of TV receivers. The receiver operates with the address 16. The only additional function, via address 14, affects D/A converters.

Option II Connect pin 18 to pin 21. This functional mode is also intended for remote controlled TV receivers via address 16. Apart from address 14 address 13 can be employed for controlling the D/A converters. The remaining addresses as well as the unused commands of partially occupied addresses stay available for further circuits connected to the data bus.

Option III Connect pin 18 to pin 24. In this variant, all functions of the receiver IC are activated via address 15. This enables parallel operation on SAA 1251 as under option I, and a further receiver IC SAA 1251 provided in a radio receiver (option III).

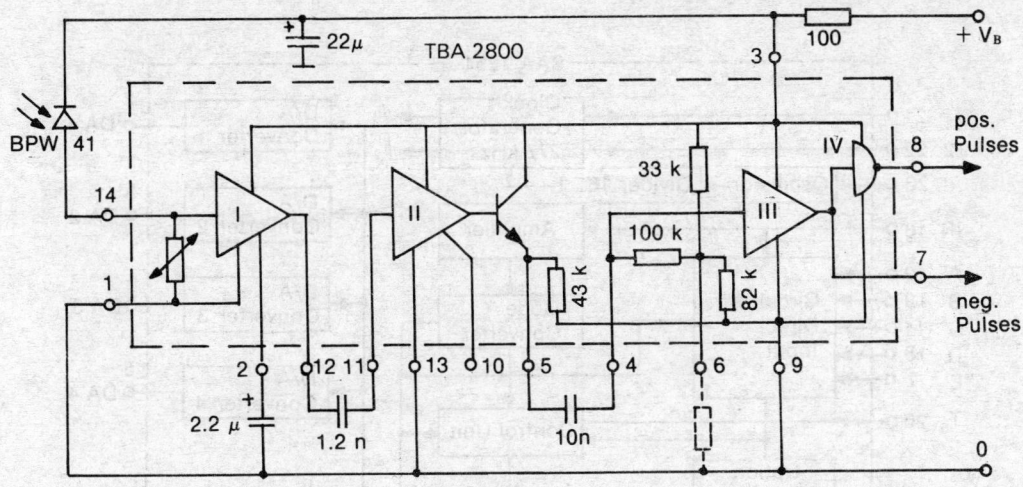

Fig. 19.15 Block diagram of TBA 2800 with associated circuit (*Courtesy:* ITT)

Option IV Connect pin 18 to pin 22. According to this option, the signals are processed in different ways, depending on the condition of a SUB flip-flop contained in the receiver IC. When the latter is initialized this flip-flop is set to OFF. This receiver will then operate with addresses 1 and 15. When this changeover has taken place, the program commands are code converted and given out by the data output DA, without effecting any change in the program location memory. The commands 2, 3, 33 and 39 are used to reset the SUB flip-flop to OFF.

Option IV is intended for the control of TV receivers with videotext, TV games and similar accessories. These accessories are treated as *subsystems*. This gives the following operating advantages. The same transmitter keys which change the program condition 'subsystem off' can be used to input numbers when the condition is 'subsystem on'. If the television receiver is switched off while operating in the subsystem mode, it returns automatically to TV operation (Subsystem off) when it is switched on again. It is therefore not necessary to provide functional mode indication.

Regardless of address and operation mode, all signals transmitted via the infrared range are outputted by the data output DA.

The block diagram of the receiver IC SAA 1251 is shown in Fig. 19.16.

The oscillator needs only a crystal (e.g. 4 MHz) between the pin 23 and V_{ss} to generate the clock that controls the 16:1 divider. The divider output is connected to a *clock generator* which produces clock pulses θ and ϕ (277 kHz) of opposite phase. These clock pulses serve for the IC and synchronise all the additional integrated circuits connected to. For this purpose they are inverted and taken to the clock outputs $\theta^{\sim\sim}$ and $\phi^{\sim\sim}$. This inversion is useful because in the additional ICs the clock pulses have to be refreshed by inverting amplifiers, where the original polarity is restored.

The pulse signal from the external preamplifier should be capacitively coupled to pin 16. This signal is transferred to a *code converter* via an internal amplifier. The converter changes the pulse space modulated information into AM information which in turn is fed to the data output register.

In the quiescent state, the data output DA is at V_{ss} (H) potential. Each serial 10-bit instruction word is preceded by and followed by an L-bit. If several H-bits or several L-bits follow in succession, the output remains unchanged from one bit to the next, until a bit of the opposite polarity is output (NRZ). Also direct input signals are taken to the data bus. Depending on the functional mode, the instruction is preceded by either address 16 or address 15. The address code is shown in Table 19.3.

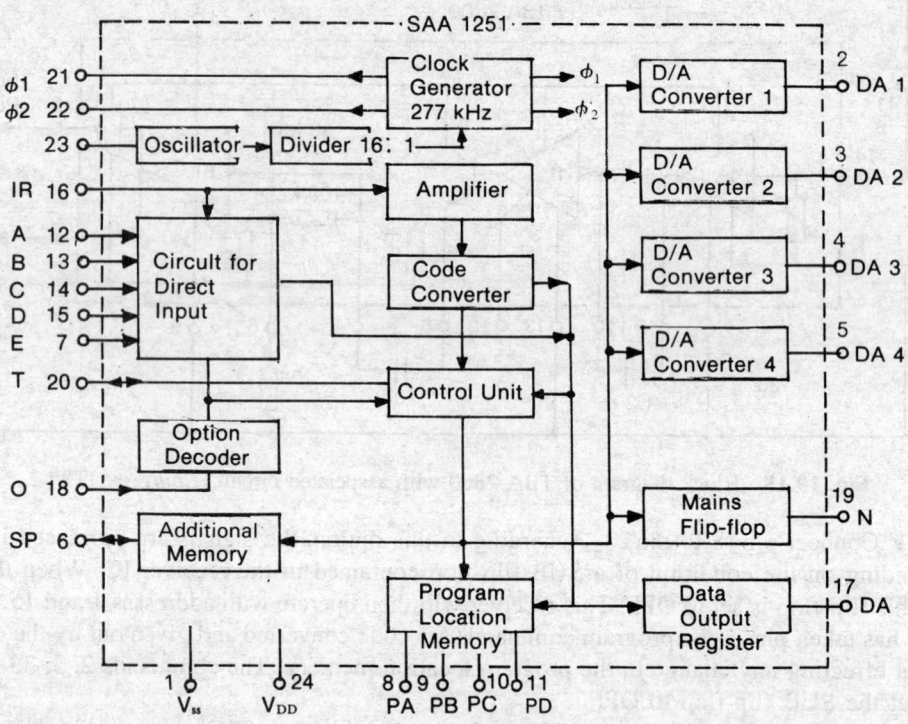

Fig. 19.16 Block diagram of the receiver IC SAA 1251 (*Courtesy:* ITT)

Table 19.3 Address code

Address Number	Code a	b	c	d
1	H	H	H	H
2	L	H	H	H
3	H	L	H	H
4	L	L	H	H
5	H	H	L	H
6	L	H	L	H
7	H	L	L	H
8	L	L	L	H
9	H	H	H	L
10	L	H	H	L
11	H	L	H	L
12	L	L	H	L
13	H	H	L	L
14	L	H	L	L
15	H	L	L	L
16	L	L	L	L

In this table, the H level corresponds to the binary 0 at the output of the transmitter SAA 1250 and the level L corresponds to binary 1. The *option decoder* recognizes the chosen functional modes and supplies corresponding data to the control unit and to the circuit for direct input. The control unit determines the entire signal processing sequence.

Pin 20 serves as input during testing of SAA 1251 and as output for controlling tuning circuit SAA 1121. In order to measure the leakage current of the output transistors, pin 20 must be connected to V_{SS}. When V_{SS} is disconnected, SAA 1251 attains the initial condition. Commands 5 and 6 produce an H-pulse of duration 36 μs and 144 μs respectively at pin 20, periodically at intervals of approximately 130 ms. Pin 20 should be otherwise open.

The mains output N is connected to the mains flip-flop which may be switched into the 'Receiver On' state in four ways:

— by the command 3 'Mains On'
— by any of the program commands 17 to 32
— by command 8, sequential program change
— via the mains output if the latter is connected to V_{SS} for at least 10 μs

'Receiver Off' takes place by means of the command 2 'Mains Off'. The switching of the mains flip-flop by means of the commands 2, 3, 8 and 17 to 32 always takes place with a delay of approximately 0.7 s, in order to avoid accidental switching as a result of the remote control keys beings briefly touched inadvertently.

An additional memory has been provided. This may be set and reset by two commands. It is initialized (L at pin 6) when the supply voltage is applied. This memory flip-flop can be set to L or level via this output. The desired potential has to last for a minimum of 10 μs at pin 6.

The four *D/A converters* supply rectangular voltages of approximately 17.3 kHz whose pulse/interval ratio is variable in 63 steps. Owing to the comparatively high frequency, the formation of dc control voltage by means of simple filter elements of small time constant is feasible. Nor does the suppression of interference caused by rectangular voltages pose any problems. A permanent command for varying analog information causes one step to occur every 130 ms. The whole range is traversed within approximately 9 seconds. Overshoot of the final position in either direction is excluded.

When VDD is applied, the D/A converters 1, 2 and 3 are set to mid-position, and the D/A converter 4 is set to a pulse/interval ratio of approximately 1 : 2 (pulse = H). As a result of the 'initialize' command, D/A converters 1, 2 and 3 are set to mid-position (pulse/interval ratio of 1 : 1). D/A converter 4 can be switched by means of command 7 (mute) to zero (output transistor blocked) and reset to the previously set value by using commands 3, 47 and 48. With every command, D/A converter 4 is switched to L (muted) for 320 ms. The program data are currently available in binary coded form at the four outputs of the program location memory, PA to PD. When VDD is applied the memory is initialized to program 1. Table 19.4 shows the code at four program outputs PA to PD.

Program selection may be direct or sequential. In the 'mains off' condition it is also possible to switch on the circuit by using the sequential program change command. In that case the program will be stepped forward when the command has been interrupted for at least 0.32 seconds. As soon as the command is issued once more, the program will be switched forward immediately. A continuous command causes one further step to be followed for every 0.7 seconds. Program stepping can only take place in one and the same direction.

For the purpose of emergency operation of remote controlled sets, and for operating sets which are not remote controlled, a circuit for direct input signals is provided, which has priority over IR signals.

The *direct inputs* may be actuated either by mechanical means or electronic tough buttons. In the latter case a 50 Hz hum voltage of large amplitude may be superposed on the input potential, formed by the

Table 19.4   Code at the outputs PA to PD (ITT)

Program Number	Code PA	PB	PC	PD
1	L	L	L	L
2	H	L	L	L
3	L	H	L	L
4	H	H	L	L
5	L	L	H	L
6	H	L	H	L
7	L	H	H	L
8	H	H	H	L
9	L	L	L	H
10	H	L	L	H
11	L	H	L	H
12	H	H	L	H
13	L	L	H	H
14	H	L	H	H
15	L	H	H	H
16	H	H	H	H

potential division between the fixed resistors and skin resistance. The SAA also accepts such cyclic input signals.

In order to prevent commands from being triggered by individual *stray pulses* acting upon the direct inputs, the first pulse is used to set one or more flip-flops at the appropriate inputs, and a counter is started. After the expiry of one 50 Hz cycle (20 ms), all flip-flops are reset. After another 20 ms a test is made to reveal whether at least one of the flip-flops has been set once more. If this is the case, resetting takes place once again, and upon the expiry of another 20 ms, a further test is made to ascertain whether or not another flip-flop has been set. Only if both tests prove positive will the command be executed, in accordance with the data available at the final test.

The directly given commands must be entered according to the command list (Table 19.5) in coded form as five-bit words. With mechanical contacts coding can take place by means of multiple contacts. With electronic touch buttons diodes may be preferred.

19.5   INTEGRATING CIRCUITS FOR ECONOMY TV RECEIVERS

The high component density and complexity of circuits in a TV receiver justifies extensive use of integrated circuits to replace complex circuit blocks in a TV receiver. Use of integrated circuits reduces the component count and assembly costs. Production line adjustments are reduced and the increased reliability minimizes service after sales cost also. The mean time between failures (MTBF) is considerably improved. Therefore, the use of integrated circuits is on the rise in the TV industry. As the VLSI designs integrate more and more functions on a single chip, the number of IC chips required for a TV receiver has gone down to one or two, integrating most of the small signal processing functions. The dream of manufacturing colour and black-and-white receiver sets using a compact PCB incorporating a couple of chips has already come true.

Table 19.5 Command code of the infrared receiver IC SAA 1251 (ITT)

Command No.	IR Code a b c d e f	Direct A B C D E	Data-bus a b c d e f	IR Receiver IC SA 1251 Options I, II and III	IR Receiver IC SAA 1251 Option IV, Address 16 Subsystem off	Subsystem on
1	0 0 0 0 0 0		H H H H H			
2	1 0 0 0 0 0	L H H H H	L H H H H	Mains off	Mains off	Mains off/Subs. off
3	0 1 0 0 0 0	H L H H H	H L H H H	Mains on/Sound on	Mains on/Sound on	Mains on/Subs. off
4	1 1 0 0 0 0	L L H H H	L L H H H	Initialisation D/A Convert.	Initialisation D/A Convert.	
5	0 0 1 0 0 0	H H L H H	H H L H H H	FT+	FT+	FT+
6	1 0 1 0 0 0	L H L H H	L H L H H H	FT-	FT-	FT-
7	0 1 1 0 0 0	H L L H H	H L L H H H	Mute	Mute	Mute
8	1 1 1 0 0 0	L L L H H	L L L H H H	Sequ. Progr. Ch./Mains on	Sequ. Progr. Ch./Mains on	
9	0 0 0 1 0 0		H H L H H			
10	1 0 0 1 0 0		L H H L H H			
11	0 1 0 1 0 0		H L H L H H			
12	1 1 0 1 0 0		L L H L H H			
13	0 0 1 1 0 0		H H L L H H			
14	1 0 1 1 0 0		L H L L H H			
15	0 1 1 1 0 0		H L L L H H			
16	1 1 1 1 0 0		L L L L H H			
17	0 0 0 0 1 0	H H H H L	H H H H L H	Program 1/Mains on	Program 1/Mains on	
18	1 0 0 0 1 0	L H H H L	L H H H L H	Program 2/Mains on	Program 2/Mains on	
19	0 1 0 0 1 0	H L H H L	H L H H L H	Program 3/Mains on	Program 3/Mains on	
20	1 1 0 0 1 0	L L H H L	L L H H L H	Program 4/Mains on	Program 4/Mains on	
21	0 0 1 0 1 0	H H L H L	H H L H L H	Program 5/Mains on	Program 5/Mains on	
22	1 0 1 0 1 0	L H L H L	L H L H L H	Program 6/Mains on	Program 6/Mains on	
23	0 1 1 0 1 0	H L L H L	H L L H L H	Program 7/Mains on	Program 7/Mains on	
24	1 1 1 0 1 0	L L L H L	L L L H L H	Program 8/Mains on	Program 8/Mains on	
25	0 0 0 1 1 0	H H H L L	H H H L L H	Program 9/Mains on	Program 9/Mains on	
26	1 0 0 1 1 0	L H H L L	L H H L L H	Program 10/Mains on	Program 10/Mains on	
27	0 1 0 1 1 0	H L H L L	H L H L L H	Program 11/Mains on	Program 11/Mains on	
28	1 1 0 1 1 0	L L H L L	L L H L L H	Program 12/Mains on	Program 12/Mains on	
29	0 0 1 1 1 0	H H L L L	H H L L L H	Program 13/Mains on	Program 13/Mains on	
30	1 0 1 1 1 0	L H L L L	L H L L L H	Program 14/Mains on	Program 14/Mains on	
31	0 1 1 1 1 0	H L L L L	H L L L L H	Program 15/Mains on	Program 15/Mains on	
32	1 1 1 1 1 0	L L L L L	L L L L L H	Program 16/Mains on	Program 16/Mains on	
33	0 0 0 0 0 1		H H H H L			Subsystem off
34	1 0 0 0 0 1		L H H H H L	Additional Memory II	Additional Memory II	
35	0 1 0 0 0 1		H L H H H L	Additional Memory L	Additional Memory L	
36	1 1 0 0 0 1		L L H H H L			
37	0 0 1 0 0 1		H H L H H L			Subsystem off
38	1 0 1 0 0 1		L H L H H L			
39	0 1 1 0 0 1		H L L H H L			
40	1 1 1 0 0 1		L L L H H L			
41	0 0 0 1 0 1	H H H L H	H H H L H L	DA1+	DA1+	DA1+
42	1 0 0 1 0 1	L H H L H	L H H L H L	DA1-	DA1-	DA1-
43	0 1 0 1 0 1	H L H L H	H L H L H L	DA2+	DA2+	DA2+
44	1 1 0 1 0 1	L L H L H	L L H L H L	DA2-	DA2-	DA2-
45	0 0 1 1 0 1	H H L L H	H H L L H L	DA3+	DA3+	DA3+
46	1 0 1 1 0 1	L H L L H	L H L L H L	DA3-	DA3-	DA3-
47	0 1 1 1 0 1	H L L L H	H L L L H L	DA4+/Sound on	DA4+/Sound on	DA4+/Sound on
48	1 1 1 1 0 1	L L L L H	L L L L H L	DA4-/Sound on	DA4-/Sound on	DA4-/Sound on
49	0 0 0 0 1 1		H H H H L L			
50	1 0 0 0 1 1		L H H H L L			
51	0 1 0 0 1 1		H L H H L L			
52	1 1 0 0 1 1		L L H H L L			
53	0 0 1 0 1 1		H H L H L L			
54	1 0 1 0 1 1		L H L H L L			
55	0 1 1 0 1 1		H L L H L L			
56	1 1 1 0 1 1		L L L H L L			
57	0 0 0 1 1 1		H H H L L L			Subsystem 1 on
58	1 0 0 1 1 1		L H H L L L			Subsystem 2 on
59	0 1 0 1 1 1		H L H L L L			Subsystem 3 on
60	1 1 0 1 1 1		L L H L L L			Subsystem 4 on
61	0 0 1 1 1 1		H H L L L L			Subsystem 5 on
62	1 0 1 1 1 1		L H L L L L			Subsystem 6 on
63	0 1 1 1 1 1		H L L L L L			Subsystem 7 on
64	1 1 1 1 1 1		L L L L L L			Subsystem 8 on

Combination ICS for Economy TV Receivers

Efforts to reduce the number of integrated circuit chips required for a TV receiver have continued, the aim being to manufacture a *single-chip TV receiver* set. The forerunners of this have been the small-signal combination ICs including the series 4500 developed by Philips. These ICs contain all small signal processing functions in one chip, requiring only a tuner and simple power stages to complete the circuitry. The IC *TDA 4503* includes a vision IF amplifier with synchronous demodulator and AGC detector with tuner output and fully synchronised vertical and horizontal drive outputs for a black-and-white receiver. *TDA 4501 and TDA 4505* are TV sub-system circuits intended for use in colour TV receivers. For a complete colour TV receiver, only a tuner, colour decoder and output stages have to be added.

This single-IC concept reduces drastically the number of peripheral components and the adjustments, providing excellent reliability at lower cost. Furthermore, it has provided an opportunity to improve performance by making use of information obtained in the IF part to control functions in the sync section and vice versa. For example:

— The AFC can be gated by pulses derived from the horizontal oscillator.
— The AFC can be switched by the coincidence detector if the transmission is not detected, thus ensuring reliable and accurate tuning.
— The time constant of the first PLL in the horizontal sync can be optimised for the characteristics of the incoming signal (weak, strong or VCR) as the AGC detects this information.

For higher-performance sets, separate ICs for vision IF (TDA 8341, TDA 2549) and for sync processing (TDA 2579) are advised. This allows modular construction where display is expected to be a separate high-performance module.

IC TDA 4503 for B/W Receiver

TDA 4503 combines all small-signal functions, except the tuner, which are required in a black-and-white receiver. For a complete black-and-white receiver, only the output stages for video, sound, horizontal and vertical deflection, and the tuner, have to be added. The internal circuit blocks are shown in Fig. 19.17, and are explained in the following.

— Vision IF amplifier with synchronous demodulator, operating with symmetrical inputs at pins 8 and 9.
— Demodulator and AFC share an external reference tuned circuit at pins 20, 21.
— IF AGC control range 60 dB, tuner AGC for *pnp/npn* RF stages
— $3V_{pp}$ video preamplifier output for an rms IF input of 70 uV.
— SIF signal coupled to sound circuit by BP filter at pin 15 amplifier-limiter, synchronous demodulator with reference tuned circuit at pin 13, 80 dB range dc volume control at pin 11, audio output 320 mV rms with delta f = 7.5 kHz.
— Transmitter identification and mute
— Horizontal synchronisation
— Horizontal oscillator and drive
— Vertical synchronisation internal integrating network
— Vertical oscillator and drive, feedback from deflection coils

Functional Description of TDA 4503

If amplifier, demodulator and A.F.C.   The IF amplifier operated with symmetrical input at pins 8 and 9 and has an input impedance suitable for SAW filter application. The amplifier sensitivity gives a peak-

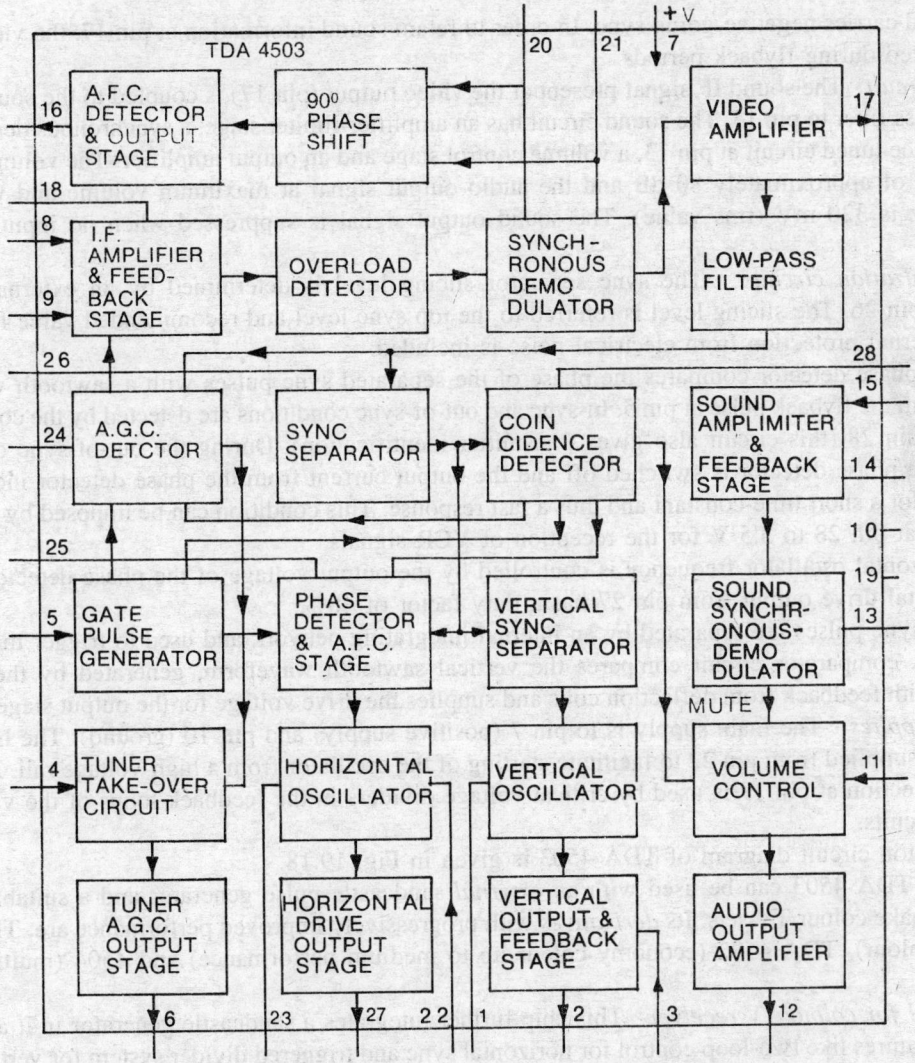

Fig. 19.17 Block diagram of IC TDA 4503 (*Courtesy:* ELCOMA PHILIPS)

to-peak output voltage of 3V for an rms input of 70 μV. The demodulator and the AFC circuit share an external reference tuned circuit (pins 20 & 21) and an internal RC network provides the phase-shifting necessary for AFC operation. The AFC circuit provides a control voltage output with a (typical) swing of 9V from pin 16 (V_p = 10.5 V).

AGC circuit Gating of the AGC detector is performed to reduce sensitivity of the IF amplifier to external electrical noise. The AGC time constant is provided by an RC-network connected to pin 24. The typical gain control range of the IF amplifier is 60 dB. Tuner AGC voltage is supplied from pin 6 and is suitable for tuners with *pnp* or *npn* RF stages. The sense of the AGC (to increase in a positive or negative direction) and the point of the tuner take-over are presented by the voltage level at pin 4 [V_4 = 3.5V (typically) for positive AGC; V_4 = 8 V (typically) for negative AGC].

Video amplifier The video signal output from pin 17 has a peak-to-peak value of 3V (top sync level

= 1.5 V) and carries negative-going sync. In order to retain sound information at pin 17, the video signal is not blanked during flyback periods.

Sound circuit The sound IF signal present at the video output (pin 17) is coupled to the sound circuit by a bandpass filter to pin 15. The sound circuit has an amplifier-limiter stage, a synchronous demodulator with reference tuned circuit at pin 13, a volume control stage and an output amplifier. The volume control has a range of approximately 80 dB and the audio output signal at maximum volume and with delta $f = 7.5$ kHz is 320 mV (rms value). The sound output signal is suppressed when no input signal is detected.

Synchronization circuits The sync separator slicing level is determined by an external resistor network at pin 26. The slicing level is referred to the top sync level and recommended value for slicing is 30%. Internal protection from electrical noise is included.

A gated phase detector compares the phase of the separated sync pulses with a sawtooth waveform obtained from the flyback pulse at pin 5. In-sync and out-of-sync conditions are detected by the coincidence detector at pin 28 (this circuit also gives transmitter identification). During the out-of-sync condition, gating of the phase detector is switched off and the output current from the phase detector increases to give a detector a short time-constant and thus a fast response. This condition can be imposed by clamping the voltage at pin 28 to 3.5 V for the reception of VCR signals.

The horizontal oscillator frequency is controlled by the output voltage of the phase detector circuit. The horizontal drive output from pin 27 has a duty factor of 40%.

Vertical sync pulses are separated by an internal integrating network and used to trigger the vertical oscillator. A comparator circuit compares the vertical sawtooth waveform, generated by the vertical oscillator, with feedback from deflection coils and supplies the drive voltage for the output stage at pin 2.

Power supplies The main supply is to pin 7 (positive supply) and pin 10 (ground). The horizontal oscillator is supplied from pin 22 to facilitate starting of the oscillator from a high-voltage rail. A special ground connection at pin 19 is used by critical voltage dividers in the feedback loops of the vision and sound IF circuits.

The application circuit diagram of TDA 4503 is given in Fig. 19.18.

The chip TDA 4503 can be used with an *external* sandcastle pulse generator and a suitable colour decoder to make colour TV sets. Its *derivatives* with progressively improved performance are: TDA 4505 (economy colour), TDA 4502 (economy colour up to medium performance) and 4504 (multistandard derivative).

IC TDA 4501 for colour TV receiver This chip further integrates a sandcastle generator in it and some additional features like two-loop control for horizontal sync and triggered divider system for vertical sync and sawtooth generation giving automatic amplitude adjustment for 50 or 60 Hz working.

IC TDA 4505 for colour TV receiver TDA 4505 is a TV subsystem for economy colour TV receiver sets. For a complete colour TV receiver, only a tuner, colour decoder and output stages are required. This chip has special sync features like two-loop horizontal synchronisation, extra time constant switches in H-AFC detector, and triggered divider system for vertical synchronisation. An external connection between the video amplifier and the sync separator is provided. This enables synchronisation of the IC from external source like teletext, VCR and video camera input. The block diagram of the IC is given in Fig. 19.19, and its application circuit in Fig. 19.19.

Functional Description of TDA 4505

If amplifier, demodulator and AFC The IF amplifier has a symmetrical input (pins 8 and 9). The synchronous demodulator and the AFC circuit share an external reference tuned circuit (pins 20 and 21). An internal RC-network provides the necessary phase-sift for AFC operation. The AFC circuit is gated by an internally

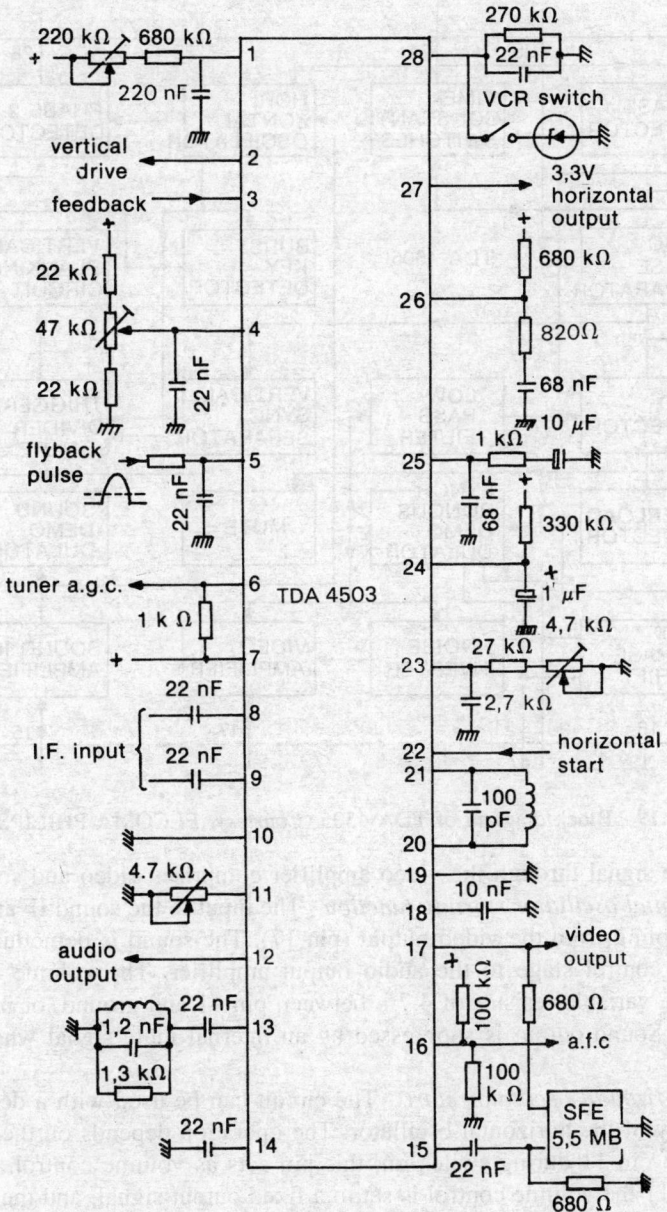

Fig. 19.18 Application circuit of TDA 4503 (*Courtesy:* ELCOMA PHILIPS)

generated gating pulse. As a result, the AFC output voltage contains no video information. The AFC circuit provides a control voltage output with a swing greater than 10V at pin 18.

AGC circuit An AGC detector is gated to reduce sensitivity of the IF amplifier to external electrical noise. The AGC time constant is provided by an RC-circuit connected to pin 19. The point of tuner takeover is preset by the voltage level at pin 1.

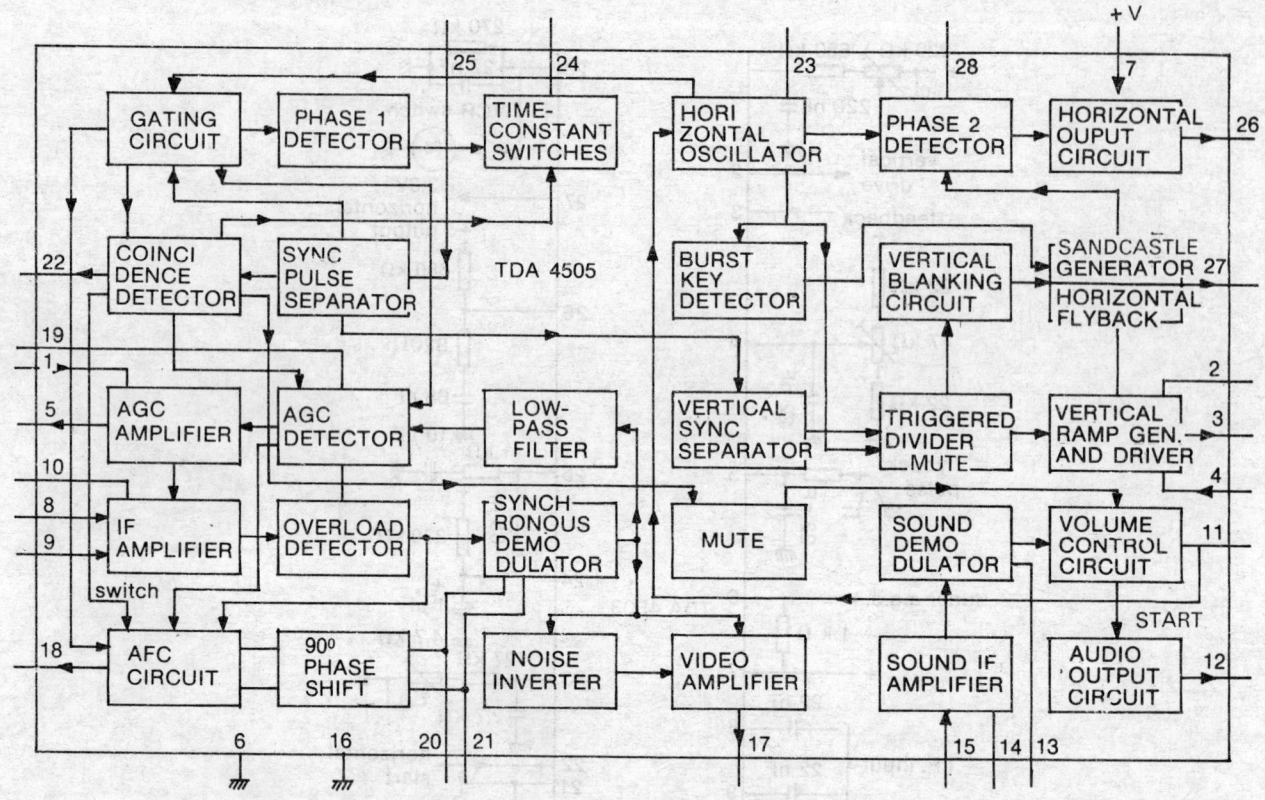

Fig. 19.19 Block diagram of TDA 4505 (*Courtesy:* ELCOMA PHILIPS)

Video amplifier The signal through the video amplifier comprises video and sound information.
Sound circuit and horizontal oscillator starting function The input to the sound IF amplifier is obtained by a bandpass filter coupling from the video output (pin 17). The sound is demodulated an passed via a dual-function volume control stage to the audio output amplifier. The volume control function is obtained by connecting a variable resistor of 4.7 k between pin 11 and ground, or by supplying pin 11 with a variable voltage. Sound output is suppressed by an internal mute signal when no TV signal is identified.

dc volume control/Horizontal oscillator start The circuit can be used with a dc volume control or with a starting possibility of the horizontal oscillator. The operation depends on the application. When no current is supplied to pin 11 during switch-on, this pin acts as volume control. When a current of 6 mA is supplied to pin 11 the volume control is set to a fixed output signal, and the IC generates drive pulses for horizontal deflection. The main supply of the IC can then be derived from the horizontal deflection.

Horizontal synchronisation The video input signal (positive video) is connected to pin 25.

The horizontal synchronisation has two control loops. This has been introduced to generate a sandcastle pulse. Using the oscillator sawtooth facilitates accurate timing of the burst key pulse. Therefore, the phase of this sawtooth must have a fixed relationship to the sync pulse, which is achieved by use of a second loop.

Horizontal phase detector The circuit has the following operating conditions.

(a) Strong input signal, synchronised or not synchronised.

(The input signal condition is obtained from the AGC circuit, the in-sync/out-of-sync from the coincidence detector.) In this condition the time constant is optimal for VCR playback, i.e. fast time constant during the vertical retrace (to be able to correct head-errors of the VCR) an such a time constant during scan that fluctuations of the sync are corrected. The phase detector is not gated.

(b) Weak signal.

In this condition the time constant is doubled compared to condition (a). Also the phase detector is gated when the oscillator is synchronised. This ensures a stable display which is not disturbed by the pulse in the video signal.

(c) Not synchronised (weak signal). In this condition the time constant during scan and vertical retrace are the same as during scan in condition (a).

Vertical sync pulse   The vertical sync pulse integrator is not disturbed when the vertical sync pulses have a width of only 10 μs with a separation of 22 μs. Such vertical sync pulses are generated by video tapes with anti-copy guard.

Vertical divider system . The TDA 4505 embodies a synchronised divider system for generating the vertical sawtooth at pin 2. The divider system has an internal frequency doubling circuit, which allows the horizontal oscillator to operate at its normal line frequency. One line period equals two clock pulses.

Use of the divider system avoids the requirement for vertical frequency adjustment. The divider has a discriminator window for automatic switching from the 60- to the 50 Hz mode. When the trigger pulse comes before line 576 the 60-Hz mode is selected, otherwise the 50 Hz mode is selected. The divider system operates with two different divider reset windows for maximum interference/disturbance protection. The windows are activated via an up/down counter. The counter increases its counter value by 1 each time the separated vertical sync pulse is within the search window. When not within the search window this value is decreased by 1. The operating modes of the divider system are as follows:

Mode A   Large (search) window (divider ratio between 488 and 722)
This mode is valid for the following conditions:

— The divider is looking for a new transmitter
— Divider ratio found—not within the narrow window limits
— Non-standard TV-signal condition detected while a double or enlarged vertical sync pulse is found after the internally generated anti-topflutter pulse has ended. This means a vertical sync pulse width > 8 clock pulses (50 Hz); > 10 clock pulses (60 Hz).
— Usually this mode is activated for video tape recorders operating in the feature trick mode.
— Up/down counter value of the divider system operating in the narrow window mode drops below count 10.

Mode B   Narrow window (divider ratio between 522 to 528 for 60 Hz, and 622 to 628 for 50 Hz)
The divider system switches over to this mode when the up/down counter has reached its maximum value of 15 approved vertical sync pulses. When the divider operates in this mode and a vertical sync pulse is missing within the window, the divider is reset at the end of window and the counter value is decreased by 1. At a counter value below 10 the divider system switches over the large window mode. The divider system also generates an anti-topflutter pulse which inhibits the phase 1 detector during the vertical sync pulse. The pulse width is dependent on the divider mode. In Mode A the start is generated by reset of the divider. In Mode B the anti-topflutter pulse starts at the beginning of the first equalising pulse.

The anti-topflutter ends at count 10 for the 50 Hz mode and count 12 for the 60 Hz mode. The vertical blanking pulse is also generated via the divider system. The start is by reset of the divider while the blanking pulse width is 34 (17 lines) for the 60 Hz mode and at count 42 (21 lines) for the 50 Hz mode. The vertical blanking pulse at the sandcastle output (pin 27) is generated by adding anti-topflutter pulse to the blanking pulse, Thus the vertical blanking pulse starts at the beginning of the first equilizing pulse when the divider operates in Mode B. The length of the vertical blanking in this condition is 21 lines in the 60 Hz mode and 25 lines in the 50 Hz mode.

Application when external video signals require synchronization The input to the sync separator is externally available via pin 25. For normal application the video output signal at pin 17 is ac-coupled to this input as shown in Fig. 19.10. It is possible to interrupt this connection and drive the sync separator from other sources such as:

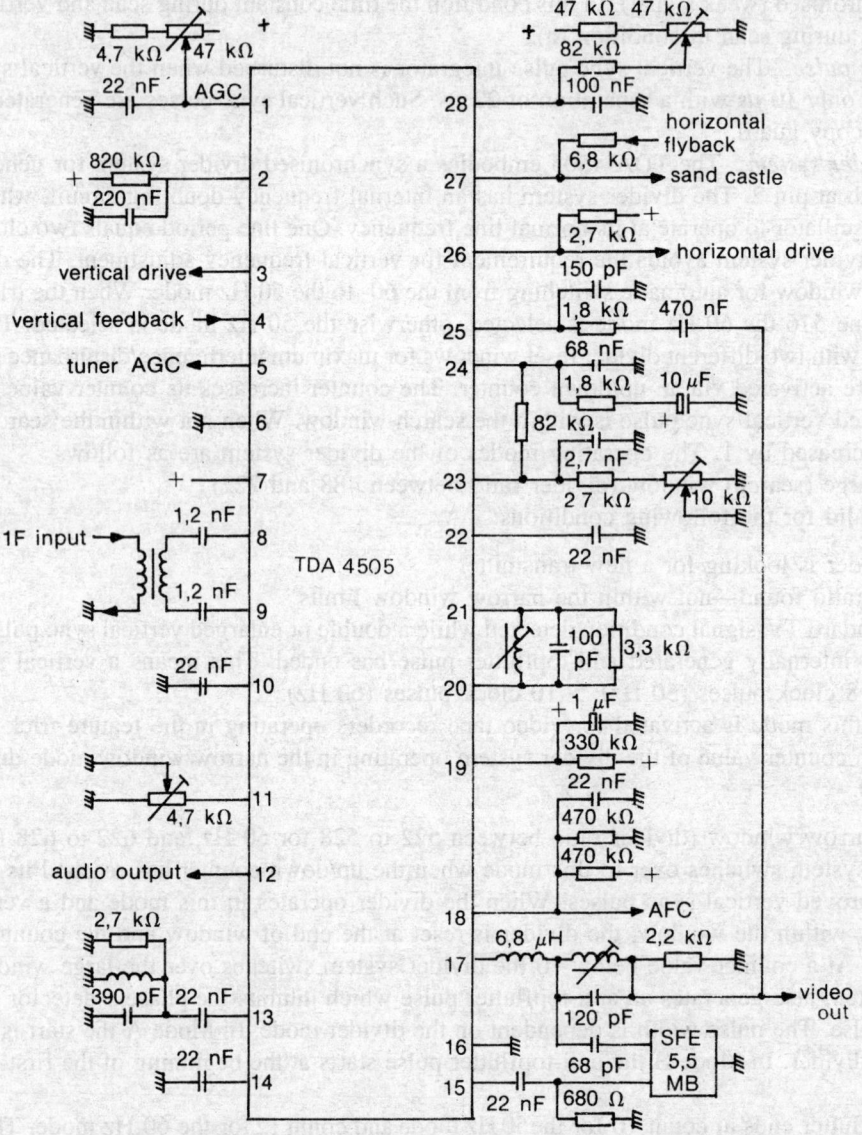

Fig. 19.20 Application circuit of TDA 4505 (*Courtesy:* ELCOMA PHILIPS)

— A teletext decoder in serial mode.

— An external audio signal via a peri-television connector.

When a teletext decoder is applied the IF amplifier and synchronization circuit are operating in the same phase which allows various connections between the two parts (i.e. AGC gating). When external signals are applied to the sync separator the connections between the two parts must be interrupted. This can be achieved by connecting pin 22 to ground, which results in the following conditions:

— AGC detector is not gated

— AFC circuit is active

— Mute circuit is not active—sound channel remains switched on

— Phase detector 1 has an optimal time constant for external video sources.

REVIEW QUESTIONS

19.1 Explain why transformerless designs were common in TV receiver circuits. How were the low dc voltages required for transistor stages obtained in such circuits for hybrid sets?

19.2 How is a solid state receiver designed to operate from a transformerless supply from ac mains? State the advantages and disadvantages of this type of supply.

19.3 State the advantages of employing SMPS in TV sets. What are the different types of basic circuits? Discuss the operation of a typical circuit used in TV sets.

19.4 What are the merits of using SMPS IC? Describe a circuit to explain block diagram of a typical IC.

19.5 What are the methods of remote control of a TV receiver? Describe the advantages of infrared remote control and discuss the technique, giving a block schematic diagram of the arrangement.

19.6 How are the remote commands encoded in the transmitter IC? Discuss the various operational modes.

19.7 State the advantages of using integrated circuits in a TV receiver. Illustrate by discussing some typical ICs.

19.8 Explain the principle of synchronous detection as employed for video detection in modern video IF ICs. What are the advantages, compared to the conventional envelope detector?

MULTIPLE-CHOICE QUESTIONS

19.1 Flyback SMPS is used in TV receivers because:
   (a) it gives better regulation
   (b) it does not require a filter choke
   (c) it is more suitable for multiple voltage outputs
   (d) (b) and (c).

19.2 Pick the wrong statement:
   (a) SMPS reduces loss in regulator circuits.
   (b) It can regulate over a very wide input voltage range.
   (c) SMPS automatically provides isolation from the mains circuit.
   (d) SMPS provides ripple-free supply and interference-free operation.

19.3 Synchronous detection:
   (a) is useless in black and white reception
   (b) reduces beats generated due to nonlinear distortion
   (c) requires two quadrature modulated carriers
   (d) (b) and (c).

20

Colour TV IC Receiver Circuits

Modern colour TV receiver designs rely heavily on integrated circuits developed for the various functional subsystems. The video IF subsystems and some of the synchronisation ICs have been already discussed earlier. The chroma decoder ICs, the RGB video output amplifier, the deflection sync processor IC horizontal output stage and EHT details are discussed in this chapter.

20.1 CHROMA DECODER IC SUBSYSTEMS

The decoding and matrixing of the Y, U and V signals in order to obtain the R, G and B, is now done in a modern receiver by a single IC chip, generally referred to as chroma processor or chroma decoder. The chroma decoder receives a detected composite colour video signal from the video IF demodulator IC subsystem. The luma and chroma signals are separated by *LC* or *SAW type comb filters* having appropriate band pass and frequency rejection. The luma and chroma signals are fed to separate processing chains inside the ICs. The IC decoder processes the video for the following:

Luma chain: Preamplification
Delay compensation
Luma amplification
Black level clamping and brightness control
Contrast control

Chroma chain: Chroma amplification with
Automatic colour control (ACC)
R–Y and B–Y Chroma components separation by
Delay line matrix
CSC reference oscillator and 90 deg phase shifting
Reference oscillator phase locking by
Gated burst phase detector
Burst phase identification for the PAL switch
Colour killer to bias off the chroma during B/W reception
R–Y and B–Y demodulation

The basic concepts in PAL decoding of composite colour video signal were discussed in Chs 5 and 11. The design blocks like ACC, and PAL switching circuit in the different ICs may vary in some details. Typical of these are the 28-pin ICs TDA 3560/3561A, and the modified TDA 3562A, that has dual standard features in that it works with NTSC signals also. TDA 3564 is a decoder for NTSC standards while TDA 3565 is a PAL decoder compact 18-pin package, with dc operated picture controls and low impedance RGB outputs for use with a 14-inch picture tube. Sets with teletext require a colour decoder with RGB inputs. TDA 3560/61A have this facility, and are pin-aligned with PAL decoder TDA 3567. Pin alignment allows the same PCB to accept either teletext compatibility or non-teletext decoder. By dual piercing of the PCB and appropriate interconnections between the two sets of holes, either the 18-pin TDA 3565 or the 28-pin TDA 3560/3561A can be fitted. This facilitates production of the set for teletext without extra cost. TDA 3566 is also a dual standard PAL/NTSC decoder, a further development of TDA 3562A with additional features and improvements. TDA 3590A is a SECAM processor while TDA 3592A transcoder helps convert SECAM signals into PAL signals.

20.2 COLOUR DECODER IC TDA 3561A

The block schematic of the 28 lead DIL plastic packaged TDA 3561A is given in Fig. 20.1. The IC combines all the functions required for identification and demodulation of PAL signals. The IC further contains a luminance amplifier, an RGB matrix and amplifier. These amplifiers give out video signals up to 5 V_{pp}, enabling direct drive to the transistor video output stages. The IC has separate inputs for external data insertion, analog as well as digital, which can be used for text display systems like teletext, channel number display, etc., signal clamping and blanking for the RGB outputs. In addition to these functions provided in TDA 3560, the 3561A includes the following features:

— The peak white limiter is only active during the time the 9.3V level at the output is exceeded. The start of the limiting action is delayed by one line period, thus avoiding peak white limiting by test patterns which have abrupt transitions from colour to white signals.
— Brightness control is obtained by inserting a variable pulse in the luminance channel. Then the ratio of the bright variation, and the signal amplitude at the RGB outputs, will be identical and independent of the difference in the gain of the three channels. Discolouring due to adjustment of contrast and brightness is thus avoided.
— Suppression of internal RGB signals when switching to external input and vice versa is improved.
— Suppression of the residual 4.43 MHz signal in the RGB output is also improved.
— Non-synchronous external RGB signals do not disturb the black level of the internal signals.
— Cascaded stages in the demodulators and burst phase detector minimize the radiation of the colour demodulator inputs.
— RGB outputs have higher current capacity.

Luminance Chain

The delayed Y signal is ac coupled through a 22-nF capacitor to the input pin 10 to permit clamping within the IC at the line rate to an internal reference level within the IC. A 1kΩ delay line can be applied because the luminance impedance is made very high. For *brightness control*, the black level of the RGB outputs can be set by the voltage at the pin 11, by the resistance-potentiometer arrangement to around 3 V. The input signal should be 0.45 V_{pp}, which is amplified in the controlled amplifier to obtain black-white output of 5 V at nominal contrast. The *contrast control* range is 20 dB, set by the control voltage change from + 2 to + 4 V at pin 6. If one or more output signals surpass the level of 9 V, the *peak white limiter* circuit becomes active and reduces the output signals via the contrast control by discharging the capacitor C_2 across pin 7 via internal current sink. The luminance signal Y is thence fed to the R, G and B matrices.

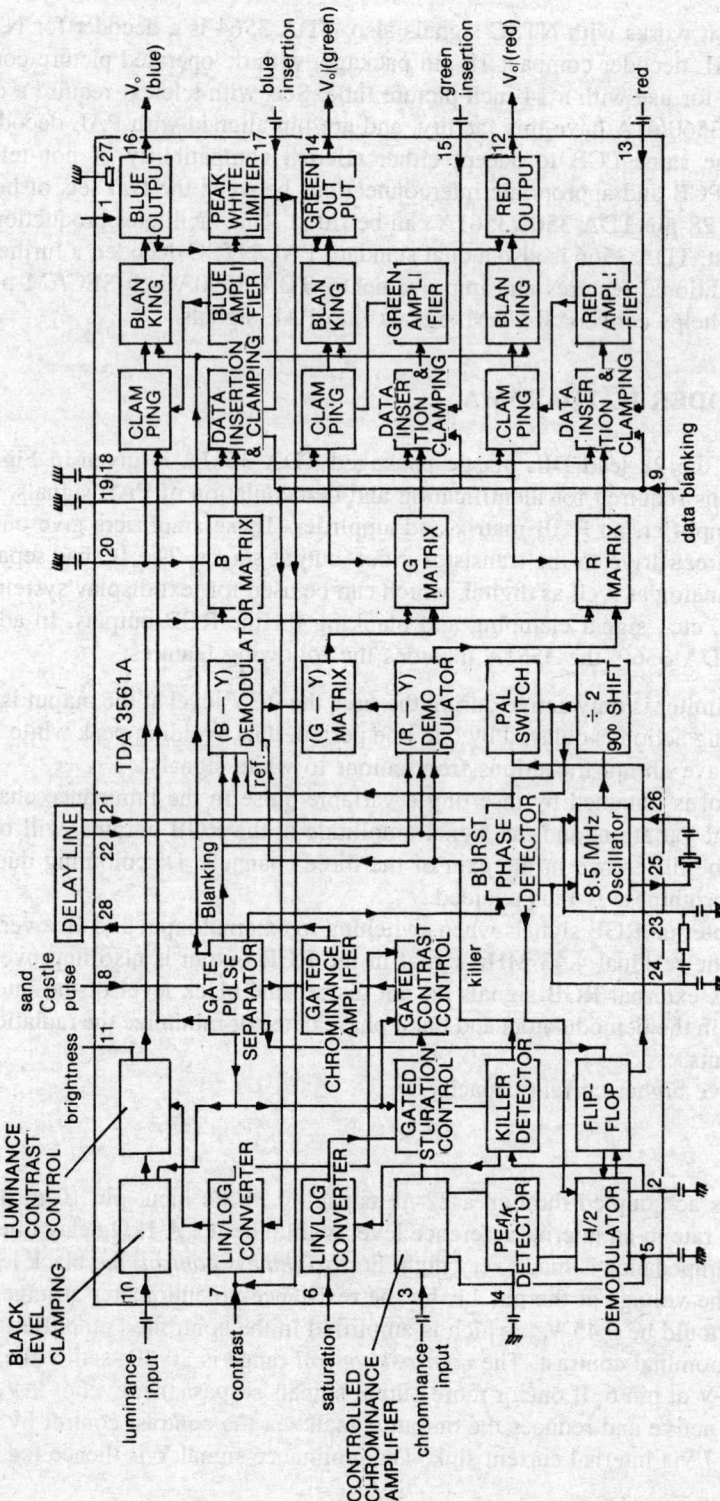

Fig. 20.1 Block diagram of TDA 3561A *(Courtesy: ELCOMA PHILIPS)*

Chrominance Chain

The chrominance signals sourced through a high pass filter comprise a base band spectrum of 3.3–5.4 MHz. They are given to the controlled chroma amplifier for ACC function to maintain the chroma level constant, by means of potential derived from the peak detector across a 330-nF capacitor at pin 4. The amplifier feeds chroma signal to the gated saturation control stage which varies gain in excess of 50 dB, under control from a dc voltage from + 2 to + 4 V supplied to pin 6, for *colour saturation* adjustment. The gating mechanism provides for an extra gain of about 12 dB, only during the forward scan of 52 μs, while passing the burst during the blanking interval, at a constant level. This prevents the saturation control from affecting the burst amplitude.

When the *colour killer* is active, the saturation voltage is reduced to a low level if the resistance of the external control network is sufficiently high. When this pin is connected to the + 12 V power supply, through the service switch, the colour killer circuit is overruled so that the colour signal is visible on the screen. In this way it is possible to set the reference oscillator frequency without using a frequency counter. The reference voltage ACC detector circuit needs external decoupling by a capacitor of 330 nF at pin 4, where the voltage is 4.9 V.

PAL Dematrix

The chroma signal along with the burst signal is available at the output pin 28. The burst-to-chroma ratio here is identical to that at the input for normal control settings. The chroma signal's fed to the external delay line as shown in the application circuit of Fig. 20.3. The modern delay line consists of a special glass block fitted with an input piezoelectric transducer that converts electrical signal into mechanical ultrasonic vibrations that travel along a defined path within the glass block, as shown in Fig. 20.2.

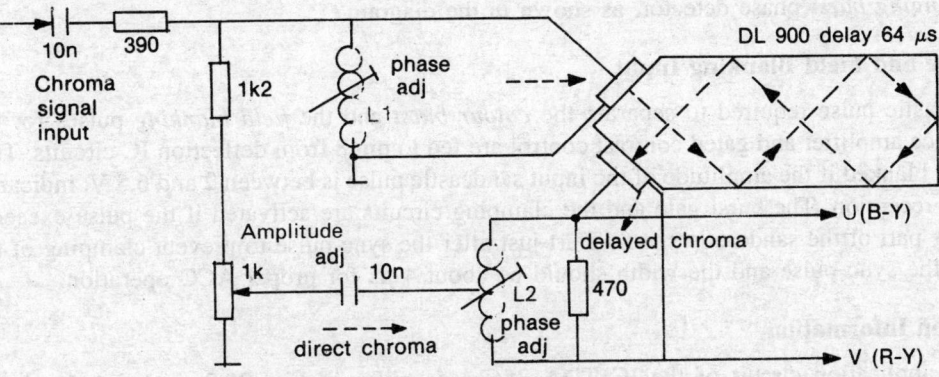

Fig. 20.2   Ultrasonic delay line and matrix arrangement

The edge surfaces of the block are arranged to fold the path back through a series of reflections to reach another transducer which converts the mechanical vibrations into output electrical signals delayed by one TV line period. The direct signal as fed to the input and the delayed signal are added and subtracted by means of the associated centre-tapped matrix arrangement to separate the U (B–Y) and V (R–Y) modulated subcarriers at the output ends.

Chroma Demodulators

The B–Y and R–Y subcarrier modulated signals separated by the delay line matrix, are brought back to the IC pins 21, 22 and are passed on to the B–Y and R–Y synchronous demodulations. These give out B–Y and R–Y video outputs only when the signal to be demodulated is fed simultaneously with the reference signals in correct frequency and phase as the suppressed subcarrier. A synchronous modulator can be regarded as a *sampling* circuit, where the circuit is switched on by the peaks of the reference oscillator for very short durations. Any misphasing between the chrominance signals and the reference oscillator leads to wrong colour output.

The colour burst signal required for phase correction of the reference oscillator is extracted from the delay line outputs, and is also fed to the *burst phase detector*, which extracts the burst available at pins 23 and 24. At these pins the burst is filtered and used to control the reference oscillator. An adequate catching range is obtained with the time constants chosen in the application circuit.

Reference Oscillator Chain

The reference oscillator in this design uses a crystal at exactly *twice* the colour subcarrier frequency, viz. $2 \times 4.433619 = 8.8672238$ MHz, connected across pins 25 and 26 in series with a trimming capacitor. Twice the frequency is used to achieve a 90° phase shift in the square wave in the digital way. The R–Y and B–Y chroma demodulators need two quadratured phased switching waveforms. Ordinarily this would require some kind of 90° phase shifting network. For a digital waveform this can be achieved simply by means of two flip-flops used as *binary dividers*, one triggered on the positive edge, while the other on the negative edge of the square wave. The triggering frequency of 8.86 MHz given to one flip-flop is inverted to obtain another waveform of the same frequency but 180° out of phase. This triggers the second flip-flop to give output, at 4.43 MHz with 90° phase shift with respect to the first. This is given to the R–Y modulator through the *PAL switch* which reverses the phase of the carrier under control of the *swinging burst* phase detector, as shown in the diagram.

Sandcastle and Field Blanking Input

The sandcastle pulse required to separate the *colour burst* and the *field blanking* pulses for the gated chrominance amplifier and gated contrast control are fed to pin 8 from deflection IC circuits. The output signals are blanked if the amplitude of the input sandcastle pulse is between 2 and 6.5 V, indicating black and white reception. The burst gate and the clamping circuits are activated if the pulse exceeds 7.5 V. The higher part of the sandcastle should start just after the sync pulse to prevent clamping of the video signal on the sync pulse and the width should be about 4 μs for proper ACC operation.

Application Information

A typical application circuit of the IC TDA 3561A is given in Fig. 20.3, and the circuit functions associated with the remaining *pins* are explained below.

1. *Power supply:*   The IC gives good operation in a supply voltage range of 8 to 13.2 V, over which the signals and current levels have a linear dependency. For typical operation, a + 12 V supply is used and fed to pin 1 with respect to the ground pin 27. The IC draws a current of typically 85 mA.

2. *Control voltage identification:*   Pin 2 requires a peak detection capacitor of 330 nF for the H/2 identification demodulator.

12, 14, 16. *RGB output:*   Circuits are identical giving output signals of 5.25 V at the pins 12, 14, 16 for nominal input of 0.45 V and contrast setting. The black level of the three outputs must have the same value. The blanking level at the outputs is 2.1 V. The peak value is limited to 9.3 V.

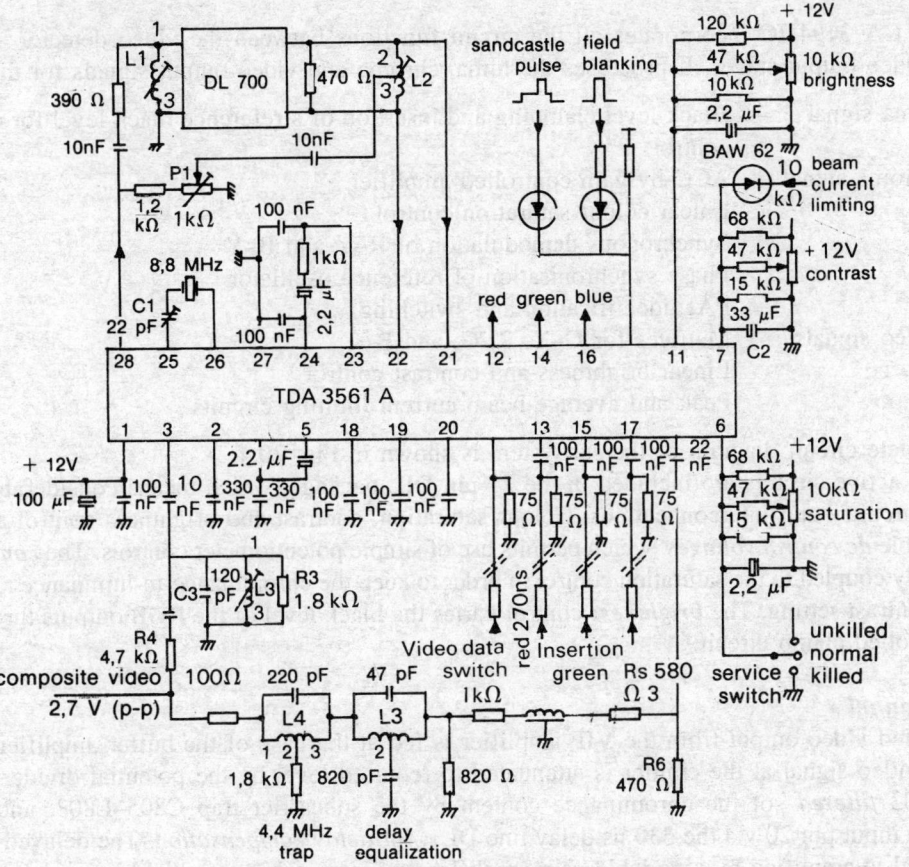

Fig. 20.3 Application circuit of TDA 3561A

$L_1 = 10.7\ \mu H$ phase delay line

$L_2 = 10.7\ \mu H$,

$L_3 = 10.7\ \mu H$, chroma input filter

$L_4 = 5.6\ \mu H$,

$L_5 = 66.1\ \mu H$

P_1 = amplitude of direct chroma signal

18, 19, 20. *Black level clamp capacitors:* The black level clamp capacitors for the three channels are connected to pins 18, 19 and 20. The value of each capacitor should be about 100 nF.

13, 15, 17. *External RGB inputs:* External signals must be ac coupled to the inputs at pins 13, 15, 17 via a coupling capacitor of about 100 nF. For 5 V output, the input signal required is 1 V, at a source impedance not exceeding 150 Ω.

9. *Video data switching* The external RGB signals are supplied to the output amplifiers by activating the insertion circuit by an input pulse between 1 V and 2 V, at pin 9. In this condition the internal signals are supplied to the output amplifiers. For normal operation, pin 9 should be connected to the negative ground pin 27.

20.3   LUMA/CHROMA PROCESSING IC CA 3194

The BEL CA 3194 IC incorporates all the circuit functions between the video detector and the high voltage video output stages. It processes the luma, chroma and video output signals for the following:

(a) Luma signal   —   Black level clamping and insertion of a reference black level for cut-off point control
(b) Chroma signal —   ACC by gain controlled amplifier.
Linear colour saturation control
Synchronous demodulation of R–Y and B–Y
Phase synchronisation of reference oscillator
PAL identification and switching
(c) Video signals —   Matrices for G–Y, R, G, and B
Linear brightness and contrast control
Peak and average beam current limiting circuits

The complete circuit diagram of the subsystem is shown in Fig. 20.4.

All the active circuits are included in the 24-pin DIL package, which means considerable saving in external circuitry and interconnections. Colour saturation, contrast and brightness control are achieved with variable *dc control voltages* which permits use of simple potentiometer controls. The *contrast* control is internally coupled to the saturation control in order to keep the chrominance-to-luminance ratio constant for any contrast setting. The *brightness* control varies the black level of the RGB outputs through a burst gate controlled clamp circuit.

Luma Channel

The detected video output from the VIF amplifier is fed at the base of the buffer amplifier Q801. The buffered video signal at the emitter is attenuated to required level by the potential divider comprising R802–R803 *filtered* of the chrominance content by the subcarrier trap C805–L803, and fed to the luminance input pin 20 via the 330 ns delay line DL1, for *delay compensation*. The delayed and filtered luma signal is amplified by a variable gain amplifier whose gain is controlled by the dc voltage at the contrast pin 22. In order to keep the RGB output black levels at a constant value, the dc voltage is automatically adjusted by the et al. closed loop formed by the luma amplifier, brightness comparator, G–Y matrix and G output blocks.

The amplitude controlled luma signal is then added to the R–Y, G–Y and B–Y signals in the *matrix* blocks to give the R, G, B signals through emitter followers to pins 18, 17 and 16. The available output level is 4 V black to white, with a nominal black level of 6.3 V. These RGB signals are internally blanked and clamped to black level by the horizontal and vertical blanking signals obtained from the *sandcastle decoder*, which gets the sandcastle pulse input from the deflection system at pin 13. The decoder separates the horizontal blanking, vertical blanking and the also the burst gate pulses which are required by the burst phase circuits.

Chroma Channel

The video output available at the emitter of Q801 is fed to the chrominance input at pin 4, through the *chroma band pass* filter comprising C803, L802, C807. The band pass filter tuned to the subcarrier of 4.43 MHz by varying L892 and is designed for a bandwidth of 1.6 MHz. The chroma signal including the burst is input to pin 4 and kept at a constant level by the first variable gain chroma amplifier stage and the *ACC* closed loop circuit. The burst is then separated from the chroma signal by the burst gate obtained from the sandcastle decoder. The separated burst is then processed in a quadrature phase detector followed by the sample and hold circuits of the automatic colour amplitude and phase closed loop control circuits.

The chroma signal passes through two variable gain stages of which one is used for colour saturation setting and the other for automatic tracking to contrast. The two stages are also used for the *colour killing* function. The processed chroma is then available at pin 2 through an emitter follower buffer stage at low output impedance across the load resistors R835–R836.

Dematrix
The chroma signal at pin 2 is led through the matching network R837-L897-C824, to the 1–H PAL delay line DL802. The coil L805 receives both delayed and the direct undelayed signals to carry out the add and subtract functions on the chroma signals of two successive lines. As a result of the addition/subtraction, modulated U or (B–Y) signal is available at the top of the resistor R825 connected to pin 14, and modulated +/– V or +/– (R–Y) signal at the bottom connected to pin 15. The separated U and +/– V signals are demodulated in the two synchronous quadrature phase detectors, which are supplied internally the reference carriers in relevant quadrature phases.

Reference Oscillator
A 4.43 MHz crystal in series with frequency trimming network is connected between the VCO output pins 11, and 0° input pin 10 of the IC. The crystal controlled VCO is phase locked with the burst and provides a 0° reference subcarrier at pin 10. By connecting inductance L896 between the pins 9 and 10, a 90° phase-shifted subcarrier is obtained at pin 9. As the 0 and 90° subcarrier signals are generated externally, adjustments of neutral colours (white balance) for achromatic signals can be made precisely. The subcarrier is applied to the +/– V detector through a line-by-line PAL flip-flop driven inverter, and to the U detector through the buffered quadrature phase equaliser. The U and V signals are deweighted to obtain the B–Y and R–Y, which are matrixed to form the G–Y.

20.4 HORIZONTAL/VERTICAL DIGITAL SYNC-SUBSYSTEM BEL CA 3223

The BEL CA 3223 is a combination IC that incorporates digital count down technique to produce synchronised deflection drive signals. This *gen-lock system* similar to that used in cameras and studio equipment provides excellent interlacing and noise-free synchronisation without the need for H and V hold controls. The important functions performed by the system are:

(a) Generation of both the horizontal and vertical drive signals phase locked to the sync pulses.
(b) Phase comparison of the line oscillator output and the line flyback pulses in order to compensate for unwanted phase shifts introduced in the line output stage.
(c) Generation of sandcastle pulse required for processing the luma and chroma signals.
(d) Preamplification of the vertical scan signal.
(e) Facility for acceptance of non-standard sync signals generated by equipment such as VCR.

The complete circuit diagram of the deflection subsystem is given in Fig. 20.5.

Sync Separator
The composite colour video signal obtained from the VIF subsystem CA 7611 is processed by the transistor sync separator stage consisting of input RC noise suppression network, transistor Q402, integrator

R415–C407 and differentiator R418–C408. The separated vertical and horizontal sync signals are applied to pin 24 and pin 1 respectively for further processing.

Horizontal Sync Processor

The horizontal synchronising system essentially consists of a reference voltage controlled oscillator (VCO) operating at a frequency of $16 \times f_H$, and *two phase locked loops* to accommodate the conflicting requirements of the deflection output and synchronising circuits. Narrow bandwidth is desirable for good noise performance of synchronising circuits, while wide bandwidth is required to minimize horizontal output switching delay variations. The two loop approach separates the horizontal output circuitry from the synchronising loop so that the two bandwidths can be independently optimized. The block schematic of the horizontal processor portion is shown in Fig. 20.6.

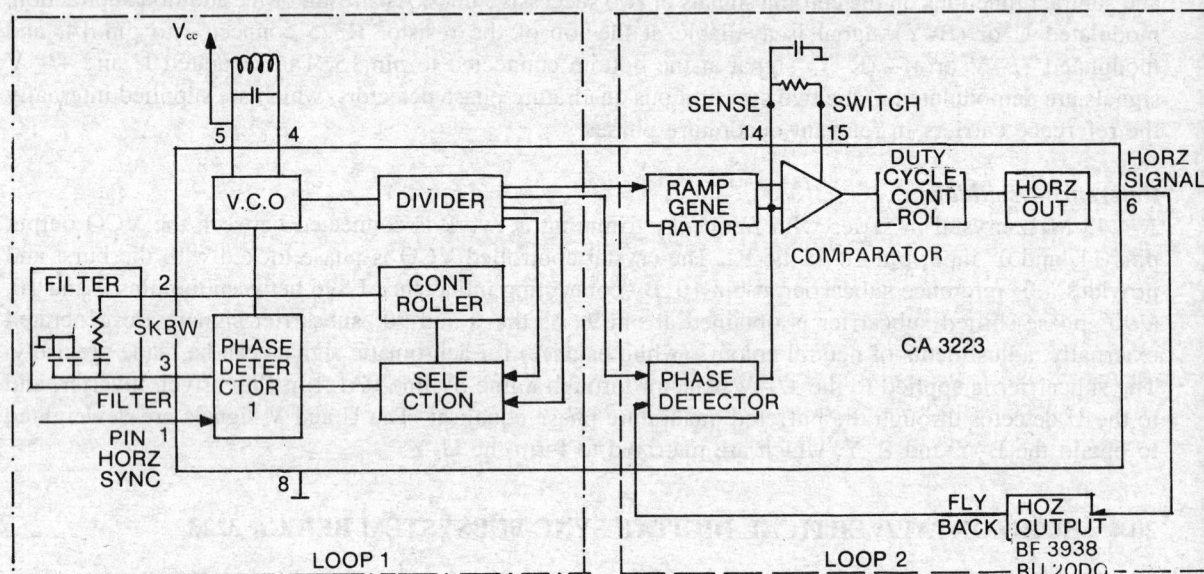

Fig. 20.6 Block schematic of the horizontal processor (*Courtesy:* BEL)

The first loop consists of the phase detector, compensating filter, VCO, synchronous frequency divider, frequency selector and controller. In the phase detector, the line oscillator output received through the selection circuit is compared with the sync pulses received at pin 1, to produce an error voltage. This is smoothened by the dual time constant RC filter network connected across pins 2 and 3, and is then applied to the VCO to correct its frequency.

The second loop consists of the line output stage, another phase detector, ramp generator and a comparator sense switch for duty cycle control, that produces an error signal for correcting picture position.

VCO

The frequency of the VCO operating at $16\,f_H$ is determined by the L402–C416 network connected across pins 4 and 5 of the IC. The output of the amplifier A_1 feeds a 90° phase shifter and then to a phase splitter that produces two components PH_1 and PH_2 that are 180° out of phase with each other. PH_1 and PH_2 are fed through gain controlled amplifiers A_2 and A_3 to a summing point and then back to the tank circuit.

Depending on which phase of the summed signal is fed back to the tank, the tank impedance is paralleled by inductive or capacitive impedance. The VCO resonant frequency therefore increases or decreases according to the dc voltage correction applied.

The VCO frequency of $16 \times f_H$ is counted down by a four-stage counter to produce f_H, $2f_H$, $4f_H$, $8f_H$ that are fed to the phase control loops as shown. Loop 1 synchronizes VCO with the input sync pulses at pin 1. The synchronised output is then corrected for horizontal output switching delay variations with the help of the loop 2 and the phase shifter comparator. It is then made available at output pin 16 through buffering for driving the horizontal output.

Vertical Sync Processor

The vertical sync processor consists of the vertical count down, ramp generator and the blanking circuits. The vertical count down circuit uses the integrated V sync input at pin 24, and the clock signals from loop 1 divider to generate a synchronised output for driving the vertical ramp generator. Use of the vertical count down (Gen-lock) system facilitates synchronisation to both standard and non-standard signals, eliminates the need for vertical hold control, improves interlace performance and enhances noise immunity. The vertical ramp generator output is available at pin 21 for further processing in the vertical output stage. The count down system also generates control signals for the blanking generator, whose output pulses are available at pin 23. The subsystem contains an internal shunt regulator maintaining internal supply rail at a constant 10 V. There is also a provision for pulse width modulated driver, which provides control signals for regulating the receiver dc power supply.

Sandcastle Generator

Transistors Q401 and Q403 form the sandcastle generator circuit deriving its input from the horizontal flyback and vertical blanking signals. The sandcastle pulse available at the collector of Q403 is applied to luma/chroma processor CA 3194, for black clamping and burst gating.

20.5 HORIZONTAL OUTPUT STAGE AND EHT

The horizontal output stage in the colour TV receiver performs the following functions:

(a) It provides the deflection current for the horizontal deflection coil of the yoke.

(b) It provides flyback pulses for blanking, AFC, AGC and other gating functions.

(c) It provides auxiliary voltage of about 600 V for grid 2 of the CPT and for dc supply for the vertical output stage.

(d) It generates the EHT voltage of 25 kV for the accelerating anode and 7 kV for the focusing anode of the CPT.

(e) It also supplies other auxiliary voltages required by other circuits in the receiver and filament power of the CPT.

The circuit diagram of a typical horizontal output stage for a colour IC TV receiver is given in Fig. 20.7.

Drive Stage

The synchronised output from the deflection IC CA 3223 is applied to the base of the horizontal driver transistor BF 393. The transistor switches over from cutoff to saturation whenever the pulse applied to its base goes positive, and in turn provides the necessary drive to the output transistor. The primary of the driver transistor is damped by D501, R502, C501, R503 and C502, to provide the necessary pulse

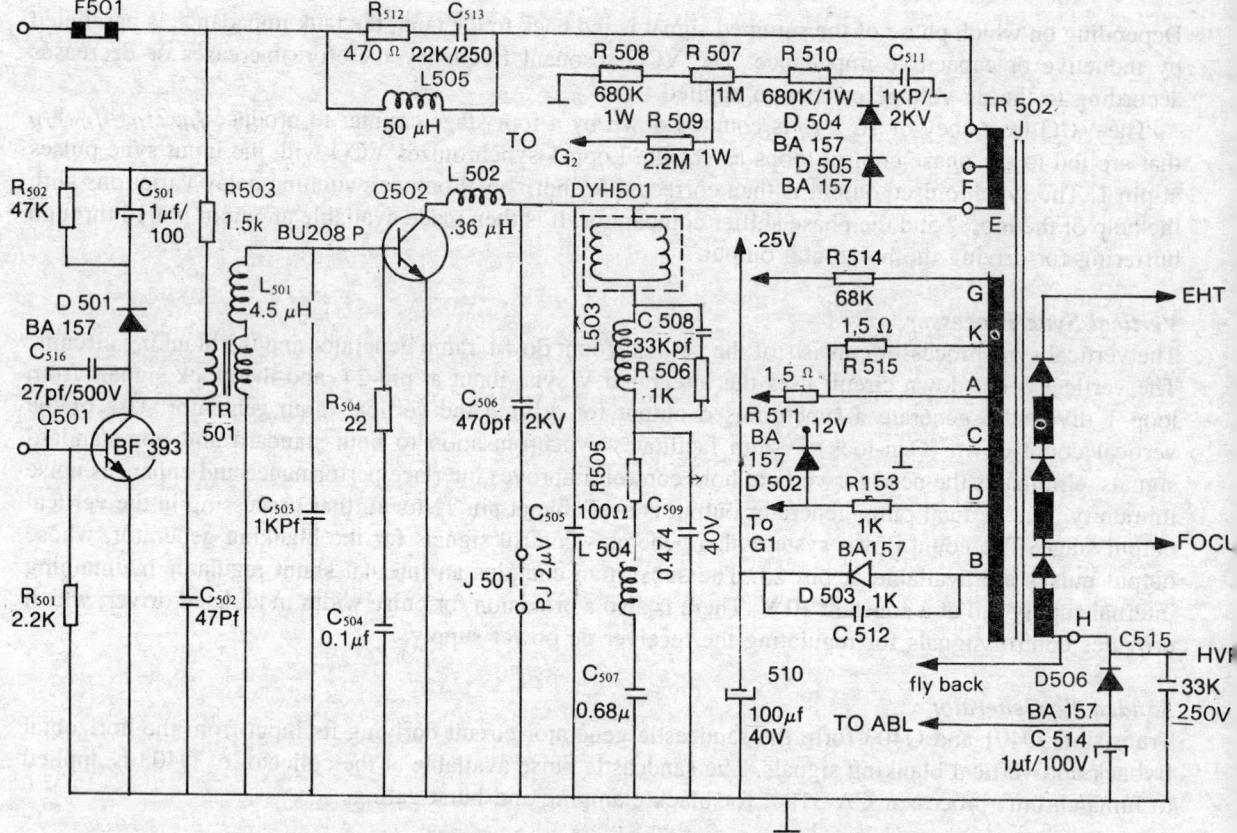

Fig. 20.7   Circuit diagram of horizontal output stage in a colour TV receiver (*Courtesy:* BEL)

shaping and avoid spikes associated with inductive load during switching. Small capacitors C516 and C502 provide the roll-offs at high frequencies, preventing parasitic oscillations. The drive pulse from the secondary of the transformer is then applied to the base of the horizontal output transistor BU 208D, through L501, C503, R504 and C505 which provide further damping to this pulse. L501 provides the fall in the base current of the output transistor necessary to reduce the transient high dissipation during flyback. The power supply is fed to the transistor through L505.

Output Stage

The output transistor drives the deflection coil and the flyback transformer generating the EHT and other voltages. The stage is supplied 105 V dc from the SMPS, through the fuse F_1, parallel resonance circuit L605–C513–R512, and inductance L_2. The fuse is mounted on top of the transistor sensing its overheating. The resonant circuit blocks the signal frequencies preventing coupling into the driver stage via the supply line. The transistor is in the cut-off region until the input pulse switches it on and drives into saturation. During this ON period, current is drawn from the deflection coil DYH501 through the transistor. This current causes a linear rise in the magnetic field causing deflection of the electron beam from the centre to the extreme right.

The energy for the yoke is initially taken from the power supply during the forward scan when the transistor is turned on and the primary portion J-E of the EHT transformer stores the energy. During flyback this energy is transferred to the deflection coil via 0.47 μF series capacitor C509. When the drive pulse switches off the transistor by negative base current, retrace is initiated. The deflection current *swings sinusoidally* to zero, charging in this process the 8.2 nF capacitor C505 to a peak voltage. After this instant the capacitor discharges into the coil with the current now swinging in the opposite direction. These sinusoidal charge and discharge periods of the capacitor C505, resonating with the deflection coil for half cycle, constitute the flyback period of the raster, *typically* 8 μs. At the end of the flyback the beam is brought back to the left extreme of the raster. Forward scan starts again when the damper diode internal to the bi-directional switching transistor BU 208D is forward biased by the swinging voltage across the capacitor C505.

L504, R505, C507 and C509 are used for S-correction due to nonlinearities in deflection currents due to fringe magnetic fields, while L503 is the nonlinearity coil providing correction for coil resistance drop.

Generation of Auxiliary Supplies
The flyback pulse at the pin *E* of flyback transformer TR2 is rectified by diodes D504 and D505, to supply the screen G2 of the CPT through potential divider arrangement of R507 to R510. The voltage across B and C of FBT TR502 is scan rectified and, to derive *25 V dc for supplying to* the vertical output stage, the voltage at pin D of TR2 is applied to grid G1 of the CPT through R513. Voltage across pins C and K of TR502 provides *heater power* to the CPT through R514–515 to set the required value. *EHT* of 25 kV obtained across the secondary of the diode split transformer windings, is connected directly to the anode of the CPT. *Focus* anode voltage of 7 kV is also derived from the first tap of the diode split winding and applied through the focus potentiometer mounted on the CPT neck board. The ground point of the secondary winding is used to sense the beam current for average beam current limiting (*ABL*) functions and for vertical tracking purposes. This is done by connecting the pin H of the EHT to ground through C515. The voltage across this capacitor, which varies with the beam current, is fed back to CPT drive circuits to achieve average brightness limiting/vertical tracking functions.

20.6 VIDEO OUTPUT AMPLIFIER AND COLOUR PICTURE BIASING

The video output stage in a colour TV receiver has to perform the following functions:

(a) Separate amplification of the video R, G and B signals available from the chroma decoder to drive the respective cathodes of the CPT.
(b) Blanking of the retrace lines.
(c) Provision of halftone and monochrome adjustments for the RGB drives for correct while balance near cutoff and peak whites.

The circuit diagram of a typical video output amplifier and picture tube biasing is given in Fig. 20.8.
The video output stage consists of three identical video amplifiers driving the RGB cathodes of the CPT. The inputs to these amplifiers are obtained from the decoded R, G and B outputs at the terminals 18, 17 and 16 of the chroma decoder IC CA 3194. The amplifiers are designed with the high voltage high frequency transistors of type BF 393, employing series peaking for high frequency compensation. The three R, G and B amplifiers are designed to be almost identical in frequency response by identical electrical circuits and component layout.

The V_{cc} of 150 V required for the three stages in common emitter configuration is obtained from the SMPS fed through a common decoupling filter L702–C708. The common *emitter bias* for all the three transistors is provided from the emitter follower Q802 (shown in Fig. 20.4) and is bypassed by 4.7 μF

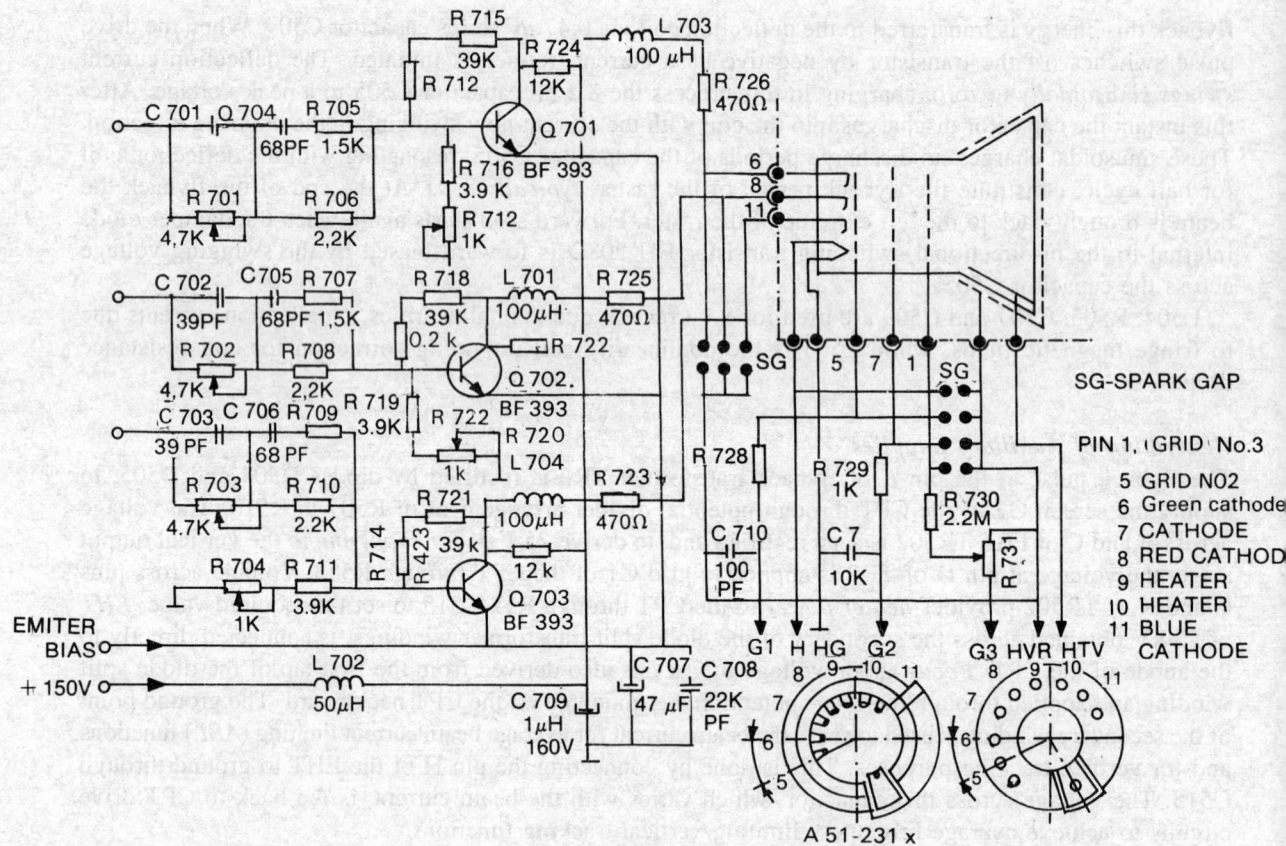

Fig. 20.8 Circuit diagram of video output amplifier and picture tube biasing (*Courtesy:* BEL)

and 22 nF capacitors C707–C708, at low as well as high frequencies. The resistors R712–R715 connected between the collector and base provide *dc negative feedback* to improve the dc bias stability, and make the amplifier performance independent of device parameters. The dc collector voltage which determines the cutoff of the CPT, is set for each beam separately by adjusting the 1k resistors in the base-to-ground lead. This enables easy adjustment of the cutoff voltage of the three guns to obtain standard white at low beam currents, constituting the *half tone/black level adjustment.*

The output at the collector load resistor of 12 k is coupled to the CPT cathodes via the 100-μH *series peaking* coils to improve HF response. The 470 Ω resistor in series connection to the cathodes help in the fast breakdown of the spark gap in the initial stages of the *flash over,* limits the current into the transistor and isolates the video stage from the short circuit produced by the spark gap. The 39 pF capacitors along with other RC coupling network in the input lead improve the step response of the amplifier. The voltage gain obtained is over 24 dB, and the rise time better than 0.18 μs. The 4.7 k resistor pots permit the gain/slope adjustment of individual R, G and B amplifiers providing *monochrome/white level adjustment* for the CPT.

CPT circuits Horizontal blanking is applied to the grid 1 (pin 5 of CPT) from the FBT through R728. The grid 2/screen grid voltage (500 V) is applied to pin 7 of the CPT, from the FBT set by R507 as shown in Fig. 20.7. The focus anode voltage (7 kV) and EHT voltage (25 kV) are also supplied from the FBT circuits. The focus voltage is set by R731 for optimum focus at the centre and edges of the CPT.

The high voltage return (HVR) return path is provided by the grounding braid connected to the external aquadag coating of the CPT. The degaussing coil is mounted on the CPT cone for degaussing stray magnetic fields.

20.7 P²CCD TECHNOLOGY FOR INTEGRATED COLOUR PROCESSING

Although most of the active functions in conventional colour transfer of processors are performed by a single IC, a considerable number of *peripheral components* are required, which include chrominance trap, luma delay line, chroma band pass, PAL delay line matrix, chroma crystal circuit, and so on. Several *manual adjustments* are still required, viz. chroma trap, chroma bandpass, PAL delay line matrix—phase and amplitude, reference oscillator and white point adjustments.

In computer-controlled TV, these adjustments have been integrated and eliminated/automated. Vg_2 adjustment can also be automated. Integration of delay lines and filters in colour processing circuitry is achieved by use of P²CCD (*profile peristaltic charge coupled device*) technology. Basically this is a kind of N-MOS technology in which dynamic, time-discrete analog shift registers can be made. An array of evenly spaced electrodes is used to move or store analog charge packets in the material under the electrodes. The electrodes are driven by a four-phase clock. Due to the doping concentration profile, as shown in Fig. 20.9, the charge packets are at a depth in the material, at which optimum use of the electric field is made for the transfer from one electrode to the next. P²CCD technology therefore has a large dynamic range, a high transfer efficiency with high speed and better frequency characteristics than conventional delay lines.

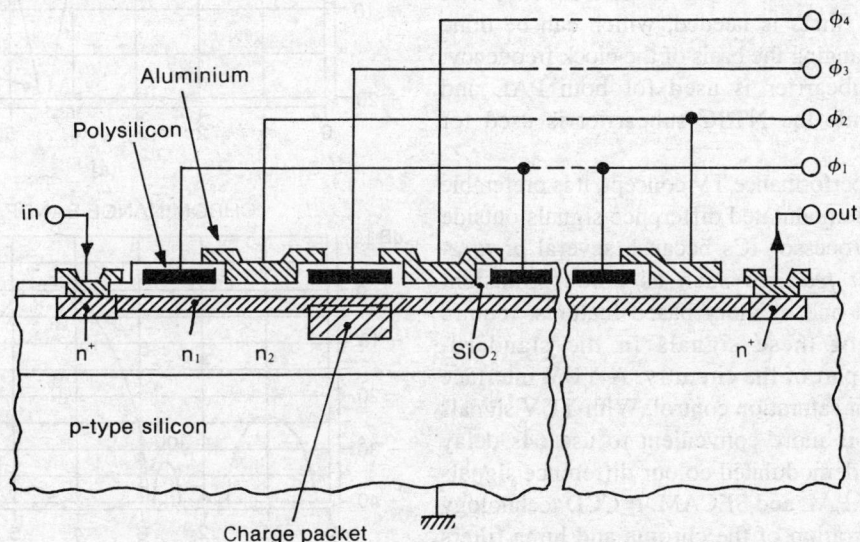

Fig. 20.9   Cross-section of P²CCD delay line (*Courtesy:* PHILIPS)

Since we can now integrate delay lines, we can also fabricate filters in the same process. The basic principle of operation of these filters is shown in Fig. 20.10.

The factors a_n are realized by the relative widths of the delay lines t_n. Summation takes place by merging the charge packets coming from the delay lines. Thus the processing between the filter input and the output takes place in the charge domain, i.e. there are no transitions from charge packets to voltages and vice versa. This solution avoids the need for complex and expensive summing stages, requiring many inputs and a wide bandwidth.

These P²CCD filters and delay lines cause an *extra delay*, due, for example, to the transitions at inputs and outputs from the voltage to the charge domain and vice versa. A compensation for these delays is incorporated in the ICs concerned.

All the characteristics of the P²CCD filters and delay lines are essentially determined by the geometries and the clock frequency. Thus, if a suitable clock is provided, no adjustments are needed. For the colour-processor filters and delay lines fabricated in P²CCD technology, the PAL/NTSC subcarrier frequency is used as the basis for the clock.

These filters are of the *non-recursive* type, and have linear phase characteristics. The characteristics of P²CCD PAL chroma trap filter in luminance channel and the chrominance bandpass filter are shown in Figs 20.11(a) and (b) respectively.

For NTSC, only a scaling of the frequency axis for both the filters is needed, which can be done simply by changing the basis of the clock frequency: The PAL subcarrier is used for both PAL and SECAM, while the NTSC subcarrier is used for NTSC.

In a high-performance TV concept, it is preferable to have the demodulated difference signals outside the colour processor ICs because several *picture-enhancement features* such as colour transient improvement and memory based features, require processing of these signals in the standard-independent part of the circuitry. A YUV interface is required for saturation control. With YUV signals available, it is more convenient to use 64s delay lines for the demodulated colour difference signals in the case of PAL and SECAM. P²CCD technology enables integration of the chroma and luma filters on the same chip, with no adjustments needed for the filters and delay lines.

EXAMPLE OF A FILTER STRUCTURE WITH DEALY—LINE SECTIONS

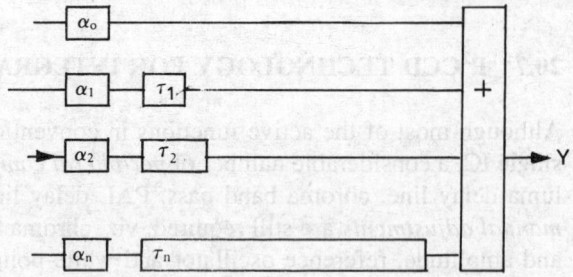

Fig. 20.10 Filter structure with delay line sections

LUMINANCE CHANNEL RESPONSE

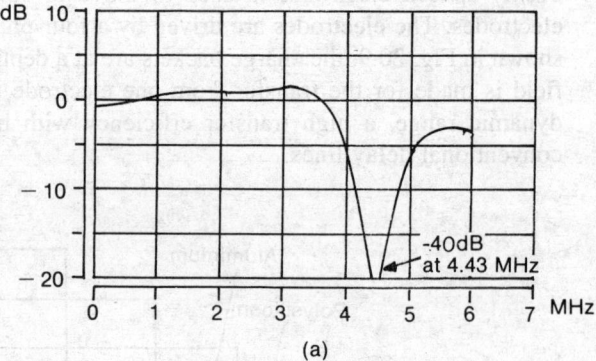

(a)

CHROMINANCE BANDPASS

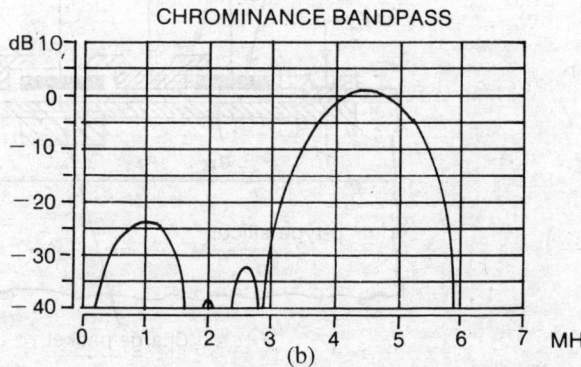

(b)

Fig. 20.11 (a) Luminance channel response of P²CCD trap filter (b) Chrominance bandpass response of P²CCD bandpass

Filter/Delay Line Combination IC TDA 8450

Apart from the functions mentioned above, this IC contains a CVBS switch, which can be I²C-bus controlled via a computer-controlled colour processor. A combination block diagram of the TDA 8450 filter/delay and a PAL/NTSC colour processor IC is given in Fig. 20.12.

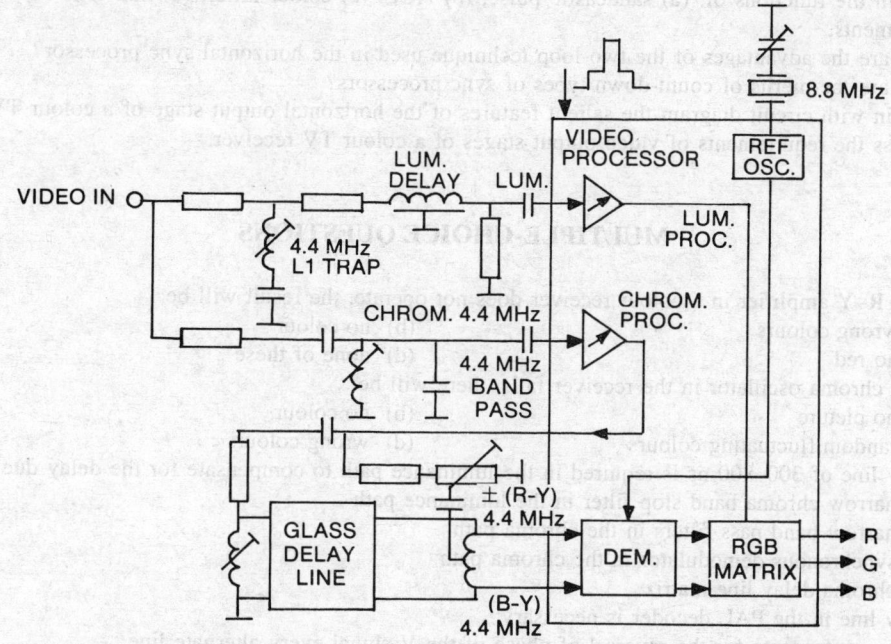

Fig. 20.12 Block diagram of filter/delay IC with PAL/NTSC processor CCTV

It also has a built-in delay for composite video, compensating for the extra delays introduced in the P²CCD filters. The output is used for synchronisation and SECAM decoding, having a normal bell-filter. The IC is driven by an on-chip clock generator, which supplies two clock frequencies derived from the subcarrier. The filters, which must process wide frequency bands, use a frequency of $4f_{sc}$, while the colour delay lines use the subcarrier frequency, f_{sc}. Since the clock signals remain within the chip, there is no interference with other parts of the set.

The colour processor adds a dc voltage to the subcarrier to enable identification of PAL/SECAM or NTSC. The delay lines are switched off for NTSC, while the filters are automatically corrected for PAL/SECAM or NTSC. Demodulation takes place before the delay lines. The YUV signals are now available at the outputs of the TDA 8450 and are further processed in the colour processor. A computer-controlled high performance PAL/NTSC processor is available in the IC TDA 8461, which contains all the active functions required for PAL/NTSC decoding and a video control section with RGB-insertion, e.g. for Teletext. A Memory feature module requires two separate sandcastle pulses for correct timing of the decoder part and the display part respectively.

REVIEW QUESTIONS

20.1 State the colour processing functions carried out by a typical chroma decoder IC.

20.2 Explain the principle and working of a delay line used in the PAL decoder. How does the delay line matrix separate the colour difference signals?

20.3 How is the colour subcarrier burst used to phase lock the reference oscillator in the receiver? How are the quadrature-phased colour subcarriers obtained?

20.4 Explain the functions of: (a) sandcastle pulse, (b) ACC, (c) colour killer, (d) half tone and monochrome adjustments.

20.5 What are the advantages of the two-loop technique used in the horizontal sync processor?

20.6 What are the merits of count-down types of sync processors?

20.7 Explain with circuit diagram the salient features of the horizontal output stage of a colour TV receiver.

20.8 Discuss the requirements of video output stages of a colour TV receiver.

MULTIPLE-CHOICE QUESTIONS

20.1 If the R–Y amplifier in a colour receiver does not operate, the result will be:
    (a) wrong colours   (b) no colour
    (c) no red   (d) none of these

20.2 If the chroma oscillator in the receiver fails, there will be:
    (a) no picture   (b) no colour
    (c) random fluctuating colours   (d) wrong colours

20.3 Delay line of 300–800 ns is required in the luminance path to compensate for the delay due to:
    (a) narrow chroma band stop filter in the luminance path
    (b) narrow band pass filters in the chroma path
    (c) synchronous demodulator in the chroma path
    (d) chroma delay line matrix.

20.4 Delay line in the PAL decoder is necessary:
    (a) to compensate for the reversal of phase of the V signal every alternate line
    (b) to average out the effect of phase reversal of V electronically
    (c) to get rid of the hanover effect
    (d) to equalise the signal delays in the decoder.

21

Testing, Alignment and Servicing of Television Receivers

A television receiver is an equipment of a rather complex circuitry, consisting of a number of subassemblies. For proper functioning of the receiver, all these must be carefully aligned and adjusted for good quality picture and sound reproduction. The adjustments have to be done with the help of special test instruments. In this chapter, the common TV test instruments are explained bringing out their operational features, and use in TV testing and alignment work. This is followed by a discussion on trouble shooting techniques in servicing a TV receiver. Diagnostic procedures are indicated to localize the faults and methods of isolating them.

21.1 TESTING AND ALIGNMENT OF TELEVISION RECEIVERS

The testing and alignment during production or servicing of a monochrome television receiver (special factors for colour receiver are discussed in Sec. 21.1) include mainly the following adjustments:

(i) *Power supply adjustments:*
    (a) Mains voltage taps, fuses.
    (b) Heater current, if required.
    (c) DC Supply voltage controls for $B+$, V_{cc}, etc.
(ii) *Picture tube adjustments:*
    (a) Brightness, high voltage (EHT).
    (b) Focus voltage tap.
    (c) Contrast
(iii) *Raster adjustments:*
    (a) Yoke clamp.
    (b) Width and height.
    (c) Centring.
    (d) Raster correction magnets.
    (e) Horizontal and vertical hold controls.
    (f) Horizontal and vertical linearity controls.
(iv) *AGC adjustments*
(v) *Frequency response alignments:*
    (a) RF tuner.

(b) Video IF amplifier.

(c) Sound IF amplifier and FM detector.

(d) Video amplifier.

For doing these adjustments properly, a number of test instruments are required to align the various stages of a TV receiver. These include:

(i) A good high sensitivity multimeter or preferably a general purpose electronic volt-ohmmeter or multimeter,

(ii) A cathode ray oscilloscope,

(iii) A wide band sweep generator,

(iv) A marker RF generator, and

(v) A pattern generator with RF modulation.

Servicing instruments that combine two or more of the above measuring facilities are also commonly available. These are:

(i) Wobbuloscope which includes a sweep generator, a marker oscillator and an oscilloscope.

(ii) Universal TV test sets which supply video signal for various geometrical patterns on the screen and also RF modulation of these on the standard television channels, along with an electronic volt-ohmmeter to measure ac/dc voltages and resistance values.

If the oscilloscope available is not of a high sensitivity type with a detector probe for HF voltage measurement, additional (i) electronic ac millivoltmeter for AF measurement, and (ii) RF millivoltmeter for RF measurement may be useful in testing the receiver.

The important operational features of the test instruments and the way they are used, are discussed in the following sections.

21.2 MULTIMETER AND ELECTRONIC VOLTMETERS

Multimeter A multimeter is a handy instrument for measuring currents, voltages and resistances. Some desirable features in a good multimeter are as follows: (i) Adequate ac current ranges along with the usually existing dc ranges. This is useful in monitoring the mains supply current drawn. (ii) A good sensitivity on the voltage ranges; 20,000 Ω/V or greater is readily available. High sensitivity is desirable for avoiding loading of the measured circuit. The shunting impedance of the meter is given by the sensitivity multiplied by the voltage range used. This must be much greater than the circuit impedance across which the voltage is measured. The sensitivity on the ac voltage ranges is sometimes much less than that on the dc ranges. When measuring voltages across high impedances, for example in the AGC circuits, multimeter readings can be erroneous. In such places electronic voltmeter should be used. (iii) The resistance measuring capability of a multimeter is also usually limited, from 1 Ω to 100 kΩ, unless a higher voltage battery (9 V/22 V) is provided for in the meter.

A multimeter can measure as voltages up to about 50 kHz only. Beyond this frequency, lead reactance problems and the limitations of the meterbridge rectifier affect the accuracy of measurement considerably.

A high voltage probe of 20 or 30 kV full scale deflection (f.s.d.) to measure the EHT voltage of the picture tube anode is a desirable accessory. This consists of a very high series resistance of 50 to 100 MΩ of high voltage type encased in a good quality insulator probe.

Electronic Volt-ohmmeter or Multimeter

These are particularly necessary for measurement of dc voltages across high impedances and ac voltages at higher frequencies. A VTVM or a FET voltmeter has a higher input impedance so that its loading effect is much less. These are generally of two types.

(i) *Rectifier-amplifier type* This has a basic double triode or a FET pair dc amplifier in bridge configuration with a meter in the output circuit. The amplifier input impedance is very high, above 10 MΩ so that the Ω/V sensitivity is very high even on lowest voltage ranges of 0.3 or 1 V f.s.d.

The ac voltages are measured on such a general purpose electronic voltmeter by means of a rectifier probe that rectifies the ac voltage before feeding it to the dc amplifier. This reduces the input capacitance loading and lead inductance effects. The lowest f.s.d. range is around 1 V measuring HF voltages up to several MHz.

Electronic voltmeters usually incorporate facility to measure currents and resistances by converting these parameters into voltage drops, when they are called electronic volt-ohm meters or multimeters.

(ii) *Amplifier-rectifier type voltmeter* This consists of a wide band gain stabilized ac amplifier followed by a rectifier-meter circuit. Its input impedance is typically 1 MΩ in parallel with 30 pF and is capable of ac millivolts or microvolts measurement in about 10 Hz to 5 MHz range with the limitation of the shunting effect of the input capacitance. It is useful in measuring low AF voltages and low video frequency voltages in low impedance circuits only, so far as the shunting effect due to the capacitance can be neglected.

A variation of the rectifier-amplifier type voltmeters is the *RF millivoltmeter* which is basically a dc microvoltmeter incorporating a stable high gain dc amplifier that can measure dc millivolts, on a meter at the output. An RF detector probe with suitable calibration enables measurement of millivolts in the HF and VHF ranges. The diode detector probe has a low input capacitance of about 4 to 5 pF and input resistance of about 40 kΩ and can be used from 100 kHz to 200 MHz or for higher frequency signals. This is quite useful in measuring RF/IF signals in the receiver and the antenna signals. It should be remembered that these indicate peak readings calibrated as rms and are responsive to all ac signals available or applied at its input terminals. Being wide band types, they respond to stray pick-ups and noise voltages in the range also. It must be checked whether the indicated output is due to the wanted signal only.

For measuring very low RF signals *selective microvoltmeters* have to be used. These are tunable to the frequency of measurement so that the unwanted signals and noise are rejected. These are actually in the form of calibrated sensitive receivers. Modified versions of these with an antenna dipole accessory can be used as *field strength meters* (FSM).

21.3 CATHODE RAY OSCILLOSCOPE

In testing and alignment of a TV receiver an oscilloscope is an indispensable instrument which is useful not only for observing the waveforms in the video and deflection circuits but also for the sweep alignment of frequency response of RF/IF stages of a TV receiver.

The video or the deflection circuit waveforms are of pulsed repetitive type containing a wide frequency spectrum in the video range. It is not possible to get a correct idea about it with a multimeter or electronic voltmeter. Observations of their shapes, peak-to-peak values and periodicities can be made on an oscilloscope only. The shape and magnitude of the waveforms in the video and deflection stages can provide useful clues to faults in a television receiver.

A very wide range of oscilloscopes are available to suit the applications and one's budget. The important blocks of a cathode ray oscilloscope are indicated in Fig. 21.1 to bring out its features and performance parameters.

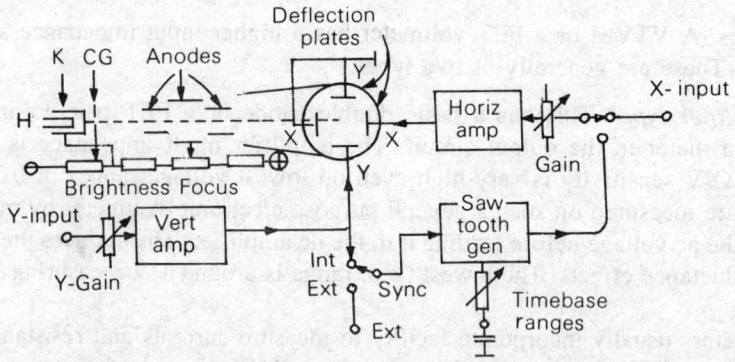

Fig. 21.1 Block diagram of a cathode ray oscilloscope

Low cost service type oscilloscopes have vertical and horizontal amplifiers that are ac coupled and uncalibrated. The vertical bandwidth is from a few Hz to a few MHz, typically from 50 Hz to 4 MHz with a sensitivity of 50 to 100 mV/cm. This may be modified by a switch to give a higher sensitivity of around 5 to 10 mV/cm at a lower bandwidth of 0.4 MHz. The time base in such oscilloscopes is a simple sawtooth generator in the range 20 Hz to 50 kHz with synchronisation from the vertical amplifier. These oscilloscopes have generally a built-in calibrating voltage to measure the waveform observed by comparison. A convenient time-base calibration is possible by intensity modulation of the beam by giving a known sine wave or pulse signal from an external source to the *z*-input of the oscilloscope. The waveform is then seen as a dashed waveform, with the dashes periodic at the modulating frequency. The periodic dashes are very convenient in measuring the pulsed waveform parameters. The horizontal amplifier of these oscilloscopes is also ac coupled with a lesser bandwidth (200 kHz) and sensitivity (200 mV/cm typically). The CRT screen size of 75 mm is usually adequate but larger screen size up to 125 mm or more can provide a greater viewing area convenient for alignment work.

Somewhat costlier oscilloscopes have dc calibrated vertical amplifiers and triggered-mode calibrated time-bases. The amplifier bandwidth is usually from dc to 5 or 10 MHz with a sensitivity of 5 to 20 mV/cm. The time-base has calibrated ranges from about 1 μs/cm to 0.1 s/cm with an external input facility for the horizontal amplifier, which is also direct coupled. In some oscilloscopes, there is provision for TV line or frame triggering in the time-base for quick locking and observation of line and frame signals.

21.4 SWEEP GENERATOR

In a TV receiver, the frequency response alignment of the RF/1F circuits cannot be done by just an RF signal generator and an output voltmeter. It is too cumbersome to plot the frequency response of these amplifiers by manually varying the input frequency and observing the output over the wide frequency range. Moreover, it is impossible to align it as per desired shape quickly since one cannot see the effect of the adjustments directly on the response characteristic.

A set-up to visually observe the frequency response as an *x-y* plot with frequency as the *x* or time-axis, is an essential measurement facility required in bandpass alignment of wide band circuits. This can be done with the aid of a sweep generator and an oscilloscope. A sweep generator is an RF generator, the frequency of which is varied or swept over the alignment range periodically by means of a sweep voltage. This sweep voltage is simultaneously given to the *x*-deflection plates of an oscilloscope to deflect the beam from left to right and establish a frequency axis horizontally on the screen. The sweep generator has manually variable frequency ranges to cover the frequencies required for alignment. At each setting, the frequency is swept electronically over an adjustable sweep width range, by variation of *L* or *C* of the RF oscillator. The sweep voltage may be a sawtooth voltage or a sine wave voltage at the mains frequency of 50 Hz. The swept frequency output must remain constant over the sweep width used, as shown in Fig. 21.2.

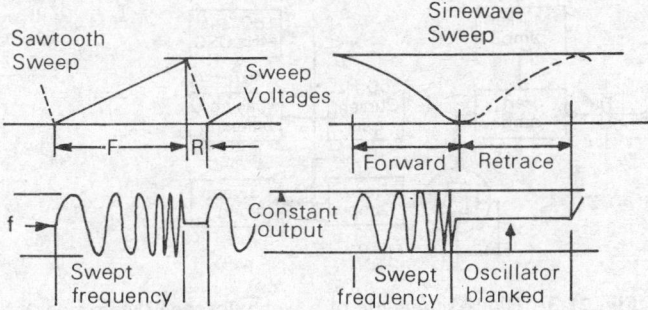

Fig. 21.2   Sweep generator output

The sweep voltage is available with a phase adjustment so that the relative phase between the voltage used for sweeping the RF oscillator and the voltage available for *x*-deflection can be varied such that the two can be synchronised. The deflection from left to right and the sweep of the RF generator frequency from minimum to maximum must start at the same instant.

21.5   MARKER RF SIGNAL GENERATOR

An RF signal generator is useful as a marker generator to identify the frequencies on the swept frequency scale. When the marker RF signal is mixed with the swept frequency output it produces a low frequency beat on the sweep response curve at the point where the sweep frequency and the RF marker frequency coincide. The low frequency beat is made distinctly visible by passing the mixed output through a low-pass filter. Variation of the marker frequency will be seen as a beat riding over the swept frequency response curve.

The RF signal generator used as a marker generator is not without application in alignment work also. When sweep generator is not at hand the RF generator can be used to make a fairly satisfactory alignment by peaking of response at important points and aligning the traps for minima. Amplitude modulated RF signal produces dark bands across the screen if the signal frequency lies in the bandpass of the amplifier. Fixed frequency markers at 1 MHz, 10 MHz and their harmonics produced from crystal oscillators may be used for more accurate calibration of the sweeps. The intermediate frequencies are then read off by interpolation.

21.6   TELEVISION WOBBULOSCOPE

A TV wobbuloscope is an instrument with a combined service facility of a wobbulator or sweep generator, a cathode ray oscilloscope and a marker generator. It is, therefore, most convenient for bandpass response alignment of RF/IF and video stages of a TV receiver. The schematic and operational features of a typical wobbuloscope shown in Fig. 21.3 are briefly discussed.

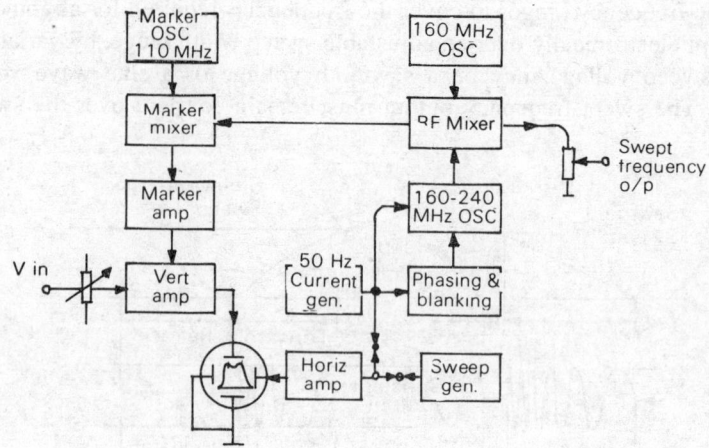

Fig. 21.3 Block schematic of a wobbuloscope (Metrimpex)

(i) Wobbulator The wobbulator provides a swept frequency output to cover frequencies from 1 MHz to 240 MHz in two or three bands. The sweep generator consists of a basic sweep oscillator continuously tunable over the band III range, from 160 MHz to 240 MHz. The manually adjustable frequency can be electronically swept or wobbled by variation of the oscillator coil inductance at a 50 Hz rate. This is done by altering the coil ferrite core saturation with the help of variation of its magnetic circuit by a 50 Hz current generator. The sweep width can be altered from 0.5 MHz to 15 MHz centered at the tuned frequency. The magnetic sweep variation technique gives a fairly constant output over the full range.

In order to obtain swept frequencies on the low side to cover the video frequencies, the IF and band I frequencies, the sweep oscillator output of the 160 MHz to 240 MHz is mixed with a 160 MHz fixed frequency oscillator output to produce difference frequency sweeps having the same sweep widths up to 15 MHz. The low frequency sweep range is thus 0 to 80 MHz. The sweep output is available through an attenuator at a constant level up to 100 mV across a 75 Ω unbalanced impedance.

(ii) Oscilloscope The oscilloscope has a vertical amplifier which has a bandwidth of about 20 Hz to 0.5 MHz so that the sweep and video waveforms can be observed on it with a little loss of HF details, besides the swept frequency response. It has a sensitivity of about 50 mV/cm adjustable with an attenuator with decade ranges. The time-base generator furnishes sawtooth sweeps of 20 Hz to 50 kHz in continuous and step variation, with facility of internal synchronisation. The horizontal amplifier is fed the time-base generator output or the 50 Hz sine wave sweeping voltage from the sweep oscillator section, in the wobbulator. The amplified signal is fed to the horizontal deflection plates of the cathode ray tube.

The oscilloscope used can be a general purpose oscilloscope to observe the video and deflection circuit waveforms in a TV receiver, or to observe the frequency response of the RF and IF amplifiers. In the sweep mode, the 50 Hz current generator is given to the *x*-deflection to establish a frequency axis, as the same voltage is varying the frequency of the sweep oscillator. For obtaining the frequency response of an amplifier, the wobbulator output is fed to the amplifier and the detected output from the same is given to the vertical amplifier input.

The sweep oscillator is blanked during the retrace by a blanking stage displaying a reference line only and a phase shifting circuit adjusts the phase of the 50 Hz sine wave signal so that the sweeping of the oscillator by the sweep signal from the blanking stage coincides with the forward scan on the screen, with minimum frequency to the left and the maximum to the right.

(iii) Markers Marker signals are useful to identify the frequencies in the displayed frequency response, as they produce a low frequency beat of the difference frequency which coincides with the marker. In the wobbuloscope

1 MHz or 10 MHz crystal oscillator frequencies and their harmonics are mixed with the sweep output from mixer and the beat marker is introduced in the vertical amplifier path to be observed on the sweep frequency response curve. Their amplitude is adjustable.

External markers can also be introduced by feeding them into the marker mixer to have any variable frequency marker on the display. This allows the distance between characteristic points on the displayed curve to be measured more conveniently.

Measurements Accessories

The following accessories shown in Fig. 21.4 are very essential for sweep alignment and oscilloscopic measurement in a TV receiver.

(i) Balun This is a 75 : 300 Ω impedance transformer providing transformation of 75 Ω unbalanced impedance of the wobbulator into 300 Ω balanced one for connection to the 300 Ω input of a receiver. The impedance transformation may be done by the minimum loss matching pad also shown in Fig. 21.4(i). This introduces, however, a loss of about 11.5 dB.

(ii) RF detector probe This is necessary for observing the swept RF output on the oscilloscope. It consists usually of a shunt diode detector with a low pass filter, as shown.

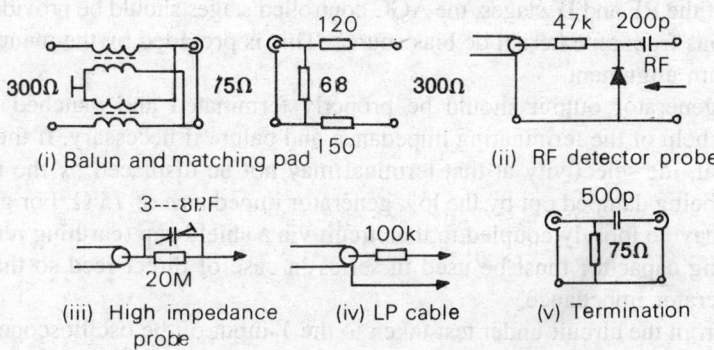

(i) Balun and matching pad   (ii) RF detector probe

(iii) High impedance probe   (iv) LP cable   (v) Termination

Fig 21.4 Wobbuloscope accessories (Metrimpex)

(iii) High-ohmic/attenuator probe This enables the input impedance of the oscilloscope to be increased, by a factor of 10 or 20. A 20 : 1 attenuator probe increases the input resistance from 1 MΩ to 21 MΩ, and reduces the input capacitance from 50 pF to about 5 pF. This reduces the loading and detuning effect on the circuit under test. The 3–8 pF capacitor serves to compensate the attenuators for transient response.

(iv) Low-pass filter cable This is useful in taking the swept frequency signal from the detection point in the receiver while rejecting stray pick-ups at high frequencies or line deflection circuits. The 100 kΩ resistance of the lead and the capacitance of the cable form a low-pass filter that by-passes high frequency pick-ups.

(v) Terminating impedance This 75 Ω resistance serves to terminate the wobbulator output for impedance match when required. A blocking capacitor blocks the dc voltage in the receiver circuits, where the wobbulator output is connected.

Applications of Wobbuloscope
Swept frequency alignment of a two-port bandpass network or selective circuit requires the set-up shown in Fig. 21.5.

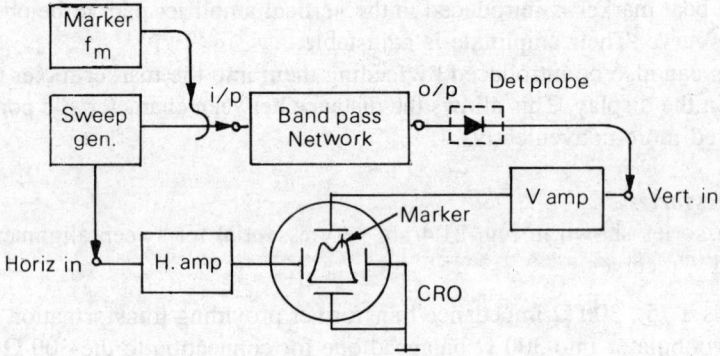

Fig. 21.5 Swept frequency alignment set-up

It is necessary to remember the following points in preparing the alignment set-ups in a television receiver:

(i) In alignment of the RF and IF stages, the AGC controlled stages should be provided the recommended value of the AGC bias from an external dc bias source. This is provided by the manufacturer or designer to ensure an optimum alignment.

(ii) The sweep generator output should be properly terminated and matched to the circuit input impedance with the help of the terminating impedance and balun, if necessary. If the output is fed to the grid or base terminal, the selectivity at that terminal may not be displaced by the response, the circuit at there feed points being damped out by the low generator impedance of 75 Ω. For guarding against this, the sweep voltage may be loosely coupled to the circuit via a shield cap (earthing removed) for the tube. An isolating-blocking capacitor must be used in series in case of direct feed so that the dc bias is not shunted by the generator impedance.

(iii) The output from the circuit under test taken to the *Y*-input of the oscilloscope should be detected. The RF detector probe must be used for this purpose. The detector probe is not necessary when the output is taken after the detector circuit.

(iv) If there is stray external pick-up by the output probe going to the oscilloscope, low-pass filter probe should be used to reject the pick-up.

(v) The RF detector probe should be connected to the recommended measuring point specified by the manufacturer so that its loading effect is minimized.

(vi) The marker amplitude should be minimum so that it does not affect the response curve. If external markers are used by coupling them along with the sweep generator to the input, they should be only loosely coupled.

Television Receiver Alignments
The sweep alignment set-ups for RF tuner, video IF, video stages, sound IF and overall receiver frequency response characteristics are shown in Fig. 21.6.

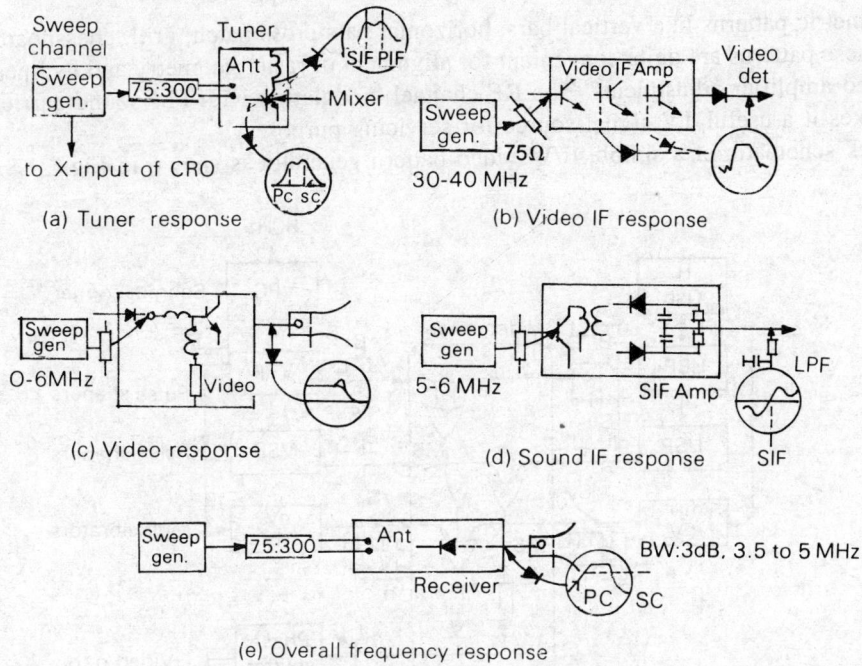

(a) Tuner response

(b) Video IF response

(c) Video response

(d) Sound IF response

(e) Overall frequency response

Fig. 21.6 TV receiver alignments

Impedance Matching

A set-up for indication of matching of antenna impedance or receiver input impedance with a cable with the help of standing wave display on a wobbuloscope is shown in Fig. 21.7. The cable length l for displaying n crests of standing waves is given by:

$$l = \frac{150 \times \text{cable velocity factor} \times n}{\text{sweep width}}$$

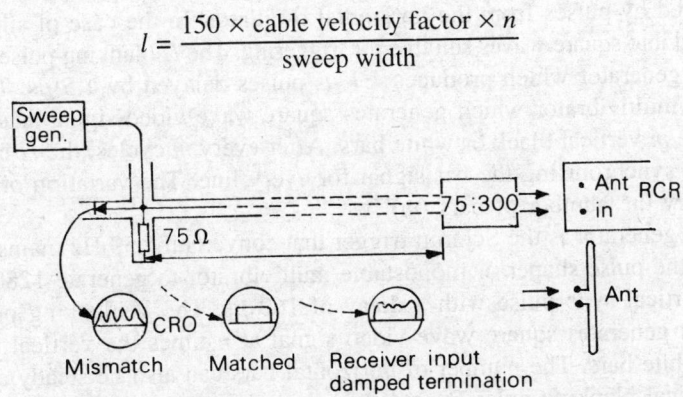

Fig. 21.7 Impedance matching—VSWR check-up with wobbuloscope

21.7 VIDEO PATTERN GENERATORS

Pattern generators provide video signals direct and with RF modulation on the standard TV channels so that it can be used for testing and alignment of TV receivers. The video signal is designed to produce simple geometric patterns like vertical bars, horizontal bars-cross hatch, grill, chessboard and gradation patterns. These patterns are quite convenient for alignment of raster geometry and its linearities, and also for the video amplifier adjustments. The RF channel modulation with FM sound carrier facility along with it, makes it a useful TV signal source for servicing purposes.

The block schematic of a simple B/W video pattern generator is shown in Fig. 21.8.

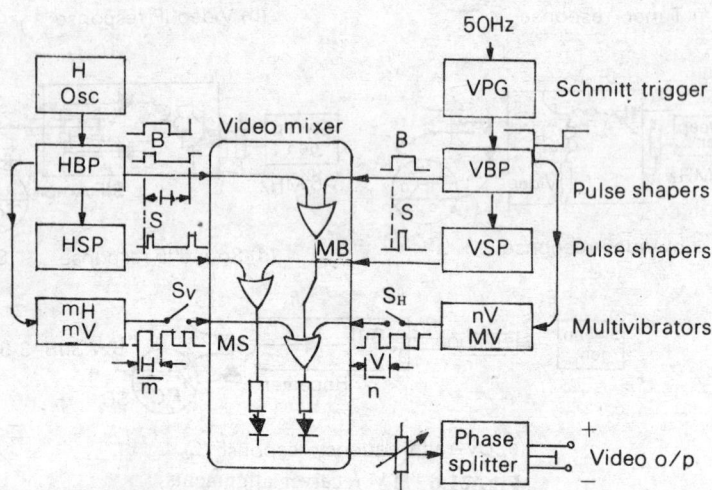

Fig. 21.8   Block diagram of a video pattern generator

A stable 15,625 Hz sine wave *LC* oscillator or a pulse generator serves as the basic horizontal or line frequency source and the 50 Hz mains frequency controls the vertical frequency source. As shown in the block diagram, the 12 μs horizontal blanking pulses are generated by a pulse shaper or monostable multivibrator triggered by pulses from the horizontal oscillator. In the case of sine wave oscillator, the sine waves are shaped into square waves suitable for triggering. The *H* blanking pulses control the triggering of the *H* sync pulse generator which produces 4.7 μs pulses delayed by 1.5 μs. The *H* blanking pulses also control the *mH* multivibrator which generates square wave video signal at *m* times the horizontal frequency to produce *m* vertical black or white bars. After every *m* cycles, the *H* blanking pulses trigger the multivibrator for synchronising the bar signal for every line. The variation of frequency of the *mH* multivibrator can vary the number of bars (*m*).

The vertical pulse generator is the Schmitt trigger that converts the 50 Hz mains voltage into a square wave. This controls the pulse shaper or monostable multivibrator to generate 1280 μs vertical blanking pulse and 160 μs vertical sync pulse with a delay of 160 μs. The *V* blanking pulses also control the *nV* multivibrator that generates square wave video signal at *n* times the vertical frequency to produce *n* vertical black or white bars. The number of horizontal bars can also be steady as the multivibrator is triggered by the vertical blanking pulse for synchronisation with the field.

The sync and blanking pulse outputs and suitable bar-multivibrator outputs are combined together in the video mixer to produce the composite video signal to give required patterns. The *H* and *V* blanking pulses as well as the video signals from *mH* and *nV* multivibrators have the same levels changing between black and white. Hence it is convenient to mix these together first. The *H* and *V* sync pulses have same levels changing between blacker-than-black and white. These may also be mixed together with suitable OR logic gate and then combined with the mixed blanking and video signal in a level proportional transmission gate like a diode-resistance OR gate to produce a composite video signal with 7 : 3 picture-sync ratio.

Provision of switches in the signal paths of the mH and nV multivibrator and their manipulation can give various patterns. If both switches are off, there will be a blank white raster. With mH switch *on*, there will be vertical bars, while nV switch being *on* will produce horizontal bars. With both switches *on*, a cross hatch pattern will be produced. These patterns and the chessboard pattern can be obtained easily by logic gating of the outputs from the mH and nV multivibrators as shown below:

$$(\text{Black level} = \text{logic } 1)$$
$$(\text{White level} = \text{logic } 0)$$

Pattern	*Output Logic Function*	
(i) Blank white raster	0	
(ii) Vertical bars	mH	only
(iii) Horizontal bars	nV	only
(iv) Cross-hatch of black bars	$mH + nV$ or $\overline{mH \cdot nV}$	
(v) Cross-hatch of white bars	$mH \cdot nV$ or $\overline{mH + nV}$	
(vi) Chessboard	$mH \cdot nV + \overline{mH} \cdot \overline{nV}$ or $\overline{mH} \cdot nV + mH \cdot \overline{nV}$	

Here black corresponds to logic level 1 and $\overline{mH}$ and $\overline{nV}$ are complements of mH and nV respectively. The logic functions can be implemented by available NAND or NOR gates also. Gradation pattern requires a video waveform in the form of a staircase generator. This can be formed by a diode capacitor pump circuit followed by a discharge circuit that discharges the capacitor after it has charged in steps from the mH or nV multivibrator, depending on whether horizontal gradation is desired or vertical. The discharging circuit can then be triggered at the horizontal or vertical rate.

Linear staircase steps can be obtained with the help of a binary or decade counter, the output from the stages of which feed into an output point through suitable weighting resistors. The output level then changes in steps as the count advances due to the mH or nV pulse train, progressively.

The multivibrators, pulse shapers and the logic gates may employ discrete transistor circuits or digital ICs or a combination of these. The actual implementation of a pattern generator may be optimized depending upon the type of hardware used.

21.8 TELEVISION TEST CHARTS

For rapid evaluation of the picture quality and help in adjustments of the television system performance, various optical test charts are used. These include Resolution Test charts that evaluate the picture sharpness, Linearity Test charts that test geometrical distortions and linearity, and the Gray-scale or Gamma charts that are useful in testing the tonal set-up, i.e. contrast capability of the system.

Universal Test Resolution Charts
These charts, which have a number of patterns combined together to rapidly evaluate the linearity, resolution, gray scale performance, etc. of the television systems, are commonly used in different forms in various countries. The RMA/RETMA/EIA standard transmitter test chart or SMPTE test chart is widely used for this purpose. The RMA/RETMA resolution test chart is shown in Fig. 21.9(a).

The chart has a large white circular area at the centre against a gray background and four smaller

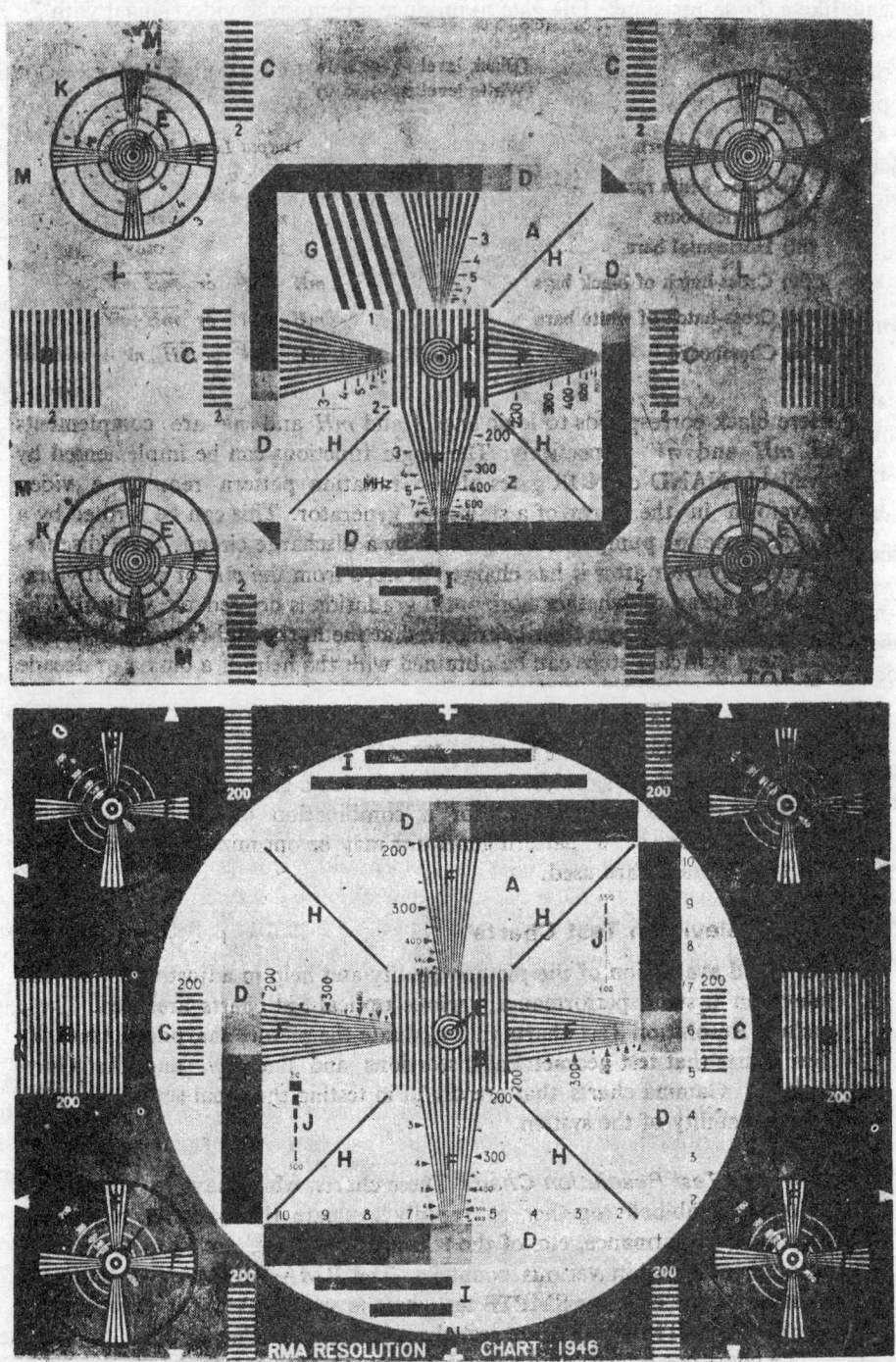

Fig. 21.9 Universal test resolution charts: (a) RMA/RETMA chart, (b) Continental test chart derived from RETMA charts

circles at the four corners. These circles are filled with different geometrical patterns and stripes, which help evaluation of the picture quality, resolution and distortions in the picture at a quick glance. This is necessary for a general check-out during maintenance of a TV system or adjustments in a TV receiver. The purpose of the various geometrical patterns in the chart is explained below:

The chart has some variations. Its later version, the RETMA/EIA resolution test chart (1956) has eight wedges that are narrower, but have a slower taper to give improved readability of the resolution lines. The continental version of the RMA chart includes in addition, a set of five slanting bars in the top left portion of the white circle in place of the diagonal line, as shown in Fig. 21.9(b).

(a) *The large white circle at the centre* indicates by its shape the distortion in the horizontal and vertical linearity as also the aspect ratio. The circle must be exactly in the centre of the screen. Because of the human eye capability of critical assessment of nonlinearity of a circle, any deviation from reasonable limits of distortion becomes quickly apparent.

(b) *The broad vertical stripe-boxes* in the centre and on the left and right hand sides of the picture have a resolution of 200 lines. These enable adjustments in the horizontal linearity of the picture. The horizontal dimensions of all of them must be same and the width of all the vertical stripes or lines should be equal.

(c) *The two narrow horizontal stripe-boxes* inside the white circle near the gradation bars and similar stripe boxes near the top and bottom edges of the picture enable evaluation of the vertical linearity. Their dimensions and the width of the stripes should be uniformly equal.

(d) *Grayscale or gradation bars* the four vertically and horizontally arranged gradation bars evaluate the gamma, the brightness transfer characteristics, by the number of discernible gray scale squares out of the ten squares which have a gray scale contrast ratio of 30 : 1. These are useful in proper adjustments of the brightness and contrast of the picture.

Brightness and contrast controls should be adjusted together. At normal contrast, the *brightness* should allow brightness to be varied smoothly from complete blacking out, to the appearance of flicker and defocusing of the bright spots due to excessive brightness. The screen glow should change uniformly over the screen surface. The *contrast* control should allow the contrast to be varied from a barely visible image to an excessively black one. After achieving optimum brightness and contrast in the picture at least 6 to 8 gradations should be discernible for satisfactory reception and proper tonal rendition of the television picture. At excessive contrast, the image tones are lost, while at excessive brightness fine details are lost and a flicker may be evident.

(e) *The five concentric circles at the centre* serve to evaluate the beam sharpness or focusing at the centre and at the edges. The focus adjustment should be such that the diameter of the central spot is smallest with the surrounding circles clearly visible. Any distortion or elliptical shape manifests a fault in the focusing or deflection system.

(f) *Resolution wedges* the tapered vertical and horizontal wedges of black and white stripes inside the square formed by the gradation bars evaluate the vertical and horizontal resolution. The figures near the tapered wedges indicate the corresponding resolution lines and the corresponding equivalent bandwidth of the television system. The resolution of the picture is given by the section of the wedge where the clearly discernible lines meet the gray mixed-up zone.

The *vertical resolution* is indicated by the horizontal wedges. It depends upon the number of blanked lines, optical and the electron beam limitations indicated in the Kell factor of the system. With a good focus adjustment in the picture tube besides the camera tube-lens system adjustments vertical resolution of 400 to 500 lines can be obtained in practice.

The *horizontal resolution* of the system is indicated by the vertical wedges. It depends upon the bandwidth of the system. In a TV receiver, the bandwidth depends upon the bandpass response of the

RF, video IF and the video amplifier stages. The effective bandwidth is affected by the setting of the picture carrier on the Nyquist flank of the video IF response curve due to the tuning adjustment. Larger the bandwidth, larger is the horizontal resolution provided the phase response is also proper; but if the picture carrier lies deep down the required 50% level of the maximum response, excessive phase errors make the picture 'plastic' in appearance with apparent ringing at the edges or borders in the picture. Optimum tuning is usually achieved by first tuning the control on one side till the sound bars tend to appear, and then tuning back a little until the bars disappear.

The number of resolution lines and the corresponding bandwidth or video response of the system is marked along the vertical wedges. The resolution is referred to the corresponding number of horizontal scanning lines. For example, resolution lines of 200 correspond to a video bandwidth of 2.5 MHz, 300 lines correspond to 3.75 MHz, 400 lines to 5 MHz.

Resolution wedges are also provided in the four corner circles to give an indication of the resolution in the respective areas.

The horizontal wedges also indicate whether the interlacing is proper, by fanning or Moiré effect at the areas where the lines tend to merge.

(g) *A group of slanting bars* inside the white circle, near the upper wedge serve as a reference of 1 MHz frequency for oscilloscopic evaluation of the waveform due to the wedges. It is not useful for any service adjustments.

(h) *The diagonal lines* in the white circle check the interlacing. The lines should be thin and uniform. With poor interlacing, the lines become thick and steps appear in them when the odd and even field lines pair. Periodic variation in the distance between the lines appears as serrated projections on the diagonal lines. This also causes the tapered ends of the horizontal wedges to bend upwards and downwards showing a fanning effect.

(i) *Black horizontal bars* inside the central white circle above and below the horizontal gradation bars check the low frequency response and phase errors at low and medium frequencies, in the form of 'streaking' forming white tails or gray elongations to the right of the dark bands depending on the loss or boost of the low frequency response.

(j) The *single resolution lines* near the vertical gradation bars indicate the frequency of ringing if the video response is underdamped.

(k) The *concentric corner circles* evaluate the geometry and resolution of the picture at the corners. The geometrical distortion at the corners and edges is corrected by pincushion magnets on the yoke, and if necessary by the linearity coil.

(l) The *uniform gray surface background* for the white circle is useful in observing noise impulse interference in the picture.

(m) The outer corners of the eight *white small triangles* at the edges of the chart from the limits of the raster for a picture tube with a rectangular format. These may not be fully visible on tubes with rounded corners.

(n) The *crosses* at the middle of the edges of the chart are used to align the optical system of large screen projection receivers.

Gray Scale Chart

This chart, used for adjustment of gamma or tonal set-up of television cameras, has 10 linear or logarithmic increment steps in a contrast range of about 30:1, providing 10 gradations bars in horizontal direction. Logarithmic chart is most useful in camera alignment because the oscilloscopic display of the signal after gamma correction is staircase waveform with equal increment steps.

Linearity Test Chart

This is used for checking the aspect ratio and geometric distortions in the raster. The chart contains a set of small triangles defining the limits of the aspect ratio and a set of small circles, the relative diameter of which depends upon the limits of distortion allowed in a given system.

All these charts are usually sent over the television camera optically. In place of the pictorial charts, sophisticated *electronic test patterns* generated by electronic circuitry are often more convenient in oscilloscopic evaluation of the circuit performance. A typical electronic test pattern includes sine waves of 1, 2, 3, 4, 5 MHz, $\sin^2$ pulse and bar signal, and gray scale of 10 steps, arranged to be sent over the various portions in each picture scan with the help of electronic pulse circuitry.

Measurement of Geometrical Distortion

A cross-hatch grating or dot pattern generator can be conveniently used for measurement of nonlinearity in the picture reproduced on the monitor or receiver screen. It generates a series of accurate narrow pulses that are synchronised to the vertical and horizontal scanning frequencies. When these are fed directly into the monitor or receiver, they appear as narrow vertical and horizontal lines, their number depending upon their frequency. Any nonlinearity in deflection shows as deviation from equi-distance spacing. A transparent overlay of the correct size may be laid against the screen to read off the nonlinearity deviations. Alternatively, EIA linearity chart of 4:3 aspect ratio with small circles along a set of cross-hatch lines may be projected optically on the screen to superpose the two optically. The deviations of the dots from the circle centres can give a measure of the nonlinearity, as explained below.

Camera Linearity

A method to isolate the nonlinearity of the camera deflection system from the viewing monitor or receiver is to superpose the cross-hatch lines or dots (obtained by feeding the electrical output of the generator to the monitor) on the linearity chart with small circles along a set of cross-hatch lines produced by the camera system optically. The two patterns superposed, are equally affected by monitor nonlinearity and hence the camera nonlinearities can be corrected. The intersection of the grating or cross-hatch bars form convenient reference for the small circles. The deviation of the crossings from the small circles of well defined diameters can quickly indicate the per cent nonlinearity. The cross-hatch grating generator can produce dots at the intersection of the grating lines, providing a dot pattern. This can be done by suitable logic gating of the video waveform pulse trains in the generator. As shown in Fig. 21.10, the small circles have an inner radius of 1% of the picture height, and an outer radius of 2% of the picture height, the total circle diameter being 4% of the picture height. The camera deflection circuitry can be adjusted for providing best linearity on the monitor to get all dots as near the centre as possible. The diameter of the dot should of course be less than 1% of the picture height. Then if the dot falls within the inner white area of the circle, the nonlinearity and geometrical distortion lies within ± 1%. If they fall just outside of the circle outer diameter, the distortion is ± 3%.

21.9 SERVICING OF TELEVISION RECEIVERS

Television receiver is a fairly complex electronic system as compared to a radio receiver. Servicing or fault finding of a TV receiver is, in general, a somewhat more difficult job, often requiring the help of the more sophisticated instruments discussed in the earlier sections.

Most troubles or faults in a TV receiver arise out of faulty alignment or defective components, viz. tubes. transistors, resistors, coils or capacitors showing opens, shorts, leakages or changes on values due to over-loading, over-heating or mechanical defects like loose contacts, breaking or dry soldering of

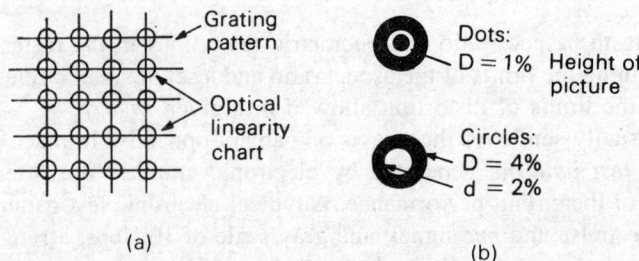

Fig. 21.10 Camera linearity check: (a) Grating pattern superposed on linearity chart of small circles, (b) Dot pattern superposed on linearity chart of small circles

leads, etc. Tubes are, of course, the most common cause of the trouble and may be checked. first, since this can be tried by quick replacement. Particularly fault-prone are the components dealing with larger currents and voltages, viz. those in the power output stages of sound, video and deflection drive and power supply circuits.

Indiscriminate replacement of components like diodes, transistors, deflection coil units, condensers, etc. may cause more harm than good by damage to printed circuit wiring and hence is ill-advised unless some indications lead to the specific faulty components. Really tricky faults are the intermittant faults due to loose contacts due to dry solder, etc. which may often prove a wild goose chase when trouble shooting them. The skill of a servicing engineer lies in isolating the faulty component by a proper diagnostic procedure in minimum amount of time. While a systematic diagnostic procedure using TV signal tracing, waveform measurements on oscilloscope and multimeter checks would ultimately lead one to the faulty adjustment or component, one should try to develop rules of the thumb and intuitive skill to pin point the fault with minimum use of equipment and keen observation of the indications.

In a TV receiver, the indication of the fault or malfunction in the set is available on the picture tube screen in addition to the indication through the loudspeaker. The indication of faults in a TV receiver is available in three ways, viz. sound, raster and picture. Typical indications are:

(i) *Sound*:   Normal, weak, distorted, hum or buzz superposed, no sound.

(ii) *Raster*:   No raster, no brightness, horizontal line, vertical line, distorted or compressed raster, foldover; trapezoidal, nonlinear, off-centre, tilted raster.

(iii) *Picture:* No picture, weak or washed out picture, excessive contrast, low brightness, blooming, hum bar, sound bars.

Loss of sync, picture rolls, tears, scrambled picture, drifts vertically and horizontally, jittery picture, negative picture, picture bending vertically.

Snowy picture, pattern interference, Moiré patterns, smeared picture, loss of detail, plastic picture.

Ghosts, venetian blind effect, black and white dots and dashes or streaks in picture.

Fault Localizing Procedure

The various possible symptoms or indications in the sound, raster and picture can provide clues to isolation of the particular section, block or stage in the TV receiver which contains the fault. One has only to remember the block diagram and the signal separation points carefully, as shown in Fig. 21.11.

From the indications of break in the path or any abnormality in output of sound, video signal, sync pulses, *H* and *V* drive waveforms, AGC voltage, etc. observed on a multimeter or an oscilloscope, one may draw logical conclusions to localize the faulty section in the receiver.

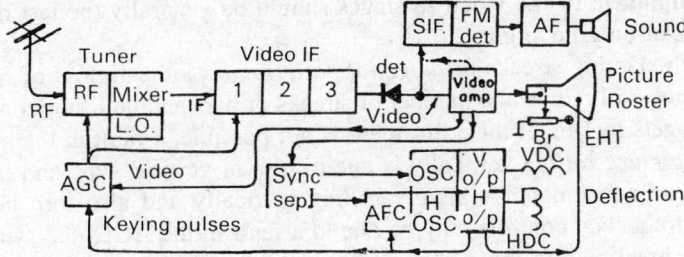

Fig. 21.11 Block diagram of TV receiver with signal flow paths

Simple indication of a working TV receiver is the appearance of *raster and snow,* black and white spots randomly appearing on the screen if the contrast and brightness controls are turned fully on, and a *hissing noise* from the loudspeaker if the volume is turned on. If the receiver sensitivity is moderate, snow may not appear on the screen, but dashes of black and white stray lines due to static interference from car ignition, etc. would appear if the antenna is connected or brushed at the aerial terminals, when TV signal is not available.

If there is *no raster* and no *sound* available from the receiver, the power supply circuit that feeds the heaters of the valves and the $B+$, V_{cc} supplies need a check. Fuses in the supply leads may be checked. If the mains fuse is blown, a series lamp of a comparable wattage may be temporarily included in the mains supply circuit to guard against frequent blowing of the fuse or damage to other components due to a short somewhere in the mains circuit, for example, a short in the RFI suppressor condensers or in the rectifier circuit due to filter condenser short, heavy leakage, etc. A check on the valves not lighting or overheating may be made to trace out any abnormality in the supply voltages due to a faulty valve or other component.

If there is *no raster* but *sound* is O.K. indicated by the typical hissing noise in the speaker, the RF-IF stages and the sound section are in order. The fault may be in the deflection system, dc conditions in the video amplifier, picture tube connections or the picture tube itself. The horizontal oscillator is essential for EHT generation without which the raster cannot appear. Any distortion in the raster height and width or nonlinearity of the raster is traced to the deflection system in the receiver. A bright horizontal line indicates fault in the vertical deflection stages or the deflection coil. A bright vertical line indicates fault in the horizontal deflection coil.

Once the *raster* appears and is O.K., it is necessary to trouble shoot in the presence of TV signal from a station or a pattern generator that can provide video and RF modulated signal on a standard channel. If there is *no picture* and also *no sound*, even after adequate TV signal from a pattern generator or TV antenna connected to the aerial terminals and the set properly tuned, the tuner and the video IF amplifier stages and the detector are suspect. This can be confirmed by measuring the detector output voltage, which should measure more than 1 or 2 V dc, on a multimeter. The detected video voltage may be monitored on an oscilloscope also. If video modulated IF signal is available, the fault can be isolated between the RF tuner and the video IF stages. If this is not available, a simple amplitude modulated RF signal in the video IF frequency range may be used to test the video IF amplifier. When this is fed to the IF amplifier, the modulating signal will produce a number of bars in the raster, depending on the modulating frequency. At this stage, it is worthwhile checking the AGC voltage and its pre-set control, which may have been disturbed to bias off and reduce the gain of the RF-IF stages. A *weak picture* or *excessively contrasted* picture with disturbed synchronisation, or *negative* picture is often ascribed to the

AGC circuit fault or adjustment. If AGC voltage is not abnormal stagewise check on the RF and IF stages may be made. The alignment of the video IF stages should be generally the last resort as misalignment in normal operation can only be slight.

A *torn picture* with slanting streaks or *diagonal pattern bars* indicate loss of *H* sync. Adjustment of the horizontal frequency will reduce the number of streaks or bars and their horizontality until the picture becomes evident and gets locked. If the adjustment is not possible, a fault in the horizontal oscillator or AFC is indicated. A *picture rolling* vertically is due to loss of vertical sync and indicates a fault in the vertical oscillator or its adjustment. A picture rolling vertically and also torn horizontally, failing to respond to *H* and *V* frequency controls may be due to a fault in the AGC, the sync separator, the noise inverter or the video signal path to these circuits. The *picture drifts* horizontally and vertically for want of sync pulses of adequate amplitude at the output of the sync separator.

A *raster bent* vertically indicates presence of hum in the horizontal AFC, horizontal oscillator or the sync separator due to 50 Hz mains pick-up or ripple in the rectified supplies. A 50 Hz hum causes a single cycle of the bend while a 100 Hz ripple from full-wave rectifier supply causes a double bend. The pick-up may occur due to heater-cathode leakage or increased ripple in the rectified supply caused by defective filtering components. This may also happen when the mains voltage falls excessively and the regulated supplies can no longer function properly. The vertical bending in this case is accompanied by a hum in the audio channel and horizontal *hum bars* in the picture. Hum introduced in the RF-IF stages due to heater-cathode leakage, etc. may cause *hum modulation* of the signal. This can cause the width of the raster to be modulated by the hum bend, in presence of the picture signal.

Sound bars in picture indicates mistuning of the channel (fine tuning) or misalignment of sound IF trap in the video IF amplifier. If the picture is O.K. while sound only is absent, the sound section must be investigated. The audio drive and the loudspeaker circuit may be tested first. If these are O.K., the FM detector and the sound IF amplifier may contain some fault or misalignment.

Tracing of the Faulty Component

Once the fault has been localized to a particular section or stage of the TV receiver, tracing of the faulty component can be done by inspection of any abnormally heated or damaged component in the section or by voltage and resistance check of the circuit for abnormal measurements which may point to the faulty component or device. The valves or other plug-in components may be replaced by new ones if it helps. The socket connections of such components as also the socket connectors of plug-in card modules should be checked for loose or dirty contacts. The circuit voltages should be checked against the typical voltages specified by the manufacturer. High dc voltages indicate open circuits. Low voltages indicate shorted or leaky capacitors or improperly biased or overloaded stages. A resistance check between various points will lead to the actual open or short circuits. While making this check, all the resistance paths between the test points should be borne in mind, particularly the forward or reverse resistance of diode or transistor that may shunt the related component test points. Care should be taken to switch off the set and discharge all the capacitors concerned before making the resistance check. The suspected component should be disconnected at least at one end to confirm short circuit or leakage resistance.

The valves indicating low currents and gains may be checked for low emission. Valves taking excessive current and indicating over-heating by excessive glow may be gassy. Faulty transistors may be isolated by voltage and resistance check in circuit is far as possible. A useful check on the transistor is to short the base emitter momentarily, which should cut-off the transistor and raise the collector voltage towards the supply value if there is a resistance load or a decoupling resistance in the V_{cc} supply lead.

21.10   ALIGNMENT AND TESTING OF COLOUR TV RECEIVERS

Alignment and testing of a colour TV receiver requires a good colour TV pattern generator in addition to the standard instruments like a CRO, Sweep generator, DMM with EHT probe, RF attenuator, AF Power meter, Distortion analyzer and a power supply. Low cost service type colour TV pattern generator provide just the standard colour bars and a few geometric patterns. Professional models are more complex and provide for additional patterns and signals that enable PAL delay line and demodulator adjustments, gray scale with multiburst, teletext etc.

Colour Bar Pattern Generator

A Colour bar pattern is useful in adjustment of the chroma circuits of TV receivers. Complex colour pattern generators provide a variety of additional patterns like DEM pattern for PAL delay line and demodulator adjustments, gray scale with multiburst, besides the geometric patterns for linearity, cross hatch and dot patterns for convergence adjustments.

A modern colour pattern generator like the *Philips 5515* is a microprocessor controlled pattern generator with RAM and PROM memory, a combined RAM/port for input and output operations and CITAC for adjustment and control of the VCO in the RF unit. A *master oscillator* generates the reference frequencies for PAL/SECAM or NTSC, that clock the sync pulse generator. The *sync pulse generator IC* locks the subcarrier, generates the H and V pulses and their subharmonic pulses which control all the circuits in the instrument. The horizontal elements and the vertical elements in all the *B/W* digital patterns are generated in the line pulse generator and the subsequent gate circuitry. All the information for the colour patterns, e.g. saturation steps, vector information are stored in the *pattern PROM.* The pattern *control register* controlled by the data, clock and strobe lines selects the actual pattern out of the PROM. The luminance data is fed to the video output path while the chrominance data are further processed in the PAL/NTSC and SECAM encoder.

The *PAL/NTSC encoder* has a VCO to generate the colour subcarrier frequency as per the TV system required, phase locked to the SPG signals. The chroma data consisting of the colour components, the saturation step signals and the vector information together with PAL/NTSC burst are applied to the UV matrix and the quadrature modulator as per PAL/NTSC encoding requirements. In the SECAM mode, the colour bar and grey scale signals are weighted, summed and then fed to the encoder which produces sequential, FM colour signals as required.

The *multiburst generator* comprises a counter, summing amplifier, current source for a triangle generator, a sine shaper and a start/stop circuit. The circuit delivers 8 packages of sine waves. A counter with subsequent summing amplifier delivers a staircase signal which is converted into packages of sinewaves with stepwise increase in the *frequencies from 0.8 to 4.8 MHz.* Control signals for circle and multiburst signal generation are fed from the SPG.

Applications of the Test Patterns

The various test patterns provided in a typical colour TV pattern generator like Philips PM 5515 can be used for alignment of a TV monitor, receiver or VCR as follows:

1. *Circle* on a black background is used to check the overall linearity and geometry. White circle on black is useful to check framing, while black circle on white is more suitable for checking reflections.

2. *Centre cross/Border lines* provide checks for centering the TV screens, deflection linearity and pincushion correction.

3. *White pattern 100%* with colour burst are used to check colour purity, and adjust the maximum beam current.

4. *Dot pattern* provides check for static convergence and focus. All dots should be pure white. Presence of coloured dots indicate the need for adjustment of the convergence magnets and focusing if necessary.

5. *Crosshatch/Center incation* with 12×17 lines (13×17 for PAL-M and NTSC) is used for checking and aligning dynamic and corner convergence, pin cushion correction, E/W-N/S correction in 110 deg colour TV receiver.

6. *Checkerboard pattern* of 8×6 squares provide a visual standard for basic picture tube alignments like centring, focus, *H* and *V* deflection, linearity, framing, aspect ratio. Bandwidth can be guaged by B/W transitions.

7. *Grayscale* is a full screen linear staircase signal with 8 equal steps from black to white that can used to locate faulty linearity of a video amplifier or grayscale setting. A colour in the gradation bars indicates the need of gun adjustments.

8. *Multiburst* contains 8 full screen vertical bars of resolution lines in the frequency ranges 0.8, 1.8, 2.8, 3.0, 3.8 and 4.8 MHz. They provide direct check on the video bandwidth as well as resolution.

9. *VCR* is used to check the bandwidth, linearity, amplitude sensitivity and AGC of the chroma amplifiers in the VCRs, by means of a combined test pattern consisting of:

— horizontal 100% white bar covering 1/6 field for exact level adjustment
— 8 bars of resolution of which 2.8–3.0–3.2–3.4 MHz are used to align the HP filter for maximum resolution of the VCR bandwidth.
— 8 steps of decreasing linear levels of saturation from 100% to 0% to check the chroma amplifier linearity and colour AGC circuitry.
— the bottom section consists out of a black horizontal bar with a moving white field to check VCR on moving pictures.

10. *Colour bar* pattern contains the standard vertical colour bars, providing overall check on the colour performance, including checks on burst keying, subcarrier regeneration, RGB amplifiers, delay colour against B/W signal and saturation check.

11. *DEM pattern* Demodulator is a combined test pattern which is divided into 4 horizontal sections. The signal content depends on the selected TV system.

For *PAL*, the *first section* consists of two horizontal bars in a line. Bar one contains R–Y and B–Y information with G–Y as zero. The adjacent bar to the right is a reference bar with no colour information, only 50% luminance or Y signal. The *second section* consists of 4 coloured squares with colour information according to the Fig. 21.12.

The 1st and 2nd squares are PAL-coded. This section indicates the proper functioning of the colour demodulator section. The third section consists of 4 squares which are colour coded but should not show any colour on a well aligned colour TV receiver or monitor, all the 4 squares should be gray.

Both R–Y squares are NTSC coded—the R–Y signal does not change direction 180° each line. The burst signal is PAL coded and ensures normal operation of the PAL switch in a colour receiver. The 3rd and 4th square contain only B–Y signal information, alternating 180° each following line.

The *third section* consists of 4 squares colour coded for *Delay line check* for the alignment of the 64 μs chrominance delay in amplitude and phase. All the 4 squares should be gray. The squares 1 and 2 are NTSC coded, while the squares 3 and 4 contain only B–Y signal information, alternating 180° phase every alternate line. *Hannover bars/Venetian blinds* in the picture indicate the need for these adjustment. By observing where this effect appears, it is possible to distinguish between the amplitude and phase faults. As the R–Y signal in the square 1 and 2 are NTSC coded, the delay line and PAL switch should

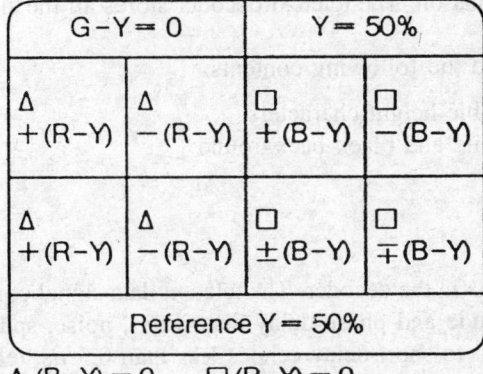

G−Y = 0		Y = 50%	
Δ +(R−Y)	Δ −(R−Y)	☐ +(B−Y)	☐ −(B−Y)
Δ +(R−Y)	Δ −(R−Y)	☐ ±(B−Y)	☐ ∓(B−Y)
Reference Y = 50%			

Δ (B−Y) = 0   ☐ (R−Y) = 0

Fig. 21.12   PAL DEM pattern in PM 5515 (PHILIPS)

eliminate all R−Y information. This is because this information in successive lines of the first two squares is being subtracted.

If an *amplitude error* occurs between the direct and delayed signals, the subtracter output of the delay line will produce R−Y information in square 1 and 2. The action of the PAL switch will cause the information to be inverted on alternate lines to give the Hannover bar effect. If a *phase error* occurs between the direct and delayed signal, the Hannover bars will show up in the squares 3 and 4. They will also appear in the yellowish horizontal bar of the upper left section of this test pattern.

Demodulator check   This pattern is useful for checking faults in the chroma demodulators of the colour TV. The subcarrier should be applied to the R−Y and B−Y demodulators in the correct phase, otherwise there will be colour in all the 4 squares.

If the subcarrier fed to a demodulator has correct phase, the R−Y demodulator will demodulate only the R−Y information and the B−Y unit will detect only the B−Y information. But if the subcarrier has a phase error, this would result in the R−Y information passing the B−Y demodulator showing up colour in the 3rd and 4th square. Similarly, R−Y demodulator could receive B−Y information and this would be seen as colour information in the 1st and 2nd square. It will be thus clear that in the event of a *phase fault* in general, the phase error in both the demodulators outputs incorrect information showing up as *colours in all the 4 squares*.

The *fourth section* at the bottom contains reference bar with 50% luminance *Y* signal only, for PAL B, D, G, H, I, N.

12. *Purity* can be checked by the three primary colour patterns available. The electron beams should strike only one set of phosphor dots or stripes on the screen. The red pattern is primarily used for the purity adjustment of the central beam in PIL tubes, followed by the adjustments of the outer beams with green and blue patterns.

These patterns can also be used to check *interference* between chroma and *sound carrier*. The patterns have 75% saturation setting and hence can be used to align the writing current of the VCRs.

Teletext   PM 5515-T and PM 5515-TX provide selection of 5 pages of teletext with special contents for teletext decoder testing as well as 'wall paper' test pattern which gives out a composite video waveform with a *pseudo-random teletext data* pattern on every line. Teletext is a technique of sending text information to TV receiver during the vertical blanking interval without disturbing the normal television program. One row of a page is sent with one line blanking so that for each page of 25 rows of 40 characters each needs 24 TV lines. It takes some 0.25 seconds to transmit a page if 2 lines per frame, viz. lines 22 and 335 are used to carry the information. The teletext decoder stores all the signals associated with the selected page in the decoder RAM.

The five *teletext pages* can be generated to provided the following contents:

p. 100 index, decoder text, concealed information, double height characters
p. 101 character set, clock cracker, graphics, flash, white and black background
p. 102 news flash
p. 103 clock cracker
p. 104 VIDEO TEXT

The 'wallpaper' pattern is suitable for sync measurements in the decoder. The teletext data signal consists of high speed pulses and transients sensitive to amplitude and phase delay distortions, noise, spurious pulses and ghosts. While the TV signal is more tolerant to short delay echoes less than 0.5 μs, teletext is affected by long delay and short delay echoes. Positioning of TV antenna needs attention and fine tuning of the TV receiver should be within +/– 50 kHz. Teletext decoder like Philips SAA 5020/5030/ 5041/5051 requires the following adjustments with the aid of the signal, as per specific decoder.

1. Frame synchronisation of the timing chain.
2. Oscillator frequency.
3. Clock coil adjustment to form a clean clock. An 'eye diagram' is viewed using the 'wallpaper' test pattern.
4. Auto-interlace to get noninterlaced scanning.
5. Amplitude of the teletext characters—intensity.

Vectorscope

A vectorscope is a specially designed oscilloscope to display on its screen, the chrominance signals in a vectorially, with their *amplitudes and phase angles in a polar plot.* A precise chroma decoder is used to derive the B–Y and R–Y signals and represent them along the x and y axes. The colour bar signals are displayed vectorially to check their amplitudes and phases. The screen is calibrated for the chroma vector amplitudes in IEEE units and the phase angle in degrees. It provides precise checks on the amplitudes and phases of colour bar signals. A vectorscope can be set up to lock in the decoder on the burst signal of one source while displaying the colour bar vector signals from other sources to compare the phases and match them. A colour bar display on a vector display is shown in Fig. 21.13.

The colour vectors are seen as bright spots because the beam stays in those positions for the duration of the respective colour bars. During the change over from one colour bar to another the display may show off the rapid amplitude and phase changes by curved line loops.

Swept Frequency Aligrment of Tuner and VIF Stages

The general methods for RF–IF and video bandpass alignments using a wobbuloscope and the adjustments of deflection circuits have been discussed earlier. The special factors dealing with a colour TV receiver are briefly dealt with here. The measurements are generally made at an ambient temperature of 25° +/– 3°C, after a warm-up period of 15 minutes. Impedance matching pads are inserted at the input wherever necessary. A 1C:1 attenuator probe should be used at the CRO input to prevent circuit loading when observing waveforms at high impedance points in the video circuits.

Testing and Alignment of the tuner Testing of a tuner involves the following measurements:

(a) Power gain
(b) Bandwidth

Fig. 21.13   Vectorscope display of colour bar display

  (c) AGC
  (d) Noise figure
  (e) VSWR
  (f) Oscillator radiation

As it is difficult to measure all parameters, power gain, bandwidth and AGC are checked for acceptance tests. All other parameters are checked occasionally to check on the quality of the tuners on a random sample basis.

Power gain: For the sake of measurement the input and output impedances are made equal and the voltage gain measured to obtain the power gain. The AGC is disconnected and set to recommended value externally. The sweep input at the desired channel is fed directly to the detector probe to set a reference line. The tuner with matching pads if necessary for impedance matching, are now inserted and the sweep attenuator set such that the peak coincides with the reference line. The power/voltage gain is given by the difference between the attenuator settings. The small attenuation of the matching pads found by direct connection between the pads, should be added to this to obtain the exact gain.

Bandwidth: This can be measured on a wobbuloscope for the 3 dB points by means of the markers.

AGC characteristics: The tuner is energized with required power supply and appropriate vision carrier from a signal generator is fed to it through an RF attenuator. The output is monitored on an RF millivoltmeter. The AGC voltage is set for maximum gain. The input signal is set at − 50 dB level and the output noted.

The input is then increased gradually and output level is maintained at the same level by adjusting the AGC voltage. The input signal level for which the output cannot be controlled by variation of the AGC voltage is noted. The difference of the two input levels is the AGC range.

Noise figure: The tuner is properly tuned for the required channel to give correct IF output at 38.9 MHz. A noise generator (R & S/Elina) is connected to the input and the output monitored on a sensitive wide band oscilloscope through a detector probe. The noise generator is kept off, and the input is terminated by its own impedance. The AGC is adjusted for maximum gain. The noise generator is switched on and the noise input level gradually increased until the noise power output is doubled. The increase in noise generator output (log ratio) is the noise figure. Some noise generators (e.g. Elina make) are calibrated for the noise figure. Alternatively, noise meters may be used to monitor the output which give direct noise figure reading.

Alignment of a Colour TV Receiver
Alignment of a modern Colour TV receiver typically involve the following adjustments. These can be carried out by use of a colour pattern generator and a CRO.

1. *dc supply voltages*: The voltages supplied by the SMPS are set to predetermined values by setting the duty cycle adjustment potentiometer, usually done with colour bar signal applied and the receiver brightness and contrast set to minimum.

The dc supply voltages at other *test points* supplying to the various subsystems like tuner, VIF subsystem, vertical stage, AF section etc. are checked as per manufacturer's instructions.

2. *VIF alignment*:
(a) *Detector coil*: The input circuit to the base VIF preamplifier is disconnected and a staircase or bar pattern modulated IF from the TV pattern generator is fed to the base through a 10 nF capacitor. The synchronous detector coil is aligned for minimum black to white amplitude at the VIF subsystem output monitored on a CRO. This corresponds to the optimum efficiency point of the detector.
(b) *AFT coil*: Recommended (7.5 V) bias is applied to the AGC terminal of the VIF module. AFT switch is closed and with no signal applied to the input, recommended 6.5 V are applied to the AFT pin of the module. The switch is then opened, external IF-AGC bias is removed and 38.9 MHz VIF is fed at the input. The AFT coil is adjusted to get the same recommended voltage 6.5 V at the AFT pin.
(c) *Tuner-VIF*: The tuner is then connected to the VIF module, recommended AGC bias of 7.5 V is applied and RF channel modulated with multiburst pattern fed to the tuner input. The tuner IF coil and the input circuit coil are aligned to obtain good bandwidth as observed in the multiburst video output on the CRO.

RF AGC delay pot is adjusted for tuner AGC begins to rise to take-over at an aerial input of 1 mV and above.

3. *Focus and EHT adjustment*: With the crosshatch pattern displayed at normal brightness and contrast, the focus potentiometer as adjusted to provide best overall focus.

EHT Regulation: At zero beam current the EHT is adjusted not to exceed 25 kV. Variations in the EHT voltage at zero and maximum beam current should not be greater than 15%.

4. *Horizontal and vertical frequencies*: A crosshatch plus circle pattern is fed to the receiver. The video signal input to the synchronisation IC is shorted to ground temporarily and the free running frequencies of the horizontal and vertical oscillators are adjusted to get a stationary or slowly floating picture.

5. *Picture geometry*: With the crosshatch pattern on the screen, the horizontal *linearity coil* is adjusted for best linearity and horizontal centering is set by *position* phase shift control associated with the synchronisation IC. *Width* may be set if provided for in the circuit. The *vertical height and linearity* controls in the vertical drive circuit are adjusted for correct raster size and linearity.

6. *SIF coil*: The discriminator coil of sound IF is set for largest undistorted AF output.

7. *SIF trap*: The sound IF series trap in the feed line to the chroma IC is adjusted for maximum SIF level. A 10:1 attenuator probe should be used to avoid loading due to CRO.

8. *Colour subcarrier trap*: This shunt trap at the chroma subcarrier frequency in the feed to the chroma input to the decoder is adjusted to give for minimum amplitude as monitored on the CRO.

9. *Chroma bandpass*: The chroma bandpass circuit at the chroma input pin of the decoder is set to give maximum amplitude of the chroma waveform observed on a CRO, indicating maximum saturation in the colour bars.

10. *Reference subcarrier*: The free running frequency of the crystal controlled oscillator is adjusted by the trimming capacitor in series with the crystal for minimum colour beat frequency on the screen, after making the colour killer and the burst phase detector ineffective. The saturation control pin is shorted to dc supply voltage to disable the colour killer and the input from the delay line matrix to the burst phase detector is shorted to make it ineffective.

11. *PAL delay line matrix*: The DEM test pattern containing V, +/– U and (G–Y = 0) signals is applied. With brightness at minimum, saturation at maximum and contrast at acceptable level, the following adjustments are made.

(a) *Amplitude adjustment*: The delay line amplitude control at the input of the PAL delay line is adjusted so that the Hannover bars in the + V part of the pattern disappears. This adjustment makes the amplitudes of direct signal and delayed signal equal.

(b) *Phase adjustment*: The delay line phase controlled by the inductance at the *input* of the delay line, is adjusted to remove Hannover bars from +/– U and (G–Y) part of the pattern.

(c) *Burst phase adjustment*: The subcarrier inputs to (R–Y) and (B–Y) demodulators are ensured quadrature phase by adjustment of the inductance across the *output* of the delay line and associated trimmer capacitor if provided. The + V and +/– U parts of the pattern should be neutral gray.

12. *Black level (half tone) and monochrome adjustments*: These six controls are set for correct reproduction of black and white picture at high lights and low lights and thereby also ensure correct balance of colours. The procedure is followed as per manufacturer's instructions. Typical sequence is on the following lines:

Black level adjustments: The drive presets, the black level presets and the saturation control are all adjusted to minimum resistance positions. All the black level presets are adjusted to maximum (earthside stop), and all the drive presets to maximum amplification. The colour saturation pin is temporarily shorted to earth. The brightness control is adjusted to obtain 6.3 V at the brightness pin of the decoder. The screen grid voltage Vg_2 is set so that the screen just illuminates. The corresponding black level controls until the screen appears 'white'. This may be done more closely by disabling the vertical oscillator and observing the horizontal line. After the adjustment at least one control must be on the earthside stop. The Vg_2 control is turned back until screen just goes dark. The earth short at the decoder is removed.

Monochrome adjustments: The brightness and contrast controls and all the colour drive presets are set to maximum. The presets are then adjusted during a standard black and white transmission, for a 'white' (6500 K) picture free from colour stains or aprons. Grayscale tracking for low to high light is checked for a black and white picture. Should any colour predominate, the alignment is not proper and is repeated.

Production Testing

When a TV receiver is productionized it has to undergo tests which are classified under three groups:

(a) *Type approval tests:* This implies exhaustive testing and the receiver is subjected to all the ISI tests for approval.

(b) *Acceptance tests:* These consist of 13 of the type approval tests, and a certain number of the TV receivers have to comply with the ISI specifications during these tests. These test procedures have been comprehensively given in the ISI standards (Refer to publication IS-4545-1968 of the Indian Standards Institution for details).

(c) *Routine tests:* These tests have been identified by the ISI in such a way that the batch subjected to these routine tests will pass through the acceptance test under an agreed sample plan.

Functional Tests

Six functional tests are conducted on each of the TV receiver before they pass the testing stage. The measurements are conducted on at least one channel each in the band I and band III. The receiver is fed the RF modulated signal from the TV pattern generator through an RF attenuator and the output measured by a CRO and a distortion analyzer. The six functional tests are:

1. Sync limited sensitivity: The sync limited sensitivity of a TV receiver is the minimum input signal required by the receiver, for which the synchronisation just loses control, and the picture quality becomes unacceptable. The receiver is fine tuned to the channel under test to setup a standard picture with a standard pattern. The RF input signal is attenuated gradually in steps and the level in dBm, at which the picture loses synchronisation is taken as the sync limited sensitivity. During measurement, the receiver controls except the fine tuning may be readjusted to obtain a stable picture.

2. Noise limited sensitivity: The noise limited sensitivity of a TV receiver is the input required for producing a standard video output with a S/N ratio of 30 dB. The S/N ratio is expressed as the ratio of the peak-to-peak black to white video signal, to rms noise at the picture tube cathode. An RF channel signal with 50% modulation is applied to the receiver input terminals. The receiver is tuned to the selected channel and set for standard video output (typically 40 Vpp), observed on a CRO. The RF input at is now decreased in steps and the peak to peak noise measured on the CRO. The input level at which the P-P noise is 1/5th of the video output gives the noise limited sensitivity of the receiver for that channel. This is because the rms noise is 1/6th of P-P noise and hence the ratio corresponds to 30 dB.

3. Colour sensitivity: The colour sensitivity of a colour TV receiver is the level of the input signal applied to the receiver at which the colour decoding circuits cease to operate resulting in only black and white picture. Using the colour bar pattern the receiver controls are set for a standard image. The input signal is then reduced in steps, each time resetting the controls for optimum performance. The level at which the picture colour quality is unacceptable or the receiver reverts to monochrome operation.

4. AGC range: AGC range is the range of input signal variations required for 6 dB variations in the output. Ratio of the maximum usable input signal to the input level at which the video output with reference to the video level of maximum usable input signal, decreases by 6 dB, gives the AGC range of the receiver.

5. *Maximum usable audio output (MUAO)*: MUAO is the output power available at the loudspeaker for a standard input signal of – 50 dBm with 1 kHz modulating signal and +/– 15 kHz deviation for sound carrier. The audio output power is adjusted with the volume control for a maximum of 10% distortion. The corresponding power output is the maximum usable audio output.

6. *Synchronising range*: The synchronising range is the frequency range over which the sync signals are able to control the frequencies of the vertical and horizontal time base circuits. The frequencies of the time base are measured in corresponding free running condition by electronic counter to find the hold and lock-in ranges. *Lock-in* range is the frequency range between two points at which the synchronization is achieved, while the *hold range* is the frequency range between the two points at which synchronisation is lost.

The receiver is adjusted for normal operation with a standard TV signal and sync controls manipulated to observe locking of picture. As a sync control is rotated from one extreme end to the other, a point is reached where the time base (vertical or horizontal whichever is controlled) first locks in to give a stable picture. As the control is further rotated, another point is reached where the synchronisation is lost. The sync control is rotated in the opposite direction to get two more similar points (which do not coincide with the previous points). The four points are noted and the corresponding free running frequencies by electronic counter.

Soaking Test and Life Characteristics

In the soaking test the receiver after undergoing alignment and testing, is kept ON for over 24 hours. The test reveals early failures in the so-called the infant mortality period or debugging period. In a large sample of production, the failure rate considered over the life span of the equipment is much larger in the initial period. This may be due to faulty manufacture—poor soldering or jointing, poor seals, cracks, surface contamination etc., transportation or installation errors. If after the soak test the receiver shows up failure or malfunctioning the same is debugged and the receiver checked and realigned before delivery to the customer.

During the normal *useful life span* of the receiver the failure rate is low and relatively constant. It is characterized by stress related failures, caused by excessive temperature and voltage levels, humidity, vibrations, shock also contributing to it. At the end of useful life span is the *wearout period* where the failure rate starts increasing rapidly due to wearout of the components and devices. The wearout failures are due primarily to deterioration of design strengths of the devices, as a consequence of continuous operation and exposure to environmental fluctuations.

21.11   TROUBLE SHOOTING COLOUR TV RECEIVERS

Broad principles of trouble shooting a faulty receiver outlined in section 21.9, for a monochrome TV receiver apply equally well to a colour TV receiver. Locating faulty component or integrated circuit is facilitated by the measurements and waveforms monitoring at standard test points in the diagrams provided by the manufacturer. The operating conditions of the various ICs should at hand to assess any faulty IC. Alignments of the various stages should not be normally disturbed without the data about alignment information.

A colour receiver must first operate in the monochrome, provided convergence or purity adjustments on the CPT yoke have not been disturbed. If *no colours* are coming in the picture, fine tuning/AFT adjustment should be checked. The chroma decoder should then be checked for the voltages, sandcastle pulse and realigned if necessary. If there is a colour tint on black and white picture, monochrome adjustments in the video drive amplifier should be done. The colour amplifier video transistors should be checked if one of the colour is missing. Specific *tinted colours* or appearance of *Hannover bars* in the picture indicate improper amplitude and phase adjustments in the colour decoder matrix. Intermittent colour can be due to nonsynchronised colour subcarrier, requiring check up in the APC and ACC circuits. It should be remembered that antenna input signal bordering on the colour sensitivity of the receiver can also be the cause of intermittent colour.

21.12   FAULT LOCALIZATION TABLE

The following table lists typical fault situations in the form of indications on the raster, picture and sound, in a TV receiver from which the defective stage may be deduced and the fault localized.

Table 21.1   Fault Localization Table

Raster	Picture	Sound	Defective stages
No raster	No picture	No sound	Power supply circuit Filament supply circuit DC rectifier circuit
No raster	No picture	Sound O.K.	Video output circuit Picture tube, bias circuit & socket connections Horizontal drive, EHT circuit
Horizontal line	No picture	Sound O.K.	Vertical deflection stages Vertical deflection coil
Vertical line	No picture	Sound O.K.	Horizontal deflection coil
Raster O.K.	No picture	Sound O.K.	Video amplifier Picture tube
Raster O.K.	No picture or weak picture	No sound	AGC stage-adjustment Tuner stages gain Video IF stages gain Antenna circuit
Raster O.K.	Picture O.K.	No sound	AF amplifier, loudspeaker Sound IF, demodulator Sound IF alignment
Raster O.K.	Picture torn slanting streaks diagonal bars	—	Horizontal oscillator Horizontal sync, AFC
Raster O.K.	Picture rolling	—	Vertical oscillator
Raster O.K.	Picture drifts torn & rolling	—	AGC stage, video stage Sync separator, Noise inverter
Raster distorted	Picture O.K.	—	Deflection stages
Vert. compressed	—	—	Vert. height, linearity Vert. deflection drive

(Contd.)

Raster	Picture	Sound	Defective stages
Horiz. compressed	—	—	Horiz. width, linearity
			Horiz. deflection drive
Tilted raster	—	—	Yoke clamp
Vert. foldover	—	—	Vert. drive & amplifier
			Vert. linearity
Horiz. foldover	—	—	Horiz. width & linearity
			Horiz. oscillator drive & amplifier
Keystoning	—	—	Deflection coils partial short-circuit
Intermittant raster	—	—	EHT, Horiz. oscillator, picture tube-loose connection
Raster bent vertically	humbar	—	Mains leakage, ripple on sync in sync separator
			Video output heater-cathode short
Raster edge pull with ripple	Horiz. hum bar on picture	sound with hum	Low supply voltage
			faulty filter condenser
do	do	Sound O.K.	Heater cathode short in picture tube
			Hum in video due to pick-up or ripple in $B+$ supply, decoupling filter
Raster enlarged	Picture O.K.		Horiz. output stage
	Picture blooming		Low EHT, EHT regulation circuit
			HV rectifier tube
Raster O.K.	Picture jittery	—	Horiz. oscillator
			Noise inverter
			Ext. impulsive interference, RFI suppressor, antenna, feeder
Raster O.K.	Excessive contrast with disturbed sync		Excessive signal
			AGC stage-adjustment
			Gassy video amplifier or picture tube
			Supply voltage high
Raster O.K.	Negative picture		AGC stage
			Video signal overload
			Picture tube
			Detector diode polarity
Raster O.K.	Excessive brightness	—	Video amplifier
			Picture tube bias circuit
			EHT adjustment
Raster O.K.	Snowy picture weak picture	Sound O.K.	AGC stage-adjustment
			Weak signal, antenna defect
Raster O.K.	Diagonal lines in picture	Sound O.K.	Oscillations in RF amp.
			RF interference pick-up
Raster O.K.	Thin closely spaced vert./slanting lines Moire patterns	Sound O.K.	Sound trap
Raster O.K.	Sound bars	Sound O.K.	Mistuning of tuner-oscillator
			Video IF amp. misalignment
Raster O.K.	White vertical bar/ fold at centre	Sound O.K.	Horizontal drive
			Booster circuit
Raster O.K.	A few vertical bars on LHS	Sound O.K.	Horizontal output stage damper circuit

Raster	Picture	Sound	Defective stages
Raster O.K.	Dark picture with trailing smear	Sound O.K.	Video stage, video IF response
Raster O.K.	Loss of detail	Sound O.K.	Video stages, HF peaking coils Video IF response
Raster O.K.	Multiple bars/ghost lines on LHS of sharp vert. line contours	Sound O.K.	Video stages peaking coil damping resistances Overpeaked video IF

REVIEW QUESTIONS

21.1 Discuss giving the block schematic the facilities provided for in a wobbuloscope and explain its use in alignment of RF tuner, video IF amplifier and sound IF amplifier.

21.2 Explain how markers are produced in the swept frequency response curve.

21.3 Draw the block diagram of a service pattern generator to explain how the sync and blanking signals are generated and combined with the pattern signal. How are the different patterns—horizontal bars, vertical bars, cross-hatch and chessboard obtained?

21.4 How are the following defects in a receiver indicated on a cross-hatch pattern?
   (a) Poor HF response
   (b) Peaky HF response
   (c) Poor LF response
   (d) Poor LF phase response.

21.5 What are the checks for the following in the Test Resolution chart?
   (a) Vertical resolution
   (b) Horizontal resolution
   (c) LF response
   (d) Interlace
   (e) Raster size
   (f) Horizontal and vertical linearity.

21.6 Explain the likely fault or the faulty stage in a TV receiver, indicated in the following symptoms.
   (a) Picture snowy
   (b) Picture torn into slanting bars
   (c) Excessive picture contrast
   (d) Excessive brightness and poor contrast
   (e) Height compressed
   (f) Width compressed
   (g) Foldover at the centre of the picture
   (h) Foldover on the left hand side of the picture
   (i) No picture, raster on but sound O.K.
   (j) Vertical edges wavy, hum in sound
   (k) A bright spot appears at the centre on switching off.

21.7 What will be the result of the following faults on the raster, picture and sound?
   (a) Video IF supply voltage very low
   (b) AGC voltage excessive
   (c) AGC voltage too low
   (d) Fine tuning not proper
   (e) H oscillator drifted away
   (f) H oscillator failed
   (g) Video output tube gassy
   (h) Booster diode gassy
   (i) EHT rectifer diode (solid state) connected with wrong polarity
   (j) Video detector diode connected with polarity
   (k) Peaking coil lead open with its damper resistance in circuit
   (l) Sound trap in the video misaligned
   (m) Sound trap in the video IF amplifier misaligned.

MULTIPLE-CHOICE QUESTIONS

21.1 Linearity test chart provides test for:
(a) camera linearity
(b) monitor linearity
(c) overall linearity
(d) none of these

21.2 Pick out the false one:
(a) vertical wedges in the test chart indicates the vertical resolution in the picture
(b) streaking and smear is caused by phase distortion in the frequencies below 100 kHz
(c) horizontal resolution of 360 lines corresponds to video frequency response of 4 MHz
(d) diagonal lines in the chart provide test for good interlacing.

21.3 Multiple outlines in the picture edges are indicative of:
(a) excessive bandwidth
(b) low gain at high frequencies
(c) excessive gain at high frequencies
(d) a and c

21.4 No raster but sound O.K. in a TV receiver, indicates fault in:
(a) video stages
(b) sync separator
(c) horizontal drive stages
(d) horizontal output stage

21.5 A snowy picture with contrast is caused by fault in:
(a) the AGC circuit
(b) antenna circuit
(c) the tuning circuit
(d) b and c

21.6 Picture size increases when brightness is increased. This may be due to fault in:
(a) power supply rectifier
(b) EHT circuit
(c) CRT biasing circuit
(d) a and b

21.7 Intermittant colour in picture is caused due to:
(a) fault in ACC and APC circuit
(b) weak TV signal at the antenna input
(c) loose connection in the video output drive circuit
(d) a and b

22

CCTV, VCR and Video Disc Systems

Design and constructional aspects of CCTV systems are discussed in this chapter. Block schematics of the systems are described to highlight the important features of video processing, sync generator and video monitors. Operation of a portable video tape recorder is also given, as it forms a common accessory to modern educational CCTV systems.

22.1   CCTV SYSTEM DESIGN FEATURES

Closed circuit TV systems are extensively used for industrial applications, security and surveillance, education and training, public information displays, and so on. Closed circuit TV systems used in some of these applications are preferably characterized by reduction in size complexity and cost. Their performance need not be of the professional grade broadcast standards. Generally a vidicon camera tube (1/2 in. – 1 in.) is used although in some industrial and medical X-ray applications, plumbicon, or in high light situations, multidiode silicon vidicon is used in modern cameras. Solid state circuitry is now common in the camera units and the display monitors as far as possible. Considerable simplification is possible by a random interlace system and less rigorous sync signal waveform, referred to as 'industrial sync'.

Random Interlace

Systems with random interlace have no master sync pulse generator for controlling the horizontal and the vertical scanning. The horizontal scanning and the vertical scanning are driven by independent sources. In standard systems, the vertical rate is obtained from a master oscillator from which the vertical as well as the horizontal rate is controlled. The vertical rate is derived from the horizontal rate through a chain of dividers. Introduction of equalising pulses and the serrations in the vertical sync pulses involve somewhat complex circuitry and elaborate pulse shaping. This is avoided in simpler industrial grade systems in what is known as 'Random interlace' systems, since in such systems, the horizontal and vertical deflection drive signals have no well defined phase relationship as such, though the two are in the particular ratio of 15625:50. Since vertical frequency is 50 Hz, which is the same as the readily available mains frequency, it is used with some wave shaping to trigger the vertical oscillator directly, while the horizontal oscillator is a crystal controlled astable multivibrator. The positioning of the two fields may not remain fixed and the line spacing may vary in a random fashion, which gives the name *random interlace*.

Another simplification resorted to in some industrial television systems, is to do away with separate sync pulses and use blanking pulses of levels up to the sync tip level for synchronisation. In practice, even with master oscillator controlled horizontal and vertical sync, with the vertical rate derived from countdown of the horizontal rate, stable interlace is difficult to achieve. During the vertical blanking, the lack of horizontal sync may drift the horizontal oscillator away before it is pulled back by the horizontal sync pulses into synchronisation again. This causes a curved type of picture shift at the top, what is called monitor pulling. The picture may also show a horizontal jitter at the top left appearing as flag waving.

In construction of CCTV cameras, two types of arrangements are possible as discussed below:

(i) Self-contained monoblock cameras The camera tube and all associated circuitry are contained in one single unit generating the composite video signal at the standard level, for feeding it to the video monitor distribution system. The deflection unit scanning drive, video processing circuitry, sync and blanking mixer may all be contained in it, or there can be a provision for external genlock pulses drive from a remote located master sync pulse generator (SPG).

(ii) Remote operation type cameras In this type of camera construction, the bulky part of the camera circuitry like power supplies, deflections drives, protection circuitry may be contained in a separate control unit linked with the camera head unit by a multicore cable of the desired length which may extend to hundreds of metres. This makes the camera head unit very compact and can then be placed in more rigorous environments of temperature, vibrations, humidity, underwater locations, etc. The camera head unit contains a vidicon tube with video signal amplifier and decoupling filter circuitry. This unit is usually without any controls so that it can be mounted or located at the ceiling, wall or a convenient place that may not be accessible readily. Motorized remote control may be used for pan and tilt control of the camera viewing axis. The optical lens system of some cameras may also have motorized mechanism for remote control of focus and aperture. The camera control unit can be located at a convenient place along with a viewing monitor.

Synchronisation Systems

When several cameras are used for monitoring scenes at different locations, all of them have to be synchronised by a common SPG signal which is looped through the various cameras and terminated finally in the cable impedance of 75 Ω, as shown in Fig. 22.1.

At each camera, the sync pulse signal is sampled through a high impedance circuit that provides sync pulses to the camera sync and blanking pulse shapers and the deflection drive circuits, without affecting the loop.

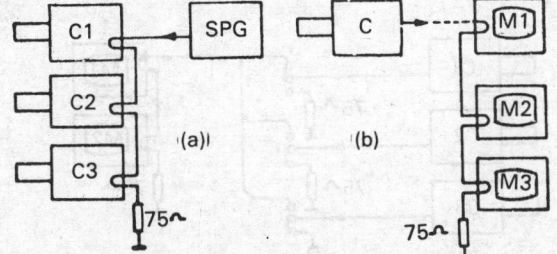

Fig. 22.1 Loop through techniques: (a) Sync pulses looped through cameras, (b) Video signal looped through monitors

Signal Distribution

When the signal from one camera is to be displayed on a number of video monitors, is it taken over a coaxial cable and 'looped through' each monitor, and terminated at the end by the cable impedance as shown in Fig. 22.2. When the monitors are located at far off distances, the output from camera may undergo considerable loss. It is, therefore, amplified by a video distribution amplifier to provide for multiple distribution of the video signal to a number of monitors. The signal may be supplied to each monitor individually or looped through a group of close-by monitors as shown in Fig. 22.2.

The signal undergoes attenuation as it propagates over the cable length, the loss increasing at higher frequencies due to the cable capacitance and increased loss at high frequencies. Whether this should be compensated by equalising amplifiers or not, depends upon the resolution quality requirements. A high quality cable with lower loss may be found more economical than adding an equalising system to a lower cost cable that has higher loss.

Long coaxial cable lengths between drive distribution amplifier and receiving monitors may cause hum to appear in the video signal by formation of earth loops via the cable shield conductor. The hum is best avoided by running a separate ground between the two ends. Whatever hum is introduced in the signal, may be eliminated by a clamping amplifier that clamps the sync pulse tips to a reference voltage. Differential amplifier technique

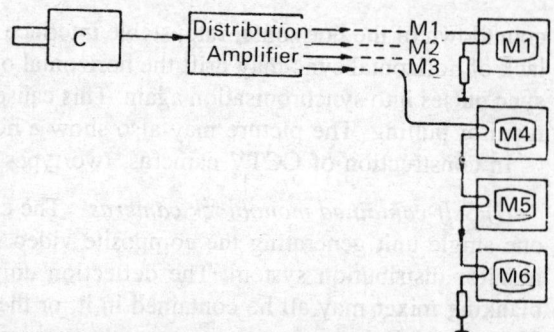

Fig. 22.2 Signal distribution by video distribution amplifier

may also be used to reject the hum which is a common mode (i.e. in phase) signal to the differential input terminals of the amplifier and is, therefore, suppressed.

Camera Selector Switch

When a number of video cameras are used for monitoring at different locations it is desired to have a choice of signals from the different cameras or other video sources, a video switching facility is necessary. In CCTV systems for industrial and surveillance applications, a simple mechanical camera selector switch is usually satisfactory, as switching is not frequent and momentary disturbances in the picture during switching can be tolerated. The switcher arrangement should connect the desired camera to the output line and terminate the unused camera outputs with proper cable impedance as shown in Fig. 22.3(a). This can be easily done by change-over switches operated by push buttons, with mechanical interlocking so that only one switch can be operated at a time. Mechanical switcher is simple, of low cost and does not require power but it is slow operating and causes switching disturbances in the monitors.

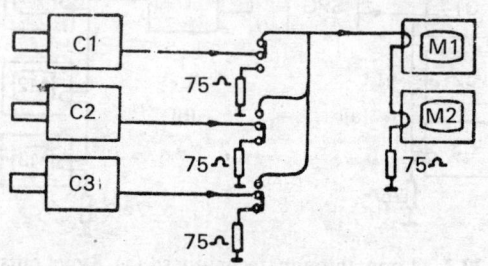

Fig. 22.3 (a) Push button camera selector switch

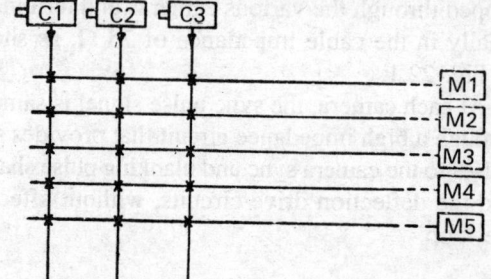

Fig. 22.3 (b) A 3 × 5 relay crossbar video switcher

Electromechanical Switcher

In an electromechanical switcher or *relay crossbar* as it is called, reed relays are used at the cross points of the matrix formed by camera output lines and the monitor lines as shown in Fig. 22.3(b).

The relays can be operated by remote control lines. Reed relays have fast operating time around 1 ms, reliability and long life. Relay switchers using them are, therefore, fast operating and can be arranged to enable switching during the vertical blanking interval to avoid the visible switching disturbance on the monitor. The relay at the cross-points of the camera cables and the monitors can be located in a central position and the relays can be controlled by remote control switches. The camera signal need not be routed to each destination but needs to be taken to the centrally located relay switcher. This effects considerable saving in the cable length loss and capital cost.

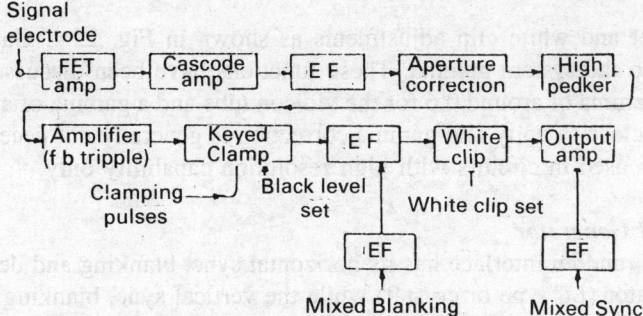

Fig. 22.4 Video processing of camera signal

Isolation amplifiers are necessary in the distribution lines to prevent transient changes and level changes caused by switching, from appearing at other monitors.

RF Modulation

In order to enable reception on commercial receivers, RF modulation facility is often provided to supply video signal from the camera system, on a standard broadcast channel via a cable. The cable again may be looped through the receiving stations. High impedance capacitive or resistive coupling to the cable or a strip-line directional coupler supplies the RF signal to the receiver, if necessary through booster amplifiers to make-up for the cable loss.

22.2 VIDEO CAMERA SIGNAL PROCESSING FEATURES

Simple monochrome closed circuit TV systems commonly employ 1 in. vidicon camera tubes. With the compact circuitry that is now possible for video signal processing and sync generation using solid state devices and integrated circuits, many modern monoblock TV systems have high grade performance features, with standard interlace scanning. Resolution better than 400 lines at 60% modulation and geometric distortions below 1 to 2% are possible. A 3 dB video bandwidth of 6, 8 or 10 MHz is aimed at depending on the system resolution. Modern vidicon tubes have a limiting resolution of 600 to 700 lines giving a life expectancy of 5000 to 10,000 hours. They are conveniently adapted for automatic light sensitivity control or compensation (ALC) to operate satisfactory over a light flux ratio of 2500 or more, without iris control.

Automatic Light Flux Compensation

This serves to regulate the target voltage and hence the sensitivity of vidicon. For this purpose, rectified picture signal provides a dc voltage which is amplified and added to the bias voltage adjustment of the target. Decrease in the signal amplitude due to decrease in face plate illumination is made to increase the target voltage to increase the sensitivity and *vice versa*.

Automatic Iris Control

In case of excessive light fluctuation which cannot be compensated any longer by automatic light flux control, automatic iris control is provided in some camera lens systems, that readjusts the iris by a drive motor controlled by the signal voltage.

Video processing of the vidicon signal involves amplification usually with a low noise FET input stage, aperture correction, and high peaking for improved resolution, dc restoration by keyed clamping circuit, black level set and white clip adjustments as shown in Fig. 22.4. Gamma correction may be introduced to match to the system gamma. These functions have been discussed in Sec. 5.9.

With the average gamma of around 0.6 for the vidicon tube and a gamma of about 2 for a picture tube, the system gamma is close to unity and gamma correction is generally not necessary in CCTV systems. Aperture correction is used in circuits with high resolution capability only.

Synchronising Signal Generator

A CCTV system with random interlace has its horizontal sync, blanking and deflection drive controlled by a horizontal oscillator (*LC* type or crystal) while the vertical sync, blanking and deflection drive are controlled by the 50 Hz mains frequency. The technique is similar to the sync and blanking pulse generation in a pattern generator discussed in Sec. 21.7, with the *H* and *V* pulse also controlling the deflection drive circuits for the deflection unit of the vidicon.

A standard interlaced CCTV system of the professional grade or studio type has its sync and deflection drive controlled by a sync pulse generator (SPG). In the sync pulse generator the basic timing is obtained from (i) a crystal controlled source at 2*H*, (ii) an external 2*H* source, or (iii) 50 Hz mains locking circuit, as shown in Fig. 22.5.

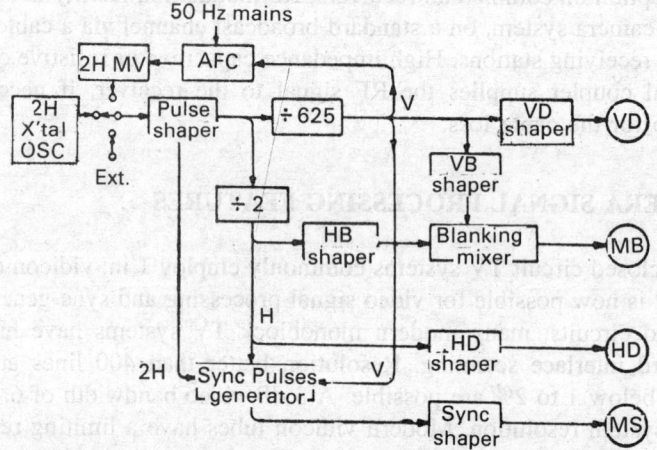

Fig. 22.5 Block diagram of a sync signal generator torr

The crystal oscillator frequency at 2*H*, viz. 31,250 Hz produces the basic timing for the equalising pulses, horizontal sync and blanking and vertical sync and blanking through suitable frequency dividers. In the mains frequency lock mode, the 2*H* multivibrator is AFC controlled by a phase detector circuit that compares the 50 Hz square wave derived from the 2*H* multivibrator through the divider chain, with the mains derived 50 Hz square wave to generate an error voltage that corrects the frequency of the 2*H* multivibrator to synchronise it with the mains frequency.

The *V* pulses obtained through the 625 : 1 frequency divider are shaped suitably to drive the vertical deflection and the blanking mixer. The *H* pulses obtained from a 2 : 1 frequency divider flip-flop drive the horizontal deflection and the blanking mixer through suitable pulse shapers, as shown in the block diagram of Fig. 22.5.

The 2*H*, *H* and *V* pulses are used for developing the vertical sync pulses with serrations, the equalising pulses and the horizontal sync pulses are then mixed with them.

The circuitry of a modern SPG employs mostly transistors or ICs. The basic building blocks required are frequency dividers, pulse shapers or stretchers, delay circuits, pulse adders and logic gates. In transistorized circuits the frequency dividers utilize flip flops as binary counters with appropriate feedback technique for obtaining the correct count down from the master oscillator at $2H$. The pulse shapers or delay circuits are monostable multivibrators or box-car circuits. A box-car circuit is a simple saturated transistor switching circuit, the base circuit, R-C timing of which determines the off state of the normally on transistor, after it is driven into off state by an input pulse at the rising or the falling edge depending upon the *pnp* or *npn* transistor circuit.

The HB and VB pulses are obtained with the help of pulse shapers driven by the H and V timing pulses. The HS pulses are obtained by a delay and a pulse shaper to provide the required front porch and the sync width. Obtaining the vertical sync signal containing five pre-equalising pulses, five serrated vertical sync pulses and the five post-equalising pulses is somewhat complex, but can also be simply realized by generation of a $2^1/_2H$ pulse, another $2^1/_2H$ delayed pulse of the same width and a $7^1/_2H$ pulse, all controlled by the V timing pulses. These pulses, used for gating the $2H$ rate timing pulses can along with suitable pulse shapers, readily produce the VS pulses which can be combined with the H sync to produce the MS pulses. In transistorized circuits, the vertical sync generator may be formed by a pulse shaper driven by $2H$ rate pulses, which can be made responsive to $2^1/_2H$ pulses such that it produces the narrow equalizing pulses during the first and the last of the three $2^1/_2H$ blocks while in the central block of $2^1/_2H$, it produces wider five vertical sync pulses, with the serration gaps in between. This can be achieved by altering the R-C timing of the pulse shaper by shunting the R with a transistor switch selectively during, the $2^1/_2H$ blocks.

22.3 SINGLE TUBE COLOUR CAMERA

Colour Dissector Systems

From the early days of colour television search for a methods of obtaining red, green and blue signals from a single camera tube has continued, to make the cameras compact, portable. Single tube cameras avoid the problems of registration of colours and equalisation of the signals by providing three matched sets of circuits. A number of colour dissector systems have been developed to separate the primary colours using a single pickup tube, with target split into tiny areas or stripes from which the primary colours could be read sequentially. A Sony system employs *vertical stripe filters* with or red, green and blue stripes, while a JVC system uses a sequence of green, cyan and white (clear) stripes. Some Panasonic and JVC cameras make use of *crossed or angular stripes* of yellow and cyan, or yellow and magenta combinations. In another system, the vertical stripes filter comprises stripes in a sequence of green, cyan and white (clear) colours. Each system has its own specific readout decoding to obtain the primary colour signals.

Trinicon

In this colour saticon pickup tube developed by Sony Corporation, a *built-in colour filter* consisting of sets of Red, Green and Blue stripes of uniform pitch is used to disassemble an object into an optical image of three primary colours. The 2/3 in. Saticon Mixed Field (SMF) Trinicon tube give a high performance close to that of three tube cameras. However, compared to three tube cameras in which all the light is diverted by the dichronic prism surfaces to the three colour sections, the incoming light is partially absorbed in the filters, limiting its use in higher illuminations.

If the electron beam scans the charge image formed on the target, due to light passing through the colour filters, the R, G and B colour signals are obtained one after the other from the target, by generating an electronic index signal at the target. This signal is used to demodulate the phase of the carrier frequency generated by the stripe filter. The *index signal* is produced by a set of electrodes which modulate the target voltage, line by line, and this in consequence modulates the video signal. The index signal is recovered by a *one line comb filter*. Fig. 22.6(a) illustrates how the signal is formed in the tube.

During phase separation, the colour signals are determined time elapsed from the starting of the scanning. The frequency component of each colour signal is the same, though the colour cannot be obtained directly. An *indexing signal* used as a phase reference to decode the multiplexed colour signal.

The *structure of target* for phase separation by indexing is shown in Fig. 22.6(b). The stripe filter consists of repetitive sets of red, blue and green primary colour stripes and is located near the face plate. It is arranged inside the tube face plate with a set of transparent conductive Nesa stripes behind a sheet of insulating glass layer as shown followed by the photoconductive target plate.

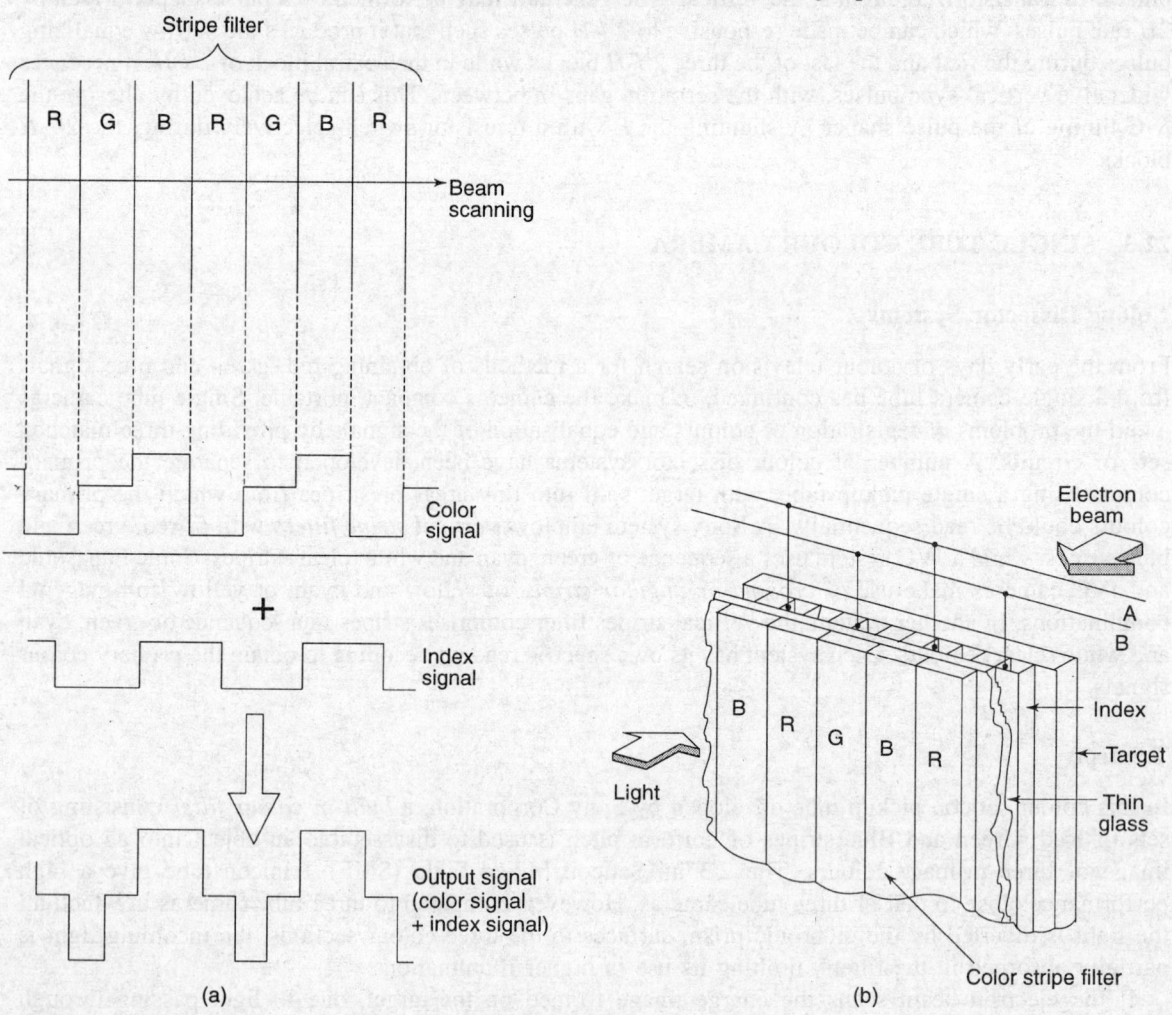

Fig. 22.6 (a) Phase separation by indexing, (b) Structure of target (*Courtesy:* Sony Corp.)

The *Nesa* film is composed of transparent conductive *comb-shaped stripes* as shown in Fig. 22.7. A set of R, G and B stripes corresponds to the indices A and B. An offset pulse (square wave) which is reversed every horizontal scanning period, as well as the DC target voltage *Vt*, is fed to indexing electrodes A and B through the drive transformer.

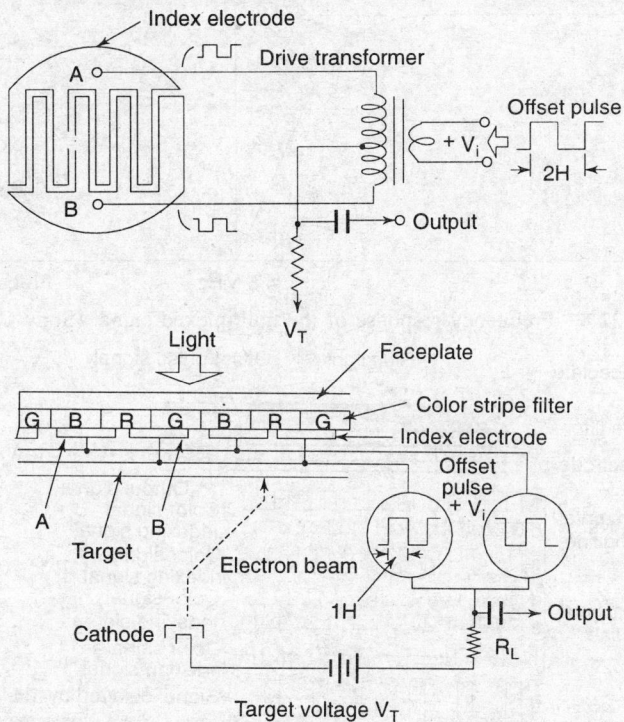

Fig. 22.7 Indexing electrodes and the comb shaped stripes (Sony Corp.)

The frequency of the signal from saticon depends on the number electrodes *A* and *B*, and the horizontal scanning rate of the electron beam. The number of pairs of indexing electrodes is determined so that frequency is the resolution limit viz. 4.5 MHz. The pitches of the indexing electrodes and the stripe filter are the same. If a small dark current flows through saticon the indexing signal can be satisfactorily obtained.

The optical image obtained on the target through the lens system, is optically colour modulated via the colour stripe filter. Phase modulation is accomplished by three primary colours as the beam scans it. A colour subcarrier frequency signal, which has amplitude proportional to the light intensity and the indexing signal are multiplexed and obtained by the indexing electrodes A and B. The frequency of the multiplexed signal is 4.5 MHz, and is the same as that of the indexing signal, as illustrated in Fig. 22.8. The advantage of the method is that high resolution can be obtained because the high band's range can be expanded as seen in the illustration.

Dissection Technique

The technique of dissection of the colour signal from the subcarrier frequency is shown in Fig. 22.9(a). If green monochrome light is incident on the target only green light can pass through the stripe filter during nth horizontal scanning. At that time, indexing signal 'b' and the colour signals are multiplexed,

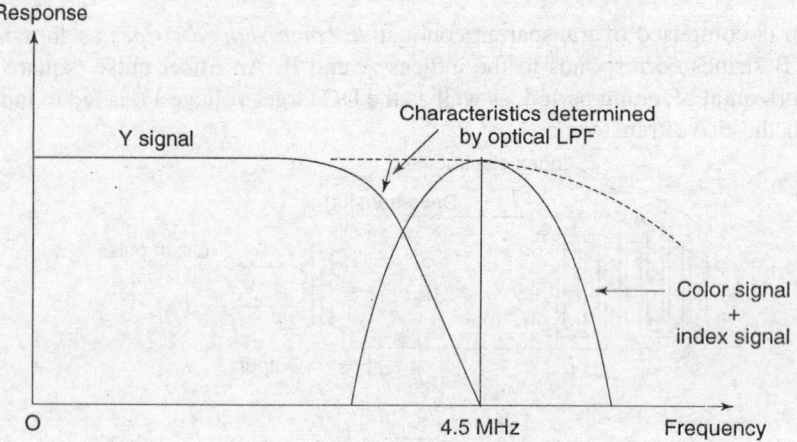

Fig. 22.8 Frequency response of the multiplexed signal (Sony Corp)

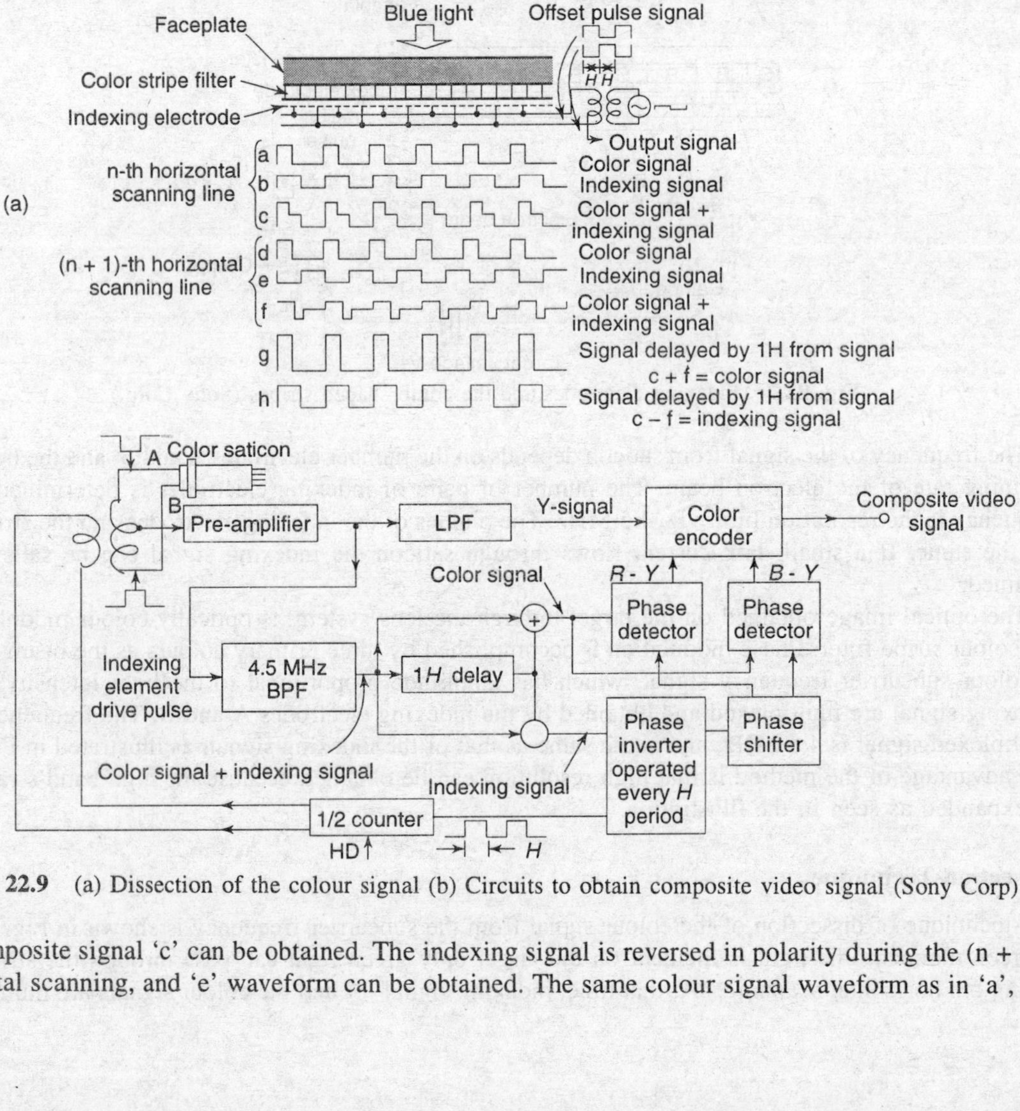

Fig. 22.9 (a) Dissection of the colour signal (b) Circuits to obtain composite video signal (Sony Corp)

and composite signal 'c' can be obtained. The indexing signal is reversed in polarity during the (n + 1)th horizontal scanning, and 'e' waveform can be obtained. The same colour signal waveform as in 'a', that is 'd', is generated and composite signal 'f' can be obtained. The actual signal contains the Y signal component, and RF signal components 'c' and 'f' are multiplied at the output. The signal thus obtained is then fed to further circuits shown in Fig. 22.9(b).

The Y signal component is obtained by LPF, and the 'colour signal + indexing signal' is obtained by the BPF centered at 4.5 MHz. Signal 'c' which is delayed by one H during the n-th horizontal scanning, and signal 'f', which is generated during the (n + 1)-th scanning, are added together to obtain colour signal 'g'. When 'f' is subtracted from 'c' indexing signal 'h' is obtained. The indexing signal is reversed in polarity every other horizontal line and the phase inverter circuit is used to arrange the indexing signal so that it is the same polarity every horizontal scanning. The phase inverter output must be, however, synchronised with the original indexing signal. In order to obtain the composite colour signal, the indexed signal, which has the same polarity every horizontal scanning, is fed through the phase shifter to the phase detector.

Video Camera Signal Processing

The small amplitude of pick-up tube signal is passed through a transformer, where the index signal is superposed onto the pickup tube output and fed into a low noise JFET input stage of a preamplifier via a series peaking coil. This reduces the effect of stray parallel capacitance and improves chroma frequency components and the S/N ratio in chroma band. Negative feedback is used to improve and adjust the LF response in the 1–2 MHz range.

The luma signal is processed for the following:

— AGC setting
— HF limiter
— horizontal detail generator for aperture correction
— vertical detail generator for aperture correction
— pedestal, Y-gamma correction and black clip
— negative/positive inversion, set up and white clip
— low light level attention
— automatic iris control
— chroma-gamma correction

The horizontal detail signal is obtained by subtracting the 150 ns delayed signal from the sum of the non-delayed and 300 delayed signals as shown in Fig. 22.10.

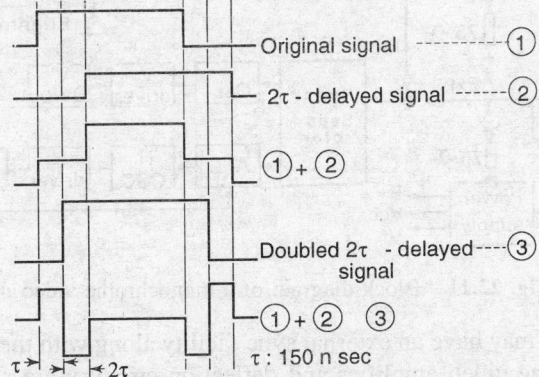

Fig. 22.10 Horizontal detail generator (Sony Corp.)

The 300 ns delayed signal is obtained by delaying the video signal twice with the help of a delay line reflected signal. This gives improvement in transient changes and aperture correction is effected.

The processing further provides for chroma signal processing and chroma encoding, shading correction for luma and chroma, automatic white balancing with the help of window gate generator, burst and sync mixing, fade in/out, colour bar generator and other facilities. Electret condenser microphone permits simultaneous audio recording.

22.4 VIDEO MONITORS

A video monitor reproduces the picture directly from the composite video signal. It does not contain any RF-IF and associated stages and is usually designed with better reproduction, qualities in respect of resolution, bandwidth, geometrical distortion, linearity, etc. The monitor may or may not have audio reproduction circuitry associated with it. Depending upon the quality of reproduction the monitors are sometimes graded into class A or class B monitors.

Bandwidth and Resolution

In professional grade CCTV broadcast systems, the video monitor resolution is high, better than 500 lines. High resolution requires well focused beam and large bandwidth with good phase response. Bandwidths of 6, 8, 10 MHz or higher are used.

Deflection Linearity

Smaller angles of deflection gives less geometrical distortion and defocusing effects near edges and corners. Hence for better linearity 90° deflection picture tube may be preferred in high grade class A monitors in place of the common 110° deflection tube in class B monitors.

Input Sensitivity

Video monitors are generally designed to operate satisfactory with 1 V_{pp} video signal with loop through link and a termination of 75 Ω for the cable supplying the video signal, if it is not looped through.

Block diagram of a video monitor is shown in Fig. 22.11.

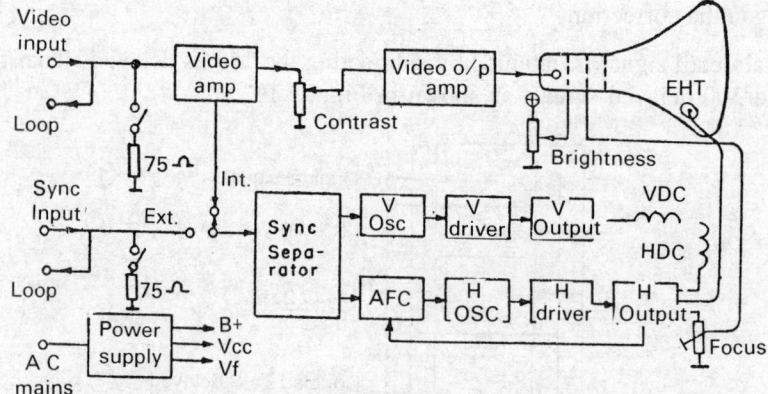

Fig. 22.11 Block diagram of a monochrome video monitor

The deflection circuits may have an external sync facility along with the normal internal sync obtained from the video signal. The video amplifier and deflection circuitry are similar to that in a normal TV receiver but for the higher grade performance generally aimed at.

22.5 VIDEO CASSETTE RECORDERS

The helical scan recorders were introduced in 1960s, The difficulty in handling streamer tape by unskilled people and multiplicity of models from different manufacturer prevented creation of a popular standard. It was in the 1970s when Sony and Philips set up new standards for video cassette recorders to open up new markets in the domestic and semiprofessional field. Philips aiming at the educational and domestic market, brought out a series of models to increase the play time to over 3 hours. Sony U-matic system became popular to set for the first time, a standard for the *3/4 inch* cassette recorders in the semiprofessional field, while in the domestic market a new range of *1/2 inch* cassette recorders were introduced extending the play time to over 3 hours to accommodate feature films.

With the widening domestic, educational and industrial market, a tussle has been going on amongst the manufacturers to make their standards popular. Although Philips had some initial success, Sony Betamax system and the JVC VHS system became the more popular in the domestic market, with periodic improvements in their performance and play time features.

Video cassette recorders use for *oading convenience* plastic containers inside which the reservoir and the take up spools are sealed. These include the—*matic, Betamax and VHS* as the major systems currently most popular the world over, and the Philips Video 2000 system, also common in the European market. There are other systems with fixed head LVR (longitudinal video Recording) technique, like the BASF system, the Toshiba system which have been marketed with very limited success. The Japanese Victor Corporation's 1/2-in VHS is very widely accepted as the *Home video* standard, while the 3/4-in U-matic is a higher quality professional standard for industrial and educational use. The compact 8 mm video camcorders have recently entered the fray.

The *Betamax* format has been adopted as interchangeable format by Aiwa, General Corporation, Sanyo, NEC, Toshiba Ampex, Zenith, and others. The *VHS* format VCRs are manufactured by JVC, Matsushita, Mitsubishi, Sharp, Akai, Panasonic, Hitachi, General Electric, Magnavox, RCA and others in the VHS group. Both these formats have undergone many improvements since they were first introduced and adopted by different manufacturers with minor constructional and design differences.

The domestic video market has been predominated by the VHS, and judging by the number of machines in the field and the program software produced, has now stabilized into a universal standard.

22.6 U-MATIC SYSTEM SMPTE TYPE E 3/4-IN FORMAT

Sony corporation introduced 3/4 in. tape cassette *U-matic* recorders in 1972, of acceptable picture quality at relatively low tape running speed of 95.3 mm/s (3.752 in/s), which gained quick acceptance for use in the industrial and educational and even broadcast applications. Compared to the earlier 1-in helical machines it had the advantages of low cost, small size and low weight for broadcast as well as industrial users. The U-matic system used a lower speed to increase the play time but guard band was kept to reduce cross talk between video adjacent tracks, as shown in the Fig. 22.12(a). The head positions on the U-matic tape transport is shown in Fig. 22.12(b).

The tape speed of 3.75 in/s (9.53 cm/s) is controlled by the capstan servo fixed by use of synchronous motor, while the head drum is controlled by a braking type servo system which takes its prerecorded speed control pulses from the control track.

U-matic Low Band Recorder

The original low band version of the U-matic machine gave 1 hour play time, resolution of 300 lines for monochrome and 240 for colour, SNR better than 40 dB and two audio tracks.

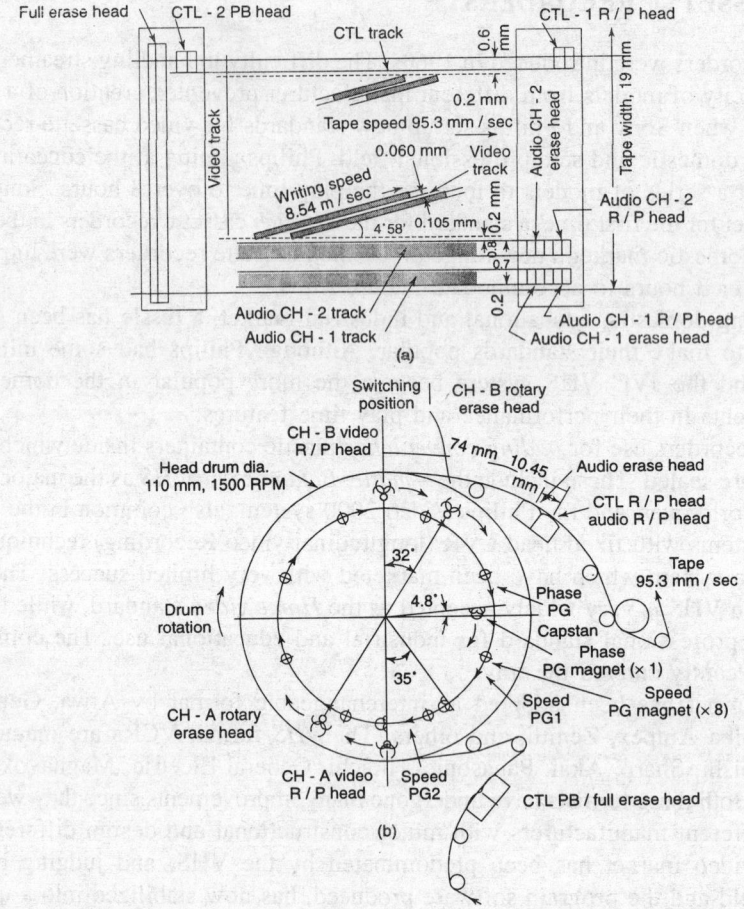

Fig. 22.12   (a) U-matic tape format (b) U-matic head positions (Sony Corp)

The low band U-matic system uses bandwidth less than 2 MHz including 0.5 MHz for the colour using the colour under technique. The luminance signal is recorded as frequency modulated signal (3.8 MHz-sync tip to 5.4 MHz-peak white), while the colour signal separately modulates 685 kHz which in turn amplitude modulates the FM luminance signal.

Popularity of this machine in industry, TV program distribution and later in electronic news gathering (ENG) for TV broadcasting, led to the development of the high band U-matic recorder and editing system for high quality and flexibility, using the same cassette but not compatible.

U-matic High Band Recorder

For obtaining improved performance of broadcast quality in the Sony BVU (Broadcast Version U-matic) range of VCRs, the FM frequencies are increased by 1 MHz, to higher frequencies of 4.8 MHz to 6.4 MHz. The colour under subcarrier employed, is increased to 924 kHz. The increase in carrier frequencies reduce interference from sidebands and the bandwidth of the recorded signals can be increased to improve the picture quality.

Superior performance has been possible in the U-matic machines by use of special *cobalt doped magnetic tapes,* that allow use of *higher FM carrier* frequencies. These special (KSP) tapes produce improved picture quality when used with SP BVU recorders which can operate in SP U-matic or the conventional high band mode. SP U-matic mode use FM frequencies of 5.6 MHz to 7.2 MHz, permitting larger colour under bandwidth and luminance sidebands band. The special KSP tape cassettes automatically switch over to the SP U-matic mode during recording or playback. SP technology maintains compatibility with High Band U-matics because the luminance signal deviation (1.6 MHz) and chroma subcarrier frequency (924 kHz) is unchanged.

Latest BVU machines incorporate *dynamic track following,* and hence can produce broadcast quality pictures over a wide range of speeds from still frame to thrice the normal speed, with the use of associated time base corrector. Studio based BVU machines have comprehensive *editing facilities* built in, including high speed or slow speed picture search at select speeds in forward or reverse mode. SP U-matic mode also provides for a Dolby type-C *noise reduction* system for the audio. The new generation SP U-matic machine like VO-9600P shown in Fig. 22.13, can operate in any of the three modes: SP, High band or low band.

Dolby noise reduction systems: High frequency noise heard as a hiss from tape, in the range of 2 to 10 kHz is more annoying to a listener than the low frequency noises that may usually be masked in the background noise of a living room, which has a quiet noise level of about 40 dB. Lack of high

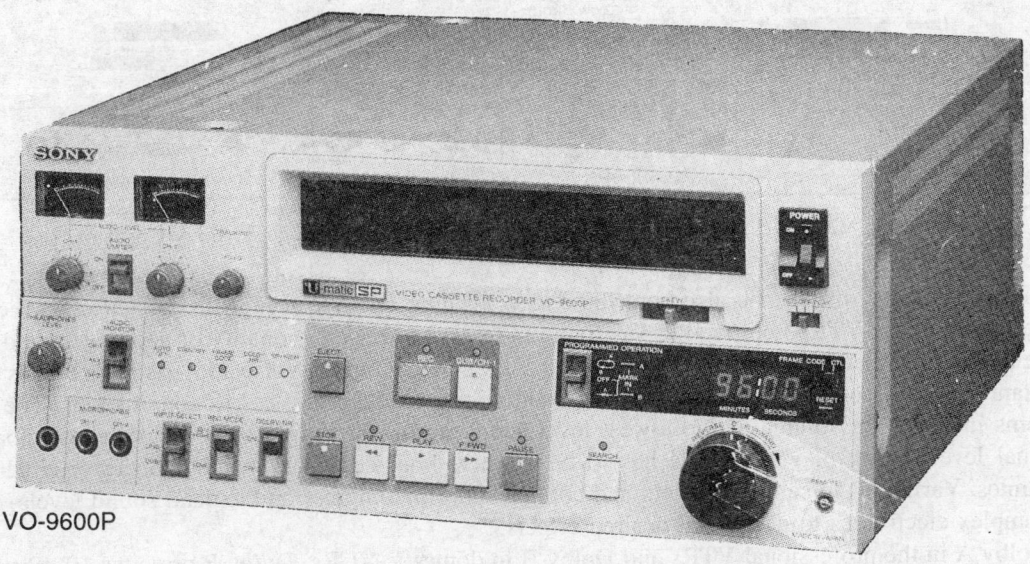

VO-9600P

Features

- High Quality Video
  —SP U-matic Technology
  —New Circuitry
  —New Videocassettes
  —SP, High Band,
  Low Band Operation
- High Quality Audio
  —Dolby™ Noise Reduction
  ——New Sendust Head
- Frame Code
  —Computer Interfacing
  —Accurate Random Access
  —Sophisticated Program Operations

Fig. 22.13   (a) New generation SP U-matic VO 9600P machine

RECORDING COMPATIBILITY

Diagram 1

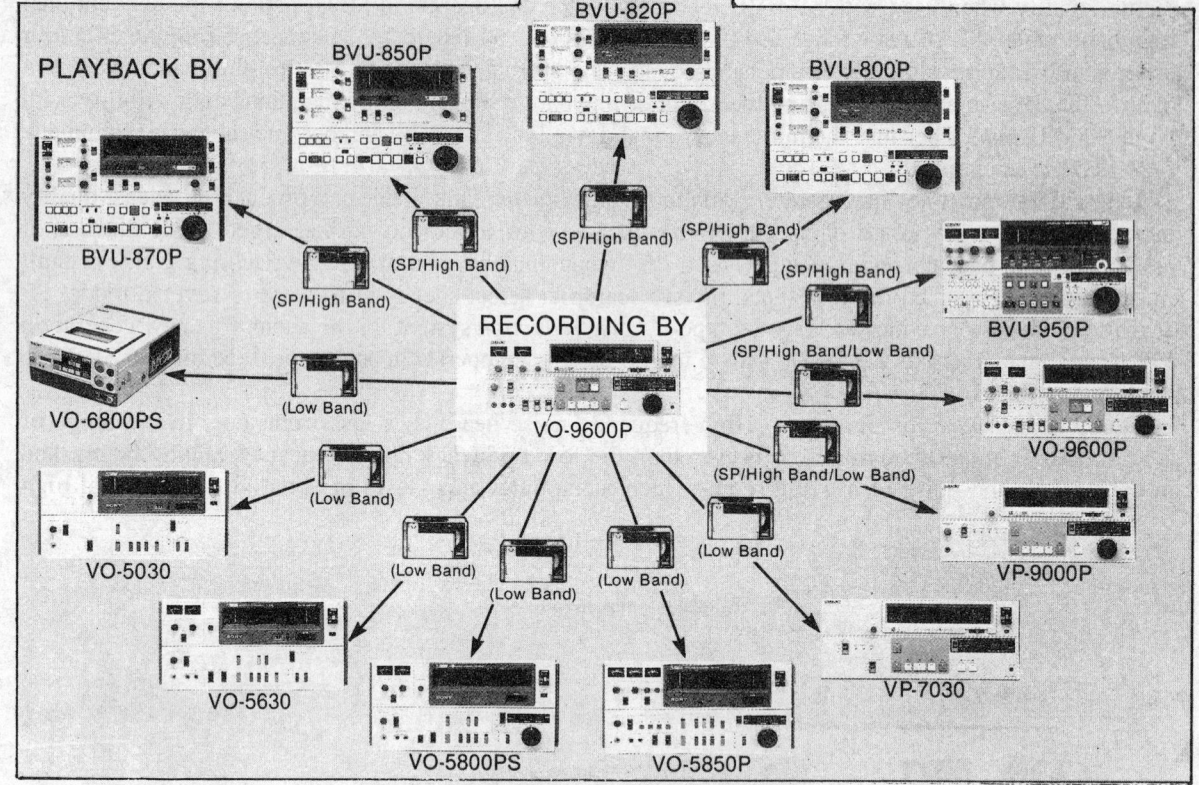

Fig. 22.13 (b) Recording compatibility of VO 9600P (*Courtesy:* Sony Corp)

frequency response of the tape also contribute to worsening of SNR at high frequencies. As the video performance of the VTRs improved it was found necessary to raise the quality of audio also, and Dolby noise reduction systems were incorporated by the VTR manufacturers. Noise and hiss from tape is almost constant while the sound level varies greatly, resulting in poor S/N ratio at low sound levels. The Dolby systems increase the volume of the lowest level sound during recording on the tape, and restore it to original level during playback. This has to be done without adding distortion to the original sound dynamics. Variations in selective filters and companders (to compress and expand sound levels) results in complex electronics to obtain the desired effects.

Dolby A in the professional VTRs and Dolby B in domestic VCRs. *Dolby B* provides 10 dB of noise reduction above about 4 kHz. A sliding band technique used, where the companding band slides up in the frequency in response to music frequencies, permitting full volume variations. Later versions of recorders employed Dolby-C and Dolby-SR (spectral recording) processes. *Dolby C*-type system brings down the tape noise by 20 dB above about 1 kHz, to a level below the common living room noise. *Dolby SR* is designed for professional master recordings providing a very low noise level and large dynamic range.

The main characteristics of the U-matic format are given in the Table 22.1, along with those of the Betamax and VHS.

Table 22.1 Characteristics of popular VCR formats

	U-matic Low band	Beta I/II/III	VHS I/II/III
SMPTE type	E	G	H
Width in.	0.75	0.5	0.5
mm	19	12.65	12.65
Head drum diam, mm	110	75	62
Head gap width um	0.6	0.6	0.3
Gap Azimuth tilt deg.	15	7	6
Video track width, um	85	58.5/29.2/19.5	58.5/29.2/19.5
Track angle deg	4.7	5	6
Head-to-tape speed, m/s	10.4	7.0	5.8
Audio track width, mm	0.8	1.05	1.0
Tape speed, cm/s	9.53	4/2/1.33	3.34/1.67/1.12
FM carrier, MHz, sync	3.8	3.5	3.4
White	5.4	4.8	4.4
chrom. subcarrier, kHz	685/688	685/688	629
Cassette size mm	$186 \times 123 \times 31$	$156 \times 96 \times 25$	$162 \times 104 \times 25$
Playing time, min	60	60/120/180	120/240/360
Tape thickness, um	27	20	20

22.7 COLOUR UNDER SYSTEM

In VCR machines, the *time base errors* caused by the mechanical factors like rubbing friction of the tape against stationary guides, can cause severe frequency and phase errors in the reproduced signal, including the colour severe frequency and phase errors in the reproduced signal, including the colour burst. This would result in loss of colour sync and produce *random colour bands,* besides disturbances in picture synchronisation. To overcome this problem the chrominance signals band limited to 0.5 MHz around the 4.43 MHz colour subcarrier, are down converted to lower frequencies between 129 kHz to 1.129 MHz, (188 kHz to 1.188 MHz in Beta) during recording, and are translated back to the original values during playback. This technique called *colour under technique* is illustrated in Fig. 22.14.

The chrominance signals are separated from video by the chroma bandpass amplifier and are *heterodyned with a* 5.059 MHz signal from a CW oscillator in VHS, (5.118 MHz in Betamax) to produce a difference carrier frequency of (5.059 – 4.43 =) 0.629 MHz in VHS, (5.118 – 4.43 = 0.688 MHz in Betamax) modulated with chrominance signals of bandwidth +/– 500 KHz, as shown in the Fig. 22.13.

The luminance signal frequency modulates an RF carrier VCO (for FM recording), to set sync tips at 3.4 MHz and peak white at 4.4 MHz in VHS, (3.8 to 5.2 MHz in Beta). This FM signal added with the *down converted* 129 kHz to 1.129 MHz VHS chroma signal, (188 kHz to 1.188 MHz in Beta) is recorded on the tape. The high frequency *FM signal acts like HF bias* for the chroma signal as in the audio recording to reduce amplitude distortion on the tape. A bandwidth of 2 MHz is used for the FM including 500 kHz for the chrominance signal. A *lower subcarrier of* 629 kHz is modulated by the chrominance signal by chrominance bandwidth of 500 kHz, which further amplitude modulates the FM luminance signal.

During playback, the 629 kHz (688 kHz in Beta) subcarrier beats with a 5.059 MHz CW signal (5.12 MHz in Beta) to produce a difference of 4.43 MHz colour subcarrier. The 5.059 MHz CW signal is generated by conversion from the 629 kHz VCO frequency and phase locked to the burst from the colour signal with the help of the locally generated 4.43 MHz carrier. The CW generator is a multiple of the

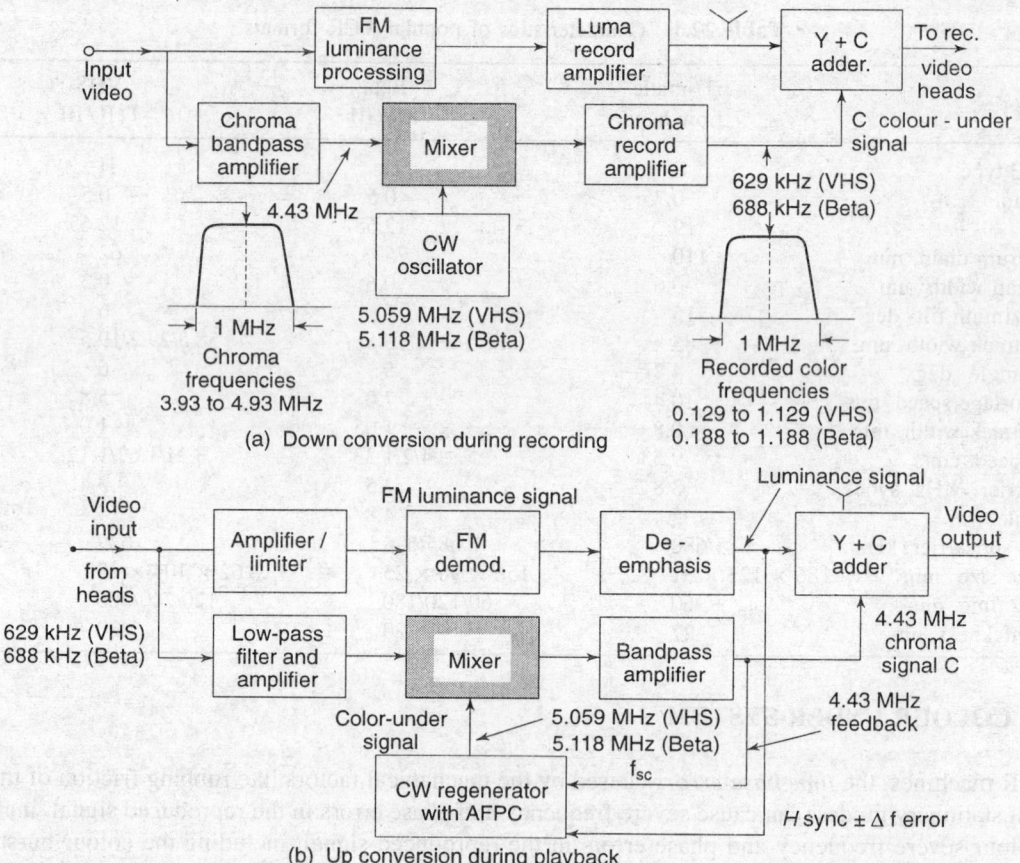

Fig. 22.14 Colour under-technique (a) Down conversion during recording, (b) Up conversion during playback

horizontal sync frequency carrying along with it, the time base errors +/– 4. During the *heterodyning* process the timing *errors cancel* out. This eliminates the time base phase errors in the chroma signal introduced by the tape transport mechanism.

22.8 HOME VIDEO FORMATS—BETAMAX AND VHS

In these formats the tape running speed was reduced substantially lower than the U-matic format, 1.57 in/s in Betamax and 1.3 in/s in VHS, the video track width and the control track width was reduced. For tape economy, the guard band between the video tracks was also dispensed with, to permit recording time of about two hours to accommodate a full length feature film on a single cassette. This tape economy has been further enhanced to 5 to 9 hours by further reduction in tape speeds, making them rapidly acceptable for the home video market. The tape usage in the home video recorder is much smaller, about 20% of the U-matic recorder. VHS tapes are generally designed by play time T-xxx, xxx indicating the time in minutes, while the Betamax tapes are usually designated by Length L-***, *** indicating the tape length in ft.

The home VCRs use *helical scan* technique, with alpha wrap of the tape around the head drum scanner for a little over 180°, the *two heads* crossing the tape at an angle of about 7° in Betamax and 6° in the VHS, from one edge of the tape to the other. The audio and control tracks are located on the edges longitudinally.

Betamax 1/2-in Format SMPTE G

Sony corporation introduced the first 1/2 in. compact cassette recorder as the betamax system in 1975, later adopted as interchangeable format by Aiwa, General Corporation, Sanyo, NEC, Toshiba Ampex, Zenith, and others. Betamax machines use 0.6 μm head gap whereas the VHS system employs 0.3 μm gap. It has a U-type tape threading system over a 74.5 mm drum scanner. using a cassette of 6.1 × 3.9 × 1 in. the Beta I/II/IIIformats marketed in 1975/77/79 could give play time of 1.7/3.3/5 hours, at the tape speeds of 1.57/3.3/5 in. per second (4/2/1.34 cm/s).

Beta Threading The Betamax machines employ Omega-wrap (also called U-wrap) for its tape threading. The tape gets loaded wraps around the head drum almost fully in omega shape, as soon as the cassette is inserted, and is ready for immediate recording or replay. The tape remains in contact with the video heads during fast forward and rewind modes also. One can always see the video rushing through, during these operations, and one remains at the same spot during these operations as well as the play mode.

VHS 1/2-in Format SMPTE H

Japanese Victor Corporation (JVC) introduced the 1/2 in. tape compact cassette recorder in 1976. VCRs in this format are manufactured by JVC, Matsushita, Mitsubishi, Sharp, Akai, Panasonic, Hitachi, General Electric, Magnavox, RCA and others in the VHS group. Lowering of the tape speed and writing video-tracks without guard band extended the play time. In VHS recorders the head drum size is decreased to 62 mm to reduce their weight. But this reduces tape speed, and hence the high frequency performance. To compensate for the high frequency loss the head gap is smaller, 0.3 μm. Using cassette size of 7.4 × 4.1 × 1 × 1 in. the VHS SP/LP/EP or I/II/III formats (marketed in 1976/77/79) provide play times of 2/4/6 hours at the tape speeds of 1.31/0.6/0.44 in. per sec (3.33/1.67/1.12 cm/s). With super long play (SLP) tapes, the playtime can be as high as 9 hours can be obtained.

VHS Threading VHS machines use a simple M-type threading (also called parallel loading system) with two movable loading arms, one for the supply side and the other for the take-up side. Loading time is reduced to about 2 seconds.

Both the supply and take up spools have motors besides the capstan driven by the capstan motor. A pinch roller presses against the capstan, which pulls the tape with the correct speed. The tape is kept in proper tension by the tension arm, as the tape is guided by four guide poles to move on to the heads and the drum, as shown in the Fig. 22.16. After the erase head which can be supplied with ac current to erase the tape, the tape wraps around the video head drum so that the two head on the drum scan the tape in turn. Slanting pins extracting the tape from the cassette move across the head drum to load the tape around it.

As seen in the threading, less tape is pulled out of the cassette, loading is fast taking a fraction of second and there is less chance of tape spillage and tangles. Loading of the tape on the drum takes place only when record or replay controls are operated, taking a little time for the actual operation. As fast forward and rewind operations are performed with the tape inside the cassette, less damage is likely to the tape, which means more life for the VHS tapes. The video can be seen during fast forward or rewind mode with playback switch on.

22.9   TILTED AZIMUTH RECORDING

In the home video systems two video heads are mounted at 180° to each other on the periphery of the rotating drum assembly, each head remaining in contact for one-half of revolution. The two interlaced fields of a complete picture are generally recorded in one complete rotation of the drum assembly. This simplifies stop motion action.

Due to elimination of the guard band intertrack cross talk is produced by the scanning heads during replay. This crosstalk is reduced by the *tilting azimuth* of the video head-gaps, from their right angled position with the video track, by 6° (7° in Betamax) as shown in Fig. 22.15.

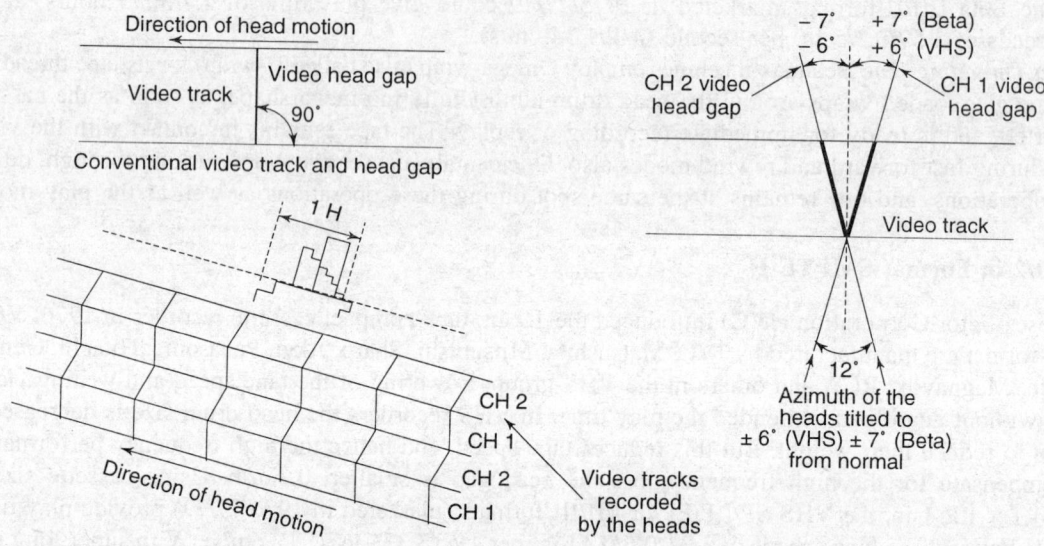

Fig. 22.15 Tilted Azimuth recording

The tilt of 6 or 7° in both the heads is in the opposite direction, so that the total *azimuth difference between the tracks* is 12° for the VHS (14° for Betamax). The tilt in the heads is fixed in the manufacturing process. The azimuth tilt in effect increases the gap size and hence, if a head plays the signal from tracks recorded by the other head, *crosstalk is ineffective* at high frequencies over 1 MHz. As the azimuth difference between the tracks in Betamax is more than the VHS the track isolation in Betamax is a little better. Azimuth loss with respect to head gap angle and recorded frequencies is shown in Fig. 17. The crosstalk caused by frequency modulated luminance signals in the range of 3.4 MHz (sync tip) to 4.4 MHz (peak white), (3.5 to 5.2 MHz in Beta) is almost eliminated. Crosstalk at lower frequencies which include the down converted chrominance signals cannot be completely eliminated.

For eliminating chroma crosstalk at low frequencies also, *Phase-shift Colour Recording* is used, in which the chrominance signals of the two heads are phase shifted 90° while recording, and on replay the signals are added to the signals of the preceding 2 lines by using a 2H delay line. The crosstalk component gets eliminated because of the phase difference, while the signal components are almost doubled. Both the VHS and Betamax systems use this technique of two line repetition sequence to eliminate the effect of crosstalk.

22.10   VHS CASSETTE RECORDER SCHEMATIC

The innovation of VHS cassette recorder lies in recording and reproducing colour video signal without phase errors and crosstalk. In the VHS machine, one head records the colour subcarrier information modulated in the colour-under scheme at about 629 kHz in the normal phase, while the other head records

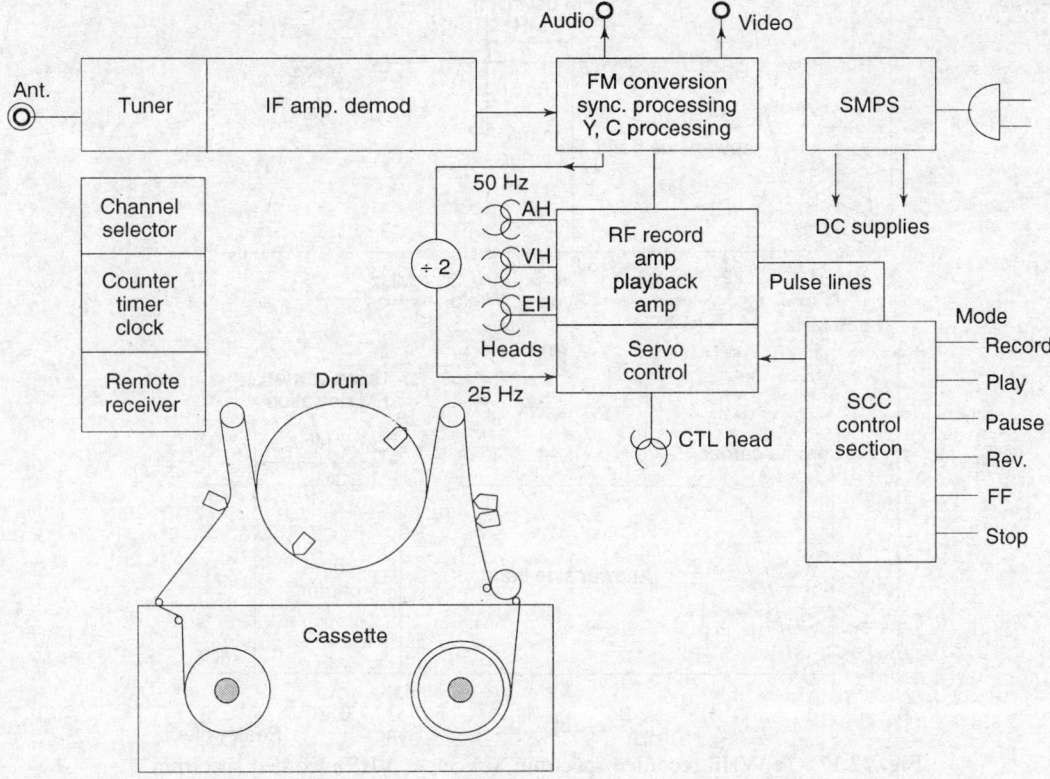

Fig. 22.16 Block schematic of VHS cassette recorder

it with a phase lag of 90° Because of the PAL switching of the V component every alternate line and the 90° phase delay introduced by the alternate scanning head in combination produce a 2-line scanning pattern of recording and replaying signals. The replay chroma signal when added to the preceding 2-line signal through a 2-line delay doubles the main in phase chroma signals, while the antiphase crosstalk is cancelled out.

In most VCRs a tuner and Video IF subsystem is provided built in to enable recording direct from antenna, often with built-in timer for recording programs as per preset timings. The video and audio from the VIF subsystem or direct from other sources is processed and given to the recording heads. The 50 Hz line sync pulses control the head drum servo during recording.

Super VHS Format

There have been remarkable improvements in picture quality of recent VHS machines. Sharper pictures are obtained in the HQ system by sharp overshoots and raising of the white clipper levels by 20%. *Detail enhancing system* and line noise cancelling system with a recursive filter are incorporated to improve picture quality. Recently the JVC has introduced the Super VHS format giving improved picture resolution of 400 lines. In the S-VHS, *higher FM carrier* is used allowing a carrier deviation from 5.4 MHz for sync tip to 7 MHz for the peak white, as against the 3.8 to 4.8 MHz in the standard VHS, as shown in Fig. 22.17.

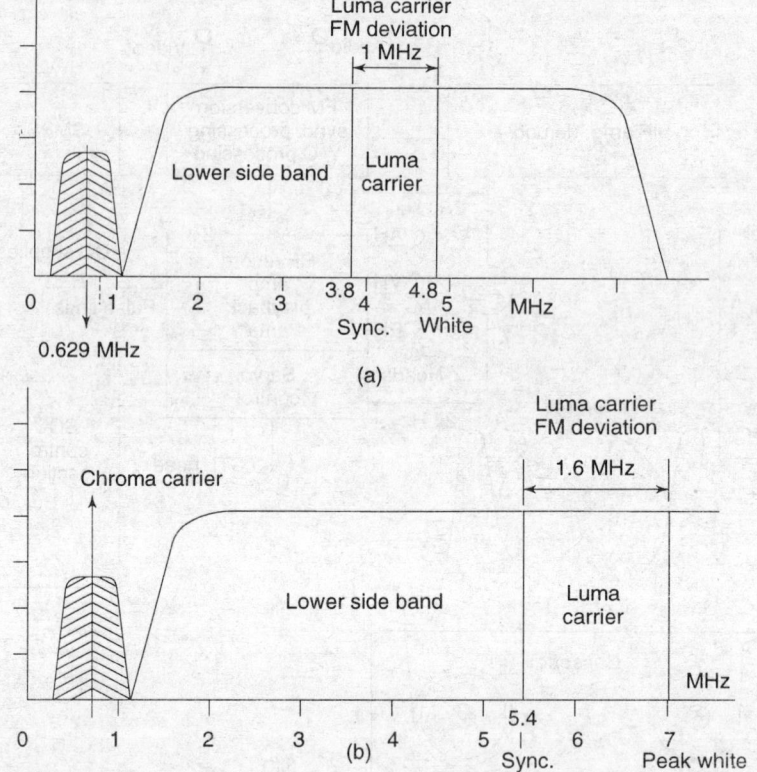

Fig. 22.17 (a) VHS recorded spectrum, (b) Super VHS recorded spectrum

This permits the luminance response of up to 5 MHz equivalent to 400 lines resolution, which is almost twice that of the standard VHS system. The wider deviation of 1.6 MHz in the new system, against the 1 MHz deviation in the old, also improves the SNR to some extent.

In the S-VHS recorder, the band width of the chroma colour-under signal centered around 629 kHz. As the FM carrier is shifted higher up, the FM sidebands of the luminance signals and the sidebands of the colour-under signal do not overlap, improving the colour reproduction. Separate luminance and chrominance signal inputs or outputs have been provided in addition to the composite colourplexed video and RF signal, permitting better picture. A tape of high coersivity, over 800 oersteds ensures a better quality of the recorded picture.

22.11 VIDEO-2000 SYSTEM

This 1/2 in. tape cassette system was introduced by Philips in 1980, using a flip over cassette for recording from both ends of the tape. As one half of the tape width is used for recording from each side, the cassette enables 8 hours recording, 4 hours each way, at a linear speed of 2.44 cm/s and the head-track interface speed of 5.08 m/s. As in the VHS two head helical scan with slanted azimuth technique is used.

The Philips system however incorporates *dynamic track following* (DTF) similar to AST in the 'C' format 1 in. professional helical VTRs, in which a piezoelectric support for each head automatically adjusts the head position by servo control so that each video scan is accurately registered. Control track is not required. Auxiliary track is provided near the middle of the tape to carry the cueing signals while audio track is kept near the outer edges of the tape as shown in the track arrangement of Fig. 22.18. With each video track scan corresponding to one field, stop-motion and fast/slow-motion is simple to implement.

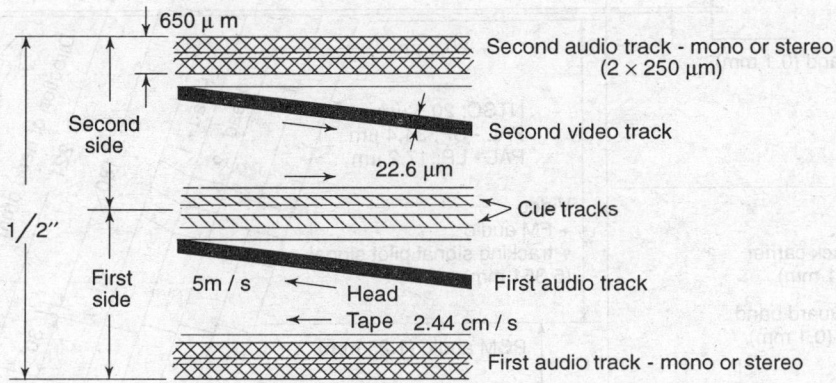

Fig. 22.18   Video track arrangement in V-2000

22.12   FIXED HEAD VTRs

While the helical scan VCRs have been very widely accepted by the consumers, industry efforts in development of longitudinal video recorders (LVR) are continuing to develop *light weight* and less expensive than the rotary head VCRs. BASF and Toshiba systems are examples of this trend.

BASF LVR System

The tape speed in this fixed head LVR system is 4 m/s, so that the 8 mm wide 600 m tape records for $2^1/_2$ min. one way, reversing automatically at the end of the track in a split 0.1 second. The single head (gap 0.3 μm) is shifted laterally to record composite video and audio signals on adjacent tracks (80 and 68 μm). 72 such tracks are provided on the 8 mm tape, giving a total running time of 3 hours. The various functions of tape transport and video electronics are microprocessor controlled.

Toshiba LVR System

In this system 1/2 in. tape length of 100 m is wound in the form of a continuous loop around a single reel. The tape speed is 6 m/s, giving a recording time of 16.7 seconds on a single run. The head shifts laterally to the next track by means a pulsed motor within a matter of 20 ms, so that the 220 tracks across the half inch tape give a recording time of 1 hour.

22.13   8 MM VIDEO

With the development of magnetic high density recording technique and *superior tapes* in recent years, Over 127 leading video-related companies have agreed on one standard 8 mm video format as shown in Fig. 22.18(a). Conventional full erase obtained by a stationary head which erases the full width of the tape, leaving blank space before the new recording, as shown in Fig. 22.19(b).

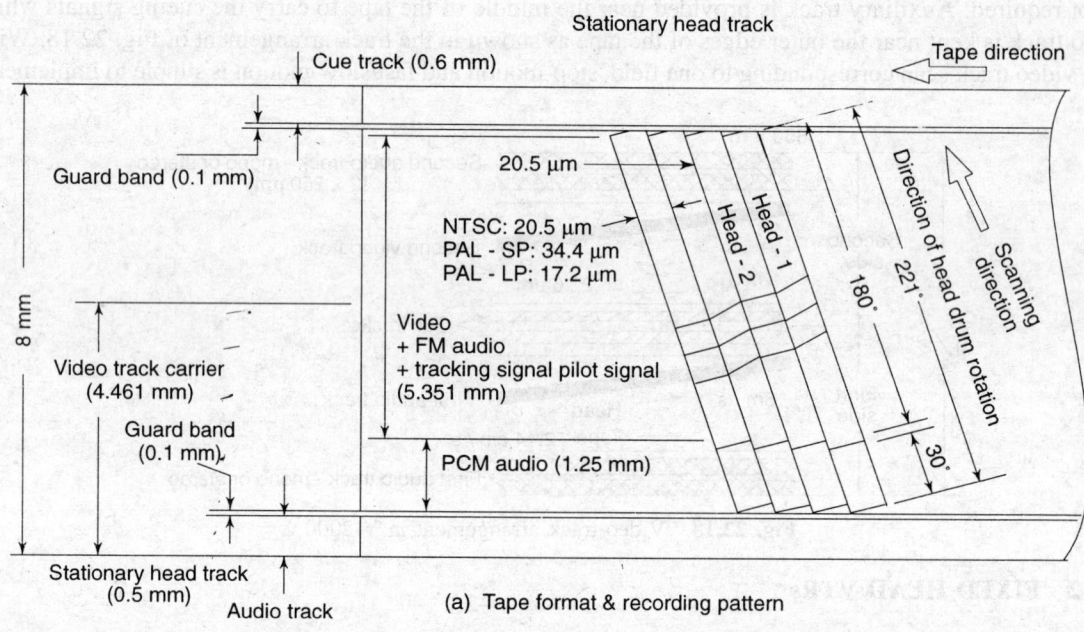

(a) Tape format & recording pattern

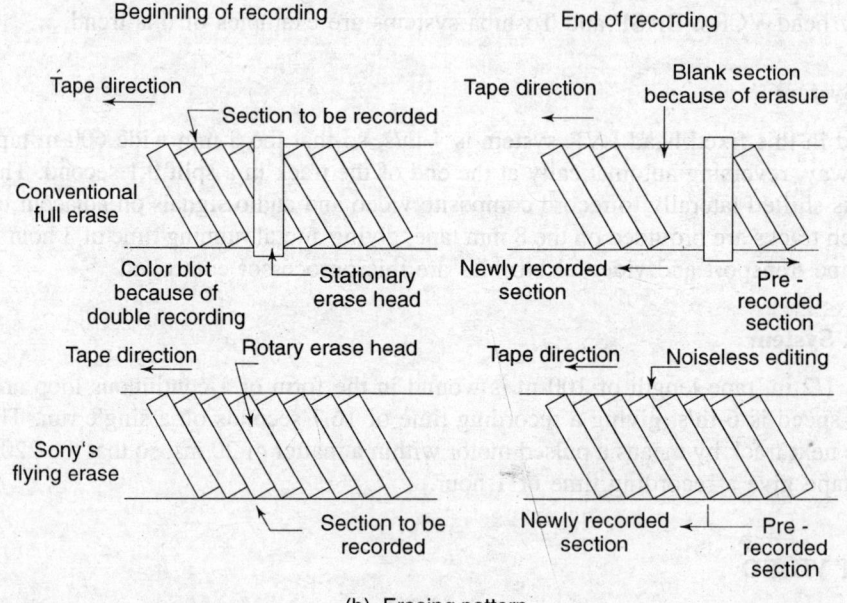

(b) Erasing pattern

Fig. 22.19   (a) Tape format and recording pattern of 8 mm video, (b) Erasing patterns in conventional and Sony's FE. (Sony Corp.)

Noisefree scene *transitions and insertions* equal in quality to those produced by professional studio equipment are delivered by Video 8's *Flying Erase* (FE) head. Mounted directly on the rotating drum the FE head has erasing pattern directly aligned to the diagonal path of the recording heads. This allows *noiseless interval recording and titles* or scenes to be placed on exactly on the tracks erased by the FE head.

The video 8 models use 8 mm tape in $95 \times 62.5 \times 15$ mm cassette (3.8 mm tape is used in the $102 \times 63 \times 12$ mm audio cassette), and operate in the standard mode at 2 cm/s and in the long play mode at half the speed, sacrificing some quality. Play time from 60 min (75 m of 13 μm tape) to 180 min (112 m of 10 μm tape) are possible. The luma signal is recorded in the standard FM technique, and the chroma is separated and recorded as separate low end spectrum by the colour under technique, as shown in Fig. 22.20.

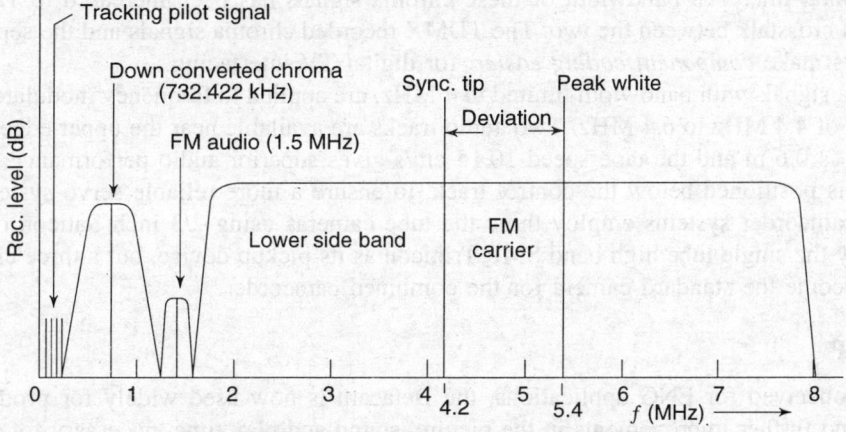

Fig. 22.20   Signal spectrum of 8 mm video

Slanted azimuth recording is used with the azimuth angle of +/– 10° to eliminate crosstalk between adjacent tracks without guard band in between. There is no control track, *automatic track finding* (ATF) system is used, which allows simpler mechanical system. Each track has a different frequency pilot tone recorded on it by the rotating video heads. They occupy the bottom end of the signal spectrum. Two tracks are available on the outer edges of the tape, one used for audio and the other for a cue track for indexing.

Audio can be recorded as *standard AFM signal* or optionally as a *PCM* signal also. The standard audio channel uses 1.5 MHz as FM carrier with a deviation of +/– 100 kHz. An optional feature is digital audio recording using 8-bit PCM, with stereo facility. Non-linear quantization is used to improve dynamic range. After passing through a noise reduction circuit and nonlinear compression by 50%, the audio signal is sampled at 31.25 kHz and converted first into 10-bit and then into 8-bits words. The tape wrap angle has to be increased from 180° to 221° to provide for the necessary recording space for PCM sound.

ATF system records 4 different *pilot tones* on every four adjacent tracks. During playback the tones are recovered by the recording heads from their own track and from the adjacent track. The ATF system monitors the recovered levels and adjusts the tape speed and drum rotation such that the level of the main pilot signal and the adjacent track signal maintain a constant frequency difference relation.

Dynamic track following (DTF) system as discussed Sec. 6.9, is used, where the heads are mounted on ceramic bimorph plates. The video heads follow the track accurately as the piezoelectric plates bend in either direction whenever the tracking deviates on slow speed or stop motion.

22.14 CAMCORDERS

Because of improved quality in small size, a compact combination of video camera and recorder (camcorder) is now increasingly used for ENG, in place of the age-old 8 mm cine film camera.

Betacam

Sony has the Betacam system using 1/2 inch tape cassette with performance better than the U-matic. The system uses two adjacent video tracks one recording *time compressed chrominance* signals and the other recording the luminance signal, by means of separate adjacently mounted *two video heads*. Compressed colour difference signals R-Y and B-Y are time division *multiplexed* and accommodated sequentially in each horizontal line. The bandwidth of these chroma signals has been increased to 1.5 MHz, without problems of crosstalk between the two. The TDMX recorded chroma signals and the separately recorded luma signals make *component coding easier*, for digital TV interfacing.

The luma signals with bandwidth limited to 4 MHz, are applied to frequency modulator producing FM frequencies of 4.4 MHz to 6.4 MHz. Two audio tracks are available near the upper edge of the tape. The track width is 0.6 m and the tape speed 10.15 cm/s gives superior audio performance. A separate *time code track* is positioned below the control track, to ensure a more reliable servo system.

Earlier camcorder systems employ three the tube cameras using 2/3 inch Saticon or plumbicon, or alternatively the single tube high band SMF Trinicon as its pickup device, but a three chip CCD camera has now become the standard camera for the combined camcorder.

Betacam SP

Although conceived for ENG applications, the Betacam is now used widely for production of studio programs and further improvements in the picture, sound and play time, necessary for multi-generation transcripts, have been desirable. This was aimed at in Betacam SP (superior performance). Use of *metal particle SP tape* with finer particle structure allows use of higher carrier frequencies higher bandwidths for chroma as well as luma signals. The superior performance tape allows 7 dB better SNR (luma > 47 dB, chroma > 50 dB), maintaining necessary picture quality in *multi-generation operations*. In addition to the two conventional longitudinal audio channels, two high quality *Audio FM* (AFM) channels are provided giving superior audio performance than in the standard Betacam. AFM entails that the frequency modulated audio is simultaneously recorded with the video information by means of *two rotary heads* on the drum.

The studio use Betacam SP VCR Sony BVP-75P is designed to provide over 90/100 (L cassette) minutes of playing time and also accepts the conventional 36 minute cassette of oxide or metal particle tape for Betacam SP. With Dynamic Tracking (DT) it can provide broadcast quality pictures from – 1 to + 2 times normal speed of 10.15 cm/s. It has built-in *expert features* like two machine editing capability with variable memory, TBC, LTC/VITC/user bit generator and reader, character generator providing 'Burnt-in' time code output, self-diagnostics etc.

8 mm Camera Recorders

Compact versatile 8 mm CCD camcorders are now available in wide range models from the video manufacturers. Sony has ultra-compact aim-and-shoot CCD-M8 Handycam, to the high performance CCD V-100 Video 8 Pro, shown in Fig. 22.21.

The CCD V-100 Video 8 Pro incorporates 2/3 in. CCD image sensor with 291,000 pixels, requiring normal illumination of 300 lux. It can handle a wide illumination range of 17–100,000 lux. It has

Fig. 22.21 CCD 8 mm video camera recorders (*Courtesy:* Sony Corp)
CCD image sensor PHOTO literature

advanced features such as 7-colour titler, white fader, edit, interval REC with Flying Erase head, built-in character generator and calendar in memory, AFM/PCM hi-Fi sound recording capability, through the Camera Lens (TCL) Auto/manual Focusing, Macro function for close-ups. The TCL digital autofocus is shown in Fig. 22.21. The microcomputer based system adjusts focus at the centre of the main lens, parallax-free.

22.15 VIDEO DISC SYSTEMS

Video disc systems for domestic use were developed in early 70s by Telefunken-Decca system and later by Philips (laservision), RCA (Selectavision) and EMI-JVC (VHD system). The Teldec disc used mechanical stylus and groove system. The RCA developed the *capacitive electronic disc* (CED) system with hill and dale pits on the surface of V-grooved PVC disc. The signals are generated by variations in the capacitance between the conductive disc surface and metallic electrode attached to the stylus while running in the spiral groove. JVC demonstrated in 1978, *very high density* (VHD) disc system also based on the capacitive stylus. Unlike the RCA CED disc, VHD disc is grooveless. The video, and two channels of audio information are recorded as rectangular depressions along a single spiral of pits, in the surface of a flat PVC electroconductive disc. A stylus tracking signal of two different alternate low frequencies is recorded on either side of the video-audio track. During replay the stylus senses the tracking signal and adjusts its position by means of servo, accordingly. Both these CED and VHD discs are easily damaged by dust and handling and must be kept in a *plastic holder* called a 'caddy'.

Optical reflective video disc   Optical video disc system was developed independently by both Philips in Netherland and MCA in the USA, initially known as *video long player* (VLP). The two companies joined forces to combine the proprietary technologies to produce an optical disc in which the video and audio signals are recorded as tiny rectangular pits in moulded in the plastic substrate. The surface of the disc is coated with thin metal layer making it reflective and a transparent acrylic coating is further applied on this layer to protect the information. The system was marketed under the brand name *LaserVision*.

The pits are approximately 0.4 μm wide and 0.1 μm deep, while the length and spacing varies between 0.5 μm to 1.5 μm depending on the FM modulating signals as shown in the coding system of Fig. 22.22 Colour-under processing is not necessary as the recording bandwidth is adequate. This is achieved using

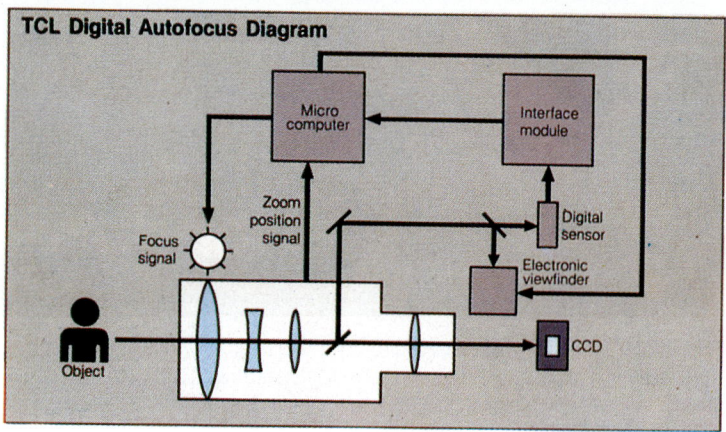

Fig. 22.22   TCL digital autofocus schematic diagram (Sony Corp.)

a system of direct video frequency modulation shown in the Fig. 22.23(a). The composite video signal is directly frequency modulated on to a carrier of 7.5 MHz. This gives an ideal sideband disposition and ensures minimum interference between luminance and chrominance components. The two audio signals are modulated on the subcarrier at 684 kHz and 1066 kHz. The audio subcarriers are then applied as PWM signals to the video frequency modulation signal. The result in a simplified form is shown in the Fig. 22.23(b). The PWM waveform is used to produce the pits in the recording, the length and spacing of which is an analog code for the complete video and audio signal. The control data is inserted in the vertical blanking intervals.

Relatively simple single frequency demodulator circuitry is needed for reproduction. The signal is recovered by scanning the disc by a low power He-Ne laser beam as seen in Fig. 22.24. The laser beam shines from below the disc and is focussed below the surface coating into the pits, so that dust on the surface has little effect on the output. The light reflected from the disc surface is modulated by the variations in the beam caused by the pits in the surface. A photodiode sensor converts the varying light into electrical signals. The laser beam has to be accurately aligned and focused along the recorded track. For this, a *tri-beam radial tracking* system is used. A special diffraction grating located at the exit of the

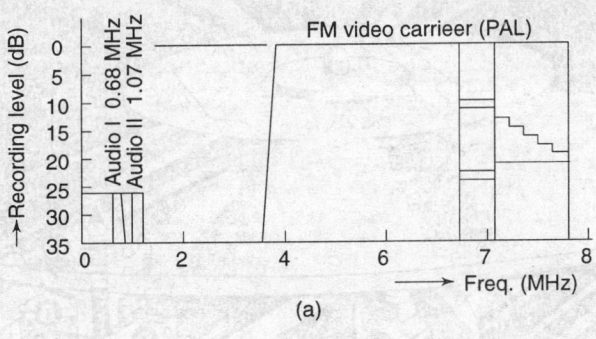

(a)

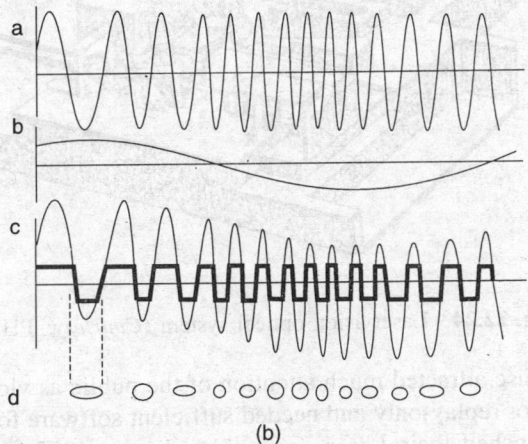

(b)

Fig. 22.23 (a) Frequency spectrum of Laservision recording, the CVBS (Luma and chroma) signal is directly modulated on 7.5 MHz carrier with two sound channels on subcarrier, (b) Video frequency demodulation followed by audio PWM produces clipped waveform, which is used to produce the variable pits on the track (*Courtesy:* PHILIPS)

laser generates two first-order beams in addition to the main scanning zero order beam. The two subsidiary beams are imaged partially on the track and partially on the spaces on either side of the scanned track, to read the edges of the pits. If the beam is off the track, each of the subsidiary beams produces unequal signals on its own detector, thus producing an error correction signal to actuate a pivoted mirror to maintain radial beam tracking. The beam focus is detected by sensing the returned beam by four photodiodes, which must give equal signal. Defocused elliptical beam gives unequal signals that are used for focus control.

Two additional servos are associated with the rotational speed and time base correction for the colour signal, during playback signal processing. *Active Play discs* run at a constant angular velocity, each rotation containing two fields, and gives a playtime of 36 minutes per side of 30 cm disc. The machine is capable of freeze frame or slow motion and frame numbering facilities. These are indispensable for the active and interactive programs that are a special feature of Laservision. In *Long Play discs* the playtime has been increased to 60 minutes per side, by obtaining a constant linear velocity for the tracking beam, the speed varying from 570 to 1500 rpm, in PAL.

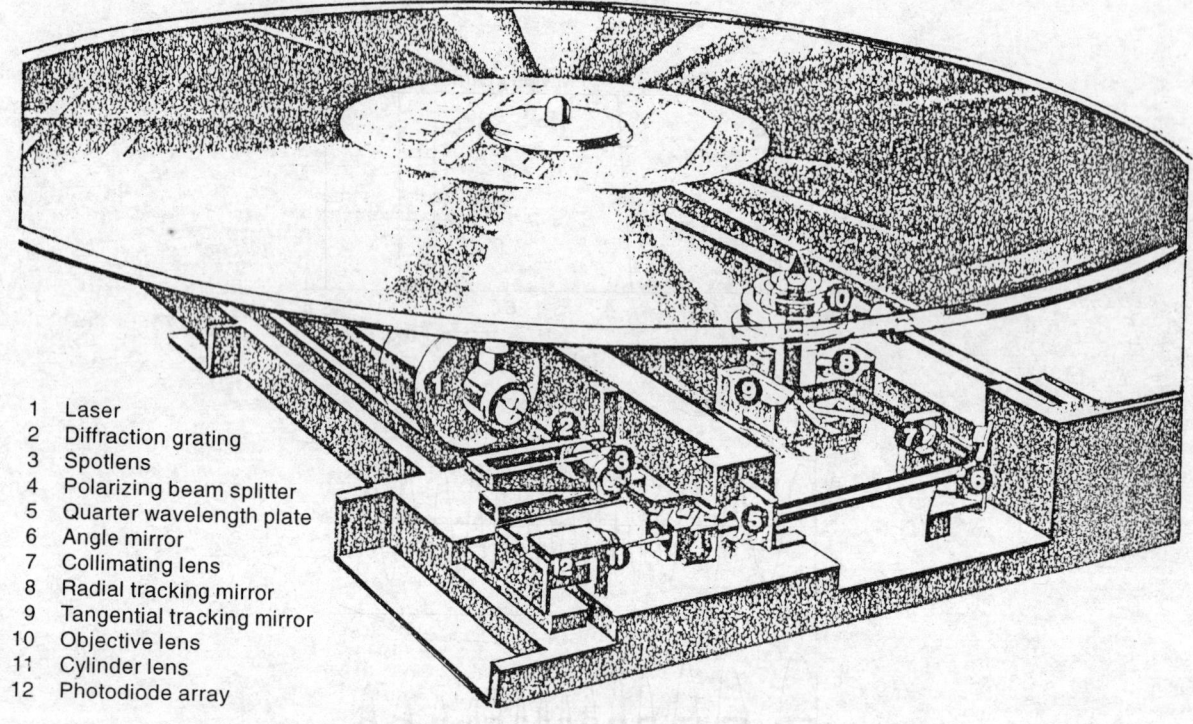

1 Laser
2 Diffraction grating
3 Spotlens
4 Polarizing beam splitter
5 Quarter wavelength plate
6 Angle mirror
7 Collimating lens
8 Radial tracking mirror
9 Tangential tracking mirror
10 Objective lens
11 Cylinder lens
12 Photodiode array

Fig. 22.24 Laservision optical system (*Courtesy:* PHILIPS)

Over many years video disc attracted much attention of the public as video parallel to the music disc records. But systems were for replay-only and needed sufficient software for the public to buy them. By the time Philips finally launched their *Laser vision* video disc in 1982, the VHS and Betamax video cassette systems were well established in the domestic market and preferred by the public as re-usable media with the freedom to record our own programs. While one can listen to music and enjoy replaying the records several times, the same cannot be said for video programs. The need for video disc in broadcasting has been removed by the B and C format recorders.

HDTV Video Disc Player

Sony corporation have demonstrated in 1989 IEEE Consumer Electronics show, a prototype video disc player, for the HDTV 1125-lines/60-Hz format, *without using bandwidth compression* techniques, which can degrade moving objects in picture. The player can give play time of 8 minutes per side in the constant angular mode, or 15 minutes replay per side in constant linear velocity mode. Wide band HDTV signals require higher angular velocity. Instead of this, Sony chose to increase the number of recording channels by developing a *two beam master disc cutting* in an optimum signal format and a *two channel optical pickup*. Precise adjustment of the two beams used for cutting the master disc is ensured by monitoring them with a CCD camera. A photoresistive krypton laser is used, that has long life. The signal consists of 12 MHz luminance and line-sequential chrominance components time division multiplexed to avoid mutual interference. The two channel pick-up employs a high sensitivity wide band photodetector and a tracking error signal detection that reduces interchannel distortion.

22.16   INTERACTIVE VIDEO SYSTEMS

The video disc is however now finding application in large screen presentations and interactive video systems, where replay is frequent and display is interactive. The recording density on the video disc is highest possible with the present state of technology. It is therefore suited for bulk storage of information, requiring fast access and frequent replay, being free from the problems of dust, rough handling, etc. With video disc systems, features like multiscreen displays, slow motion or stop motion displays in either direction with video disc are more easily manipulated by computers by digital system interfaces, than is possible with VCRs.

   Interaction with the viewer is most important in training and educational video systems. Special discs can be produced to react to response from the student, through keyboard of light pen as interfaces. The system can sequence the lesson presentation depending on the response, to repeat necessary portions and correct the student wherever required in a *programmed learning* sequence. It is also possible to overlay text or graphics from a computer onto the video display from the disc, by suitable genlock interface. Future teaching method may depend much upon interactive video in the field where shortage of teachers may continue and student interaction is vital.

   On a new type of *Laser Vision ROM disc,* 648 MBytes of data in addition to 108,000 frames of high quality pictures can be stored. The LV-ROM disc drive is compatible with normal discs and is designed to interface with a microcomputer via the standard SCSI interface, with random access time of less than 1 second.

   The audio compact disc can be used for data storage as ROM to store over 500 Mbytes (200,000 pages of text) on a single side of a 12 cm disc by optical laser technique introduced by Philips. The bulk storage CD-ROM may have significant impact on large data bases and information systems allowing the program producer to combine text, graphs, pictures and sound. CD-ROM data is impressed into the disc substrate as a series of *pits of variable length*. A binary 1 is represented by a land/pit or pit/land transition and the number of 0's is defined by the path length between transition. The data is read by laser focused onto the reflective surface, through the transparent substrate. In the absence of a pit, light is reflected strongly while in the presence of a pit, it is scattered with reduced level of reflected light, providing binary output information.

   The CD-ROM is designed to provide *maximum bit density* as far as the resolution permits, which depend on the aperture of the focusing lens, and ultimately by the laser beam wavelength. For 780 nm laser beam, the data density cannot be beyond 1 μm. The system has to be self clocking, with error correcting mechanism. Optical disc has storage capacity with a density around 10 Mbits/mm squared, an order of magnitude higher than the magnetic tape, with the present technology.

22.17   STILL PICTURE CAMERAS

Recently electronic still cameras have been introduced to store still pictures magnetically on microfloppy disks. As the IC memory sizes shrink, *IC memory storage* is now competing with the slow magnetic storage. The CCD and IC memory technologies have been put together by Toshiba to produce electronic still camera, that store several still pictures on a 85.6 × 54 mm IC card. This produces picture information that can be easily processed and used directly as input to computer and communication applications. Prototypes demonstrated have a horizontal resolution 400 lines. The *card camera,* equipped with auto focus and auto exposure control, produces image on the CCD sensor that has some 400,000 (775 × 494) pixels along with on-chip RGB filters. A schematic of the system is shown in Fig. 22.25.

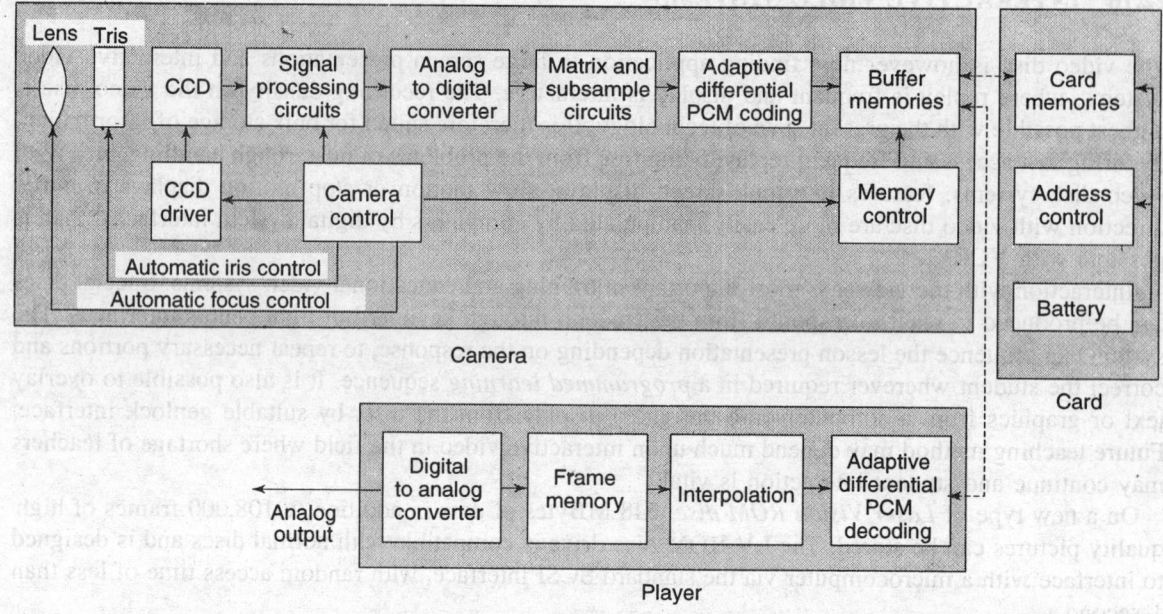

Fig. 22.25 Schematic diagram of IC card camera system (Toshiba) (*Courtesy:* IEEE N Y)

The CCD RGB outputs are amplified, corrected for white balance and gamma correction in the preprocessing circuits and then converted into 8-bit digital data. The digital RGB data are converted into luminance Y and chrominance (R-Y and B-Y) data. The full digital data of the full picture frame is first reduced to 3 M bits by *subsampling techniques*. This is further reduced to 1.5 M bits by halving each 8-bit sample, by an *adaptive differential PCM encoder* and the finally stored in 8 M bit buffer memories.

The 20 M bit memory on the 85.6 × 54 mm card is made up of 20 chips of 1 M bit static RAMs. The player reads out the digital data on the card, converts it into analog TV signals through adaptive differential PCM decoder and interpolation circuits as shown.

REVIEW QUESTIONS

22.1 What are the simplifications possible in the circuit features of industrial CCTV systems?

22.2 Discuss the different design variations in the construction of industrial or educational CCTV systems.

22.3 How is the signal distributed to monitors at various locations? How does the length of cable affect the signal distribution? Discuss the problems and the way they are overcome.

22.4 Name the different types of video switchers. Discuss their operating features and relative advantages giving applications.

22.5 Develop a gating circuitry to obtain vertical sync pulses with equalising pulses for interlaced scanning using the basic timing pulses at 2H, H and V rates.

22.6 Explain the terms: Automatic light compensation, Vertical interval switcher, Loop through connection, Hi-peaker circuit, Cable compensation.

22.7 In what way does a video monitor differ from a standard TV receiver?

22.8 Explain the circuits for processing of video signal for recording it on a video tape, and for playing back the recorded signal.

22.9 Draw the block diagram of a head drum servo to explain how its speed is locked to the camera sync, with the help relevant waveforms.

22.10 Explain the technique of colour dissection in single tube colour cameras. Give the construction of the target arrangement.

22.11 Explain how indexing signal is used to decode the colours to form the CCVS.

22.12 Review the development of video cassette recorders from the U-matic to the 8 mm video, highlighting features that made compact systems possible.

22.13 Explain how the colour under technique is employed to eliminate time base phasing errors in colour reproduction.

22.14 Bring out the special features of (a) Video-2000 system, (b) Fixed head VCRs, (c) Betacam system, (d) Super-VHS.

22.15 Explain the modern methods of video recording on a disc. What are the factors that have not given it an edge over the VCRs?

22.16 What is interactive video? Discuss the merits of the technology in programmed learning.

MULTIPLE-CHOICE QUESTIONS

22.1 How many fields are recorded on one slant track of VHS recorder:
    (a) 1   (b) 2
    (c) 4   (d) 8

22.2 Crosstalk between adjacent tracks is reduced in VHS recorders by:
    (a) automatic scan tracking
    (b) tilting azimuth of the two heads in opposite direction
    (c) by 90 deg phase shift recording of the chroma signals
    (d) b and c

22.3 In colour-under technique,
    (a) chroma signals are band limited
    (b) chroma signals are down-converted
    (c) chroma and luma signals are down converted
    (d) a and b

22.4 Pick out the false one. Broadcast version U-matic recorder employs:
    (a) high band recording   (b) special recording tape
    (c) increased guard band   (d) dynamic track following

22.5 High band U-matic recorders use FM carrier frequencies:
    (a) 3.8 MHz to 5.4 MHz   (b) 4.8 MHz to 6.4 MHz
    (c) 5.6 MHz to 7.2 MHz   (d) 3.4 MHz to 4.4 MHz

22.6 Guard band is absent in:
    (a) U-matic format   (b) Betamax format
    (c) VHS format   (d) b and d

22.7 Pick out the incorrect statement:
    (a) In optical video disc recording recorded pits on the disc vary in depth
    (b) The length and spacing of the pits is changed with the video signal
    (c) Optical video discs are used as replay systems only
    (d) Optical video discs do not need colour-under processing of the video signal.

22.8 Good multigeneration recording operations are possible with:
    (a) Low band U-matic   (b) SP U-matic
    (c) Betacam SP   (d) a, b and d
    (e) b and c

23

Cable Television and Direct Broadcast Satellite Systems

The last few years have seen the emergence of a large demand for conditional access to choice television programs, and more generally to any service available through cable networks or satellites. The cable and satellite system operators offer a *wide variety of choice TV programs* some of them are for subscribers only or pay-TV service. Smart 'key-card' technologies are coming in for these type of controlled access systems for both coaxial cable video communication networks and satellite broadcasting. The MAC packet signals have been proposed, which can encode such facility along with enhanced quality TV broadcast via satellites. These are discussed at some length in chapter 26. The D2-MAC/packet standard evolved by the European Broadcasting Union, has extended to EUROMAC Consortium of firms. Satellite system operators have formulated through this Consortium, specifications for open paid services for the imminent TV-broadcasting/distribution satellites.

A new dimension to televiewing is added by the technology of *interactive television* whereby the viewer can choose any one of the cameras covering programs like the sport events, from different angles and close ups. The interactive technology enables the viewer to select the shots of his choice from his remote control, as if sitting in the control room of the studios! This flexibility is possible through the cable TV circuits.

23.1 CATV SYSTEMS AND CHANNELS

Cable Systems

Cable TV began in the USA as community antenna television (CATV) systems for the benefit of communities in rural areas, that were out of range of broadcast transmissions and those living in shadow zones. Programs received by elaborate CATV antennas, microwave links or later by satellite links, were supplied to the members of the communities via cable TV systems. Today, cable television has developed far beyond small communities into large cable or pay-TV networks, even in areas where good broadcast reception is possible. The cable TV networks provide can offer a wide choice of program material free from external interferences and ghost images due to reflections from high rise buildings in large cities.

Program Choice: The video tape recorder and studio equipment costs have come down considerably so that local program generation is easy, and can be a ready source for feeding popular programs into the cable TV network. On cable TV networks, there is scope for two way interactive communication with computer databases to provide services like videotext. HDTV can also be realized in some cable networks as the full cable spectrum is available to accommodate the large bandwidth required by it.

The cable TV operators can provide programs originating from a variety of sources that may include 'Off-air' terrestrial or satellite reception, local VCR fed or studio generated broadcast exclusive programs of *wide choice* educational and entertainment programs not obtainable except over the cable. Systems used in the 80s have 50 to 100 channel capacity with interactive capabilities. This has generated a great interest and demand from the viewing public and financial incentives for the cable operators with potential for local advertising. Growth of cable TV has been very fast and by 1988 itself over 50% of household viewers in many countries subscribe to cable TV.

Cable Channels

Standard TV channels are often used for cable TV, which can be directly connected to the tuner input for necessary selection. Other special cable TV channels employ the mid-band range between Band I (low band) and Band III (high band), avoiding the FM broadcast band II, and the superband range above band III. The CCIR allocation in the mid-band includes channels S1 to S10 from 104 to 174 MHz in the midband, and S11 to S20 from 230 to 300 MHz in the superband. For receiving these special channels separate converter units are required to convert any of these to a standard TV channel, usually channel 2 or 3 of band I. The newer systems are capable of operating in the range 5 to 400 MHz, excluding the frequencies used for aircraft navigation and communication bands which are prohibited.

Some systems operate on the standard TV channels but operating slightly off the assigned frequencies, but in the fine tuning range. The chosen frequencies are harmonically related as multiples of the channel width (HRC system), which simplifies the frequency synthesis in the cable converter.

23.2 CABLE TYPES AND NETWORKS

RF-coaxial cable has been the transmission conduit for distributing the CATV signals, while the glass-fibre optic cable is slowly emerging as a viable alternative for wide band data communication, that could include digital TV signals. The fibre optic technique use infra-red light as the carrier, usually in pulse code modulated (PCM) mode, and requires more complex terminal equipment, though it has advantages of small size cable, immunity to electrical interference, wide band high data rate capability and so on.

Coaxial Cables

Currently coaxial cable is well established for CATV use, and has the advantages of easy interfacing with existing equipment, lower initial costs. Larger diameter (20 to 25 mm) coaxial cable giving lower attenuation (approx. 1 dB per 30 m, at 400 MHz) is used on the main trunk route. The trunk cable has inside aluminum tubing, a thick copper-clad aluminum conductor centrally supported by polyethylene foam or spacers at regular intervals. Protective polyethylene jacket and steel armor is used for waterproofing and strength, so that the cable can be mounted on poles, or laid underground or underwater. Thinner cable can be used for branch lines, while the subscriber drop line is generally the RG-59U 1/2 in. flexible copper braid coaxial cable. The nominal cable impedance is generally 75 Ω.

Fibre Optic Cables

Instead of the bulky and heavy copper and aluminum cables fibre optic cables, no thicker than human hair become available in late 70s, carrying vast amount of data using light as the modulated carrier.

Glass-fibre optic cable consists of a thin glass fibre of 50–100 μm diameter, enclosed inside a low refractive index cladding of about 30 μm thickness and uses light energy as carrier for the data. The modulated light is launched into and guided along the thin glass fibres via a process of *total internal reflections* from the inner surfaces of the fibre walls interfacing with the cladding due to sudden change in the refractive index of the so called *step-index fibre*, as shown in Fig. 23.1.

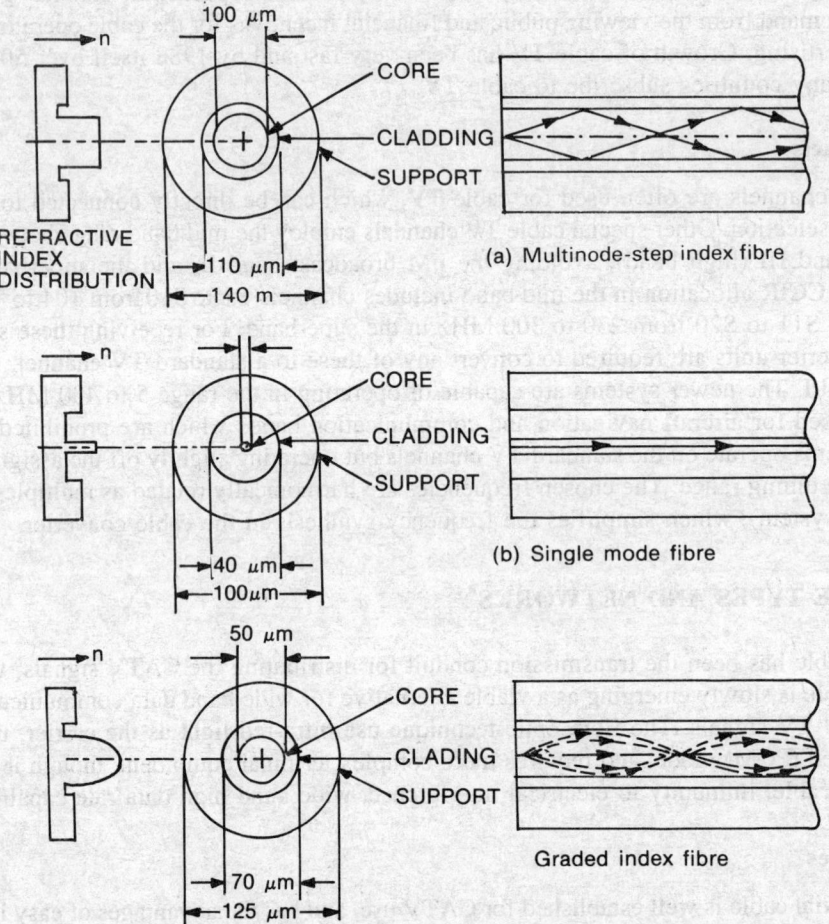

Fig. 23.1 Fibre optic cable transmission modes

Inside the fibre light waves can travel at several discrete angles with the axis, resulting in different modes of the fibre. The light source e.g. an LED couples light into the fibre at different angles. Lower the angle of light path with the axis, lower the number of reflections with shorter path between reflections. This mode is refered as lower order transmission. Larger reflection angle gives longer path and higher order transmission modes. This distributes the optical energy in different modes which propagate along the fibre and are combined on the detector at the far end. We get thus, a *multimode transmission* in step-index fibres with relatively *large core diameter*. The problem then is that these modes travel different distances from source to detector and hence a transmitted narrow pulse reaches the detector at different times via different modes resulting in pulse broadening, called *intermodal dispersion*. This phenomenon requires that the transmitted pulses be separated widely, lest they overlap each other. This separation, however *reduces the bit rate* of the system defeating the purpose of using optical fibres.

One way to avoid this is to use very thin fibre core propagating only the axial ray. In a *single mode fibre,* the core diameter is made very small, less than 10 μm so that only the axially launched ray propagates along the axis as in a single mode. This implies that the light source must couple light into a *very thin core;* and the alignment requirement for coupling between the source and core or two fibre cores is very severe. A compromise is made in a graded index multimode fibre which has approx. 50 μm diameter core, but the core refractive index varies with the radial distance.

In the *graded-index fibre*, the refractive index along the radius is gradually varied to give a graded-index profile as shown in Fig. 23.1. The light rays entering at an angle with the axis bend around gradually, as shown in the figure. The modes that have to travel through longer paths encounter smaller refractive index near the cladding, travelling faster in the regions away from the axis. If the refractive index is so designed that the *time delays for different modes* propagating in the fibre are *equalised*, multimode dispersion is eliminated. It also reduces the reflection loss as at the boundary of step-index fibres, but cost of fabrication is higher.

The sending device is an LED, or for larger distances IR LED laser diode encapsulated with the fibre for coupling maximum light into the fibre cable. Typically, 1–3 mW of the IR radiant energy is available from the laser diode at about 860 nm. The receiving end device is a PIN photodiode or an avalanche photodiode (APD) for long-haul reception, coupled intimately to the fibre end.

Optical fibre technique has several *advantages* in terms of data handling capabilities, viz. wide spectrum for high speed data in the small size thin fibres, greater transmission efficiency, immunity from electrical interference and data security, although the terminal equipment is more complex and costly. Earlier FOC links operated at 850 nm, having attenuation around 3 dB/km. The next generation FOC links are built around 1300 nm, giving a loss below 0.5 dB/km. These provide very high bandwidth links with 30–40 km length. Sources and detectors in this region are more expensive. Recently, there is a greater interest in the 1550 nm system which offers very low attenuation 0.1 dB/km, with the possibility of repeater distances over 300 km. Because of greater transmission efficiency, repeater spacing can be very large and hence for very high data rates up to 140 MB/s, it is an attractive proposition for trunk routes to the local distribution points, leaving the branch lines to the subs multitaps for the coaxial cables.

Presently Fibre optic cables are used for transporting programs over trunk routes or specific links. Their use in subscriber distribution network will not be quite justified until telephone, data and video information are transmitted at acceptable investment level per subscriber level. In the longer perspective, the wideband fibre optic network has much appeal as the cost of optical fibres and associated components show a steady downward trend. The cost effectiveness of fibre optic system will increase if optimum network topology is combined with large scale integration of components and functions required for value added networks services.

Subscriber Network

For cable distribution to the subscribers, traditionally the *tree and branch network* has been used, with trunk lines branching into different regions, with further branching into twigs to feed to the subscriber taps. An alternative system is the *switched star network,* where the available programs are piped to a point central to the region covered. Each subscriber gets interactively the desired TV program from star network, like telephones in the PSDN. The cable TV system consists of the trunk and distribution system taking the signals from the head end, where all the signals originate, to the subscriber television sets through subscriber drops. The drop may connect directly into the set if the standard TV channels are used, or may require a converter, or a descrambler for special channels like pay-TV channels. In the interactive systems, the subscriber is allowed to interact with the program source through bi-directional repeater network and get videotext information or special channels. For this, additional equipment like a videotext decoder and cable modem for data transmission, is needed.

Bi-Directional Networks

Most of the bi-directional CATV networks single cable use frequency division for upstream and downstream communication, typically 5 to 35 MHz for upstream and 50 MHz and higher for downstream transmission. Each line repeater is split into two paths with *band splitter filters* at the input and output. Each path includes a plug-in equaliser, an attenuator and a hybrid amplifier module to take care of the upstream and downstream signals separately. The terminals equipment can be of transmitter or *transponder* type. Transmitter type terminal serve for monitoring functions only, such as pay-TV or home security messages for the head end where a scanning receiver polls the subscriber transmitters in a CATV system. The date rates are slow, less than 3 kB/s. A transponder terminal comprises a transmitter, a receiver and address recognition circuitry, and can be *polled by the computer* at the head end through a common addressing channel to send data at pretty high rates, 256 kB/s typically.

23.3 HEAD END PROCESSORS

Head End

All the signals carried on the cable TV network originate at the head end, where they are frequency multiplexed and combined for transmission over the cable. The signals are received off the air, from terrestrial microwave links and from TV satellites by means of *elaborate antenna systems*. For reception of distant stations, directional aerials of high gain are installed on tall towers, located in areas of low noise and minimum interference. *Low noise booster* preamplifiers are used near the pickup aerials for weak signals. Interfering strong adjacent channels are rejected through suitable filters if necessary. For satellite TV reception, television receive only (TVRO) earth stations with suitable size parabolic dish antenna and low noise front end converter are used, also located for minimum interference from terrestrial microwave links.

Head End Processors

At the head end the various signals obtained, are processed to bring the output to a constant level, rejecting the out of band signals and then multiplexing them on the cable for simultaneous transmission, by converting each to different cable channels. The multiplexing is done by modulating the programs on suitable cable channel carriers. Processing of off-the-air TV signals is done by a heterodyne processor or a demodulator-modulator pairs. A *heterodyne processor* shown in Fig. 23.2, converts the incoming signal to the IF carriers. The video and audio IF carriers are amplified filtered and level controlled independently, before they are recombined and heterodyned—upconverted to the desired output channel. Two *pilot carriers* are generally transmitted near the ends of the cable channels spectrum to provide reference levels for the wide band trunk amplifier distribution system. All the signals are grouped together frequency multiplexed on different cable channels in the spectrum, in *combining networks* to that feed into a common cable.

In a *demodulator-modulator pair,* the video and audio are first obtained in a high performance TV receiver and then modulated on a low power TV transmitter also employing IF modulation technique. The demodulated video and audio baseband signals first modulate IF carriers, 38.9 MHz for picture and

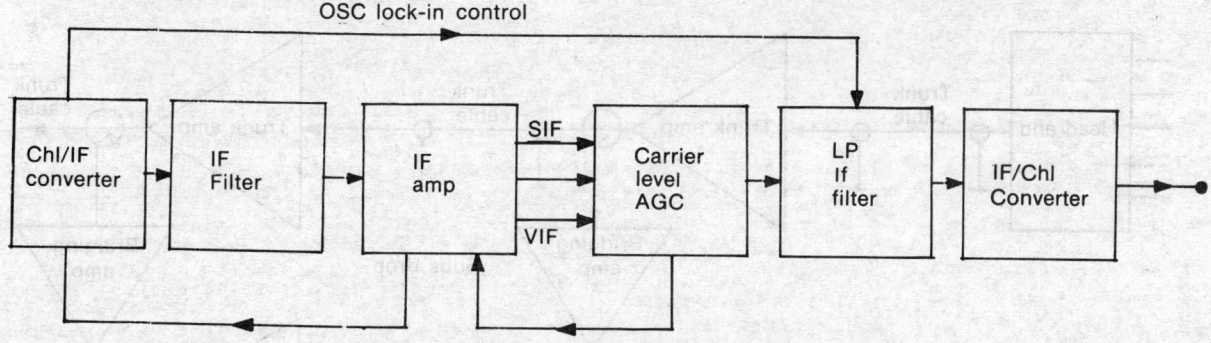

Fig. 23.2 Schematic of a heterodyne processor

33.4 MHz for the sound. The IF modulators for all inputs are thus identical, while suitable unconverter places these IF signals on the desired cable TV channels. This method provides better selectivity and signal level control besides flexibility in switching of signal sources.

23.4 TRUNK AND CABLE DISTRIBUTION SYSTEM

The multiplexed channels from the up-converters are combined into the trunk cable system connecting to a number of distribution points. Long haul trunk cable systems carrying a large group of cable channels are referred to as *supertrunks*. **Trunk amplifiers** with equalisers are used at regular intervals in the trunk system to overcome the losses in the cable, which increase towards the high end of the spectrum. These trunk line repeaters providing a gain of about 20 dB, compensate the corresponding loss of a cable run of about 600 m, depending on the type of cable, and maintain the signal level of 1 to 3 mV. Minimum input for TASO *excellent grade picture* is 1 mV (0 dB mV).

Automatic gain control (AGC) and *automatic slope control* (ASC) is incorporated in the amplifiers to maintain the constant levels flat within +/– 0.25 dB, over the whole cable spectrum with the help of the two pilot carriers. The maximum level allowable in a cable amplifier is usually below 32 dBmV to keep down the cross modulation of the picture signals to below 96 dB, typically. The amplifiers are supplied power through the cable, which results in a hum modulation which has to be kept below – 40 dB relative to signal carrier. Although the noise figure of a hybrid trunk amplifier is typically around 7 dB, the overall noise figure of a trunk station, with equalisers, filters etc. may add up to about 10 dB. Trunk amplifier cascade up to 64 repeaters are possible, depending on the channel spectrum, modulation method, etc. For long distance trunks, FM is preferred to minimize nonlinear distortion at the cost of bandwidth. The schematic of a trunk distribution system is given in the Fig. 23.3(a).

Fibre Optic Trunk Cable Links

An immediate economic motivation for using fibre in CATV distribution is to cut down on the need for repeaters and improve signal quality to customers on the network. With copper coaxial distribution repeaters are required every 500 meters or so. Fibre link can carry the optical signal up to 20 km. Optical repeaters are used to extend the trunk distance range to tens of km. Within existing CATV distribution network, the number of repeaters can be reduced and signal made less vulnerable to trunk amplifier failures by providing *fibre optic backbone* for long haul drop locations. The system can be used as a link between a master head end and smaller 'hub' distribution system. Fibre optic transport systems are also

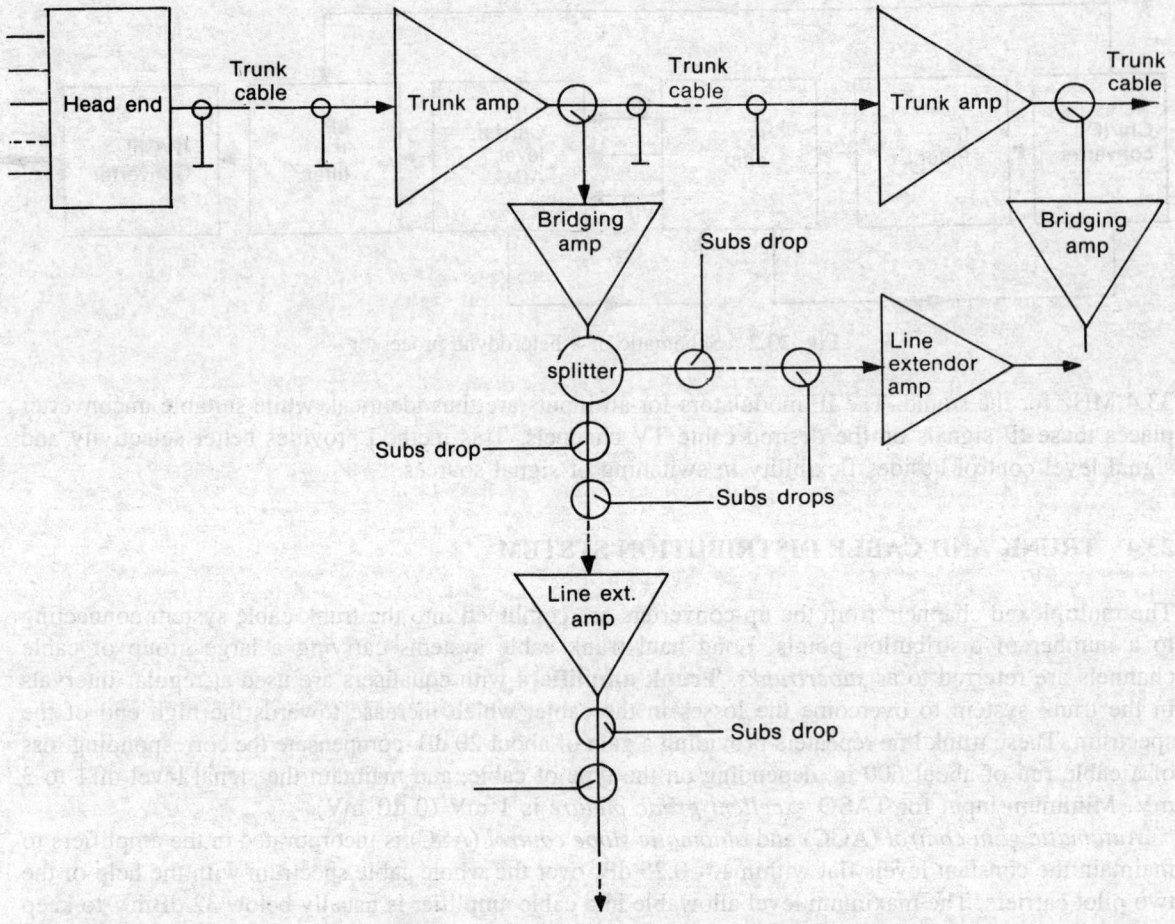

Fig. 23.3 Trunk and cable distribution system

employed on CATV trunk lines and as *wide band remote link for TVRO* earth stations feeding to CATV head-end processors.

In a multichannel fibre optic transport system including *drop and retransmit locations*, it is possible to send the down converted wide band FM IF signal on the fibre optic system giving distinct improvement in S/N ratios over the copper cable systems. For sending more channels, the baseband video inputs are frequency modulated and frequency division multiplexed by suitably upconverting each signal. The combined multichannel FM signal is fed to the laser diode transmitter that couples the optically to the fibre cable. At the receiver end an APD converts the optical signal back into electrical FM signal. This is demultiplexed and demodulated into the video baseband. S/N ratio over 50 dB can be maintained on the optical fibre cables carrying wide band multichannel signal.

Fibre to home feeding CATV channels and other value added services are expected to be soon a reality. These services include integrated phone, home security, fire and medical aid. Several regional Bell Operating Companies (RBOCs) in the USA, have field trials by now and are expecting to scale up operations in the mid-1990s. These potential services, the advent of HDTV demanding increased bandwidth not available via coax, increased penetration rate of CATV in resistential market are driving forces in favour of the wide band fibre optic technology. Critical to the CATV fibre industry are low cost solutions for volume production of taps and couplers along with innovative multiplexing techniques in the optical and electronic domain.

Drop Location

At drop locations, the fibre optic link drops the signals as it is received by the APD receiver as electrical signal, demultiplexed, and demodulated into the basebands, before it proceeds to the laser transmitter and the fibre cable ahead. The baseband signals are available for the local distribution. No channel switching or insertion of new local signal into the hub is intended in a drop location.

Retransmit Location

At these locations, new channels can be added into the fibre system which may not fully utilized. The received signals from the APD receiver are demultiplexed, and passed on to the multiplexer along with new channels, modulated and added into the unused bandwidth. Some of the demultiplexed channels may be demodulated and dropped at the location for local distribution, and some of the new channels may be replaced in place of the dropped channels.

Satellite TVRO Links

When the satellite TVRO station is not very near to the head-end processer fibre optic cable link is an attractive proposition as a wide band transport system. The FM IF signal from the front end converters can be directly multiplexed on to the optical link, and received and demultiplexed at the head-end as FM IF signals for further processing and distribution.

Bridging Amplifiers

Bridging amplifier is used for feeding a *branch lines* from the main trunk, distributing the signal to subscriber drops in a particular area. A bridging amplifier, sometimes housed together with the trunk amplifier, provides a gain of about 20 to 40 dB, to feed the signals to the subscriber cables through directions couplers and signal splitters designed to introduce isolation from reflections or interferences coming from subscriber sets. The signal splitters and the couplers or the *tap-offs* are fabricated using transmission line techniques or ferrite core transformers.

Impedance matching of the coaxial cables is important to prevent reflections and ghost effects. For dividing, redirecting, tapping the signal from cables, or terminating of CATV signals, well designed standardized passive components are used. These include: (i) signal splitters, (ii) multitaps, (iii) directional couplers, (iv) matching transformers, (v) terminators. *Signal splitters* divides the input power into 2 or more paths. A 2-way splitter introduces a loss of 3.5 dB on each leg. This can be used to further split the signal into 3, 4 or more paths when additional 3.5 dB loss will be incurred in each successive splitter. *Multitaps* are units placed on the cables to enable portion of the signal to be sent to the subscriber. 1, 2, 4, or 8 outlets are provided from the cable by specially designed junction boxes (tap-offs) that are freely accessible. Individual outlets may be connected or disconnected without disturbing other outlets. *Directional coupler* provides one way tap off from the cable, restricting any signal flow in the opposite direction (tap isolation). A directional coupler introduces insertion loss of about 1 dB. 75 Ω to 300 Ω Balun *matching transformers* are used match with 300 Ω the antenna input of a TV receiver, if necessary. 75 Ω resistance connectors are used as *terminators* for the ends of the cables.

Line Extender Amplifiers

When the line run from the bridging amplifier to the drop lines is very long, line extender amplifiers providing a gain of about 20 to 40 dB, are inserted in the branch line to compensate for the cable loss in it.

23.5 SCRAMBLING AND CONDITIONAL ACCESS SYSTEMS

Traps

While cable TV offers some basic service for a minimum fee, special premium programs are offered only to those subscribers who pay extra premium. Two techniques are possible to authorize the subscribers. One is the *trap method* implemented either by inserting an interfering carrier in the pay channel and notching it out by a suitable sharp filter at the subscriber (positive notch method), or by introducing a sharp notch filter on the picture carrier of the premium channel, installed outside at the drops of the not authorized subscribers (negative trap method). These methods are easy to tamper and hence scrambling technique is now more commonly employed for conditional access to a channel.

Scrambling

A common method of scrambling is to *suppress the sync* signals at the transmitting head end. The picture without H and V sync cannot lock into a steady picture, usually seen as rolling picture and diagonal bars, with overloaded contrast due to loss of AGC bias also. For descrambling a pilot carrier containing the sync timing information is transmitted on a separate carrier or within the base band of the channel. In the descrambler located at the subscriber's converter, the pilot signal is used to reinsert the sync signals to the video.

Another method offering greater security with greater complexity is the *base band scrambling,* where polarity of randomly selected lines is inverted. The scrambled line code, transmitted in the video band is decoded in the descrambler to restore the inverted lines. This requires demodulations and remodulation after separating the code.

More sophisticated scrambling is possible by digital processing of the video and audio and introducing complex encryption and technique in the process and smart key-card for controlled access.

The problem of signal security in cable TV can also be solved by *off-premise* converters which allow only authorized channels into the subscriber's home. For this, *addressable converters* are now available, which include a tunable converter and address recognition circuitry. Each converter can be addressed and controlled from the head end by a computer so that specific channels can be turned off or on to the subscriber.

Conditional Access

The functional features like equipment compatibility, access, support, user system interface, telephone communication module for conditional access to television, depend on the characteristics of the supporting network and the transmission standard (PAL/SECAM or D2-MAC). The technique uses for scrambling, and the procedures for injecting *encrypted messages* pertaining to the conditional access have an impact on the scramblers and injectors upstream and descrambler modules or terminals at the users. Mass produced *smart key-card* with 'buried' key components, have been adopted by the operators and manufacturers and are expected to reduce costs. Both in the cable and satellite systems three different operational scenarios can be projected, viz. access to a given channel for a fixed duration, pay per view with preselected choice of a channel in advance, or pay per view without preselection.

23.6 DIRECT BROADCASTING SATELLITES

Geostationary Orbit

As indicated in Section 7.12, satellites orbiting at a height of about 36,000 km from the earth, at an orbital speed of about 3 km/s (11,000 km/h) can act as a geostationary satellite, when the centrifugal force acting on the satellite just balances the gravitational pull of the earth.

If M is the mass of the earth, m is the mass of the satellite, r, the radius of the orbit, and G, the gravitational constant, we have the centrifugal force,

$$\frac{mv^2}{r} = G\,\frac{Mm}{r^2} \quad \text{the gravitational pull,}$$

which gives

$$v = \sqrt{GM/r} \tag{23.1}$$

Also, the orbital period T of the satellite should be 24 hours.

Hence,

$$T = \frac{2\pi \cdot r}{v} = 24 \times 3600 \text{ seconds.}$$

Substituting for v from the above equation, and the values for the mass of earth $M = 5.974 \times 10^{24}$ kg, and $G = 6.6672 \times 10^{-11}$, gives the orbital radius of a synchronous satellite as 42164 km. Deducting the radius of earth equal to 6378 km, the distance from earth surface will be 35786 km.

A geostationary satellite must have an *equatorial orbit*, in order to prevent apparent North-south oscillation due to the tilt in the axis of the earth with respect to the solar orbit of earth. Any drifts from its earth-relative position and altitude have to be corrected by small on-board gas jet motors. Such geostationary satellites having powerful transponders, can beam concentrated signal to desired areas and act as direct broadcasting satellites (DBS). Geostationary orbit in the plane of the equator implies solar eclipse twice a year around the equinoxes around 21st of March and 21st of September, when the satellite passes into the shadow of the earth. The duration of the eclipse lasting for some 52 days in a year rises to a maximum of 72 minutes, around the midnight. If the satellite is kept on the west of its footprint, the eclipse time can be delayed, 4 minutes per degree, thus requiring over 9 deg west for (72/2 = 36 minutes) for delaying the eclipse to beyond midnight, a necessity for viewing the 'late night movies' without a midnight 'black-out'!

In a communication link, the carrier to noise *C/N ratio* determines its performance. In DBS, this depends on the EIRP, rain attenuation and the *figure of merit* (G/T) of the earth receiving terminal. It is improved as EIRP is increased, or the service coverage area is reduced while increasing the transmitting antenna gain. The figure of merit G/T, is the ratio of the *antenna gain* in dB, to the *effective noise temperature* of the receiver. One can use a large dish with a high gain with a poor input amplifier, or alternatively do with a smaller dish used with a high performance low noise front end. WARC '77 formulated DBS plans on the basis of G/T ratios of 6 dB/K. Developments in low noise RF GaAs FET amplifiers have now enabled to achieve G/T of 10 dB/K in domestic installations with a 0.9 m dish.

Footprints

As the satellite radio beam is aimed towards the earth, it illuminates on the earth an *oval service area,* called the 'footprint'. Because of slant illumination of the earth by the equatorial satellite, this is actually an egg-shaped area with the sharper side pointing towards the pole. The size of the footprint depends on how greatly the beam spreads on to the surface of the earth intercepted by it. The foot print for contours of 3 dB or half power beam width are usually considered. The beamwidth planning depends on the angle of incidence of the beam on the earth or the angle elevation of the satellite. It can be directly controlled by the size of the on-board parabolic antenna. Present day launchers can carry antennas of around 3 m, giving a minimum beamwidth of about 0.6°. With difficulties in *accurate station-keeping*, it is prudent to allow for a margin of around 0.1° when planning the footprint to cover a country. Some satellites employ additional antennas to emit spot beams that cover regions beyond the normal oval shape. The *slant range* of a satellite involves calculation of the distances from the *bore sight point* of the beam, covered by the semi-beamwidth angle, considering the geometry of footprint.

The *free space loss* depends on the path length d which is related to the angle of elevation. The radio waves undergo attenuation loss due to scattering and absorption in the lower layers of atmosphere and by rain, clouds etc. The *atmospheric loss* depends on the length of path through atmosphere, and naturally increases with lower angles of elevation. The loss also varies according to the atmospheric fluctuating with respect time. Hence maximum attenuation loss values, encountered for 99% or 99.9% time, during which satellite broadcasts are received are considered, depending on the degree of reliability sought. For *99% reliability*, the attenuation in the 12 GHz band is found to increase e.g. from 1.5 dB at 45° angle of elevation, to 6.8 dB at 5° and for 99.9% reliability it increases to 4.8 dB and 14 dB resp.

Let us consider the power budget for a typical satellite-earth link as shown in Fig. 23.4.

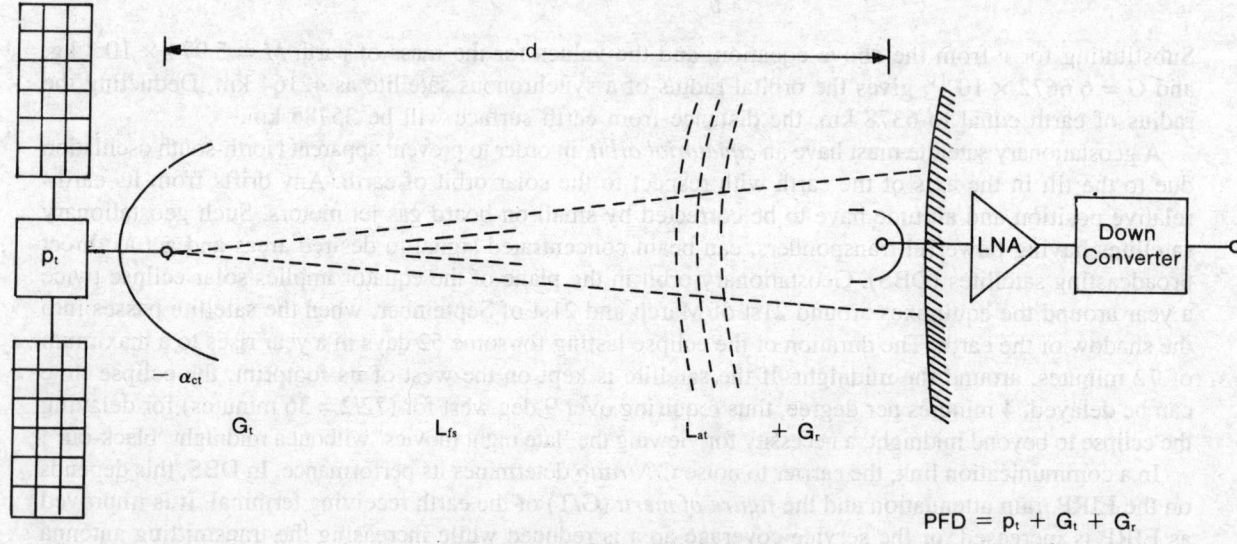

Fig. 23.4 Satellite-earth link budget

At the satellite transponder, a power amplifier feeds power Pt to the transmitting antenna of maximum directive gain Gt. The maximum radiated power EIRP from the antenna is *Pt· Gt*.

$$\text{EIRP} = Pt + Gt$$

As this power propagates towards the earth it spreads into space and encounters the so called free space loss. The spreading factor is given by $1/4\ \pi d^2$. The power density along the direction of maximum radiation is

$$\text{PFD} = Pt \cdot Gt/4\pi d^2$$

When a parabolic dish receiving antenna is positioned to collect maximum power from the radiated power, the total power intercepted and received is given by

$$Pr = \text{PFD} \cdot \text{Aeff}$$

where Aeff is the effective dish area or aperture, ($= \eta \times A$, efficiency coefficient η, accounting for the dish coupling loss)

$$Pr = \text{PFD} \times \text{Aeff}, \text{ where Aff} = \eta \times A.$$

The *power gain* of an antenna (in the direction of maximum directivity) with respect to isotropic antenna is given by the basic relation:

$$G = \frac{4\pi \cdot \text{Aeff}}{\lambda^2} \tag{23.2}$$

where λ = received wavelength, or

$$\text{Aeff} = \frac{G\lambda^2}{4\pi} \tag{23.3}$$

If Gr is the maximum directivity gain of the receiving antenna,

$$Pr/Pt = Gt \cdot Gr \ (\lambda/4\pi d)^2 \tag{23.4}$$

This can be expressed conveniently in dB to give

$$Pr/Pt = (Gt)\text{dB} + (Gr)\text{dB} + 10 \log (\lambda/4\pi d)^2 \tag{23.5}$$

The last term represents the spreading loss known as free space propagation loss. Thus the free space propagation *magnitude of loss* in dB between isotropic antennas is given by

$$
\begin{aligned}
Lfs &= 10 \log Pt/Pr = 10 \log (4\pi d/\lambda)^2 \\
&= 20 \log (4\pi \cdot d/\lambda) \\
&= 20 \log (4000\pi \cdot f \cdot d/0.3), \text{ where } d \text{ is in km and } f \text{ is in GHz} \\
&= 32.45 + 60 + 20 (\log f + \log d) \text{ dB} \\
&= 92.45 + 20 \log f + 20 \log d
\end{aligned}
\tag{23.6}
$$

For geostationary satellite having footprint away from the equator the path length d, is related to its height h (= 35786 km) from the equator, earth radius R (6378 km) and the angle of elevation β, by simple geometric relation

$$(R + h)^2 = R^2 + d^2 + 2Rd \cos (90 + \beta) \tag{23.7a}$$

i.e.

$$(42164)^2 = 6378^2 + d^2 + 12756\, d \cos (90 + \beta) \tag{23.7b}$$

where β is the angle of elevation of the satellite with the horizon.

The received power in dBW is the sum of the following factors expressed in dB: EIRP, free space propagation loss, feed-line loss, receiving antenna gain, plus any other losses such as polarization mismatch, antenna pointing error, atmospheric loss etc. as applicable.

Beam Width

The radiation pattern from a parabolic dish can be calculated from the equation:

$$E\phi = J1\left(\frac{\pi \cdot D}{\lambda} \sin \phi\right) \times \frac{2\lambda}{\pi \cdot D \cdot \sin \phi} \tag{23.8}$$

where J1 = Bessel function of first order,
  D = diameter of the parabolic dish,
  ϕ = angle of direction with respect to the principle axis of the antenna aperture.

The expression (argument) within the bracket is evaluated and the Bessel function obtained for the argument, from Bessel function tables/graphs. The values of the argument for which the Bessel function becomes zero, will be found to be 3.8, 10.3 and 13.5. The angle of the radiation pattern where the first null occurs is given by

$$J1\left(\frac{\pi \cdot D}{\lambda} \sin \phi\right) = J1(3.8) \tag{23.9}$$

which gives

sin ϕ = 3.8 $\lambda/\pi \cdot D$ = 1.22 λ/D, and hence
  ϕ = 1.22 λ/D radians, or

$$\phi = 70(\lambda/D) \text{ degrees} \tag{23.10}$$

The main lobe of circular dish lies within the angle ϕ between the first nulls on either side and is given by twice this angle 0.
  The 3 dB beamwidth of the main lobe is given by the half lobe angle

$$\phi \text{ 3dB} = 58 \ \lambda/D \tag{23.11}$$

It may be observed that the antenna gain is inversely proportional to square of the beam width. That is, a decrease of the beam width by a factor of 2 obtained by doubling the diameter of the dish increases the antenna gain by a factor of 4 (6 dB).

Example Consider a satellite antenna fed transmitter power of 200 W, at 12 GHz. The antenna gain is 43 dB (20,000), the transmitter antenna coupling loss Lt is 2 dB. If the distance between the satellite and the receiving antenna is 36,000 km, and the atmospheric loss is 1.5 dB, find the signal power available from the receiving parabolic dish antenna of 0.9 m diameter and efficiency of 50%.

Pt = 100 × 2W = (20 + 3) = 23 dBW

Satellite EIRP = $Pt + Gt - Lt$ = 23 + 43 − 2 = 64 dBW
For 12 GHz and distance of 36,000 km, the loss is given by substituting values in equation 23.6,

  Lfs = 92.45 + 21.58 + 91
    = 205 dB

Signal power flux density *PFD* available to *isotropic* receiving antenna on earth is

  PFD = EIRP − Lfs − Latm
    = 64 − 205 − 1.5 = − 142.5 dBW/m^2

For a circular dish antenna of diameter $D = 0.9$ m, and $\eta = 50\%$

$$Aeff = \eta \cdot A = \eta\pi D^2/4,$$
$$= 0.5 \times \pi \times 0.81/4$$
$$= 0.32 \text{ sqm}$$

At 12 GHz, the wavelength in meters is

$$\lambda_m = 300/f_{MHz} = 0.3/f_{GHz}$$
$$= 0.3/12 = 1/40 \text{ m}$$

The receiver antenna gain for the above values is given by:

$$Gr = \frac{4\pi \cdot Aeff}{(\lambda)^2} = 38 \text{ dB}$$

Power received at the antenna terminals is given by

$$Pr = PFD + Gr \text{ in dB units}$$
$$= -141.5 + 38 = -103.5 \text{ dBW}$$

Noise Figure Considerations

In DBS, the minimum power flux density depends on the figure of merit G/T of the receiving station, defined as the ratio of the receiving antenna gain to the effective noise temperature of the receiving terminal (also see section 12.2). This includes *natural noise* as stellar noise, sky noise, receiver noise and *manmade noise* like RF interference, cross talk, quantization noise, mains hum etc. entering the receiving bandwidth of the system. In the lower fractions of 1 GHz, the atmospheric noise predominates. As we approach 1 GHz, the main source of noise is the galactic noise. In the microwaves range where $hf << kT$, the amount of noise received by a ground based receiver from a black body like the earth, at a temperature $T(K)$, is given by Plank's law, which can be approximated to $P_n = kTB$. Noise emanating from solar and celestial bodies is most commonly expressed as *noise temperature*, in terms of this relation. Above 0.1 GHz, the galactic noise temperature of a few thousand K at 0.1 GHz falls off rapidly to a few K, at 1 GHz.

For considering the effects of various noise sources the concept of noise temperature is most convenient. For a receiving installation the total system noise output Pn is the sum of the noise contributions from the antenna, RF, mixer and IF unit gains and noise temperatures.

$$Pn = Gk \, Ts \, B = (GrGmGi \cdot Ta + GrGmGi \cdot Tr + GmGi \cdot Tm + Gi \cdot Ti) \, kB.$$

Hence the equivalent system temperature is given by

$$Ts = (Ta + Tr + Tm/Gr + Ti/GrGm) \tag{23.12}$$

In terms of the noise figure F, the equivalent noise temperature,

$$Te = To \, (F - 1), \text{ by equation (10.6)}$$

G/T Ratio

For a receiving station, the G/T ratio is defined as the ratio of its receiving antenna gain to the effective noise temperature the system including effects of the sky noise, receiver noise coupling loss etc. This ratio is often refered to as 'sensitivity or threshold' of the receiving system. This G/T ratio is the figure of merit of the system. It can be expressed as

$$\frac{G}{T} = \frac{\alpha \beta Gr}{aTa + (1 - \alpha) To + (F - 1) To}$$

(23.13)

where To = abs. temp 290 K
Ta = eff. noise temp.
F = rec. noise figure
α = coupling loss
β = pointing and polarization loss

The ratio give a flexibility to the designer to chose between a small dish aerial with a low noise front end or a large aerial with a higher noise figure front end. The WARC 77 adopted a G/T of 6 dB/K for earth reception terminals with the then available technology. Present GaAs MOSFET technology has raised this figure to 10 dB/K.

C/N Ratio

of the carrier power received to system noise, C/N is given by

$C/N = Prx/kTsB$, therefore

$$C/N = \frac{Pt \cdot Gt \cdot Gr}{4\pi \cdot d^2} \times \frac{\lambda^2}{4\pi} \times \frac{1}{(kTsB)}$$

(23.14)

$= $ constant $\times Gr/Ts$,   for a satellite system.

This indicates that C/N is proportional to Gr/Ts, remaining terms being constants for a particular DBS system. The G/T ratio is the figure of merit used for evaluating the noise performance of a receiving installation. It may be noted that while the FM bandwidth of the satellite signal is around 27 MHz, the baseband is 6 to 7 MHz. The system noise is thus reduced in lieu of the reduced bandwidth and pre-emphasis. The S/N at the output of the FM demodulator is better than the C/N before demodulation. The improvement ratio is given by

$$\frac{S/N}{C/N} = \frac{B}{f_m} \times 1.5 \left(\frac{f_{max}}{f_m}\right)^2, \qquad \begin{array}{l} \text{where } B = \text{FM bandwidth} \\ f_{max} = \text{max. deviation} \\ f_m = \text{modulating frequency} \end{array}$$

$$= \frac{B}{f_m} \times 1.5 \, m^2$$

(23.15)

$= 3m^2(m + 1)$, because $B = 2fm(m + 1)$

In FM system, the signal cannot be allowed to fall below the FM threshold, below which the signal to noise ratio rapidly degrades. This occurs in the conventional FM discriminators when the RF signal power to noise power ratio is around 10 dB. The required video signal to noise ratio is determined from subjective visual tests and the minimum value required is 33 dB over 99% of the worst month. C/N ratio is kept above 14 dB to avoid degradation of picture due to trunkation noise, for 27 MHz bandwidth. With C/N at 14 dB, and G/T of 6 dB/K the minimum PFD for 99% of the worst month works out to $- 103.3$ dBW/m^2.

For the purposes of planning DBS systems in the 12 GHz bands, WARC 79 assumed a power flux density of -103 dBW/m^2 at the edges of the satellite footprints, 3 dB below the PFD of 100 dBW/m^2 at the centre. A PFD of -103 dBW means an available power density of 50 pW. For an 27 MHz FM system, a receiver of noise figure of 7 dB, a signal power of -104 dBW is required to provide a good acceptable picture with a video SNR of 40 dB, making allowance for a signal loss of about 0.5 dB in the antenna to receiver feed, parabolic antenna of 0.9 m and 50% efficiency (loss of 3 dB).

In present DBS, the transmitter RF power of upto 200 W is concentrated into a narrow beam by highly directional transmitting aerials at the satellite so that the EIRP is high of the order of 3000 W per channel. Tunnel diode and TWT amplifiers supply generate the required microwave power. The tubes may be *derated* to operate at lower efficiencies to improve their life to as long as 7 years. The solar cell panels with back-up batteries are here required to supply a DC power of about 350 to 400 W per channel, which means around 1800 W to operate say, 5 channels. The panels are servo controlled to keep them facing the sun all the time. The telecontrol and telemetry services on board also require additional power. Equatorial countries require larger beam widths and hence larger power to produce the same flux density, though the free space loss can be smaller.

After the Indian SITE experiment in 1976, a number of *DBS experiments* were attempted. Canada had the Hermis experiment also with US cooperation, when a satellite carrying 12 MHz transmitter was launched. The Japanese almost started a DBS service from their orbital position 110° east, but for the snag in the transponder's traveling wave tube, which relegated the service into an experiment. The European Space Agency launched an Orbital Test Satellite (OTS) 10° east, used to study the propagation and reception of the 12 GHz link planned for DBS. BSB, the British Satellite Broadcasting Consortium is to launch soon the 'Sky-television' with a capability of direct reception on home TV, using a very small panel antenna and a sensitive front end attachment that would provide a mixture of free and pay channels of high quality, using MAC packet signals.

Present DBS satellites include the Eutelsat in Europe, TV-sat D3 of FRG, UNISAT of Britain, AUSSAT of Australia, the PALAPA in Indonesia, the INSAT 1-B series by India, the ARABSAT and several others, which have made international TV directly accessible from space, provided suitable standards receiver or converters are available.

Satellite Channels

The frequencies for satellite communication as regulated by the ITU are broadly as shown in Table 23.1

Table 23.1

Band	UHF	S-band	C-band	Ku-band			K-band	Ka
f(GHz)	0.3–1	2–4	4–8	12	18		18–27	27–40
Uplink	6		6	14	17.3–17.8			29/30
Downlink	0.620–0.790	2.5	4	11/12	12.2–12.7		22.5–23	19/30
					11.7–12.5			

Details are discussed in the allocation plan of WARC in section 23.7. For the present the frequencies used are: *C band* 4–8 GHz: 6 GHz for uplink and 4 GHz for downlink; *Ku band* 12–18 GHz, 17.3 to 17.8 GHz for uplinks and 12.2 to 12.7 GHz for downlink in region 2, 11.7–12.5 for downlink in region 1 and 3). In *Ka band* 27–40 MHz, uplink frequencies in use are 27.5–31 GHz. With EIRP of thousands of Watts, it possible to use smaller dish antennas of about 0.9 m for direct reception. For downlinks, a lower frequency of 2.5 GHz in the 2–4 GHz S band has also been employed for direct broadcasting satellite, easing the front end design for domestic TV receiver. Countries that have not established terrestrial UHF band IV-V service would be able to radiate in the band V allocation of 620–790 MHz, which ease the design problems, as in the SITE experiment. The manmade noise in this band is high, compared to receiver noise, though the free space attenuation proportional to $\log f$ is lower down to about 182 dB.

The allocated channels 25–30 MHz wide, using FM are wide enough to provide for use of C-MAC/ D2-MAC/Packet digital TV signals. Adjacent channels are allocated to widely separated countries, and shared channels are given opposite directions of circular polarization, in which the electrical field polarizing angle continuously rotates through 360°, in right hand or left hand direction. Several channels of conventional 525 or 625 line video standards and MAC digital TV or HDTV standards using nearly double the number of lines and wider video baseband are possible in DBS systems.

Development Stages

Development stages of some important satellite television services as introduced chronologically are given in table 23.1.

Table 23.1 Development stages of Satellite Television services

Year	Designation	Orbit	Use
1962	Telstar	Orbiting	Fixed service, point-to-point
1963/64	Syncom I–III	Geostationary	First synchronous satellite Telecommunication/TV
1965	Intelsat I (Early Bird)	Geostationary	Fixed service, Telecommunication/TV
1975/76	ATS-6	33 East	Distribution satellite TV
1980	Intelsat V	60 East	Telecommunication/TV satellites
1984	ECS 1-F1 Eutelsat	13 East	Distribution satellite TV
1984	Anik C1	108.5 West	Canada
1984/85	BS-II a/b	110 East	DBS-NHK Japan
1984	INSAT 1B	74 East	Multipurpose-Telecom/DBS
1985	TV-sat D3	19 West	Broadcasting satellite-FRG
1985	TDF-1	19 West	Broadcasting satellite-France
1985/86	AUT-sat	19 West	Broadcasting satellite-Austria
1985/86	SUI-sat	19 West	Broadcasting satellite-Switzerl.
1986	Rainbow I/II	132/79 West	USA
1986	ABC I	130 West	USA
1987	RCA	77 West	USA
1987	Videosat I	37.5 East	France
	BS-1		Japan
1988	INSAT 1C	93 East	Multipurpose-Telecom/DBS
1990	Arabsat ??		
1990	BS-2 ???		HDTV DBS Japan
1990 ?	INSAT 1D	? East	Multipurpose-Telecom/DBS

23.7 INDIAN SATELLITE INSAT SERIES

Indian National Satellite system, INSAT has been multipurpose system for domestic long distance telecommunications, meteorological earth observations and data relay, and nationwide TV broadcasting and program distribution.

INSAT 1B Satellite

INSAT 1B launched in August 1083 is operating as the primary INSAT 1 satellite from 74° East longitude. INSAT 1C launched in July 1988 operated at half power because of failure in its power supply, is positioned 93.5° East longitude to join INSAT 1B at the end of its life. INSAT 1 satellites as shown in Fig. 23.5, have been built by Ford Aerospace Corporation, and weigh about 1200 kg at lift-off reducing to about 650 in geostationary orbit. These satellites have a solar array of roughly 11.5 sqm area, generating

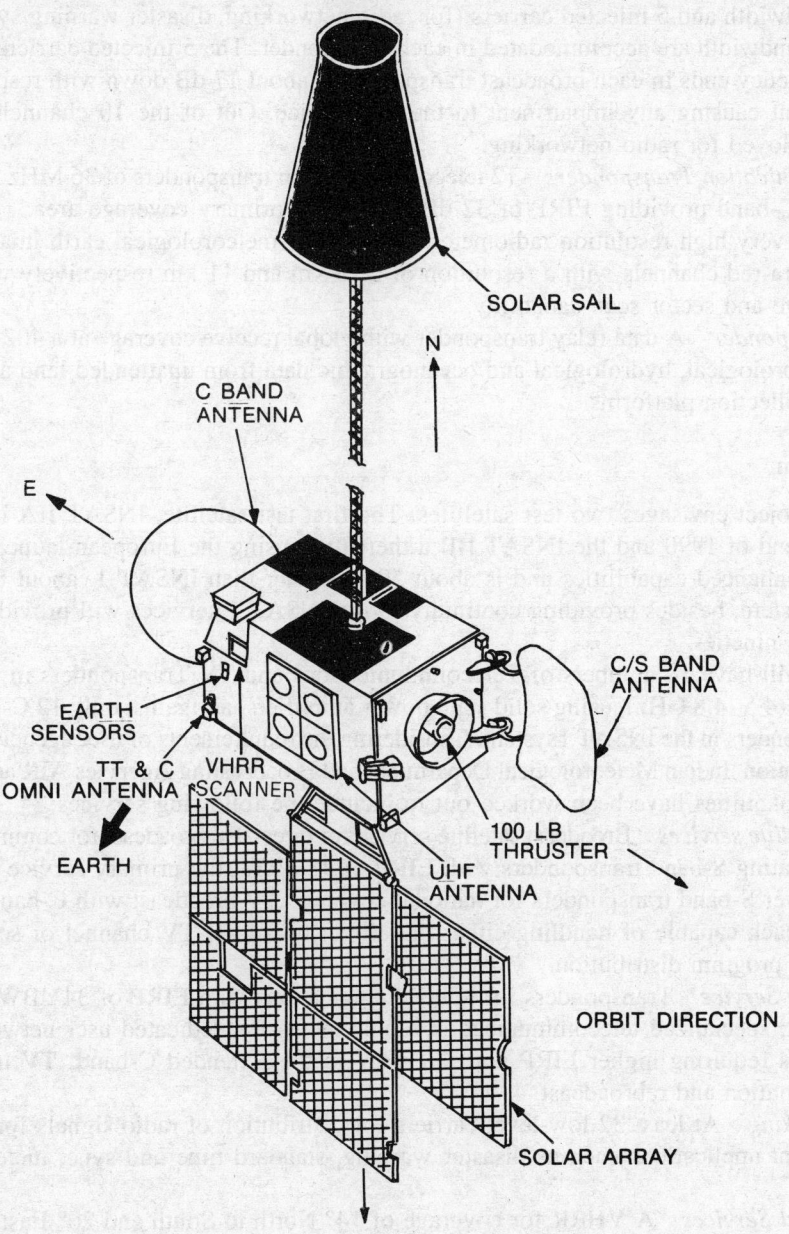

Fig. 23.5 INSAT spacecraft construction (ISRO)

a power of 900 W. These satellites are designed for a minimum of 7 years life, and carry 15 transponders providing the following capabilities:

(1) *TV Transponders* 2 High power TV broadcast transponders for national coverage operating in the C-band (5855–5935 MHz) for the uplink, and S-band (2555–2635 MHz) for the downlink, each capable of handling one direct broadcast (community reception) TV channel and several low level carriers for radio program distribution giving EIRP of 42 dBW over the primary coverage area. One TV carrier of 30 MHz FM bandwidth and 5 injected carriers (for radio networking, disaster warning system etc.) each of 170 kHz RF bandwidth are accommodated in each transponder. The 5 injected carriers are introduced at the lower frequency ends in each broadcast transponder at about 17 dB down with respect to the main TV carrier without causing any impairment to the TV service. Out of the 10 channels so derived, 5 channels are employed for radio networking.

(2) *Telecommunication Transponders* 12 telecommunication transponders of 36 MHz bandwidth each operating in the C-band providing EIRP of 32 dBW over the primary coverage area.

(3) *VHRR* A very high resolution radiometer (VHRR) for meteorological earth imaging, operating in visible and infra-red channels with a resolution of 2.75 km and 11 km respectively with half hourly full earth coverage and sector scan capability.

(4) *Data Transponder* A data relay transponder with global receive coverage at a 402.75 MHz uplink for relay of meteorological, hydrological and oceanographic data from unattended land and ocean based automatic data collection platforms.

INSAT II System

The INSAT II project envisages two test satellites. The first test satellite, INSAT IIA is scheduled for launched by the end of 1990 and the INSAT IIB a thereafter, using the European launcher Ariane. The INSAT IIA has enhanced capabilities and is about 50% heavier than INSAT 1, about 850 kg in orbit. The INSAT II system, besides providing continuity to the INSAT 1 services will provide for new class of services in the nineties.

The satellite will have 18 numbers of Telecommunications and TV Transponders in the C-band and Extended C-band (4.5–4.8 GHz), using solid state power amplifiers, as against only 12 C-band Travelling Wave tube transponders in the INSAT 1system. Considering the requirements of user agencies—Department of Telecommunication, Indian Meteorological Department and Broadcasting Agencies AIR and Doordarshan, the functional capabilities have been worked out to include the following services:

Broadcast Satellite services Broadcast satellite service for direct TV broadcast for community reception, to include 4 operating S-band transponders with EIRP of 42 dBW the primary service area. INSAT II has two high power S-band transponders for national coverage TV broadcast with C-band uplink and S-band downlink, each capable of handling either one direct broadcast TV channel or several low level carriers for radio program distribution.

Fixed Satellite Service Transponders for telecommunications with EIRP of 34 dBW in C-band and Extended C-band; specialized telecommunication requirements for dedicated user networks employing roof-top terminals requiring higher EIRP upto 36 dBW in the Extended C-band; TV transponders for nationwide distribution and rebroadcast.

Radio Networking At least 32 low-level carriers for distribution of radio signals for rebroadcast or point-to-multipoint applications such as disaster warning, standard time and sync, meteorological data distribution etc.

Meteorological Services A VHRR for coverage of 14° North to South and 20° East to West with a resolution of 2 km in the visible and 8 km in the infrared channels; transmission of processed met data via satellites; a 400 MHz uplink and 4 GHz downlink data relay transponder for access by low cost unattended data collection platforms; a disaster warning service to address selected receivers in the disaster-prone areas.

Satellite-aided Search and Rescue (SAS & R): A 406 MHz uplink and 4 GHz downlink relay electronics to provide instantaneous alert capability via geostationary satellites in this area as a part of the international COSPAS/SARSAT system.

23.8 INTERNATIONAL DIRECT BROADCAST SATELLITES

The ITU WARC held in Geneva in 1977 and later in 1985 made frequency allocations in the 12 GHz band providing channels to each country and the orbital positions for satellite transmissions to various countries. Countries who are likely to view each other's programs are grouped together and have been allocated the same orbital positions, e.g.

United Kingdom, Ireland, Spain, Portugal, Iceland	31° West
France, West Germany, Belgium, Netherland	
Luxembourg, Italy and Austria	19° West
Norway, Sweden, Finland and Denmark	5° East
Poland, East Germany, Czechoslovakia, Hungary, Rumania and Bulgaria	1° West

Larger countries like USA, USSR, China, India, Canada and Australia have been allocated several orbital positions for country wide coverage. The USA has 8 orbital positions at 61.5°, 101°, 110°, 119°, 148°, 166° and 175° West. India has its INSAT 1B orbiting at 74° East, INSAT 1C operating at 93° East.

As per WARC 77 and WARC 85 recommendations, the world countries have been grouped into *3 regions* for agreed international technical standards, though there were some differences evinced in the two conferences. Europe, USSR and Africa in the region 1, the Far East and Australia in the region 2 and America in region 3, have to hold international conferences and set up agreed standards for the respective region.

For the *regions 1 and 3*, WARC 77 divided the 800 MHz Ku band (11.7–12.5 GHz) into 40 channels, each 27 MHz wide and spaced 19.7 MHz apart. This resulted in some overlap interference which has been avoided by polarization in opposite directions. The 40 channels are grouped into 8 families with 5 frequencies spaced at 4 channel intervals. Each country in region 1 has been allotted 5 channels, while in region 3 the positions and channels made available to large countries also vary considerably.

For *region 2*, the 500 MHz band (12.2–12.7 GHz) was divided by WARC 85 into 32 channels each 24 MHz wide, also resulting in overlap that will have to be avoided by reversing direction of circular polarization at the 32 channels have 4 frequencies at 4 channel interval. The number of channels, given varies from only 4 to smaller countries while as much as 128 are available for the USA!

The *power flux densities* recommended for the outer edges of the footprints (areas of coverage) are – 103 dBW/sqm for region 1 and 3, while – 107 dBW/sqm for the region 2.

The *coding modulation standards* were not agreed upon and will vary in different regions. Presently these are NTSC in America and Japan, D-MAC for UK and D2-MAC for France and Germany. In the MAC standards C and D2 relate to the audio systems. C-MAC which was adopted as EBU standard, allows 4 stereo signals to be used for multilingual transmissions (required by scandinavian countries). This standard is not used by the French TDF-1 and the German TV-SAT satellites, which are to use D2-MAC, with duobinary coding for the audio, while UK is going for D-MAC. FM is used presently for the satellite carrier, but digital coding may soon be coming up.

Typical Television Broadcasting Satellite Data

Typical characteristic data of some international broadcasting and communications satellites and receiving system is given in Table 23.2 to give an idea of the parameters involved.

Table 23.2 Typical characteristic data of DBS systems

Parameters	TV-sat D3, TDF-1	Eutelsat I-F1 (ECS 1, spot West)
Frequency downlink	11.7 to 12.1 GHz	10.95 to 11.7 GHz
	12.1 to 12.5 GHz	
Channel number	5	2 × 6
Channel spacing	19.18 MHz	83.33 MHz
Channel width	27 MHz	83.33 MHz
Polarization	circular	linear horizontal (X) or vertical (Y)
Sense of rotation right-hand (1), left-hand (2)		
Type of Modulation	FM	FM
Transponder Power	250 W, 24 dBW	20 W, 13 dBW
Antenna gain	42 dB	32 dB
EIRP	4000 kW, 66 dBW	32 kW, 45 dBW
PFDo	– 104 dBW/sqm	– 118 dBW/sqm
Diameter of receiving antenna	0.9 m, 1.8 m	3.7 m
Gain	37 dB, 43 dB	50 dB
Figure of merit G/T of receiving equipment	6 dB/K, 14 dB/K	25 dB/K
Carrier/noise ratio C/N	>15 dB	>18 dB
Video S/N ratio (unweighted)	42 dB, 48 dB	>50 dB
Time of utilization	>99%	>99.9%
Attenuation caused by rain	not considered	3.5 dB

Satellite Sound Broadcasting

In a TV channel of the broadcasting and distribution satellites up to 16 high-quality stereo sound program channels may be accommodated in digital form. Signal-to-noise ratio of 85 to 90 dB can be achieved by quantizing with 14 or 16 bits/amplitude value, providing Compact Disc quality music programs.

23.9 DBS-TV RECEPTION

Receiving Antennas

The limiting factors for reception are set by antenna gain and noise figure of the front-end of the receiver as discussed earlier. *Parabolic dishes* are most effective for the microwave signal reception. The design considerations include gain, aperture efficiency, blocking loss, feed system suitable for polarization, side lobe spillover, mount, stearability, environmental conditions like temperature, wind etc. For the same size, the parabolic dish may be deep or shallow, implying different *focus-to-diameter ratios* (f/d). This lies between 0.25 and 0.5 in practical dish antennas. Dishes with low f/d ratio are generally designed for higher side lobe suppression, while those with high f/d ratio are better designed for high gain.

For the power flux densities specified for DBS, a diameter of 1 to 3.5 m have been used to give a noise figure of 8 to 3.5 dB. For satellite downlinks operating in the C band like in USA (3.7–4.2 GHz), or in the S band (2.5 GHz), a wire mesh type parabolic dish can be used. The dish reflector can be a *fine aluminium wire chicken mesh* or expanded mesh of 22 SWG wire suitably supported on ribs and plates. But for 12 GHz Ku band the mesh is not suitable, as the mesh has to be less than 1/10th of the wavelength for reflection of all incident flux. A solid aluminium dish is necessary for signals over 12 GHz. *Fibre glass dish* with thin conducting coating sprayed on it can form a light robust structure. The dish structure has to be strong and mounted stable to withstand wind pressures and temperature effects, to be able to track the satellite within a narrow angle of their receiving beam width.

Attempts have been going on to develop smaller *flat panel antennas*, less obstrusve than parabolic dishes. These design include phased arrays of multiple slots cut into sheets of aluminium, or arrays of printed circuit dipoles appropriately phased to provide the required beam pattern and gain. Active phased arrays with a large number of semiconductor elements assembled on flat panels to give required gain are also in the offing.

Antenna Feed

The feed has to be optimized to minimize blockage and coupling loss. Blockage loss in the *front feed* caused by the low noise feed and its supports at the *primary focus* in the beam path, can be reduced by *offset feed*, aligning the dish at an angle with the beam axis and the mounting the feed at the focus, shifted away from the beam path. An alternative arrangement is the *Cassegrain* with rear feed, when a small *hyperbolic subreflector located at the vertex* of the dish directs the received energy to the feed at the centre of the dish. These arrangements increase the cost and are used in large community installations only. *Gregorian antenna* is a advanced form, using *offset subreflector as well as the feed*, that steer clear of the beam path, and is used in ground transmitting and receiving stations.

Because of the circularly polarization, a helical antenna is most convenient in the S band. Typically for 2.6 GHz. The helix will have a pitch of $\lambda/4 = 2.9$ cm, and 2.5 turns of 12 SWG copper wire. For higher frequencies like 12 GHz band, wave guide feedhorn is more suitable.

Microwave Amplifier

A good low noise microwave preamplifier is essential before the signal is down converted. Microwave solid state techniques using devices like Gunn diodes, IMPATT devices and tunnel diodes may be used at 12 GHz in the front ends. GaAs FET amplifier provides a lower cost solution. *Strip line* planer transmission technique is most suitable for microwave frequencies. Consisting of microstrip of suitable geometry on a substrate backed by ground plane, the technique allows easy fabrication by photolithographic methods as well as control of the line parameters for filter design. *Hybrid integrated circuit FET amplifiers* with the tuning and filtering network etched onto GaAs substrate, have been designed to provide low noise figure of 2.5 to 3 dB. Front ends as a packaged as a LN converter unit with multistage IF amplifier are now mass produced to provide high gain of 27 to 30 dB.

For 12 GHz reception, *double superheterodyne technique* is used, because very high stability of local oscillator is difficult to achieve for a low IF of 70 MHz. The signal from low noise microwave amplifier is first converted into IF of 750 MHz typically, by means of a stable local oscillator (STALO). The first IF is filtered, amplified and converted into 2nd IF of 70 MHz. The second IF is amplified and filtered with the required bandwidth of about 27 MHz. After further IF amplification the FM signal is demodulated in the discriminator to obtain the video signal after filtering.

DRS for INSAT S-band Downlink

INSAT 1B DRS consists of the following:

(1) a chicken mesh parabolic antenna of about 3.6 m diameter,
(2) a low noise front end converter mounted on the antenna feed and
(3) an indoor unit comprising IF amplifier, limiter and discriminator.

Antenna

The antenna consists of an aluminium alloy mesh parabolic reflector, typically consisting of 12 main radial ribs, 12 sub-ribs extending from 2 m diameter to the outermost edge and two circumferential ribs, for ease of transportation and assembly on site. It is mounted on a tripod with the rear leg having telescopic adjustment to set the elevation angle and the base frame centrally pivoted to adjust the azimuth angle. A two and half turn helical feed made of 10 SWG Nickel plated brass wire at the prime focus provide for reception of left hand circular polarized wave. The helix is surrounded by a cone which helps to properly illuminate the parabola and increase the efficiency. The antenna system typically provides a gain of 36 dB.

Head End (Outdoor) Unit

The antenna receives television signals from the satellite at 2.575/2.615 GHz corresponding to the transponders I and II. The head end unit amplifies these RF signals (– 85 dBm) in a 3-stage Low Noise Amplifier (gain > 20 dB, bandwidth 20 MHz) and down converts them to 70 MHz IF (conversion gain 50 dB) by mixing them with crystal controlled local oscillator frequencies (2505/2545 MHz). Crystal frequencies at 104.375/106.01466 MHz are multiplied 3 times and buffered into a step recovery diode (SRD) multiplier circuit. They are multiplied here 8 times and passed through a microstrip bandpass filter to select the desired L.O. frequency of 2505/2545 MHz and suppress unwanted harmonics. The RF signal and L.O. signal are fed to mixer and IF preamplifier card. The diode mixer produces a 70 MHz signal that is preamplified to a level of around – 40 dBm, and fed via a coaxial cable to indoor unit.

Tail End (Indoor) Unit

The indoor unit processes the frequency modulated 70 MHz IF signal at nominal value of – 45 dB to produce the video and audio signals. It comprises an IF amplifier, predetection band pass filter, amplitude limiter, broad band frequency demodulator, base band amplifier, audio subcarrier demodulator, video low pass filter, de-emphasis circuits. The video and audio are given to TV receiver/monitor.

REVIEW QUESTIONS

23.1 What are the factors that have made cable TV popular, in spite of the wider TV broadcast network? Compare the merits of conventional TV broadcasts, CATV and DBS as utility services for the consumer.

23.2 Discuss the different types of cable distribution networks, the various frequency bands and relative merits and limitations. What are the special features of fibre optic cable systems?

23.3 How does a bidirectional network interact with the transmitter? Explain the methods of scrambling for pay-TV and conditional access to specific programs.

23.4 Explain the functional requirement of the head-end processor with the help of a block schematic.

23.5 Discuss the system requirements with reference to power, spectrum usage, antenna etc. for DBS to cover wide multilingual regions.

23.6 Draw the set-up for a DBS reception unit giving the details of requirements for the front-end converter.

23.7 Discuss the considerations for channel frequency allocations to large number of countries in the world for DBS.

23.8 Give reasons for the following:
   (a) DBS satellites are placed in equatorial orbits,
   (b) MAC packet encoding is most suitable for satellite TV broadcasting.

23.9 A parabolic dish of diameter 3 m receives a PFD of – 103 dBW. The antenna connected via a waveguide to a low noise amplifier of gain 18 dB and noise temperature of 20 K. Assume an antenna noise temperature of 100 K and a waveguide loss of 0.5 dB. The signal is then fed to a main amplifier giving a gain of 23 dB and noise figure of 4 dB. The output goes into a receiver having a noise figure of 8 dB. Draw the block diagram indicating the signal budget and calculate the equivalent system noise temperature. What is the C/N at the output of the system if the antenna impedance is 75 Ω and the system bandwidth is 27 MHz?

23.10 A satellite front end converter consists of a low noise amplifier, a mixer-oscillator, and an IF amplifier. The noise figures for the LN amplifier, mixer and IF amplifier are 4 dB, 6 dB and 2 dB; and the gains are 10 dB, – 3 dB and 30 dB respectively. Compute the system noise figure.

23.11 Calculate the power gain and 3 dB beam width of a 3 m parabolic dish at 3.5 GHz and at 12 GHz.

MULTIPLE-CHOICE QUESTIONS

23.1 Pick out the correct statement:
   (a) Trunk cables for cable TV are thicker because they carry larger power.
   (b) Fibre optic cables have high channel capacity but large power loss.
   (c) Cables for CATV require equalisation by automatic slope control for the amplifiers.
   (d) Weak signals of channels cause cross modulation.

23.2 Pick out the incorrect statement:
   (a) Graded index cables provide larger bandwidth.
   (b) Pilot carriers are sent for automatic gain control of the trunk amplifier.
   (c) Bridging amplifiers are used to feed the subscriber branch lines.
   (d) Encryption allows the subscriber conditional access to TV channels.

23.3 Solar eclipse period of Satellite TV broadcasts:
   (a) is constant
   (b) can be reduced
   (c) can be postponed
   (d) b and c

23.4 S/N ratio at the output of the demodulator in DBS receiver is:
   (a) same
   (b) better
   (c) worse
   (d) uncertain, than in the RF signal

23.5 If the diameter of parabolic dish is 2 m, antenna efficiency is 50%, and the received signal is at 12 GHz, the antenna gain will be:
   (a) 38 dB
   (b) 19 dB
   (c) 29 dB
   (d) 26 dB

23.6 The direction of the helix antenna depends on:
   (a) the type of feed connection
   (b) the impedance
   (c) the polarization
   (d) none

23.7 The power flux density *PFD* from a DBS depends on:
   (a) the frequency used
   (b) distance of DBS from the footprint
   (c) atmosphere in the path
   (d) gain of receiving antenna

23.8 Cassegrain antenna is a parabolic antenna:
   (a) with offset feed
   (b) with back feed
   (c) with two reflectors
   (d) all of these

23.9 A desired figure of merit of a receiving station is obtained if
   (a) a low noise front end is employed and the antenna size is decreased
   (b) a lower cost front end is used and the antenna size is increased
   (c) a low noise front end is used and antenna size is also increased
   (d) none of these

24

Broadcast Information Services

The concept of utilizing the vertical blanking interval (VBI) for carrying useful signals or information was pioneered by the British Broadcasting Corporation (BBC). The first experimental narrowband teletext service was launched by both BBC and IBA in England, and made available commercially to British public by the BBC as *Ceefax* in 1976, and by the Independent Broadcasting Authority (IBA), as Oracle in 1978. This led to considerable interest in this form of *broadcast information services* as well as computer-based *home information services* like *Prestel*, over other communication facilities like public switched telephone network (PSTN), cable TV, satellite networks, in all advanced countries. Many countries are now providing visual text or graphic information of public interest, on existing TV channels, directly on unused channels or during the vertical blanking interval of the composite video signal, without interfering with the normal TV program.

While basic teletext services are limited to a few lines in the vertical blanking interval, it is possible to have *partial or full channel teletext* transmissions allowing huge amount of data. It is also possible to use teletext equipped TV receiver to decode similar digital data from computer information centre sent over telephone lines. The impact of videotext embracing a wide ranging computer based communication services on social services, business transactions, banking, continuing education and training, entertainment and several other activities central to the modern information oriented society, can be hardly imagined. Video information data services are known by a number of names: *videotex, teletext, viewdata, videotext, teletex* and so on, and the terminology has been somewhat confusing, although used in specific type of service.

Videotex can be taken as the most general generic term for video information systems that provide simple-to-use, low-cost computer-based information services via public communication facilities including broadcast television, cable or satellites, telephone system etc. There are two forms of videotex: (1) *teletext* providing one way broadcast information, and (2) *viewdata* providing two-way switched interactive services via a telecom network. As videotex continued to be used to represent both interactive and pseudointeractive videotex, the term *videotext* was coined to represent the combination of videotex i.e. viewdata and teletext. Viewdata is thus refered to as *videotext* also, for example in Germany as *Bildschirmtext*. The term *teletex* refers a messaging system similar to telex service. *Electronic Mail* is another type of messaging service for sending one computer network system user to send personalized information to and receive information from other users of the system as each terminal has an address. Every user can periodically check his computer mailbox to see if anybody has sent a message for him. The information systems have different proprietary versions and names in different countries. There is a trend in many of the countries, to integrate the features of teletext, viewdata and other electronic services as an overall information network plan. Studies and efforts are under way to have a *broadband integrated services digital network* providing for various types of video information or programs.

24.1 VIDEOTEX INFORMATION SYSTEMS

Teletext is a *one way* service based on broadcast or cable TV system, providing a wide variety of information useful to the viewer, without recourse to the often overcrowded telecommunication switching networks. Teletext has only one way link, but is pseudointeractive in that the user can select the information to be viewed. There can be an *open-channel teletext,* when the user receives preselected information in a predetermined order. There is no interaction in any way.

A teletext provides the technique of transmitting blocks of digital data containing alphanumeric or graphic information, which the teletext decoder in a home TV receiver into a series of displays called *pages* of an electronic *magazine.* The large number of pages of video information in the form of coded digital data, are continuously transmitted in a cyclic sequence from a central information storage facility, during the vertical blanking interval of the television channel, or in an entire unused television channel.

Each magazine contains hundreds of pages containing information on such topics as news headlines, weather forecasts, programs, airline schedules, stock exchange and so on. The viewer can select on a specially adapted receiver with teletext decoder, the normal programs or teletext pages or a combination of both, by a remote control key-pad. In teletext mode, an *index page* provides the key to select the desired information and on entering the appropriate *page number* through the key-pad, the wanted page can be accessed in a matter of seconds.

By entering the identification numbers on a *key pad* accessory, the receiver *decoder circuitry* extracts from the whole of the bitstream, the block of digital data associated with the desired frame, stores and decodes it for display on the TV screen. The total numbers of frames or pages of information available has to be smaller, in order to avoid a long wait for the desired page to be captured from the continuous cyclic streams of pages sent. Teletext service has large bandwidth of the TV channel available, allowing greater speed of transmission of data.

Teletext can offer a number of *special services* like closed captioning for the deaf, variable comprehension and reading rates, conceal-reveal mode for question/answer session for educational use, alarm mode permitting important news break while viewing normal TV program. Flashing, Scrolling and animation are also possible.

Teletext Developments

There are four major systems developed for teletext. First teletext service was introduced in UK in 1972, under the proprietary trade names—**Ceefax** (see facts) of BBC and **Oracle** (Optical Reception of Announcements by Coded Line Electronics) of IBA, later in 1974 made a compatible **Teletext** to have a common digital data format in the vertical blanking interval (VBI), and enable reception with the same decoder in the receiver. This was followed by the French **Antiope** system a joint project of French television and telephone services, planned to integrate teletext and viewdata interactive services, and provide a full channel teletext information service in the extra spectrum of the discontinued 819-line TV channels. In Japan the NHK developed **Captain** (Character Pattern Telephone Access Information Network), to suit the pictorial Japanese script, combining both the teletext and viewdata requirements. **Telidon** system in Canada also combined the teletext and viewdata capabilities, with higher resolution in colour graphics and variable grayscale with a better pictorial capability, through a code system known as *picture description instructions* (PDIs), capable of operating at different resolution levels. Depending on the decoder capability, graphic images in Telidon may vary from a matrix of 60×80 pixels for simple alphanumeric display, to a high resolution matrix of 960×1280 pixels for HR colour graphics.

These four major teletext systems have been adapted in other countries in their own way. In the USA inspite of attempts and recommendations by EIA for a single teletext standard, different standards are currently in use–NALPS is a specification brought out by American National Standards Institute (ANSI) and the Canadian Standards Institute on the basis of AT & T videotex and Teledon. CEPT video presentation level protocol is another specification for viewdata only, published by the Conference Europe'ennes Postes et Te' le' communications used by the European countries. NABTS (North American Broadcasting Teletext Specification) system, supported by NBC, CBS and others, combines some features of Antiope and Telidon into a hybrid format to suit the 525/60 scan; while the World System Teletext (WST) supported by Taft, Keycom, Bonneville and others is based on the Ceefax/Oracle of UK. Both WST and NABS can employ lines 10 to 21 in the VBI, and can transmit full field teletext on active picture lines. They have similar attributes of data displays at comparable service levels of operation. NABTS is an *asynchronous* data transmission system with *seven layers architecture* for data processing and transfers. WST has a synchronous data signal at the bit rate of 5.727272 Mb/s, i.e. $364 \times fh$. WST has multilevel standards that come in 5 levels as explained later in the enhanced teletext systems. The French Antiope system has been adapted in India as Indtext.

24.2   TELETEXT SYSTEMS—CEEFAX/ORACLE

These teletext services operating in Great Britain have the same format with a transmission rate of $444 \times fh$, i.e. 6.9375 Mb/s. Each page of these services contains up to 24 rows of maximum 40 alphanumeric characters, with a special top row called the *page header*, during which an identification code number is broadcast corresponding to the page being sent. The first eight character code is used for the page number and control codes. One can key in the page number required from the remote control unit to select the required page. With the current text transmissions containing about 100 pages per magazine (while capable of containing up to 800), it takes up to   seconds on average to access a requested page. As the requested page is identified and received, it is line by line written into a page of receiver digital memory, and is continuously scanned to refresh the screen data being displayed, until a different page is selected and overwritten on the one page memory.

The digital coded data streams are inserted on the six teletext data lines 13 to 18 of the odd field 1, and 326 to 331 of the even field 2. (Initially these were lines 17, 18 and 330, 331 only 2 per field, later increased to 6 lines per field to improve the average access time to 7.5 s/page.) Test signals are introduced on lines 19, 20 and 332 and 333 on the alternate fields. These data signals can be visible as twinkling dots at the top of the picture if the raster height is reduced during TV reception.

As the teletext data is slotted into the 6 spare lines per field of 312 lines during the vertical blanking period of the program, the data transfer takes place for about $(6/312 =)$ 2% of the time. This available time is further reduced by the *framing* pulses for synchronisation, error correction pulses and other control data attributes such as colour, size and so on. The *bit rate* of the data is decided by the video bandwidth of the system. For teletext the rate is $444 \times fh$, i.e. 6.9375 MHz. To minimize the bandwidth, NRZ (Non-return to zero) format is used, with bit 1 voltage high at 66% of white level, and bit 0 voltage low at low at black level, so that if several 1 bits or several 0 bits follow in a sequence, the signal levels remains at 1 level or 0 level over the period of the sequence, reducing the 1–0 level transition rate.

Bandwidth and Eye Height

Although before transmission, the digital data signals are square wave pulses at the maximum bit rate permitted, viz. 6.9375 Mb/s, the bandwidth limitations of the TV channel affect their shape. Also, other phenomena such as reflections, ringing, ghosting in the signal path, can affect the teletext reproduction severely, while the signal strength is of lesser importance because of the digital nature of the information.

The digital teletext signals are filtered through the limited bandwidth of video transmission and reception system, where only the fundamental frequency of the bit-rate can pass through. Even the theoretical transitions of the teletext signals are not square wave, but a rounded 'raised cosine' waveform. Different frequency components of the signal also arive at slightly different times causing time jitter.

In practice therefore, every pulse in a data pattern affects to a certain extent the shape of the following pulses and the mean level of the signal varies according to the data pattern present. This effect is known as the *intersymbol interference* (ISI). As the bits overlap and blur together it is difficult to determine the bit value correctly. The extent of the ISI can be simply examined in the 'eye height' display obtained in the form of Lissajous pattern on an oscilloscope. The data signal is applied to the Y-deflection and the X-deflection set to two bit periods (one clock cycle), triggered by the teletext clock. This effectively superposes the wave shapes of the successive data pulses to produce the so-called *eye height display* as shown in Fig. 24.1(b).

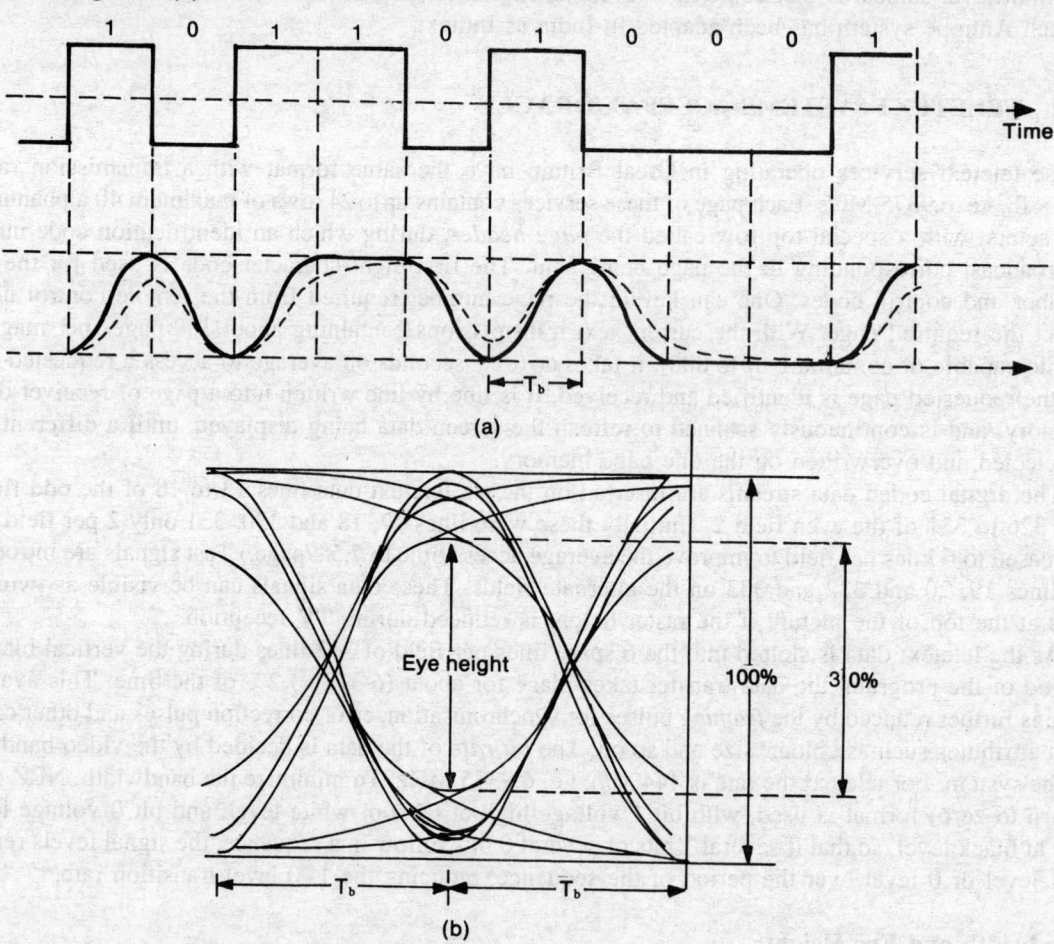

Fig. 24.1   (a) Ideal rounded waveform, (b) Eye height display

The observed waveform shows elliptical shapes like 'eyes', with the measurement between the 1 and 0 levels termed as the '*eye height*'. As the data bits are reduced in amplitude, the eye begins to close. The amount of closing reduces the difference between 1 and 0, reducing the noise margin. Sampling in the receiving circuitry should be timed at those instants at which the eye opening is the greatest. This may not be in the middle of the bit as the timings depends on the distortions in the system. The percentage opening quickly allows estimation of the ISI degradation due to non ideal Nyquist amplitude and phase characteristics of the system bandwidth. For a given bit error rate (BER) to be maintained for the most unfavourable sequence of bits and signal to noise ratio, the ideally minimum eye height is refered to as the worst eye height.

24.3 TELETEXT DATA ORGANIZATION

The basic data organization in the teletext data lines of CEEFAX/ORACLE is as shown in Fig. 24.2.

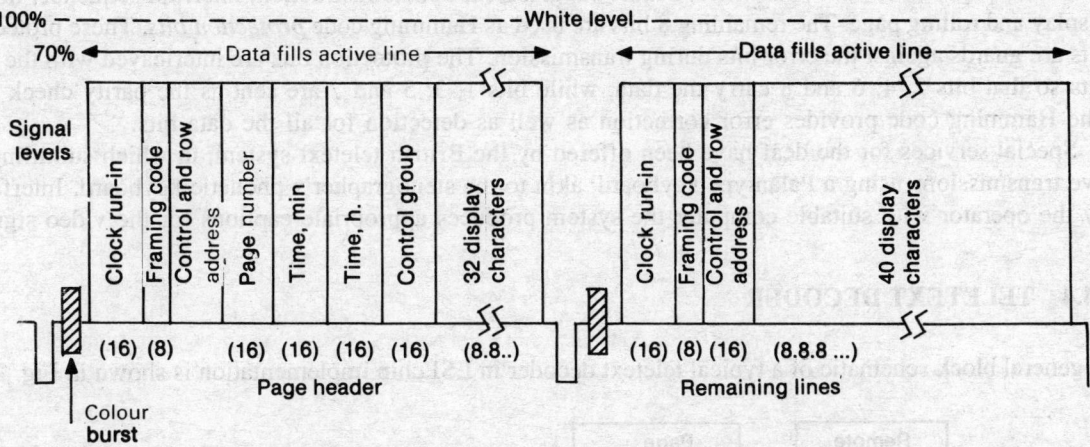

Fig. 24.2   Teletext data organization (*Courtesy:* TV Engg Handbook, McGraw-Hill Inc. NY)

Each active part of the line scan can accommodate 45 bytes of 8 pulses each. The first 5 bytes contain synchronising signals for the decoder and the remaining 40 bytes carry the display information in all the rows except the page header in which the first row contains 8 bytes for additional codes to leave only 32 bytes for display.

The *first two bytes* of alternate 1–0 sequence (1010101010101010) known as *clock run in* are used to synchronise of the receiver data clock. Since there is data transition at the end of every bit period during these two bytes, they can determine the relative phase of the local oscillator clock and lock it to the incoming data rate. The synchronisation of the local clock can be maintained throughout the rest of the line by using the logic level transitions as timing references, as the teletext coding has been arranged to make at least one logic level transition within each data byte.

Next follows the *framing code* a byte (11100100) serving data line recognition for locating the start of a line of text. The framing code was chosen so that it cannot be confused with character byte even if one bit went wrong. This is done by a shift register in which incoming data is loaded and matched with the framing code. The next two bytes contain the hundreds-digits of the *page number* and the *row number*, both with Hamming code error protection bits, header suppression and other control bits. After these 5 bytes are sent the *40 bytes of display* information each with an odd-parity bit. This may be in alphanumeric codes, graphic combination codes, or instruction codes containing attributes to specify *colour, flashing or boxing* for the following line.

Page Header

The first row forming the *page header* contains following the first 5 synchronising bytes, 8 bytes of additional header information viz. 2 bytes containing the last two digits of the page number, 4 bytes of time code carrying nominal transmission time in hours and minutes, 2 bytes for various control functions like header suppression etc. This leaves in the page header, only *32 bytes for display* information, each with an odd parity bit.

The 4 bytes of *time code* in hours and minutes, is not necessarily the actual clock time. It can identify the different editions of the same page to be sent at different times. Some decoders have the facility for storing a particular page for future display, to be recalled for comparison. The last two bytes 12 and 13, in the *control group* contain 16 bits, of which 8 bits are use to control an equal number of functions such as—erase page, news flash, sub-titles, header suppression, update instruction, interrupt sequence, inhibit display and rolling page. The remaining 8 bits are used as Hamming code *protection bits*. These protection bits are guards against the error bits during transmission. The protection bits are interleaved with the data bits so that bits 2, 4, 6 and 8 carry the data, while bits 1, 3, 5 and 7 are sent as the parity check bits. The Hamming code provides error correction as well as detection for all the data bits.

Special services for the deaf have been offered by the British teletext system, in which subtitling on live transmissions using a Palantype keyboard, akin to the stenographer's phonetic keyboard. Interfaced by the operator with suitable computer the system produces appropriate captions for the video signals.

24.4 TELETEXT DECODER

A general block schematic of a typical teletext decoder in LSI chip implementation is shown in Fig. 24.3.

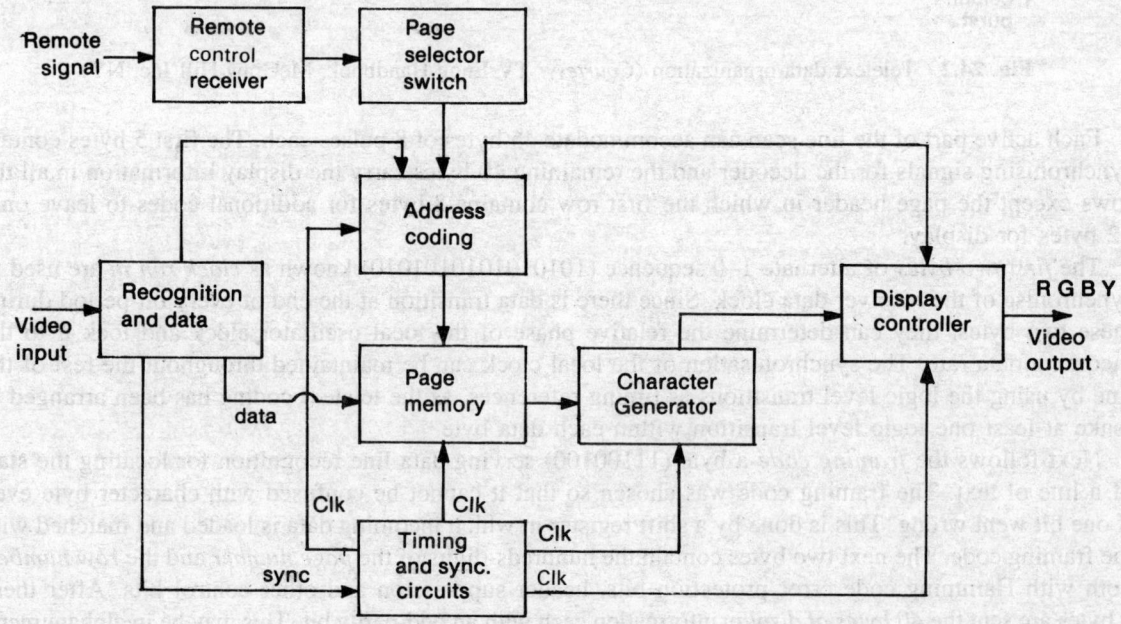

Fig. 24.3 Block diagram of teletext decoder

The decoder is implemented using four MOS N-channel ICs. Teletext Timing Chain (SAA 5020) which provides the necessary timing signals to the teletext page memory and the Character Generator (SAA 5050 series). It works in conjunction with the Video Processor Circuit (SAA 5030) and the Teletext Acquisition and Control Circuit SAA 5040 (series), maintaining synchronisation between the teletext system and the incoming video signal.

Teletext Video Processor The video processor SAA 5030 consists of a data retrieval section and a display clock generator as shown in Fig. 24.4.

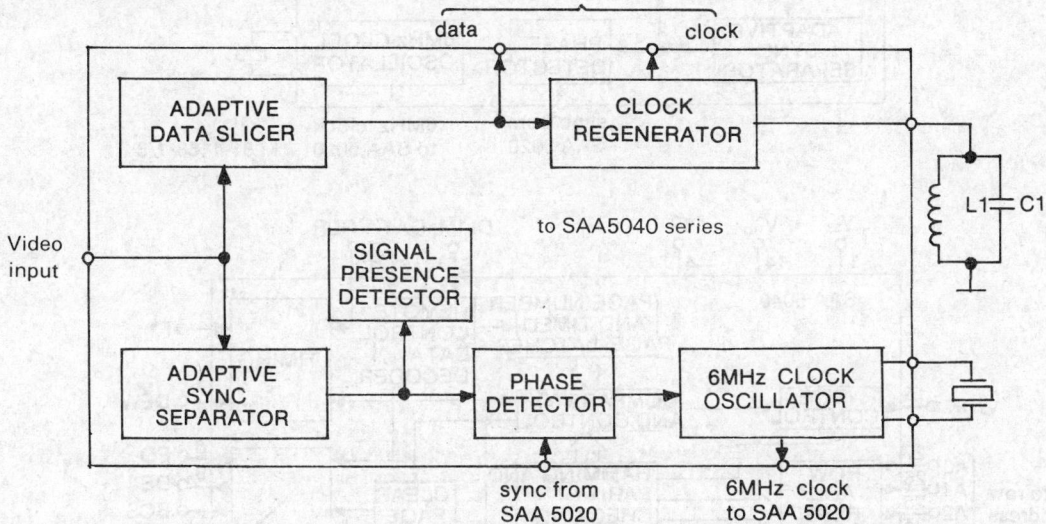

Fig. 24.4 Video processor SAA 5030 (PHILIPS)

It extracts data and data clock information from the television composite video signal (CCVS) and feeds this to the Acquisition and Control circuit IC SAA 5040. The adaptive data slicer sets the level at half the data amplitude at which the incoming is detected, thus proving some immunity to noise and interference. A clock signal is generated from the incoming data at 6.7395 MHz using the external tuned circuit L1C1.

An *adaptive sync separator* is also' provided which derives line and field sync from the input video in order to synchronise the timing chain. A 6 MHz crystal controlled phase locked oscillator is incorporated which drives the Timing Chain IC SAA 5020. This is divided in the timing chain IC SAA 5030 to give clock pulse every 64 μs. This is returned to the video processor IC, where it is compared with the incoming line sync in the phase detector to phase lock the timing chain to the transmission.

Teletext data Acquisition and Control (TAC) IC SAA 5040 processes and routes the data so that they can be written into the memory, as shown in the block diagram of Fig. 24.5.

The IC has been designed to meet the September 1976 Broadcast Teletext specifications of BBC/IBA/ BREMA. It consists of two main sections.

(a) *Data acquisition section:* The basic input to this section is the serial teletext data stream DATA from the SAA 5030 video processor IC. This data stream is clocked at 6.9375 MHz clock rate (F7) from the video processor. The incoming data stream is processed and sorted so that the page of data selected by the user is written as 7-bit parallel word into the system memory. Hamming and parity check is performed on the incoming data reduce transmission errors. WOK (write OK) indicates to the memory when valid data are received and can be written in, and the WACK (Write Address Clock) causes the

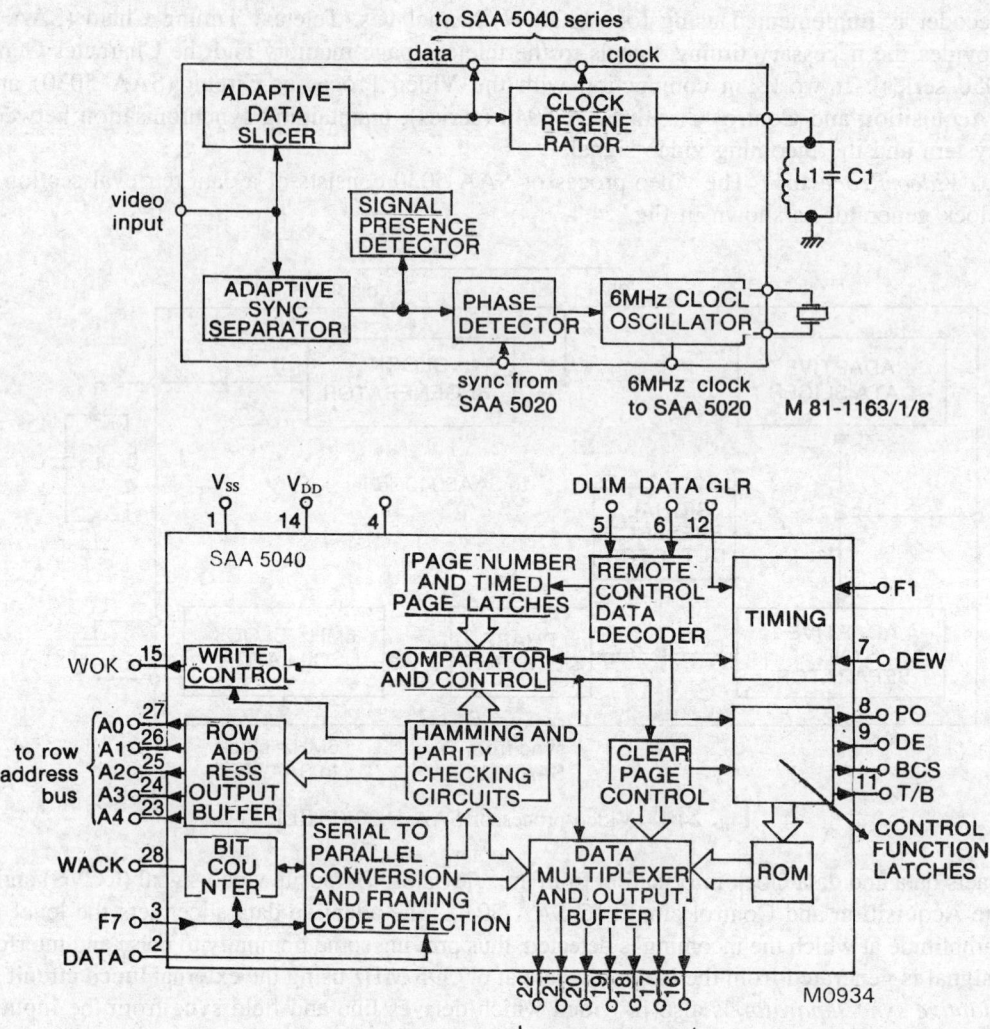

Fig. 24.5 Teletext data acquisition and control IC SAA 5040 (PHILIPS)

address counters to step on after each character. Provision is also made to process the control bits in the page header.

(b) *Control section:* The basic input to this section is the 7-bit serial data (DATA)‾‾‾‾ from the remote control decoder circuit such as IC SAA5012/SAB3012. This is clocked by the DLIM signal. The remote control commands are decoded and the control functions are stored. The control section also writes data into page memory independently of the data acquisition section. This gives on-screen display of certain user selected functions such as page numbers and program name.

The 3-state data and address outputs to the system memory are set to high impedance state if certain remote control commands are received, e.g. viewdata mode. This allows another circuit to access the memory using the same address and data. The address lines are also high impedance while the acquisition and control is not writing into the memory.

Teletext Timing Chain   IC SAA 5020 provides timing signals to the teletext page memory and to the character generator (SAA 5050 series), and maintains the synchronisation between the teletext system and the incoming video signal. Block diagram of the IC SAA 5020 is given in Fig. 24.6.

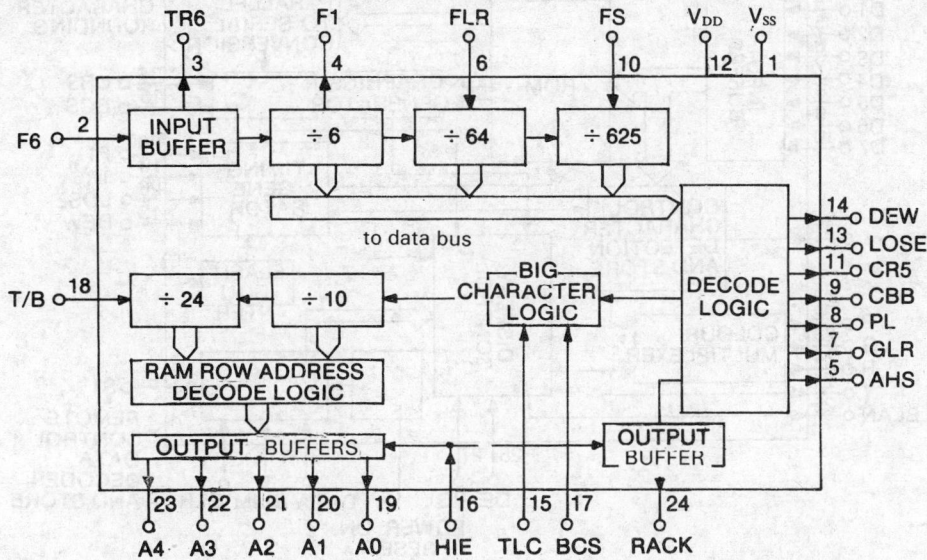

Fig. 24.6   Block diagram of the timing chain IC SAA 5020 (PHILIPS)

The basic input to SAA 5020 is the 6 MHz clock signal from the Video Processor Circuit. This clock signal is buffered and made available as output. A divide-by-6 counter produces the character rate of 1 MHz. This is followed by a divide-by-64 to produce the line rate and a further 312/313 to derive the field rate.

The line rate is also divided by 10 to clock a divide-by-24 counter for teletext memory row addresses. Logic is incorporated to enable the selection of the big character display, and enable the display of transmitted large characters. An output is provided to enable character rounding for normal height characters. A composite sync signal (AHS)~~~ is also available as an output which can be used to synchronise the display time bases.

The teletext memory consists of two $1k \times 4$ static RAMs e.g. 2114, arranged as four 32×32 matrices with each storage addressed by 10 binary address lines. The display is arranged as a 40×24 character matrix with a 5 bit row and 6 bit column address. These $(5 + 6 =)$ 11 bits of display address is converted into a 10 bit address code for selection of the address locations in the RAMs.

Block diagram of a *teletext character generator* (SAA 5050 series) is shown in Fig. 24.7.

The IC SAA 5050 incorporates a fast access character generator ROM (4.3 kbits), the logic decoding for all the teletext control characters and decoding for some of the remote control functions. The IC generates 96 alphanumeric and 64 graphic characters. In addition there are 32 control characters which determine the nature of the display.

The basic input to the IC is the character data from the teletext page memory. This is a 7 bit code. Each character code defines the *character dot matrix pattern*. The character period is 1 μs and the character dot rate is 6 MHz. The timing is derived from two external input clocks F1 (1 MHz) and TR6 (6 MHz) which are amplified and re-synchronised internally. Each character rectangle is a 6 dot wide by 10 TV

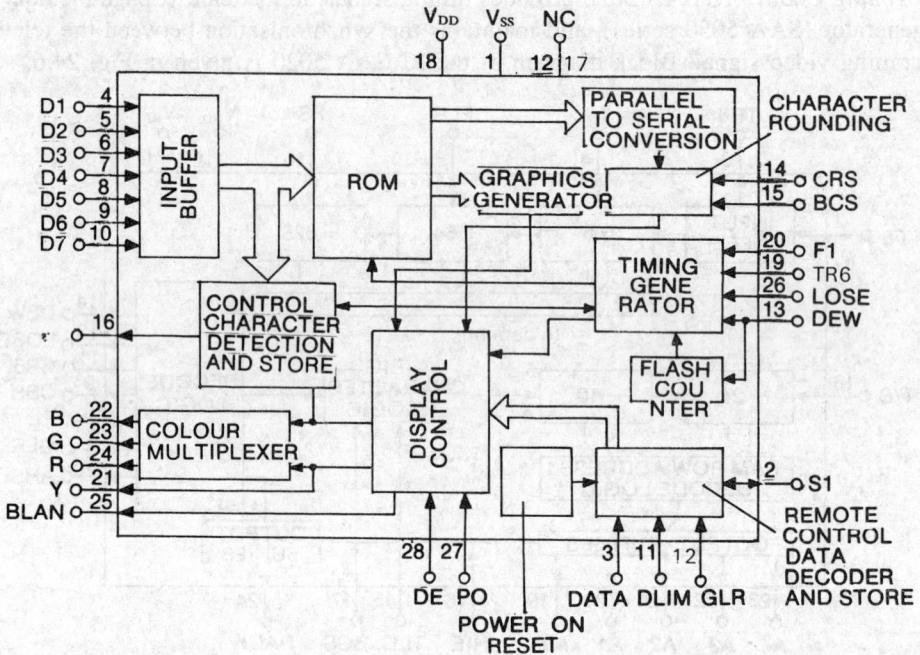

Fig. 24.7 Block diagram of teletext character generator IC SAA 5050 (PHILIPS)

lines high. One dot space is left between adjacent characters, and there is one line space left between rows. Alphanumeric characters are generated on a 5 × 9 matrix, allowing space for descending characters. Each of the 64 graphic characters is decoded to form 2 × 3 block arrangement which occupies the complete 6 × 10 dot matrix. Graphic characters may be either contiguous or separated. The alphanumeric characters are rounded, i.e. a half dot is inserted before or after a whole dot in the presence of a diagonal in the character matrix.

The *character video signals* comprise a monochrome signal and RGB signals for a colour receiver. A blanking output signal is provided to blank out the television signal under the control of PO (picture on input) and DE (display enable input) inputs and the box control characters. The monochrome data signals can be used to inlay characters into the television video. The use of the 32 characters provide information on the nature of the display, e.g. colour. These are also used to provide other facilities such as concealed display, flashing words etc. The full character set is given in Table 24.1.

24.5 ENHANCED TELETEXT SYSTEMS

In the current teletext format we can display 23 rows of up to 40 characters each (920 alphanumeric characters). In the graphic mode, a maximum of 6000 pixels are possible with limitations on colour selection, due to the need of serial attribute characters in the rows. In order to improve the teletext services while retaining the reverse compatibility with the existing decoders, enhanced teletext services of higher levels of graphic capability, employing Alpha-geometric coding have been developed. The CCIR has proposed five levels of teletext, of which the present teletext (and Prestel viewdata) is at the lowest *level 1*. This uses an alpha-mosaic system to display a basic Latin-based alphabet set of 24 rows

Fig. 24.1 Full character decoding set of SAA 5040 (PHILIPS)

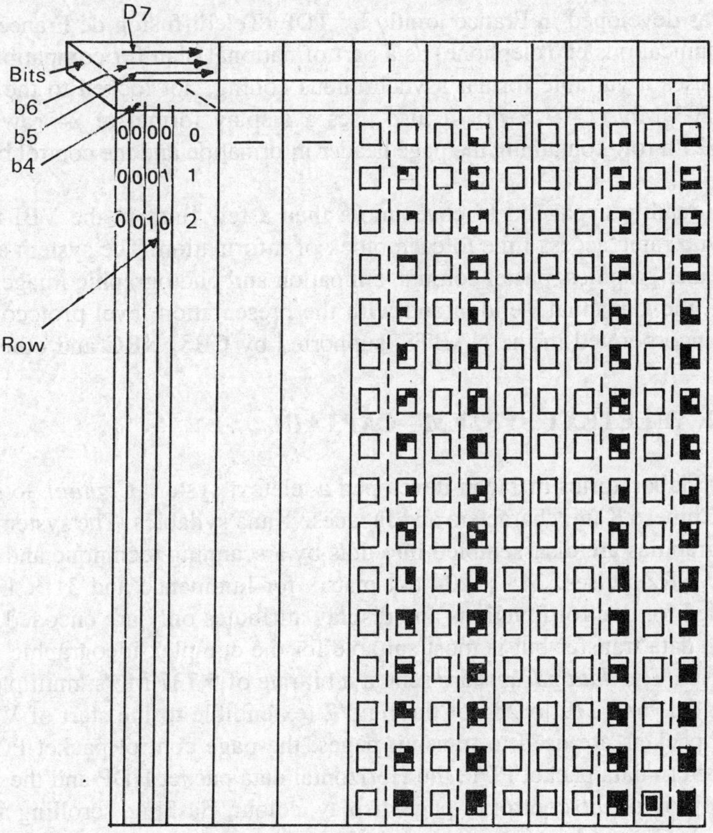

and 40 characters along with standard graphics. The display attributes include 6 colours, black and white, double height and captioning.

Level 2 called enhanced alpha-mosaic offers more versatile character set by augmenting the rectangular outlines of the 6-element mosaic with sloping boundaries giving smoother mosaics and rounded characters. It permits overwriting, addressing individual text-pixels, and permits wider range of pastel shades of colours. *Level 3* has Dynamically Redefinable Character Sets (DRCS) which extends tremendously the graphics capability. The incoming DRCS data can program the character generator at the decoder to define the foreground and background pixels from 12×10 dot matrix in each character cell, to form any special characters or graphics of high resolution. *Level 4* employs *alpha-geometric* mode making available over 75,000 pixels, each of which can be separately addressed and assigned any of the available 16 colours. Level 4 text instructions call for a great deal of data storage and some computing power. They are essentially computer programs that can be a subset of a telesoftware language. *Level 5* is known as *alpha-photographic* mode that has full grayscale and complete colour capability. This is full picture teletext with the resolution limited rather by the display devices and the transmission channel bandwidth.

24.6 ANTIOPE SYSTEM

Antiope system was developed in France jointly by TDF (Telediffusion de France) and the French PTT (Postes Telecommunications et Telephone) as a part of national plan for compatible videotex interactive services. Antiope uses a variable format asynchronous coding, not locked to the horizontal frequency. The bit rate is 6.203 Mb/s ($397 \times fh$). It also uses a display format of 24 rows of 40 alphanumeric characters each. A 25th row containing the page header information and the control bits, may be suppressed from display.

Antiope can be readily expanded to cover more than a few lines in the VBI to transmit very large amounts of data with rapic access time to each block of information. The system also has defined levels 0 to 6, to include finer graphics, pastel colours, animation and photographic images via teletext. Antiope is compatible with the Canadian Telidon and with the presentation level protocol (PLP) of the AT&T viewdata system, incorporated in the NABTS, supported by CBS, NBC and others.

24.7 JAPANESE TELETEXT SYSTEM—CAPTAIN

The NHK Research laboratories in Japan developed a teletext system *Captain*, to suit their pictographic script comprising Chinese Kanji characters and Japanese Kana syllabics. The system utilizes photographic coding. Text and graphics are disassembled into dots by a scanning technique and transmitted as dots in the form of binary NRZ pulses. 248×204 dot matrix for luminance and 31×17 for colour blocks is used. Signals needed for control functions and display attributes only are encoded for transmission. The system is slower in data transfer but is most suitable for the complex ideographic and fine line graphics of the Japanese script. The *data packets* are sent at a bit rate of 5.737 Mb/s, multiplexed on the VBI lines 16 to 21 of the field 1, and 279 to 284 of the field 2 (extendible to the start of VBI lines 10 and 273). Five types of data packets are used to transmit pages: the page control packet PCP, the colour control packet CCP, the pattern data packet PDP, the Horizontal data packet HDP and the program index packet PIP. Together they provide the control of page display, colour, flashing, scrolling and conceal functions.

At the receiver, *a frame memory* is used for display instead of the typical character generators in other teletext systems. A single page memory in the decode requires 55 kbits, including colour code memory. The display modes include text and graphics over the entire screen, superimposed text and subtitling. There is a provision of vertical scroll mode rolling the full-screen text upwards and a horizontal scroll mode, when a line text can be shifted right/left. Some Japanese teletext receivers have an eight page memory that can be programmed to store any specific information for almost instant access. A small 4 in. printer providing a *hard copy* of the teletext page is an added convenience. For more sophisticated viewdata applications, full colour printers rendering high resolution colour graphics are also available.

24.8 NABTS AND WORLD SYSTEM TELETEXT

NABTS is an asynchronous teletext transmission system derived from Antiope, compatible with Telidon and the presentation level protocol for interactive videotex developed by the AT&T in the USA. It has a 7-layer architecture with forward and reverse compatibltiy for future enhancements. The layers 1 to 4 concern the data transfer while the layers 5 to 7 concern data processing and usage as follows:

1. The *physical layer* defines the physical and electrical characteristics of the transmission medium and procedural functions in order to establish, maintain and release physical connections.
2. The *data link layer* provides a reliable data transmission link across one or several physical

connections. Its functions include error correction, sequencing and flow control in order to maintain the data integrity.

3. The *network control* layer provides routing path, switching and network access via various links to specific destinations.

4. The *Transport layer* provides a communication path, the end to end transparent virtual data circuits over the tandom network facilities.

5. The *session layer* provides the means of establishing connection and orderly exchange of data and other related control functions for a particular information service. Functions include the handling procedures like LOG-ON and LOG-OFF, security information and the like.

6. The *presentation layer* provides the means to represent information in characters, graphics etc. in a data coding format in a way that preserves its meaning. The function of the layer include terminal handling, data compression, encryption etc.

7. The *application layer* is the highest layer that provides the actual service sought by the end user. It defines and manages the interaction between the user and the server, through command protocols at this highest layer.

Designed to operate with the CCIR M 525/60 system, with NTSC coding NABTS could use lines 10 to 21 in the VBI on both the fields, or for full field data transmission, when the channel is not sending picture.

World System Teletext

WST developed in the USA, is claimed to be simpler, rugged and more widely used as it is based on the British Ceefax/Oracle enhanced teletext and is compatible with the West German standards for videotext at level 3. It has multilevel standards that are offered in 5 levels of operation. WST can employ . same VBI lines as in NABTS and can, if desired, transmit on all active lines of the full field and give precedence to emergency messages on priority to normal teletext. WST employs a synchronous signal format fixed relative to the horizontal scan rate. Multipath effects like ghosting can be minimized by a decoder associated adaptive equalising filter. WST has 5 layers of operation as in the enhanced levels of teletext in the UK.

Level 1. *Alphanumeric* format: 24 rows × 40 characters, basic Latin alphabet set and Standard graphics
Attributes: 6 colours, black and white, double height, captioning.
Compatible upwards.

Level 2. Alphanumeric display: Full multilingual character set.
Enhanced graphics.
Parallel attributes permit colour changes in the same line.
Reverse compatible.

Level 3. Includes non-Latin language alphabets like Greek, Japanese. *DRDS for fine shapes and high definition graphics.*
Digitized video images possible.
Attributes: 16 colours, 4096 shades.

Level 4. *Alpha-geometric coding* for very high resolution. Computer generated graphics with special graphic terminal. Computer Telesoftware service possible.

Level 5. Picture Teletext/*Alpha-photographic* Transmission.
Full-colour still picture transmission with mixing of text.
Compatible with videotext/viewdata systems.
Extra decoding and *HR graphics* facilities required in the TV receiver.

24.9 VIEWDATA/VIDEOTEXT SYSTEMS

Viewdata or videotext is a *two-way* information service using colour TV receiver as video display unit or an independent CRT terminal/VDU. It is essentially a value-added messaging service where the information is transmitted bidirectionally on the public subscriber network. Data from a variety of computer data bases is available to the user as modulated data via local telephone lines. The information service generally provides for data in four major areas: Finance-stock exchanges and money markets and reservation, messaging-topical news and reviews etc., microcomputing-*Telesoftware*.

A subscriber to the database can select the desired information via the keypad accessory for the adapted TV set, or keyboard of the video monitor.

The system is interactive with possibility of user to user communication. There is instant communication with the data base but it has limited simultaneous use capability. **Prestel** was the first videotex service launched in UK, and is emulated by some countries while others are developing their own, with their proprietary names—**Teletel** in France, **Teleset** in Finland, **Teledata** in Denmark, **Telidon** in Canada, **Viewtron** in USA, **Captain** in Japan. Viewdata systems use all kinds of host computers, from minicomputers to mainframes depending on the computational power required for the service and the anticipated number of users catered to, at a time.

Viewdata is fully interactive requiring two-way communication facilities via PSTN, packet switching networks. As the present PSTN are designed for analog voice signals, the digital data signals are handled by the terminals, through *modems* which translate them into modulated voice range frequencies for the telephone system. The modem may be acoustically coupled to the telephone handset or directly wired to the telephone line. Modems may operate at different speeds—commonly at 300/1200 bps, adequate to home terminal users. Higher speed modems operating at 2400/4800 bps are also available.

View Data Chip (Lucy)

LSI IC SAA 5070 from Philips is a microprocessor peripheral chip intended for use in wired viewdata communication systems. Simplified block diagram of the IC is given in Fig. 24.8. It performs most of the hardware functions of a viewdata terminal including an autodialer circuit, a 1200 band demodulator and asynchronous receiver, and a 75/1200 baud modulator and asynchronous transmitter. It includes a tape interface for recording the character codes of the pages of the text on a standard audio cassette recorder, and an IBYS receiver/transmitter. There are 2 general purpose I/O ports. One could be used as interface to a nonvolatile RAM which can store telephone numbers for autodialing and user passwords and the other could be used for display control.

Viewdata Terminals

CRT display terminals are used to display the information for viewing and editing, before taking a hard copy on paper in a printer. It can be an intelligent terminal or a microcomputer. *Intelligent terminals* are specially equipped with microprocessors and data storage memory capacity to process the data. *Microcomputers* with appropriate software are even more powerful in data storage and local processing capabilities.

In low cost home computers, the video information is modulated on a standard channel to suit antenna input of a TV receiver. A TV monitor with enhanced video bandwidth is used in professional computers and PCs to improve resolution to serve as the CRT display monitor or video display unit (VDU). The scan synchronising and video signals obtained from character generators or graphic adapters, are fed directly as the composite video input.

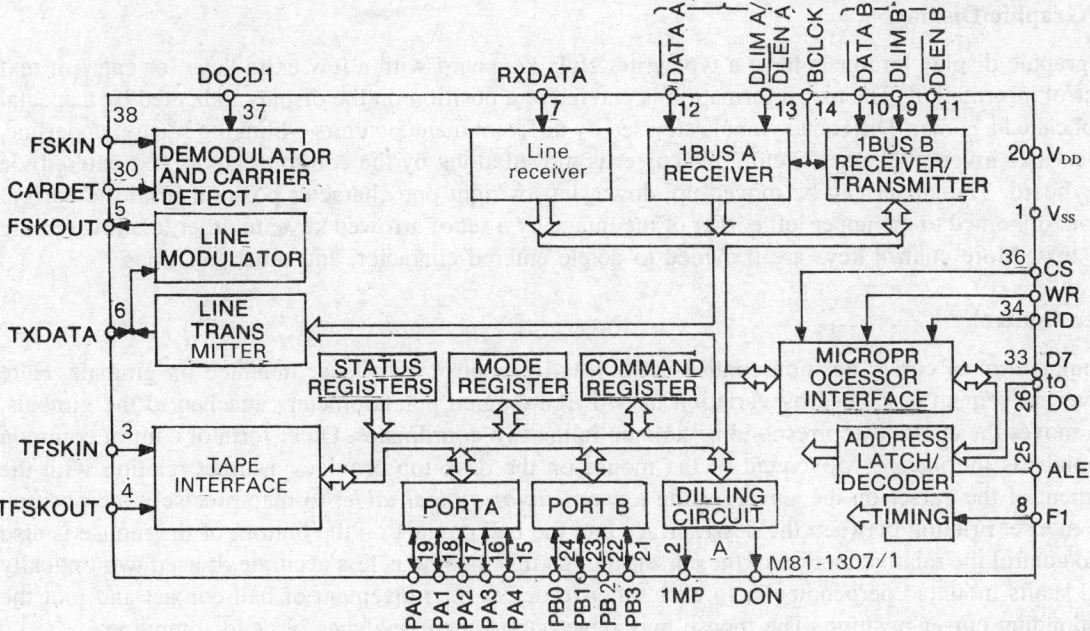

Fig. 24.8 Viewdata IC chip Lucy SAA 5070 (PHILIPS)

24.10 CRT DISPLAY TERMINAL

The Teletext data or graphic information is to be displayed on the screen of the CRT video terminal. For this, refreshing of the screen display is provided by the control logic from the information stored in a memory, called the *video RAM* or *Screen memory*. This memory may be separate or a part of the associated microcomputer which can address and alter the information stored.

For co-ordinating the position and the intensity of the beam two methods can be used. In the *raster-scan method* the beam sweeps through the raster along predetermined pattern, as in the regular TV set. Each scan line is divided into a number of small intervals called pixels or pels. As the beam scans through the different pixels, the information stored in the video RAM turns the beam intensity on or off at the respective pixels. In *vector-scan method* the beam position and the beam intensity are both simultaneously controlled by means of the stored information, drawing the graphic images like a pen writing on paper.

The stored information may be *bit-mapped* individually for each pixel, or it may be *character-mapped* (alphanumeric display) for each group of pixels, usually in a rectangular matrix array. In bit mapping, the binary memory image is represented by the pixel-pattern, while in character-mapping, a particular character is employed to access a font-pattern which is displayed on the screen under the control of a character generator circuit. Character mapped display controllers do not allow full graphics, but they require much less memory which means lower cost, and also put less burden on the software allowing quicker manipulation of the text. *Bit-mapped displays* are however, much more *flexible* allowing different fonts and characters sizes to be displayed besides providing full graphics capability. Bit-mapped displays have been costlier because of *large memory* but the decreasing cost of IC memories and increasing processor speeds have made them popular in Mcintosh-style user interfaces. The demands placed by bit mapped displays on the system software are considerable.

CRT Graphic Displays

Most graphic display terminals have a typewriter style keyboard with a few extra keys for entry of text or control information. The text information is entered at a position on the display indicated by a special symbol called a *curser*. Different symbols are used by different manufacturers—blinking square, underline, reverse video inverted character, etc. This curser is moved along by the control logic as text entered via the keyboard. The curser can be moved up, down, left or right one character position from the current position, or homed to the upper left corner of the image by a set of arrowed keys, in order to allow editing of the text. More control keys are provided to delete entered character, and other functions.

Curser Control

A popular form of curser position control is the *joystick* control by a lever mounted on gimbals. Here the lever movement is gagged by variation in two right-angled potentiometers attached to the gimbals, which moves the curser by corresponding amount in the two coordinates. Other form of control common in graphics is the *mouse*. Movement of the mouse on the desk top provides a direct relation with the movement of the curser on the screen, using a *magnetic or optical tablet* to map precisely the position of the mouse. Friction between the desk surface and the ball contact on the bottom of the mouse is also used to control the relative motion of the curser, though this method is less accurate. It used two optically sensed shafts mounted perpendicular to each other to detect the movement of ball contact and plot the corresponding curser position. The mouse may have one to three switches on it for inputting.

For display of more complex graphic information like in CAD, *light pen or mouse* is used for input of information and output display. A light pen senses any light within the field of its view, and the time taken from the start of the raster scan to the time, when the light pen signal is sensed give the position of the pen over the image. In the *vector scan* display, the centering of a special pattern under the pen is sensed, and the pattern is kept centered by its movement under control of host computer.

24.11 CRT CONTROLLERS

Character Mapped CRT Controllers

As pointed out earlier character-mapped display controllers for alphanumeric displays do not allow full graphics but are cheaper requiring less memory and enable rapid manipulation of text. Characters are formed on the screen by dot matrix of 5×7, 7×9 or 7×11 depending on the *letter quality* desired. For spacing between horizontal characters, one dot is acceptable, but two dots are preferable for legibility. For vertical spacing, two dots are essential while three dots improve legibility. Thus, for a character dot matrix of 7×11, each *character cell* including the space dots will require a 9×14 dot matrix, as shown in Fig. 24.9.

Each dot row in the cell is called a *cell line*, which will be followed by the TV scanning line. The number of characters along horizontal line will decide the minimum horizontal resolution, and the number of character rows vertically, will decide on the number of usable scan lines. Consider the standard 80 characters $\times$ 25 rows display used in computer monitors. The 25-rows display requires $25 \times 14 = 350$ usable dot-scan lines. Allowing for 25 lines for retrace blanking, a total of 375 lines is required per frame. For a non-interlaced display this requires a line scan rate of $50 \times 375 = 18.75$ kHz. The line period is then 52 μs, of which 80% viz. 40 μs is available for active video is accommodate $80 \times 9 = 720$ dots at the dot rate of 18 MHz.

Each character to be displayed requires specific dot patterns which are stored in special purpose ROM

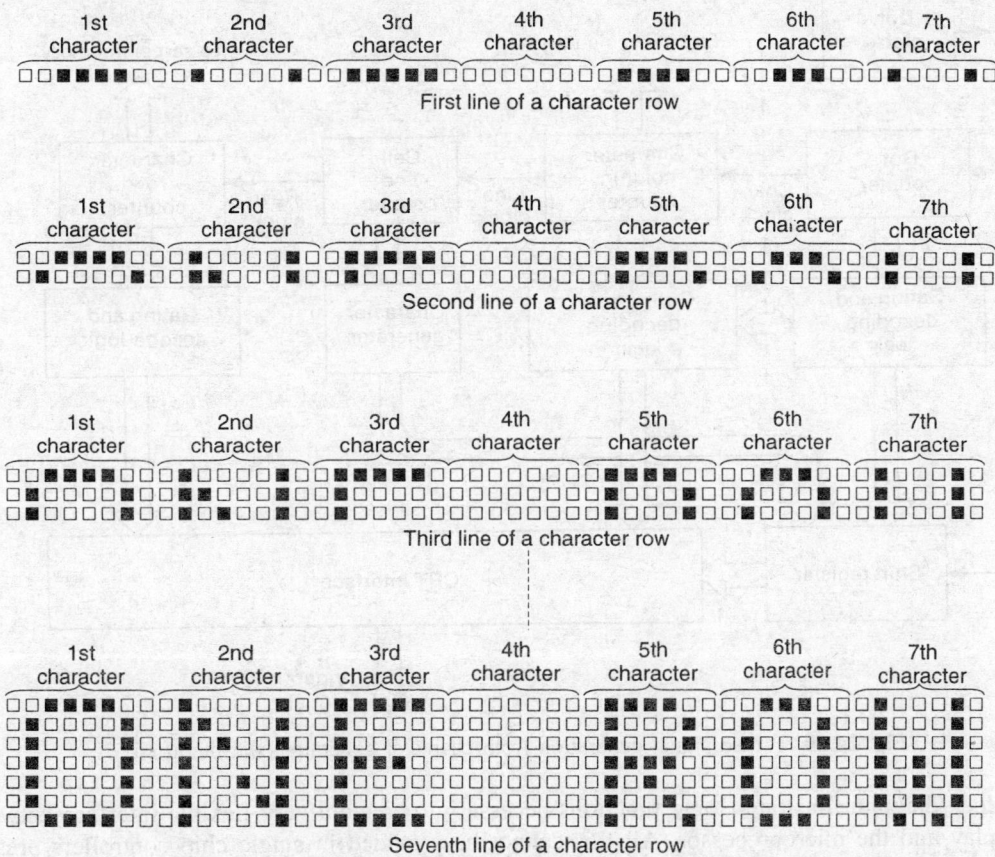

Fig. 24.9 7 × 11 character matrix in a character cell of 9 × 14 (INTEL)

called *character generator.* Each character is addressed by the ASCII address, as another part of the address specifies the cell line to be displayed. The character generator ROM outputs and loads all the bits in a cell line, into a shift register which is clocked out to provide the video signal, as indicated in Fig. 24.10.

The CRT controller has to provide appropriate *character codes and cell line addresses* to the character generator as the electron beam scans the raster, line by line. For each horizontal line, only one cell line corresponding to all the characters in a row, is required in sequence as shown in Fig. 24.9. The size of character generator memory depends on the dot matrix size and the number of characters. ASCII 7-bit code is typically used, which can be extended to another set of 128 characters by adding the 8-bit. A 7 × 11 matrix requires 4 cell line address bits for 14 cell lines, and 8 character select address lines for 256 character capability. This means a total of 12 address lines, or 4k memory.

CRT Controller Character generators are available as ROMs with standard ROMs, although custom mask-programmed EPROMs, or RAMs with system software for desired character pattern can be used for greater flexibility. A generalized logic functions that are provided in CRT controllers are shown in Fig. 24.11.

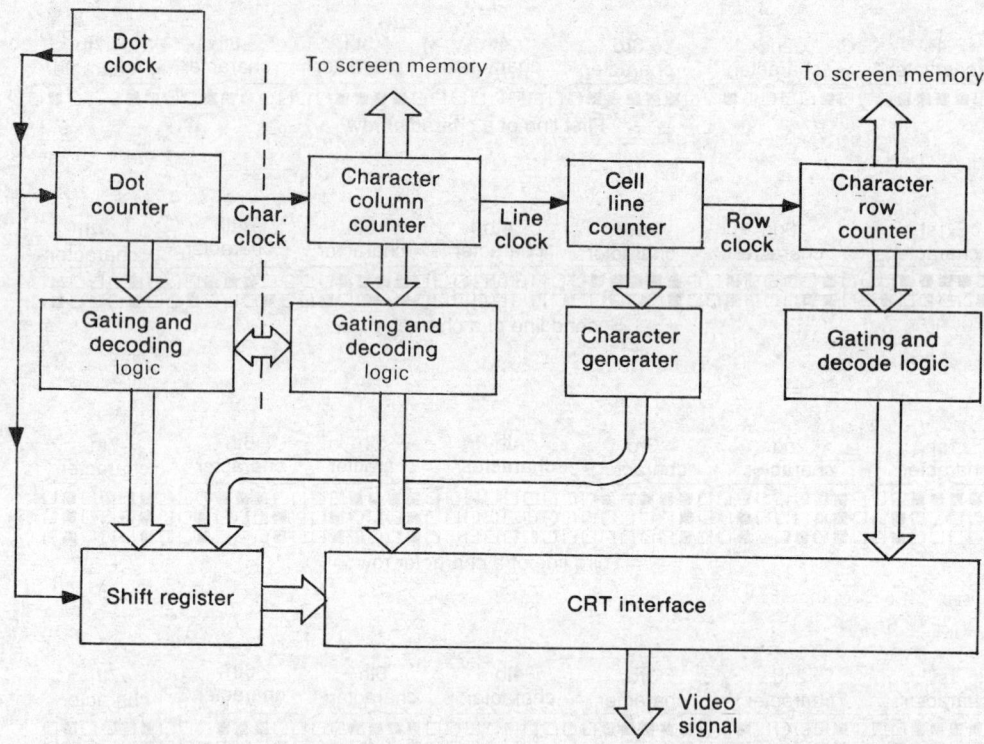

Fig. 24.10 Video signal generation from character generator (slat-modified)

The block diagram illustrates logic functions necessary to create a very flexible interface between a CRT display and the microprocessor. All this cannot be provided in single chip controllers presently available. Depending on the flexibility desired in the system, the chips incorporate the necessary logic functions. For example, the logic functions included in 6845 CRT controller are shown thick lines in the diagram. A number of CRT controllers are available, which include National Semiconductors DP 8350, Intel 8275, Motorola 6845.

Monitor Resolution and RGB Interfaces

As discussed in chapter 2, the vertical resolution of a CRT display depends on the scan lines and the horizontal resolution upon the video bandwidth which limits the number of dots that can be displayed on each line. The resolution is also affected by the sharpness of beam focus, hence good monitors employ have well focused beam with moderate deflection angle tubes. Out of the 64 μs TV line period, typically 45 μs are available for active video, the remainder being used up in blanking.

For a *horizontal resolution of 640 dots per 45 μs active line*, which means a period of about 2×70 ns for each alternate black and white dot pair, the minimum bandwidth required is over 7 MHz. For obtaining sharp edges on each dot the monitor must pass the third harmonic of the minimum, fundamental frequency and hence have a video bandwidth of over 21 MHz. Increasing vertical resolution also increases the bandwidth as the line period reduces too. High resolutions systems require bandwidths of 60 to 100 MHz.

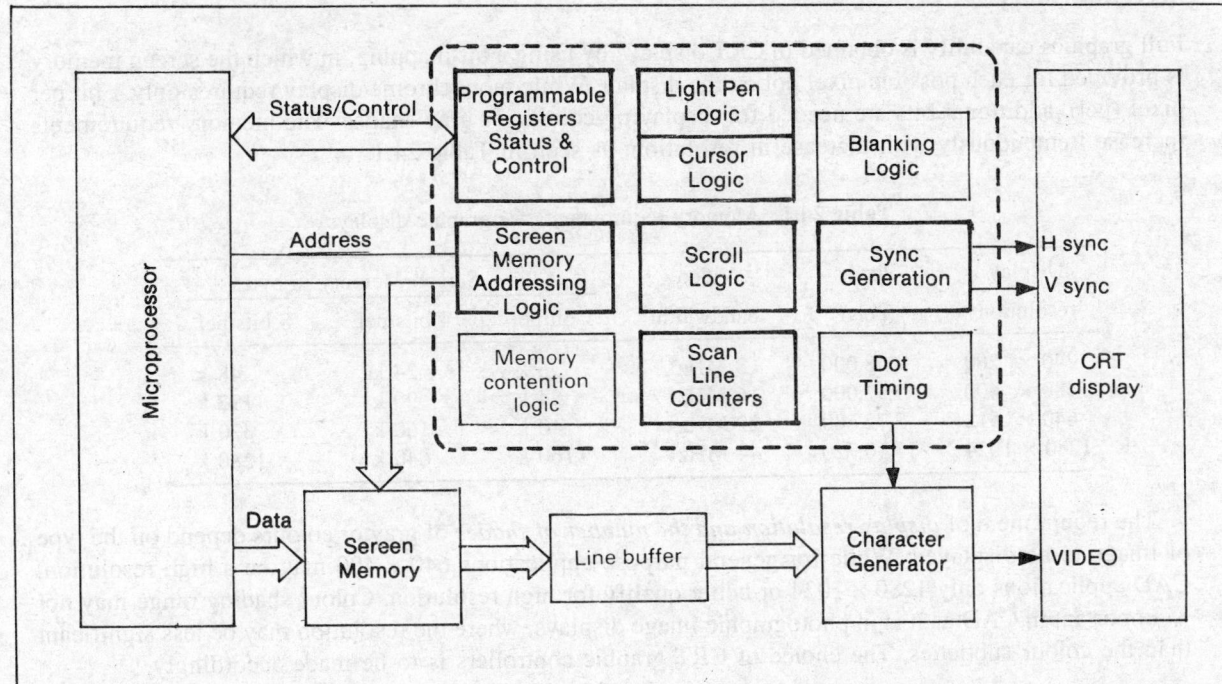

Fig. 24.11   Logic functional blocks provided in CRT controllers

Colour monitors have pixels, each comprising three RGB phosphor dots. The spacing between the dot triads called the *dot pitch* sets a physical limitation on the resolution on colour displays, regardless of the video bandwidth. High resolution colours CRTs have a dot pitch of about 0.31 mm.

, There are two types of *RGB interfaces* used for colour video monitors. In the *analog-interface* the voltage levels on RGB inputs decide the colour of dot triad, allowing an infinite range of hues to be produced, while in monitors with *TTL-interface*, each of the R, G and B input is treated as digital input, turning the corresponding colour phosphor dot on/off. Thus, the 3 bits controlling R G B inputs provide a palette of 8 colours for display. The digital TTL-interface is often extended to provide a palette of 16 colours, by adding a digital input signal which allows a high/low *binary intensity* control for the RGB primary colours. This type of interface providing 2 shades of 8 colours, is used in the CGA—colour graphics adapter (640 × 200 dots) and EGA—enhanced graphics adapter (640 × 350 dots) for the IBM PC. Generally the sync signals are fed separately, but some monitors accept these on the green video input.

As an emerging technology, *computer graphics* has assumed great importance in computer usage, especially in personal computing where the user interface is made highly interactive. *Graphic standards for PCs* have emerged to bring high quality colour graphics to personal computing. These are discussed in section 24.12. Apart from the simple choice between monochrome and colour, the graphic adapers offer the a number of options in resolution, e.g. VGA-video/virtual graphics array (640 × 480), super VGA (800 × 600), 8514A (1024 × 768). With increased resolution the number of pixels to be accessed multiply and the screen displays slow down.

Screen Memory

Full graphics capability is obtained in CRT displays by using a bit-mapping, in which the screen memory is provided for each possible pixel dot on the display. While monochrome display requires only 1 bit per pixel (pel), additional bits are needed for displaying colours or gray shades. The memory requirements increase tremendously with increase in resolution, as seen in Table 24.1.

Table 24.1 Memory requirements for graphic displays

Display resolution	Pixels (pels)	Video bandwidth	Screen Memory in bytes		
			1 bit/pel	4 bits/pel	8 bits/pel
240 × 200	48,000	8 MHz	6 k	24 k	48 k
480 × 400	1,92,000	16 MHz	24 k	90 k	192 k
640 × 512	3,72,680	22 MHz	40 k	160 k	320 k
1280 × 1024	13,10,720	44 MHz	160 k	640 k	1280 k

The requirement of *display resolution and the number of shades* of gray or colours depend on the type of image to be displayed. While for general purpose applications 640 × 400 may be a high resolution, CAD applications only 1280 × 1024 or better qualify for high resolution. Colour shading range may not so important in CAD as it is in photographic image displays, where the resolution may be less significant than the colour subtleties. The choice of CRT graphic controllers is to be made accordingly.

Graphics CRT Controllers

Memory Requirements

In graphic CRT controllers, the considerations for scanning, dot clocking, memory addressing are similar to the alphanumeric controllers. Instead of the display data coming through the character generator, the data from the screen memory is used directly for bit-mapping it to generate a graphic dot pattern. The CRT controller addresses the bits from each location from the screen memory to specify the state of successive pixels. Dynamic RAMs have been specifically designed for use as screen memories or video RAMs in *bit-mapped graphics* display. These DRAM chips have *2-ports*, one identical to a standard RAM, while the other provides a *separate serial data output* via an internal shift register, at rates upto 25 MHz. As the microprocessor and the CRT controller have separate ports for access to the RAM, and the internal shift register is loaded with several bytes at a time to output bits at a fast rate, video graphics can be generated pretty fast.

Gray Levels

For producing gray levels in the display more than 1 bit per pixel are necessary. The number of bits per pixel can be visualized as a number of *bit planes* of memory, each with 1 bit per pixel. 8 bits per pixel can provide 256 gray levels in the display, by a suitable D/A converter. But this increases the memory access time to 8 times, necessitating use of very fast memory chips as well as high speed D/A converter. One way to overcome this demand on access time is to build a wide-word memory array for the screen memory (e.g. 32 k × 64, rather than 256 k × 8, handling 4 pixels at a time) so that larger number of bits corresponding to more pixels are read and latched in each access. These are multiplexed in turn, to provide pixel outputs to the D/A Converter.

Colour Palette

Colour CRT controllers must produce *three video signals* corresponding to the colour primaries R, G and B. One bit for each of these signals allows 8 different colours, and addition of a fourth bit for luminance can extend this to 16 colours. For more colours more than one bits are used as *attributes*, along with the analog RGB monitor, employing a memory array for the colour attributes, and one D/A converter for each colour. 4 bits for each of the 3 colours (12 bits per pixel) give 2_{12}, i.e. 4096 colours which are generally adequate. High end systems use as many as 8 bits per colour, giving over 16,800,000 colours to produce true to life colour renditions! For the screen memory, wide-word memory organization with output multiplexers is essential to reduce the memory access time requirement.

Screen Memory

The screen memory requirement can be reduced by limiting the colour selection by use of a *colour lookup table*, which consists of a high speed RAM that is written by a particular application program. The lookup table RAM translates the colour number from the screen memory into the RGB data bits for the DAC. For example, if 4 bits of screen memory are used for each pixel, only 16 colours can be displayed simultaneously. Any 16 of the 4096 colours can be selected by writing the desired colour number in the RAM which can assign any 16 colours to 16 colour numbers. The screen memory requirement is reduced to one third, while only a small but fast access RAM storing 16 words of 12 bits each is required.

DAC Requirements

The D/A converter must produce new output for each pixel. In high resolution graphics displays this require high performance D/A converters. The *bandwidth* has to be typically 40 to 200 MHz and the word size 4 to 8 bits. IC chips containing *high speed* triple D/A converters are now available, some include sync input and can drive standard monitors directly. Colour palette ICs are now available which combine the colour lookup table RAM and DAC on a single chip.

24.12 STANDARD GRAPHIC DISPLAY ADAPTERS FOR PCs

Computer graphics have recently assumed great importance in computer usage. Graphic display adapters have enhanced the potential applications of personal computers where the user interface is made highly interactive. Advances in memory technology, VLSI circuits, and CRT display design have now brought high quality colour graphics within realm of the PCs. Graphic standards for the popular IBM PC compatibles have evolved to bring high quality colour graphics useful in a host of popular applications like CAD, DTP, video graphics, animation etc. in personal computing. Apart from the simple choice between monochrome and colour, the *graphic adapters* offer a number of standard options in resolution, HGA (720×350), CGA (650×200), EGA (640×348), VGA (640×480), super VGA (800×600), 8514A (1024×768).

Monochrome Display Adaptor (MDA)

The original IBM PC was sold with monochrome display and the *monochrome display adapter* (MDA). MDA had *no graphics* capabilities but has been widely adopted for text applications. The resolution of MDA (720×350) is better than than of the EGA (640×350). MDA is a good choice offering crisp monochrome text at low cost, if colour and graphics are not required.

Herculus Graphics Adapter (HGA)

Soon after the IBM PC was introduced a number of add-on options were developed by others companies.

Herculus Computer Technology, Inc. introduced a monochrome adapter for the IBM PC monochrome display that offered *graphics capability* (resolution of 720×348 using 1 bit per pixel, i.e. 8 pixels per byte) besides the MDA text mode capability. Herculus graphics adapter is based on 6845 CRT controller. Its screen memory is burdened with a nonlinear address map that is worse than the CGA map, as scan line data is splintered into 4 sections as against 2 sections for the CGA. The *Herculus graphics card* (HGA) became popular quickly, to become the standard for monochrome graphics. It was not until the introduction of the enhanced graphics adapter (EGA) that IBM offered an adapter with monochrome graphics display capability and more compatible displays having 50 Hz refresh rate, followed. Herculus graphics is *not a standard mode for EGA* or *VGA* adapters though some third party EGA products offer HGA compatible modes.

Colours Graphics Adapter (CGA)

Colour display standard was established by IBM with the introduction of the IBM colour display and *colour graphics adapter* (CGA), that included an output for driving monochrome display also, and is often used in low end colour displays. Capable of displaying 16 colours, the CGA supports 4 colour graphics and 8 colour text. But with a resolution of 640×200, the text is grainy and poorly shaped, as each character cell gets only 8×8 *pixels*. It uses only 1 bit per pixel (8 pixels per byte), 0 bit displaying black while 1 displays white. The pixel data is stored in the colour plane 0 of the 4 plane screen memory. Display data is serialized MSB taken first. Use of 6845 as CRT controller resulted in nonlinear mapped screen memory address space on all graphic modes. The complicates the drawing algorithms, because a computation is required for translating the pixel position into bit position in the screen memory.

On the original IBM PC CGA board, the processor access to the video RAM interfered with screen refresh operations resulting in *noise/snow* on the screen. Most CGA software blanks the display during refresh functions, but this shows off in *flicker* in the standards CGA operating modes. Because of its graphics capability, colours at low price tag caused CGA to be quite a success and a vast amount of installed base of CGA compatible hardware and software is currently found. However, the CGA does not find much acceptance from serious users of personal computers because of its *poor display quality*.

Enhanced Colour Displays (ECD)

For improved resolution in colour graphic displays, IBM introduced the Enhanced Graphics Adapter (EGA) with a higher resolution (640×350) with enhanced colour displays. EGA offers 64 colours instead of 16 in the CGA and the text modes are improved by an enhanced character set which uses a character cell of 6×14 *pixels*, reproducing text that almost matches the quality of the MDA monochrome text. The EGA and the ECD were designed with a certain amount of backward compatibility with the CGA and the CD. The ECD has a dual frequency display that enables its connection to a CGA and automatically operate at the CD resolution (640×200).

Monochrome graphic modes of the EGA are not compatible with the HGA. Because of the superior resolution of the HGA compared to the EGA, and a large base of the existing Herculus compatible software, some manufacturers have added an amount of Herculus compatibility modes to their EGA, that can run most of popular the Herculus software packages.

Sensing the trend towards higher resolution, NEC Corp. created their own *Multisync line of displays*, that are can operate over a wide range of horizontal and vertical sync frequencies and are therefore able to support any required range of screen resolution. Thus we have these displays providing resolution starting from the EGA (640×350) extending to higher resolutions of 640×400, 640×480, 800×600, 1024×768, 1250×1024. Multisync displays offer unlimited colour capabilities when operated as analog RGB displays. The new versions are VGA compatible.

Vector Graphic Array (VGA) Displays

The VGA was developed by IBM to improve upon the EGA resolution, as a standard for their PS/2 family of PCs. This was accompanied by a new family of VGA compatible high resolution *analog RGB display and monochrome analog display* that shows colour information as gray shades. It supports resolution of 640 × 480 pixels, displaying up to 256 colours at a lower resolution. Unlike the EGA, the VGA is not compatible with a number of displays, but offers operating modes that simulate the performance of other displays and are partially software compatible.

Like all display adapters of the PC/AT family, the EGA and FGA are dumb adapters and do not have on board processing power. The system microprocessor is responsible for all drawing operations, writing directly into the bit-mapped screen memory. Predefined graphic software environment such as Microsoft Windows, GKS can be used by the application programmers; or they may write their own graphics drawing routines.

Standard Operating Modes

IBM has defined a set of standard operating modes for the EGA and VGA, which have been used as standard video interfaces for application software. Some EGA compatible boards have however implemented device drivers to operate in nonstandard modes to emulate CGA or Herculus and other higher resolutions such as 640 × 480. The modes 0, 1, 3, 7 are for colour text; 0 and 1 for 40 column 25 row matrix on the screen, and modes 2 and 3 for 80 column 25 row display. Each character cell has 8 × 8 pixels, increased to 8 × 14 or 9 × 16 in enhanced pixel modes (marked by * or +). Modes 4, 5, 6, and 10 to 13 are for colour graphics providing different resolutions and colours.

Architecture of EGA and VGA

A block schematic of the EGA/VGA architecture is given in Fig. 24.12.

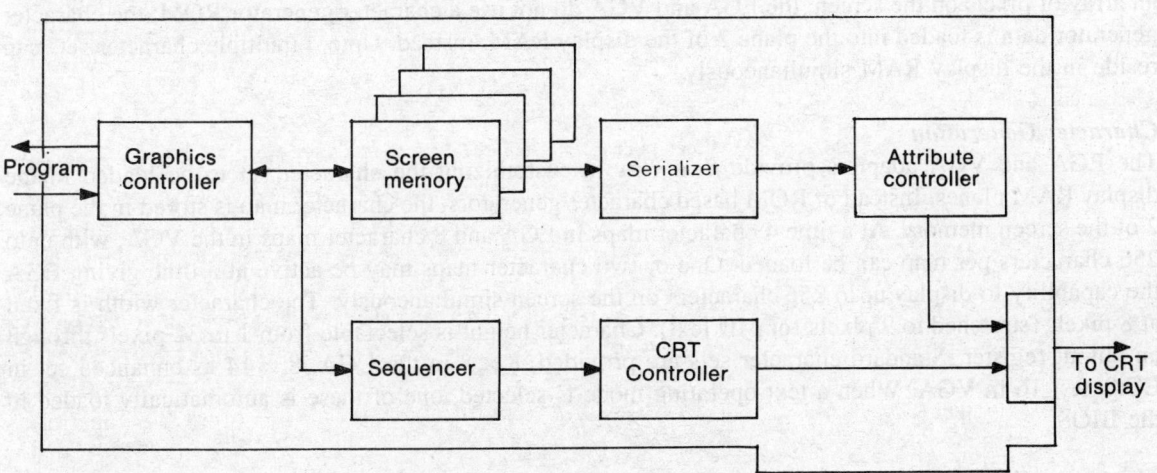

Fig. 24.12 Basic architecture of a EGA/VGA card

The basic elements in the architecture are:

— *screen memory* comprising 256 k DRAM divided into 4 colour planes.
— *graphic controller* in the data path, programmed to perform logic functions on the data, that provide hardware assistance to simplify drawing operations.

— *data serializer* that captures the parallel bit data from the screen memory and converts it into serial bit stream.

— *attribute controller* contains a colour lookup table and translates the colour information from the screen memory into colour information required for the CRT, as per the contents of the lookup table.

— *CRT controller* that generates timing signals for the display.

— *Sequencer* which controls the overall timing operations of all the functions on the card.

The EGA adapter permits the processor to access the screen memory while the display refresh is in progress, while the CGA does not. The sequencer controls access to the screen memory, interleaving screen refresh cycles with processor read/write cycles. In high resolution modes, the display requires more amount of data and hence the processor is allowed only 1 out of 5 memory cycles.

All IBM adapters following the CGA include *40 column text* modes. These modes were created to allow the text to be displayed on home television sets, which have resolution too poor to display 80 columns of text. Special adapter circuitry is required to connect an IBM compatible to a television set.

Screen Memory
The EGA/VGA cards contain 256 kbytes of memory, arranged in 4 independent colour planes of 64 k-bytes each, but in the memory space of the same processor. The settings of I/O registers determine the colour plane active at a time. The processor can write in all the 4 planes in a single memory write cycle. In the text mode displays, the display screen is divided into 25 lines with either 40 or 80 character text per line. Two bytes of screen memory are used to define each character. The 1st byte mapped at an even memory address contains the ASCII character code, and the 2nd byte called the *character attribute*, is mapped at the odd memory address, giving the colour information. Obviously, 2 kbytes of memory are needed for a 40 column page and 4 k for an 80 column page. For converting ASCII character code into an array of pixels on the screen, the EGA and VGA do not use a character generator ROM; the character generator data is loaded into the plane 2 of the display RAM, instead. Upto 4 multiple character sets can reside in the display RAM simultaneously.

Character Generation
The EGA and VGA adapters provide *flexibility* in customizing the character set to be loaded in the display RAM planes. Instead of ROM based character generators, the character map is stored in the plane 2 of the screen memory. At a time 4 character maps in EGA and 8 character maps in the VGA, with upto 256 characters per map can be loaded. One or two character maps may be active at a time giving EGA the capability to display up to 256 characters on the screen simultaneously. The character width is fixed at 8 pixels (stretched to 9 pixels for MD text). Character height is selectable from 1 to 32 pixels through an output register. Standard character sets are provided, 8×8 in the CGA, 8×14 as enhanced set in EGA, 9×16 in VGA. When a text operating mode is selected, one of these is automatically loaded in the BIOS.

Standard Text Attributes
When operating in a standard colour text mode, the attribute bits in a byte define the attributes as follows:

Bits D0-D2 for foreground colour selecting the character body colour
Bits D4-D6 for background colour selecting the character cell colour
Bit D3 as foreground intensity control, doubling colours to 16
Bit D7 as foreground blink control or background intensity control

Standard attributes used for the *foreground/background colours* are:

Attribute Bits:	000	001	010	011	100	101	110	111
Standard colours:	Black	Blue	Green	Cyan	Red	Magenta	Brown	Gray
Intensified colours:	Gray	Light Blue	Light Green	Light Cyan	Light Red	Light Magenta	Yellow	White

The interpretation of the text attributes depend on the operating mode. *Monochrome (MDA) text attributes* are similar, bits D0-D2 are used for foreground control to give normal/blanked/underlined character, and D3 for intensified character. D4-D6 are used to provide a reverse video. D7 can be used as foreground blink control or background intensity control, as controlled by the Mode Control register of the Attribute Controller. It is possible to *customize* the text attributes by reprogramming the Attribute Controller.

Attribute Controller

The attribute controller determines the colour to be displayed by use a Colour Look-up Table which translates the 4 bit data from the screen memory into 6 bit colour information for the EGA or 12 bit colour information for the VGA. The colour look-up table is initialized with colour data appropriate for that mode, during the BIOS mode select operation.

Graphics Controller

The graphic controller in its default state is *transparent*, allowing any data to be written into or read from the screen memory, without alterations. It can be programmed to assist in drawing operations by performing certain tasks of the system processor. It is possible to *quickly copy data* from one memory plane to another because the read latches can latch the data from all the four planes simultaneously, whereas the processor can read at a time data from one plane only. The data read from the screen memory by the system processor is also latched on the EGA card. This data can be logically combined with write data from the main processor during write operations. Logical functions like AND/OR/XOR are useful in adding or removing foreground display elements over the background. Write data can be rotated for performing block transfers of non-byte aligned data.

During processor read cycle, the graphic controller can perform *Colour Compare* function which is useful for drawing algorithms such as 'flood fill' where a specific colour or change in colour has to be detected. The graphics controller can compare data from all the 4 planes against a reference colour, while the main processor may interrogate only one plane at a time.

Image Processing Cards

Frame grabber and processing cards are now available that digitally capture and process the TV picture frames as fast as they occur—at 25 frames per second. Architecturally the key features include a frame memory that stores 512×512 pixels, at 6/8/12 bits each, and lookup tables of $4096 \times 6 - 12$ bits each. Some of them use CMOS gate arrays extensively, reducing the IC chip count and power consumption. A block diagram of such an image processor is shown in Fig. 24.13.

The image processing feature include brightness enhancement, comparison; addition, subtracting and averaging of the brightness value of two images; graphic overlays and display windows, etc. Applications of such PC-AT based cards include automated inspection, medical imaging, security systems, graphic design.

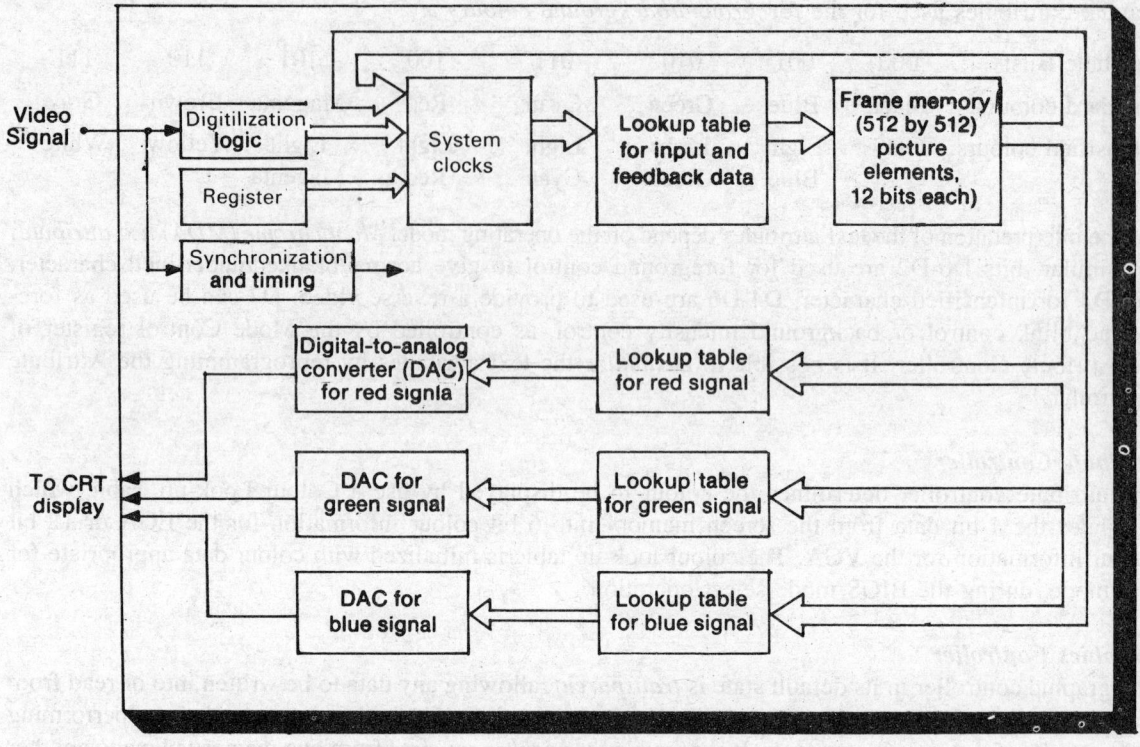

Fig. 24.13 Block schematic of an image processor card

REVIEW QUESTIONS

24.1 Discuss the significance of modern broadcast information services? What is the difference between videotext and teletext services?

24.2 Explain the basics of a teletext information system indicating how the information in the VBI is accessed sequentially.

24.3 Give details of a how the teletext data is digitally encoded into a data stream and the standard format in which data is organized in the Teletext system.

24.4 What type of digital code is used in the teletext? What is intersymbol interference?

24.5 Draw the schematic arrangement of a teletext decoder required for a TV receiver.

24.6 What are the levels proposed by CCIR to enhance the capabilities of teletext systems towards graphics mode?

24.7 Discuss the differences in bit-mapped and character-mapped displays.

24.8 A CRT display has a resolution of 640 × 480 and vertical non-interlaced scan at 50 Hz. 20% of each scan line is used for retrace blanking and 30 lines are used for vertical retrace blanking. Calculate the dot rate and minimum bandwidth. Repeat for interlaced scanning.

24.9 What are the merits and demerits of character mapped and bitmapped graphics displays?

A bit-mapped graphic system has a pixel resolution of 1280 × 1024. Calculate the screen memory requirements for

(a) monochrome, 64 gray levels

(b) colour with a palette of 256 colours

(c) colour with choice of 16 colours from a palette of 4096 colours.

24.10 Draw a block diagram of the video generation circuits for the bit-mapped displays in the question above.

24.11 Explain the following terms: Page header, clock run-in, eye height, DRDS, alpha-photographic teletext.

MULTIPLE-CHOICE QUESTIONS

24.1 Teletext is a video information service that is:
    (a) an interactive   (b) a one way only
    (c) a pseudo-interactive   (d) b and c

24.2 For teletext data NRZ coding format is used because:
    (a) it is less prone to error   (b) it is easier decoded
    (c) it requires less bandwidth   (d) none of these

24.3 Intersymbol interference is caused by:
    (a) the data pulses not being ideal
    (b) the unequal delays in arrival of the spectral component frequencies
    (c) limited bandwidth of the system

24.4 Teletext data bit 1 voltage is:
    (a) at the black level   (b) at white level
    (c) at 50% of white level   (d) at 66% of white level

24.5 Teletext data line contains:
    (a) 40 bytes of display information
    (b) 45 bytes of display information
    (c) 45 bytes with 5 bytes as synchronisation
    (d) 45 bytes with 8 bytes of coded information

24.6 Synchronisation of teletext signals is ensured by:
    (a) the crystal clock in the decoder
    (b) the clock run-in bytes sent with each line
    (c) by phase locking the teletext clock with the line sync pulses
    (d) b and c

24.7 State whether True/False
    Teletext information requires lower bandwidth because it is sent at a slow pace during the vertical blanking interval. T/F?

25

Digital Television Technology

As in other areas of electronics, digital technology has made its impact in television systems too. In television transmission system, the input signal at the camera pick up, and the output drive to the picture tube are inherently taken as analog, continuous with respect to space and time. If TV signals are digitized into two level binary sequences, several *technical advantages* are possible. However, digitization of the video signals involves high speed A/D converters and very high bit rates for transmission. With the rapid advances in microelectronic technology a wide variety of studio equipments are getting available, continuously improving and offering innovative variety of digital video effects. Digital techniques are profitably used in time base correctors, frame synchronizers, digital recording, and signal processing for improved picture in TV receivers. Because of limitations of hardware, equipments and lack of worldwide standardization, TV broadcasts continue to be analog transmissions.

The CCIR has unanimously adopted in 1982, *Recommendation 601* for encoding parameters of Digital Television for studios, which established an agreement on a *digital code compatible with both 525/60 and 625/50 standards.* There is now renewed interest in digital transmission of TV signals because of the emergence of Broadband Integrated Services Digital Networks (B-ISDN) and proliferation of digital video techniques in TV studios.

25.1 MERITS OF DIGITAL TECHNOLOGY

Even with today's analog TV systems, the receiver manufacturers have already started innovating their TV sets with digital technology. Digital TV can provides long term stability of colour picture and increased reliability, besides flexibility to adapt readily to facilities like teletext, picture-in-picture. Digital signal processing of the analog TV signals improves picture reliability, eliminates the effects of component aging over time, and can provide a ghost-free picture, so important in teletext to prevent misreading of characters.

With the VLSI technology packing over 50,000 transistors in a chip, the component count is also lower. This can reduce costs and facilitates automatic production and testing. ITT-INTERMETALL in Freiburg, formulated the digital TV concept in 1973, and worked out the applications criteria through discussions with leading TV manufacturers by 1979, and set upon the design and development of ICs required for a *real time digital TV signal processing system.* The first digital TV receiver based on the DIGIT 2000 IC system was unveiled by the ITT in West Germany, in 1983. Other manufacturers have followed the concept in different ways. Analog-digital interfaces, signal processing ICs and microcomputer chips now available at economic prices have made digital television techniques feasible and attractive, offering several additional features.

25.2 FULLY DIGITAL TELEVISION SYSTEM

Analog signals suffers degradation in transmission path due to noise, crosstalk, linear and nonlinear distortion in the circuitry. In a fully digital system, the analog signal is converted into digital signal by means of an A/D converter. The A/D converter performs three basic functions: Sampling, Quantization and Encoding.

(1) Sampling
The essentially band limited TV signal, is sampled at a rate of twice the maximum band-limit frequency, according to the sampling theorem, also known as the *Nyquist criteria*. The spectrum or the Fourier transform of a band limited signal sampled at the interval T seconds, is a periodic signal consisting of the original spectrum of the signal scaled by $1/T$ and shifted on the frequency domain by nw_0, repeated at w_0 interval given by $w_0 = 2\ \pi/T$. Such a sampling process is shown in Fig. 25.1, in the time and frequency domain.

When the original band limited signal does not have frequency components avove $w_0/2$, the components of the Fourier spectrum, do not overlap and the scaled down version of the original spectrum could be recovered by low pass filtering. Stated as the *Nyquist uniform sampling theorem* this means: A signal band limited to *fm* Herz is uniquely defined by its samples taken uniformly at $1/2\ fm$ seconds apart. If the sampling rate is less than this Nyquist frequency, the analog signal cannot be reconstructed without errors.

If the input signal is not band limited or the sampling frequency does not meet the requirement of sampling theorem, the various components of the spectrum overlap and the signal recovered in reception is not be the exact replica of the original signal. It suffers from distortion called *aliasing* effect shown in Fig. 25.1(b), and Moire patterns appear in the picture, in areas of fine detail. An example of this effect is in the backward strobing of rotary movements of wagon wheels observed in cinematography, which because of the frame repetition technique is a sampled data process rather than an analog process. Actually, it is interesting to note that by virtue of the frame scanning technique, the analog TV signal is also virtually a sampled data signal!

(2) Quantization
The sampled output, although discrete in time is continuous in amplitude, and has to the assigned a discrete numerical value as the A/D converter maps a range the input amplitude in the minimum resolution range, onto each single digital output value. Since the D/A converter in the reverse process can assign only one analog amplitude to the digital samples within this range, this introduces an impairment in the digital system, which is referred to as the *quantization error*.

(3) Encoder
The sampled numerical value is assigned a binary code along with additional coding like parity for error detection, scrambling etc., depending on the nature and error characteristics of the system handling the digital signal. In digital systems, the errors occur during transmission or storage of the signals. By introduction of some redundancy in the digital signals the some of the errors can be detected and corrected. When an error is merely detected, the system can either hide its effect or request for retransmission. The latter remedy, because of the inherent delay is not acceptable in television. Coding process with *forward error correction* has its benefits and limitations. Block coding and convolution coding are two promising approaches in this direction.

Advantages of Digital Techniques

As the digital signals consist of binary pulse train they can be made immune amplitude degradations, by

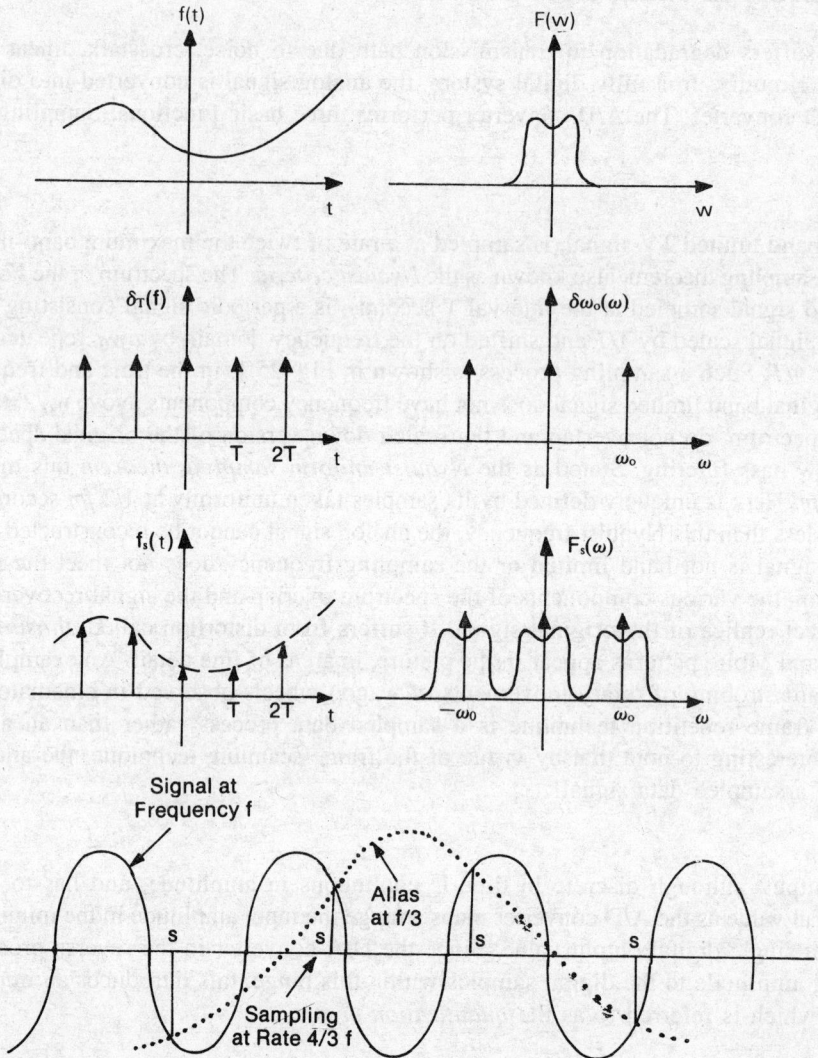

Fig. 25.1   (a) Sampling process in time frequency domain (*Courtesy:* Benson TV Engg Handbook, McGraw-Hill Inc.), (b) Aliasing effects (*Source:* Benson, TV Engg Handbook, *Courtesy:* McGraw-Hill Inc. NY)

regeneration of the pulses. In digital form, the signal can be easily stored, delayed and manipulated by means of microcomputer chips, integrated in the system. The digital signals can be time multiplexed or interleaved with other ancillary signals. Moreover, the use of forward error correcting codes can preserve integrity of the signal.

The problem of component aging on the alignment, tuning etc. and consequent instability encountered in analog systems is eliminated with digital technology which can have automatic self-tuning and adjustment routines. The digital TV system adapts more easily to additional video display functions from teletext to TV games, videotext, computer terminal etc.

25.3 DIGITAL TELEVISION SIGNALS

The digitization of TV signals can take two forms. In *composite coding*, the composite NTSC, PAL or SECAM signals are quantized, the resulting digitized signals are referred to as 'digitized video'. PAL signals can be digitally coded and decoded satisfactorily, and there are no difficulties in sampling PAL at multiples of line or colour subcarrier frequencies. SECAM presents some problems as the colour signals are not available simultaneously. Decoding the digitized composite SECAM into *Y*, *R-Y* and *B-Y* components requires conversion to analog before the components can be obtained.

In *component coding*, the luminance and chroma signals (*Y*, *U*, *V*) or the Red, Green and Blue (R, G, B) from the camera, rather than the composite video, are separately digitized, when the digitized signals are referred to as 'all-digital video'. Component coding is more attractive being independent of the colour system standards. Frame stores, noise reduction, special video effects are easier in component coding. Cross colour interference or low diagonal resolution in PAL can be eliminated in component coding using MAC signal.

The *lowest Nyquist frequency* for PAL can be seen to be $3 \times fsc$. However because of the subcarrier is a multiple of the frame frequency, the sample dots cannot form an *orthogonal structure* on the scanned raster. It is obviously possible to obtain this by sampling at $2 \times fsc$ or $4 \times fsc$. The sub-Nyquist rate of $2 \times fsc$ reduces the bit rate and reduce the bandwidth problem without significant impairment.

Because the video signals in television systems are highly structured, digitization results in *quantization errors* which may cause perceptible impairment of the picture, like contouring. There is a large amount of redundancy existing in the structured nature of the video signal which can allow bit rate reduction, provided sampling parameters are properly chosen.

Sampling Patterns

The basic television signal is one-dimensional representation of the *time varying image of two-dimensional* picture, sampled by the scanning mechanism of the camera. The analog signal in composite form or in the R, G, B component form is a sampled version of the three dimensional image. In the *sampling requirements* of digital television signal, it is therefore necessary to consider the scanning process, which has already sampled image in vertical, horizontal and time dimensions. Further sampling at the minimum required rate for digitization, results in a three dimensional sampling grid patterns. The number of scanning points have to be as low as possible without introducing aliasing. Since the line and field scanning rates are derived from the colour subcarrier, the horizontal scanning frequency is conveniently, a multiple of the same. The rectangular scanning grids can take different patterns as shown in Fig. 25.2.

The field aligned scheme shown in Fig. 25.2a, consists of a sampling grid having rectangular projection both in the spatial and horizontal-temporal direction. To *avoid aliasing errors* the sampling frequency in PAL system must exceed 10 MHz. The field offset scheme shown in Fig. 25.2b, has half sampling period offset for the interlaced field and is more efficient from considerations of spectral density. Checkerboard sampling grid of Fig. 25.2c is formed by offsetting the sampling pattern from one line to the next line by half horizontal-sampling intervals, whereas double checkerboard sampling is obtained by half sampling period offsets after every two lines, as shown in Fig. 25.2d, has better spectral properties.

Bits per Sample

Subjective studies have indicated that satisfactory picture quality is possible with quantization of 8 bits per sample. If the video has vertical blanking interval test signals 8 bits per sample are necessary to preserve the integrity of the test signals. For the primary digital coding uniformly spaced quantization

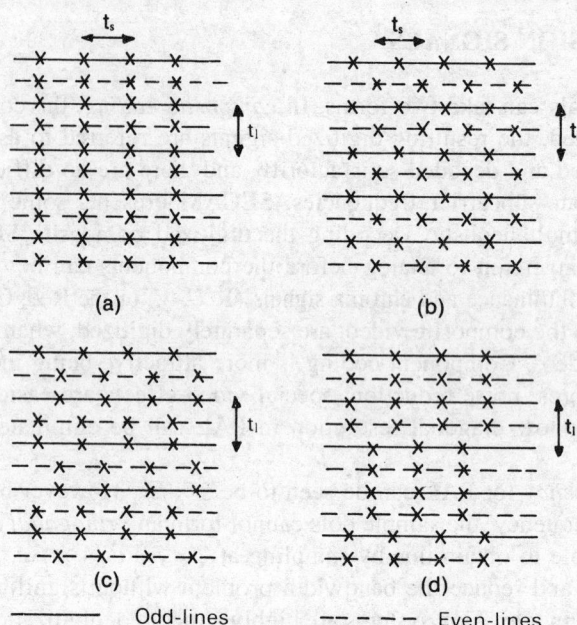

Fig. 25.2 Rectangular sampling patterns (a) Field aligned pattern, (b) Field offset pattern, (c) Checkerboard pattern, (d) Double checkerboard pattern (*Source:* Benson, TV Engg Handbook, *Courtesy:* McGraw-Hill Inc. NY)

is preferred because of simplicity in implementation and signal processing. The resulting signal is then referred to as *uniformly quantized PCM* signal. Non-uniform *tapered* quantization is used for bandwidth reduction techniques.

25.4 DIGITIZED VIDEO PARAMETERS

The important parameters of digitized video signal are listed in Table 25.1.

Table 25.1 Parameters of Digitized Video

	525/60 system	625/50 system
Active lines	485/s	575/s
Vertical sync	7.6% H	8% H
Horizontal blanking	16.0% H	18% H
Frame rate	30 Hz	25 Hz
Line Frequency	15,750 Hz	15,625 Hz
Bandwidth	4.5 MHz	5.5 MHz
Active line period	53 μs	52 μs
Subcarrier	3.579545 MHz	4.4336187 MHz
	$= (227 + 1/2)fh$	$= (284 - 1/4) fh{-}25$
Bit rate at 3 fsc	85.90908 Mbits/s	N/A
Bit rate at 4 fsc	114.54544 Mbits/s	141.8758 Mbits/s
Bit rate at 3 fsc active pels only	66.679192 Mbits/s	N/A
Bit rate at 4 fsc active pels only	88.905589 Mbits/s	114.01139 Mbits/s

(*Source:* Benson, TV Engg Handbook, McGraw-Hill Inc. NY)

In digital video standards, extendible family of compatible digital codes are recommended so that they could suit the future needs. The members of the family will have spatially static sampling patterns with two simultaneous colour difference signals spatially cosited except when the signals represents R, G, and B signals. In that case all the three samples are cosited, as shown in Fig. 25.3.

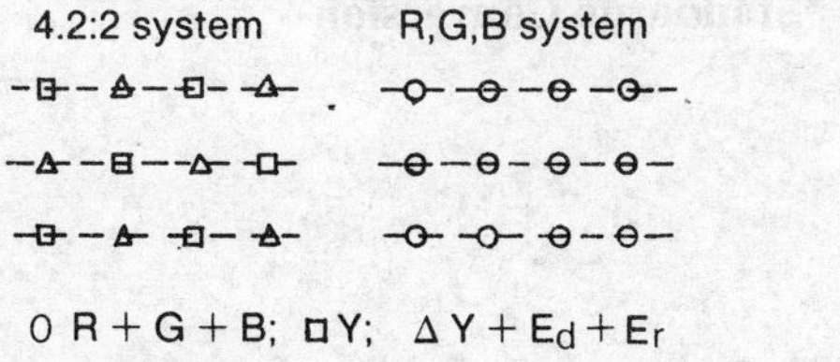

A code from such extendible family of codes is the 4 : 2 : 2 system, which has a wide support. The system is so called because the sampling frequencies adopted for it are close to 4*fsc* and 2 *fsc*. The numbers 4 : 2 : 2 denote the *sampling rate multiples of fsc* in terms of the colour subcarrier, associated with the luminance *Y* and colour difference signals UV

Fig. 25.3 Digital video sampling patterns (*Source:* Benson, TV Engg Handbook, *Courtesy:* McGraw-Hill Inc. NY)

respectively. The CCIR standards committee has produced the digital studio interface standard *CCIR 601*, based on this component coding permitting manufacturers to develop equipment having many common features of 625 and 525 line systems.

The lowest common multiple for scanning frequency which results in *static orthogonal pattern* in both 625/50 and 525/60 system, is 2.25 MHz. Integral multiple of this produces sampling frequencies of 13.5 MHz for luminance and 6.75 MHz for chrominance signals in a 4 : 2 : 2 system. The encoding patterns for this system are given in Table 25.2. A higher performance 4 : 4 : 4 system has also been proposed for digital encoding, in which the luma and chroma sampling frequencies are in the ratio 4 : 4 : 4. This system will be suitable for source equipment and high definition video signal processing.

Table 25.2 Encoding parameters for All Digital Video in TV studios. (4 : 2 : 2 component coding)

Parameters	525/60 systems	625/50 systems
Coded signals	Y, R-Y, B-Y	
Sampling frequency for Y	13.5 MHz	
Sampling frequency for R-Y and B-Y	6.75 MHz	
Number of samples per digital active line for Y	720	
Number of samples per digital active line for R-Y/B-Y	360	
Number of samples per total line for Y signals	858	864
Number of samples per total line for each R-Y/B-Y signal	429	432
Sampling structure:	Orthogonal, line, field, and frame locked R-Y, B-Y samples cosited with odd Y samples	
Quantization:	Uniformly quantized PCM for each signal	
Digital code* :	The luminance Y signal is encoded in 220 quantization levels such that the binary word 16 corresponds to the black and 235 to pack white. The colour difference signal will be encoded in 225 quantization levels symmetrically respect to the 0 signal corresponding to the digital word 128.	

Note: Digital words FF and 00 are reserved for timing reference signaling.

Pixel Oriented (line locked) Digital Signal Processing

If the video information is sampled with a frequency related to the actual line frequency, it is possible to identify the part of the display to which a sample belongs. With non-standard signals, the ratio between the line frequency and the colour burst frequency changes from line to line. With a burst locked clock, this results in a different number of samples for each line. In contrast, a sampling frequency that is an integer of the line multiple of the line frequency results in *orthogonal sampling* of the field. The processed samples at line locked frequencey stored in memories is suitable for a variety of applications such as:

*Standards Conversion

> — Field doubling, non-interlacing etc.
> Picture in picture (PIP)
> Multi-PIP
> Picture enlargement
> C-MAC decoding etc.

*Interline, Interfield or Interframe (orthogonal) Signal Processing
Temporal filtering with field memories for

> — Cross colour reduction
> — Noise reduction
> — Cross luminance reduction
> — Resolution enhancement etc.

Vertical filtering with one line memory for

> — Comb filters with one line delay

Vertical filtering with more than one line memory for

> — Vertical contour correction
> — Progressive scan

Furthermore, orthogonal sampling with line locked frequency is essential for *matrix displays* such as LCDs, plasma displays and dot matrix printers. It is also essential for good quality *character displays*. The clock frequency is determined by the CCIR studio standard, viz. 13.5 MHz.

The incoming video line frequency delivers the line lock for the entire system and decodes colour standards into line-locked samples and process the luminance component. The *synchronisation and colour decoder functions* can be combined together in a single IC, to facilitate not only the introduction of features, but also the functional coupling between sync decoding and colour decoding. The processing can be done independent of the incoming video standard.

25.5 DIGITAL VIDEO HARDWARE

The high bit rates involved in the digital video demand strict speed constraints on the digital hardware used. For logic gates and ALUs, Schottky TTL and ECL circuits can meet the speed requirements. For line memories 1 K chips are useful and can be paralleled to store (8 bits) of information. Larger IC memories are useful for frame memories, which can be slower than the logic gates if multiplexing technique is employed.

Video A/D converters form the vital component in practical realization because of the high speed requirement in lieu of the large bandwidth TV signal. Flash converters using parallel comparators for quantization of the video amplitude are only suitable. For N-bit quantization $(2^N - 1)$ comparators are required increasing the component count. The timing accuracy and jitter in switching is very critical. These have been now available in VLSI, flash converter chips operating up to 50 Megasamples per second with 8 bit/sample resolution.

For higher resolution *cascading* techniques are possible, by multiplexing the outputs from two A/D converters, fed through two comparators, to determine whether the input is more than half the total range and setting the additional MSB to 1 if the input exceeds the half range value. *D/A converters* are also implemented in VLSI technology currently available for 8 bits operating at 100 Megasamples per second.

Digital Studio Equipment

Experimental studios are already operating in some countries like France. The analog signals from cameras are digitized following wide band amplification. In order to reduce complexity of switching equipment serial data link is preferred, for the digital component coded signal, operating at over 200 MBits/s. *Digital telecine* employing CCD sensor has already discussed in section 6.6. Developments on *digital video recorder* have been going on using component coding. The large bandwidth requires high writing speeds and hence large tape consumption which are possible by reducing track width, which is possible because of the digital signal recording. Improved metal oxide tapes (19 mm, at lin. speed of 286.9 mm/s for 625/50), with coersivity of 68,000 A/m are recommended by CCIR using 4 : 2 : 2 standard coding. Only the tape format is specified by the EBU/SMPTE, leaving the manufacturers to specify the mechanics of tape wrap, heads and the drum scanner. The video and audio are recorded as bursts of data by four rotating heads laying down four tracks per revolution. The time code track are as per standard EBU code. Sony corporation has brought out the first Component Digital VCR to match the specifications of SMPTE C-format VTRs. The video and audio are recorded as bursts of data.

25.6   TRANSMISSION OF DIGITAL TV SIGNALS

Digital TV signals are transmitted via copper cables, fibre optic cables, or for over digital microwave radio for satellites relay links. Limiting factors for the capacity and distance for cable transmission are attenuation, crosstalk, echos, intersymbol interference, thermal and impulse noise. Using equalisation and shielding, a twisted pair copper cable can deliver without repeater 10 MB per second up to 1 km, while RF coaxial cables can transmit without repeaters, at the Gigabits per second range over 1 km. The range can be extended to longer distances by using regenerative repeaters.

Fibre optic cables can give very high capacities over larger distances, a span reach of tens of km, at 2 to 100 Mbit/s depending on the wavelength (typically 850 nm and 1300 nm) and the type of transmitter-detector. Use of laser LED for transmitter and avalanche photodetector (APD) provide larger distances than LED and PIN diode combination. Because of costly terminal equipment, fibre optic systems are economic over larger distances only. The digital transmission technique can be *bit-parallel or bit-serial*. *Bit-parallel* transmission is usually employed for interconnections of digital hardware within a studio complex. The interface standards are based on the 4 : 2 : 2 scheme of the studio digital codes as per CCIR Recommendation 601 given in the Table 25.2 above. For the 8-bit digital word corresponding to the 256 level PCM, the interface requires 8 conductor pairs, (D0-D7), each carrying a bit of the same significance of the multiplexed luminance and chrominance samples. The luminance is sampled at 13.5 MHz, and chrominance difference signals at 6.75 MHz. The multiplexed bit rate for each of the conducting pairs is 27 Mbit/s. An additional pair carries the 27 MHz clock along with the data consisting of NRZ binary ECL compatible output.

In place of the repetitive deterministic sync/blanking signals, special timing reference signals in the form of **FF 00 00XY** to indicate various sync and blanking information, and ancillary information signals are included, as shown in Fig. 25.4. The first three words **FF 00 00** is a unique preamble as both FF or 00 are not used for coding the video or even the ancillary signals. Table 25.4 gives the information pertaining to the type of timing reference carried by the last word XY.

Bit-Serial transmission is used for transmitting digital video over longer distances. Telecommunication networks are rapidly adapting to all digital environments, and in order to optimally utilize the transmission capabilities, the digital networks are highly structured in architecture, operating at the progressively higher hierarchical levels DS-1 through DS-4, as shown in Table 25.3.

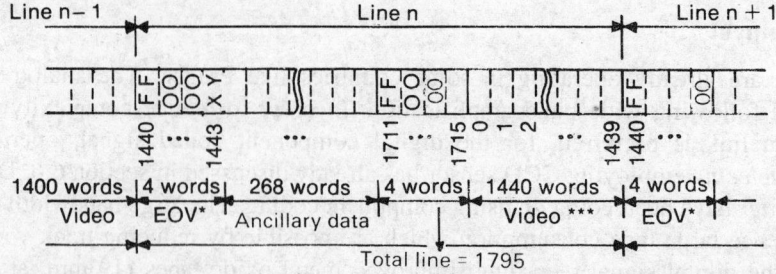

Fig. 25.4   Bit parallel transmission-timing reference signals (*Source:* Benson, TV Engg Handbook, *Courtesy:* McGraw-Hill Inc. NY)

Table 25.3   Hierarchical levels and the bit rates of digital Networks

	North America		Japan		Others	
	Mbits/s	*mij*	Mbits/s	*mij*	Mbits/s	*mij*
DS-1	1.544		1.544		2.048	
		4		4		4
DS-2	6.132		6.132		8.448	
		7		5		4
DS-3	44.736		32.064		34.368	
		6		3		4
DS-4	274.776		97.728		139.264	

(*Source:* Benson, TV Engg Handbook, *Courtesy*: McGraw-Hill Inc. NY)

Table 25.4   Timing Reference Signals in Bit-parallel transmission

Bit position	7	6	5	4	3	2	1	0
XY	1	F	V	H	P3	P2	P1	P0

F = 1 ...... even field

V = 1 ...... vertical blanking

H = 1 ...... start of horizontal blanking (EOV)

P0 to P3 (6, 3) ...... Hamming code plus 1 even parity bit

(F and V can change when H = 1)

(*Source:* Benson, TV Engg Handbook, *Courtesy*: McGraw-Hill Inc. NY)

Any data information traffic entering the transmission channels can do so only at defined multiplexed levels, or a multiple of such level. The total bit rate for the required for the digital video must be compatible with the transmission hierarchy. The only the third and fourth levels DS-3 and DS-4 have capacity to carry *broadcast quality* video signal. Because of the high cost of the high capacity DS-4 level can be considered for high definition or very high quality digital TV transmission such as the 4 : 2 : 2 component standard only. A single or multiple DS-3 level at 44.736 Mbit/s or 34.368 Mbit/s is satisfactory for digitized composite video.

In addition to the video, a number of ancillary signals such as the following have also to be accommodated in the DS-3 bit rate:

1. Digitized audio channels, sampled at 48 or 32 kHz, each encoded into 12 bit word, A-law companded PCM, including a singal parity bit.
2. Teletext videography signals transmitted on 10 to 21 lines of the vertical blanking interval.
3. Captions for the deaf encoded in ASCII 8-bit code
4. Vertical interval test signal (VITS) and vertical interval reference signal (VIRS) for monitoring the performance of the analog sections.
5. Colour burst for phase reference of CSC.
6. Error correcting codes.
7. Framing information etc.

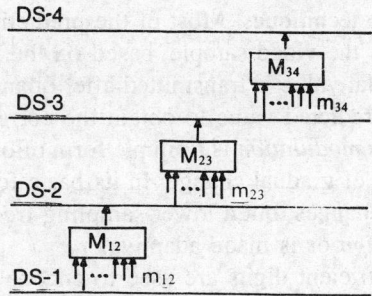

Fig. 25.5 Common carriers' multiplex and Transmission hierarchy (*Source:* Benson, TV)

25.7 BIT-RATE REDUCTION

Digital television signal, because of the high bit rates requires more than 3 times bandwidth required by analog FM circuits, displacing thousands of audio channels. It is therefore highly desirable to reduce the bandwidth and bit rate in the digital domain by redundancy techniques. With the present state of art algorithms, it is possible to reduce this to DS-3 rate with broadcast quality, while still further reduction to less than 1.5 Mbits/s is possible for applications like video teleconferencing. Some of the possible bit-reduction methods are indicated below.

Removing the Deterministic Portions

The repetitive synchronisation signals need not be sent with the video. If the locations of these pulses are transmitted, the receiving terminal can insert locally generated timing signals for synchronisation of the picture. A saving of over 6% of the total capacity is possible by removal of vertical sync-blanking interval and additional 18% is possible by removing the horizontal sync-blanking interval.

Variable Word-Length or Entropy Coding

This medhod can reduce transmission rate by assigning shorter code words for frequently occurring output values and longer codes for less frequent larger output values. This is possible in *differential pulse code modulation* (DPCM), where the resultant code has a very skewed unequal code. The line transmission rate of course becomes picture or information dependent and entropy codes must employ elastic buffers to produce uniform transmission rate. In order to handle the possible overflows of the finite length coders, suitable remedies like reduction of sampling rate or reduction of the number of quantization levels, may be employed.

Predictive Coding

This method relies on the statistical properties of the transmitter and the receiver including the subjective human element, to achieve bit rate reduction at the expense of some less significant subjective details. This can be achieved in *time domain coding* or *transform domain coding* employing fixed parameters or adaptive techniques. Most of the predictive coding techniques rely on DPCM. The DPCM predicts the value of the video sample, based on the past history. The difference error between the predicted value and actual value is transmitted after quantization and encoding. At the receiver the error signal is added to the predicted value to obtain the correct video sample.

Delta modulation is a simple form of one digit DPCM and is excellent for encoding luminance signals in areas of gradual change. In its basic form it requires a very high sampling frequency to handle large abrupt changes, but a lower sampling frequency can be used if the delta modulator incorporates *double integration* or is made adaptive.

If sufficient digits are used to encode each difference, DPCM can both handle *abrupt changes*, and avoid objectional contours in low detail areas. By using instantaneous companding technique, i.e. arranging for the quantum levels to be closer together for small differences than for large differences, the number of digits can be reduced. For example, with instantaneous companding 16 quantum levels are found sufficient for video telephone applications.

Dual mode schemes, in which two different encoding techniques are used for different areas of the picture. *DPCM* may be used for areas containing large variation in luminance while *delta modulation* is used for areas of gradual change. In colour video signal which includes chrominance subcarrier, DPCM encoder is in a continuous overloaded condition unless each prediction is based on a sample separated by an integral multiple of the subcarrier periods. If the sampling frequency is chosen to be integral multiple of the subcarrier frequency, intermodulation products can be reduced. 5-digit or 8-digit DPCM is considered sufficient for most colour television systems. Greater efficiency is possible if luminance and chrominance are encoded separately, using a lower sampling rate and transmission of chrominance on alternate lines.

Sometimes bandwidth reduction is easier done in transform domain after linear transformation. Instead of transforming the entire field of video which involves huge data, it is more practical to split the video into smaller blocks before transform coding. The transform coding techniques include Karhunen-Loeve Transform, Discrete Fourier Transform (DFT), Discrete Cosine Transform (DCT), etc. After transformation of the image, bandwidth reduction techniques like DPCM coding as in the time domain can be applied.

25.8 DIGITAL TV RECEIVERS

Since the launching of digital TV receiver by ITT INTERMETALL Freiburg, in 1984, the transition from the hitherto completely analog TV receivers to digital has started, and a sizable section of TV manufacturers have already introduced digital modes into the market to cash on the additional features like PIP, and a number of advantages the digital technology can provide. Digital technology can improve the picture and sound quality by digital signal processing, and maintain it throughout the life of the TV set, without readjustments as required in analog TV sets due to effects of component aging. It can provide many additional features like stereo sound, picture within picture, single chip teletext decoder, with improved reliability. Easier to automate, the technology also cuts production costs in components count, material handling, quality assurance, testing and alignment and with the ability to include extra features like automatic switching to any transmission system, ease of conversion to HDTV, digital technology is going to be the inevitable choice in the long run.

25.9 ITT DIGIT 2000 IC SYSTEM

This ITT INTERMETALL system consists of a series of VLSI chips each consisting of about 50,000 transistor functions. These VLSI chips and a small number of additional components simplify the digital TV receiver manufacture and improve the set performance. The advantages of the DIGIT 2000 concept are:

— small external component count, no tolerances, no drift and aging, programmable processing, operating facilities flexible determined by software, computer-controlled fully automatic alignment during set production, stereo sound or bilingual facility and adaptability to all standards: PAL, NTSC, SECAM, D2-MAC

      The digital signal processing concept also offers a number of *additional features:*

— reception of digitally transmitted TV signals via DBS (Direct Broadcast Satellite) using the D2-MAC/packet standard or via broadband integrated fibre optic network for telecommunications, like the BIGFON local network of the German postal authority
— adaptive noise reduction by correlation or non-linear response
— flicker-free picture by means of intermediate storage and readout with increased deflection speed
— automatic ghost compensation by suppression of reflections
— picture-in-picture (PIP-another channel shown at reduced size)
— improved picture quality for NTSC receiver by means of digital filter
— direct processing of digital audio signals
— Text processing of Teletext, Viewdata, Prestel, Antiope, etc.
— terminal for home computer systems

The following basic functions are served by the VLSI chips of the DIGIT 2000 system:

Central Control Unit

This is a 8-bit single chip NMOS microcomputer, intended to control the digital colour TV receiver from keyboard input. Three versions of this offer different ROM and RAM capacities, viz.

    CCU 2030 6.5 kB ROM and 120 Bytes RAM
    CCU 2050 8   kB ROM and 256 Bytes RAM
    CCU 2070 16 kB ROM and 256 Bytes RAM

Non-volatile memory EEPROMs MDA 2062 1k bits

Picture processors:

Video Codecs	VCU 2133/34
Video Processor	VPU 2203
Deflection Processors	DPU 2553/54/55
NTSC Comb Filter Processors	CVPU 2233/35/70
SECAM Chroma Processor	SPU 2220

Sound Processors

Audio A/D converters	ADC 2300/10
Audio Processor	APU 2400/70
Audio Mixing Unit	AMU 2585

Clock Generators
MCU 2600/32

Improvement ICs

Processor for flicker-free picture	PSPU 2240
Digital Colour Transient Improvement processor	DTI 2223

Progressive Scan Processor	PSP 2210/32
Video A/D Converters	VAD 2150/70
Video Memory Controller	VMC 2260
Video/Sync Processor	VSP 2850

Additional Feature ICs for special and allied functions:

Teletext Processors	TPU 2732/33/40
D2-MAC Decoder	DMA 2270
Picture-in-Picture processor	PIP 2250

Peripheral ICs

D/A and Bus Converter	MEA 2050
Tuner Interface	MEA 2901
IR Remote-Control Transmitter	SAA 1250
IR preamplifier	TBA 2800

For developing user program CCU 2000 Emulator is available, usable with external memory. The concept does not include a multistandard colour decoder, but the audio processor APU 2400 has the promise of a powerful version which will perform identification processing for mono, stereo and bilingual broadcasts and generate the sound to optimally match the loudspeakers.

Block diagram of a basic CTV system DIGIT 2000 is given in Fig. 25.6. This is a standard version of digital CTV receiver with infrared remote control, equipped with PAL and SECAM reception and Teletext.

The TV signal from the tuner is demodulated in the IF processor as usual to separate the sound and the composite video. These are fed respectively to the audio processor chip APU 2400/70, and the video codec unit VCU 2100/33 and further to the video processor unit VPU 2203.

In NTSC receivers, the screen displays only 450 lines, because of the 525 lines 25 are used up for vertical flyback blanking, and 50 cannot be seen due to 10% overscan needed. To improve the poor picture quality, a double-scan flicker free scanning systems can be implemented using PSP 2210 to display each line two times, which means about 900 lines on the screen and gives better picture impression in the NTSC receivers. To do this, the horizontal deflection is at about 32 kHz, and the needed information for the second line is derived by storing one line in a line memory and by special filtering of the video signals.

Central Control Unit

The ICs CCU 2030, CCU 2050 or CCU 2070 differing only in their ROM-RAM capacities, can be programmed during production according to customer specifications, with the help of emulator boards. Combined with peripheral hardware the CCUs offer the following facilities:

— infrared remote control
— front panel keyboard control with up to 32 commands
— tuning by frequency synthesis (PLL) and band switching
— non-volatile program storage
— LED display for channel indication
— storage of alignment information during production
— generation and recognition of various control signals
— control of digital signal processors for video, audio deflection and teletext via a serial IM (Intermetall) bus

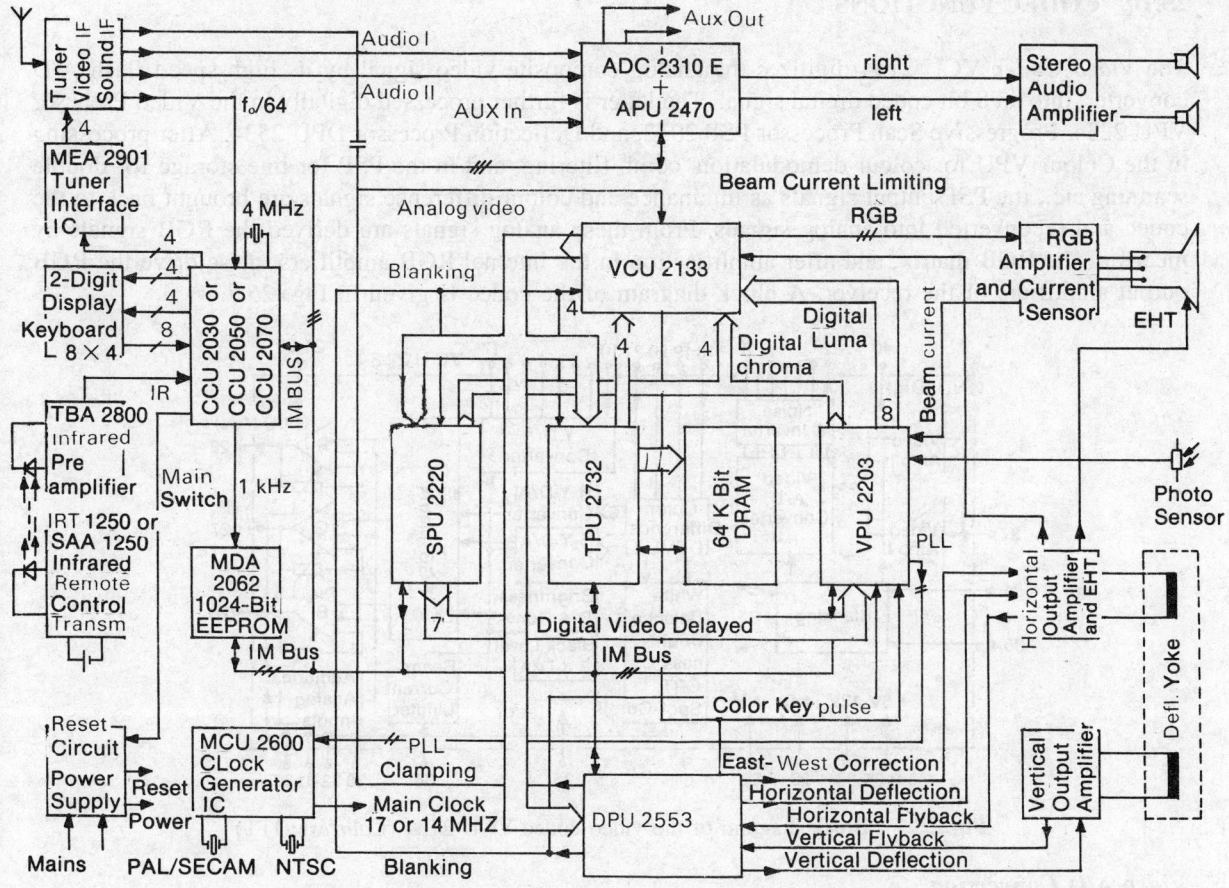

Fig. 25.6 Block diagram of a CTV receiver-the ITT-DIGIT 2000 concept, designed for PAL and SECAM (*Courtesy:* ITT)

The CCUs, produced in N-channel HMOS technology, are housed in 40-pin Dil plastic package, and contain on one chip the following functions:

— 8049 8-bit microcontroller
— remote control decoder
— Ports P2 and P3 for connecting a maximum of 32 keys and 4-digit seven-segment LED channel indication
— PLL tuner circuit for VHF and UHF
— IM bus interface for inputting and outputting control signals and for inputting alignment instructions
— crystal-controlled clock oscillator which also serves as reference for the PLL circuit
— mains flip-flop and reset circuit

The CCU 20*0 provides an efficient interface between the user and TV set. Their programmability enables different design specifications. The main function of the CCU is to process the user settings and control the digital signal processors for video, audio and deflection. It controls storage and output of factory alignment values that have been programmed during production of the TV set.

25.10 CODEC FUNCTIONS

The *Video Codec* VCU 2134 digitizes the analog composite video signal by its high speed flash A/D converter, into an 8 bit coded digital signal. The latter is further processed digitally in the Video Procesor VPU 2234, Progressive Scan Processor PSP 2032, and Deflection Processor DPU 2534. After processing in the Colour VPU for colour demodulation, comb filtering, and in the PSP for line storage for double scanning etc., the PSP output signals as luminance and colour difference signals are brought back to the codec and reconverted into analog signals. From these analog signals are derived the RGB signals by means of the RGB matrix, and after amplification in the internal RGB amplifiers, these drive the RGB output amplifiers of the receiver. A block diagram of the codec is given in Fig. 25.7.

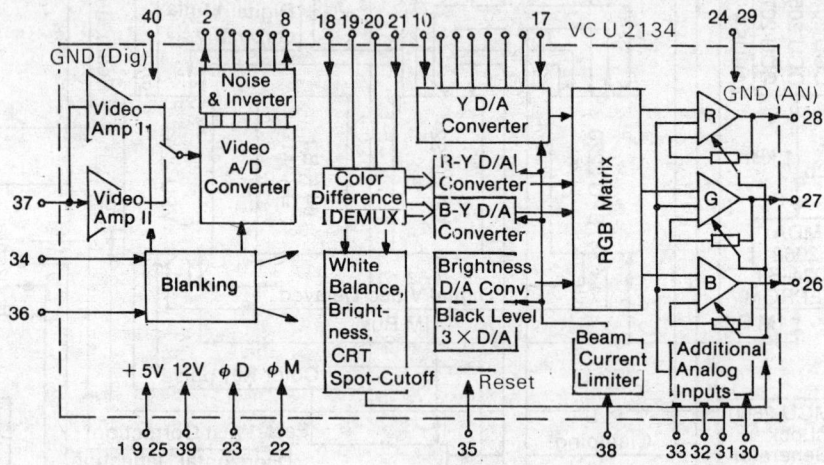

Fig. 25.7 Block diagram of the video codec VCU 2134 (*Courtesy:* ITT)

Video A/D Conversion

A flash A/D converter for n bits needs $2^n - 1$ comparators. This means that the number of comparators must be doubled if one additional bit is needed. Hence it is important to have as few bits as possible. For a slowly varying video signal, 8-bits are required. In order to achieve an 8-bit picture resolution using a 7-bit converter, a trick is used; the reference voltage of the A/D converter is changed by an amount corresponding to one half of the LSB, during alternate horizontal sweep. Thus a gray value located between two 7-bit steps is converted to the next lower value during one line and to the next higher value during the next line. The two gray values reproduced on the screen are averaged by the eye, producing the impression of gray values with 8-bit resolution.

The A/D converter sampling frequency is 17.7 MHz for PAL and 14.3 MHz for NTSC, the clock being supplied by the MCU 2632 clock generator IC. The output signal from the A/D converter is Gray-coded to eliminate spikes and gliches resulting from different comparators speeds or from the coder itself.

25.11 VIDEO PROCESSOR UNIT

The video processor unit contains the following functions:

— code converter
— chroma bandpass filter

— chroma trap with peaking facility
— contrast multiplier with limiter, for luminance signal
— all colour signal processing such as automatic colour control (ACC), colour killer, PAL identification, decoder with PAL compensation or NTSC comb filter, hue correction
— colour saturation multiplier with multiplexer for colour difference signals
— IM bus interface for the CCU
— circuitry for measuring dark current (CRT spot cutoff), white level and photocurrent, and for transferring this to the CCU.

The block diagram of the N-MOS chip VPU 2203 is shown in Fig. 25.8(a) while the combined block schematic taking into account the sequence of signal flow is shown in Fig. 25.8(b).

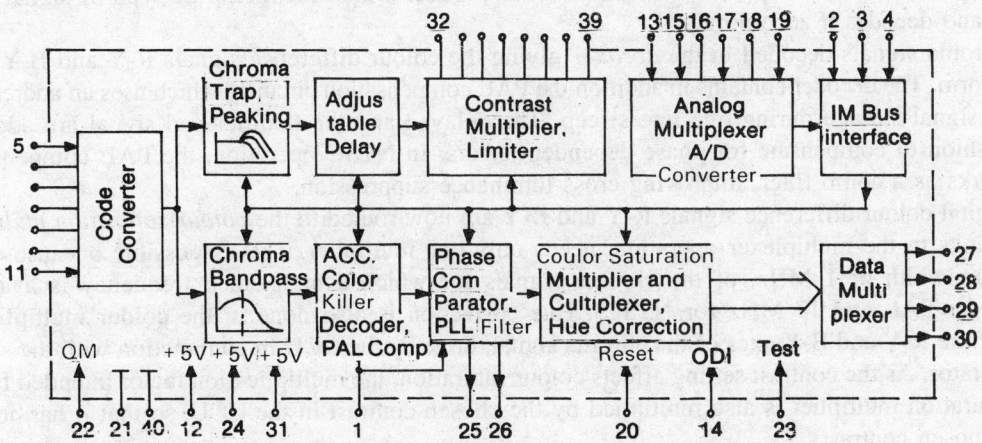

Fig. 25.8 (a) Block diagram of the video processor VPU 2203(ITT)

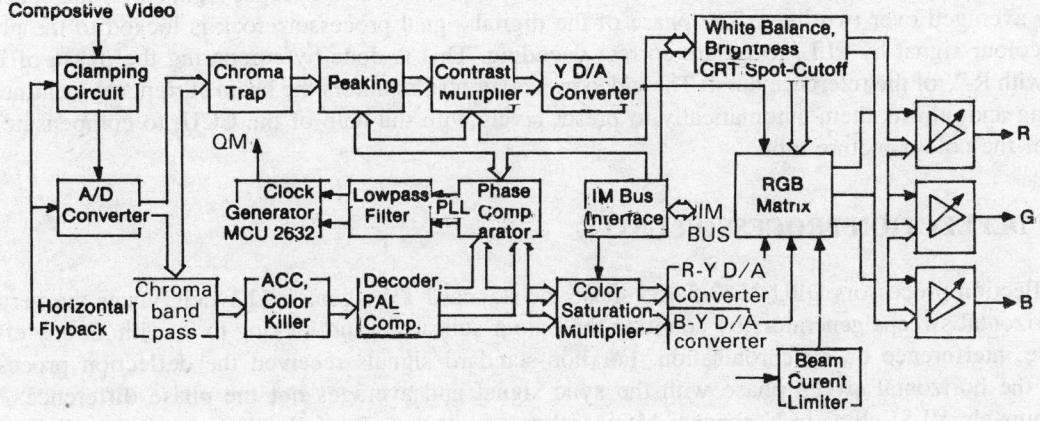

Fig. 25.8 (b) Block schematic of the signal flow in video codec and the video processor (ITT)

The code converter takes the digitized video signal from the video codec in a parallel Gray code form and converts it into simple binary signal for the luminance channel and into offset-binary-coded signal for the chrominance channel. The *luminance channel* has a chroma trap, a phase-linear digital filter which separates the luminance signal and handles the peaking so that the high frequency components of the luminance signal in the range of 3 MHz are raised to improve picture resolution and sharpness. The peaking at 3 MHz can be set by the user in eight steps, from − 3 dB to + 10 dB. (The principle of digital filtering is explained in section 25.13, in the context of audio processing.) The brightness and contrast are set by the multiplier according to the keyboard input. It is adapted to the room lighting by means of a photosensor connected to pin 17 of the VPU 2203.

The chrominance channel consists of a phase linear chroma bandpass filter, that can be switched to asymmetrical or symmetrical response and broad or narrow bandwidth by means of the CCU via the IM bus. Asymmetrical response provides correct compensation for the IF amplifier response, while the symmetrical filter curve is used for external composite video from VCR or video disc player. The chroma signal amplitude is maintained constant by the automatic colour control circuit (ACC) having a contrast range of 36 dB, and is then passed on to the colour decoder which identifies the type of signal PAL/SECAM, and decodes it accordingly.

The chroma signals decoded in the *decoder* giving the colour difference signals R-Y and B-Y, both in digital form. The decoder contains in addition the PAL compensation circuitry which uses an addressable RAM for signal storage during one line sweep. The delayed and the non-delayed signal are added in known fashion to compensate for phase dependent errors. In NTSC operation, the PAL compensation circuit works as a comb filter, improving cross luminance suppression.

The digital colour difference signals R-Y and B-Y are now routed to the *colour saturation multiplier* which, thanks to the multiplexer operation, needs only one multiplier. This is possible because of the smaller bandwidth of 1 MHz, of the chroma signals for which a high clock frequency is available (14 MHz for PAL and 17 MHz for NTSC). Hue correction is also done in the colour multiplier by rotation of the R-Y and B-Y axes of the chroma signal, done by the CCU in conjunction with the setting by the operator. As the contrast setting affects colour saturation, the multiplication factor intended for the colour saturation multiplier is also multiplied by the chosen contrast in the CCU, so that it harmonizes with the chosen contrast.

The *phase comparator* for synchronising the colour demodulator-decoder is also contained in the chroma channel. The reference colour burst is derived from the colour difference signal after the decoder, already averaged over two lines. The phase of the digital signal processor clock is locked to the phase of the colour signal by PLL, to ensure correct decoding. This is done by comparing the phase of B-Y signal with R-Y of the reference burst. The video processor also monitors the beam currents by continuous sampling and adjusts them automatically to preset levels with the help of the CCU, to compensate for aging of the colour picture tube.

25.12   DEFLECTION PROCESSOR UNIT

The deflection processors DPU 2553/54/55 sense the standard TV signals and synchronizes the vertical and horizontal sweeps generators by counting the colour subcarrier and locking to it, without any effect of noise interference on synchronisation. For non-standard signals received the deflection processor checks the horizontal sweep phase with the sync signal and averages out the phase difference. The programmable VLSI chips in N-channel MOS technology contain the following circuit functions:

— video clamping
— horizontal and vertical sync separation
— generation and synchronisation of the horizontal and vertical deflection frequencies
— horizontal output pulse of adjustable duration and phase

— east-west correction, for flat screen picture tubes
— vertical sawtooth generation including S-correction
— text display mode with increased deflection frequencies (18.7 kHz horizontal and 60 Hz vertical)
— D2-MAC operation mode

DPU 2554 incorporates in addition:

— double-scan horizontal deflection
— normal and double-scan vertical deflection

The DPU communicates via the bidirectional serial IM bus with the CCU which supplies the picture-correction alignment information stored in the EEPROM during set production, when the set is turned on. The DPU is normally clocked with a sinusoidal 17.734 MHz (PAL), 14.3 (NTSC) or 20.25 MHz (D2-MAC) clock signals from the clock generator IC. The functional diagram of the DPU 2553–5 is shown in Fig. 25.9.

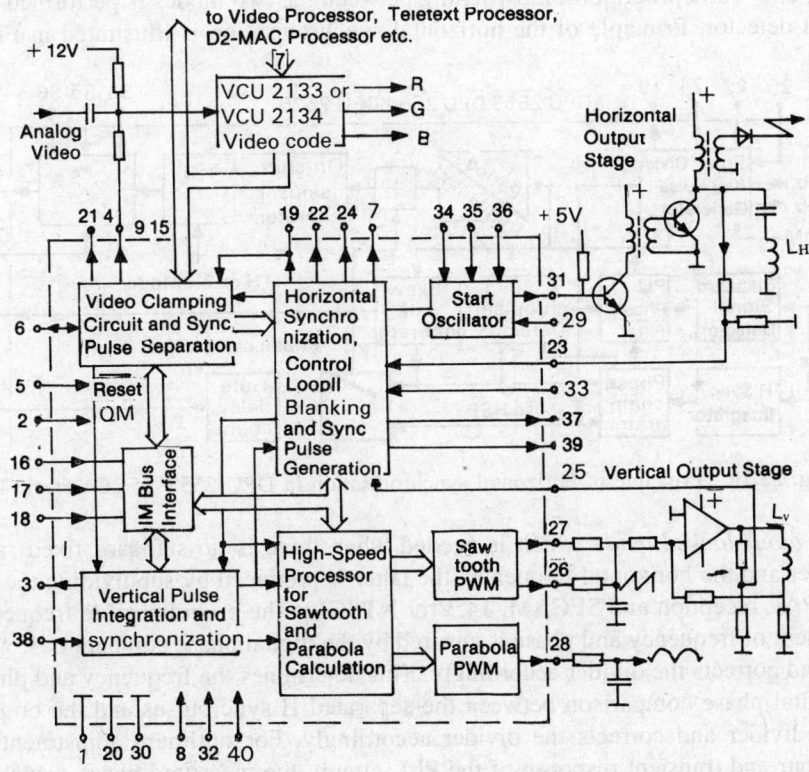

Fig. 25.9 Block diagram of the Deflection Processor DPU 2553–55 (*Courtesy:* ITT)

Video Clamping and Sync Separation

The digitized composite video signal delivered as a 7-bit parallel signal by the video codec is first noise filtered by a 1 MHz digital filter, and in order to improve the noise immunity of the clamping circuit, is additionally filtered by 0.2 MHz low pass filter before being routed to the minimum and back porch level detectors. The DPU has two different clamping outputs one of which is supplied to the codec VCU.

After the TV receiver is switched on, the video clamping circuit first of all ensures, by means of horizontal frequency current pulses from the clamping output of the DPU to the coupling capacitor of the analog video, that the video signal is optimally biased for the operating range of the A/D converter (5 to 7 V). For this the sync top level is digitally measured and set to a constant level (5.125 V) by these current pulses. The *H* and *V* pulses are now separated by a fixed separation level (5.250 V), so that the horizontal synchronisation can lock to the correct phase.

When horizontal synchronisation is achieved, the slice level for the sync pulse is set to 50% of the sync pulse amplitude by averaging the sync top and the black level. This ensures optimum pulse separation, even with small sync pulse amplitudes.

Horizontal Synchronisation

Two operating modes are provided in horizontal synchronisation depending on whether or not, the TV signal is standard PAL or NTSC, in which there is fixed ratio between colour subcarrier and horizontal frequency. In this case the operation is 'colour locked'; otherwise the operation is 'non-colour locked' e.g. with black-and-white programmes. Switching between the two modes is performed automatically by standard signal detector. Principle of the horizontal synchronisation is illustrated in Fig. 25.10.

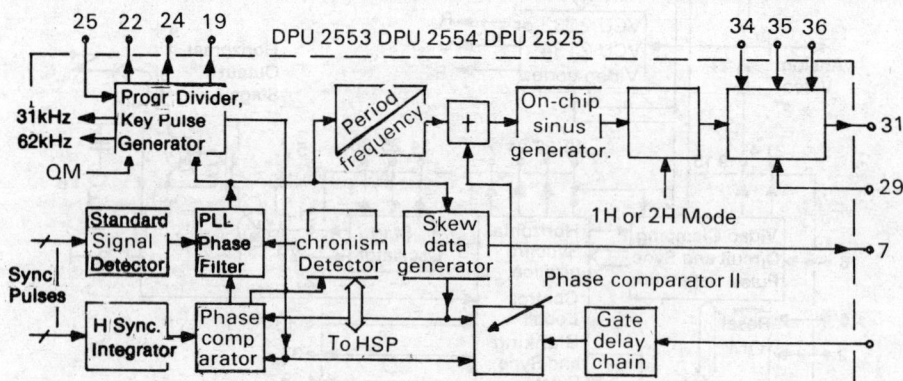

Fig. 25.10 Principle of horizontal synchronisation in DPU 2553–55 (*Courtesy:* ITT)

In the *non-colour-locked mode* which is needed when there is no standard fixed ratio between the colour subcarrier and the horizontal frequency, the latter is produced by subdividing the clock frequency (17.7 MHz in PAL reception and SECAM, 14.3 for NTSC) in the programmable frequency divider. The correct adjustment of frequency and phase is ensured by the digital phase comparator I, which determines the deviation and corrects the divider accordingly. This determines the frequency and phase deviation by means of a digital phase comparison between the separated H sync pulses and the output signal of the programmable divider and corrects the divider accordingly. For optimum adjustment of phase jitter, capture behaviour and transient response of the PLL circuit, the measured phase deviation is filtered in a digital low pass filter (PLL phase filter).

In the case of non-synchronised horizontal PLL this filter is set to wide band PLL response with pull-in range of +/– 800 Hz. If the PLL circuit is locked the PLL filter is automatically switched to narrow band response by an internal synchronism detector in order to limit the phase jitter to minimum, even in the case of weak and noisy signals.

For the purpose of equalising phase changes in the horizontal output stage due to switching response tolerances or video influence, a second phase control loop II is used in which the phase difference (deviation) between the output stage of the programmable divider and the leading edge of the horizontal flyback pulse at pin 23 is measured by means of a balanced gate delay line. This deviation from the desired phase difference and the information on the horizontal frequency derived from the phase comparator I are added together to control an on-chip digital sine wave generator (about 1 MHz), which acts a phase shifter.

By means of the phase control loop II, the horizontal output pulse at pin 31 is shifted such that the horizontal flyback pulse at pin 23 attains the desired phase position with respect to the output signal of the programmable divider, which in turn, due to phase comparator I, retains a fixed phase position with respect to the video signal.

When in the *colour locked operating mode*, after the phase has been set in the non-colour locked mode, the programmable divider is set to the standard division ratio (1135 for PAL, 910 for NTSC) and phase comparator is disconnected so that the interfering pulses and noise cannot influence the horizontal deflection. Because the phase comparator II is still connected, the phase errors of the horizontal output stage are also corrected in the colour locked operating mode.

To protect the horizontal output stage of the TV receiver, during changing the standard and for using the DPU as a low power start oscillator, an additional oscillator is provided on chip with the output connected to pin 31. It is controlled by a 4 MHz signal independent from the main clock and is powered separately.

The DPU supplies a combination of colour key pulse and undelayed horizontal blanking pulse in the form of a three level sand castle pulse. The processor also provides for switching the horizontal power stage from 15.6 kHz (PAL/NTSC) to 18.7 kHz (Text display) for improved text mode picture quality. In the D2-MAC operation mode, the programmable divider is set to divide by 1296 to generate a horizontal frequency of 15.625 kHz with the clock rate of 20.25 MHz used in the D2-MAC standard. The DPU is synchronised to this sync signal from D2-MAC decoder DMA 2270.

Vertical Synchronisation

The vertical sync pulse is derived by means of digital integration from the separated composite sync signal. The trigger pulse from the high speed processor and the vertical deflection is produced in the working counter which divides the double horizontal frequency (31 kHz) by (625 +/– 64) for PAL and SECAM, and (525 +/– 64) for NTSC. The working counter can be set to three different operating modes:

1. *Non-locked* operation using a *wide trigger window* for the divider which can be reset at counter position of 561 to 689 for PAL/SECAM or 461 to 589 for NTSC, when it can be synchronised by vertical frequencies in the range from 45 to 55 for PAL/SECAM or 54 to 66 for NTSC.

2. *Non-locked* operation using *narrow window*, in which the working counter reset range is restricted to 618 to 632 for PAL/SECAM or 518 to 532 for NTSC to improve noise immunity with non-standard signals or in VCR operation.

3. *Locked operation* in which the standard division ratio is fixed, viz. 625 : 1 for PAL/SECAM or 525 : 1 for NTSC, ensuring optimum noise immunity for all standard signals.

Audio A/D Converter

IC ADC 2300 U for multichannel sound standard in USA proposed by Zenith, or ADC 2300 J FM-FM multiplex system for Japan, or ADC 2300 or 2310 E for European system, digitizes the analog sound signals in the TV receiver with stereo sound channel.

The IC ADC 2300/10 E works with the audio processor like APU 2400 T or 2470 being controlled by the CCU. As shown in the block diagram of Fig. 25.11(a), the chip contains the following functions:

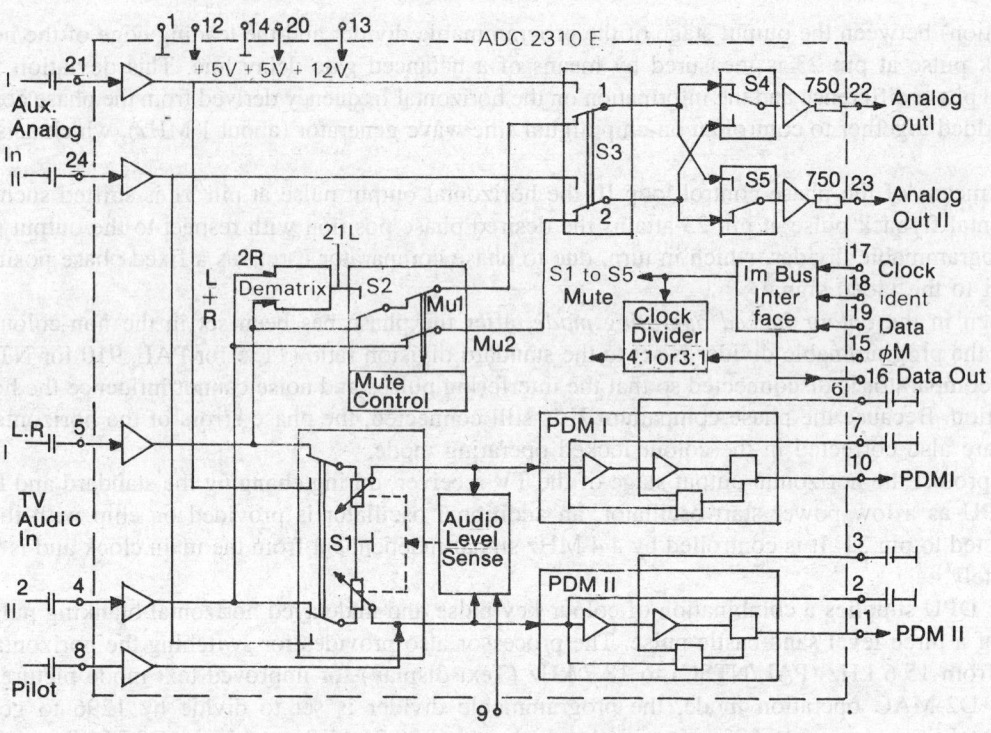

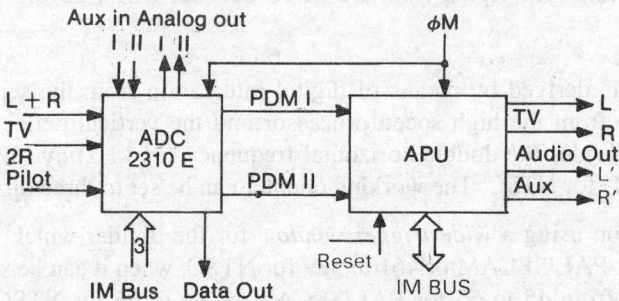

Fig. 25.11 (a) Block diagram of the A/D converter ADC 2310 E, (b) Stereo processing system comprising ADC 2300 and APU 2400 T (*Courtesy:* ITT)

— several analog input and output amplifiers, (in ADC 2310 E)
— five analog switches for selecting different signal sources,
— an analog stereo dematrix circuit,
— a level control facility,
— two pulse density modulators (PDM I and PDM II),
— IM bus interface, and muting circuit.

The two pulse density modulators PDM I and PDM II are sigma-delta modulators equipped with two feedback loops. At the outputs they supply pulse trains whose pulse density is proportional to the amplitude of the input signal. The maximum sampling and hence the maximum pulse rate is 4.7 MHz. This data is then transferred to the audio processor APU where the digital decimation filters are the input, performing the second step of the conversion process. Due to the very high sampling rate of the pulse-density modulators, no steep anti-aliasing filters are needed at the input. The digital output data of the whole converting system has a SNR comparable to that of a conventional 13 bit A/D converter.

The TV Audio inputs get their analog signals (L + R and 2R) from the stereo decoder of the TV set, whereas the Aux Analog inputs are intended to receive an audio signal from a VCR or another external source. The Analog Out I and II supply the selected signal to VCR or other equipment via the SCART connector.

25.13   AUDIO PROCESSOR

Programmable digital real-time processing of the audio signals is carried out by means of the ADC 2310 E and the audio processor APU 2400 T. Together they comprise a complete stereo sound processing system as shown in Fig. 25.11(b). The input signal is sampled and converted into 16 bit digital stream in the *Audio ADC* which employs a pulse density modulator and digital converter. A digital identification filter identifies the type of audio broadcast signal—mono, stereo or bilingual. A parallel to serial converter multiplexes the output to the audio processor to reduce pin count. The *audio processor* is designed to receive stereo and bilingual broadcasts in addition to mono. To cater to stereo and high fidelity broadcasts, the processor has two parallel channels with ability to respond to high fidelity and stereo adjustment commands from the user. The processor, controlled by software carries out the following functions for two different audio signals (stereo/bilingual).

— converting PDM signals into parallel data at audio sample rate
— DC offset suppression
— pre-emphasis
— dematrixing TV signals
— dematrixing auxiliary channels
— bass and treble controls
— stereo basewidth enlargement
— APD effect (pseudo stereo)
— balance
— volume control
— conversion of the processed digital audio signal into PWM
— automatic mode detection (mono, stereo or bilingual)

The basic functions are mask programmed and can be modified by mask programming. Software blocks and hardware interfaces of the APU 2400 T are given in Fig. 25.12.

Digital Filters
The main task of digital signal processing carried out in the APU 2400 T, is filtering. A digital filter, in a sense is essentially a *circuit* for converting a series of samples of an incoming digital signal $X(nT)$, into another series of signal samples $Y(nT)$, as outgoing digital signal, that has some more desirable properties. Since the signal samples can be in the form of binary numbers, a digital filter can be considered as a technique for converting a given input sequence of binary numbers into another desired output sequence by *numerical procedure*. Digital filtering can therefore be performed entirely by means of *software* in the form of computer program based on appropriate *algorithm*. The flexibility, stability and

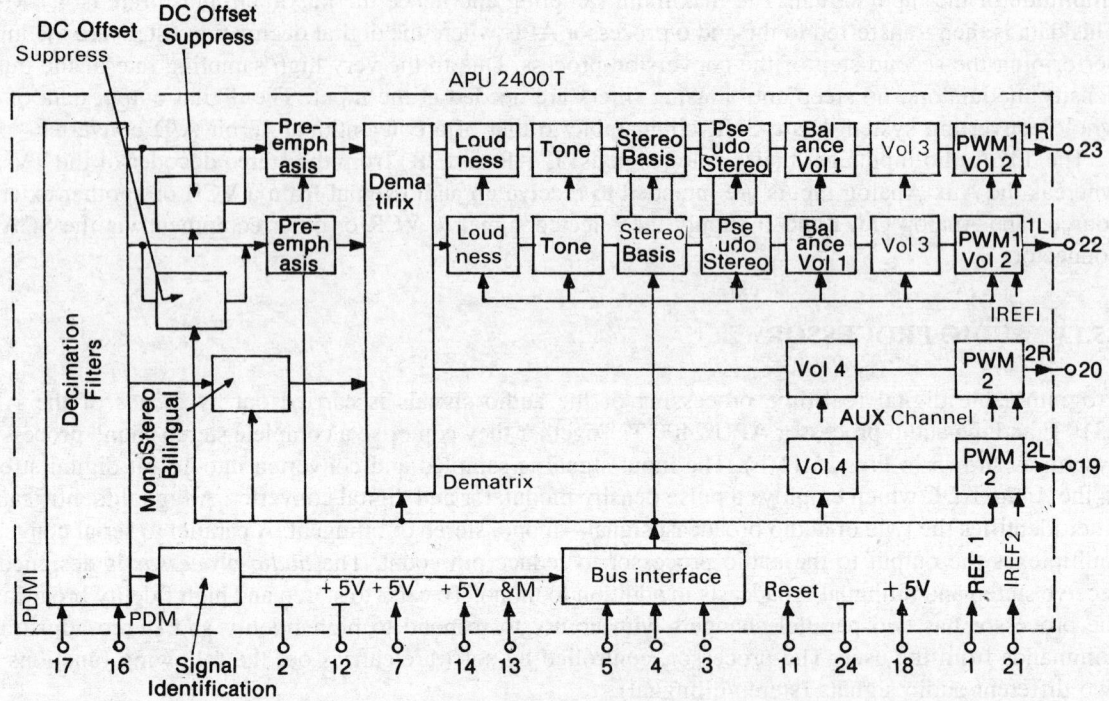

Fig. 25.12 Software blocks and hardware interface of APU 2400 T (*Courtesy:* ITT)

compatibility to digital systems, are great advantages of digital filters. However, the quantisation process, depending on the number of bits, determines the accuracy and quality of filtering. *Errors and noise* due to quantisation, and the 'rounding off' errors in the multipliers used, limit the quality of filtering.

Large scale computers or microcomputers have been used for achieving the required signal processing. With the VLSI technology, digital filters can be implemented using single chip *hardware*, incorporating three simple elements: *address, multipliers and delays (shift registers).* Hence the cost and complexity of digital filter implementation has considerably decreased. A digital filter design involves selecting and interconnecting a number of these elements and determining the multiplier coefficient values. Delays can be positive, conventionally indicated by z^{-1}. Negative delays or *advance* indicated by z, are used to look ahead to the next value in the sequence. This is possible in applications where the entire data sequence is available at the start of the processing, e.g. in image processing. The principle of digital filtering is explained by means of a simple first-order filter shown in Fig. 25.13(a).

Here the block z^{-1} represents delay by one sample clock cycle (state variable memory). The output $Y(nT)$ is calculated by:

$$Y(nT) = a \cdot X(nT) + b \cdot (nT) + c \cdot Y(nT - T) \tag{25.1}$$

The z-transformation yields

$$H(z) = \frac{a + b \cdot z^{-1}}{1 - c \cdot z^{-1}}$$

Derived from this is the amplitude characteristic

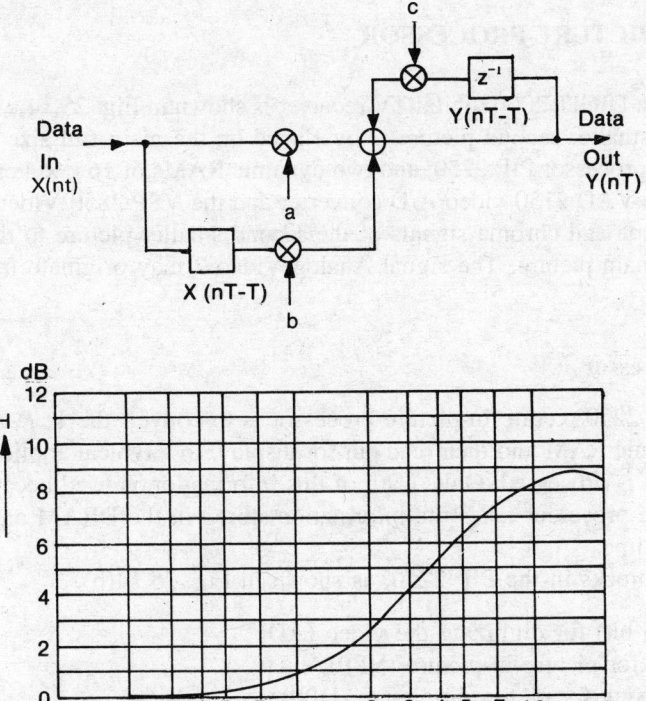

Fig. 25.13 (a) Structure of digital filter, (b) Frequency response of the calculated digital filter (*Courtesy:* ITT)

$$[H(f)] = \frac{a^2 + b^2 + 2ab \cdot \cos(2pi \cdot f/f_s)}{1 + c^2 - 2c \cdot \cos(2pi \cdot f/f_s)}$$

Assuming that the sampling rate f_s = 31.468 kHz and sampling period T_s = 31.8 μs and the filter coefficients are:

a = 2.189 147, b = − 1.510 422 and c = 0.321 274 8

Then the frequency response shown in Fig. 25.13(b) is obtained. As can be seen from equation (1), even with this simple filter three multiplications and one addition of three addends are necessary to calculate the sequence of outputs. The basic operation of such systems is *multiply and add/subtract.* For the given example three such operations and the associated data transfer must be carried out within a 31.8 μs sampling period. For the entire sound processing system about 100 such operations are required. This means that one basic operation must be executed in less than 318 ns. Special hardware is required for such high processing speeds.

An arithmetic logic unit (ALU) is used to carry out filtering. A fast multiplier is used by all the filters. The basic filter coefficients are stored in ROM, while the coefficients of the filters, controlled by the user commands from keyboard input, are stored in a RAM in the CCU. As the processor is based on sequencer ROM which controls the processing algorithm, the processor functions can be redesigned by change of the ROM providing much flexibility.

25.14 PICTURE-IN-PICTURE PROCESSOR

An advanced version of a DIGIT 2000 digital TV receiver is shown in Fig. 25.14, which features picture-in-picture. With this, a smaller second picture is overlaid on the main full size TV picture by means of the picture-in-picture processor PIP 2250, and two dynamic RAMs of 16 k × 4 each. The second video source is made up of the VAD 2150 video A/D converter and the VSP 2850 Video/Sync Processor. The latter supplies digital luma and chroma signals of the second smaller picture to the PIP, as VPU video processor does for the main picture. The signal Analog Video 2 may originate from a second tuner or from a VCR.

Picture-in-Picture Processor

The function of the PIP 2250 picture-in-picture processor is to convert the *Y*, *R-Y* and *B-Y* into a form to be stored in the dynamic RAM and then read out for display. In a typical application only 80% of the active picture dimension is processed. Only 1/9th of this information is used because 1/3 by 1/3 size of picture is generated. The processor takes the information stored in the DRAM and presents the data to the VCU at the proper time.

There are six major blocks in the PIP 2250, as shown in Fig. 25.14(b).

— A/D converter (6 bit) for digitizing the video (AD)
— video processing for picture-in-picture (VPP)
— deflection processing for picture-in-picture (DPP)
— picture input processor (PIP)
— picture output processor (POP)
— DRAM interface (DI)

The AD, VPP, DPP, and PIP comprise the input section. The inputs fed are the two composite video signals, main chassis horizontal and vertical signals, serial control bus (IM bus), external analog RGB and fast blanking, main chassis clock (14.3 or 17.7 MHz). The sampling rate for the small picture is every third pixel and every third line. Each pixel is stored as 5 bit luminance and 3 bit chrominance in the following way: 4 × 5 bit luminance and 4 × 3 bit chrominance. During four luminance pixels, one complete chrominance pixel with 6 bits for R-Y and 6 bits for B-Y is stored.

The output section is made up of POP providing analog RGB outputs for border of the small picture, PPP fast blanking, digital *Y*, R-Y and B-Y, and tristate outputs to VPU or CVPU. The DI interfaces with both these section and the DRAM.

The VPP decodes the composite video signal and generates Y, R-Y and B-Y. The DPP generates the horizontal and vertical timing signals necessary for the VPP and the PIP. The latter converts the composite video signals into a form suitable for the DRAM. The POP takes the DRAM data and displays it.

The filtering in the VPP and PIP contains a horizontal *alias* filter with peaking and a vertical *alias* filter to overcome the frequency folding caused by the sampling rate of 1 : 3 for the stored picture. Also included are interpolation filters, which convert the skew data from the DPU 2532 into time shift for data. This is necessary because the PIP 2250 uses the main chassis clock which is not line-locked for the stored picture.

To create the border around the small picture, the additional RGB inputs from VCU are used. To maintain the possibility of teletext of external RGB to be used as an input, the PIP 2250 has an internal RGB switch in the POP part as well as additional RGB inputs for external RGB signals.

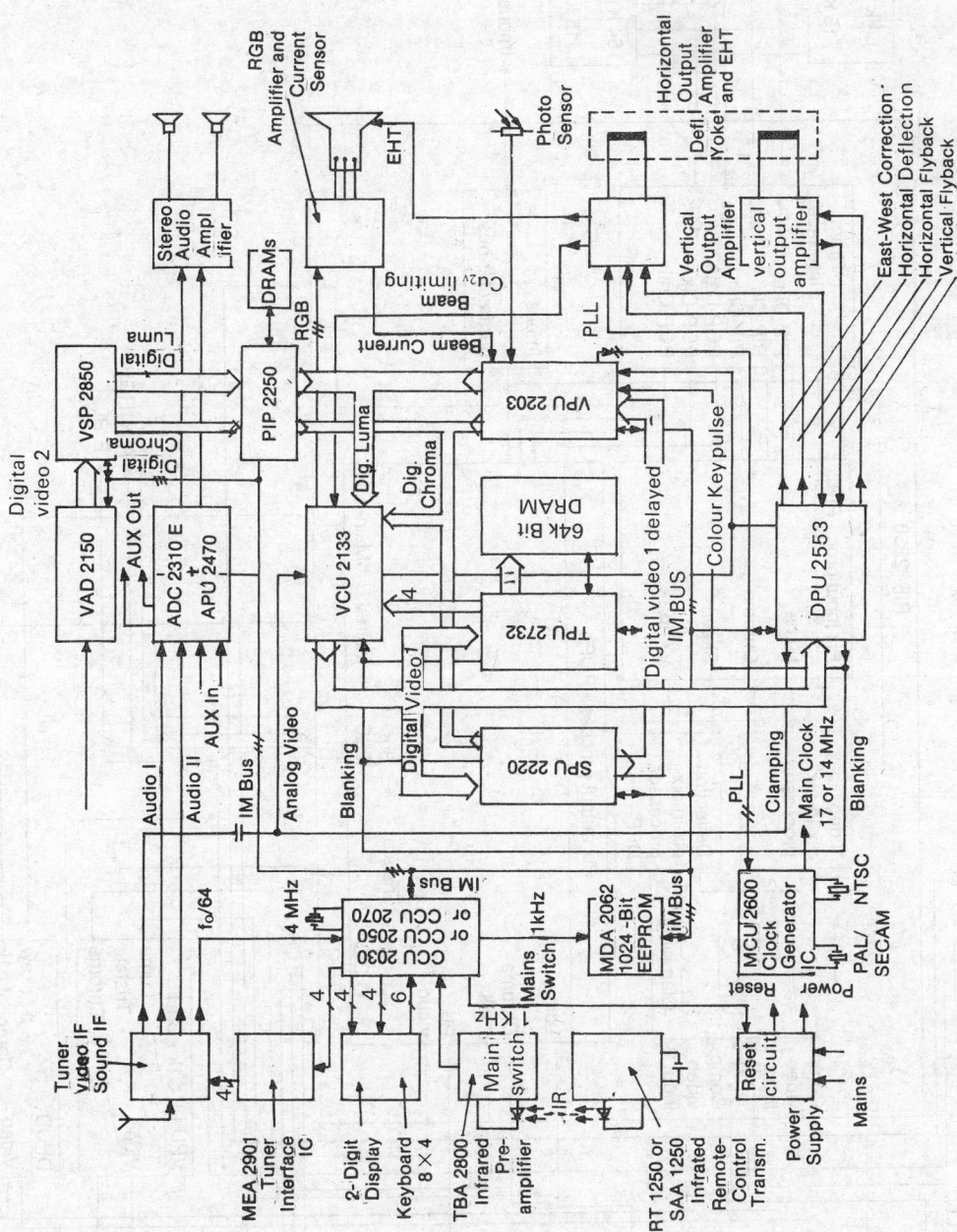

Fig. 25.14 (a) Advanced version of Digit 2000 TV receiver featuring PIP (*Courtesy*: ITT)

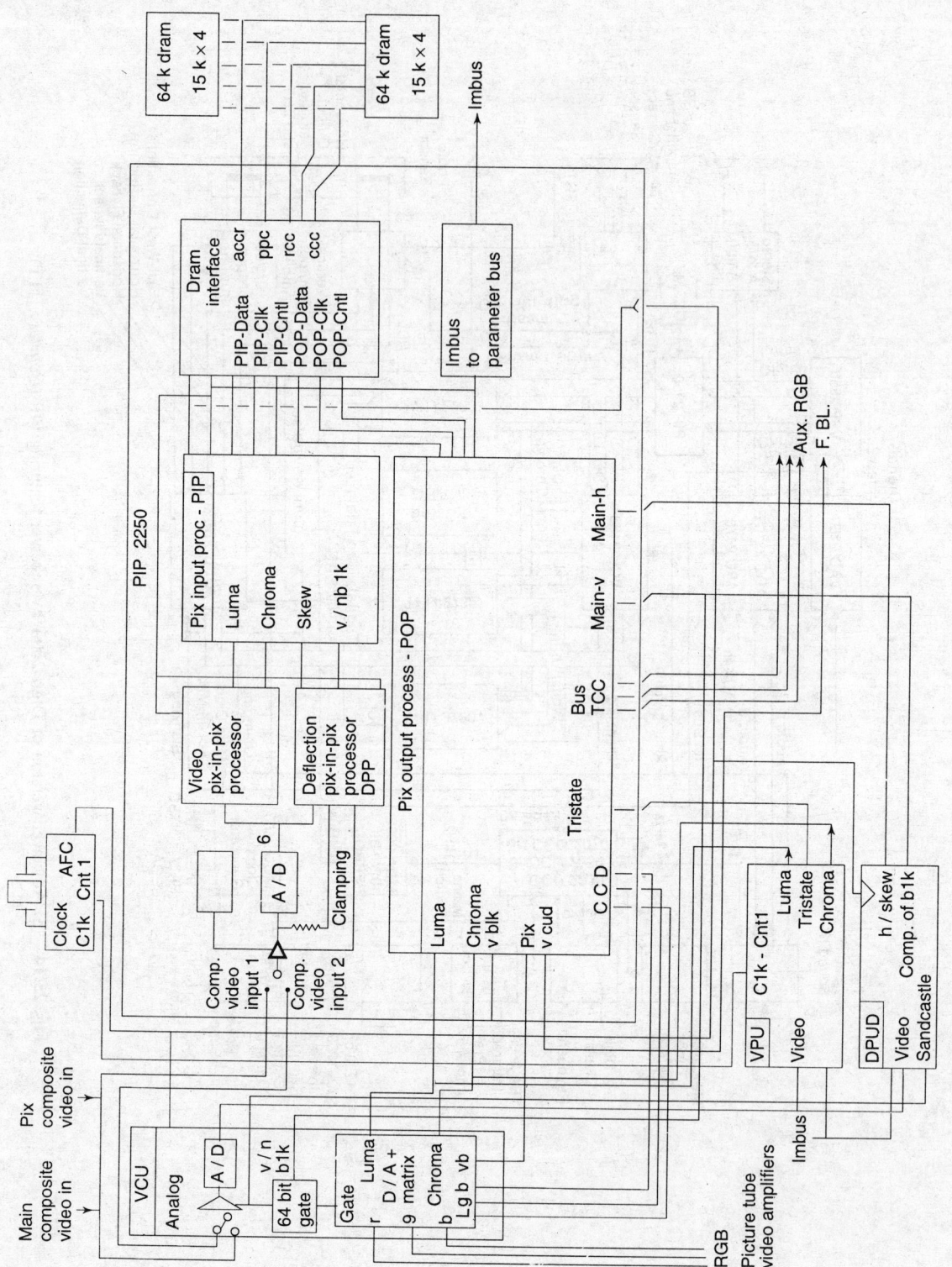

Fig. 25.14 (b) Block diagram of PIP processor IC PIP 2250 (*Courtesy: ITT*)

25.15 OTHER DIGITAL RECEIVER DESIGNS

Euro-chassis ICC 5 developed by Thompson-Brandt forms the basis of some manufacturers including Saba, Nordmende, Telefunken. This design emphasizes flexibility to suit the manufacturers. It incorporates the SCART interface which would probably be obligatory in all the latest TV sets. The chassis includes following ICs designed with the aid of CAD system:

— a power supply processor with switched end stages for horizontal and vertical deflection (by Thompson-CSF)
— a quasi-digital video processor (by Hitachi and Telefunken)
— a multi-standard colour decoder which operates alignment-free, requiring no quartz crystal and synchronised to any subcarrier (Toshiba)
— an audio processor modified version of APU 2400 (ITT)
— a C-MOS microprocessor as a CPU with an 8-bit memory IC which among other things contains the control for teletext decoder
— a two-chip teletext decoder and EUROM (Valvo) offering an option of both videotext and teletext, so that the two features can be easily retrofitted as a plug-in decoder.

Manufacturers like Grunding have digitized only the following functions: PLL tuner, remote control, stereo decoder, deflection control. Features like microprocessor-controlled tuning of all functions by means of standard PC-bus (Valvo) for all sets, and teletext decoder with auxiliary videotex decoder for which there are also advanced retrofit kits, have been very commonly incorporated. The audio and video functions have not been digitized by many, as the improvements and benefits are considered by them as few, or marginal.

This types of 'open' concepts assume that a single basic chassis is sufficient for a large number of TV models and can be adapted to meet any desired operational, reception and equipment requirements (including D2-MAC), retrofitting modules and assembly boards using special sockets.

Philips Computer-controlled TV

The present market is thus demanding an increasing diversity of colour TV receivers, ranging from simple economy sets to high performance models with *additional features and enhanced quality* reproduction of picture and sound. To meet these demands M/s Philips, Holland have also developed a new generation of TV ICs that can serve as building blocks for TV sets, ranging from economy sets to high performance computer-controlled colour TV receivers. In the computer-controlled sets, additional functions can be added and several modern techniques used to improve the picture and sound quality, within limits of today's television broadcast systems, by more fully exploiting the contents of the transmitted signal, and by special digital processing of the analog vision and sound signals.

It is possible to have extreme *functional flexibility* is possible in the architecture of the TV set, by employing integrated analog processing with digital building blocks. Functional partitioning of the ICs and the use of simple 2-wire bus to control them simplifies the board layout and wiring. Drastic reduction in the number of peripheral components is achieved by using advanced technologies to incorporate *on-chip filters and delay lines* which do not require alignment (see section 20.7). These include for example,

— analog time discrete p^2CCD filters and delay lines for colour processing
— analog gyrators for functions such as colour transient improvement

Automatic Alignment

By incorporating digital control via the Inter-IC (I^2C) bus, many other manual adjustments required due to components which are subject to aging, e.g. coils, trimpots etc., are eliminated by *automatic adjustments* (electronic screw driver technique). With computer-controlled TV, it is also relatively simple to *monitor quality* via the I2C bus during set production. Besides such normal functions as tuning, program/channel display, remote control, analog picture and sound controls, the microcontroller also controls the various alignment functions. Provided the signal processing ICs have the necessary facilities, automatic alignment with production-line computer is possible. When the alignment is complete, the production-line computer stores the alignment data in the non-volatile memory. After the power-on reset, the internal system microcontroller fetches this data and transfers it to the signal processing ICs. The set incorporates a *diagnostic plug* to which the internal bus is connected. Automatic alignment done in this way, minimizes human errors and avoids shortcomings of semi-automatic electromechanical alignment systems. The *benefits of computer control* are:

— elimination of many expensive components like potentiometers, coils and glass delay lines which are integrated on chips.
— faster and flexible system design due to microcontroller bus, new features like CCD memory applications, peri-TV connectors can be readily added.
— simpler wiring due to better partitioning by means of control bus.
— simpler PCB layout due to absence of manually adjusted components.
— faster production, testing and alignment through use of production-line computer.
— better quality monitoring.

Servicing Techniques

The flexibility of computer-controlled LSI based TV concepts allows the manufacturers to devote a larger part of its system design capability to the development of software which will become increasingly important. The software should use optimal use of the microcontroller during manufacture, normal operation and servicing of the TV set. Servicing a computer-controlled TV set is different from a conventional, as there are no trimpots to adjust. There are *four ways* to approach the servicing requirement:

— to use a dedicated service computer with minimum microcontroller hardware and alignment software.
— to use a standard PC added with the TV microcontroller bus.
— to incorporate in the set a service switch that runs a service program and stores realignment values.
— to change remote control address to a service address so that all commands with this address become service commands.

Once the digital tuning and remote control techniques are added to the basic set, a microcontroller and the I^2C-bus become available to the designer. These can then also be used for user controls such as volume and brightness, and for automatic alignment. Colour decoder IC TDA 8442 has four DACs especially for colour processing user controls, three output ports for switching, for example: internal/external RGB and PAL/NTSC.

By having broadcast *standard-independent interfaces* (YUV and baseband audio), optimum use can be mode of peri-TV connectors and additional functions; and performance improvements like colour transient improvement, memory-based functions and picture enhancements, can be easily be added. The IC TDA 4565 provides colour-transient improvement by sharpening the transients in the colour difference signals.

Synchronisation and Deflection

The deflection processor TDA 8432 in combination with the widely used count-down type TDA 2579 sync processor, reduces the number of peripheral components to simplify and automate the adjustment procedure by computer control. The processor performs the following functions:

— alignments of horizontal oscillator
— setting the correct picture position and size
— generation of the waveforms required by vertical deflection and East-West modulation of the horizontal deflection current to obtain the correct picture geometry.
— static and dynamic anti-breathing correction in the vertical and East-West modulation waveforms.

For this purpose, the sync processor IC requires following *inputs:* information about horizontal frequency, a trigger pulse for vertical scanning from the sync processor, the vertical feedback and EHT variation. The *outputs* are the control voltages for the horizontal VCO and the second horizontal PLL (phase shift), the vertical drive and the East-West modulation drive.

The *horizontal oscillator alignment* is automatic, as a DAC converts the commands from a production line computer into control voltage for the VCO. The production-line computer increments the DAC until the VCO frequency equals the synchronisation frequency. At this instant, a 'centre' bit is generated, the production-line computer stops the DAC and the DAC digital value is stored in the receiver's non-volatile memory.

The *vertical drive and raster correction* waveforms are also generated through the accurately aligned horizontal oscillator, via the *countdown circuit* in the sync processor. The deflection processor incorporates a vertical sawtooth generator which produces a linear vertical sawtooth of constant stabilized amplitude. To effect corrections in the horizontal deflection via East-West modulator, a parabola is required. This is generated by squaring the vertical sawtooth $X(t)$, to apply basic parabola correction $X^2(t)$; other corrections can be made using $X^3(t)$ and $X^4(t)$, and shifts and multiplication with DC voltages.

The multiplication factors used in the processor are made frequency-independent and bounce-free as no RC components are used as in a Miller integrater. The parabolic waveforms are produced by means of a network of *Gilbert-cell multipliers* (shown in Fig. 25.15) and adder stages, that provide the desired power series of $X(t)$. The various coefficients are controlled by the DACs; however one DAC can control more than one coefficient. This makes the height, width, parabola and corner corrections independent of the screen and simplifies the software for alignment.

High Performance Computer Controlled TV

A simplified block diagram of a high performance computer-controlled TV with PIP employing Philips ICs is shown in Fig. 25.16.

Front-end

TV signals can be selected by the synthesizer tuning system, bus controlled by CITAC with the SAB 3036, a PLL system with the TSA 5510 or a VST system with the PCFC84XXX series. The vision IF signals are amplified and demodulated in the vision IF circuit the TDA 8341 which offers increased video bandwidth.

Video Processing

Internal and external signals are selected and filtered in the TDA 8452. Its output delivers a chroma signal to the colour decoder TDA 8466, which in turn delivers colour difference signals to the integrated

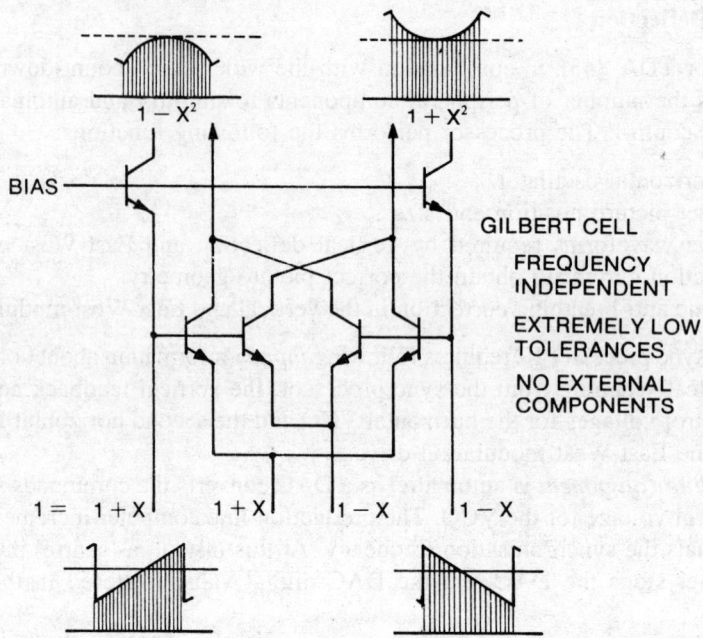

Fig. 25.15 Gilbert cell—example of squaring state (*Courtesy:* PHILIPS)

delay lines TDA 8451. The delayed colour difference signals are matrixed with the luminance signal in the standard-independent part of TDA 8466. The bus controlled gain settings in the TDA 8466 allow automatic alignment of the white point. The YUV inputs and outputs in the colour decoder TDA 8466 facilitate very simple interfacing with a feature box or a picture-in-picture module. The two sets of RGB inputs on the TDA 8466 allow simple interfacing to teletext and external RGB or On-Screen-Display signals. The RGB outputs are further processed in the integrated video output stages TDA 6100, which requires very few peripheral components. Circuitry for black current measurement has also been integrated. RGB amplifiers TDA 6100 are each in small SIL-9 package which needs no further heat sinking.

Synchronisation and Deflection
The TDA sync processor can be extended with the geometry processor TDA 8433 to permit automatic alignment of picture geometry.

Control Part
The various functional blocks in the receiver are controlled via the I^{2}C-bus by the microcontroller MAB 8461. For Teletext, a separate microcontroller can be added. The non-volatile memory PCF 8582 stores data such as favourite channels and preferred analog settings. TDA 3047 amplifies the infrared data signals received from the remote control transmitter. Options of peri-TV connectors can be provided by CVBS and sound switch, which can select between the two connectors.

Picture-In-Picture System
The block diagram of the PIP system is shown in Fig. 25.17. It can display a small (3 × 3 times compressed) picture from a second video source in any of the four corners of the main picture. Eight selectable border

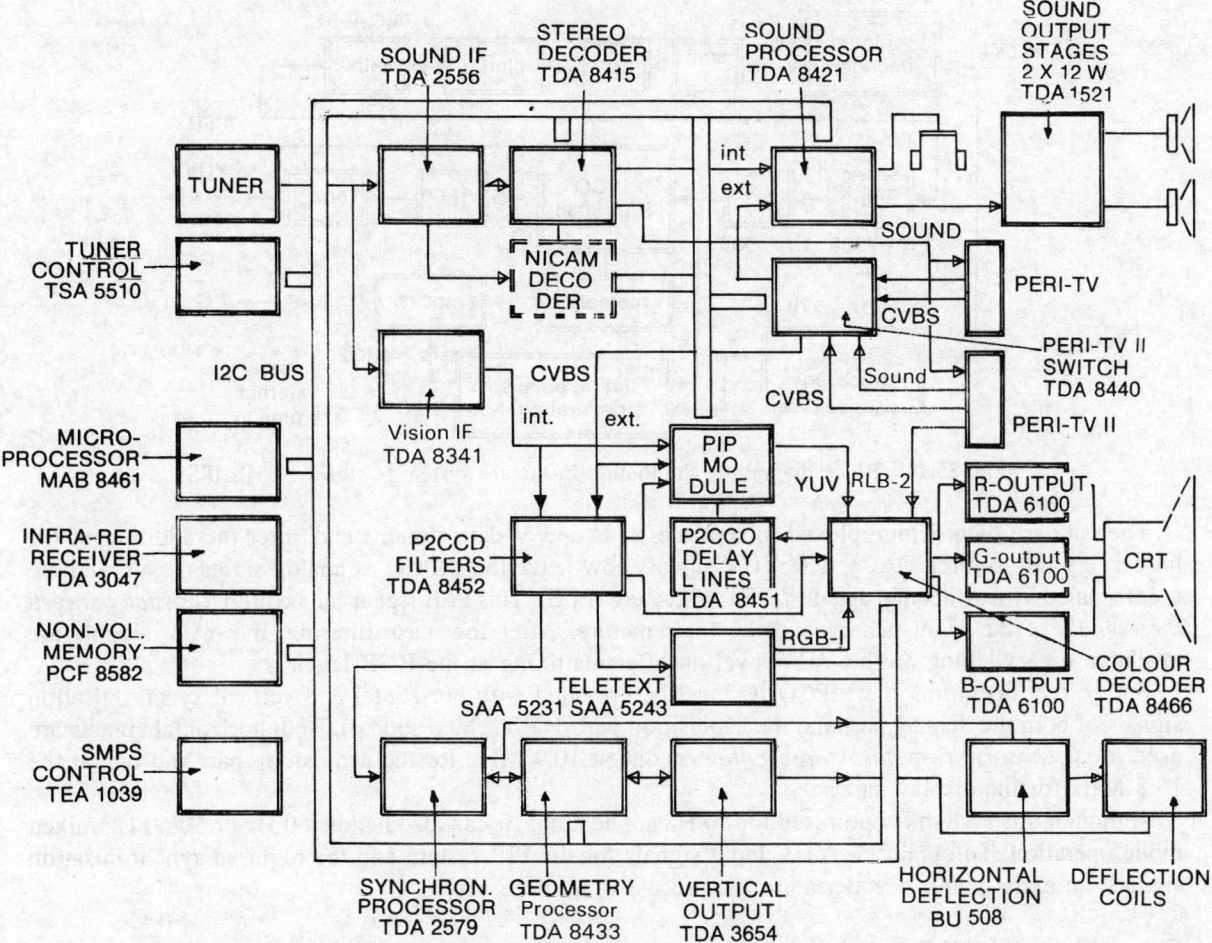

Fig. 25.16 Block diagram of a computer controlled TV with PIP (*Courtesy:* PHILIPS)

colours are provided for clearly separating the inset picture. The several features of this system, including freeze and blank picture, can be selected via the two wire I^2C bus.

Once the YUV signals are band limited, amplified and clamped to the correct levels, they are fed time-division multiplexed to the single ADC PNA 7509. Due to this time multiplexing only one ADC is needed. The 5 bit wide data stream from the ADC is then processed in a Digital Vertical Filter DVF SAA 9069, to prevent vertical aliasing components due to vertical subsampling. The output of this filter which consists of *one averaged line from every three input lines,* is then sent to the PIP controller, PIPCO SAA 9068, in a 4 : 1 : 1 format of 128Y + 32U + 32V active samples per line. Here the samples are delayed by about 560 ns, to compensate for the delay-difference in analog prefilters.

After temporal storage in the line memory (to synchronise read and write), the data are written into a 10k × 8 static RAM. At appropriate time, these data are read from the RAM and stored in a second line memory. This memory is *read at high speed* to obtain a line compression by a factor of 3. This also eases the speed requirements of the field memory.

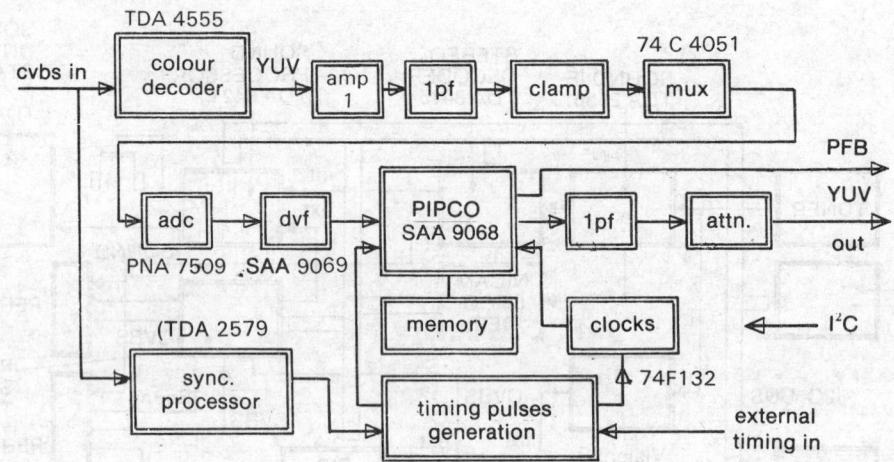

Fig. 25.17 Block diagram of the Picture-In-Picture system (*Courtesy:* PHILIPS)

The data are then demultiplexed to separate Y, U and V data streams and, after the addition of the border, are fed to three DACs. The YUV signals now leave the PIPCO as analog signals again and are accompanied by a switching signal *PIP Fast Blanking* (PFB). This PFB signal can be used to *switch between the signals* of the main picture and the inset picture. After low pass filtering, the YUV signals are available for switching on the YUV level or, after matrixing at the RGB level.

For correct operation, the PIPCO IC must be provided with horizontal and vertical synchronisation signals of both the display part and the acquisition part (= PIP video source). Both horizontal signals are used to *synchronise two start/stop oscillators,* one at 10.9 MHz for the acquisition part and one at the 15.8 MHz for the display part.

Although this system is optimised for 50 Hz applications, it can also handle 60 Hz or 50/60 Hz mixed mode operation. To obtain the YUV input signals for the PIP system and the required synchronisation signals, an extra colour decoder and sync processor are necessary.

Future Oriented TV Receiver

Today, many different types of television programs or video input options are available through terrestrial or satellite distribution. Several of the transmission standards have been in existence for some time whereas others are still being developed. There were no options for video display or sound reproduction in TV sets. Development of new transmission standards and introduction of advanced analog and digital signal processing now makes several *options* possible which can be incorporated in a computer controlled TV set for the future. The decoding options and display options are listed the Table 25.5. The block diagram of a multistandard receiver with B/G/L stereo, txt, 2 × PT-plug based on Philips ICs is shown in Fig. 25.18.

Architecture

Introduction of line and field memories will allow a diversity of options for video display features which result in line and field conversion to different scanning frequencies and have a strong impact on the television architecture. With the introduction of memories and peri-TV connector the TV receiver has to be rather designed so that it can be used as an advanced *monitor* for the display of uncoded baseband signals. This requires '*vertical*' *partitioning* between the tuning-decoding and the display section, rather

Table 25.5 TV options for decoding and display

Options for decoding		Options for display/reproduction
Video system:	PAL/B, G, H, I	
	SECAM/D, K, K1, L	Colour picture tube
	NTSC/M, N	
	Satellite MAC	LCD
Videotext:	WST	
	Antiope	Plasma displays
	Captain	
	Telidon	Digital memories (display map)
	NABTS	
Audio:	B/G stereo/dual	Stereo/multichannel
	Two carrier	
	Satellite digital	
	Hi-Fi, VCR, CD	
	US/Jap. MPX/SAP	
	French L-stereo	
Sync:	625/2 × 50	Standard 625/2 × 50, or 525/2 × 60
	525/2 × 60	Line doubling 635 × 50, or 525 × 60
		Field doubling 625/2 × 100, 525/2 × 120

than the 'horizontal' partitioning based on the nature of signals e.g. video, teletext, audio and sync sections.

The digital market has also been affected by the advent of HDTV. Digital TV could be easily converted to HDTV, by altering necessary software during manufacture. While NHK of Japan put forward a basically analog system for world standards on HDTV, a consortium of European companies Philips, Bosch, Thomson and Thorn/EMI has been working on high performance systems based on digital technology. Thus the digital features in TV are continuing to expand, and as the problems of world standards get settled, digital technology would assert in this hitherto analog field of consumer electronics in a big way. The future TV receiver would look more like a wide screen high resolution microcomputer serving as a *multifunction information terminal*, with a remote keyboard control facilities.

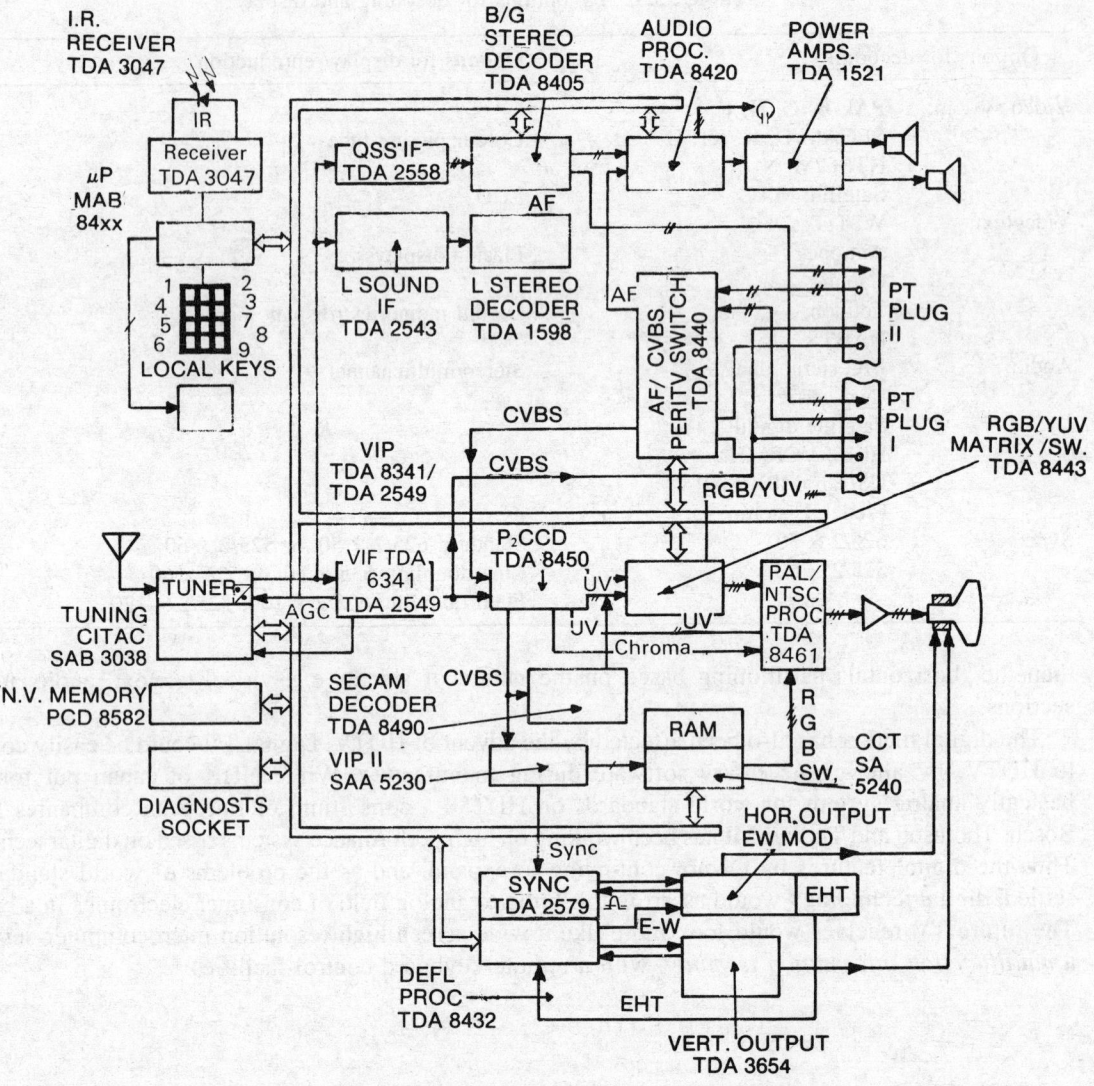

Fig. 25.18   Multistandard receiver with B/G/L stereo, txt, 2 × PT-plug (*Courtesy:* PHILIPS)

REVIEW QUESTIONS

25.1 What are the technical advantages of using digital technology in television systems?

25.2 What are the sampling requirements for conversion of the analog video signal? Discuss the sampling patterns that can be employed and their relative merits.

25.3 What are the requirements for digitization of video signal as regards the quantization, bit rate etc.? Compute the minimum bit rate for the 625/50 PAL system. Discuss the methods of bit rate reduction.

25.4 How are the digitized video signals structured for transmission in the digital services networks? Explain the different levels of digital transmission.

25.5 What is the difference between component coding and composite coding? Give the main features of CCIR Rec. 601 for digital video standards.

25.6 Explain the following terms: Aliasing effects, Sub-Nyquist sampling frequency, 4:2:2 and 4:4:4 encoding.

25.7 Discuss the suitability of bit-parallel and bit-parallel transmission. What are the ancillary signal requirements of bit serial transmission?

25.8 Draw the block diagram of a digital TV receiver giving the functions of each block.

25.9 Explain the digital signal processing carried out in *any one* of the ICs:

    (a) video processor   (b) audio processor

    (c) deflection processor   (d) PIP processor.

25.10 What are the studio equipments presently available that employing digital techniques? Explain the main features of digital VTR as per recommendations of SMPTE.

25.11 What are the benefits of computer controlled TV sets? How do the alignment and servicing techniques differ from the conventional sets?

25.12 What is line locked digital signal processing? Explain where this technique is advantageous.

25.13 What are the options available for a modern high performance TV set? How can these be kept optimally open by modifying the architecture of the receiver?

25.14 State the functions performed by deflection processor and explain how these are carried out in a digital receiver by synchronisation and deflection processors to facilitate automatic alignment?

26

Advanced Television Systems

Intensive research and development effort is presently going on towards improvements in the quality and service capabilities of television technology. Video programmes are today available from many different sources, some of which can provide far higher quality than the conventional broadcast television programs. Large screen displays are now widely used and show up the line structure is picture at closer distances. As colour receiver performance improves and large display units become common, the viewers would require higher quality signals. The recommendation for direct broadcasting in the 12 GHz band and in 22 GHz in the future, has provided the broadcasters a rare opportunity to reconsider the restructuring of the television signal for high quality satellite broadcasting. Innovations towards achieving this include the multiplexed analog components—**MAC encoding,** and **High Definition Television (HDTV)**. MAC encoding eliminates interference between chrominance and luminance components by time division multiplexing, and allows transmission of multichannel stereo sound or data using digital techniques. HDTV refers to television systems with approximately twice the horizontal and vertical resolution of the conventional system, wide aspect ratio.

NHK in Japan pioneered developments in HDTV, with 1125 lines/60 fields scanning standards by 1986, and demonstrated their HDTV MUSE transmissions during the 1988 Seoul Olympics. Regular experimental broadcasts have started in Japan in June 1989 using a single (27 MHz) channel of the Japanese direct broadcasting satellite BS-2b and NHK and MUSE-LSI based receiver manufacturers are planning broadcast service via the satellite BS-3, when it is launched in 1990. With the tremendous commercial potential, many other countries in the three corners of the globe, have plunged into develop the HDTV knowhow and a compatible system of comparable quality. With the Japanese lead of 10 years in the HDTV, it is a hard chase for Europe and the United States as there are futile attempts to evolve a worldwide standard.

EBU in Europe is making a concerted effort to increase the European level of know how in the field. Some 27 member countries of the European Broadcasting Union (EBU) have joined hands to develop the 1250 lines/50 fields HDTV system under the Project *Eureka EU95,* as one of the series of multinational research and development projects under the general title 'Eureka'. They have administered the first version of fully compatible HDTV chain, including programme material, in September 1988, as HD-MAC system. In the USA and Canada, the road to HDTV is seen by many as a *process of evolution* via Improved Definition TV (IDTV) and Extended Definition TV (EDTV). IDTV can include improvements in coding, filtering, ghost cancellation etc. in the standard NTSC 4:3 aspect ratio format. EDTV can include more improvements like increased aspect ratio and resolution that require modifications in NTSC emissions, but are NTSC receiver compatible. A number of such systems have been proposed.

The United States Advanced Television Systems Committee ATSC approved in January 1988, the *standard A/27*, Signal parameters of the 1125/60 HDTV production system, and recommend it to US state Department. The decision on this was withdrawn however, probably on the realisation that the European countries have not accepted the 60 Hz rate. When the USA willy-nilly supported Japanese system in 1985, it had an eye on studio standards and programming rather than manufacturing. European objection to acceptance of the NHK system led to crash program to develop their own compatible HDTV technology under the project Eureka 95. In May 1989, ITU-COM at Genua failed to reach a compromise on standards and as of now, it would probably take 1995 to decide on a common standard, if Europeans, Americans and the Japanese agree to share with the rest of the world the HDTV programs.

26.1   DIGITAL VIDEO AND AUDIO SIGNALS

As discussed earlier in chapter 27, the digitization of TV signals can take two forms. The composite NTSC, PAL or SECAM video signal (encoded with the colour subcarrier) can be quantized at high sampling rate to produce the *composite digital signal.* Alternatively, each of the luminance and chrominance signals *Y , U, V* or the Red, Green and Blue (R, G, B) from the camera, can be *quantized separately* to give so-called *component coding* in the form of three digital pulse streams which are combined for transmission. Because the video signals in television systems are highly structured, digitization errors can result in perceptible impairment of picture, like contouring. There exists also a large amount of redundancy in the structured nature of the video signal, which can allow bit rate reduction, provided sampling parameters are properly chosen.

In **component coding**, the sampling rate for the chroma signals U and V can be less, about one half of that required for the luminance signal, because of the smaller bandwidth for chroma. For 8 bit encoding, a total of 255 quantization levels are available, wherein the chroma signals can vary, going positive or negative with respect to the mean level of 128. Each of the components Y, U and V can be kept separate throughout the transmission and reception. They may be encoded in whatever standard form prevalent in the country or the intended area. On reception and demodulation these can be fed separately to the corresponding chroma decoder, RGB matrix and the display.

In 1981, CCIR set up a *world standard for digital component coding* of video signals in studio equipment, according to which the luminance and chrominance signals are encoded separately. The digital interface standard referred to as **CCIR 601**, enables manufacturers to develop equipment with many common features of 625/50 and 525/60 scan systems. The standard normally uses 4 : 2 : 2 ratio for the sampling frequencies. A higher performance is possible in the standard using 4 : 4 : 4 ratio. The luminance signal Y is digitized in an 8 bit A/D converter and encoded at 13.5 mega samples per second, giving output at a bit rate of $(8 \times 13.5 =) 108$ Mb/s. The colour difference signals *V* and *U* are digitized and encoded at 6.75 mega samples per second giving a bit rate of $(8 \times 6.75 =) 54$ Mb/s for each. The complete digitized signal thus requires a base bandwidth of over 200 MHz.

HDTV is the highest level of video quality at *level 1* in Television systems as well as B-ISDN with data rates around 216 Mb/s, where the image quality approximates that of the 35 mm motion picture film. The EDTV is at the *level 2*, midway between the HDTV and present day TV with distribution signal data rate of about 135 Mb/s. The *level 3* in the field of B-ISDN is for digitally coded composite video systems, with distribution rates of 32 to 42 Mb/s.

26.2   THE MULTIPLEXED ANALOG COMPONENT (MAC) SIGNALS

In the *conventional PAL/NTSC* systems the luminance and chrominance information is transmitted simultaneously for each line in interleaved form. For sound information, there is an additional carrier, and there are two separate carriers in the case of stereo sound. This creates problems of interference beats, called *cross luminance and cross colour* effects associated with areas of high colour saturation. These problems had to be avoided in the new satellite standard for direct broadcasting satellites (DBS)wherein bandwidth is less constricted than the terrestrial broadcasts in VHF/UHF bands. With this in view, new encoding systems were proposed, which included the extended PAL proposed by the BBC, MAC proposed by Brittain's Independent Broadcasting Authority (IBA). These were further developed by EBU as a family of MAC packet signals.

For studio production, it was convenient to use the PAL/NTSC format for TV production, since separate components were difficult to use because of *timing constraints*. As technology has advanced to introduce features like digital video effects, the need for higher performance in the studio has increased. MAC systems came about as a convenient way to maintain separate components without having to worry about the critical timing of the TV signal parameters by compressing the luminance and colour difference and placing them on the same line along with digitally encoded sound/data and sync signals.

In the multiplexed analog components (MAC), encoding the luminance and chrominance signals are *time-division multiplexed*, i.e. sent in time sequence as separate components. This eliminates the interaction between the luminance and chroma signals. In the conventional compatible signal PAL/NTSC systems where these are sent simultaneously, they can interact and interfere mutually, causing cross colour and cross luminance effects. The MAC components forming 8 MHz video base band are transmitted from DBS to specially equipped cable TV head-end processors or direct satellite reception terminals, as FM signals of about 24 MHz bandwidth, typically operating in the 12 GHz band.

Compression of Signals

As the colour and brightness information no longer appears simultaneously in each line, but have to be transmitted sequentially during the same line period, the two have to be time compressed. In the MAC system, the luminance and chrominance signals are individually *time compressed* so as to accommodate them both in the 64 μs line scanning period. Digital technology is cleverly employed here. The luminance and chrominance signals are separately digitized and buffer-stored. The stored line data is then *read at a faster* bit rate so as to compress it into a smaller time interval than the line scan period. This leaves the residual line scan time as a window to accommodate the chrominance and sound signals. The stored data thus read out at a faster rate is in a compressed form and is transmitted separately in sequence, the colour signal first for 17 μs followed by the luminance for 35 μs.

Scanning Frequencies

The scanning frequencies of 13.5 MHz for luminance, and 6.75 MHz for chrominance as per the digital component studio standard (CCIR Rec. 601/SMPTE RP 125) are used for writing the signals into the digital store. For time compression they are read out at a common sampling rate of 20.25 MHz. This means a *compression factor of 1.5 for the luminance and 3 for chrominance*. The digital signals are converted back into analog form, and are transmitted in frequency modulated form. At the receiving end the signals are decompressed. The bandwidth for the video signals is increased to 8.4 MHz so that no information is lost as a result of time compression.

Although the video signal remains in analog form it is *sampled at* 20.25 MHz. rate, producing (64 μ × 20.25 M =) 1296 samples per line. Each MAC line thus contains 1296 samples, of which 1114 samples are used for chroma and luma signals – 349 samples for one of the two colour difference signals transmitted on alternate lines, and 697 samples for luminance signals. The remaining period of about

10 μs in the blanking interval corresponding to 182 samples is used for sync and sound-data signals, as shown in Fig. 26.1.

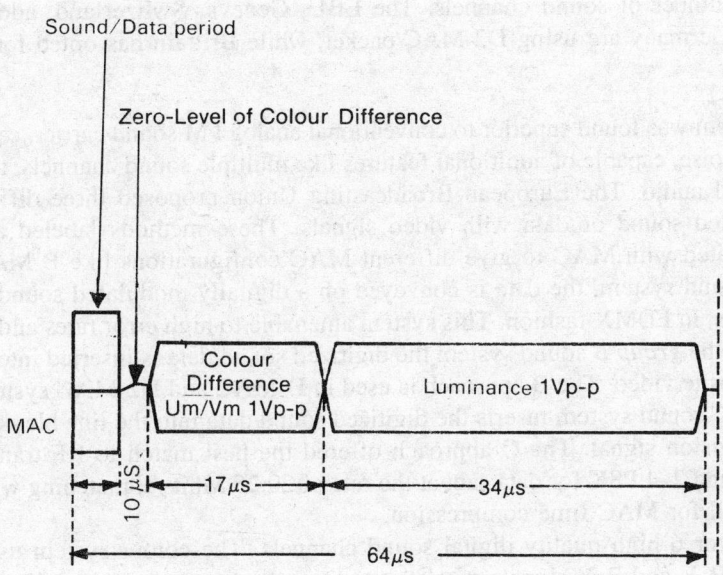

Fig. 26.1 MAC encoding format

For time compression, the digitized video signal comprising the 1296 samples per line in 64 μs (8 bits per sample), is written at the standard scanning rate into a suitable digital IC memory and is read out at a faster rate in just 35 μs, a factor of 1.5, leaving some 17 μs slot to fit in the chroma and the remaining line blanking time of 10.17 μs to carry 206 bits of digital sync, sound/data signals at 20.25 Mb/s. These bits consist of a run-in, a sync word, a colour sync and digital sound/data which can be used to provide up to 8 high quality audio channels or other data. The colour sync identifies which of the two chroma signals V and U are included in the line, as they are sent alternately. The sync word provides the scan synchronisation.

Chroma Signals

The chroma signals V and U are time compressed by a factor of 3. They are sampled at 6.78 MHz, written into memory and read at clocked rate of 20.25 MHz, before multiplexing them on a sequential basis into the time slot of *alternate lines,* similar to the SECAM technique. Thus, during the forward scan we have the V or U signals alternately occupying the first 17 μs, followed by Y signal taking 35 μs. U is transmitted on odd lines and V on the even lines of the picture; each being sent on the lines prior to the coincident luminance signal. One line delay is required at the receiving end, to store the V chroma information of a line, until the U chroma information is available from the next line for decoding the colours. The number of scan lines and the alternate transmission of the chroma components limits the useful vertical resolution. In order to prevent aliasing effects, frequencies in the chroma components above this limit are filtered out.

EBU MAC Systems

Different versions of MAC systems were developed by the European Broadcasting Union and TV receiver manufacturers, in the efforts for standardisation—B-MAC, C-MAC, D-MAC, D2-MAC, depending on the sound encoding techniques employed. The MAC systems differ slightly in their compression ratios, data rates and the number of sound channels. The EBU, Geneva, Switzerland, adopted the C-MAC/packet; France and Germany are using D2-MAC/packet, while Brittain has opted for D-MAC/packet.

Sound/Data Signals

As digital sound system was found superior to conventional analog FM sound carrier, capable of providing high fidelity, lower noise, capable of additional features like multiple sound channels, there was a general preference for digital audio. The European Broadcasting Union proposed three different methods for combining multiplexed sound or data with video signals. These methods labeled as A, B, C sound systems were associated with MAC to give different MAC configurations like B-MAC, C-MAC etc.

In the *group A* sound system, the data is conveyed on a digitally modulated sound subcarrier placed above the video range, in FDMX fashion. This system amenable to high error rates and low data capacity was not favoured. In the *group B* sound system the digitized sound/data is inserted into the line Blanking interval of the composite video. The B approach is used in D-MAC and D2-MAC systems recommended by EBU. The *group C* sound system inserts the digitized sound/data into the line blanking interval of the Carrier modulated vision signal. The C approach offered the best match to RF transmission channel, using FM for vision and 2–4 PSK for data sent at the rate of 20.25 Mbits/s, matching with the 20.25 MHz sampling rate adopted for MAC time compression.

B-MAC provides for 6 high quality digital sound channels. The colour system uses line sequential format, carrying the R-Y and B-Y signals on alternate lines. It can accommodate the wider aspect ratio of 16 : 9. B-MAC was adopted by Australia for DBS transmission in 1985. The MAC-C has potential for 8 simultaneous data channels, which may carry sound tracks or teletext pages. The *C-MAC/packet* format was originally meant for cable distribution of C-MAC/packet satellite service. It was optimized for transmission from a satellite where a relatively large bandwidth of about 27 MHz was available. The *D-MAC and D2-MAC* systems were subsets derived from C-MAC/packet format as a result of continued development work. D-MAC, an 8-channel system with FM forms the basis of the standards for DBS of UK. D2-MAC, a 4-channel system is supported by France and West Germany.

EBU MAC Systems

The family of MAC systems developed by the EBU, each associated with different type of packet data multiplex, summarized in the Table 26.1, as recommended for different uses.

Table 26.1 EBU MAC/packet family

	MAC/packet	Data burst rate	Vision modulation	Use
1.	C-MAC	20.25 Mbits/s (2–4 PSK)	FM	DBS
2.	D-MAC VSB AM	20.25 Mbits/s (duo-binary at baseband)	VSB AM	CATV
3.	D-MAC FM	20.25 Mbits/s (duo-binary at baseband)	FM	DBS
4.	D2-MAC VSB AM	10.125 Mbits/s (duo-binary at baseband)	VSB AM	narrow band CATV channels
5.	D2-MAC FM	10.125 Mbits/s (duo-binary at baseband)	FM	DBS

26.3 D2-MAC/PACKET SIGNAL

For digital sound transmission, the phase modulation of (2–4-PSK) of C-MAC was initially considered but rejected because it would have required, at the given clock rate of about 12 MHz, too large a bandwidth to suit the 7 MHz VHF or 8 MHz UHF channels of the broadband cable networks. The *duo-binary* (a higher level of binary) coding of sound signals developed in France, to reduce the bandwidth requirement for the data burst was chosen. Fig. 26.2 shows the time-division scheme of a D2-MAC video line.

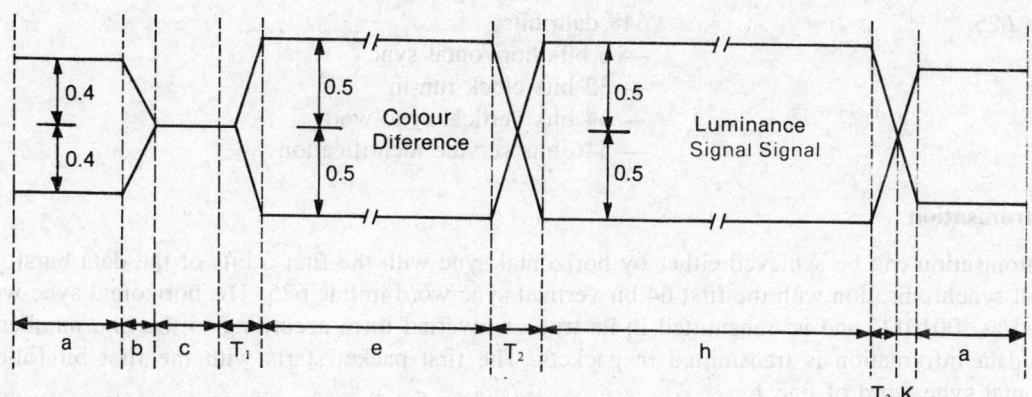

Fig. 26.2 D2-MAC Baseband signal waveform for normal unscrambled picture transmission (drawn not to scale) (ITT)

Unlike C-MAC, the time division multiplex of the analog video and digital sound/data is performed at the baseband, that is before modulation of the subcarrier. Hence the use of both AM/VSB (cable) and FM (satellite) system is possible. The duobinary coding halves the required bandwidth for a given bit rate by use of a 3-level signal instead of the binary to level signal.

Duobinary Coding

Every two adjacent bits of level 0 and 1 are combined into a new information unit, giving rise to possible pairing as 00, 11 as well as 10 and 01. Because 10 and 01 are treated the same, three possible pairings are left, each of which is assigned to a new level. In this manner a system of three levels is obtained which has only half the transmission capacity present in the case of binary coding, and it is possible to halve the clock rate, and with it the sound bandwidth. With the D for duo-binary coding and 2 for halving of the clock rate, the new satellite TV standard is referred to as **D2-MAC**.

Transmission Multiplex Characteristics of D2-MAC are as follows:

lines per picture	625
lines with data bursts	1 to 625
lines with video signal	24 to 310 and 336 to 622
lines with luminance signal	Y in each line
lines with chrominance signal	U in odd lines, V in even lines
interlace ratio	2 : 1
aspect ratio	4 : 3/(5.33 : 3)
luma compression ratio	3 : 2

chroma compression ratio	3 : 1
sampling clock frequency	20.25 MHz
instantaneous bit rate	10.125 Mbits/s
samples per line	1296
chrominance samples	349
luminance samples	697
bits per data burst	105 (6 bits horizontal. sync and 99 bits data)
line 624	105 bits + analog reference
line 625	648 data bits,
	— 6 bits horizontal sync
	— 32 bits clock run-in
	— 64 bits vertical sync word
	— 546 bits service identification

Synchronisation

Synchronisation can be achieved either by horizontal sync with the first 6 bits of the data burst, or by vertical synchronisation with the first 64 bit vertical sync word in line 625. The horizontal sync word is defined as '001011' and is transmitted in its true or inverted form according to the line number. The sound/data information is transmitted in packets. The first packet starts with the first bit following horizontal sync word of line 1.

Sound/Data Packets

As 6 bits out of 105 are spent for line sync and data code for scrambling and descrambling of special programs like the conditional access Pay-TV channels are also transmitted in the 10.17 μs interval per line, only 99 bits per line are left for sound information. This is not adequate. The 99 data bits of lines 1 to 623 are combined into 82 packets of 751 bits each, to contain either sound or data information. The remaining bits in line 623 are not used. In order to minimize the effects of multiple bit errors, the 751 bits of each packet are interleaved with a distance of 94 bits. Then, for spectrum-shaping purposes, the bits are scrambled by adding (modulo 2) a *pseudo-random binary sequence* (PRBS). The PRBS generator operates continuously, at a clock rate of 10.25 MHz and is initialized every 625 lines in synchronism with the first data bit in line 1.

As shown in Fig. 26.3, each packet consist of a header of 23 bits followed by 91 bytes of useful data:

— 23 bit header   — 10 bit address field
                                   — 2 bit continuity index
                                   — 11 bit protection suffix
— 8 bit packet type   — BC1 or BC2 (sound packet)
                                         BI1 or BI2 (data packet)
— 720 bits sound/data information

The header contains fixed addresses for the different sound and data services, which can be used for automatic configuration of the receiver. To guarantee a secure packet recognition the header is highly protected with a Golay cyclic code, which is able correct upto three errors.

The PT bytes enable the receiver to distinguish between sound and data packets and allows the precise switching process from one service to another. As there are only four different PT bytes, the Hamming distance is equal to at least 5 and allows a secure distinction of sound and data packets.

The satellite can transmit simultaneously 3 or 4 high quality sound channels of compact disc quality, or 8 common commentary-quality channels. Line 625 is used to transmit pure data only including a picture synchronising word, with the key for the decoding of the D2-MAC signal in the receiver.

D-MAC/packet Multiplex

In this *packet structure,* 624 lines of video signal are preceded by a burst of 206 bits of data. Line 625 contains entirely of 1296 bits of data. The first 7 bits (run-in bit and 6 bit sync word) at the beginning of each line serve for line synchronisation. Frame synchronisation can be derived from line-by-line pattern of the line sync words or a frame sync word of 64 bits can also be sent on the line 625. The D-MAC packet structure is shown in Fig. 26.3.

Each packet contains 91 bytes of useful data preceded by 23 bit header. The data can be assigned to any one of the different services identified by the first 10 address bits in the 23 bit header. The next 2 bits serve for continuity link to detect any loss of packet and the suffixed 11 bits serve for error protection for the address and index information providing detection upto 3 errors.

The *sound/data packet* can be transmitted in a string of 751 bits including the 23 bit header. The useful data area of 726 bits corresponds to a bit rate of 2.9848 Mbits/s. In the absence of TV picture, the switching multiplex can consist entirely of 20.25 Mbits/s data. The 623 lines of digital data divided into two sub-frames of 90 bits, each carrying 82 packets. This simplifies transcoding data into narrow channel D2-MAC.

The *sound coding* provides for high quality (40 Hz–15 kHz) mono, high quality stereo and reduced bandwidth (40 Hz–7 kHz) mono. Linear coding with 14 bits/sample and companded coding with 10 bits/sample are specified.

The D-MAC packet is planned to be very flexible and a variety of services can be offered. For information of the viewer, a menu driven *service identification system* is formed by sending data packets with address 0, the receiver then acting as a menu driven CRT terminal. As the time compressed baseband video bandwidth of the MAC is about 8.5 MHz, it is feasible to convey it within a channel of about 10.5 MHz on *cable TV* systems. It is not possible to retain the data rate of 20.25 Mbits/s 2–4 PSK which requires a full 27 MHz bandwidth. For this purpose D2-MAC/packet having half this data rate,

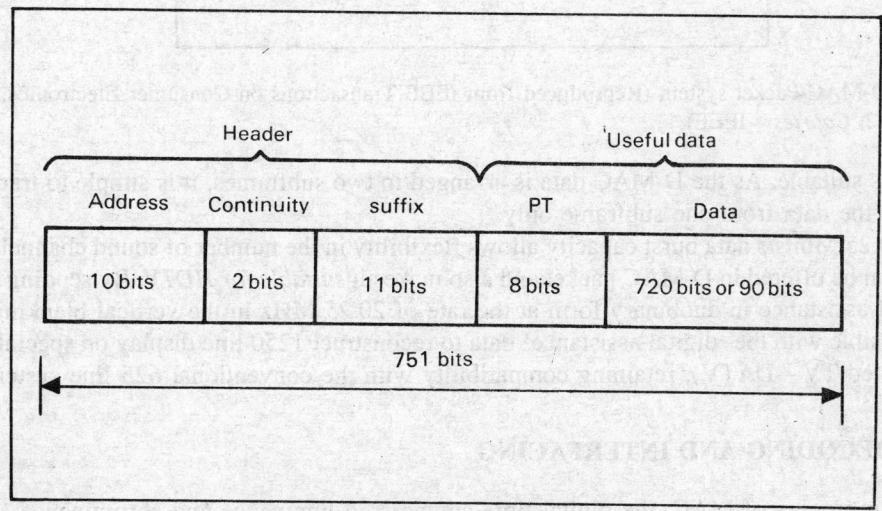

Fig. 26.3 (a) Data packet structure

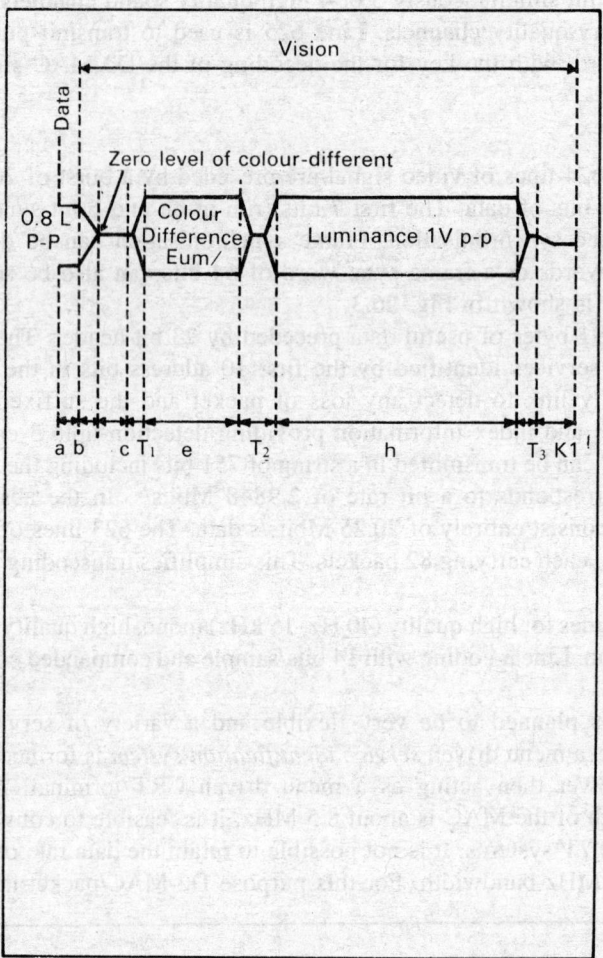

Fig. 26.3 (b) D-MAC/Packet system (Reproduced from IEEE Transactions on Consumer Electronics, Feb. 1988 p-132, *Courtesy:* IEEE)

10.125 Mbits is suitable. As the D-MAC data is arranged in two subframes, it is simple to transcode the data by taking the data from one subframe only.

The high 20.25 Mbits/s data burst capacity allows flexibility in the number of sound channels and data services that can be offered in D-MAC packet and also makes it *suitable for HDTV*. By sending additional data for digital assistance in duobinary form at the rate of 20.25 MHz in the vertical blanking interval, it could be possible with the 'digital assistance' data to reconstruct 1250 line display on special receivers (digitally assisted TV—DATV), retaining compatibility with the conventional 625 line system.

26.4 MAC DECODING AND INTERFACING

In the decoding process of MAC, the digital time compressed luminance and chrominance signals are first written into a *digital line memory*. They are read out from the memory at the normal scan rate to decompress and separate the Y signal and the line sequential *V* and *U* signals. A 64 μs delay line and an electronic switch is used to combine the sequential *V* and *U* information into simultaneous information updated every 2 lines.

Thus, we have at the output of the MAC decoder the *Y, V and U or R, G and B* signals as the base band signals. These could be fed directly to the colour picture tube drive circuitry for best results. Since in the initial stages the MAC decoder is designed as a part of the down converter add-on unit suitable for a standard TV receiver or as integral part of the head-end unit of a wide band cable TV system, the base band would have to be encoded in the PAL or SECAM format and RF modulated on a spare channel.

When the transmissions are fed into standard cable networks and converted into PAL, the large video bandwidth of D2-MAC signal has to be reduced after necessary *conversion from FM to AM* so that it is suitable for firstly for cable even in the super band 300–400 MHz, and then for PAL demodulation. This reduction in bandwidth and the deterioration in signal at each conversion affects the picture quality considerably.

26.5 ADVANTAGES OF MAC SIGNAL

MAC systems can offer higher performance than composite systems because of separation of luma and chroma signals which eliminates cross modulation, and use of full bandwidth for luma and chroma signals. Because of compression the baseband bandwidth is increased by 50% and the vertical resolution is reduced due to the sequential chroma format.

In MAC the picture synchronisation is ensured by means of the data time multiplexed data signals. The wasteful use of line and frame sync pulses occupying 30% of the signal amplitude is thus avoided, and therefore the *signal power of* MAC is greater than in PAL. This compensates to some extent the increase in noise generated in the time compression process. MAC/packet approach is highly flexible in the scope and types of services it can offer. The use of *encryption* in this packet is seen as a very important factor in commercial success of the satellite delivered Pay-TV services.

MAC signal is best suited for direct reception of satellite programs by individual and community antenna installations to maintain high quality of picture. Many of the constraints like monochrome compatibility have by now disappeared and the channel bandwidth available on *cable TV systems and satellite channels* is larger, permitting better quality. Satellite transmissions can give tremendous coverage over different countries having different TV encoding systems and languages, when MAC can provide multiple sound channels.

The component coding in which the each colour signal has separate indentity in the transmission chain gives it a flexibility in transcoding into any other format. The *programmability* feature of the MAC video and sync system has also the advantage of flexibility for future changes and improvements in the television technology. Advanced studio techniques like digital video effects for post production put on timing constraints on the signal parameters in conventional systems. With MAC, higher performance is readily possible because of separate components.

The *data structure* of MAC signals can be altered at will, by transmitting the structure map for the receiver to identify the type of service mix offered. MAC processing produces all 625 lines on each scan instead of the interlaced scan. As such, higher definition picture can be displayed. For transmission on standard 625 line TV network, the MAC signal is processed to remove alternate lines. The MAC system is adaptable into various formats having different specifications for different purposes.

26.6 HIGH DEFINITION TV SYSTEMS

The 35 mm films have been taken for the last several years, as the highest quality standard picture reproduction. Expensive TV programs have therefore been shot using cinematographic methods usually on 35 mm films in preference to electronic video recording techniques. The TV picture quality has been, with the current 625 or 525 line scanning standards, inferior compared to photographic pictures. HDTV system offers a wide screen format and a fineness of definition that is does not show up the scan line artifacts even when the viewing distance is as close as 3 times the screen height.

According to CCIR, a high definition TV is a system to allow viewing at about *3 times the picture height*, such that the system is virtually, or nearly transparent to the quality of portrayal such as motion and perception of depth in the original scene or performance, as perceived by a discerning viewer with normal visual acuity. This implies large screen display of cinema quality, using more than 1000 lines. HDTV is usually defined as system having *twice* the horizontal spatial and vertical spatial resolution of the current television systems, improved colour rendition by separate colour difference and includes one or more stereophonic channels that provide quality comparable to that of compact disk—CD player.

The NHK System

The *Japanese Broadcasting Company NHK* has been working on the development of High Definition TV system for several years since 1970. After evaluation of a wide range of psycho-physical factors affecting perception of visual images, they put forward a standard for the HDTV in 1980. The standard has been supported by many people, as a worldwide high definition studio production standard is highly desirable. The 1125/60 HDTV meets the studio production requirements with resolution comparable to, or better than 35 mm theater-films releases, which is a world standard for motion pictures.

The NHK HDTV standard was envisioned to provide a *high perception base* for the entire range of video systems needed by the information oriented society of the future. These include high definition recording and transmission of pictures, and a wide range of applications in motion picture production (electronic cinematography), video theaters, film processing and electronic high quality image printing and publishing, medical diagnostics and many other fields. The facility of immediate reproduction, review, and editing have made HDTV video systems more attractive to the motion picture producers than 35 mm films. Various techniques using HDTV technology, such as electronic picture synthesization, video effects, electronic editing and computer graphics have made a steady headway in the field of printing, to make video processing more versatile. Japan has been viewing HDTV as an element of the new information society, and are carefully planning to build up the consumer market as the regular HDTV broadcast start via BS-3a in 1990 and BS-3b in 1991.

HDTV System Considerations

HDTV broadcasting system consists of the basic *three parts:* (i) program generating *studio* equipment—cameras, VCRs, (ii) *transmitters* and transmission links—microwave, satellite, (iii) *large screen TV receivers* sets, projection TV equipment. Although new program generating equipment can be designed for the HDTV standards, the present day transmission links and the DBS do not have the capacity for very large bandwidth signal in the 12 GHz band. The new DBS planned to transmit in the 22 GHz band will have HDTV direct broadcast to suitably equipped TV receivers. Until then the HDTV signal generation is restricted to the studio and displayed directly on HDTV receivers. This itself has a *wide range applications* in motion picture production, video theaters, printing, medical diagnostics etc. There are a number of innovative systems proposed for HDTV transmission, using two channels with one compatible current channel and/or MAC approach.

HDTV Equipment

Cameras, VTRs, telecine, monitors and standards converter, special effects equipment for the NHK 1125/60 HDTV system with satisfactory performance are already available and the present efforts are to directed to improve the stability, handiness and operational ease. Small 17 mm camera tubes with high resolution capability has been developed, suitable for small compact colour camera improving camera sensitivity by a factor of 10. Amongst them is the HARP (*High Gain Avalanche Rushing Photoconductor*) tube cameras, which are ten times as sensitive as the current ones. Experimental digital VTRs with improved electronic tape-editing has been developed. High efficiency encoding for transmission and conversion systems for HDTV to NTSC/PAL have also been developed. Development of high resolution laser telecine, 127 cm rear projection displays, 16 : 9 aspect ratio 104 cm trinitron, have contributed to realization of HDTV. *Band compression MUSE* based receivers, which have been available in costly protypes, are to be montaged by 9 manufacturers for the consumer market by late 1990, but would only slowly penetrate to typical consumers by 1995. Large screen flat panel displays or projectors at reasonable cost are considered critical for development of HDTV as a consumer item.

26.7 HDTV STANDARDS AND COMPATIBILITY

CCIR report 601 adopted in 1982, provides an extensive family of digital coding standards for studio signals, upto the point of emission providing for high definition TV also. In 1985, a world standard proposed, based on the 1125/60 system developed in the NHK laboratory under the guidance of Dr Fujio. As the commercial interest in the HDTV increased in different regions in the world, the quest for greater compatibility with 50 Hz systems led to alternative proposal in 1987, for the *1250 line system using progressive scanning.* For aspect ratio increased from 4 : 3 to 16 : 5, the bandwidth of luminance signal has to be increased from 5 MHz to $(5 \times 2 \times 2 \times 16/5 \times 3/4 =)$ 26.7 MHz. Additional bandwidth of some 10 MHz is needed for the separate chrominance signal and about 0.6 MHz for hi-fi digital sound.

After the NHK proposals was put up, as the HDTV standard, several organizations started working on HDTV transmission systems adopting evolutionary approach. They focused on application of HDTV as public transmission service emphasizing on compatibility, using existing channels for increased bandwidth. In Europe, the *Eureka project* EU95 is formed in *9 project subgroups*, studying and perfecting different aspects of HDTV, developing extensive intellectual and technical resources in television industry, broadcasters, PTTs and academicians. These include BBC, IBA, British Telecom, Philips, Thomson, Bosch, Quantel, ITT Intermetall, Dortmund and Braunshweig university, Swetel ITVA and many others. Their approach is evolutionary aimed to produce a compatible standards and bring picture of HDTV quality into consumer TV equipment at reasonable cost, so that the new system finds quick public acceptance. The Groups are working to recommend the standards, assessment methodologies, studio interfaces and program material for demonstrations, aimed to cover the 1992 Barcelona Olympics in Spain, on the HDTV.

Field Rate

NHK initially proposed 1125/60, 2 : 1 interlace scanning choosing in aspect ratio of 5 : 3, as larger aspect ratio is important for the viewers' perspective and involvement in the program. The number of lines chosen was 1125, a compromise between 2 times the most common scanning line standards of 525 and 625 lines. This allowed simple relationship with the number of lines in the conventional television systems, because $1125 = 525 \times 15/7 = 625 \times 9/5$. Interlace has been accepted as a simple effective technique of bandwidth compression, achieved merely by a shift of the scanning pattern by plus-minus half line. The rate of 60 fields per second was chosen in preference to 50, because of reduced flicker and higher temporal sampling rate. The aspect ratio of 16 : 9 suggested later was preferred, in order to give more flexibility in shooting. Many interested countries having 50 Hz mains, find 60 Hz field rate unacceptable, although it is debatable whether this is really a major problem. For 60 Hz countries, lowering the 60 Hz rate to 50 Hz for higher quality HDTV system appears a contradiction in terms.

The *EBU Sub-group G5* specialists argue that the smoothness of motion may be acceptable in both 50 and 60 Hz field rates, when dynamic resolution can be adjusted by camera shuttering and display flicker can be corrected by display up-conversion. With large high contrast display, one of the problems is the large area flicker, whereby the eye perceives a flicker effect, conspicuous in the peripheral vision. Imperfections due to interlaced picture shows up as *field aliasing* distortions. The EBU studies based on subjective tests, indicate that a flicker-free display requires display field frequency of about 80 Hz and above, at high contrast levels and distances likely to be encountered in HDTV. A 50 Hz *progressive scan* system would have a better dynamic performance than 60 Hz interlaced system, because of greater distance to alias. Many post production processes suffering from the drawbacks of interlaced scanning could improve. Of course the 60 Hz progressive scan would be better still! Incidentally, one may note that with liquid crystal displays having a much longer hold time of the information received, large area flicker will be much less of a problem.

In 1987, the CCIR study group 11 discussed a dual mode production standard, firstly, to devise HDTV broadcast *transmission interface* which could link the production to all other HDTV program support media such as air, cable, VCR, etc. and secondly, to approach an agreed HDTV standard via *the digital routes*. The first obvious route was to define a set of digital parameters which should be universally adopted. A second route was to standardize on a unified system, employing a *common interface—data bus* easily transcoded to different sources and receivers. A third route could employ a *phased approach* using equipment that can be switched between *two field rates,* as per requirements of the present system. In the second phase, the rate could converge on the one in a universal format.

How to line up all other parameters than the field rates, with maximum convenience and quality, is also a major technical problem. Aligning the overall bit rate to the 1920 samples per active line, a common feature of all the HDTV proposals, is called the *Common Data Rate* or CDR approach. Aligning the number of active lines per picture is called the *Common Image Format* or CIF approach. A number of different parameter sets are under study, with their bearing on quality, convenience and cost factors. With maximum commonality in parameters such as line frequency and sampling frequency the CDR approach may provide benefits in recording and transmission. The CIF approach would include common image elements such as aspect ratio, pixels per active line, calorimetry etc. and may provide benefits from the considerations of solid state image sensors and displays. ATSC in the USA has expressed strong support to the CIF approach, whereby program exchange would be possible without final approval of all the video parameters by all nations. The CCIR may at least hope for an acceptable choice between the two formats.

The issue of *worldwide standard* for HDTV is thus complicated and the debate for compatible HDTV system is continuing in the forum of the International Radio Consultative Committee of the International Telecommunication Union (CCIR/ITU). The *CCIR study group 11* is devoted exclusively to High Definition television to recommend appropriate HDTV studio, emission and transmission standards. The group has in its latest meeting in 1989 agreed on a draft recommendation which covered the parameter values that are common to both 1125/60 and 1250/50 proposals! In the characteristic intractability of international debates, which are more related to economic rather than technical concerns, quest for common standards is indeed not easy. Everyone is convinced on the need for a single worldwide standard, but is equally convinced that the one should be most convenient to themselves.

Compatibility of HDTV standards with the existing conventional systems can be conceived in different levels and the evolution to HDTV planned accordingly. *New HDTV transmissions* may be received by the receivers system with HDTV quality on new receivers, or with a reduced quality comparable to that of the conventional system, using an add-on transcoding converter to reformat the new signal. *Compatible transmissions* have format compatible with the conventional television receivers, and is augmented by additional information for displaying full quality HDTV on new receivers, which could be reverse compatible. *Multistep systems* approach envisages improved or enhanced quality TV signals for some years not compatible with the existing standards, to be followed by HDTV.

A lower level of compatibility would need a low cost adapter/converter, while the lowest level of compatibility would require a rather costly adapter, so costly that a new system is preferable than adapting to the conventional old one. Higher the performance of the HDTV transmission system, the lower are likely to be the levels of compatibility. There is also the need for the backward compatibility, so that the new HDTV receivers are able to receive present signals.

Thus, a proper *trade-off* between higher performance of tomorrow, and higher level of today's compatibility with existing equipment is what could decide the choice of new systems like HDTV. If the broadcast HDTV falls short of the capabilities of other video distribution media like cable TV, satellite distributions direct or in conjunction with cable TV, video discs or VCR systems, these media would create their own separate HDTV standard, to cater to quality conscious videophiles. Satellite emissions are attractive to provide wide bandwidths required for HDTV. Optical disc recorder which has seen considerable progress recently, could revolutionalize video recording to deliver HDTV programs via optical fibre cables.

The main **1125/60/2:1 HDTV** system parameters are as follows:

TV scanning lines		1125
Aspect ratio		16 : 9
Interlace ratio		2 : 1
Active lines		1035
Field frequency		60 Hz
Line frequency		33750 Hz
Luminance signal	Y	20 MHz
Samples per active line		1920 for luminance
		960 for colour difference
Wide band colour signal	Cw	7.0 MHz
Narrow band colour signal	Cn	5.5 MHz

The CCIR Rec. 601 specifies 720 luminance samples during the active line and 360 samples for each of the color difference signals for the current television systems. Corresponding HDTV sampling requires $2 \times 720 \times 3/4 \times 16/9 = 1920$ samples per line for luminance and 960 for the chrominance signals. This demands a bandwidth of 30 MHz for luma and 15 MHz for chroma signals. The SMPTE has recently decided on a bandwidth of 30 MHz for luma as well as each of the chroma components. A uniform proposal detailing the various parameters of the 1125/60/2:1 system was done up as a new recommendation, through discussions amongst Japan, USA and Canada, and included in the CCIR report in 1986 as the only proposal for HDTV, the decisions on standards was kept pending till the nest plenary session of the CCIR in 1989. The proposal has been adopted as a *studio standard (240M) by the SMPTE* and domestic organizations in the USA (ATSC), Canada and Japan (BTA). But there has been no compromise on the standards issue even in 1989.

The basic characteristics of the 1125/60/2: 1 HDTV systems are given in the Table 26.1. Colometric characteristics of the system are given in Table 26.2 and digital parameters in Table 26.3.

Tabl 26.2 Colorimetric characteristics of the 1125/60/2:1 HDTV System

Item	Characteristics				
1	Chromaticity of reference primaries			x	y
			G	0.310	0.595
			R	0.630	0.340
			B	0.155	0.070
2	Chromaticity coordinates for equal primary signals (Eg=Er=Eb) (reference white)		D 65	x 0.313	y 0.329
3	Gamma correction		Electro-optical transfer characteristic (1/0.45) $L = ((V+0.1115)/1.1115)$ $L = V/4.0$ for $V > 0.0913$ for $V < 0.0913$		
4	Luminance signal		$Y' = 0.701G' + 0.087B' + 0.212R'$		
5	Color difference signals	R'–Y'	$R'-Y' = 0.701G' - 0.087B' + 0.788R'$		
		B'–Y'	$B'-Y' = 0.701G' + 0.913B' - 0.212R'$		
6	Scaling of color difference signals	$\dfrac{P'}{R}$	$\dfrac{P'}{R} = \dfrac{R'-Y'}{1.576}$		
		$\dfrac{P'}{B}$	$\dfrac{P'}{B} = \dfrac{B'-Y'}{1.826}$		

Table 26.3 Digital Parameters of the 1125/60/2:1 HDTV System

Item	Characteristics		
1	Active sample points per line	Y' / G'	1920 / 1920
		P' / R' R	960 / 1920
		P' / B' B	960 / 1920
2	Sampling frequency (MHz)	Y' / G'	74.25 / 74.25
		P' / R' R	37.125 / 74.25
		P' / B' B	37.125 / 74.25

The quest for greater compatibility with the 50 Hz systems and awakening commercial interest in HDTV in Europe and USA, led to an alternative proposal for *1250 lines 50 Hz progressive scanning* from the EBU nations at the November 1987 meeting of the CCIR, and the issue remained unresolved.

Standards Conversion

In the international exchange of programs standards conversion without affecting the quality, has been a very difficult task for broadcasters. Using motion compensation techniques, little loss of quality in conversion from 1125/60 to 625/50 systems has been demonstrated in Japan. Telecine transfer from 24 frames per second film to 1125/60, using *motion compensation* has also been demonstrated. Conversion from 1125/60 to 525/59.94 is more difficult, because the two field rates are too close. It can be achieved by using a drop frame or insert frame every 33 seconds. Where the 60 Hz field rate has to coexist with the 50 Hz, it is prudent to use an 60/50 Hz *adaptive display* in the receiver which accepts signals of both 60 and 50 Hz field rates, in combination with suitable decoders.

Colorimetry

The conventional colour TV systems used introduce colour errors and other impairments in the displayed picture, largely because of the assumptions about the colour primaries, which are not quite correct. In these systems, the derivation of colour primaries is made only in an approximate way, which violates the *constant luminance principle*. These impairments were considered acceptable in the present day systems alongside the other limitations of the systems. When introducing the HDTV system, attention can be paid to improve colour rendition by studying colorimetry aspects and amplitude transfer characteristics. It is necessary to design the new system so that it copes with the evolution of image sensor and display technology. Improved colour processing structure with increased performance and flexibility has been defined and a set of *optimized colorimetric primaries* preserving receiver compatibility specified.

The main idea of the EBU Eureka EU5 group is to specify a *standard primary (RGB) transmission interface,* based on a wide set of primaries that lie outside the locus of real surface colours. This set of primaries is obtained by matrixing the camera pickup outputs. The matrix would change according to improvements in the camera pickups. The RGB transmission interface however remains the same. The transmission primaries would then be matrixed to obtain the luminance and chrominance (colour difference) signals. A *nonlinear precorrection* would then be applied to improve the overall noise performance. After processing and distribution in the camera chain, a reciprocal arrangement would be used, leading to a second transmission primary (RGB) interface. Here any necessary *display correction* could be applied. From the second transmission interface, the picture display primaries required for a particular receiver could then be matrixed.

The proposal for **1250/50/1 : 1 HDTV production standard** from EBU is based on the following parameters:

TV scanning lines per picture	1250
Active lines	1152
Scanning	1 : 1 progressive
Aspect ratio	16 : 9
Field frequency	50 Hz
Line frequency	62.5 kHz
Samples per active lines	1920

The above parameters set refered to as *high definition progressive*—'HDP', offer ease of conversion to 50 Hz and 60 Hz interlaced conventional and HDTV systems, and have simple relation to digital standards of CCIR Rec. 601. Bandwidth reduced versions, and a digital *quincunx version* 'HDQ' is being developed to simplify digital recording. In the past, progressive scanning camera was not favoured because of poor noise performance and recordability. These problems are under study and improvements are in sight. Any HDTV system must include satisfactory HDTV standards converter and VTR-Telecine converter. High quality bit reduction techniques like the DCT studies in France and Italy, show promise for satisfactory digital recording.

The EBU countries in Europe, adopting the MAC packet family of standards have established a step-by step approach to improve the quality of the TV signal delivered to the receivers. France has D2-MAC/packet format satellite broadcasts via their TDF1 satellite. FRG will have the same format for their DBS TVSAT. ASTRA the European television satellite will carry encrypted Pay-TV channels in the MAC/packet format besides several PAL channels. The next step in this process is going to be the *HD-MAC* (section 26.9), with different system proposals for AM/VSB and FM transmissions, for which there is an intensive design and development effort in the project EU 95.

26.8 THE MUSE SYSTEMS

The 1125/60 HDTV, which provides 5 times as much information as the conventional TV, requires video bandwidth of 20 to 30 MHz. For transmission in the 12 GHz band for DBS, some form of bandwidth reduction is essential. The NHK solution to this is the *multiple subnyquist sampling encoding* (MUSE) which produces bandwidth compression by using a 4:1 subsampling technique. The basic principle is to use a phase alternating sub-nyquist sampling pattern over a 4 field sequence and motion compensated interframe interpolation . There are 1125 lines with 1440 horizontal pixels. The band limited luminance and colour difference signals with high definition are converted into a digital form by sampling at the rate of 64.8 MHz, enough to retain high resolution. The luminance and chrominance signals are sent separate in time compressed multiplexing (TCM) form with time compression of four for the line sequential chrominance signals. For suppression of aliasing two different prefilter are introduced, one for stationary portions and the other for moving portions. Depending on the motion detected, the outputs are combined before subsampling the pattern pixel by pixel basis. The subsampling pattern is shown in Fig. 26.4(a).

Only 1 out of every 4 samples is transmitted during each field, requiring a sequence of 4 fields for each sample. During each field 373 samples are transmitted, with minimum vertical spacing of 1/1035th of the picture height and horizontal spacing of 1/1500th of the width. Stationary areas are reconstructed by temporal interpolation of samples from the 4 fields. The receiver is equipped with 4 field stores, and the missing samples are interpolated between the sample pixels received, to re-form the original picture. For moving areas, the picture has to be reconstructed by using samples within one field only. This narrows down the spatial frequency response as shown in Fig. 26.4(b), but it is not much noticed. But for the movement of the whole area, such as due to camera pan and tilt, the loss of resolution can be noticeable. Motion detectors are used in the encoder to compensate for it.

In the original MUSE system called *MUSE-E (Emission)*, designed for satellite broadcasting, the wide aspect ratio 20 MHz HDTV signal is time compressed into 8.1 MHz. Luminance and chrominance signals are sent separate in the MAC format, suitable for satellite broadcasting, using frequency modulation. The information is transmitted to the receiver as a digital signal along with digital stereo sound. NHK also proposed a three level signal for sync signal. The sync timing is then more precisely carried by the zero crossing between the negative and positive pulses. For interstudio applications and field production, a wider bandwidth 16.2 MHz MUSE called *MUSE-T (Transmission)* has been demonstrated.

Three further variations of MUSE have been proposed for terrestrial broadcasting. *MUSE-6* is 6 MHz NTSC compatible system, where the wide aspect ratio is obtained by masking the top and bottom of the picture. 1125 line HDTV signal is converted to 750 lines for broadcasting, which is stepped up to 1125 line signal at the receiver. While the lower vertical frequency components of the 750 line signal are carried on the 345 active lines of the NTSC 525 line signal. The higher vertical frequency

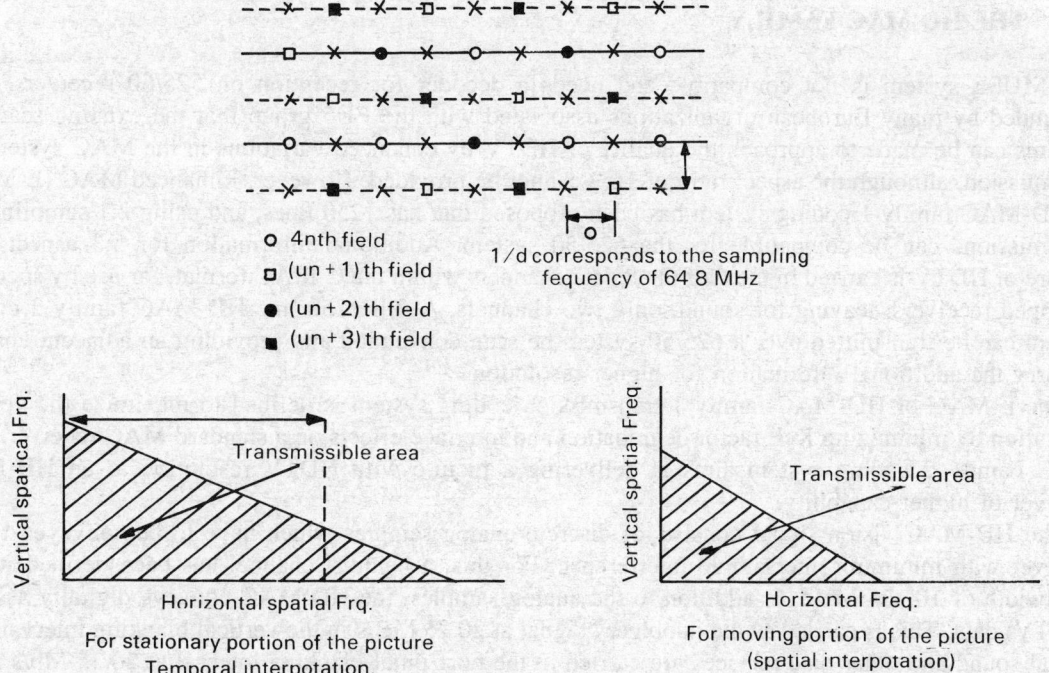

O 4nth field
□ (un + 1) th field
● (un + 2) th field
■ (un + 3) th field

1/d corresponds to the sampling fequency of 64 8 MHz

Transmissible area

Vertical spatical Frq.

Horizontal spatial Frq.

For stationary portion of the picture
Temporal interpotation.

Transmissible area

Vertical spatial Freq.

Horizontal Freq.

For moving portion of the picture
(spatial interpotation)

Fig. 26.4 (a) MUSE subsampling pattern, (b) Spacial frequency response. (Reproduced from IEEE Trans. on Consumer Electronics Feb. 1988 p.31, *Courtesy:* IEEE)

components and horizontal chrominance components are carried during the top and bottom mask portions of the signal. Horizontal luminance resolution is increased in still pictures by folding spectrum from 3.9 MHz 5.8 MHz and the spectrum from 5.8 MHz to 7.7 MHz to overlap the 1.9 MHz to 3.9 MHz region.

MUSE-9 system is an enhanced version, built on MUSE-6, with an augmentation signal in a second channel. The augmentation signal carries a baseband of 2.1 MHz, configured as 525 line VSB AM signal in the 3 MHz channel. This improves the dynamic resolution during movements, from 50% to 75% of the static resolution. Besides regular audio channels, two channels of digital stereo sound are transmitted. *Narrow MUSE* a 6 MHz non-compatible system, is third variation broadcast in *simulcast mode* (broadcast simultaneously—NTSC format in one channel and HDTV format in a second channel).

In North America, several organizations have made proposals for Advanced Television (ATV) systems, for terrestrial, cable and satellite broadcasting, some of them based on MUSE. These are under examination and valuation, with no consensus yet on the technical standards. Several manufacturer members of the HDTV 1125/60 group have been exhibiting equipment and systems applying the SMPTE 240M 1125/60 HDTV standard for studio organisation and program exchange, building clout for the standard. Zenith Electronics Corp. displayed recently its *Spectrum Compatible* (SC-HDTV) system on large projection TV system. The Zenith system can compress 30 MHz video information into 6 MHz US standard TV channel. It transmits the low frequency part of the TV signal digitally at low power, and so makes it possible for broadcasters to use channels currently unused because of interference to adjacent channels.

26.9 THE HD-MAC FAMILY

The MUSE system is not compatible and needs a decoder for reception on 525/60 receivers. It is contended by many European organizations associated with the EU5 group that the existing scanning systems can be made to approach the quality of HDTV by enhanced variations in the MAC system for transmission, although the aspect ratio of 16:9 cannot be provided. However, Enhanced MAC (E-MAC) or HD-MAC family 1 coding system has been proposed that has 1250 lines, and using 2:1 sampling the transmissions can be compatible for the 625/50 system. Additional information for 5:3 aspect ratio picture of HDTV is carried in the 6 of the 8 data channels within the C-MAC format, for use by specially equipped receivers, leaving for sound, only two channels. A full 1250 line HD-MAC family 2 coding system can be transmitted over a 625/50 system be scan conversion and providing an adjacent channel to carry the additional information for higher resolution.

The E-MAC or HD-MAC family 1 transmission coding system is defined to maximize the vertical resolution by minimizing Kell factor degradation and interlace effects on a standard MAC receiver. HD-MAC Family 2 coding system aims at delivering a picture with HDTV resolution to an HD-MAC receiver of higher capability.

The HD-MAC vision signal consists of discrete analog samples which have to be conveyed to the receiver with minimum intersample interference. For this, a Nyquist channel has been defined with a bandwidth of 10.125 MHz. In addition to the analog samples, the HD-MAC conveys digitally assisted (DATV) data. This is carried as the duobinary signal at 20.25 MB/s in the vertical blanking interval. The digital sound and other data services are carried in the horizontal blanking interval at 20.25 Mb/s in D-Mac or 10.125 Mb/s in D2-MAC.

Encoding Strategies

For transmission of HDTV signal through a MAC channel, bandwidth compression by a factor of 4 must be achieved. The nature of the initial filtering or subsampling procedure at the encoding side, and the subsequent interpolation at the decoding side, must be adapted to the local motion content of the picture. An efficient coding procedure has to be identified that operates block-by-block in the picture, characterized by mode or branch switching according to the motion content of the block. The non-interlaced signal, processed in a motion adaptive progressive scan converter, has an orthogonal scanning pattern at 10 MHz is filtered with a two dimensional diagonal LP filter, enabling an alias free resampling with quincunx pattern of 54 MHz and 1250 lines. The basic subsampling process contains 4 field sequences. This signal is further processed in a bandwidth reduction encoder/decoder, which is capable of processing at 4 times the sampling rate, which is 13.5 MHz for luminance.

The information on the filtering procedure adopted on the encoding side is transmitted to the receiver through *digitally assisted television* (DATV) channel. Suitable coding algorithms for this purpose and the complexity of the hardware implementation for each have been assessed using high definition test sequences. A trade-off between the quality of the HD-MAC picture and of the MAC compatible picture is necessary in making distinction in the pre- and post-filtering and the type of motion adaptive signal processing.

Conventional television can give picture quality close to that of 16 mm film. The information content in HDTV is about 5 times that in a conventional television, giving picture quality corresponding to that of the 35 mm film cinema. With further progress of applications to the print media, the next technical improvement may have to include about 2250 scanning lines. Just as 70 mm film cinema appeared to improve 35 mm cinema into *cinemascope* technique, stereoscopic TV is now being studied to pursue the feeling of being present in the happenings on the screen; that brings us to the efforts towards 3-D TV.

26.10   3D-STEREOSCOPIC TELEVISION TECHNIQUES

As in the cinematographic motion picture reproduction, scientists have been working over different methods to provide the third dimension to the TV picture by stereoscopic effects. We perceive the three dimensional depth because of binocular viewing, the left and right eyes seeing the same object with slightly different angles, so that the two images produces on our retina are slightly different. Our brain processes and interprets the two *overlapping 2-D images* as a three dimensional object with the depth and distance estimated from the difference in the two images. The obvious solution to 3-D TV would be to employ a two camera system to obtain the binocular view and corresponding signals. If these signals are separately transmitted and used to produce separate images in a common format, a 3-D reconstruction would be realized.

The two images could be based on *two complementary colours*, say yellow for left camera/eye and blue for the right camera/eye. These two images can then be produced in the respective colours in a shadow mask tube having yellow and blue phosphor dots or stripes. When viewed through corresponding filter glasses for the left and right eye, the two binocular images are superposed to produce a stereoscopic black and white image. If the two colours chosen are from the standard primaries, the method can be made simpler by using two standard phosphors of a shadow mask tube by cutting off the beam for the third phosphor. However, the resultant picture will be in a colour, additive of the two primary colours.

In another scheme polarizing films are used to isolate the left and right eye images. Binocular images are obtained from two separate cameras and are displayed on separate monitor screens fitted with films, polarized at right angles to each other. When optically arranged to be viewed superposed via partial to get the stereo vision 3-dimensional effect. Stereoscopic systems should normally incorporates the stereophonic sound in it so that the binaural as well as binocular effect gives a real three dimensional effect.

Abdy 3-D System

One simple adaptation process to achieve a 3D illusion in a 2-D colour TV signal, was invented in 1982 by Rolt Canes in West Germany. Under this system, the television signals containing the red information bits are delayed by some 600 ns with the help of special electronic circuitry. This produces a double image being formed on the TV screen which, when viewed through special glasses enable the viewer to perceive the third dimension of depth. The process patented by Abdy, a company in West Germany is popularity known as *Abdy 3-D* system and was introduced commercially in 1984 by a number of TV manufacturers as an innovation in their sets. The viewer has to wear special filter glasses so that the two eyes see slightly different images, the red version delayed with respect to the normal one, and creating an illusion of depth in the picture.

Holographic techniques have been used for viewing still photographs in laser light with 3-D effect. It is expected that this technique will be exploited for producing 3-D effects in television images also.

REVIEW QUESTIONS

26.1   What was need of MAC encoding? Explain the general format of MAC signals for transmitting colour TV signals.

26.2   Discuss the different MAC systems, indicating the variations in each. What is duo-binary coding? Illustrate by encoding and decoding schematic. How does duobinary coding scheme reduces the bandwidth requirement.

26.3   Explain the composition of D-MAC or D2-MAC packet signal in details and indicate whether these can be conveyed on cable TV and how?

26.4 What is the need of HDTV? Review briefly the developments made so far to evolve HDTV and the standards.

26.5 Discuss the NHK system and the MUSE system for digital encoding of the signals.

26.6 Deduce the bandwidth requirement of HDTV system using 1125 lines and aspect ratio of 16 : 9, scanning at 60 Hz 2 : 1 interlace.

26.7 Discuss the compatibility problems in HDTV. How far does the NHK system meet this requirement?

26.8 Indicate other compatible HDTV proposals and their merits in compatibility.

26.9 Draw the schematic for producing 3-D effect employing
   (a) two colours or
   (b) two axes polarized films for viewing the binocular images separately.

26.10 Explain the following techniques:
   (a) HD-MAC   (b) ABDY system
   (c) Duobinary coding   (d) DATV.

MULTIPLE-CHOICE QUESTIONS

26.1 MUSE is a method of:
   (a) bandwidth compression   (b) digital encoding
   (c) analog multiplexing   (d) a and b

26.2 MAC encoding improves quality of picture
   (a) by bandwidth compression
   (b) by digitizing analog signals
   (c) by time-multiplexing the luminance and chrominance signals
   (d) b and c

26.3 Common Bit rate for HDTV may give benefits in
   (a) adapting to pickup devices   (b) transmission of signals
   (c) recording of signals   (d) b and c

26.4 MUSE stands for
   (a) multiplexed sub-Normal encoding   (b) Multiplexed sub-Nyquist encoding
   (c) Multiple sub-Nyquist encoding   (d) none of these

26.5 In HDTV the line structure must not show up, when viewed at a distance greater than
   (a) 2 meters   (b) 5 times the height of screen
   (c) in projection TV   (d) 3 times the height of the screen.

26.6 In HDTV the sampling rate is
   (a) 960 samples   (b) 1920
   (c) 1125   (d) 1250, samples per active line.

Bibliography

BOEKHORST A. and STOLK, J., *Television Deflection Systems* (Philips Technical Library, Eindhoven), 1962.

ENNES, HAROLD E., *Television Broadcasting Equipment, Systems and Operating Fundamentals* (W. Foulsham & Co., Slough, Bucks, England), 1971.

FINK, D.G., *Electronics Engineers Handbook;* (McGraw-Hill Inc., New York), 1975.

FINK, D.G., *Television Engineers Handbook* (McGraw-Hill Inc., New York), 1957.

GLASFORD, G.M., *Fundamental of Television Engineering* (McGraw-Hill Inc., New York), 1955.

GROB, BERNARD, *Basic Television Principles and Servicing* (McGraw-Hill Inc., New York), 1975.

GUNTHER, FELLBAUM, *Fernsehservice Handbuch* (Franzis Verlag, München), 1971.

HANSEN, GERALD E., *Introduction to Solid State Television Systems* (Prentice-Hall Inc., New Jersey), 1969.

HETERSCHEID, W., *Designing Transistor I.F. Amplifiers* (Philips Technical Library, Eindhoven), 1966.

HETERSCHEID, W., *Transistor Bandpass Amplifiers* (Philips Technical Library, Eindhoven), 1964.

HILL, M.T. and EVANS, B.G., *Transmission Systems* (George Allen, London), 1973.

HOLM, W.A., *Colour Television Explained* (Philips Technical Library, Eindhoven), 1968.

HOLM, W.A., *How Television Works* (Philips Technical Library, Eindhoven).

HOWARD, W. SAMS Editorial Staff, *Photofact Television Course* (W. Foulsham Sams & Co., Slough, Bucks, England), 1968.

HUTSON, G.H., *Television Receiver Theory* (Edward Arnold Publishers, London), 1966.

INDIAN STANDARDS, *Method of Measurement on Receivers for Monochrome Television Broadcast Transmissions*—IS: 4545–1968 (Indian Standards Institution (ISI), New Delhi.)

INDIAN STANDARDS, *Methods of Measurement of Radiations from Television Receivers*—IS: 4546–1968 (ISI, New Delhi).

INDIAN STANDARDS, *Performance Requirements of Receivers for Monochrome Television Broadcast Transmissions*—IS: 4547-1978 (ISI, New Delhi).

INDIAN STANDARDS, *Specifications for Limits for Electromagnetic Interference* (first revision), IS: 6842-1977 (ISI, New Delhi).

KENNEDY, GEORGE, *Electronic Communication Systems* (McGraw-Hill Inc., New York), 1970.

KING, GORDON, *The Practical Aerial Handbook* (Odhams Book Ltd., London), 1967.

KIVER, M.S., *Television Simplified* (Litton Educational Publishing Inc.), 1973.

KOUBECK, M., *Fernsehempfanostechnik Schwartzweiss und Farbe* (Franzis Verlag, München), 1969.

KRIVOSHIV, M. and DVORKOVICH, V., *Measurements in Television Channels* (Mir Publishers, Moscow), 1977.

LIMANN, OTTO., *Fernsehtechnik ohne Ballast* (Franzis Verlag, München), 1973.

MENDE, HERBERT G., *Practische Antennenbau* (Franzis Verlag, München), 1977.

REED, C.R.G., *Principles of Colour Television Systems* (Sir Isaac Pitman & Sons Ltd. London), 1969.

SAMOYLOV, V.F. and KHROMOY, B.P., *Television* (Mir Publishers, Moscow), 1977.

SIMS, H.V., *Principles of PAL Colour Television and Related Systems* (Illife Books Ltd., London), 1969.

SJOBBEMA, D., *Aerials—TV and FM Receiving Aerials* (N.V. Gloeilampenfabrieken, Eindhoven), 1964.

SONY, CORPORATION, *Instruction Manual for Monochrome Video Tape Recorder* AV-3420CE (Sony Corporation, Tokyo).

Sony, Corporation, *Video Cassette Recorder—Transistor Circuits and Applications* (4)—*Video Cassette Instruction Booklet* (Sony Service Training School, Tokyo).

Texas Instruments Electronic Series, *Circuit Design for Audio, AM/FM and TV* (McGraw-Hill Inc., New York), 1967.

Weaver, L.E., *Television Measurements* (Peter Peregrims Ltd., London), 1971.

Westman, H.P., *Reference Data for Radio Engineers* (Howards W. Sams and Co. Inc., New York), 1970.

Zwaraber, H. and Kaufmann, R., *Practische Aufbau and Prüfung von Antennenanlagen* (Dr Alfred Hütig Verlag, Heidelberg), 1973.

Article References

1. BINGLEY, F.J., "A Half Century of Television Reception", *Proc. IRE*, **50** (No. 5), (May, 1962), 799–804.
2. STONE, R.F., "A Practical Narrowband Television System: Sampledot", *IEEE Trans.,* **BC-22** (No. 2), (June, 1976), 21–52.
3. DEUTSCH, S., "Visual Display Using Pseudorandom dot scan", *IEEE Trans.,* **Com-22** (Jan., 1973), 65–75.
4. WILMOTTE, R.M., "Technical Frontiers of Television" *IEEE Trans.,* **BC-22** (No. 3), (Sep., 1976), 73–80.
5. WILMOTTE, R.M., "TV Look-ahead", *IEEE Spectrum*, (Feb., 1976), 34–39.
6. OWEN CLURE, H., "Television Broadcasting", *Proc. IRE*, **50** (No. 5), (May, 1962), 818–824.
6. (a) TSUKOMOTO, K., OGAWA, O. and TAMAI, S., "Technical aspects of the Japanese Broadcasting Satellite Experiments, IEEE, Transactions on Broadcasting, Vol. **B-A24**, No. 4, (Dec. 1978)", pp. 81–91.
7. BARBE, D.F., "Imaging Devices using CCD's", *Proc. IEEE*, 63 (No. 1), (Jan., 1975), 38–67.
8. KOUBECK, M., "Interplex—A new Versatile Full Resolution Single Tube Colour TV Camera System", *IEEE Trans.,* **BC-22** (No. 3), (Sep., 1976), 80–85.
9. CCIR, *Characteristics of Television Systems*, Report, Doc. XIII, Plenary Assembly, Geneva, 1974.
10. RAO, V.V., "TV Camera Tube for Monochrome, Electronics Today", (Sib Publications, Bombay), 7 (No. 12), (Dec., 1974), 15–27.
11. ELCOMA DIVISION, Application Lab., "Television Camera Tubes, Electronic Application News", (Philips India Ltd., Bombay), **12** (No. 1 and 2), (Jan./Feb. and Mar./Apr., 1975).
12. DOLLEKAMP, J., "One-inch diameter Plumbicon Camera Tube", *Mullard Tech. Commun.,* **11** (No. 109), (Jan. 1971), 196–200.
13. SCHUT, TH. G. and WEIJLAND, W.P., "30 mm Plumbicon Camera Tube with Fibre Optics Face-plate, Anti-comet Tail Gun and Light Pipe", *Mullard Tech. Commun.*, **11** (No. 109), (Jan., 1971), 186–195.
14. WENTWORTH, JOHN W., "The Technology of Program Production and Recording", *Proc. IRE*, **50** (No. 5), (May, 1962), 830–836.
15. SUBRAMANYAM, P.K., "Trends in Television Recording", *Electronics Today*, (Sib Publications, Bombay), **7** (No. 12), (Dec., 1974), 28–31.
16. MURUGESAN, M., "TV Modulation at Intermediate Frequency", *Electronics Today*, (Sib Publications, Bombay), **7** (No. 11), (Nov., 1974), 33–48.
17. DEVANATHAN, R. and PARAMESWARAN, N., "International Television via Intelsat System", *Electronic Today*, (Sib Publications, Bombay), **7** (No. 11), (Nov., 1974), 19–24.
18. KRISHNAMURTHY, S., "Satellite Instructional Television Experiment", *Electronics Today*, (Sib Publications, Bombay), **7** (No. 11), (Nov., 1974), 45–47.
19. RAO, B.S., RAMAIAH and SWAMINATHAN, V.L., "Television Receiver for Direct Reception from Broadcasting Satellite", *J. Inst. Electron. Telecommun. Engrs.*, New Delhi, **9** (No. 8), (Aug., 1973), 453–458.

20. RAO, B.S., ARORA, O.P. and RAO, G.T., "Site Selection Considerations for Transmitter for Limited Rebroadcast Applications in Satellite Instructional Television Experiment (Site)", *J. Inst. Electron. Telecommun. Engrs.*, New Delhi, **20** (No. 9), (Sep., 1974), 471–476.

21. RAO, B.S. and ARORA, O.P., "Earth Station for Television Reception in SITE", *J. Inst. Electron. Telecommun. Engrs.*, New Delhi, **19** (No. 8), (Aug., 1973), 438–443.

22. MOHANVELU, A.S. and RAO, B.S., "Basic Considerations in the Planning of Broadcasting Satellite System", *J. Inst. Electron. Telecommun. Engrs.*, New Delhi, **22** (No. 6), (June, 1976), 347–355.

23. SINGH, M.D., BHATNAGAR, P.S. and DUBEY, P.C., "Development of channelcut Antenna for Band III", *J. Inst. Electron. Telecommun. Engrs.*, New Delhi, **21** (No. 10), (Oct., 1975), 551–555.

24. BHATNAGAR, P.S., DUBEY, P.C., SINGH, M.D. and RANGOLE, P.K., "Development of Multichannel Antenna for TV reception", *J. Inst. Electron. Telecommun. Engrs.*, New Delhi, **21** (No. 11), (Nov., 1975), 625–628.

25. JOHNSON, K.G. and ENRIQUEZ, E.A., "Investigation of Surface Educational Television Distribution", *IEEE Trans.*, **BC-22** (No. 2) (June, 1976), 45–52.

26. MUKHARJEE, S.K., "Booster Amplifiers for Television Receivers", *J. Inst. Electron. Telecommun. Engrs.*, New Delhi, **21** (No. 1), (Jan., 1975), 17–20.

27. RANGOLE, P.K., SACHAN, S.B.L. and DUBEY, P.C., "TV reception in Fringe Areas", *J. Inst. Electron. Telecommun. Engrs.*, New Delhi, **22** (No. 10), (Oct., 1976), 645–648.

28. MURUGESAN, M., "Design Considerations for Transistor Line Deflection Circuits", *Electronics Today,* (Sib Publications, Bomday), **7** (No. 11), (Nov., 1974), 57–71.

29. ABOUFADEL, B., "Transistorized Horizontal Output Stage", *IEEE Trans.*, **BTR-12** (No. 1), (1966), 15–22.

30. BOULTON P.G. and SKELTON, D., "Transistor Line Time Base for 110° Monochrome Receivers", *Mullard Techn. Commun.*, **12** (No. 112), (Oct., 1971), 38–43.

31. MURUGESAN, M., "Reactance Stage for Line Oscillator in Transistor TV Receivers". *J. Inst. Electron. Telecommun. Engrs.*, New Delhi, **21** (No. 11), (Nov., 1975), 619–621.

32. ELCOMA DIVISION, Application Lab., "Applications of Signal Processing ICs TBA 890/900", *Electronic Application News.* (Philips India Ltd., Bombay), **9** (No. 6), (Nov./Dec., 1972), 18–31.

33. ELCOMA DIVISION, Application Lab., "A New Range of Linear Integrated Circuits-III", *Electronic Application News.* (Philips India Ltd., Bombay), **12** (No. 1 and 2), (Jan./Feb., and Mar./Apr., 1975), 34–41.

34. TV Games, 1876 Wescon Professional Program, 1976, Western Electronics Show and Convention: KENNETH, D., LISTON (Jr.) and SLURASKI, Video Games ISI Flexibility Magnovox Co., Fort Wayne, In.
KAM LI, Technical Aspects of Video Games, Signetics Corp. Sunnyvale, Ca.

Index